The Index of SOLVENTS

The Index of SOLVENTS

An International Guide to 1700 Products by Trade Name, Chemical, Application, and Manufacturer

Compiled by

Michael and Irene Ash

Gower

Published by
Gower Publishing Limited
Gower House
Croft Road
Aldershot
Hants GU11 3HR
England

Gower
Old Post Road
Brookfield
Vermont 05036
U.S.A.

British Library Of Cataloguing in Publication Data
The index of solvents: an international guide to 1700 products by trade name, chemical, application, and manufacturer
1.
I. Ash, Michael II. Ash, Irene, 1948-
0000

ISBN 0-566-07884-8

Library of Congress Cataloging-in-Publication Data
Ash, Michael.
The index of solvents : an international guide to more than 1700 products by trade name, chemical, application, and manufacturer / compiled by Michael and Irene Ash.
p. cm
Includes index.
ISBN 0-566-07884-8 (cloth)
1. Solvents. I. Ash, Irene. II. Title.
TP247.5.A77 1996
660´.29´29462--dc20 96-29187
CIP

Typeset in Palermo by Synapse Information Resources, Inc.
Printed and bound in Great Britain by Hartnolls Limited, Bodmin, Cornwall

Contents

Preface

This reference work describes more than 1700 trade name products and chemicals that are used as solvents in the chemical industry and in research. The worldwide solvents market is estimated at over 30 billion pounds per year. Solvents are used in a wide range of industries and applications including: metal cleaning and degreasing, dry-cleaning operations, automotive and aviation fuel additives, paints, varnishes, lacquers, paint removers, plastics and rubber products, adhesives, textiles, printing inks, pharmaceuticals and food processing.

Solvents include a variety of commonly used chemical groups such as alcohols (amyl, butyl, ethyl, isopropyl, etc.), esters (amyl, butyl, and ethyl acetates), glycol ethers, ketones (acetone, methyl ethyl ketone), chlorinated hydrocarbons (ethyl and methylene chloride, etc.), petroleum-derived hydrocarbons (benzene, toluene, xylene).

In the current competitive marketplace, key personnel in the areas of industrial chemistry must find ways to meet new standards for volatile organic compound (VOC) emissions and hazardous air pollutants (HAPs). This task is becoming more daunting since the increasing number of regulatory requirements have imposed additional restrictions on some of the most effective and widely used organic solvents. It is, therefore, more difficult for formulators, suppliers, and other solvent users to meet VOC regulations, product performance requirements, and corporate profitability.

The extensive information presented here is gathered from more than 700 worldwide manufacturers, distributors, trade magazines, reference books, and chemical databases. It functions as a single source for decision-making in formulating products, purchasing solvents, and understanding the safety issues by providing current information on chemical composition, properties, function and application, toxicology, environmental impact and transportation restrictions for both trade name and generic solvents that are manufactured worldwide.

This reference contains comprehensive information on a broad range of solvents available from major chemical manufacturers and serves to expedite the material selection process for chemists, formulators and purchasing agents while providing important toxicological and environmental information for industrial hygienists, safety officers, and researchers. Trade name products are cross-referenced by chemical composition, application, CAS and EINECS numbers. Generic chemicals that function as solvents are also included along with their manufacturers and distributors.

The book is divided into four sections:

Part I—*Trade Name Reference* contains more than 1000 alphabetical entries of trade name solvents. Each entry references its manufacturer, chemical composition, associated CAS and EINECS identifying numbers, general properties, applications and functions, toxicology, and compliance and regulatory information as provided by the manufacturer and other sources.

Part II—*Chemical Dictionary/Cross-Reference* contains more than 700 alphabetical entries of chemicals that function as solvents and/or are components of trade name solvents. Each chemical entry lists the trade name products that are equivalent to the chemical compound or contains that chemical compound as a product's chemical constituent. Also included in the entry are: CAS (Chemical Abstract Service) numbers, EINECS (European Inventory of Existing Commercial Chemical Substances) numbers, synonyms, molecular and empirical formulas, definition, classification, general properties, uses, regulatory, toxicological, and precaution information wherever possible. Synonyms are thoroughly cross-referenced back to the main entry.

Part III—*Application Cross-Reference* contains an alphabetical listing of major solvent application categories. Approximately 30 categories are included, e.g., adhesives, agriculture, electronics manufacturing, food and food processing, industrial and consumer cleaning, metalworking, minerals and mining, paints and coatings, etc. Each application category entry is followed by an alphabetical listing of the trade name products and chemicals that find use in these application areas.

Part IV—*Manufacturers Directory* contains detailed contact information for the manufacturers of the more than 1700 trade name products and generic chemicals that are referenced in this handbook. Wherever possible telephone, telefax, and telex numbers, toll-free 800 numbers, and complete mailing addresses are included for each manufacturer.

The *Appendix* contains the following cross-references:

CAS Number to Trade Name Cross-Reference orders many trade names found in Part I by identifying CAS numbers; it should be noted that trade names contain more than one chemical component and the associated CAS numbers in this section refer to each trade name product's primary chemical component.

CAS Number to Chemical Cross-Reference orders chemical compounds found in Part II by CAS numbers.

EINECS Number to Trade Name Cross-Reference orders many trade names found in Part I by identifying EINECS numbers that refer to each trade name product's primary chemical component.

EINECS Number to Chemical Cross-Reference orders chemical compounds found in Part II by EINECS numbers.

This book is the culmination of many months of research, investigation of product sources and trade journals, and sorting through a variety of technical data sheets and brochures acquired through personal contacts and correspondences with major chemical manufacturers worldwide. We are especially grateful to Roberta Dakan for her skills in chemical information database management. Her tireless efforts have been instrumental in the production of this reference.

M. & I. Ash

NOTE:

The information contained in this reference is accurate to the best of our knowledge; however, no liability will be assumed by the publisher or the authors for the correctness or comprehensiveness of such information. The determination of the suitability of these products for prospective use is the responsibility of the user. It is herewith recommended that those who plan to use any of the products referenced seek the manufacturers instructions for the handling of that chemical.

Abbreviations

ABS	acrylonitrile-butadiene-styrene
abs.	absolute
absorp.	absorption
ACGIH	American Conference of Governmental Industrial Hygienists
act.	active
ADI	acceptable daily intake (FAO/WHO)
adsorp.	adsorption
agric.	agricultural
agrichem.	agrichemical(s)
agrochem.	agrochemical
a.i.	active ingredient
alc.	alcohol
Am., Amer.	American
amts.	amounts
anhyd.	anhydrous
APHA	American Public Health Association
applic(s).	application(s)
aq.	aqueous
ASA	acrylic-styrene-acrylonitrile
atm	atmosphere
at.wt.	atomic weight
aux.	auxiliary
avail.	available
avg.	average
a.w.	atomic weight
BATF	Bureau of Alcohol, Tobacco, and Firearms (U.S.)
BGA	Federal Republic of Germany Health Dept. certification
BHA	butylated hydroxyanisole
BHT	butylated hydroxytoluene
biochem.	biochemical
biodeg.	biodegradable
bldg.	building
blk.	black
BOD	biological oxygen demand
BP	British Pharmacopeia
b.p.	boiling point
BR	butadiene rubbers, polybutadienes
B&R	Ball & Ring
br., brn.	brown
brnsh.	brownish
BS	British Standards
B/S	butadiene/styrene
BSS	British Standard Sieve
Btu	British thermal unit
B.U.	Brabender units (viscosity)
byprod.	byproduct(s)
C	degrees Centigrade
CAA	Clean Air Act
cap.	capillary
CAS	Chemical Abstracts Service
CC	closed cup
cc	cubic centimeter(s)
CCl_4	carbon tetrachloride

CD	completely denatured
CDA	completely denatured alcohol
CEL	corporate exposure limit
CERCLA	Comprehensive Environmental Response, Compensation, & Liability Act (U.S.)
CFC	chlorofluorocarbon
CFR	Code of Federal Regulations (U.S.)
ch.	chapter
char.	characteristic, characterized
chel.	chelation
chem(s).	chemical(s)
CI	Color Index
CIIR	chlorobutyl rubber
CIR	Cosmetic Ingredient Review
cks	centistoke(s)
CL	ceiling concentration
cl	clear
cm	centimeter(s)
cm^3	cubic centimeter(s)
CMC	carboxymethylcellulose
CMC	critical Micelle concentration
c.m.p.	capillary melting point
CNS	central nervous system
CO	carbon monoxide
COC	Cleveland Open Cup
COD	chemical oxygen demand
coeff.	coefficient
compat.	compatible
compd(s).	compound(s)
compr.	compression
conc(s).	concentrated, concentration
conduct.	Conductive, conductivity
const.	constant
contg.	containing
cosolv.	cosolvent
CP	Canadian Pharmacopeia
cp	centipoise(s)
cps	centipoise(s)
CPVC	chlorinated polyvinyl chloride
CR	chloroprene rubber, polychloroprene
cryst.	crystalline, crystallization
cs	centistoke(s)
cSt	centistoke(s)
ctks	centistoke(s)
cwt	hundred weight
dc	direct current
DEA	diethanolamide, diethanolamine
dec.	decomposes
decomp.	decomposition
DEG	diethylene glycol
deliq.	deliquescent
dens.	density
deriv(s).	derivative(s)
descrip.	description
dg	decigram(s)

DI	deionized
diam.	diameter
dielec.	dielectric
dil.	dilute
disp.	dispersible, dispersion
dissip.	dissipation
dist.	distilled
distrib.	distributor
distort.	distortion
dk.	dark
DOP	dioctyl phthalate
DOT	Department of Transportation (U.S.)
DW	distilled water, deionized water
eb, EB	electron beam
EC	European Community
EC50	environmental concentration, 50%
EDTA	ethylenediamine tetraacetic acid
e.g.	for example
EINECS	European Inventory of Existing Commercial Chemical Substances
elec.	electrical
ELINCS	European List of Notified Chemical Substances
elong.	elongation
EMI	electromagnetic interference
EMS	electromagnetic shielding
EO	ethylene oxide
EP	European Pharmacopoeia
EPA	Environmental Protection Agency (U.S.)
EPDM	ethylene-propylene-diene rubber, ethylene-propylene terpolymer
EPR	ethylene-propylene rubber
EPS	expandable polystyrene
equip.	equipment
equiv.	equivalent
ESD	electrostatic discharge
ESP	electrostatic protection
esp.	especially
EU	European Union
Eur.Ph.	European Pharmacopeia
EVA	ethylene vinyl acetate
exc.	excellent
F	degrees Fahrenheit
FA	fatty acid
FAO	Food and Agriculture Organization (United Nations)
FCC	Food Chemicals Codex
FDA	Food and Drug Administration (U.S.)
FD&C	Foods, Drugs, and Cosmetics
FEMA	Flavor and Extract Manufacturers' Association (U.S.)
FEP	fluorinated ethylene propylene
FG	food grade
FIFRA	Federal Insecticide, Fungicide, and Rodenticide Act (U.S. EPA)
fl	fluid
flamm.	flammable, flammability
flex.	flexural
f.p.	freezing point
FR-ABS	flame retardant ABS
FRP	fiberglass-reinforced plastics

F-T	Fischer-Tropsch
ft	foot, feet
f.w.	formula weight
G	giga
g	gram(s)
gal	gallon(s)
G-H	Gardner-Holdt
GI	gastro-intestinal
glac.	glacial
gr.	gravity
gran.	granules, granular
GRAS	generally regarded as safe
grn.	green
GRP	glass-reinforced plastics, glass-reinforced polyester
GVS	Gardner varnish scale (color)
h	hour(s)
HALS	hindered amine light stabilizer
HAP	hazardous air pollutant
HC	hydrocarbon
HCl	hydrochloride, hydrochloric acid
HDPE	high-density polyethylene
Hg	mercury
HIPS	high-impact polystyrene
HLB	hydrophilic lipophilic balance
HMIS	Hazardous Material Information System
hr	hour(s)
hyd.	hydroxyl
hydrog.	hydrogenated
Hz	hertz
IARC	International Agency for Research on Cancer (United Nations)
i.b.p.	initial boiling point
I&I	industrial and institutional
IIR	isobutylene-isoprene rubber
IM	intramuscular
immisc.	immiscible
in.	inch(es)
inc.	increases, increased
INCI	International Nomenclature Cosmetic Ingredient
incl.	including
incompat.	incompatible
ing.	ingestion
ingred(s).	ingredient(s)
inh.	inhalation
inj.	injection
inorg.	inorganic
insol.	insoluble
Int'l.	International
IP	intraperitoneal
IPA	isopropyl alcohol
IPM	isopropyl myristate
IPP	isopropyl palmitate
IR	isoprene rubber (synthetic), polyisoprene
IU	International Unit
IV	intravenous
J	joule

JCID	Japanese Cosmetic Ingredients Dictionary
JP	Japanese Pharmacopoeia
JSCI	Japanese Standard of Cosmetic Ingredients
JSFA	Japan Standards for Food Additives
k	kilo
KB	Kauri-Butanol
kg	kilogram(s)
KTPP	potassium tripolyphosphate
KU	Krebs units
l	liter(s)
lb	pound(s)
LC50	lethal concentration 50%
LCLo	lethal concentration low
LD0	lethal dose 0%
LD50	lethal dose 50%
LDLo	lowest published lethal dose
LDPE	low-density polyethylene
LED	light-emitting diode
lg.	large
liq.	liquid
LLDPE	linear low-density polyethylene
lt.	light
Ltd.	Limited
LVP	low vapor pressure
M	mega
M	mole
m	milli
m	meter(s)
m-	meta
manuf.	manufacturer
max.	maximum
mbar	millibar
MEA	monoethanolamine, monoethanolamide
mech.	mechanial
med.	medium
MEK	methyl ethyl ketone
mfg.	manufacture
mg	milligram(s)
MIBK	methyl isobutyl ketone
microcryst.	microcrystalline
microgran.	microgranules, microgranular
MID	Meat Inspection Division (USDA)
MIL	Military Specifications
mil	$^1/_{1000}$th inch
min	minute(s)
min.	mineral
min.	minimum
MIPA	monoisopropanolamine, monoisopropanolamide
misc.	miscible, miscellaneous
mixt(s).	mixture(s)
ml	milliliter(s)
mm	millimeter(s)
MMW-HDPE	medium molecular weight high density polyethylene
mN	millinewton(s)
mo, mos	month(s)

mod.	moderately
mod.	modulus
monocl.	monoclinic
m.p.	melting point
mPa·s	millipascal-second(s)
mus	mouse
m.w.	molecular weight
N	normal
nat.	natural
NBR	nitrile rubber, nitrile-butadiene rubber
NC	nitrocellulose
need.	needles
neut.	neutral, neutralized
NF	National Formulary
NFPA	National Fire Protection Association
ng	nanogram
nm	nanometer
no.	number
nonalc.	nonalcoholic
nonaq.	nonaqueous
nonbiodeg.	nonbiodegradable
nonflamm.	nonflammable
nonyel.	nonyellowing
NR	natural rubber, isoprene rubber (natural)
NSF	National Sanitation Foundation
NSF	National Standards Foundation
NV	nonvolatiles
o-	ortho
OBPA	oxybisphenoxarsine
OC	open cup
ODC	ozone-depleting compound
ODP	ozone-depletion potential
OEL	occupational exposure limit
OEM	original equipment manufacturer
OMS	odorless mineral spirits
org.	organic
OSHA	Occupational Safety and Health Administration (U.S.)
o/w	oil-in-water
oz	ounce
p-	para
Pa	Pascal
PBT	polybutylene terephthalate
pbw	parts by weight
PC	polycarbonate
PCMX	p-chloro-m-xylenol
PCP	Pest Control Product Act, Canada, 1972
PCTFE	polychlorotrifluoroethylene
PE	polyethylene
PEEK	polyetheretherketone
PEG	polyethylene glycol
PEK	polyetherketone
PEL	permissible exposure level
percut.	percutaneous
PES	polyether sulfone
PET	polyethylene terephthalate

petrol.	petroleum
PG	propylene glycol
pH	hydrogen-ion concentration
Ph.	Pharmacopoeia
pharm.	pharmaceutical
Ph.Eur.	European Pharmacopoeia
phr	parts per hundred of rubber or resin
PIB	polyisobutylene
pkg.	packaging
PM, P-M	Pensky-Martens
PMCC	Pensky-Martens closed cup
PMMA	polymethyl methacrylate
PMOC	Pensky-Martens open cup
PO	propylene oxide
POE	polyoxyethylene, polyoxyethylated
polyunsat.	polyunsaturated
POM	polyoxymethylene
POP	polyoxypropylene, polyoxypropylated
powd.	powder
PP	polypropylene
ppb	parts per billion
ppt	parts per trillion
PPE	polyphenylene ether
PPG	polypropylene glycol
pph	parts per hundred (percent)
ppm	parts per million
PPO	polyphenylene oxide
PPS	polyphenylene sulfide
pract.	practically
prep(s).	preparation(s)
prod.	product(s), production
props.	properties
ps	poise
PS	polystyrene
psi	pounds per square inch
psia	pounds per square inch absolute
psig	pounds per square inch gauge
pt.	point
Pt-Co	platinum-cobalt
PTFE	polytetrafluoroethylene
PTMEG	polytetramethylene ether glycol
PU	polyurethane
PUR	polyurethane
PVA	polyvinyl alcohol
PVAc	polyvinyl acetate
PVAL	polyvinyl alcohol
PVB	polyvinyl butyral
PVC	polyvinyl chloride
PVDC	polyvinylidene chloride
PVDF	polyvinylidene fluoride
PVP	polyvinylpyrrolidone
qt	quart
quat.	quaternary
R&B	Ring & Ball
rbt	rabbit

RCRA	Resource Conservation and Recovery Act (40 CFR §261)
rdsh.	reddish
rec.	recommended
ref.	refractive
reg.	register, registry
regs.	regulations
rep.	represents
resist.	resistance, resistant, resistivity
resp.	respectively
RFI	radio frequency interference
r.h.	relative humidity
rhomb.	rhombic
RIM	reaction injection molded/molding
RQ	reportable quantity
R.T.	room temperature
RTECS	Registry of Toxic Effects of Chemical Substances
RTV	room temperature vulcanizing
s	second(s)
s-	secondary
SAN	styrene-acrylonitrile
sapon.	saponification
SARA	Superfund Amendments and Reauthorization Act (U.S.)
sat.	saturated
S/B	styrene/butadiene
SBR	styrene/butadiene rubber
SBS	styrene-butadiene-styrene
SD	specially denatured
SDA	specially denatured alcohol
SE	self-emulsifying
SEBS	styrene-ethylene/butylene-styrene
sec.	secondary
semicryst.	semicrystalline
semiliq.	semiliquid
semisyn.	semisynthetic
sl.	slight, slightly
sm.	small
soften.	softening
sol.	soluble, solubility
solid.	solidification
sol'n.	solution
solv(s).	solvent(s)
sp.	specific
spec.	specification, specialty
spp.	non-specified species
SSU	Saybolt Universal Seconds
std.	standard
STEL	short term exposure limit
Stod.	Stoddard solvent
str.	strength
subcut.	subcutaneous
subl.	sublimes
surf.	surface
SUS	Saybolt Universal Seconds
susp.	suspension
syn.	synthetic

t	tertiary
TCC	Tag closed cup
TCLo	toxic concentration low
TDLo	toxic dose low
TDS	total dissolved solids
TEA	triethanolamine, triethanolamide
tech.	technical
temp.	temperature
tens.	tensile , tension
tert	tertiary
THF	tetrahydrofuran
thru	through
TIPA	triisopropanolamine
TKPP	tetrapotassium pyrophosphate
TLV	Threshold Limit Value
TOC	Tag open cup
TPE	thermoplastic elastomer
TSCA	Toxic Substances Control Act
tsp	teaspoon
TWA	time weighted average
TWC	time weighted concentration
typ.	typical
UF	urea formaldehyde
UHF	ultra high frequency
UL	Underwriter's Laboratory
UN No.	United Nations Substance Identification Number (transport)
unsat.	unsaturated
USDA	U.S. Department of Agriculture
USP	Unites States Pharmacopeia
uv, UV	ultraviolet
V	volt
VA	vinyl acetate
VAE	vinyl acetate ethylene
VC	vinyl chloride
VdC, VDC	vinylidene chloride
veg.	vegetable
visc.	viscous, viscosity
VM&P	Varnish Makers and Painters
VOC	volatile organic compounds
vol.	volume
v/v	volume by volume
wh.	white
WHO	World Health Organization (United Nations)
wks	weeks
w/o	water-in-oil
wt.	weight
w/v	weight by volume
w/w	weight by weight
yel.	yellow
ylsh.	yellowish
yr	year
#	number
%	percent
±	plus or minus
<	less than

$>$	greater than
$\leq$	less than or equal to
$\geq$	greater than or equal to
@	at
α	alpha
β	beta
δ, Δ	delta
ε	epsilon
γ	gamma
ω	omega
μ	micron, micrometer
μg	microgram
$\approx$	approximately equal to

Part I
Trade Name Reference

A

Acetamide MEA [Witco/Organics]
Chem. Descrip.: Acetamide MEA
CAS 142-26-7; EINECS 205-530-8
Uses: Solvent, humectant in cosmetic formulations
Properties: Liq.; 100% or 70% aq. sol'n.; nonionic

Acetol® 1706 [Henkel/Cospha; Henkel Canada]
Chem. Descrip.: Cetyl acetate, acetylated lanolin alcohol
Uses: Water repellent; strongly hydrophobic emollient, penetrant, lubricant, and cosolvent used in hair prods., creams, lotions, bath preps., suntan preps., and baby prods.
Properties: Lt. yel. liq.; sol. in IPM, min., castor, and olive oil; water-insol.

Acetulan® [Amerchol]
Chem. Descrip.: Cetyl acetate, acetylated lanolin alcohol
Uses: Binder for pressed powds.; emollient, plasticizer, cosolvent, NV and sebum solvent for personal care prods.; lubricant for clay, talc, and starch; stabilizer for lanolin; solubilizer in aerosols; penetrant and spreading agent
Properties: Pale yel. thin oily liq.; odorless; sol. in ethanol, min., castor and veg. oil, IPM, IPP, IPA, silicone, butyl stearate, sulfonated castor oil, ethyl acetate, and org. solv.; insol. in water; sp.gr. 0.850-0.880; visc. 10 cps; acid no. 1.0 max.; sapon. no. 180-200; hyd. 8.0 max.; pH neutral

Acintene® AC-P [Arizona]
Chem. Descrip.: 32% α-Pinene, 19% dipentene, 13% 1,4-cineole, 9% terpinolene, 8% camphene, 7% 1,8-cineole
Uses: Terpene solvent is nontoxic replacement for petroleum-based hydrocarbon solvents; solvent and antistripping agent for paints, solvent in mfg. of varnishes, component of household and industrial cleaners and polishes; printing ink solvent as flow promoter, solvent in reclaiming rubber, solvent and paraffin wax inhibitor in oil field production chems., replacement for chlorinated solvents, and component of degreasing compds.
Properties: APHA 10 color; mild pine-like odor; misc. in alcohols; insol. in water; sp.gr. 0.860 (15.5 C); dens. 7.16 lb/gal (15 C); i.b.p. 157 C; flash pt. (CC) 35 F; ref. index 1.4680; KB value 94

Acintene® B Beta Pinene [Arizona]
Chem. Descrip.: 76.3% β-pinene, 11% dipentene, 7.5% α-pinene
Uses: Intermediate in mfg. of syn. resins
Properties: APHA 5 color; sp.gr. 0.8708; dens. 7.19 lb/gal; i.b.p. 165-167 C; flash pt. (CC) 100 F; ref. index 1.4774

Acintene® DP 738 [Arizona]
Chem. Descrip.: 24% Dipentene, 17% α-pinene, 16% 1,4-cineole, 12% terpinolene, 11% 1,8-cineole, 7% camphene, 7% α-terpinene
Uses: Solvent and antistripping agent for paints, solvent in mfg. of varnishes, component of household and industrial cleaners and polishes, extender for pine oil disinfectant cleaners; printing ink solvent as flow promoter, solvent in

reclaiming rubber, solvent and paraffin wax inhibitor in oil field production chems., replacement for chlorinated solvents, replacement for limonene as solvent, and component of degreasing compds.

Properties: Colorless clear liq.; strong pine-like odor; misc. in alcohols; insol. in water; sp.gr. 0.870 (15.5 C); dens. 7.24 lb/gal (15 C); i.b.p. 170 C; flash pt. (CC) 49 F; ref. index 1.4680; KB value 103

Acintene® LS 160 A [Arizona]

Uses: Solvent for paints, replacement for chlorinated hydrocarbons, substitute for d-limonene, component of household and industrial cleaners and polishes, paint stripping formulations, solvent in mfg. of varnishes; wood penetrant and carrier oil, printing ink solvent as flow promoter, substitute for d-limonene, component of degreasing compds.

Properties: Gardner 1 color; citrus-like odor; misc. in alcohols; insol. in water; sp.gr. 0.860 (15.5 C); dens. 7.19 lb/gal (15 C); flash pt. (CC) 35 F; KB value 68

Acintene® LS 160 B [Arizona]

Uses: Solvent for paints, replacement for chlorinated hydrocarbons, substitute for d-limonene, component of household and industrial cleaners and polishes, paint stripping formulations, solvent in mfg. of varnishes; wood penetrant and carrier oil; flow promoter and solvent for printing inks

Properties: Gardner 1 color; citrus-like odor; misc. in alcohols; insol. in water; sp.gr. 0.860 (15.5 C); dens. 7.16 lb/gal (15 C); flash pt. (CC) 37 F; KB value 83

Acintene® LS 160 D [Arizona]

Uses: Solvent for paints, replacement for chlorinated hydrocarbons, substitute for d-limonene, component of household and industrial cleaners and polishes, paint stripping formulations, solvent in mfg. of varnishes; wood penetrant and carrier oil, printing ink solvent as flow promoter, substitute for d-limonene, component of degreasing compds., extender for pine oil in disinfectant cleaners

Properties: Gardner 1 color; citrus-like top note odor, turpentine-like odor base; misc. in alcohols; insol. in water; sp.gr. 0.880 (15.5 C); dens. 7.33 lb/gal (15 C); flash pt. (CC) 49 F; KB value 100

Acintene® P-AC Alpha-Pinene [Arizona]

Chem. Descrip.: 81% α-pinene and 14% camphene

Uses: Solvent intermediate in mfg. of pine oils and camphene, monomer for production of polyterpene resins and terpene-phenolic resins, intermediate in mfg. of wide variety of terpene-based aromatic compds.

Properties: APHA 20 color; mild turpentine-like odor; misc. in alcohols; insol. in water; sp.gr. 0.865 (15.5 C); dens. 7.19 lb/gal (15 C); i.b.p. 157 C; flash pt. (CC) 35 F; ref. index 1.4670; KB value 62; < 0.1% moisture

Actrel® 1111L [Exxon]

Chem. Descrip.: Aliphatic hydrocarbon with oxygenated hydrocarbons

Uses: Cleaner effective for oils and greases, OEM and maintenance applics.; oil-sol. metalworking (cutting) oils, quenching oils, electro-discharge machining oils; rust preventative oils, drawing/stamping oils, rolling/forming oils, silicone oils; water-sol. metalworking fluids, honing or lapping oils, buffing compds., carbonized oils and residues, lubricating greases, waxes, dye penetrants; no ODP, no HAPS, no SARA reportables

Properties: Very low odor; dens. 0.81 kg/l; flash pt. (CC) 114 C; surf. tens. 27 dynes/cm

Toxicology: OEL 300 ppm; very low acute, oral, and inhalation toxicity

Actrel® 1138L [Exxon]

Chem. Descrip.: Aliphatic hydrocarbon with oxygenated hydrocarbons

Uses: Cleaner effective for oils and greases, OEM and maintenance applics.; oil-sol. metalworking (cutting) oils, quenching oils, electro-discharge machining oils; rust preventative oils, drawing/stamping oils, rolling/forming oils, silicone oils; water-sol. metalworking fluids, honing or lapping oils, buffing compds., carbonized oils and residues, lubricating greases, waxes, dye penetrants; no ODP, no HAPS, no SARA reportables

Properties: Moderate odor; dens. 0.77 kg/l; flash pt. (CC) 42 C; surf. tens. 24 dynes/cm

Toxicology: OEL 300 ppm; very low acute, oral, and inhalation toxicity

Actrel® 1160L [Exxon]

Chem. Descrip.: Aliphatic hydrocarbon with oxygenated hydrocarbons

Uses: Cleaner effective for oils and greases, OEM and maintenance applics.; oil-sol. metalworking (cutting) oils, quenching oils, electro-discharge machining oils; rust preventative oils, drawing/stamping oils, rolling/forming oils, silicone oils; water-sol. metalworking fluids, honing or lapping oils, buffing compds., carbonized oils and residues, lubricating greases, waxes, dye penetrants; no ODP, no HAPS, no SARA reportables

Properties: Sl. odor; dens. 0.79 kg/l; flash pt. (CC) 79 C; surf. tens. 25 dynes/cm

Toxicology: OEL 300 ppm; very low acute, oral, and inhalation toxicity

Actrel® 1171L [Exxon]

Chem. Descrip.: Aliphatic hydrocarbon with oxygenated hydrocarbons

Uses: Cleaner effective for oils and greases, OEM and maintenance applics.; oil-sol. metalworking (cutting) oils, quenching oils, electro-discharge machining oils; rust preventative oils, drawing/stamping oils, rolling/forming oils, silicone oils; water-sol. metalworking fluids, honing or lapping oils, buffing compds., carbonized oils and residues, lubricating greases, waxes, dye penetrants; no ODP, no HAPS, no SARA reportables

Properties: Sl. odor; dens. 0.80 kg/l; flash pt. (CC) 79 C; surf. tens. 25 dynes/cm

Toxicology: OEL 300 ppm; very low acute, oral, and inhalation toxicity

Actrel® 1960L [Exxon]

Chem. Descrip.: Aliphatic hydrocarbon with oxygenated hydrocarbons

Uses: Cleaner effective for oils and greases, OEM and maintenance applics.; oil-sol. metalworking (cutting) oils, quenching oils, electro-discharge machining oils; rust preventative oils, drawing/stamping oils, rolling/forming oils, silicone oils; water-sol. metalworking fluids, honing or lapping oils, buffing compds., carbonized oils and residues, lubricating greases, waxes, dye penetrants; no ODP, no HAPS, no SARA reportables

Properties: Strong odor; dens. 0.80 kg/l; flash pt. (CC) 61 C; surf. tens. 25 dynes/cm

Toxicology: Very low acute, oral, and inhalation toxicity

Actrel® 1971L [Exxon]

Chem. Descrip.: Aliphatic hydrocarbon with oxygenated hydrocarbons

Uses: Cleaner effective for oils and greases, OEM and maintenance applics.; oil-sol. metalworking (cutting) oils, quenching oils, electro-discharge machining oils; rust preventative oils, drawing/stamping oils, rolling/forming oils, silicone oils; water-sol. metalworking fluids, honing or lapping oils, buffing compds., carbonized oils and residues, lubricating greases, waxes, dye penetrants; no ODP, no HAPS, no SARA reportables

Properties: Strong odor; dens. 0.80 kg/l; flash pt. (CC) 79 C; surf. tens. 26

dynes/cm

Toxicology: Very low acute, oral, and inhalation toxicity

Actrel® 3338L [Exxon]

Chem. Descrip.: Aliphatic hydrocarbon with oxygenated hydrocarbons

Uses: Cleaner effective for oils and greases, OEM and maintenance applics.; oil-sol. metalworking (cutting) oils, quenching oils, electro-discharge machining oils; rust preventative oils, drawing/stamping oils, rolling/forming oils, silicone oils; water-sol. metalworking fluids, honing or lapping oils, buffing compds., carbonized oils and residues, lubricating greases, waxes, dye penetrants; no ODP, no HAPS, no SARA reportables

Properties: Very low odor; dens. 0.75 kg/l; flash pt. (CC) 40 C; surf. tens. 24 dynes/cm

Toxicology: OEL 300 ppm; very low acute, oral, and inhalation toxicity

Actrel® 3349L [Exxon]

Chem. Descrip.: Aliphatic hydrocarbon with oxygenated hydrocarbons

Uses: Cleaner effective for oils and greases, OEM and maintenance applics.; oil-sol. metalworking (cutting) oils, quenching oils, electro-discharge machining oils; rust preventative oils, drawing/stamping oils, rolling/forming oils, silicone oils; water-sol. metalworking fluids, honing or lapping oils, buffing compds., carbonized oils and residues, lubricating greases, waxes, dye penetrants; no ODP, no HAPS, no SARA reportables

Properties: Very low odor; dens. 0.76 kg/L; flash pt. (CC) 53 C; surf. tens. 25 dynes/cm

Toxicology: OEL 300 ppm; very low acute, oral, and inhalation toxicity

Actrel® 3360L [Exxon]

Chem. Descrip.: Aliphatic hydrocarbon with oxygenated hydrocarbons

Uses: Cleaner effective for oils and greases, OEM and maintenance applics.; oil-sol. metalworking (cutting) oils, quenching oils, electro-discharge machining oils; rust preventative oils, drawing/stamping oils, rolling/forming oils, silicone oils; water-sol. metalworking fluids, honing or lapping oils, buffing compds., carbonized oils and residues, lubricating greases, waxes, dye penetrants; no ODP, no HAPS, no SARA reportables

Properties: Essentially odorless; dens. 0.77 kg/l; flash pt. (CC) 61 C; surf. tens. 26 dynes/cm

Toxicology: OEL 300 ppm; very low acute, oral, and inhalation toxicity

Actrel® 4493L [Exxon]

Chem. Descrip.: Aliphatic hydrocarbon with oxygenated hydrocarbons

Uses: Cleaner effective for oils and greases, OEM and maintenance applics.; oil-sol. metalworking (cutting) oils, quenching oils, electro-discharge machining oils; rust preventative oils, drawing/stamping oils, rolling/forming oils, silicone oils; water-sol. metalworking fluids, honing or lapping oils, buffing compds., carbonized oils and residues, lubricating greases, waxes, dye penetrants; no ODP, no HAPS, no SARA reportables

Properties: Essentially odorless; dens. 0.76 kg/l; flash pt. (CC) 93 C; surf. tens. 26 dynes/cm

Toxicology: OEL 300 ppm; very low acute, oral, and inhalation toxicity

Adol® 42 [Witco Corp.]

Chem. Descrip.: Tallow alcohol

EINECS 308-980-4

Uses: Emulsifier, lubricant, foam control agent, cosolvent, plasticizer, stabilizer, emollient, intermediate; for metal rolling oils and lubricants, gas scrubbing, printing inks, auto antifreeze, textile dyeing and finishing, emulsion systems, min. processing, oil field chemicals, fabric softeners

Properties: Lovibond 5Y/0.5R color;

sol. in fatty alcohols, IPA, benzene, trichlorethylene, acetone, turpentine, VMP naphtha, kerosene, lt. min. oil; m.w. 258; sp.gr. 0.814 (60/25 C); cloud pt. 36 C max.; sapon. no. 3.5 max.; 100% conc.; nonionic

Adol® 52 NF [Witco Corp.]
Chem. Descrip.: Cetyl alcohol
CAS 36653-82-4; EINECS 253-149-0
Uses: Coemulsifier, lubricant, foam control agent, cosolvent, plasticizer, stabilizer, emollient, intermediate; for metal lubricants, inks, textiles, emulsions, paper, cosmetics, mineral processing, oil field chemicals, fabric softeners
Properties: Lovibond 5Y/0.5R max.; sol. in fatty alcohols, IPA, benzene, trichloroethylene, acetone, turpentine, VMP naphtha, kerosene, lt. min. oil; m.w. 247; sp.gr. 0.815 (60/25 C); m.p. 45-50 C; acid no. 1.0 max.; sapon. no. 3.0 max.; nonionic
Unverified

Adol® 60 [Witco Corp.]
Chem. Descrip.: Behenyl alcohol
CAS 661-19-8; EINECS 211-546-6
Uses: Lubricant, foam control agent, cosolvent, plasticizer, emollient
Properties: Lovibond 10Y/1.0R; m.w. 316; sp. gr. 0.79 (100/25 C); acid no. 1.5 max.; iodine no. 6 max.; sapon. no. 3 max.
Discontinued prod.

Adol® 66 [Witco Corp.]
Chem. Descrip.: Isostearyl alcohol
CAS 27458-93-1; EINECS 248-470-8
Uses: Coemulsifier, lubricant, foam control agent, cosolvent, plasticizer, stabilizer, emollient, intermediate; for metal lubricants, inks, textiles, emulsions, paper, mineral processing, oil field chemicals, fabric softeners; emollient for skin care prods.; as replacement for oleyl alcohol
Properties: Lovibond 5Y/0.5R max. liq.; sol. in fatty alcohols, IPA, benzene, trichlorethylene, acetone, turpentine, VMP naphtha, kerosene, lt. min. oil; m.w. 295; sp.gr. 0.861; cloud pt. 8 C max.; acid no. 1.0 max.; sapon. no. 2.0 max.; 100% conc.; nonionic

Adol® 80 [Witco Corp.]
Chem. Descrip.: Oleyl alcohol
CAS 143-28-2; EINECS 205-597-3
Uses: Coemulsifier, lubricant, foam control agent, cosolvent, plasticizer, stabilizer, emollient, intermediate; for metal rolling oils and lubricants, gas scrubbing, printing inks, auto antifreeze, textile dyeing and finishing, emulsion systems, paper pulping, cosmetics, cationic surfactants, min. processing, oil field chemicals, fabric softeners
Properties: Lovibond 5Y/0.5R liq.; m.w. 263; sol. in fatty alcohols, IPA, benzene, trichlorethylene, acetone, turpentine, VMP naphtha, kerosene, lt. min. oil; sp.gr. 0.840; cloud pt. 13 C max.; sapon. no. 3.0 max.; 100% conc.; nonionic

Adol® 85 NF [Witco Corp.]
Chem. Descrip.: Oleyl alcohol NF
CAS 143-28-2; EINECS 205-597-3
Uses: Emulsifier, lubricant, foam control agent, cosolvent, plasticizer, emollient for personal care prods.
Properties: Lovibond 3Y/0.3R liq.; m.w. 268; sp.gr. 0.84; acid no. 1.0; iodine no. 85-95; sapon no. 2 max.; cloud pt. 10 C max.; 100% conc.; nonionic

Adol® 520 NF [Witco Corp.]
Chem. Descrip.: Cetyl alcohol
CAS 36653-82-4; EINECS 253-149-0
Uses: Coemulsifier, lubricant, foam control agent, cosolvent, plasticizer, stabilizer, emollient, intermediate; for metal lubricants, inks, textiles, emulsions, paper, cosmetics, mineral processing, oil field chemicals, fabric softeners; emollient, emulsion stabilizer, visc. modifier, conditioner for skin care prods.; opacifier for creams and lotions
Properties: APHA 50 max. flake; sol. in fatty alcohols, IPA, benzene, trichlorethylene, acetone, turpentine, VMP naphtha, kerosene, lt. min. oil; m.w. 246; sp.gr. 0.815 (60/25 C); m.p. 45-50 C; solid pt. 48-53 C; acid no. 1.0 max.; sapon. no. 2.0 max.; 100% conc.; nonionic
Discontinued prod.

Adol® 620 NF [Witco Corp.]
Chem. Descrip.: Stearyl alcohol
CAS 112-92-5; EINECS 204-017-6
Uses: Coemulsifier, lubricant, foam control agent, cosolvent, plasticizer, stabilizer, emollient, intermediate; for metal lubricants, inks, textiles, emulsions, paper, cosmetics, mineral processing, oil field chemicals, fabric softeners; emollient, emulsion stabilizer, visc. modifier, conditioner for skin care prods.; opacifier for creams and lotions
Properties: APHA 50 max. flake; sol. in fatty alcohols, IPA, benzene, trichlorethylene, acetone, turpentine, VMP naphtha, kerosene, lt. min. oil; m.w. 267; sp.gr. 0.817 (60/25 C); m.p. 55-60 C; acid no. 2.0 max.; sapon. no. 2.0 max.; 100% conc.; nonionic
Discontinued prod.

Adol® 640 [Witco Corp.]
Chem. Descrip.: Cetearyl alcohol
Uses: Emulsifier, lubricant, foam control agent, cosolvent, plasticizer, emollient
Properties: APHA 40 max. flake; m.w. 253; sp.gr. 0.82 (50/50 C); m.p. 43-46 C; acid no. 1.0 max.; iodine no. 1.5 max.; sapon. no. 1.0 max.; 100% conc.; nonionic
Discontinued prod.

Agent #5032 [Polymer Research Corp. of Am.]
Uses: Antitarnishing agent for pigment roller printing; solvent, lubricant, antistat, conditioner for textile fibers
Properties: Water-wh. clear liq.; dens. 9.16 lb/gal (55 C); 25% act. in water
Discontinued prod.

Agnosol 3 [Croda Chem. Ltd]
Chem. Descrip.: PEG-3 lanolate
CAS 68459-50-7
Uses: Emulsifier, solvent
Discontinued prod.

Agnosol 5 [Croda Chem. Ltd]
Chem. Descrip.: PEG-5 lanolate
CAS 68459-50-7
Uses: Emulsifier, solvent
Unverified

Agrisol PX401 [Akcros UK]
Chem. Descrip.: Aromatic alkoxylates
Uses: Cosolvent, penetrant, flow promoter for agrochemical toxicants
Properties: Clear liq., faint odor; insol. in water; sp.gr. 1.098; visc. 27 cs; pour pt. -15 C; flash pt. (COC) > 100 C; pH 7.0 (1% aq.); 100% act.; nonionic
Toxicology: Damaging to eyes

Agrisol PX413 [Akcros UK]
Chem. Descrip.: Aromatic alkoxylate
Uses: Cosolvent, penetrant, flow promoter for formulation of agrochemical toxicants
Properties: Clear liq., faint odor; sol. in water; sp.gr. 1.121; visc. 64 cs; pour pt. -6 C; flash pt. (COC) > 150 C; pH 6.4 (1% aq.); 100% act.; nonionic

Agrisynth® BLO [ISP]
Chem. Descrip.: γ-Butyrolactone
CAS 96-48-0; EINECS 202-509-5
Uses: Solvent for PAN, PS, fluorinated hydrocarbons, cellulose triacetate, shellac; used in paint removers, petrol. processing, hectograph process, specialty inks; intermediate for aliphatic and cyclic compds.; reaction and diluent solvent for pesticides; used in dyeing of acetate; wetting agent for cellulose acetate films, fibers, solvent welding of plastic films in adhesive applics.

Agsol Ex BLO [ISP]
Chem. Descrip.: γ-Butyrolactone
CAS 96-48-0; EINECS 202-509-5
Uses: Solvent for PAN, PS, fluorinated hydrocarbons, cellulose triacetate, shellac; used in paint removers, petrol. processing, hectograph process, specialty inks; intermediate for aliphatic and cyclic compds.; reaction and diluent solvent for pesticides; used in dyeing of acetate; wetting agent for cellulose acetate films, fibers, solvent welding of plastic films in adhesive applics.

Albrite® TBP [Albright & Wilson Am.]
Chem. Descrip.: Tributyl phosphate
CAS 126-73-8; EINECS 204-800-2
Uses: Plasticizer for CR, CPE, Hypalon elastomers, PVC, PVAc, cellulosics, PS, ABS, ink systems;

lubricant additive; antifoam; ink solvent; paints and coatings additive; low odor

Aldo® MCT KFG [Lonza]
Chem. Descrip.: Caprylic/capric triglyceride
CAS 73398-61-5; EINECS 265-724-3
Uses: Dietary supplement; dense source of calories; used to produce energy; emollient in personal care prods.; carrier or suspension medium for oil-sol. antibiotics and drugs; lubricant for gel capsules, confectionery coatings, antisticking agent; carrier for spray coatings; solvent, diluent, and flavor fixative
Regulatory: Kosher
Properties: Clear liq., neutral odor and taste; acid no. 0.10 max.; iodine no. 1.0 max.; sapon. no. 335-360; hyd. no. 10 max.; 0.1% max. moisture

Alkasurf® OD-410 [Rhône-Poulenc Surf. Canada]
Chem. Descrip.: Phenol ethoxylate (1-1.5 EO)
Uses: Solvent for vinyl, phenolic, polyester, alkyd, cellulose acetate, nitrocellulose, and other resins; used in hot caustic paint strippers, solv.-based cleaners and inks; nonionic

Ameroxol® OE-2 [Amerchol; Amerchol Europe]
Chem. Descrip.: Oleth-2
CAS 9004-98-2
Uses: Solubilizer, emulsifier, dispersant, stabilizer, lipophilic cosolvent for creams and lotions, shampoos, and detergents, fluid and gelled transparent emulsions, fragrance prods., and aerosols
Properties: Pale straw-colored clear liq.; bland odor; sol. in min. oil, isopropyl esters, anhyd. ethanol; HLB 5.0; acid no. 0.2 max.; sapon. no. 2 max.; pH 4.5-7.0; 100% conc.; nonionic
Discontinued prod.

Amgard® TBPO4 [Albright & Wilson Am.]
Chem. Descrip.: Tributyl phosphate
CAS 126-73-8; EINECS 204-800-2
Uses: Flame retardant plasticizer in plastics and coatings; solvent extractant; in hydraulic fluids because of its low flamm. props.; 11.6% P

Antarox® P-400 [Rhône-Poulenc Surf. & Spec.]
Chem. Descrip.: PPG
Uses: Component in hydraulic fluids; solvent and cosolvent for cosmetics, pharmaceuticals, industrial applics.; chem. intermediates
Properties: Liq.; pour pt. < -25 C; 100% conc.; nonionic

Antarox® P-700 [Rhône-Poulenc Surf. & Spec.]
Chem. Descrip.: PPG
Uses: Component in hydraulic fluids; solvent and cosolvent for cosmetics, pharmaceuticals, industrial applics.; chem. intermediates
Properties: Liq.; pour pt. < -25 C; 100% conc.; nonionic

Antiblaze® DMMP [Albright & Wilson Am.]
Chem. Descrip.: Dimethyl methylphosphonate
CAS 756-79-6; EINECS 212-052-3
Uses: Flame retardant for rigid urethane foams, unsat. polyester resins, and water-borne latexes; also as a solvent
Properties: Liq.; 25% P

Arcal™ AAP [ARCO]
Chem. Descrip.: Allyl alcohol propoxylate
CAS 9042-19-7
Uses: Functional intermediate for modification of polyester epoxy, silicone, and vinyl resins; for UV/EB or peroxide cure resins, silicone caulks and surfactants, vinyl coatings, polyurethane coatings, adhesive resins, durable floor and furniture coatings; can formulate VOC compliant coatings
Properties: APHA < 50 color; sol. 14% in water; m.w. 150 ± 10; dens. 0.933 g/cc; visc. 5 cps; b.p. 156 C-190 C; hyd. no. 370-380 mg KOH/g; flash pt. (Seta) 146 F; < 0.5% water
Toxicology: LD50 (oral) 1.1 g/kg; LD50 (skin) > 2 g/kg; severe eye irritant
Precaution: Combustible; incompat.

with strong acids and oxidants
Storage: Store under nitrogen to preserve prod. purity

Arconate® 1000 [ARCO]
Chem. Descrip.: Propylene carbonate
CAS 108-32-7; EINECS 203-572-1
Uses: Solvent with high boiling pt., low toxicity, broad range of applics.; reactive diluent for woodbinders, urethane foams and coatings, foundry sand binders, in textile and syn. fiber industry, natural gas treating; lubricant in cosmetics
Properties: Amber liq.; mild ether-like odor; m.w. 102.09; sp.gr. 1.203-1.210 (20/20 C); dens. 10 lb/gal (20 C); visc. 2.4 cps; vapor pressure 0.03 mm Hg (20 C); f.p. -50 C; i.b.p. 170 C; decomp. pt. 320 F; flash pt. (Seta) 226 C; ref. index 1.419; pH 4.5-6.5 (10% aq.); 99% min. assay

Arconate® HP [ARCO]
Chem. Descrip.: Propylene carbonate
CAS 108-32-7; EINECS 203-572-1
Uses: High purity solvent with high boiling pt., low toxicity, broad range of applics.; reactive diluent for woodbinders, urethane foams and coatings, foundry sand binders, in textile and syn. fiber industry, natural gas treating; lubricant in cosmetics
Properties: Crystal clear, mild odor; sp.gr. 1.203-1.210 (20/20 C); dens. 10.1 lb/gal (20 C); visc. 4 cs; vapor pressure 0.02 mm Hg (20 C); f.p. -50 C; b.p. 242 C; flash pt. (TOC) 132 C; ref. index 1.419; pH 6.5-7.5 (10% aq.); 99.6% min. assay

Arconate® PC [ARCO]
Chem. Descrip.: Propylene carbonate
CAS 108-32-7; EINECS 203-572-1
Uses: Solvent with high boiling pt., low toxicity, broad range of applics.; reactive diluent for woodbinders, urethane foams/coatings/elastomers, foundry sand binders, in textile and syn. fiber industry, natural gas treating, petrol. fraction extraction; in surf. coatings, electrolytes, lubricants for cosmetics; also as accelerator, reactive intermediate, compatibilizer, dispersant, wetting agent, or scavenger in resin systems
Regulatory: Exempt from Rule 66
Properties: Sol. 8% in water (20 C); m.w. 102.07; sp.gr. 1.203 (20/20 C); dens. 10.0 lb/gal (20 C); visc. 6.6 cs (0 F); vapor pressure < 0.1 mm Hg (20 C); f.p. -50 C; b.p. 242 C (760 mm); flash pt. (TOC) 270 F; ref. index 1.421 (20 C)
Toxicology: Pract. nontoxic based on oral and dermal acute testing; moderate eye irritant, sl. skin irritant
Precaution: May decompose in strong acid or base

Arcosolv® DPM [ARCO]
Chem. Descrip.: Dipropylene glycol monomethyl ether
CAS 34590-94-8
Uses: Solvent for coatings, cleaners, inks, agric. prods., cosmetics, chemical intermediate applics., epoxy laminates, adhesives, floor polish, fuel additives, oil field, mining, and electronic chemicals
Properties: APHA 15 max. liq.; mild, pleasant char. odor; sol. 100% in water (20 C); misc. with a number of org. solvs.; m.w. 148.2; sp.gr. 0.950-0.953; dens. 7.91 lb/gal; visc. 3.6 cstk; vapor pressure 0.17 mm Hg (20 C); f.p. -80 C; b.p. 188.3 (760 mm); flash pt. (Seta) 167 F; ref. index 1.422; surf. tens. 28.2 dynes/cm; sp. heat 0.54 cal/g/°C
Precaution: Combustible

Arcosolv® DPM Acetate [ARCO]
Chem. Descrip.: Dipropylene glycol methyl ether acetate
CAS 88917-22-0
Uses: Solvent where a slow evaporating nonhydroxylic solvent is required; effective in coatings; as a coalescent in waterborne emulsion systems; also for cleaners, inks, epoxy laminates, agric., adhesives, floor polish, fuel additives, oil field, mining; electronic chemicals
Properties: APHA 15 max. color; sol. 12.3% in water (20 C); m.w. 190.2; sp.gr. 0.972; dens. 8.14 lb/gal; visc. 2.2 cstk; vapor pressure < 0.10 mm Hg (20 C); b.p. 205 C (760 mm Hg);

ref. index 1.4142; flash pt. (Seta) 186 F; surf. tens. 28.3 dynes/cm

Arcosolv® DPNB [ARCO]

Chem. Descrip.: Dipropylene glycol n-butyl ether

CAS 29911-28-2

Uses: Slow evaporating solvent with good solvency for coating resins, cleaners, cosmetics, agric., electronics, ink, textile and adhesives prods.; low volatility; HAPS-free; SARA nonreportable

Properties: APHA 20 max. color; mild odor; sol. 5% in water (20 C); sp.gr. 0.908-0.915; dens. 7.6 lb/gal; visc. 4.8 cstk; vapor pressure 0.03 mm Hg (77 F); b.p. 215 C (760 mm); flash pt. (Seta) 212 F; ref. index 1.42; surf. tens. 29 dynes/cm

Arcosolv® DPNP [ARCO]

Chem. Descrip.: Dipropylene glycol n-propyl ether

CAS 29911-27-1

Uses: Solvent with reduced causticity and performance features for a range of industrial, institutional, and consumer cleaners; HAPS-free; SARA nonreportable

Properties: Sol. 18% in water (20 C); m.w. 176.3; sp.gr. 0.922 (20 C); dens. 7.7 lb/gal; visc. 4.4 cstk; vapor pressure 0.05 mm Hg (77 F); b.p. 212 C; flash pt.(Seta) 190 F; ref. index 1.422; surf. tens. 25.8 dynes/cm

Arcosolv® DPTB [ARCO]

Chem. Descrip.: Dipropylene glycol t-butyl ether

CAS 132739-31-2

Uses: Slow evaporating solvent with good solvency for coating resins; HAPS-free; SARA nonreportable

Properties: APHA 20 max. color; mild odor; sol. 12% in water (20 C); sp.gr. 0.890-0.190; dens. 7.6 lb/gal; visc. 4.9 cstk; vapor pressure 0.08 mm Hg (77 F); b.p. 200 C (760 mm); flash pt. (Seta) 190 F; ref. index 1.42; surf. tens. 26 dynes/cm

Precaution: Ether cleavage can occur in acidic formulations; stable in neutral and alkaline conditions

Arcosolv® PE [ARCO]

Chem. Descrip.: Propylene glycol monoethyl ether

CAS 52125-53-8

Uses: Fast evaporating solvent for coatings, cleaners, cosmetics, agric., electronics, ink, textile and adhesive prods.; HAPS-free; SARA nonreportable

Properties: APHA 15 max. color; mild pleasant ether-like odor; sol. 100% in water; misc. with a number of solvs.; m.w. 104.1; sp.gr. 0.890-0.905; dens. 7.5 lb/gal; visc. 2.08 cstk; vapor pressure 4.40 mm Hg (20 C); b.p. 148 C (760 mm); flash pt. (Seta) 92 F; ref. index 1.405; surf. tens. 29.7 dynes/cm

Toxicology: Low toxicity

Arcosolv® PE Acetate [ARCO]

Chem. Descrip.: Propylene glycol ethyl ether acetate

CAS 98516-30-4

Uses: Highly misc. solvent with broad solvency for coatings, cleaners, agric., cosmetics, electronics, ink, textile and adhesive prods.

Properties: APHA 15 max. color; mild odor; sol. 10% in water (20 C); m.w. 146; sp.gr. 0.942-0.948; dens. 7.5 lb/gal; visc. 1.3 cstk; vapor pressure 1.5 mm Hg (68 F); b.p. 158 C (760 mm); flash pt. (Seta) 129 F; ref. index 1.40; surf. tens. 26.3 dynes/cm

Toxicology: Low toxicity

Arcosolv® PM [ARCO]

Chem. Descrip.: Propylene glycol methyl ether

CAS 107-98-2; EINECS 203-539-1

Uses: Solvent for coatings, cleaners, inks, agric. prods., cosmetics, chemical intermediate applics., epoxy laminates, adhesives, floor polish, fuel additives, oil field, mining, and electronic chemicals

Properties: APHA 10 max. liq.; mild, pleasant char. odor; sol. 100% in water @ 20 C; misc. with a number of org. solvs.; m.w. 90.1; sp.gr. 0.918-0.921; dens. 7.65 lb/gal; visc. 1.8 cstk; vapor pressure 8.1 mm Hg (20 C); f.p. -95 C; b.p. 120.1 C (760 mm); ref. index 1.404; flash pt. (Seta) 89 F; surf. tens. 26.5 dynes/cm; sp. heat 0.57 cal/g/°C

Precaution: Flamm.

Arcosolv® PM Acetate [ARCO]

Chem. Descrip.: Propylene glycol methyl ether acetate

CAS 108-65-6; EINECS 203-603-9

Uses: Slow-evaporating solvent with good solvency for many commonly used coating resins, e.g., acrylics, NC, and urethanes; used in lacquers, water-based paints

Properties: APHA 10 max. liq.; mild, ester-like odor; sol. 18.5% in water (20 C); m.w. 132.2; sp.gr. 0.963-0.966; dens. 8.02 lb/gal; visc. 1.1 cstk; vapor pressure 3.7 mm Hg (20 C); b.p. 146.0 C (760 mm); flash pt. (Seta) 114 F; ref. index 1.400; surf. tens. 27.4 dynes/cm

Precaution: Combustible

Arcosolv® PNB [ARCO]

Chem. Descrip.: Propylene glycol n-butyl ether

CAS 5131-66-8; EINECS 225-878-4

Uses: Solvent with good coupling and solvency for coating resins, cleaners, cosmetics, agric., electronics, ink, textile and adhesives prods.; HAPS-free; SARA nonreportable

Properties: APHA 15 max. color; low ether-like odor; sol. 6.4% in water (20 C); sp.gr. 0.874-0.881; dens. 7.3 lb/gal; visc. 3.5 cstk; vapor pressure 0.92 mm Hg (77 F); b.p. 160 C (760 mm); flash pt. (Seta) 138 F; ref. index 1.42; surf. tens. 26 dynes/cm

Precaution: Combustible

Arcosolv® PNP [ARCO]

Chem. Descrip.: Propylene glycol n-propyl ether

CAS 1569-01-3

Uses: Solvent with exc. coalescing props. and low volatility for coatings, cleaners, cosmetics, agric., electronics, ink, textile and adhesives prods.; HAPS-free; SARA nonreportable

Properties: APHA 10 max. color; low odor; sol. 100% in water; m.w. 118.2; sp.gr. 0.880-0.890; dens. 7.4 lb/gal; visc. 2.3 cstk; vapor pressure 2.5 mm Hg (77 F); b.p. 140 C (760 mm); flash pt. (Seta) 119 F; ref. index 1.410; surf. tens. 27 dynes/cm

Precaution: Flamm.

Arcosolv® PTB [ARCO]

Chem. Descrip.: Propylene glycol mono-t-butyl ether

CAS 57018-52-7

Uses: Solvent offering a blend of hydrophobicity and hydrophilicity for coatings, cleaners, electronic, and ink applics.; strong coupling ability; used in lt.-duty and hard-surf. cleaners, water-reducible polyester and alkyd resin prod.; chemical intermediate in the synthesis of monomeric and polymeric prods.; cosolvent

Properties: APHA 15 max. liq.; sol. 17.5% in water (20 C); m.w. 132.2; sp.gr. 0.870-0.874; visc. 3.8 cstk; vapor pressure 1.9 mm Hg (20 C); f.p. -56 C; b.p. 151 C (760 mm Hg); flash pt. (Seta) 113 F; ref. index 1.4116; surf. tens. 24.2 dynes/cm

Toxicology: LD50 (oral) 3771 mg/kg, (dermal) > 2.0 g/kg; sl. toxic orally; mildly irritating to skin, severe-but-reversible eye irritant in pure form

Precaution: Ether cleavage can occur in acidic formulations; stable in neutral and alkaline conditions

Arcosolv® TPM [ARCO]

Chem. Descrip.: PPG-3 methyl ether

CAS 25498-49-1

Uses: Slow evaporating solvent for coatings, cleaners, inks, agric. prods., cosmetics, chemical intermediate applics., epoxy laminates, adhesives, floor polish, fuel additives, oil field, mining, and electronic chemicals

Properties: APHA 15 max. liq.; mild, pleasant char. odor; sol. 100% in water (20 C); misc. with a number of org. solvs.; m.w. 206.3; sp.gr. 0.962-0.965; dens. 8.03 lb/gal; visc. 5.8 cstk; vapor pressure 0.03 mm Hg (20 C); f.p. -79 C; b.p. 242.4 C (760 mm); ref. index 1.430; flash pt. (Seta) 232 F; surf. tens. 29.0 dynes/cm; sp. heat 0.51 cal/g/°C

Arcosolv® TPNB [ARCO]

Chem. Descrip.: Tripropylene glycol n-butyl ether

Uses: Versatile and low toxicity solvent for use in coatings systems such as aerosol, automotive, industrial maintenance, metal containers, plastic substrates, architectural, coil, machinery & equip., metal furniture & fixtures, wood furniture

Properties: Sol. 3% in water (20 C); sp.gr. 0.934; vapor pressure < 0.01 mm Hg (20 C); b.p. 276 C; flash pt. (Seta) 276 F; surf. tens. 29.9 dynes/cm

Storage: Avoid contact with air when storing for extended periods; do not allow contact with copper or copper alloys, aluminum or aluminum alloys, or galvanized steel

Argobase 125T [Westbrook Lanolin]

Chem. Descrip.: Lanolin alcohol, min. oil, and octyldodecanol with 150 ppm BHT (antioxidant)

Uses: Liq. absorption base; enhances appearance and elegance of cosmetics and pharmaceuticals, skin and hair care prods.; stabilizer for w/o or o/w emulsions, liq. make-up, pigmented systems; plasticizer in hair sprays; solvent in aerosol prods.

Properties: Almost colorless to pale yel. clear oily liq., faint char. odor; sol. in min. oil, IPA, IPM; partly sol. in ethanol; insol. in water; sp.gr. 0.84-0.87; HLB 8.0; acid no. 1 max.; iodine no. 12 max.; sapon.no. 3 max.; hyd. no. 11-18; flash pt. > 100 C; ref. index 1.46-1.476

Toxicology: Extremely low toxicity; LD50 (oral, rat) > max. pract. dose of 16 g/kg; nonirritating to eyes and skin

Environmental: Nonhazardous

Precaution: Nonflamm., but will burn if strongly heated

Storage: 2 yrs shelf life when stored in well-closed containers under good conditions; avoid prolonged heating above 80 C; if cloudiness occurs, gently warm and stir

Arlamol® E [ICI Spec. Chems.; ICI Surf. Am.; ICI Surf. Belgium]

Chem. Descrip.: PPG-15 stearyl ether with preservatives

CAS 25231-21-4

Uses: Emollient, solvent for personal care prods.

Properties: Colorless oily liq.; sol. in min. oil, isopropyl esters, cottonseed oil, ethanol, IPA, hexadecyl alcohol; sp.gr. 0.95; visc. 80 cps; pour pt. typ. < 0 C; 100% act.; nonionic

Unverified

Arlasolve® DMI [ICI Spec. Chems.; ICI Surf. Am.; ICI Surf. UK]

Chem. Descrip.: Dimethyl isosorbide

CAS 5306-85-4; EINECS 226-159-8

Uses: Surfactant, emollient, solubilizer for personal care industry; solvent for cosmetics and pharmaceuticals

Properties: Colorless liq.; sol. in water, alcohol, cottonseed oil, propylene glycol; visc. 6 cps; nonionic

Arlox FE [Aquaness]

Uses: Aq. reactive scale solvent that rapidly dissolves iron sulfide and calcium carbonate scale from wellbore, downhole equip., and formation; evolves carbon dioxide gas, providing agitation, and resulting in faster scale removal; sol. calcium chelate and sol. sulfate salt are reaction prods.; acidizing not necessary following use; may be used as preflush in sour wells ahead of corrosion inhibitor batch treatments

Use Level: 50-100 ppm (initially); 5-15 ppm (continuous)

Properties: Liq.; sol. in water, insol. in isopropanol, heavy aromatic solv.; dens. 11.49 lb/gal; pour pt. -20 F; flash pt. (TCC) > 200 F; pH 3.5-4.5

Aromatic 100 [Exxon]

Chem. Descrip.: Aromatic hydrocarbon fluid

CAS 64742-95-6

Uses: Solvent, diluent for coatings, printing inks, agrochems.; very low HAPS; slow evaporation rates for good flow and leveling; high elec. resist.; good solvency; cosolvent as coalescer for acrylic emulsions; acrylic polymerization solvent; azeotrope

Regulatory: FDA 40CFR§ 180.1001 (c)(d)(e)

Properties: Bright clear appearance; m.w. 120; sp.gr. 0.868-0.878 (15.6/15.6 C); dens. 7.27 lb/gal (60 F); visc. 0.8 cP; vapor pressure 4.4 mm Hg (20 C); i.b.p. 154 C min.; flash pt. 42 C min.; ref. index 1.4993 (20 C); surf. tens. 30 dynes/cm; KB value 94; 99.8% aromatic
Toxicology: TLV/OEL 50 ppm

Aromatic 150 [Exxon]
Chem. Descrip.: Aromatic hydrocarbon fluid
CAS 64742-94-5
Uses: Solvent, diluent for coatings, printing inks, agrochems.; very low HAPS; slow evaporation rates for good flow and leveling; high elec. resist.; good solvency; cosolvent as coalescer for acrylic emulsions; acrylic polymerization solvent; azeotrope
Regulatory: FDA 40CFR §180.1001(b)
Properties: Dens. 7.45 lb/gal (60 F); visc. 1.2 cP; vapor pressure 0.5 mm Hg (20 C); b.p. 184-205 C; flash pt. 151 F; surf. tens. 31.4 dynes/cm; KB value 97; 99.8% aromatic
Toxicology: TLV/OEL 100 ppm

Aromatic 150 [Texaco]
Chem. Descrip.: 150° Flash aromatic naphtha with narrow distillation range
Uses: High solvency, high flash pt. for paint and protective coatings, herbicide and pesticide carrier, syn. resin mfg., degreasing applics.
Properties: Sp.gr. 0.8927-0.9100 (60 F); i.b.p. 355 F min.; flash pt. 145 F min.; 96% aromatic content

Aromatic 200 [Exxon]
Chem. Descrip.: Aromatic hydrocarbon fluid
CAS 64742-94-5
Uses: Solvent, diluent for coatings, agrochems.; slow evaporation rates for good flow and leveling; high elec. resist.; good solvency; cosolvent as coalescer for acrylic emulsions; acrylic polymerization solvent; azeotrope; active tail solvent for most polyester coil coatings; controls sagging and blistering during high temp. curing; complies with criteria for LVP solvent
Regulatory: FDA 40CFR §180.1001(b)
Properties: Dens. 8.33 lb/gal (60 F); visc. 2.6 cP; vapor pressure < 0.1 mm Hg (20 C); b.p. 231-277 C; flash pt. 219 F; surf. tens. 38 dynes/cm; KB value 101; 99.8% aromatic
Toxicology: TLV/OEL 100 ppm

Arosolve 9D5R [Arol Chem. Prods.]
Chem. Descrip.: Trichlorobenzene, butyl benzoate
Uses: High exhaustion carrier for use on polyester and blends where extremely effective swelling, penetration and leveling are required
Properties: Lt. yel. clear liq.; sp.gr. 1.25 g/cc; dens. 10.43 lb/gal; pH 8.0 ± 0.3; nonionic
Discontinued prod.

Arosolve 570-HF [Arol Chem. Prods.]
Chem. Descrip.: Aromatic solvs./detergent blend
Uses: Non-red label solvent scouring agent, detergent for textiles; roller cleaner, tar remover
Properties: Water-wh. to pale yel. thin liq., solv.-type odor; disp. in water; sp.gr. 0.88 g/cc; dens. 7.39 lb/gal; flash pt. (COC) 55 C; pH 8.5 ± 0.5; 100% act.; anionic/nonionic

Arosolve 570 Special [Arol Chem. Prods.]
Chem. Descrip.: Aromatic solvs./detergent blend
Uses: Solvent scouring agent for textiles; roller cleaner; tar remover
Properties: Water-wh. to pale yel. thin liq., solv.-type odor; disp. in water; dens. 7.39 lb/gal; flash pt. (COC) 117 F; pH 8.5 ± 0.5; 100% act.; anionic/nonionic
Precaution: Combustible

Arosolve B-950 [Arol Chem. Prods.]
Chem. Descrip.: Chlorinated solvs./emulsifiers blend
Uses: Solvent scour for removal of grease, graphite and oil stains on polyester, nylon and other syn. and natural fibers; degreaser and tar remover; stable to most acids, alkalies, metal salts
Use Level: 2-6% ow
Properties: Lt. straw liq.; pH 7.5 (0.5%); 100% act.; anionic/nonionic

Arosolve CON [Arol Chem. Prods.]
Chem. Descrip.: Chlorinated aromatic solvs./alcohols/emulsifiers blend
Uses: Scouring agent for natural and syn. fibers for batch processing in becks or jigs; removes oil, sizes, and waxes for subsequent dyeing
Use Level: 1-3% ow
Properties: Yel. to amber clear liq., chlorinated aromatic-type odor; sol. in water to wh. emulsion; sp.gr. 1.028 g/cc; dens. 8.58 lb/gal; pH 6.6-7.8; anionic/nonionic

Arosolve DBM [Arol Chem. Prods.]
Chem. Descrip.: Aromatic hydrocarbon/emulsifiers blend
Uses: Low-toxicity carrier auxiliary producing level dyeings on syn. and natural fibers
Use Level: 6-15%
Properties: Pale yel. to amber thin liq.; sol. in water forming stable blue-wh. emulion; sp.gr. 0.94 g/cc; dens. 7.84 lb/gal; flash pt. (COC) 78 C; pH 8.0 ± 0.5 (5%); anionic/nonionic

Arosolve DMI-F [Arol Chem. Prods.]
Chem. Descrip.: Solvs./detergents blend
Uses: High foaming solvent scouring agent and machine cleaner for textile industry
Properties: Water-wh. to pale straw clear thin liq., mild solv.-type odor; disp. in warm water; pH 8-9; 100% act.

Arosolve FX-6 [Arol Chem. Prods.]
Chem. Descrip.: Surfactant/solv. blend
Uses: One-bath desize, scour, and dyeing auxiliary for syn., natural and blended fabrics; anti-redeposition aid for soil and sizings
Use Level: 2-4% ow
Properties: Yel. clear liq., mild aromatic solv.-type odor; sp.gr. 0.935 g/cc; dens. 7.8 lb/gal; flash pt. (COC) 106 F; pH 7.0 ± 0.2

Arosolve IWS [Arol Chem. Prods.]
Uses: Solvent scouring agent for removal of soils, oils and sizes from cellulosics and syn. fabrics and their blends; wetting agent with good high caustic stability
Properties: Water-wh. sl. visc. liq., sl. aromatic odor; sol. in water; sp.gr. 0.925; dens. 7.72 lb/gal; pH 7.2 ± 0.3; nonionic/anionic

Arosolve MN-LF [Arol Chem. Prods.]
Chem. Descrip.: Emulsified aromatic naphthas
CAS 1321-94-4
Uses: Low foaming polyester dye carrier for use in jets and other high-pressure dyeing equip.; produces level and bright shades on polyester
Properties: Pale yel. to amber liq., mild aromatic odor; sp.gr. 1.002 g/cc; dens. 8.36 lb/gal; flash pt. (PMCC) 190 F; pH 8.0-8.3; 82% volatile

Arosolve MRC-A [Arol Chem. Prods.]
Chem. Descrip.: Aromatic and oxygenated solvs./detergents blend
Uses: Roller cleaner
Properties: Pale yel. clear thin liq., mild solv. and ammonia odor; disp. in water; 95+% act.; nonionic/anionic

Arosolve MRC-HF [Arol Chem. Prods.]
Chem. Descrip.: Aromatic and oxygenated solvs./detergents blend
Uses: Roller cleaner
Properties: Pale yel. clear thin liq., mild solv. odor; disp. in water; 95+% act.; nonionic/anionic

Arosolve RCB [Arol Chem. Prods.]
Chem. Descrip.: Low-foaming emulsifiers/aromatic petrol. distillate/biphenyl blend
Uses: Solvent carrier for use on syn. fabrics and their blends; low foaming for high shear jets
Properties: Lt. yel. to pale amber liq., mild naphthenic odor; sol. in warm water to stable blue-wh. emulsion; dens. 8.15 lb/gal; flash pt. (PMCC) 130 F; pH 6-7; anionic/nonionic
Precaution: Combustible

Arosolve SH-180 LF [Arol Chem. Prods.]
Chem. Descrip.: Solvs./low foaming detergent blend
Uses: Scouring agent and equip. cleaner for rubber rollers; tar remover; dye dispersant; stable to inorg. builders, alkalies, and aqua ammonia

Properties: Water-wh. to pale yel. thin liq., mild odor; misc. with water; sp.gr. 0.93; flash pt. (COC) 204 F; pH 7.7 ± 0.3; 95+% act.

Arosolve TMB [Arol Chem. Prods.]
Chem. Descrip.: Chlorinated/aromatic hydrocarbons/aromatic ester blend
Uses: Polyester dye carrier for pressure jet procedures
Properties: Lt. amber clear thin liq., mild chlorinated odor; disp. in warm water forming stable low foaming emulsions; pH 7-8

Arosolve VNO-LF [Arol Chem. Prods.]
Uses: Polyester dye carrier imparting levelness and depth of shade to syn. fabrics dyed with disperse dyes; low foaming
Properties: Pale yel. to lt. amber thin liq., mild aromatic odor; sp.gr. 1.05 g/cc; dens. 8.76 lb/gal; flash pt. 115 F; pH 6.6
Precaution: Combustible

Arosolve XNF-1 [Arol Chem. Prods.]
Chem. Descrip.: Solvs./low foaming detergent blend
Uses: Low foaming pressure jet solvent scour for difficult grease, graphite and oil stains on polyester, nylon and other syn. and natural fibers; general degreaser and tar remover; stable to most acids and alkalies
Use Level: 2-6% ow
Properties: Clear liq.; disp. in warm water; sp.gr. 0.91; dens. 7.6 lb/gal; pH 7.5-8.0; 95+% act.; nonionic

Axol® E 61 [Goldschmidt AG]
Chem. Descrip.: Acetylated hydrog. lard glyceride
CAS 8029-91-2
Uses: Food emulsifier, lubricant, solvent, plasticizer and coating material for foodstuffs (meat prods., cheese, dried fruit, nuts), and cosmetics; can be combined with hydrophilic emulsifiers to increase whipping props. in shortenings/topping
Regulatory: EU E472a, FAO/WHO compliance
Properties: Ivory waxy slabs, faint char. odor; misc. with fats and oils; insol. in water; m.p. 34-38 C; HLB 2-3; iodine no. 3 max.; sapon. no. 310-340; hyd. no. 75-95; flash pt. > 100 C; 100% conc.; nonionic
Toxicology: Nonhazardous

Axol® E 66 [Goldschmidt AG]
Chem. Descrip.: Acetic acid ester of mono/diglycerides
CAS 68990-55-6
Uses: Lubricant, solubilizer, solvent, and plasticizer for foodstuffs and cosmetics; solubilizer for spice oils, essential oils, antioxidants; plasticizer for fats
Regulatory: EU E472a compliance
Properties: Pale yel. liq., faint char. odor; misc. with fats and oils; insol. in water; sp.gr. 0.900; HLB 2-3; iodine no. ≈ 30; sapon. no. 360-390; hyd. no. 20 max.; cloud pt. 10-15 C; flash pt. > 100 C; 100% conc.; nonionic
Toxicology: Nonhazardous

B

Bernel® Ester DOM [Bernel]
Chem. Descrip.: Dioctyl maleate
Uses: Emollient, oxybenzone solubilizer, cleanser and wax solvent; noncomedogenic; imparts shine in hair conditioners
Regulatory: FDA, EU, and Japanese compliances
Properties: Colorless liq.; sol. in oil

Berol 02 [Akzo Nobel Surf. Chem. AB]
Chem. Descrip.: Nonylphenol ethoxylate
CAS 68412-54-4
Uses: Emulsifier; solvent cleaner; emulsion polymerization
Properties: Hazen < 200 clear liq.; sol. in ethanol, xylene, trichloroethylene, lt. fuel oil; disp. in water, paraffin oil, wh. spirit; sp.gr. 1.04; visc. 350 cps; HLB 10.9; cloud pt. 60-70 C; flash pt. > 100 C; pour pt. < 0 C; surf. tens. 31 dynes/cm; pH 6-7 (1% aq.); surf. tens. 31 mN/m (0.1%); Draves wetting 1.6 g/l (25 s); Ross-Miles foam 20 mm (initial, 0.05%, 50 C); 100% conc.; nonionic
Environmental: Biodeg.

BLO® [ISP]
Chem. Descrip.: γ-Butyrolactone
CAS 96-48-0; EINECS 202-509-5
Uses: Solvent for PAN, PS, fluorinated hydrocarbons, cellulose triacetate, shellac; used in paint removers, petrol. processing, hectograph process, specialty inks; intermediate for aliphatic and cyclic compds.; reaction and diluent solvent for pesticides; used in dyeing of acetate; wetting agent for cellulose acetate films, fibers, solvent welding of plastic films in adhesive applics.; APHA 40 clear liq.; sol. in acetone, benzene, CCl_4, ethyl ether, methanol, chlorobenzene, water; m.w. 86; sp.gr. 1.124; visc. 1.7 cp; f.p. -44 C; b.p. 204 C; flash pt. (TCC) 93 C; fire pt. 99 C; ref. index 1.435; pH 4.5 (10% aq.); sp. heat 0.40 cal/g/°C; dielec. const. 39 (20 C)
Toxicology: LD50 (oral, rat) 1.5 cc/kg; severe eye irritant; nonprimary skin irritant, nonsensitizer; noncarcinogenic

Butoxyethoxy Propasol [Union Carbide]
Uses: Solvent for coatings
Properties: Sol. in water (20 C); m.w. 176.26; sp.gr. 0.938 (20/20 C); dens. 7.81 lb/gal (20 C); b.p. 230.0 C; flash pt. (CC) 197 F

Butoxyl® [Hoechst AG]
Chem. Descrip.: 3-Methoxy-n-butyl acetate
CAS 4435-53-4; EINECS 224-644-9
UN No.: 2708
Uses: Solvent for coatings, esp. brush-applied paints; also for rubber roller-applied paints, spray paints, stoving enamels, high-solids paints; solvent in isocyanate- and epoxy-contg. systems
Properties: Colorless liq.; faint odor; misc. with common org. solvs.; very low sol. in water; sp.gr. 0.954-0.956 (20 C); visc. 1.43 mPa·s (20 C); vapor pressure 1.46 mbar (20 C); f.p. < -80 C; b.p. 169-173 C; acid no. 0.05 max.; flash pt. 62 C; ref. index 1.408-1.410 (20 C); sp. heat 1.93 kJ/kg·K (20 C); dielec. const. 8.0 (20 C); 98.5% min.act.
Toxicology: LD50 (oral, rat) 4210 mg/kg; irritant; may cause narcotic ef-

fects in high concs.
Environmental: Good biological eliminability
Storage: Virtually unlimited storage life when stored in cool, dark place, preventing contact with air and water

Butyl Carbitol® [Union Carbide]
Chem. Descrip.: Butoxydiglycol
CAS 112-34-5; EINECS 203-961-6
Uses: Solvent, coalescing aid for water-disp., water-sol., and high-solids coatings
Properties: Sol. in water (20 C); m.w. 162.23; sp.gr. 0.954 (20/20 C); dens. 7.96 lb/gal (20 C); visc. 4.2 cP (20 C); vapor pressure 0.01 mm Hg (20 C); b.p. 230.6 C; flash pt. (CC) 214 F

Butyl Carbitol® Acetate [Union Carbide]
Chem. Descrip.: Diethylene glycol butyl ether acetate
CAS 124-17-4; EINECS 204-685-9
Uses: Solvent for high-solids coatings, NC emulsions; coalescing aid for water-disp. coatings
Properties: Sol. 6.5% in water (20 C); m.w. 204.27; sp.gr. 0.980 (20/20 C); dens. 8.16 lb/gal (20 C); visc. 3.6 cP (20 C); vapor pressure 0.04 mm Hg (20 C); b.p. 246.7 C; flash pt. (CC) 221 F

Butyl Cellosolve® [Union Carbide]
Chem. Descrip.: Butoxyethanol
CAS 111-76-2; EINECS 203-905-0
Uses: Solvent for high-solids coatings (NC, butyrate, ethyl cellulose); coalescing aid for water-disp. and water-sol. coatings
Properties: Sol. in water (20 C); m.w. 118.18; sp.gr. 0.902 (20/20 C); dens. 7.51 lb/gal (20 C); visc. 3.8 cP (20 C); vapor pressure 0.6 mm Hg (20 C); b.p. 171.2 C; flash pt. (CC) 150 F

Butyl Cellosolve® Acetate [Union Carbide]
Chem. Descrip.: Butoxyethanol acetate
CAS 112-07-2
Uses: Solvent for high-solids coatings (NC, butyrate, ethyl cellulose acrylic); coalescing aid for water-disp. coatings
Properties: Sol. 1.5% in water (20 C); m.w. 160.21; sp.gr. 0.942 (20/20 C); dens. 7.84 lb/gal (20 C); visc. 1.8 cP (20 C); vapor pressure 0.9 mm Hg (20 C); b.p. 192.3 C; flash pt. (CC) 165 F

Butyl Diglyme [Ferro/Grant]
Chem. Descrip.: Diethylene glycol dibutyl ether
CAS 112-73-2; EINECS 204-001-9
Uses: Solvent which tends to solvate cations; used in electrochemistry, polymer and boron chemistry; physical processes such as gas absorption, extraction, stabilization; used in industrial prods. such as fuels, lubricants, textiles, pharmaceuticals, pesticides; solvent, drug carrier, stabilizer in pharmaceuticals
Properties: Colorless clear liq.; mild nonresidual odor; sol. 0.3% in water; misc. with ethanol, acetone, benzene, diethyl ether, octane; m.w. 218.34; sp.gr. 0.8814; dens. 7.36 lb/gal; visc. 2.4 cP; f.p. -60.2 C; b.p. 256 C; flash pt. (CC) 118 C; ref. index 1.4235; pH neutral; surf. tens. 27 dynes/cm (20 C); sp. heat 0.495 cal/g/°C; 98.5% min. purity
Environmental: LD50 3900 mg/kg; low to mod. acute toxicity; chronic exposure may cause reproductive effects
Precaution: Slowly biodeg.

Butyl Di-Icinol [ICI Australia]
Chem. Descrip.: 2-(2-Butoxyethoxy)-ethanol
CAS 112-34-5; EINECS 203-961-6
Uses: Solvent for use in protective coatings, inks, cleaning prods., agric. chems.; aids wetting, penetration, and soil removal; coupling solvent
Unverified

Butyl Dioxitol® [Shell]
Chem. Descrip.: Butoxydiglycol
CAS 112-34-5; EINECS 203-961-6
Uses: Solvent for nitrocellulose, phenolics, Epon resins, alkyds, and acrylics, oils, and dyes; used in lacquers and inks which require slow evaporating solvent; lacquer and enamel formulations requiring improved flow-out and gloss; coupling solvent in preparation of spe-

cialized cleaning sol'ns. and cutting oils; coalescing agent in latex paints; component of brake fluids in automotive industry
Properties: Colorless liq.; mild odor; water-misc.; sp.gr. 0.949-0.952; 0.10% max. water

Butyl Dipropasol [Union Carbide]
Uses: Solvent, coalescing aid for water-disp. and water-sol. coatings
Properties: Sol. 3% in water (20 C); m.w. 190.30; sp.gr. 0.915 (20/20 C); dens. 7.63 lb/gal (20 C); b.p. 227.9 C; flash pt. (CC) 230 F

Butyl Icinol [ICI Australia]
Chem. Descrip.: 2-Butoxyethanol
CAS 111-76-1; EINECS 203-905-0
Uses: Solvent for lacquers, enamels, water-borne systems, inks, industrial cleaners, waterless hand cleaners
Properties: Sp.gr. 0.901-0.904; visc. 6.4 cps; b.p. 166 C; flash pt. (PMCC) 68 C; ref. index 1.419
Unverified

Butyl Oxitol® [Shell]
Chem. Descrip.: Butoxyethanol
CAS 111-76-2; EINECS 203-905-0
Uses: Solvent for nitrocellulose, alkyds, phenolics, acrylics, Epon resins, nat. resins, dyes, waxes, oils, and org. materials; used in surf. coating formulations such as lacquers and enamels where it imparts gloss and improved leveling chars; coupling agent in cleaners and cutting oils
Properties: Colorless liq.; mild odor; water-misc.; sp.gr. 0.898-0.901; 0.20% max. water

Butyl Propasol [Union Carbide]
Chem. Descrip.: Propylene glycol t-butyl ether
CAS 57018-52-7
Uses: Coalescing aid, solvent for water-disp. and water-sol. coatings
Properties: Sol. 6.4% in water (20 C); m.w. 132.20; sp.gr. 0.880 (20/20 C); dens. 7.37 lb/gal (20 C); b.p. 170.2 C; flash pt. (CC) 138 F

BVA ND-205 Solv. [Ridge Tech.]
Chem. Descrip.: Highly refined naphthalene-depleted aromatic hydrocarbon solv.
CAS 64742-94-5
Uses: Solvent for textile, agric., resin, industrial degreasing, and fuel oil additive applics.; exc. solvency
Regulatory: EPA clearance
Properties: Pale yel. clear liq., typ. aromatic odor; sp.gr. 0.99; dens. 8.24 lb/gal; visc. 6.1 cs (32 F); i.b.p. 440 F; pour pt. -10 F; flash pt. (TCC) 215 F; > 98% total aromatics

C

Capmul® MCM [ABITEC]
Chem. Descrip.: Glyceryl caprylate/caprate
Uses: Solvent, dispersant, emulsifier, solubilizer, vehicle/carrier, penetrant with bacteriostatic effects for cosmetics, pharmaceuticals, nutrition; used in aerosols, clinical nutrition, coating, delivery/absorp. enhancement, dermatologicals, microemulsions, nutritional/sports supplements, suppositories
Properties: Lovibond 4R max. liq. to semisolid; sol. in oil and alcohol; HLB 5.5-6.0; acid no. 2.5 max.; iodine no. 2 max.; 55% min. alpha mono

Capmul® MCM-90 [ABITEC]
Chem. Descrip.: Glyceryl caprylate/caprate
Uses: W/o emulsifier, solvent for cosmetics, pharmaceuticals
Properties: Liq.; acid no. 2.5 max.; iodine no. 2 max.; 80+% alpha mono

Capmul® POE-O [ABITEC]
Chem. Descrip.: Polysorbate 80
CAS 9005-65-6
Uses: Food emulsifier for frozen desserts; solubilizer for oils into water systems; used for flavors, fragrances, vitamins, pharmaceuticals, solvent
Properties: Gardner 6 max. liq.; water-sol.; HLB 15.0; acid no. 2 max.; sapon. no. 45-55; hyd. no. 65-80; 100% conc.; nonionic
Discontinued prod.

Captex® 200 [ABITEC]
Chem. Descrip.: Propylene glycol dicaprylate/dicaprate; conforms to hydrog. veg. oil NF
CAS 68583-51-7
Uses: Carrier, coupler, solvent for flavors, fragrance oil, sol. colorants, vitamins, pharmaceuticals, medicinals, cosmetics, foods; emollient for creams, lotions, makeup
Properties: APHA 100 liq.; sol. in alcohol, oils, hydrocarbons, ketones; visc. 7-13 mPa·s; acid no. 0.1 max.; iodine no. 1.0 max.; sapon. no. 315-335; cloud pt. < -20 C; ref. index 1.4393

Captex® 300 [ABITEC]
Chem. Descrip.: Caprylic/capric triglyceride
CAS 65381-09-1; EINECS 265-724-3
Uses: Solvent for colors and perfumes; emollient, moisturizer in cosmetics, toiletries, pharmaceuticals; plasticizer; nutritional applics.
Properties: Gardner 2 max. liq., bland odor and flavor; sol. in alcohol, oils, hydrocarbons, ketones; visc. 24-30 mPa·s; acid no. 0.1 max.; iodine no. 0.5; sapon. no. 335-350; cloud pt. < -5 C; ref. index 1.4481

Captex® 350 [ABITEC]
Chem. Descrip.: Caprylic/capric/lauric triglyceride; conforms to hydrog. veg. oil NF
CAS 68991-68-4
Uses: Emollient, solvent, carrier, fixative, and extender for pharmaceutical, nutritional, and cosmetic applics.
Properties: Gardner 2 max. liq., bland odor and flavor; visc. 36-42 mPa·s; acid no. 0.1 max.; iodine no. 1.52 max.; sapon. no. 290-310; cloud pt. 0 C max.; ref. index 1.4582

Captex® 810A [ABITEC]

Chem. Descrip.: Caprylic/capric/linoleic triglyceride; conforms to hydrog. veg. oil NF
CAS 67701-28-4
Uses: Emollient, solvent, fixative, and extender in pharmaceutical, nutritional, and cosmetic applics. incl. lipsticks, lip glosses, makeup bases, bath oils, aftershave lotions, aerosols; carrier for flavors and fragrances; avail. as developmental prod.
Properties: Gardner 2 max. liq.; acid no. 0.1 max.; iodine no. 25; sapon. no. 307-320

Captex® 810B [ABITEC]
Chem. Descrip.: Caprylic/capric/linoleic triglyceride
CAS 67701-28-4
Uses: Emollient, solvent, fixative, and extender in pharmaceutical, nutritional, and cosmetic applics. incl. lipsticks, lip glosses, makeup bases, bath oils, aftershave lotions, aerosols; carrier for flavors and fragrances
Properties: Gardner 2 max. color; visc. 30 cps; acid no. 0.1 max.; iodine no. 55; sapon. no. 280-296; cloud pt. 10 C
Discontinued prod.

Captex® 810C [ABITEC]
Chem. Descrip.: Caprylic/capric/linoleic triglyceride
CAS 67701-28-4
Uses: Emollient, solvent, fixative, and extender in pharmaceutical, nutritional, and cosmetic applics. incl. lipsticks, lip glosses, makeup bases, bath oils, aftershave lotions, aerosols; carrier for flavors and fragrances
Properties: Gardner 2 max. color; acid no. 0.1 max.; iodine no. 75; sapon. no. 257-275
Discontinued prod.

Captex® 810D [ABITEC]
Chem. Descrip.: Caprylic/capric/linoleic triglyceride; conforms to hydrog. veg. oil NF
CAS 67701-28-4
Uses: Emollient, solvent, fixative, and extender in pharmaceutical, nutritional, and cosmetic applics. incl. lipsticks, lip glosses, makeup bases, bath oils, aftershave lotions, aerosols; carrier for flavors and fragrances; avail. as developmental prod.
Properties: Gardner 2 max. liq.; acid no. 0.1 max.; iodine no. 85; sapon. no. 235-253

Captex® 910A [ABITEC]
Chem. Descrip.: Caprylic/capric/oleic triglyceride
CAS 67701-28-4
Uses: Emollient, solvent, fixative, and extender in pharmaceutical, nutritional, and cosmetic applics. incl. lipsticks, lip glosses, makeup bases, bath oils, aftershave lotions, aerosols; carrier for flavors and fragrances
Properties: Gardner 2 max. color; acid no. 0.1 max.; iodine no. 19; sapon. no. 304-318
Discontinued prod.

Captex® 910B [ABITEC]
Chem. Descrip.: Caprylic/capric/oleic triglyceride
CAS 67701-28-4
Uses: Emollient, solvent, fixative, and extender in pharmaceutical, nutritional, and cosmetic applics. incl. lipsticks, lip glosses, makeup bases, bath oils, aftershave lotions, aerosols; carrier for flavors and fragrances
Properties: Gardner 2 max. color; acid no. 0.1 max.; iodine no. 35; sapon. no. 280-296
Discontinued prod.

Captex® 910C [ABITEC]
Chem. Descrip.: Caprylic/capric/oleic triglyceride
CAS 67701-28-4
Uses: Emollient, solvent, fixative, and extender in pharmaceutical, nutritional, and cosmetic applics. incl. lipsticks, lip glosses, makeup bases, bath oils, aftershave lotions, aerosols; carrier for flavors and fragrances
Properties: Gardner 2 max. color; acid no. 0.1 max.; iodine no. 48; sapon. no. 260-275
Discontinued prod.

Captex® 910D [ABITEC]

Chem. Descrip.: Caprylic/capric/oleic triglyceride
CAS 67701-28-4
Uses: Emollient, solvent, fixative, and extender in pharmaceutical, nutritional, and cosmetic applics. incl. lipsticks, lip glosses, makeup bases, bath oils, aftershave lotions, aerosols; carrier for flavors and fragrances
Properties: Gardner 2 max. color; acid no. 0.1 max.; iodine no. 60; sapon. no. 238-250
Discontinued prod.

Captex® 8227 [ABITEC]
Chem. Descrip.: Triundecanoin
CAS 13552-80-2; EINECS 236-935-8
Uses: Emollient, solvent, fixative, and extender in pharmaceutical, nutritional, and cosmetic applics. incl. lipsticks, lip glosses, makeup bases, bath oils, aftershave lotions, aerosols; carrier for flavors and fragrances
Properties: APHA 150 solid; acid no. 0.5 max.; iodine no. 5 max.; sapon. no. 270-290; cloud pt. 21 C

Carbitol® [Union Carbide]
Chem. Descrip.: Ethoxydiglycol
CAS 111-90-0; EINECS 203-919-7
Uses: Solvent

Carbitol® PM-600 [Union Carbide]
Uses: Solvent, coalescing aid for water-disp. and water-sol. coatings
Properties: Sol. in water; sp.gr. 1.027 (20/20 C); dens. 8.55 lb/gal (20 C); b.p. 195.0 C; flash pt. (CC) 204 F

Carbitol® Acetate [Union Carbide]
Chem. Descrip.: Ethoxydiglycol acetate
CAS 112-15-2; EINECS 203-940-1
Uses: Solvent

Carbitol® Low Gravity [Union Carbide]
Chem. Descrip.: Ethoxydiglycol
CAS 111-90-0; EINECS 203-919-7
Uses: Solvent for waterborne coatings
Properties: Sol. in water (20 C); m.w. 134.18; sp.gr. 0.990 (20/20 C); dens. 8.25 lb/gal (20 C); visc. 4.4 cP (20 C); vapor pressure 0.08 mm Hg (20 C); b.p. 202.7 C; flash pt. (CC) 182 F

Carbopol® 940 [BFGoodrich]
Chem. Descrip.: Carbomer 940
Uses: Emulsifier, thickener, stabilizer, suspending agent used in cosmetics, topical pharmaceuticals, for die-casting and forging lubricants, thixotropic paints; solvent thickening with or without neutralizing
Regulatory: NF, BP, JSPI compliance; FDA 21CFR §175.105, 175.300, 175.320, 176.170, 176.180, 176.200, 177.1210, 177.2260; 40CFR § 180.1001
Properties: Wh. fluffy powd.; sol. in water, polar solvs., many nonpolar solvs. blends; sp.gr. 1.41; bulk dens. 208 kg/m^3; visc. 40,000-60,000 cps (0.5%); pH 2.5-3.0 (1% aq. disp.); 100% conc.; anionic
Toxicology: Nonirritating to skin and eyes
Environmental: Low aquatic toxicity; BOD=0, nonbiodeg.; removed with the biomass in typ. wastewater treatment
Storage: 2 yr min. shelf life stored in sealed containers, protected from moisture and extreme temps.; hygroscopic

Carbopol® 980 [BFGoodrich]
Chem. Descrip.: Carbomer 940
Uses: Efficient solvent thickener for cosmetics, topical pharmaceuticals, with or without neutralizing; for sparkling clear water or hydroalcoholic topical gels, aq. or solv. systems
Regulatory: NF, DAB, JSPI compliance; FDA 21CFR §175.105, 175.300, 175.320, 176.170, 176.180, 176.200, 177.1210, 177.2260; 40CFR §180.1001
Properties: Wh. fluffy powd.; sp.gr. 1.41; bulk dens. 208 kg/m^3; visc. 13,000-30,000 cps (0.2%), 40,000-60,000 cps (0.5%); ; pH 2.5-3.0 (1% aq. disp.)
Toxicology: Sl. irritating to skin; minimal eye irritant
Storage: 2 yr min. shelf life stored in sealed containers, protected from moisture and extreme temps.; hygroscopic

Carbowax® PEG 300 [Union Carbide]
Chem. Descrip.: PEG-6
CAS 25322-68-3; EINECS 220-045-1

Uses: Coupling agent, solvent, vehicle, humectant, lubricant, binder, base; used in adhesives, agric., ceramics, chem. intermediates, cosmetics/toiletries, electroplating/electropolishing, food processing, household prods., lubricants, metal fabrication; paints, paper, pharmaceuticals, printing, rubber and elastomers, textiles, wood processing

Properties: Water-wh. visc. liq.; sol. in water, alcohols, glycerin, glycols; m.w. 285-315; sp.gr. 1.1250; dens. 9.38 lb/gal; visc. 5.8 cSt (99 C); f.p. -15 to -8 C; hyd. no. 356-394; flash pt. (PMCC) > 180 C; ref. index 1.463; pH 4.5-7.5 (5% aq.); surf. tens. 44.5 dynes/cm

Carbowax® Sentry® PEG 300 [Union Carbide]

Chem. Descrip.: PEG-6, FCC grade
CAS 25322-68-3; EINECS 220-045-1

Uses: Coupling agent, solvent, vehicle, humectant, lubricant, binder, base, bodying agent, dispersant, conditioner for cosmetics/toiletries (antiperspirants, creams/lotions, hand cleaners, toothpaste, makeup, shampoo, shave cream, sunscreens), food processing (citrus coatings, defoamer, pkg. component), pharmaceuticals (coating, binder, plasticizer, and lubricant in tablets), as glycerin replacement

Regulatory: FDA approved

Properties: Water-wh. visc. liq.; sol. in water, alcohols, glycerin, glycols; m.w. 285-315; sp.gr. 1.1250; dens. 9.38 lb/gal; visc. 5.8 cSt (99 C); f.p. -15 to -8 C; hyd. no. 356-394; flash pt. (PMCC) > 180 C; ref. index 1.463; pH 4.5-7.5 (5% aq.); surf. tens. 44.5 dynes/cm

Carbowax® Sentry® PEG 400 [Union Carbide]

Chem. Descrip.: PEG-8, FCC grade
CAS 25322-68-3; EINECS 225-856-4

Uses: Coupling agent, solvent, vehicle, humectant, lubricant, binder, base for cosmetics/toiletries (bath oil, cologne, creams/lotions, toothpaste, hair dressing, lipsticks, makeup, makeup removers, nail polish remover, shampoo, shave cream, soap, sunscree; food processing (citrus coatings, bodying agent, dispersant, defoamer, pkg. component), pharmaceuticals (coating, binder, plasticizer, and lubricant in tablets)

Regulatory: FDA approved

Properties: Water-wh. visc. liq.; sol. in water, methanol, ethanol, acetone, trichloroethylene, Cellosolve®, Carbitol®, dibutyl phthalate, toluene; m.w. 380-420; sp.gr. 1.1254; dens. 9.39 lb/gal; visc. 7.3 cSt (99 C); f.p. 4-8 C; hyd. no. 267-295; flash pt. (PMCC) > 180 C; ref. index 1.465; pH 4.5-7.5 (5% aq.); surf. tens. 44.5 dynes/cm

Carbowax® Sentry® PEG 540 Blend [Union Carbide]

Chem. Descrip.: PEG-6 and PEG-32, FCC grade

Uses: Coupling agent, solvent, vehicle, humectant, lubricant, binder, base for cosmetics/toiletries (hair dressings, hand cleaners, ointments, antiperspirant, creams and lotions, shaving preps., sunscreens); food processing (citrus coatings, bodying agent, dispersant, defoamer, pkg. component), pharmaceuticals (coating, binder, plasticizer, and lubricant in tablets)

Regulatory: FDA approved

Properties: Wh. soft waxy solid; sol. in methylene chloride, 73% in water, 50% in trichloroethylene, 48% in methanol; m.w. 500-600; sp.gr. 1.0930; dens. 9.17 lb/gal (55 C); visc. 15.1 cSt (99 C); m.p. 38-41 C; hyd. no. 187-224; flash pt. (PMCC) > 180 C; pH 4.5-7.5 (5% aq.)

Carbowax® Sentry® PEG 600 [Union Carbide]

Chem. Descrip.: PEG-12, FCC grade
CAS 25322-68-3; EINECS 229-859-1

Uses: Coupling agent, solvent, vehicle, humectant, lubricant, binder, base for cosmetics/toiletries (antiperspirants, colognes, creams/lotions, dentifrices, lipsticks, makeup, shampoo, shave

creams, sunscreens); food processing (citrus coatings, bodying agent, dispersant, defoamer, pkg. component), pharmaceuticals (coating, binder, plasticizer, and lubricant in tablets)
Regulatory: FDA approved
Properties: Water-wh. visc. liq.; sol. in water, alcohols, glycols; m.w. 570-630; sp.gr. 1.1257; dens. 9.40 lb/gal; visc. 10.8 cSt (99 C); f.p. 20-25 C; hyd. no. 178-197; flash pt. (PMCC) > 180 C; ref. index 1.46; pH 4.5-7.5 (5% aq.); surf. tens. 44.5 dynes/cm

Carbowax® Sentry® PEG 900 [Union Carbide]
Chem. Descrip.: PEG-20, FCC grade
CAS 25322-68-3; EINECS 203-989-9
Uses: Coupling agent, solvent, vehicle, humectant, lubricant, binder, base for cosmetics/toiletries (antiperspirants, creams/lotions, shave creams); food processing (citrus coatings, bodying agent, dispersant, defoamer, pkg. component), pharmaceuticals (coating, binder, plasticizer, and lubricant in tablets)
Regulatory: FDA approved
Properties: Wh. soft waxy solid; sol. 86% in water; m.w. 855-945; sp.gr. 1.0927 (60 C); dens. 9.16 lb/gal (55 C); visc. 15.3 cSt (99 C); m.p. 32-36 C; hyd. no. 119-131; flash pt. (PMCC) > 180 C; pH 4.5-7.5 (5% aq.)

Carbowax® Sentry® PEG 1000 [Union Carbide]
Chem. Descrip.: PEG-20, FCC grade
CAS 25322-68-3; EINECS 203-989-9
Uses: Coupling agent, solvent, vehicle, humectant, lubricant, binder, base for cosmetics/toiletries (antiperspirants, creams/lotions, ointments, shave creams); food processing (citrus coatings, bodying agent, dispersant, defoamer, pkg. component), pharmaceuticals (coating, binder, plasticizer, and lubricant in tablets)
Regulatory: FDA approved
Properties: Wh. soft waxy solid; sol. 80% in water; m.w. 950-1050; sp.gr. 1.0926 (60 C); dens. 9.16 lb/gal (55 C); visc. 17.2 cSt (99 C); m.p. 37-40 C; hyd. no. 107-118; flash pt. (PMCC) > 180 C; pH 4.5-7.5 (5% aq.)

Carbowax® Sentry® PEG 1450 [Union Carbide]
Chem. Descrip.: PEG-32, FCC grade
CAS 25322-68-3; EINECS 203-989-9
Uses: Coupling agent, solvent, vehicle, humectant, lubricant, binder, base for cosmetics/toiletries (antiperspirants, creams/lotions, ointments, shave creams); food processing (citrus coatings, bodying agent, dispersant, defoamer, pkg. component), pharmaceuticals (coating, binder, plasticizer, and lubricant in tablets)
Regulatory: FDA approved
Properties: Wh. soft waxy solid or flake; sol. 72% in water; m.w. 1300-1600; sp.gr. 1.0919 (60 C); dens. 9.17 lb/gal (55 C); bulk dens. 30 lb/ft^3 (flake); visc. 26.5 cSt (99 C); m.p. 43-46 C; hyd. no. 70-86; flash pt. (PMCC) > 180 C; pH 4.5-7.5 (5% aq.)

Carbowax® Sentry® PEG 3350 [Union Carbide]
Chem. Descrip.: PEG-75, FCC grade
CAS 25322-68-3; EINECS 203-989-9
Uses: Coupling agent, solvent, vehicle, humectant, lubricant, binder, base for cosmetics/toiletries (antiperspirants, bath powds., creams/lotions, toothpaste, hair dressing, makeup, nail polish remover, soap); food processing (citrus coatings, bodying agent, dispersant, defoamer, pkg. component), pharmaceuticals (coating, binder, plasticizer, and lubricant in tablets)
Regulatory: FDA approved
Properties: Wh. hard waxy flake or powd.; sol. 67% in water; m.w. 3000-3700; sp.gr. 1.0926 (60 C); dens. 8.94 lb/gal (80 C); bulk dens. 30 lb/ft^3 (flake), 40 lb/ft^3 (powd.); visc. 90.8 cSt (99 C); m.p. 54-58 C; hyd. no. 30-37; flash pt. (PMCC) > 180 C; pH 4.5-7.5 (5% aq.)

Carbowax® Sentry® PEG 4600 [Union

Carbide]
Chem. Descrip.: PEG-100, FCC grade
CAS 25322-68-3; EINECS 203-989-9
Uses: Coupling agent, solvent, vehicle, humectant, lubricant, binder, base for cosmetics/toiletries (antiperspirants, creams/lotions); food processing (citrus coatings, bodying agent, dispersant, defoamer, pkg. component), pharmaceuticals (coating, binder, plasticizer, and lubricant in tablets)
Regulatory: FDA approved
Properties: Wh. hard waxy flake or powd.; sol. 65% in water; m.w. 4400-4800; sp.gr. 1.0926 (60 C); dens. 8.95 lb/gal (80 C); bulk dens. 30 lb/ft^3 (flake), 40 lb/ft^3 (powd.); visc. 184 cSt (99 C); m.p. 57-61 C; hyd. no. 23-26; flash pt. PMCC) > 180 C; pH 4.5-7.5 (5% aq.)

Carbowax® Sentry® PEG 8000 [Union Carbide]
Chem. Descrip.: PEG-150, FCC grade
CAS 25322-68-3; EINECS 203-989-9
Uses: Coupling agent, solvent, vehicle, humectant, lubricant, binder, base for cosmetics/toiletries (antiperspirants, creams/lotions, hair dressing, nail polish remover); food processing (citrus coatings, bodying agent, dispersant, defoamer, pkg. component), pharmaceuticals (coating, binder, plasticizer, and lubricant in tablets)
Regulatory: FDA approved
Properties: Wh. hard waxy flake or powd.; sol. 63% in water; m.w. 7000-9000; sp.gr. 1.0845 (60 C); dens. 8.96 lb/gal (80 C); bulk dens. 30 lb/ft^3 (flake), 40 lb/ft^3 (powd.); visc. 822 cSt (99 C); m.p. 60-63 C; hyd. no. 13-16; flash pt. (PMCC) > 180 C; pH 4.5-7.5 (5% aq.)

Cardolite® NC-507 [Cardolite]
Chem. Descrip.: 3-(n-Pentadecyl) phenol
CAS 501-24-6; 3158-56-3
Uses: Starting raw material for surfactants, antioxidants, anticorrosives; lubricant additive; cosolvent for insecticides, germicides; resin modifier
Properties: Tan waxy solid; sol. in org. solvs. incl. aliphatic hydrocarbons; m.w. 304; sp.gr. > 1.0; m.p. 44-48 C (760 mm); b.p. 243-249 C (10 mm); ref. index 1.494

Cardolite® NC-510 [Cardolite]
Chem. Descrip.: 3-(n-Pentadecyl) phenol
CAS 501-24-6
Uses: Starting raw material for surfactants; coupling agent for pigments and dyes; cosolvent for insecticides, germicides; resin modifier for phenolic-aldehyde, polyester, PC polymers; also in photographic industry
Properties: Sl. pink to wh. waxy solid; sol. in org. solvs. incl. aliphatic hydrocarbons; m.w. 304; sp.gr. > 1.0; m.p. 49-51 C (760 mm); b.p. 190-195 C (1 mm); ref. index 1.4750

Cardolite® NC-511 [Cardolite]
Chem. Descrip.: 3-(n-Penta-8′-decenyl) phenol
CAS 8007-24-7
Uses: Starting raw material for surfactants, antioxidants, anticorrosives; lubricant additive; cosolvent for insecticides; in drilling muds; resin modifier
Properties: Brn. to amber oily liq.; sol. in org. solvs. incl. aliphatic hydrocarbons; m.w. 302; sp.gr. 0.930 (20 C); b.p. 223-227 C (10 mm); ref. index 1.5112

Cellosolve® [Union Carbide]
Chem. Descrip.: Ethoxyethanol
CAS 110-80-5; EINECS 203-804-1
Uses: Solvent for coatings (NC, ethyl acetate, butyrate, acrylic, water-sol.)
Properties: Sol. in water (20 C); m.w. 90.12; sp.gr. 0.931 (20/20 C); dens. 7.74 lb/gal (20 C); visc. 2.3 cP (20 C); vapor pressure 4.1 mm Hg (20 C); b.p. 135.6 C; flash pt. (CC) 108 F

Cellosolve® Acetate [Union Carbide]
Chem. Descrip.: Ethoxyethanol acetate
CAS 111-15-9; EINECS 203-839-2
Uses: Solvent for coatings (NC, cellulose acetate, ethyl cellulose, butyrate, acrylic, urethane)
Properties: Sol. 22.9% in water (20 C); m.w. 132.16; sp.gr. 0.975 (20/20 C);

dens. 8.10 lb/gal (20 C); visc. 1.1 cP (20 C); vapor pressure 2.0 mm Hg (20 C); b.p. 156.3 C; flash pt. (CC) 126 F

Ceraphyl® 28 [ISP Van Dyk]
Chem. Descrip.: Cetyl lactate
CAS 35274-05-6; EINECS 252-478-7
Uses: Lubricant and emollient for cosmetics and topical pharmaceuticals; binder for pressed powds.; imparts sheen and silkiness to skin and hair; reduces tack in deodorant sticks; solvent for dyes in lipstick formulas
Properties: Wh. solid, faint char. odor; sol. @ 5% in min. oil, peanut oil, IPM, oleyl alcohol, and 95% ethanol; partly sol. in water, propylene glycol; sp.gr. 0.893-0.905; HLB 13-15; acid no. 2 max.; sapon. no. 174-189
Toxicology: LD50 (oral, rat) > 20 ml/kg; nonirritating to eyes and skin

Ceraphyl® 140 [ISP Van Dyk]
Chem. Descrip.: Decyl oleate
CAS 59231-34-4; EINECS 222-981-6
Uses: Emollient, cosolvent, slip agent, lubricant for cosmetic/pharmaceutical creams, lotions, moisturizers, skin care prods., bath oils, liq. makeup; binder for pressed powds.; pigment dispersant
Properties: Wh. to straw-colored liq., char. mild odor; sol. @ 5% in peanut oil, 95% ethanol, IPM, oleyl alcohol; insol. in water; sp.gr. 0.858-0.864; acid no. 5 max.; iodine no. 55-65; sapon. no. 130-145; ref. index 1.4540-1.4560; 100% act.; nonionic
Toxicology: LD50 (oral, rat) > 40 ml/kg; sl. eye irritant; very sl. skin irritant; nonsensitizing

Ceraphyl® 140-A [ISP Van Dyk]
Chem. Descrip.: Isodecyl oleate
CAS 59231-34-4; EINECS 261-673-6
Uses: Emollient, cosolvent, and solubilizer for cosmetic systems, topical pharmaceuticals, liq. makeup; binder for pressed powd.; wetting agent for iron oxides; cleansing agent for emulsions; lighter, drier skin feel than Ceraphyl 140
Properties: Wh. to straw-colored liq., char. mild odor; sol. @ 5% in peanut oil, 95% ethanol, IPM, oleyl alcohol; insol. in water; sp.gr. 0.858-0.864; acid no. 5 max.; iodine no. 50-65; sapon. no. 130-145; ref. index 1.4540-1.4560; 100% act.; nonionic
Toxicology: LD50 (oral, rat) > 40 ml/kg; nonirritating to eyes; mild skin irritant; nonsensitizing

Ceraphyl® ICA [ISP Van Dyk]
Chem. Descrip.: Isocetyl alcohol
CAS 36311-34-9; EINECS 252-964-9
Uses: Nongreasy emollient for creams and lotions, pigment dispersant, binder for pressed powds., hair and skin conditioner; reduces tack in deodorant sticks; solvent for lipstick dyes; carrier and extender for flavor and fragrances; ideal for oil-free cosmetics and high-pH applics.; topical pharmaceuticals
Properties: Colorless clear liq., low odor; sol. @ 5% in peanut and min. oils, 95% ethanol, IPM, oleyl alcohol, castor oil, cyclomethicone; insol. in water; m.w. 242; sp.gr. 0.830-0.840; HLB 12-14; acid no. 5 max.; iodine no. 10 max.; sapon. no. 10 max.; hyd. no. 195-230
Toxicology: LD50 (oral, rat) > 5 g/kg; mild eye and skin irritant; noncomedogenic

Cetiol® B [Henkel/Cospha; Henkel Canada; Henkel KGaA/Cospha]
Chem. Descrip.: Dibutyl adipate
CAS 105-99-7; EINECS 203-350-4
Uses: Emollient; oily component for skin oils, massage oils, sun protection oils, pharmaceuticals, bathing preps., day creams, liq. emulsions; good penetration on skin; solvent for lipoid-sol. substances; superfatting agent and plasticizer for hair sprays
Properties: Colorless oily liq., faint intrinsic odor; sp.gr. 0.958-0.962; visc. 5-7 mPa·s; solid. pt. < -30 C; acid no. 0.5 max.; iodine no. 1 max.; sapon. no. 420-440; hyd. no. 1 max.; cloud pt. < -25 C; flash pt. 150 C min.; ref. index 1.434-1.437; 100% act.
Storage: 1 yr storage if in sealed origi-

nal containers at temps. below 40 C, protected from moisture
Discontinued prod.

Chemax E-200 MO [Chemax]
Chem. Descrip.: PEG-5 oleate
CAS 9004-96-0
Uses: Emulsifier for min. and fatty oils, solvent; degreaser, dispersant, detergent, lubricant; metal, textile, cosmetic, plastisol formulations
Properties: Liq.; oil-sol.; sapon. no. 118; HLB 8.3; 100% conc.; nonionic

Chemax E-1000 MO [Chemax]
Chem. Descrip.: PEG-20 oleate
CAS 9004-96-0
Uses: Emulsifier for min. and fatty oils, solvent; degreaser, dispersant, detergent, lubricant
Properties: Solid; 100% conc.; nonionic

CHP® [ISP]
Chem. Descrip.: N-Cyclohexyl pyrrolidone
CAS 6837-24-7; EINECS 229-919-7
Uses: Solvent for plating engineering plastics, esp. EMS applics.; dyeing of aromatic amide fibers; photo resists; gas separation; elec. insulating coatings; chem. synthesis; absorp. of sulfur dioxide and nitric acid; aromatic extraction; high temp. polyimide processing aid; superior resist. to hydrolysis
Properties: Liq.; water-sol.; m.w. 167; sp.gr. 1.026; visc. 11.5 cps; f.p. 12 C; b.p. 284 C; flash pt. (CC) 293 F; ref. index 1.4950; 98% min. purity
Toxicology: LD50 (oral, rat) 0.37 g/kg; primary skin irritant; may cause permanent eye damage
Environmental: Biodeg.

CHT Intensol TH-B [Catawba-Charlab]
Chem. Descrip.: Proprietary
Uses: Dyestuff solvent and machine cleaner
Properties: Straw-colored sl. hazy liq.; sl. odor; infinitely sol. in water; sp.gr. 1.01; b.p. 212 F; flash pt. (TCC) > 200 F
Toxicology: Could irritate or burn eyes; prolonged/repeated contact may cause skin irritation; inh. of very high concs. may cause headache, giddiness, mental confusion, nausea
Environmental: COD 367 mg O_2/g; BOD_5 72 mg O_2/g
Storage: Store in cool area

Cithrol 2DL [Croda Universal Ltd]
Chem. Descrip.: PEG-2 dilaurate
CAS 9005-02-1
Uses: Dispersant, w/o emulsifier, wetting agent and cosolvent; used in cosmetic and industrial applics.
Properties: Paste; HLB 6.0; 97% conc.; nonionic

Cithrol 2DO [Croda Universal Ltd]
Chem. Descrip.: PEG-4 dioleate
CAS 9005-07-6
Uses: Dispersant, w/o emulsifier, wetting agent and cosolvent; used in cosmetic and industrial applics.
Properties: Liq.; 97% conc.; nonionic

Cithrol 4DL [Croda Universal Ltd]
Chem. Descrip.: PEG-8 dilaurate
CAS 9005-02-1
Uses: Dispersant, w/o emulsifier, wetting agent and cosolvent; used in cosmetic and industrial applics.
Properties: Liq.; HLB 10.2; 97% conc.; nonionic

Cithrol 4DO [Croda Universal Ltd]
Chem. Descrip.: PEG-8 dioleate
CAS 9005-07-6
Uses: Dispersant, w/o emulsifier, wetting agent and cosolvent; used in cosmetic and industrial applics.
Properties: Liq.; HLB 8.3; 97% conc.; nonionic

Cithrol 6DL [Croda Universal Ltd]
Chem. Descrip.: PEG-12 dilaurate
CAS 9005-02-1
Uses: Dispersant, w/o emulsifier, wetting agent and cosolvent; used in cosmetic and industrial applics.
Properties: Liq.; HLB 12.0; 97% conc.; nonionic

Cithrol 6DO [Croda Universal Ltd]
Chem. Descrip.: PEG-12 dioleate
CAS 9005-07-6; EINECS 288-459-5
Uses: Dispersant, w/o emulsifier, wetting agent and cosolvent; used in cosmetic and industrial applics.
Properties: Liq.; HLB 10.4; 97% conc.; nonionic

Cithrol 10DL [Croda Universal Ltd]
Chem. Descrip.: PEG-20 dilaurate

CAS 9005-02-1
Uses: Dispersant, w/o emulsifier, wetting agent and cosolvent; used in cosmetic and industrial applics.
Properties: Paste; HLB 14.7; 97% conc.; nonionic

Cithrol 10DO [Croda Universal Ltd]
Chem. Descrip.: PEG-20 dioleate
CAS 9005-07-6
Uses: Dispersant, w/o emulsifier, wetting agent and cosolvent; used in cosmetic and industrial applics.
Properties: Liq.; HLB 13.2; 97% conc.; nonionic

Citroflex® 1 [Morflex]
Chem. Descrip.: Trimethyl citrate
CAS 1587-20-8
Uses: Environmentally friendly solvent offering safety, low volatility, high flash pt.; easily emulsified; compat. with nonaq. systems; used as actives in household, industrial, and institutional cleaners; solvent, film-former, and plasticizer for inks, adhesives, and coatings
Properties: APHA 100 max. solid; sol. 8.2 g/100 ml in water, < 4 g/100 ml in toluene; insol. in heptane; m.w. 234; m.p. 76 C; acid. no. 2.6; flash pt. (COC) 345 F; > 97% assay

Citroflex® 2 [Morflex]
Chem. Descrip.: Triethyl citrate
CAS 77-93-0; EINECS 201-070-7
Uses: Environmentally friendly plasticizer for cellulose acetate, cellulose acetate butyrate, cellulose nitrate, chlorinated rubber, ethyl cellulose, PVAc, PVB, PVC, PVP, PVDC, PET, poly(acrylate/methacrylate), dammar, ester gums; enhances grease resist.; lightfastness makes it desirable for lacquer formulations; fragrance carrier in perfumes, deodorants, shampoos, detergents; solvent in household, I&I cleaners; film-former, plasticizer, solvent for inks, adhesives, coatings
Regulatory: FDA 21CFR §175.300, 175.320, 175.380, 175.390, 176.170, 177.1210, 181.27, 182.1911, 184.1911; kosher
Properties: Clear liq., essentially odorless; sol. in toluene, 6.5 g/100 ml in water; insol. in heptane; m.w. 276.3; sp.gr. 1.135-1.139; visc. 35 cps; pour pt. -45 C; flash pt. (COC) 155 C; ref. index 1.440-1.442; 99% assay
Toxicology: LD50 (oral, rat) 7 cc/kg, (IP, mice) 1750 mg/kg
Precaution: Incompat. with oxidizing materials

Citroflex® 4 [Morflex]
Chem. Descrip.: Tri-n-butyl citrate
CAS 77-94-1; EINECS 201-071-2
Uses: Environmentally friendly plasticizer for cellulose nitrate, ethyl cellulose, chlorinated rubber, cellulosic coatings, PVAc, PVB, PVC, PVP, PVDC, poly(acrylate/methacrylate); defoamer in proteinaceous sol'ns.; improved lt. stability in cellulose acetate; solvent for household, I&I cleaners; solvent, film-former, plasticizer in inks, adhesives, coatings
Regulatory: FDA 21CFR §175.105; kosher
Properties: Clear liq., essentially odorless; sol. in toluene, heptane, < 0.1 g/100 ml water; m.w. 360; sp.gr. 1.037-1.045; visc. 32 cps; vapor pressure 1 mm Hg (170 C); pour pt. -62 C; flash pt. (COC) 185 C; ref. index 1.443-1.445; 99% assay
Toxicology: LD50 (oral, rat) > 30 ml/kg, (IP, mice) 2900 mg/kg; avoid breathing vapors or mist
Precaution: Incompat. with oxidizing materials; combustion produces CO, CO_2
Storage: Store in well ventilated area in tightly closed containers; protect from moisture

Citroflex® 6 [Morflex]
Chem. Descrip.: Trihexyl citrate
CAS 16544-70-0
Uses: Environmentally friendly solvent offering safety, low volatility, high flash pt.; easily emulsified; compat. with nonaq. systems; used as actives in household, industrial, and institutional cleaners; solvent, film-former, and plasticizer for inks, adhesives, and coatings
Properties: APHA 100 max. liq.; sol. < 0.1 g/100 ml in water; infinite sol.

in toluene and heptane; m.w. 444; sp.gr. 0.996; visc. 40 cps; vapor pressure 8.3 x 10^{-6} mm Hg (20 C); pour pt. -57 C; acid. no. 0.02; flash pt. (COC) 400 F; ref. index 1.449; surf. tens. 29.11 dynes/cm; > 99% assay

Citroflex® A-1 [Morflex]
Chem. Descrip.: Acetyltrimethyl citrate
CAS 146340-22-9
Uses: Environmentally friendly solvent offering safety, low volatility, high flash pt.; easily emulsified; compat. with nonaq. systems; used as actives in household, industrial, and institutional cleaners; solvent, film-former, and plasticizer for inks, adhesives, and coatings
Properties: APHA 50 max. liq.; sol. 5.5 g/100 ml in water; sol. in toluene; insol. in heptane; m.w. 276; sp.gr. 1.2324; visc. 366 cps; vapor pressure 0.16 mm Hg (20 C); pour pt. -2 C; acid. no. 0.02; flash pt. (COC) 311 F; ref. index 1.4440; surf. tens. 38.01 dynes/cm; > 98.5% assay

Citroflex® A-2 [Morflex]
Chem. Descrip.: Acetyltriethyl citrate
CAS 77-89-4; EINECS 201-066-5
Uses: Environmentally friendly plasticizer for more polar resins such as cellulosics, polyacrylates, PVAc, polyvinyl butyral, PVC, PVP, PVDC, chlorinated rubber; lightfastness for lacquer formulations; plasticizer in food wraps; solvent in household, I&I cleaners; solvent, film-former, plasticizer for inks, adhesives, coatings
Regulatory: FDA 21CFR §175.105, 175.300, 175.320, 175.380, 176.170, 177.1210, 178.3910; kosher
Properties: Clear liq., essentially odorless; sol. in toluene, 0.72 g/100 ml water; m.w. 318.3; sp.gr. 1.135-1.139; visc. 54 cps; pour pt. -43 C; flash pt. (COC) 188 C; ref. index 1.432-1.441; 99% assay
Toxicology: LD50 (oral, rat) 7 cc/kg, (IP, mice) 1150 mg/kg; sl. but transient eye irritant; nonirritating to skin

Citroflex® A-4 [Morflex]
Chem. Descrip.: Acetyl tri-n-butyl citrate
CAS 77-90-7; EINECS 201-067-0
Uses: Environmentally friendly plasticizer for PVC, PVDC, esp. food films, medical articles; also suitable for cellulose nitrate, chlorinated rubber, ethyl cellulose, PVAc, PVB, PVP, PU, poly(acrylate/methacrylate); sol'n. coatings on paperboard and foil; vinyl toys; solvent in household, I&I cleaners; solvent, film-former, and plasticizer for inks, adhesives, coatings
Regulatory: FDA 21CFR §172.515, 175.105, 175.300, 175.320, 175.380, 175.390, 176.170, 177.1210, 178.3910, 181.27; kosher
Properties: Clear liq., essentially odorless; sol. in toluene, heptane, < 0.1 g/100 ml water; m.w. 402.5; sp.gr. 1.045-1.055; visc. 33 cps; pour pt. –59 C; flash pt. (COC) 104 C; ref. index 1.441-1.4425; 99% min. assay
Toxicology: LD50 (oral, rat) 30 cc/kg, (IP, mice) 4000 mg/kg; sl. but transient eye irritant; nonirritating to skin

CNC Sol BD [CNC Int'l.]
Chem. Descrip.: Solv./emulsifier blend
Uses: For removal of waxes, greases, oils, and greige-mill dirt from cotton, rayon, and blends
Unverified

CNC Sol UE [Sybron]
Chem. Descrip.: Aliphatic and aromatic solvs., terpenes, low titer soaps, and nonionic biodeg. detergents
Uses: Detergent/solvent for removal of oil, dirt, waxes, graphite, for textile scouring
Properties: Amber liq.; dissolves in hot or cold water
Unverified

CNC Sol XN [CNC Int'l.]
Chem. Descrip.: Solv./emulsifier with alkaline buidlers, nonionic detergents
Uses: Solvent/emulsifier blend for removal of waxes, oils, and greige mill dirt from cotton, rayon, and blends

Unverified

CNC Sol XNN #11 [CNC Int'l.]
Chem. Descrip.: Solv./emulsifier blend
Uses: For removal of waxes, greases, oils, and greige-mill dirt from cotton, rayon, and blends
Unverified

CNC Solv 809 [Sybron]
Chem. Descrip.: Solv./surfactant blend
Uses: For deinking and repulping resin coated mixed waste furnishes; prevents redeposition of agglomerated inks and binders
Unverified

Comperlan® COD [Henkel KGaA/Cospha; Henkel/Pulcra]
Chem. Descrip.: Cocamide DEA
CAS 61791-31-9; EINECS 263-163-9
Uses: Foam and visc. increasing agent with emulsifying properties; conditioner for shampoos and other cosmetic and pharmaceutical applics.; solvent
Properties: Yel. clear liq.; solid. pt. -5 to 5 C; pH 9.5-10.5 (2%); 85% conc.
Unverified

Consoscour M [Consos]
Chem. Descrip.: Emulsifiable solv.
Uses: Solvent scour for removing oils, wax and grease from goods prior to dyeing
Properties: Liq.; 100% conc.; nonionic

Consoscour P [Consos]
Chem. Descrip.: Aromatic solv. mixture with emulsifier and surfactants
Uses: Solvent-based cleaner for textile printing and dyeing equip.; solvent scour
Properties: Liq.; 100% conc.; anionic

Consoscour TEK [Consos]
Chem. Descrip.: Solvent/emulsifier blend
Uses: Low temp. solvent scour for polyester and nylon knit goods; removes heavy oil and wax deposits; low foam
Properties: Liq.; nonionic
Environmental: Biodeg.
Unverified

Corexit CL8594 [Exxon]
Chem. Descrip.: Ethoxylated alkylphenol and ethoxylated alcohol
Uses: Heavy-duty hydrocarbon solvent-based cleaner and degreaser that solubilizes heavy oils, asphaltic materials and grease deposits to permit their removal by a water or solvent flush
Properties: Liq.; 21% conc.; nonionic

Corial® [BASF AG]
Uses: Special org. solvent for leather finishes

Corona Lanolin [Croda Chem. Ltd]
Chem. Descrip.: Anhyd. lanolin
CAS 8006-54-0; EINECS 232-348-6
Uses: Conditioning emollient, moisturizing agent, cosolvent, plasticizer, w/o emulsifier, superfatting agent, wetting/dispersing agent, used in cosmetics, as pharmaceutical base
Properties: Paste; oil-sol.; 100% conc.; nonionic

Coronet Lanolin [Croda Chem. Ltd]
Chem. Descrip.: Super refined cosmetic grade lanolin
CAS 8006-54-0; EINECS 232-348-6
Uses: Conditioning emollient, moisturizing agent, cosolvent, plasticizer, w/o emulsifier, superfatting agent, wetting/dispersing agent, used as pharmaceutical base; minimum odor, color, batch variation in this grade; slightly higher m.p. than USP and BP standards
Properties: Paste; 100% conc.; nonionic

Cosmetol® X [CasChem]
Chem. Descrip.: Unrefined castor oil USP, with antioxidant
CAS 8001-79-4; EINECS 232-293-8
Uses: Deodorized specially refined grade of castor oil containing a food grade antioxidant; emollient, pigment wetter, cosolvent, lubricant for cosmetics, makeup, antiperspirant sticks
Properties: Pour pt. -23 C; acid no. 2; iodine no. 86; sapon. no. 180; hyd. no. 164

Cremophor® EL [BASF; BASF AG]
Chem. Descrip.: PEG-35 castor oil
CAS 61791-12-6
Uses: Solubilizer, emulsifier used for essential oils, pharmaceuticals,

cosmetics, veterinary medicine; solvent for steroids and fat-sol. vitamins
Properties: Pale yel. liq.; sol. in water, ethanol, propanol, ethyl acetate, chloroform, CCl_4, benzene, toluene, and xylene; sp.gr. 1.05-1.06; visc. 700-850 cps; HLB 12-14; acid no. < 2; sapon no. 65-70; ref. index 1.471; pH 6-8 (10% aq.); 100% act.; nonionic
Toxicology: May cause anaphylactic reactions

Crill 1 [Croda Food Prods. Ltd; Croda Universal Ltd]
Chem. Descrip.: Sorbitan laurate
CAS 1338-39-2
Uses: Emulsifier, pigment dispersant, cosolvent, wetting agent, antifoam, visc. reducer, mold release, antiblock agent, corrosion inhibitor, lubricant, antistat; used for cosmetics, food and food pkg., insecticides and herbicides, leather treatment; metalworking fluids, oil slick dispersing, paints and inks, pharmaceuticals, plastics, polishes, textiles; antifoam for syrups, calf starters
Regulatory: UK clearance
Properties: Pale yel. clear visc. liq.; sol. in ethanol, oleyl alcohol, min. oil; HLB 8.6; acid no. 4-7; sapon. no. 160-175; hyd. no. 330-358; 98% conc.; nonionic
Storage: Store under cool, dry conditions

Crill 2 [Croda Universal Ltd]
Chem. Descrip.: Sorbitan palmitate
CAS 26266-57-9; EINECS 247-568-8
Uses: Emulsifier, pigment dispersant, cosolvent, wetting agent, antifoam, visc. reducer, mold release, antiblock agent, corrosion inhibitor, lubricant, antistat; used for cosmetics, food and food pkg., insecticides and herbicides, leather treatment; metalworking fluids, oil slick dispersing, paints and inks, pharmaceuticals, plastics, polishes, textiles; antifoam for syrups, calf starters
Properties: Pale tan hard waxy solid; partially sol. in propylene glycol, ethyl and oleyl alcohols, olive oil, oleic acid; m.p. 46 C; HLB 6.7; sapon. no. 140-150; 98% conc.; nonionic

Crillet 6 [Croda Inc; Croda Chem. Ltd]
Chem. Descrip.: PEG-20 sorbitan isostearate
CAS 66794-58-9
Uses: O/w emulsifier, solubilizer for fragrances and perfumes; for creams, lotions, ointments; improved resistance to oxidation; o/w emulsifier, solvent, and wetting agent for topical pharmaceuticals
Use Level: 0.5-5%
Properties: Clear yel. liq.; sol. in water, ethyl, isopropyl, and oleyl alcohols, oleic acid, xylene, trichlorethylene; HLB 14.9; acid no. 2 max.; sapon. no. 40-50; hyd. no. 65-85; surf. tens. 38.6 dynes/cm (0.1%); 97% conc.; nonionic

Crodamol AB [Croda Surf. Ltd]
Chem. Descrip.: C12-15 alkyl benzoate
CAS 68411-27-8; EINECS 270-112-4
Uses: Fine dry emollient, vehicle; solvent and perfume fixative; reduces greasy/oily texture and promotes spreading; for antiperspirants, skin care prods., makeup, sunscreens
Properties: Colorless to pale straw liq.; sol. in castor oil, corn oil, min. oil, oleyl alcohol, ethanol; sp.gr. 0.92; visc. 13.7 cst
Unverified

Crodamol CAP [Croda Inc; Croda Surf. Ltd]
Chem. Descrip.: Cetearyl octanoate and isopropyl myristate
Uses: Emollient simulating properties of preen gland oil; provides water repellency, promotes spreading, reduces stickiness and occlusivity, imparts smooth afterfeel to skin care and makeup preps.; solvent and nonoily emollient for pharmaceuticals (burn creams, acne creams and lotions, antibiotic ointments)
Use Level: 5-10%
Properties: Colorless clear liq., low odor; sol. in castor oil, corn oil, min. oil, oleyl alcohol, ethanol, IPA;

sp.gr. 0.848-0.853; visc. 11.9 cst; cloud pt. 0 C
Toxicology: LD50 (oral, rat) 13.6 g/kg; moderate skin irritant; nonirritating to eyes

Crodamol DA [Croda Surf. Ltd]
Chem. Descrip.: Diisopropyl adipate
CAS 6938-94-9; EINECS 248-299-9
Uses: Solvent for aromatic oils and perfumes; superfatting agent for aq. alcoholic lotions; solvent for lipstick dyestuffs; used for aftershave lotions, colognes, skin fresheners; plasticizer in nail varnish and hair spray; emollient in skin and bath preps.
Properties: Water-wh. liq., almost odorless; sol. in castor oil, corn oil, min. oil, oleyl alcohol, ethanol; sp.gr. 0.950-0.962; visc. 3.9 cst; f.p. –1 C; acid no. 0.5 max.; sapon. no. 480-500; ref. index 1.422-1.424
Unverified

Crodamol GTCC [Croda Inc; Croda Food Prods. Ltd.; Croda Surf. Ltd]
Chem. Descrip.: Glyceryl tricaprylate/caprate
Uses: Imparts fine, soft emollience to skin; solvent, vehicle and diluent for active ingreds.; substitute for min. hydrocarbon oils in cosmetics and pharmaceuticals; exc. dermatological props.; bakery lubricant, release agent, glazing agent for sugar confectionery, flour confectionery; solvent carrier for flavors and fragrances; prevents excessive dehydration in dried fruit; hypoallergenic baby food formulations
Use Level: 2-15%
Properties: Colorless to pale straw clear liq.; sol. in castor oil, corn oil, min. oil, oleyl alcohol, ethanol; sp.gr. 0.950; visc. 30 cst; acid no. 0.2 max.; iodine no. 1 max.; sapon. no. 325-345; 100% conc.; nonionic
Storage: Store under cool, dry conditions

Crodamol IPL [Croda Chem. Ltd; Croda Surf. Ltd]
Chem. Descrip.: Isopropyl laurate
CAS 10233-13-3; EINECS 233-560-1
Uses: Emollient, plasticizer, and cosolvent for cosmetics
Properties: Colorless liq.; sol. in castor oil, corn oil, min. oil, oleyl alcohol, ethanol; sp.gr. 0.845-0.852; visc. 3.7 cst
Unverified

Crodamol IPM [Croda Inc; Croda Chem. Ltd; Croda Surf. Ltd]
Chem. Descrip.: IPM, perfumery grade
CAS 110-27-0; EINECS 203-751-4
Uses: Spreading agent, emollient, cosolvent, and vehicle for cosmetic raw materials
Properties: Water wh. liq.; odorless.; misc. with veg. and min. oils; sp.gr. 0.847-0.853; visc. 5.3 cst; acid no. 2 max.; ref. index 1.432-1.434; 92% act.

Crodamol IPP [Croda Chem. Ltd; Croda Surf. Ltd]
Chem. Descrip.: Isopropyl palmitate
CAS 142-91-6; EINECS 205-571-1
Uses: Emollient, spreading agent, vehicle, and cosolvent for cosmetic applics.
Properties: Colorless liq.; sol. in castor oil, corn oil, min. oil, oleyl alcohol, ethanol; sp.gr. 0.850-0.855; visc. 7.7 cst
Unverified

Crodamol PMP [Croda Inc; Croda Surf. Ltd]
Chem. Descrip.: PPG-2 myristyl ether propionate
Uses: Emollient, lubricant with dry, lt. greaseless feel; coupling agent; solvent for sunscreen actives; for bath oils, creams, moisturizers, emulsions, pharmaceuticals (burn creams, acne creams and lotions, antibiotic ointments)
Use Level: 5-20%
Properties: Gardner 1 max. colorless clear liq., very mild char. odor; sol. (1%) in min. oil, IPM, oleyl alcohol, ethanol/water, lanolin, cetyl alcohol; sp.gr. 0.870-0.880; visc. 8.9 cst; cloud pt. -5 C; acid no. 0.5 max.; iodine no. 1.0 max.; sapon. no. 140-155
Toxicology: LD50 (oral, rat) > 5 g/kg; minimal skin irritant, nonirritating to eyes; esp. low in comedogenic potential

Crystal® O [CasChem]
Chem. Descrip.: Refined castor oil USP
CAS 8001-79-4; EINECS 232-293-8
Uses: Emollient, pigment wetter, co-solvent, lubricant for cosmetics, makeup, antiperspirant sticks
Regulatory: FDA approval
Properties: Gardner 1- color; sp.gr. 0.959; visc. 7.5 stokes; pour pt. -23 C; acid no. 2; iodine no. 86; sapon. no. 180; hyd. no. 164

Crystal® Crown [CasChem]
Chem. Descrip.: Refined castor oil USP, with antioxidant
CAS 8001-79-4; EINECS 232-293-8
Uses: Emollient, pigment wetter, co-solvent, lubricant for cosmetics, makeup, antiperspirant sticks
Regulatory: FDA approval
Properties: Gardner 1- color; sp.gr. 0.959; visc. 7.5 stokes; pour pt. -23 C; acid no. 2; iodine no. 86; sapon. no. 180; hyd. no. 164

D

Dalpad A [Dow]
Chem. Descrip.: Aromatic-based glycol ether
Uses: Solvent, emulsifier, coupling agent providing improved surf. wetting, soil penetration in household, commercial and industrial cleaning prods.; coalescent in adhesives, water-based architectural and industrial coatings; plasticizer for cellulosics; solvent for printing inks; carrier solvent for textile dyes; intermediate for soap/perfume fixatives; antimicrobial in cosmetics and fragrances; paint removers; ingred. in metal corrosion protection slurries
Properties: Sol. 2.3 ml/100 ml in water; m.w. 138.2; sp.gr. 1.104; dens. 9.2 lg/bal; vapor pressure 0.007 mm Hg; f.p. 51 F; b.p. 245.6 C; flash pt. (Seta) 250 F; surf. tens. 42 dynes/cm
Toxicology: Can cause mod. eye irritation and corneal injury; prolonged/repeated exposure may cause hemolysis
Environmental: Pract. nontoxic to fish and aquatic invertebrate; high index of biodeg.

DB [Occidental]
Chem. Descrip.: Butoxydiglycol
CAS 112-34-5; EINECS 203-961-6
Uses: Solvent

DB [ARCO]
Chem. Descrip.: Diethylene glycol butyl ether
CAS 112-34-5; EINECS 203-961-6
Uses: Solvent for coatings and cleaning applics.; HAPS listed; SARA reportable
Properties: 100% sol. in water @ 20 C; m.w. 162.23; sp.gr. 0.955 (20/20 C); dens. 7.94 lb/gal; visc. 4.7 cps; vapor pressure 0.02 mm Hg (20 C); b.p. 230 C; HLB 7.7; flash pt. (COC) 232 F; ref. index 1.4316; surf. tens. 30.0 dynes/cm; sp. heat 0.54 cal/g/°C

Dba [ARCO]
Chem. Descrip.: Diethylene glycol butyl ether acetate
CAS 124-17-4; EINECS 204-685-9
Uses: Solvent for coatings and cleaning applics.; HAPS listed; SARA reportable
Properties: 6% sol. in water @ 20 C; m.w. 204.27; sp.gr. 0.980 (20/20 C); dens. 8.16 lb/gal; visc. 3.2 cps; vapor pressure 0.04 mm Hg (20 C); b.p. 235 C; HLB 7.7; flash pt. (COC) 240 F; ref. index 1.4239; surf. tens. 30 dynes/cm

DB Acetate [Occidental]
Chem. Descrip.: Diethylene glycol butyl ether acetate
CAS 124-17-4; EINECS 204-685-9
Uses: Solvent

DBE [DuPont Nylon]
Chem. Descrip.: Dibasic ester mixt. (60% dimethyl glutarate, 15% dimethyl adipate, 24% dimethyl succinate)
Uses: Solvent for coatings, cleaners, inks, textile lubricants, urethane prod.; plasticizer for flexible thermoset polyester; polymer intermediate for polyester polyols for urethanes, wet-str. paper resins, polyester resins; specialty chemical intermediate
Properties: Clear colorless liq., mild sweet odor; m.w. 159; sol. in alcohols, ketones, ethers, most hydro-

carbons; sl. sol. in water and higher paraffinic hydrocarbons; sp.gr. 1.092; dens. 9.09 lb/gal; visc. 2.4 cSt; f.p. -20 C; b.p. 196-225 C; acid no. 0.3 max.; flash pt. (TCC) 100 C; surf. tens. 35.6 dynes/cm; 99% min. esters

Toxicology: LD50 (oral, rat) 8191 mg/kg, (dermal, rabbit) > 2250 mg/kg; mild to severe skin irritant, moderate eye irritant; may cause nose and throat irritation and blurry vision

Precaution: Do not mix with strong oxidants, acids, or alkalies

Storage: Store in well ventilated place in tightly closed container

DBE-2, -2SPG [DuPont Nylon]

Chem. Descrip.: Dibasic ester mixt. (76% dimethyl glutarate, 24% dimethyl adipate, 0.3% dimethyl succinate)

Uses: Solvent for coatings, cleaners, inks, textile lubricants, urethane prod.; plasticizer for flexible thermoset polyester; polymer intermediate for polyester polyols for urethanes, wet-str. paper resins, polyester resins; specialty chemical intermediate

Properties: Clear colorless liq., mild odor; m.w. 163; sol. in alcohols, ketones, ethers, most hydrocarbons; sl. sol. in water and higher paraffinic hydrocarbons; sp.gr. 1.081; dens. 9.00 lb/gal; visc. 2.5 cSt; f.p. -13 C; b.p. 210-225 C; acid no. 1 max.; flash pt. (TCC) 104 C.; 99% min. esters

Toxicology: May cause eye irritation

DBE-3 [DuPont Nylon]

Chem. Descrip.: Dibasic ester mixt. (89% dimethyl adipate, 10% dimethyl glutarate, 0.2% dimethyl succinate)

Uses: Solvent for coatings, cleaners, inks, textile lubricants, urethane prod.; plasticizer for flexible thermoset polyester; polymer intermediate for polyester polyols for urethanes, wet-str. paper resins, polyester resins; specialty chemical intermediate

Properties: Clear colorless liq., mild odor; m.w. 173; sol. in alcohols, ketones, ethers, most hydrocarbons; sl. sol. in water and higher paraffinic hydrocarbons; sp.gr. 1.068; dens. 8.89 lb/gal; visc. 2.5 cSt; f.p. 8 C; b.p. 215-225 C; acid no. 1 max.; flash pt. (TCC) 102 C; 99% min. esters

Toxicology: May cause eye irritation

DBE-4 [DuPont Nylon]

Chem. Descrip.: Dimethyl succinate
CAS 106-65-0; EINECS 203-419-9

Uses: Solvent for coatings, cleaners, inks, textile lubricants, urethane prod.; plasticizer for flexible thermoset polyester; polymer intermediate for polyester polyols for urethanes, wet-str. paper resins, polyester resins; specialty chemical intermediate

Properties: Clear colorless liq., mild odor; m.w. 146; sol. in alcohols, ketones, ethers, most hydrocarbons; sl. sol. in water and higher paraffinic hydrocarbons; sp.gr. 1.121; dens. 9.33 lb/gal; visc. 2.5 cSt; f.p. 19 C; b.p. 196 C; flash pt. (TCC) 94 C; 98.5% min. esters

Toxicology: May cause eye irritation

DBE-5 [DuPont Nylon]

Chem. Descrip.: Dimethyl glutarate
CAS 1119-40-0; EINECS 214-277-2

Uses: Solvent for coatings, cleaners, inks, textile lubricants, urethane prod.; plasticizer for flexible thermoset polyester; polymer intermediate for polyester polyols for urethanes, wet-str. paper resins, polyester resins; specialty chemical intermediate

Properties: Clear colorless liq., mild odor; m.w. 160; sol. in alcohols, ketones, ethers, most hydrocarbons; sl. sol. in water and higher paraffinic hydrocarbons; sp.gr. 1.091; dens. 9.08 lb/gal; visc. 2.5 cSt; f.p. -37 C; b.p. 210-215 C; acid no. 1 max.; flash pt. (TCC) 107 C; 99% min. esters

Toxicology: May cause eye irritation

DBE-6 [DuPont Nylon]

Chem. Descrip.: Dimethyl adipate
CAS 627-93-0; EINECS 211-020-6

Uses: Solvent for coatings, cleaners,

inks, textile lubricants, urethane prod.; plasticizer for flexible thermoset polyester; polymer intermediate for polyester polyols for urethanes, wet-str. paper resins, polyester resins; specialty chemical intermediate

Properties: Clear colorless liq., mild odor; m.w. 174; sol. in alcohols, ketones, ethers, most hydrocarbons; sl. sol. in water and higher paraffinic hydrocarbons; sp.gr. 1.064; dens. 8.86 lb/gal; visc. 2.5 cSt; f.p. 10 C; b.p. 227-230 C; acid no. 1 max.; flash pt. (TCC) 113 C; 99% min. esters

Toxicology: May cause eye irritation

DBE-9 [DuPont Nylon]

Chem. Descrip.: Dibasic ester mixt. (66% dimethyl glutarate, 33% dimethyl succinate, 0.2% dimethyl adipate)

Uses: Solvent for coatings, cleaners, inks, textile lubricants, urethane prod.; plasticizer; polymer intermediate for polyester polyols for urethanes, wet-str. paper resins, polyester resins; specialty chemical intermediate

Properties: Clear colorless liq., mild odor; m.w. 156; sol. in alcohols, ketones, ethers, most hydrocarbons; sl. sol. in water and higher paraffinic hydrocarbons; sp.gr. 1.099; dens. 9.15 lb/gal; visc. 2.4 cSt; f.p. -10 C; b.p. 196-215 C; flash pt. (TCC) 94 C; 99% min. esters

Toxicology: May cause eye irritation

DBE-IB [DuPont Nylon]

Chem. Descrip.: 15% Diisobutyl adipate, 58% diisobutyl glutarate, 27% diisobutyl succinate mixt.

Uses: Solvent used in waterborne and solvent coatings, as coalescing aid; solvent for printing ink, cleaning; textile lubricant

Properties: Misc. with most org. solvs.; sol. in water < 0.1% wt.; m.w. 242 (avg.); sp.gr. 0.955 g/ml; dens. 7.97 lb/gal; visc. 18.8 cSt; vapor pressure 0.01 mm Hg (20 C); f.p.≈-55 C; b.p. 275-295 C; flash pt. (TCC) 133 C; ref. index 1.427-1.428; surf. tens. 27.2 dynes/cm

Environmental: Readily biodeg.

Storage: Not classified as flamm.; store and handle using mild steel or aluminium containers and equip.; stable at normal storage conditions of temp. and humidity

DE [Occidental]

Chem. Descrip.: Ethoxydiglycol

CAS 111-90-0; EINECS 203-919-7

Uses: Solvent

DE [ARCO]

Chem. Descrip.: Diethylene glycol ethyl ether

CAS 111-90-0; EINECS 203-919-7

Uses: Solvent for coatings and cleaning applics.; HAPS listed; SARA reportable

Properties: 100% sol. in water @ 20 C; m.w. 134.17; sp.gr. 0.990 (20/20 C); dens. 8.25 lb/gal; visc. 4.5 cps; vapor pressure 0.12 mm Hg (20 C); b.p. 198 C; HLB 8.6; flash pt. (TCC) 195 F; ref. index 1.4620; surf. tens. 32.2 dynes/cm; sp. heat 0.55 cal/g/°C

DEA [ARCO]

Chem. Descrip.: Diethylene glycol ethyl ether acetate

CAS 112-15-2; EINECS 203-940-1

Uses: Solvent for coatings and cleaning applics.; HAPS listed; SARA reportable

Properties: 100% sol. in water @ 20 C; m.w. 176.21; sp.gr. 1.012 (20/20 C); dens. 8.42 lb/gal; visc. 2.8 cps; vapor pressure 0.05 mm Hg (20 C); b.p. 214 C; HLB 8.7; flash pt. (TCC) 225 F; ref. index 1.4220; surf. tens. 31.7 dynes/cm

DE Acetate [Occidental]

Chem. Descrip.: Diethylene glycol ethyl ether acetate

Uses: Solvent

Unverified

Decalin [DuPont]

Chem. Descrip.: Decahydronaphthalene

CAS 91-17-8; EINECS 202-046-9

Uses: Solvent and stabilizer for shoe creams and floor waxes; solvent in paint and lacquers, oils, resins, rubber, and asphalt

Properties: Colorless to pale yel. liq.; m.w. 138.2; f.p. -45 C; b.p. 186 C;

negligible sol. in water; sp.gr. 0.876; dens. 7.39 lb/gal; flash pt. 57 C (TCC); 97.0% min. conc.

Decap [Dynaloy]
Chem. Descrip.: DMSO-based
CAS 67-68-5; EINECS 200-664-3
Uses: Solvent for electronic depotting, deflashing, and decapsulation of epoxy castings, transfer moldings, and for general adhesive and polymer removal; used hot (150 C); nonselective

Degreez [Alzo]
Chem. Descrip.: Methyl oleate, methyl stearate, methyl palmitate, methyl laurate, methyl myristate
Uses: Degreasing agent for removing oil stains from metal, textiles, etc.
Properties: Yel. liq., mild typ. almost fruity odor; disp. in water; sp.gr. 0.87; b.p. > 400 F (760 mm); f.p. 18 F; cloud pt. 27 F; flash pt. (COC) 259 F
Toxicology: Nontoxic, nonirritating; LD50 (rat, acute oral) > 20 g/kg; mild transient eye irritation
Environmental: Biodeg.

Delios® C [Grünau GmbH]
Chem. Descrip.: Capric acid triglyceride
Uses: Solvent for flavors and oil-sol. food additives; surf. treating agent for dried fruits, confectionery, dietary foodstuffs
Properties: Wh. cryst. mass, neutral odor, sl. to coconut taste; m.p. 30 C; acid no. 0.1 max.; iodine no. 0.5 max.; sapon. no. 300; hyd. no. 5 max.
Storage: Store under cool, dry conditions in tightly closed containers below 25 C; 18 mos storage life in original pkg.

Delios® S [Grünau GmbH]
Chem. Descrip.: Caprylic/capric acid triglyceride
CAS 65381-09-1; EINECS 265-724-3
Uses: Solvent for flavors and oil-sol. food additives; very good oxidation stability; for dietetic prods., sport foods; glazing and release agent for dried fruit and confectionery; lubricant, release agent for baked goods; antidusting agent; visc. reducer; also avail. as kosher grade (Delios SU)
Properties: Colorless to sl. ylsh. liq., neutral odor, sl. coconut taste; misc. with ethanol; visc. 27 mPa·s; cloud pt. 0 C max.; acid no. 0.1 max.; iodine no. 0.5 max.; sapon. no. 335-350; hyd. no. 5 max.; 70% caprylic acid, 30% capric acid
Storage: Store in tightly closed containers @ 0-30 C; avoid contact with water; 18 mos storage life in original pkg.

Delios® V [Grünau GmbH]
Chem. Descrip.: Caprylic/capric acid triglyceride
CAS 65381-09-1; EINECS 265-724-3
Uses: Solvent for flavors and oil-sol. food additives; exc. oxidation stability; for dietetic prods., sport foods; glazing and release agent for dried fruit and confectionery; lubricant for cutting machines; antidusting agent; visc. reducer
Properties: Colorless to sl. yel. liq., neutral odor, sl. to coconut taste; misc. with ethanol; visc. 30 mPa·s; cloud pt. 0 C max.; acid no. 0.1 max.; iodine no. 0.5 max.; sapon. no. 330-345; hyd. no. 5 max.; 60% caprylic acid, 40% capric acid
Storage: Store in tightly closed containers @ 0-30 C; avoid contact with water; 18 mos storage life in original pkg.

Dermol EB [Fabriquimica]
Chem. Descrip.: Butyl stearate
CAS 123-95-5; EINECS 204-666-5
Uses: Emollient, dermatophilic additive; plasticizer and solvent for pigments/lipsticks
Properties: Acid no. < 1; iodine no. < 1; sapon. no. 168-176
Unverified

Dermol ICSA [Alzo]
Chem. Descrip.: Isocetyl salicylate
CAS 138208-68-1
Uses: Emollient; solvent plasticizer for benzophenone in sunscreen formulations, waxes, resins, polymers
Properties: Pale yel. clear liq., sl. typ. odor; sp.gr. 1.02; acid no. 3; sapon. no. 145-160; flash pt. (PMCC) > 200 F; ref. index 1.4865 ± 0.005; 100%

conc.; nonionic

Detersol HF [Finetex]
Chem. Descrip.: Hydrocarbon blend, emulsified
Uses: High flash pt. solvent scour for textiles; low foam
Properties: Liq.; 100% conc.; nonionic
Discontinued prod.

Detersol X-66 [Finetex]
Chem. Descrip.: Solvent/surfactant blend
Uses: General-purpose solvent scour
Properties: Liq.; 100% conc.; nonionic
Discontinued prod.

DF-2000™ [Exxon]
Chem. Descrip.: Syn. high purity hydrocarbon solvent
Uses: Safe, effective dry-cleaning fluid which acts as replacement for Stoddard solvent; good cleaning power, low residue, and low corrosion; rapid drying; high flash pt.; mild on sensitive fabrics, trims, sequins, etc.; virtually HAPS-free
Properties: Water-wh. clear appearance; essentially odorless; low water-sol.; sp.gr. 0.77 (15.6/15.6 C); vapor pressure 0.8 mm Hg (68 F); flash pt. 147 F; KB value 27
Toxicology: OEL 300 ppm

Diamond Quality® [CasChem]
Chem. Descrip.: Unrefined castor oil USP
CAS 8001-79-4; EINECS 232-293-8
Uses: Emollient, pigment wetter, cosolvent, lubricant for cosmetics, makeup, antiperspirant sticks
Properties: Pour pt. -23 C; acid no. 2; iodine no. 86; sapon. no. 180; hyd. no. 164
Discontinued prod.

Diglycolamine® Agent (DGA®) [Huntsman]
Chem. Descrip.: 2-(2-Aminoethoxy) ethanol
CAS 929-06-6; EINECS 213-195-4
Uses: Solvent for removal of CO_2 or H_2S from gases, for recovery of aromatics from refinery streams, for prep. of foam stabilizers, wetting agents, emulsifiers, and condensation polymers
Properties: Clear colorless slightly visc. liq.; mild amine odor; misc. with water, alcohols, aromatic hydrocarbons, ethyl ether; m.w. 105.14; sp.gr. 1.0572; dens. 8.7 lb/gal; b.p. 221 C; f.p. -12.5 C; flash pt. (PMCC) 255 F; ref. index 1.4598; 98% min. act.

Diglyme [Ivax]
Chem. Descrip.: Dimethoxydiglycol
CAS 111-96-6; EINECS 203-924-4
Discontinued prod.

Diglyme [Ferro/Grant]
Chem. Descrip.: Diethylene glycol dimethyl ether
CAS 111-96-6; EINECS 203-924-4
Uses: Solvent which tends to solvate cations; used in electrochemistry, polymer, and boron chemistry; physical processes such as gas absorption, extraction, stabilization; used in industrial prods. such as fuels, lubricants, textiles, pesticides, coatings; solvent for pharmaceutical applics.
Properties: Colorless clear; ethereal, nonresidual odor; water-sol.; misc. with ethanol, acetone, benzene, diethyl ether, octane; m.w. 134.17; sp.gr. 0.9451; dens. 7.88 lb/gal; visc. 2.0 cP; vapor pressure 2 mm Hg (20 C); f.p. -64 C; b.p. 162 C; flash pt. (CC) 57 C; ref. index 1.4078; pH neutral; surf. tens. 27 dynes/cm (20 C); sp. heat 0.403 cal/g/°C; 99.6% min. purity
Toxicology: LD50 4670 mg/kg; low to mod. acute toxicity; avoid exposure to vapors, skin contact; TLV/8h TWA: 5 ppm; STEL 25 ppm; chronic exposure may cause reproductive effects
Environmental: Slowly biodeg.

Dional 11, 113 [Hoechst Celanese]
Chem. Descrip.: Fluorinated hydrocarbon
Uses: Solvent for dry-cleaning
Discontinued prod.

Dioxitol-High Gravity [Shell]
Chem. Descrip.: Diethylene glycol monoethyl ether, ethylene glycol
Uses: Solvent for natural and syn. resins, dyes, fats used as component of heavy duty automotive brake fluids; in paint industry in nonaq. stains, and nongrain raising

stains for furniture and wood; yarn and cloth conditioner in textile industry; dye solvent where deep penetration and bright shades are obtained; component of industrial cleaners; coupling agent; mfg. of esters; intermediate in mfg. of plasticizers, solvents
Properties: Colorless liq.; water-misc.; sp.gr. 1.021-1.027; 72% Dioxitol-Low Gravity, 28% ethylene glycol
Discontinued prod.

Dioxitol-Low Gravity [Shell]
Chem. Descrip.: Diethylene glycol monoethyl ether
CAS 111-90-0; EINECS 203-919-7
Uses: Solvent for resins, dyes, fats, in auto brake fluids, paint industry, industrial cleaners, mfg. of esters; intermediate in mfg. of plasticizers; yarn and cloth conditioner in textiles; dye solvent where deep penetration and bright shades are obtained
Properties: Colorless liq.; water-misc.; sp.gr. 0.986-0.990
Discontinued prod.

Dipentene No. 122 [Hercules]
Chem. Descrip.: Dipentene
CAS 138-86-3; EINECS 205-341-0
Uses: Solvent

Dizene [PPG Ind.]
Chem. Descrip.: Orthodichlorobenzene
Uses: Aromatic used in industrial plants, stockyard pens, restaurants, hotels, hospitals, camps, parks, land-fill areas, cesspools, septic tanks, grease traps, dry wells, and soil; solvent for oil and grease
Properties: Ylsh. clear liq.; strong pleasant odor; dens. 10.63 lb/gal; flash pt. 150 F (TCC); 57% min. conc.
Unverified

DM [Occidental]
Chem. Descrip.: Methoxydiglycol
CAS 111-77-3; EINECS 203-906-6
Uses: Solvent
Unverified

DM [ARCO]
Chem. Descrip.: Diethylene glycol methyl ether
CAS 111-77-3; EINECS 203-906-6
Uses: Solvent for coatings and cleaning applics.; HAPS listed; SARA reportable
Properties: 100% sol. in water @ 20 C; m.w. 120.15; sp.gr. 1.023 (20/20 C); dens. 8.51 lb/gal; visc. 3.9 cps; vapor pressure 0.2 mm Hg (20 C); b.p. 191 C; HLB 9.2; flash pt. (TCC) 191 F; ref. index 1.4628; surf. tens. 34.8 dynes/cm; sp. heat 0.54 cal/g/°C

DMSO [Elf Atochem N. Am.]
Chem. Descrip.: Dimethyl sulfoxide
CAS 67-68-5; EINECS 200-664-3
Uses: Aprotic solvent used as reaction medium in mfg. of pesticides, pharmaceuticals, dyes, inks; solvent in polymers, electronics, refining; chem. intermediate for pesticides

Dow 15-200 [Dow]
Chem. Descrip.: PPG-24 glycereth-24
CAS 9082-00-2
Uses: Provides solvency, low odor, low irritation, and low toxicity to personal care prods. and cosmetics
Properties: Liq.; m.w. 2600; sp.gr. 1.063; dens. 8.81 lb/gal; visc. 420 cSt; pour pt. -40 C; flash pt. (PMCC) > 450 F; ref. index 1.460; sp. heat 0.470 cal/g/°C

Dow 112-2 [Dow]
Chem. Descrip.: PPG-66 glycereth-12
CAS 9082-00-2
Uses: Provides solvency, low odor, low irritation, and low toxicity to personal care prods. and cosmetics
Properties: Liq.; m.w. 4900; sp.gr. 1.028; dens. 8.56 lb/gal; visc. 1000 cSt; pour pt. -18 C; flash pt. (PMCC) 455 F; ref. index 1.455; surf. tens. 33.7 dynes/cm (100 ppm); sp. heat 0.430 cal/g/°C

Dow B100-1000 [Dow]
Chem. Descrip.: Polybutylene glycol
Uses: Provides solvency, low odor, low irritation, and low toxicity to personal care prods. and cosmetics
Properties: Liq.; sol. in ethanol, IPM, lt. min. oil, cyclomethicone, sunscreens, lactic acid; partly sol. in water, lanolin, stearyl alcohol; m.w. 1000; dens. 8.10 lb/gal; visc. 86.9 cSt (40 C); pour pt. -26 C; flash

pt. (PMCC) > 350 F; ref. index 1.4529

Dow B100-2000 [Dow]

Chem. Descrip.: Polybutylene glycol

Uses: Provides solvency, low odor, low irritation, and low toxicity to personal care prods. and cosmetics

Properties: Liq.; m.w. 2000; dens. 8.09 lb/gal; visc. 209.7 cSt (40 C); pour pt. -26 C; flash pt. (PMCC) > 350 F; ref. index 1.4540

Dow B100-4800 [Dow]

Chem. Descrip.: Polybutylene glycol

Uses: Provides solvency, low odor, low irritation, and low toxicity to personal care prods. and cosmetics

Properties: Liq.; m.w. 4800; dens. 8.10 lb/gal; visc. 731.4 cSt (40 C); pour pt. -21 C; flash pt. (PMCC) > 350 F; ref. index 1.4551

Dow E200 [Dow]

Chem. Descrip.: PEG-4

CAS 25322-68-3; EINECS 203-989-9

Uses: Carrier, modifier with exc. solvency, lubricity for personal care prods. (makeup, bath preps., toothpaste), foam control agents, syn. lubricants, rubber and plastics, metalworking fluids, chem. intermediates; controls cure rates in clay-contg. rubber formulations; improves flexibility, imparts antistatic props. to thermoplastics; as heat transfer fluids in plastic thermoforming; mold release

Properties: Liq.; m.w. 200; sp.gr. 1.124; dens. 9.35 lb/gal; visc. 40 cSt; f.p. supercools; flash pt. (PMCC) 340 F; ref. index 1.459; sp. heat 0.524 cal/g/°C

Dow E300 [Dow]

Chem. Descrip.: PEG-6

CAS 25322-68-3; EINECS 203-989-9

Uses: Carrier, modifier with exc. solvency, lubricity for personal care prods. (makeup, bath preps., toothpaste), foam control agents, syn. lubricants, rubber and plastics, metalworking fluids, chem. intermediates; controls cure rates in clay-contg. rubber formulations; improves flexibility, imparts antistatic props. to thermoplastics; as heat transfer fluids in plastic thermoforming; mold release

Properties: Liq.; m.w. 300; sp.gr. 1.125; dens. 9.36 lb/gal; visc. 69 cSt; f.p. –10 C; flash pt. (PMCC) > 400 F; ref. index 1.463; sp. heat 0.508 cal/g/°C

Dow E300 NF [Dow]

Chem. Descrip.: PEG-6 NF

CAS 25322-68-3; EINECS 220-045-1

Uses: Used in foods, personal care prods. (makeup, bath preps., toothpaste) and pharmaceuticals (carrier for ointments for antiseptics and other medicaments, tablet coatings, base for suppositories, solvent, suspending agent in liq. preps., gelatin capsule

Properties: Clear visc. liq.; sol. in water; m.w. 300; sp.gr. 1.125; dens. 9.36 lb/gal; f.p. -10 C; visc. 69 cSt; flash pt. (PMCC) > 400 F; ref. index 1.463; sp. heat 0.508 cal/g/°C

Toxicology: Extremely low acute oral toxicity; little or no irritation to eyes and skin; not recommended for use on abraded skin; may generate irritating vapors on decomp.

Dow E400 [Dow]

Chem. Descrip.: PEG-8

CAS 25322-68-3; EINECS 203-989-9

Uses: Carrier, modifier with exc. solvency, lubricity for personal care prods. (makeup, bath preps., toothpaste), foam control agents, syn. lubricants, rubber and plastics, metalworking fluids, chem. intermediates; controls cure rates in clay-contg. rubber formulations; improves flexibility, imparts antistatic props. to thermoplastics; as heat transfer fluids in plastic thermoforming; mold release

Properties: Liq.; m.w. 400; sp.gr. 1.125; dens. 9.36 lb/gal; f.p. 6 C; visc. 90 cSt; flash pt. (PMCC) > 450 F; ref. index 1.465; sp. heat 0.498 cal/g/°C

Dow E400 NF [Dow]

Chem. Descrip.: PEG-8 NF

CAS 25322-68-3; EINECS 225-856-4

Uses: Used in foods and pharmaceuticals (carrier for ointments for antiseptics and other medicaments, plasticizer in tablet coatings, base for suppositories, carrier, solvent, suspending agent in liq. preps., vehicle in gelatin capsules)

Properties: Clear visc. liq.; m.w. 400; sp.gr. 1.125; dens. 9.36 lb/gal; f.p. 6 C; visc. 90 cSt; flash pt. (PMCC) > 450 F; ref. index 1.465; sp. heat 0.498 cal/g/°C

Toxicology: Extremely low acute oral toxicity; little or no irritation to eyes and skin; not recommended for use on abraded skin; may generate irritating vapors on decomp.

Dow E600 [Dow]

Chem. Descrip.: PEG-12

CAS 25322-68-3; EINECS 203-989-9

Uses: Carrier, modifier with exc. solvency, lubricity for personal care prods. (makeup, bath preps., toothpaste), foam control agents, syn. lubricants, rubber and plastics, metalworking fluids, chem. intermediates; controls cure rates in clay-contg. rubber formulations; improves flexibility, imparts antistatic props. to thermoplastics; as heat transfer fluids in plastic thermoforming; mold release

Properties: Liq.; sol. in water, ethanol, cyclomethicone, sunscreens, lactic acid; m.w. 600; sp.gr. 1.126; dens. 9.37 lb/gal; f.p. 22 C; visc. 131 cSt; flash pt. (PMCC) > 450 F; ref. index 1.466; sp. heat 0.490 cal/g/°C

Dow E600 NF [Dow]

Chem. Descrip.: PEG-12 NF

CAS 25322-68-3; EINECS 229-859-1

Uses: Used in foods and pharmaceuticals (carrier for ointments for antiseptics and other medicaments, plasticizer in tablet coatings, base for suppositories, carrier, solvent, suspending agent in liq. preps., vehicle in gelatin capsules)

Properties: Clear visc. liq.; sol. in water, ethanol, cyclomethicone, sunscreens, lactic acid; m.w. 600; sp.gr. 1.126; dens. 9.37 lb/gal; f.p. 22 C; visc. 131 cSt; flash pt. (PMCC) > 450 F; ref. index 1.466; sp. heat 0.490 cal/g/°C

Toxicology: Extremely low acute oral toxicity; little or no irritation to eyes and skin; not recommended for use on abraded skin; may generate irritating vapors on decomp.

Dow E900 [Dow]

CAS 25322-68-3; EINECS 203-989-9

Uses: Carrier, modifier with exc. solvency, lubricity for personal care prods. (makeup, bath preps., toothpaste), foam control agents, syn. lubricants, rubber and plastics, metalworking fluids, chem. intermediates; controls cure rates in clay-contg. rubber formulations; improves flexibility, imparts antistatic props. to thermoplastics; as heat transfer fluids in plastic thermoforming; mold release

Properties: Wh. waxy solid; m.w. 900; sp.gr. 1.204; f.p. 34 C; visc. 100 cSt (100 F); flash pt. (PMCC) > 450 F

Dow E900 NF [Dow]

Chem. Descrip.: PEG NF

CAS 25322-68-3

Uses: Used in foods and pharmaceuticals (carrier for ointments for antiseptics and other medicaments, plasticizer in tablet coatings, base for suppositories, carrier, solvent, suspending agent in liq. preps., vehicle in gelatin capsules)

Properties: Wh. waxy solid; m.w. 900; sp.gr. 1.204; f.p. 34 C; visc. 100 cSt (100 F); flash pt. (PMCC) > 450 F

Toxicology: Extremely low acute oral toxicity; little or no irritation to eyes and skin; not recommended for use on abraded skin; may generate irritating vapors on decomp.

Dow E1000 [Dow]

Chem. Descrip.: PEG-20

CAS 25322-68-3; EINECS 203-989-9

Uses: Carrier, modifier with exc. solvency, lubricity for personal care prods. (makeup, bath preps., toothpaste), foam control agents, syn. lubricants, rubber and plastics, metalworking fluids, chem. intermediates; controls cure rates in clay-contg. rubber formulations; improves flexibility, imparts antistatic props. to thermoplastics; as heat transfer fluids in plastic thermoforming; mold release

Properties: Wh. waxy solid; m.w. 1000; sp.gr. 1.214; f.p. 37 C; visc. 18 cSt

(210 F); flash pt. (PMCC) > 450 F

Dow E1000 NF [Dow]

Chem. Descrip.: PEG-20 NF

CAS 25322-68-3; EINECS 203-989-9

Uses: Used in foods and pharmaceuticals (carrier for ointments for antiseptics and other medicaments, plasticizer in tablet film coatings, base for suppositories, carrier, solvent, suspending agent in liq. preps., vehicle for actives in gelatin capsules)

Properties: Wh. waxy solid; m.w. 1000; sp.gr. 1.214; f.p. 37 C; visc. 18 cSt (210 F); flash pt. (PMCC) > 450 F

Toxicology: Extremely low acute oral toxicity; little or no irritation to eyes and skin; not recommended for use on abraded skin; may generate irritating vapors on decomp.

Dow E1450 [Dow]

Chem. Descrip.: PEG-32

CAS 25322-68-3; EINECS 203-989-9

Uses: Carrier, modifier with exc. solvency, lubricity for personal care prods. (makeup, bath preps., toothpaste), foam control agents, syn. lubricants, rubber and plastics, metalworking fluids, chem. intermediates; controls cure rates in clay-contg. rubber formulations; improves flexibility, imparts antistatic props. to thermoplastics; as heat transfer fluids in plastic thermoforming; mold release

Properties: Wh. waxy solid; m.w. 1450; sp.gr. 1.214; f.p. 44 C; visc. 29 cSt (210 F); flash pt. (PMCC) > 450 F

Dow E1450 NF [Dow]

Chem. Descrip.: PEG-32 NF

CAS 25322-68-3; EINECS 203-989-9

Uses: Used in foods and pharmaceuticals (carrier for ointments for antiseptics and other medicaments, plasticizer in tablet film coatings, base for suppositories, carrier, solvent, suspending agent in liq. preps., vehicle for actives in gelatin capsules)

Properties: Wh. waxy solid; m.w. 1450; sp.gr. 1.214; f.p. 44 C; visc. 29 cSt (210 F); flash pt. (PMCC) > 450 F

Toxicology: Extremely low acute oral toxicity; little or no irritation to eyes and skin; not recommended for use on abraded skin; may generate irritating vapors on decomp.

Dow E3350 [Dow]

Chem. Descrip.: PEG-75

CAS 25322-68-3; EINECS 203-989-9

Uses: Carrier, modifier with exc. solvency, lubricity for personal care prods. (makeup, bath preps., toothpaste), foam control agents, syn. lubricants, rubber and plastics, metalworking fluids, chem. intermediates; controls cure rates in clay-contg. rubber formulations; improves flexibility, imparts antistatic props. to thermoplastics; as heat transfer fluids in plastic thermoforming; mold release

Properties: Wh. waxy solid, pract. odorless; sol. > 100 g/100 g in water; m.w. 3350; sp.gr. 1.224; visc. 93 cSt (210 F); f.p. 54 C; b.p. dec.; flash pt. (PMCC) > 232 C; pH 4.5-7.5 (5% aq.)

Dow E3350 NF [Dow]

Chem. Descrip.: PEG-75 NF

CAS 25322-68-3; EINECS 203-989-9

Uses: Used in foods and pharmaceuticals (carrier for ointments for antiseptics and other medicaments, plasticizer in tablet film coatings, base for suppositories, carrier, solvent, suspending agent in liq. preps., vehicle for actives in gelatin capsules)

Properties: Wh. waxy solid, pract. odorless; sol. > 100 g/100 g in water; m.w. 3350; sp.gr. 1.224; visc. 93 cSt (210 F); f.p. 54 C; b.p. dec.; flash pt. (PMCC) > 232 C; pH 4.5-7.5 (5% aq.)

Toxicology: Single dose oral toxicity believed to be very low; may cause sl. transient eye irritation; avoid prolonged/repeated contact with abraded skin

Precaution: Dusts may present explosive hazard; incompat. with oxidizers, conc. min. acids

Dow E4500 [Dow]

Chem. Descrip.: PEG-100

CAS 25322-68-3; EINECS 203-989-9

Uses: Carrier, modifier with exc. solvency, lubricity for personal care prods. (makeup, bath preps., toothpaste), foam control agents, syn. lubricants, rubber and plastics, metalworking fluids, chem. intermediates; controls cure rates in clay-contg. rubber formulations; improves flexibility, imparts antistatic props. to thermoplastics; as heat transfer fluids in plastic thermoforming; mold release

Properties: Wh. waxy solid, sl. polyether odor; colorless visc. liq. above 136 F; sol. > 100 g/100 g in water; m.w. 4500; sp.gr. 1.224; visc. 180 cSt (210 F); f.p. 58 C; b.p. dec.; flash pt. (PMCC) > 232 C; pH 4.5-7.5 (5% aq.)

Dow E4500 NF [Dow]

Chem. Descrip.: PEG-100 NF

CAS 25322-68-3; EINECS 203-989-9

Uses: Used in foods and pharmaceuticals (carrier for ointments for antiseptics and other medicaments, plasticizer in tablet film coatings, base for suppositories, carrier, solvent, suspending agent in liq. preps., vehicle for actives in gelatin capsules)

Properties: Wh. waxy solid, sl. polyether odor; colorless visc. liq. above 136 F; sol. > 100 g/100 g in water; m.w. 4500; sp.gr. 1.224; visc. 180 cSt (210 F); f.p. 58 C; b.p. dec.; flash pt. (PMCC) > 232 C; pH 4.5-7.5 (5% aq.)

Toxicology: LD50 (oral, rat) > 50,000 mg/kg; extremely low oral toxicity; may cause sl. transient eye irritation; avoid prolonged/repeated exposure on abraded skin; contact with heated material may cause thermal burns

Precaution: Supports combustion; do not breathe smoke; incompat. with oxidizers and strong acids

Dow E8000 [Dow]

Chem. Descrip.: PEG-150

CAS 25322-68-3; EINECS 203-989-9

Uses: Carrier, modifier with exc. solvency, lubricity for personal care prods. (makeup, bath preps., toothpaste), foam control agents, syn. lubricants, rubber and plastics, metalworking fluids, chem. intermediates; controls cure rates in clay-contg. rubber formulations; improves flexibility, imparts antistatic props. to thermoplastics; as heat transfer fluids in plastic thermoforming; mold release

Properties: Wh. waxy solid, clear liq. above 65 C, pract. odorless; sol. > 100 g/100 g in water; m.w. 8000; sp.gr. 1.224; f.p. 60 C; visc. 800 cSt (210 F); flash pt. (PMCC) > 260 C; pH 4.5-7.5 (5% aq.)

Dow E8000 NF [Dow]

Chem. Descrip.: PEG-150 NF

CAS 25322-68-3; EINECS 203-989-9

Uses: Used in foods and pharmaceuticals (carrier for ointments for antiseptics and other medicaments, plasticizer in tablet film coatings, base for suppositories, carrier, solvent, suspending agent in liq. preps., vehicle for actives in gelatin capsules)

Properties: Wh. waxy solid, clear liq. above 65 C, pract. odorless; sol. > 100 g/100 g in water; m.w. 8000; sp.gr. 1.224; f.p. 60 C; visc. 800 cSt (210 F); flash pt. (PMCC) > 260 C; pH 4.5-7.5 (5% aq.); > 99% act.

Toxicology: LD50 (oral, rat) > 50 g/kg (low oral toxicity); may cause sl. transient temporary eye irritation; avoid prolonged/repeated exposure on abraded skin

Precaution: Dusts may present explosive hazard; incompat. with oxidizers, conc. min. acids

Dow EP530 [Dow]

Chem. Descrip.: Poloxamer-181

CAS 53637-25-5

Uses: Provides solvency, low odor, low irritation, and low toxicity to personal care prods. and cosmetics; defoamer in scald baths for poultry defeathering and hog dehairing machines; defoamer for food-grade papers and coatings and sanitizing sol'ns. for food processing equip.

Regulatory: 21CFR §172.808, 176.200, 176.210, 178.1010

Properties: Liq.; m.w. 2000; sp.gr. 1.017; dens. 8.46 lb/gal; visc. 321 cSt; pour pt. -32 C; flash pt. (PMCC)

> 420 F; ref. index 1.452; surf. tens. 42.6 dynes/cm (100 ppm)

Dow L910 [Dow]
Chem. Descrip.: PPG-14 butyl ether
CAS 9003-13-8
Uses: Provides solvency, low odor, low irritation, and low toxicity to personal care prods. and cosmetics
Properties: Liq.; m.w. 910; sp.gr. 0.9833; dens. 8.23 lb/gal; visc. 83 cSt; pour pt. -43 C; flash pt. (PMCC) 345 F; ref. index 1.444

Dow L1150 [Dow]
Chem. Descrip.: PPG-18 butyl ether
CAS 9003-13-8
Uses: Provides solvency, low odor, low irritation, and low toxicity to personal care prods. and cosmetics
Properties: Liq.; m.w. 1150; sp.gr. 0.9888; dens. 8.28 lb/gal; visc. 115 cSt; pour pt. -40 C; flash pt. (PMCC) > 400 F; ref. index 1.446

Dow MPEG350 [Dow]
Chem. Descrip.: PEG-6 methyl ether
CAS 9004-74-4
Uses: Provides solvency, low odor, low irritation, and low toxicity to personal care prods. and cosmetics
Properties: Liq.; m.w. 350; sp.gr. 1.097; dens. 9.14 lb/gal; visc. 27 cSt; f.p. 0 C; flash pt. (PMCC) > 350 F; ref. index 1.455
Unverified

Dow MPEG550 [Dow]
Chem. Descrip.: PEG-10 methyl ether
CAS 9004-74-4
Uses: Provides solvency, low odor, low irritation, and low toxicity to personal care prods. and cosmetics
Properties: Liq.; m.w. 550; sp.gr. 1.101; dens. 9.17 lb/gal; visc. 56 cSt; f.p. 20 C; flash pt. (PMCC) > 450 F; ref. index 1.461
Unverified

Dow MPEG750 [Dow]
Chem. Descrip.: PEG-16 methyl ether
CAS 9004-74-4
Uses: Provides solvency, low odor, low irritation, and low toxicity to personal care prods. and cosmetics
Properties: Solid; m.w. 750; sp.gr. 1.102 (50 C); dens. 9.04 lb/gal (50 C); visc. 53 cSt (100 F); f.p. 30 C; flash pt. (PMCC) > 450 F; ref. index 1.463
Unverified

Dow P425 [Dow]
Chem. Descrip.: PPG-9
CAS 25322-69-4
Uses: Provides solvency, low odor, low irritation, and low toxicity to personal care prods. and cosmetics
Properties: Liq.; m.w. 425; sp.gr. 1.007; dens. 8.39 lb/gal; visc. 70 cSt; pour pt. -45 C; flash pt. (PMCC) 330 F; ref. index 1.447; sp. heat 0.477 cal/g/°C

Dow P1000TB [Dow]
Chem. Descrip.: PPG-17
CAS 25322-69-4
Uses: Provides solvency, low odor, low irritation, and low toxicity to personal care prods. and cosmetics; defoamer for food-grade coatings and paper prods.
Regulatory: 21CFR §176.200, 176.210
Properties: Liq.; m.w. 1000; sp.gr. 1.005; dens. 8.38 lb/gal; visc. 162 cSt; pour pt. -25 C; flash pt. (PMCC) > 360 F; ref. index 1.448; surf. tens. 40.7 dynes/cm (100 ppm); sp. heat 0.461 cal/g/°C

Dow P1200 [Dow]
Chem. Descrip.: PPG-20
CAS 25322-69-4
Uses: Provides solvency, low odor, low irritation, and low toxicity to personal care prods. and cosmetics; boiler water additive for prep. of steam that will contact food; defoamer for beet sugar and yeast processing; defoamer for food-grade coatings and paper prods.
Regulatory: 21CFR 173.310, 173.340 (a)(3), 176.200, 176.210
Properties: Liq.; m.w. 1200; sp.gr. 1.007; dens. 8.38 lb/gal; visc. 175 cSt; pour pt. -40 C; flash pt. (PMCC) 345 F; ref. index 1.448; surf. tens. 43.3 dynes/cm (100 ppm); sp. heat 0.459 cal/g/°C

Dow P2000 [Dow]
Chem. Descrip.: PPG-26
CAS 25322-69-4
Uses: Provides solvency, low odor, low irritation, and low toxicity to personal care prods. and cosmetics; boiler water additive for prep.

of steam that will contact food; defoamer for beet sugar and yeast processing; defoamer for food-grade coatings and paper prods.
Regulatory: 21CFR 173.310, 173.340 (a)(3), 176.200, 176.210
Properties: Liq.; sol. in ethanol, IPM, cyclomethicone, sunscreens, lactic acid; partly sol. in water, lt. min. oil; m.w. 2000; sp.gr. 1.002; dens. 8.34 lb/gal; visc. 300 cSt; pour pt. -30 C; flash pt. (PMCC) 390 F; ref. index 1.449; surf. tens. 39.6 dynes/cm (100 ppm); sp. heat 0.452 cal/g/°C

Dow P4000 [Dow]
Chem. Descrip.: PPG-30
CAS 25322-69-4
Uses: Provides solvency, low odor, low irritation, and low toxicity to personal care prods. and cosmetics; boiler water additive for prep. of steam that will contact food; defoamer for food-grade paper coatings
Regulatory: 21CFR §173.310, 176.200
Properties: Liq.; m.w. 4000; sp.gr. 1.005; dens. 8.36 lb/gal; visc. 800 cSt; pour pt. -26 C; flash pt. (PMCC) 365 F; ref. index 1.450; surf. tens. 39.6 dynes/cm (100 ppm)

Dow PT250 [Dow]
Chem. Descrip.: PPG glyceryl ether polyglycol
CAS 25791-96-2
Uses: Provides solvency, low odor, low irritation, and low toxicity to personal care prods. and cosmetics
Properties: Liq.; m.w. 250; sp.gr. 1.091; dens. 9.07 lb/gal; visc. 850 cSt; pour pt. -18 C; flash pt. (PMCC) 390 F; ref. index 1.459

Dow PT700 [Dow]
Chem. Descrip.: PPG-10 glyceryl ether
CAS 25791-96-2
Uses: Provides solvency, low odor, low irritation, and low toxicity to personal care prods. and cosmetics
Properties: Liq.; m.w. 700; sp.gr. 1.033; dens. 8.59 lb/gal; visc. 236 cSt; pour pt. -32 C; flash pt. (PMCC) 500 F; ref. index 1.453; sp. heat 0.442 cal/g/°C

Dow PT3000 [Dow]
Chem. Descrip.: PPG-55 glyceryl ether
CAS 25791-96-2
Uses: Provides solvency, low odor, low irritation, and low toxicity to personal care prods. and cosmetics
Properties: Liq.; m.w. 3000; sp.gr. 1.010; dens. 8.40 lb/gal; visc. 480 cSt; pour pt. -27 C; flash pt. (PMCC) 430 F; ref. index 1.451; sp. heat 0.445 cal/g/°C

Dowanol® DB [Dow]
Chem. Descrip.: Diethylene glycol n-butyl ether
CAS 112-34-5; EINECS 203-961-6
Uses: Solvent, coupling agent providing improved surf. wetting, soil penetration in household, commercial and industrial cleaning prods.; coupling solvent for resins and dyes in water-based printing ink, textile dyeing and printing; coalescent in water-based architectural and industrial coatings, latex adhesives; deactivator, stabilizer for agric. pesticides
Properties: Sol. in water; m.w. 162.2; sp.gr. 0.952; dens. 7.94 lb/gal; vapor pressure 0.06 mm Hg; f.p. -90 F; b.p. 230 C; flash pt. (PMCC) 226 F; surf. tens. 30 dynes/cm
Toxicology: May cause sl. hemoglobinuria

Dowanol® DM [Dow]
Chem. Descrip.: Methoxydiglycol
CAS 111-77-3; EINECS 203-906-6
Uses: Solvent, coupling agent providing improved surf. wetting, soil penetration in household, commercial and industrial cleaning prods.; solv.-based printing inks, textile printing pastes; vat dyeing fabrics; rust removers, aluminum brighteners; paint and varnish removers; deactivator and stabilizer for agric. pesticides; anti-icing additive for jet fuels
Properties: Sol. in water; m.w. 120.1; sp.gr. 1.021; dens. 8.47 lb/gal; vapor pressure 0.25 mm Hg; f.p. -92 F; b.p. 194.1 C; flash pt. (Seta CC) 197 F; surf. tens. 34.8 dynes/cm

Dowanol® DPM [Dow; Dow Europe]
Chem. Descrip.: Dipropylene glycol methyl ether

CAS 34590-94-8
Uses: Solvent, coupling agent providing improved surf. wetting, soil penetration in household, commercial and industrial cleaning prods., grease/paint removers, metal cleaners; stabilizer for agric. herbicides; solvent, coupling agent for water- and solv.-based coatings, printing inks, vat dyeing fabrics; solvent, coupling agent, and emollient in cosmetic formulations
Properties: Sol. in water; m.w. 148.2; sp.gr. 0.951; dens. 7.91 lb/gal; vapor pressure 0.55 mm Hg; f.p. -117 F; b.p. 184 C; flash pt. (Seta) 167 F; surf. tens. 28.8 dynes/cm
Toxicology: ACGIH TLV 100 ppm; no effects at any level

Dowanol® DPMA [Dow]
Chem. Descrip.: PPG-2 methyl ether acetate
CAS 88917-22-0
Uses: Solvent, coupling agent providing improved surf. wetting, soil penetration in household, commercial and industrial cleaning prods.; solvent for solv.-based coatings and printing inks; coalescent for architectural water-borne coatings
Properties: Sol. 19.4 ml/100 ml water; m.w. 190.2; sp.grf. 0.976; dens. 8.12 lb/gal; vapor pressure < 1.0 mm Hg (20 C); f.p. < -67 F; b.p. 209.3 C; flash pt. (TCC) 186 F; surf. tens. 28.6 dynes/cm

Dowanol® DPnB [Dow; Dow Europe]
Chem. Descrip.: Dipropylene glycol n-butyl ether
CAS 29911-28-2
Uses: Solvent, coupling agent providing improved surf. wetting, soil penetration in household, commercial and industrial cleaning prods., grease/paint removers, metal cleaners, hard surf. cleaners; effective surfactant props.; coalescent in water-borne latex coatings; solvent for solv.-based coatings; chem. intermediate for prod. of epoxides, acid ester derivs., solvents, and plasticizers
Properties: Sol. 5 g/100 g water; m.w. 190.3; sp.gr. 0.9; dens. 7.56 lb/gal; visc. 4.9 cps; vapor pressure 0.06 mm Hg (20 C); f.p. -75 C; b.p. 229 C; flash pt. (Seta) 212 F; ref. index 1.425; surf. tens. 28.8 dynes/cm; sp. heat 0.59 cal/g/°C
Toxicology: May cause skin irritation, liver effects
Precaution: Flamm. limits in air 0.6 (145 C) to 20.4 (180 C)

Dowanol® DPnP [Dow]
Chem. Descrip.: Dipropylene glycol n-propyl ether
CAS 29911-27-1
Uses: Coalescent in water-borne latex coatings; active solvent for solv.-borne coatings
Properties: Colorless liq.; low odor; mod. water sol.; m.w. 176.2; sp.gr. 0.922; dens. 7.70 lb/gal; visc. 4.3 cSt; vapor pressure 0.12 mm Hg; b.p. 212.0 C; flash pt. (Seta) 190.0 F; surf. tens. 27.6 dynes/cm

Dowanol® EB [Dow]
Chem. Descrip.: Ethylene glycol n-butyl ether
CAS 111-76-2; EINECS 203-905-0
Uses: Solvent, coupling agent providing improved surf. wetting, soil penetration in household, commercial and industrial cleaning prods., rust removers, disinfectants; solvent for solv.-based coatings, printing inks; agric. pesticides; coupling agent, coalescent for water-based industrial and architectural coatings
Properties: Sol. in water; m.w. 118.2; sp.gr. 0.897; dens. 7.49 lb/gal; vapor pressure 0.88 mm Hg; f.p. -125 F; b.p. 171.1 C; flash pt. (TCC) 150 F; surf. tens. 27.4 dynes/cm
Toxicology: May cause blood effects and toxicity by inhalation

Dowanol® EPh [Dow]
Chem. Descrip.: Ethylene glycol phenyl ether
CAS 122-99-6; EINECS 204-589-7
Uses: Solvent, emulsifier, coupling agent providing improved surf. wetting, soil penetration in household, commercial and industrial cleaning prods.; coalescent in adhesives, water-based architectural

and industrial coatings; plasticizer for cellulosics; solvent for printing inks; carrier solvent for textile dyes; intermediate for soap/perfume fixatives; antimicrobial in cosmetics and fragrances; paint removers; ingred. in metal corrosion protection slurries

Properties: Colorless liq., rose-like odor; sol. 2.3 ml/100 ml in water; m.w. 138.2; sp.gr. 1.104; dens. 9.2 lg/bal; vapor pressure 0.007 mm Hg; f.p. 51 F; b.p. 245.6 C; flash pt. (Seta) 250 F; surf. tens. 42 dynes/cm

Toxicology: May cause minor skin effects; no evidence of systemic toxicity by dermal exposure; may cause kidney and other organ effects by oral exposure; may cause mod. eye irritation and corneal injury

Environmental: Pract. nontoxic to fish and aquatic invertebrate; high index of biodeg.

Dowanol® PM [Dow; Dow Europe]

Chem. Descrip.: Propylene glycol methyl ether

CAS 107-98-2; EINECS 203-539-1

Uses: Solvent, coupling agent providing improved surf. wetting, soil penetration in household, commercial and industrial cleaning prods., rust removers, hard surf. cleaners; carrier solvent for writing pen inks; solvent for agric. pesticides; deactivator and emollient for livestock pesticides; solvent, coupling agent for solv.-based coatings, printing inks

Properties: Sol. in water; m.w. 90.1; sp.gr. 0.919; dens. 7.65 lb/gal; vapor pressure 12.5 mm Hg; f.p. -142 F; b.p. 120.1 C; flash pt. (Seta) 90 F; surf. tens. 27.7 dynes/cm

Toxicology: ACGIH TLV 100 ppm; may cause CNS depression, liver effects

Dowanol® PMA [Dow]

Chem. Descrip.: Propylene glycol methyl ether acetate

CAS 108-65-6; EINECS 203-603-9

Uses: Solvent, coupling agent providing improved surf. wetting, soil penetration in household, commercial and industrial cleaning prods.; solvent for solv.-based coatings and printing inks

Properties: Sol. 20.5 ml/100 ml water; m.w. 132.2; sp.gr. 0.966; dens. 8.03 lb/gal; vapor pressure 3.7 mm Hg (20 C); f.p. < -67 F; b.p. 145.8 C; flash pt. (Seta CC) 114 F; surf. tens. 26.4 dynes/cm

Toxicology: May cause respiratory tract irritation, mild liver effects at high doses

Dowanol® PnB [Dow; Dow Europe]

Chem. Descrip.: Propylene glycol n-butyl ether

CAS 5131-66-8; EINECS 225-878-4

Uses: Solvent, coupling agent providing improved surf. wetting, soil penetration in household, commercial and industrial cleaning prods., grease/paint removers, metal cleaners, hard surf. cleaners; effective surfactant props.; solvent, coupling agent, coalescent in water-reducible and water-borne coatings; active solvent in solv.-based coatings; chem. intermediate for prod. of epoxides, acid ester derivs., solvents, and plasticizers

Properties: Low odor; sol. 6.4 g/100 g water; m.w. 132.2; sp.gr. 0.9; dens. 7.31 lb/gal; visc. 3.1 cps; vapor pressure 0.6 mm Hg (20 C); f.p. < –85 C; b.p. 170 C; flash pt. (TCC) 138 F; ref. index 1.416; surf. tens. 26.3 dynes/cm; sp. heat 0.63 cal/g/°C

Toxicology: May cause skin irritation, liver and kidney effects

Precaution: Flamm. limits in air 1.1 (80 C) to 8.4 (145 C)

Dowanol® PnP [Dow]

Chem. Descrip.: Propylene glycol n-propyl ether

CAS 1569-01-3

Uses: Effective coalescent in water-borne latex coatings; coupling co-solvent in water-reducible aerosol paint formulations; coupling agent, solvent in household and industrial cleaners, grease/paint removers, metal cleaners, hard surface cleaners

Properties: Colorless liq.; low odor; completely water sol.; m.w. 118.2; sp.gr. 0.885; dens. 7.38 lb/gal; visc. 2.2 cSt; vapor pressure 1.7 mm Hg; b.p. 149.8 C; flash pt. (Seta) 119.0 F; surf. tens. 25.9 dynes/cm

Dowanol® PPh [Dow]

Chem. Descrip.: Propylene glycol phenyl ether

CAS 770-35-4

Uses: Latex coalescent in water-based architectural and industrial coatings; carrier solvent for textile dyes; solvent for inks, textile printing pastes; paint removers; coalescent for latex adhesives

Properties: Sol. 1.1 ml/100 ml in water; m.w. 152.2; sp.gr. 1.063; dens. 8.8 lb/gal; vapor pressure < 0.01 mm Hg (20 C); f.p. 55 F; b.p. 242.7 C; flash pt. (TCC) 240 F; surf. tens. 38.1 dynes/cm

Toxicology: No evidence of systemic toxicity; may cause mild transient dermal irritation

Dowanol® TBH [Dow]

Chem. Descrip.: Triethylene glycol butyl ether

CAS 143-22-6; EINECS 205-592-6

Uses: Solvent

Properties: Infinite sol. in water; m.w. 231.2 (avg.); sp.gr. 0.996; dens. 8.3 lb/gal; visc. 9.2 cSt; vapor pressure < 0.01 mm Hg; b.p. 283 C; flash pt. (PMCC) 285 F; surf. tens. 31.4 dynes/cm (20 C)

Dowanol® TMH [Dow]

Chem. Descrip.: Triethylene glycol methyl ether/highers

CAS 112-50-5; EINECS 203-978-9

Uses: Solvent

Properties: Infinite sol. in water; m.w. 173 (avg.); sp.gr. 1.054; dens. 8.8 lb/gal; visc. 7.0 cks; vapor pressure < 0.01 mm Hg; b.p. 232 C; flash pt. (PMCC) 255 F; surf. tens. 39.1 dynes/cm (20 C)

Dowanol® TPM [Dow]

Chem. Descrip.: PPG-3 methyl ether

CAS 25498-49-1

Uses: Solvent, coupling agent providing improved surf. wetting, soil penetration in household, commercial and industrial cleaning prods.; rust, paint, and varnish removers; penetrating oils; stamp pad and pen inks

Properties: Sol. in water; m.w. 206.3; sp.gr. 0.965; dens. 8.03 lb/gal; vapor pressure 0.02 mm Hg; f.p. -110 F; b.p. 242.4 C; flash pt. (Seta) 232 F; surf. tens. 30 dynes/cm

Toxicology: Mild skin irritant; may cause narcosis at lower doses, mortality at high dose

Dowanol® TPnB [Dow]

Chem. Descrip.: Tripropylene glycol n-butyl ether

CAS 55934-93-5

Uses: Solvent for heavy duty cleaners, possible additive for floor wax strippers, plasticizer in selected resin systems, coupling agent and grease penetrant in oven cleaners, and solvent for inks in ballpoint and felt tip pens, stamp pads; textile printing pastes; suitable for indoor decorative paints

Properties: Colorless low visc. liq.; pract. no odor; sol. 3.0 g/100 g in water; m.w. 248.4; sp.gr. 0.932; dens. 7.77 lb/gal; visc. 7.3 cks; vapor pressure < 0.01 mm Hg (20 C); b.p. 276 C; f.p. -75 C; flash pt. (PMCC) 255 F; ref. index 1.433; surf. tens. 29.9 dynes/cm

DP [ARCO]

Chem. Descrip.: Diethylene glycol propyl ether

CAS 6881-94-3

Uses: Solvent for coatings and cleaning applics.; HAPS listed; SARA reportable

Properties: 100% sol. in water @ 20 C; m.w. 148.20; sp.gr. 0.963 (20/20 C); dens. 8.04 lb/gal; visc. 4.1 cps; vapor pressure 0.03 mm Hg (20 C); b.p. 202 C; HLB 8.2; flash pt. (TCC) 200 F; ref. index 1.4290; surf. tens. 32.3 dynes/cm; sp. heat 0.55 cal/g/°C

Drakeol® 7 [Penreco]

Chem. Descrip.: Lt. min. oil NF

CAS 8042-47-5

Uses: Carrier, base ingred. in ointments, lotions, baby oils, sun tan lotions, makeup; solvent and emollient in creams, waterless hand cleaners; protective coating for foods, eggs, fruits, and vegetables; divider oil, mold release lubricant for food indus; food pkg. materials; meat packing; pigment dispersant, lubricant for plastics; topical pharmaceutical ointments; gelatin capsule lubricant

Regulatory: FDA 21CFR §172.878, 178.3620, 573.680

Properties: Sp.gr. 0.828-0.843; dens. 6.94-7.08 lb/gal; visc. 10.8-13.6 cSt (40 C); pour pt. -9 C; flash pt. 177 C; ref. index 1.4632

DX 3641 Naphtha [Exxon]

Chem. Descrip.: Hydrocarbon solv.

CAS 64742-47-8

Uses: Solvent

Properties: Sp.gr. 0.79 (15.6/15.6 C); dens. 6.60 lb/gal (60 F); b.p. 187-210 C; flash pt. (TCC) 64 C; surf. tens. 25.1 dynes/cm; KB value 33; 49% paraffin, 45% cycloparaffin, 6% aromatics

Toxicology: OEL 200 ppm

Dynacerin® 660 [Hüls Am.; Hüls AG]

Chem. Descrip.: Oleyl erucate

CAS 17673-56-2; EINECS 241-654-9

Uses: Emollient, lubricant, dispersant, solubilizer for cosmetics/pharmaceuticals; carrier/vehicle and solvent for topical ointments, creams, lotions; jojoba oil substitute; oxidation-stable

Properties: Ylsh. liq. wax; sol. in alcohol, min. oil, acetone; acid no. 1 max.; iodine no. 95 max.; sapon. no. 95-110; cloud pt. 15 C

Dynaflush [Dynaloy]

Uses: Nonflamm., noncorrosive, noncarcinogenic, non-ozone depleting solvent for flushing and cleaning residues from urethane processing equip. at R.T.; can be used hot (95 C) for faster results; cleans residues from mixing and metering equip., nozzles, spray guns, RIM molds

Properties: Clear liq.; sp.gr. 1.06; b.p. 202 C; flash pt. (CC) 191 F; pH 4-6

Dynasolve 100 [Dynaloy]

Chem. Descrip.: Dimethylformamide-based solv.

CAS 68-12-2; EINECS 200-679-5

Uses: Nonselective solvent for removal of adhesives, polymers, and fiber optics coatings; electronic depotting, decapsulation on cured acrylics, epoxies, and urethanes

Dynasolve 150 [Dynaloy]

Chem. Descrip.: Tetrahydrofuran-based solv.

CAS 109-99-9; EINECS 203-726-8

Uses: Flamm. solvent for electronic conformal coating removal; for R.T. use only

Dynasolve 165 [Dynaloy]

Chem. Descrip.: Blend of methylene chloride-based solvs. and org. chems.

Uses: R.T. decapsulation solvent for dissolving cast epoxy resins; for general adhesive and polymer removal, paint stripping, conformal coating removal, depotting, deflashing

Dynasolve 180 [Dynaloy]

Chem. Descrip.: N-Methyl pyrrolidone base

Uses: Solvent for removal of adhesives and inks from rollers, general polymer removal, paint stripping; suitable for cured epoxies; also for electronic depotting, deflashing, decapsulation, deinking applics., surf. prep. and photoresist stripping; use hot (105-107 C); not selective

Dynasolve 185 [Dynaloy]

Chem. Descrip.: N-Methyl pyrrolidone base

Uses: Solvent for removal of adhesives and inks from rollers, general polymer removal, paint stripping, removal of fiber optic coatings; suitable for cured epoxies; also for electronic depotting, deflashing, decapsulation, deinking applics.; surf. prep. and photoresist stripping; use hot (105 C); not selective; modified version of Dynasolve 180

Dynasolve 185 SG [Dynaloy]

Chem. Descrip.: N-Methyl pyrrolidone base

Uses: Electronic grade solvent for depotting, deflashing, decapsulation, and deinking applics., surf. prep., photo resist stripping

Dynasolve 190 [Dynaloy]

Chem. Descrip.: Methanol base
Uses: Solvent for electronic depotting, conformal coating removal, cable assembly repair, elec. connector repair; suitable for cured urethanes and silicones; for R.T. use only; stronger than Uresolve Blue but less selective
Precaution: Flamm.

Dynasolve 200 [Dynaloy]
Chem. Descrip.: Methylene chloride-based
CAS 75-09-2; EINECS 200-838-9
Uses: Selective solvent effective at R.T.; used for polysulfide removal from aircraft fuel tanks; adhesives, polymers, inks

Dynasolve 210 [Dynaloy]
Chem. Descrip.: Methylene chloride-based
CAS 75-09-2; EINECS 200-838-9
Uses: Solvent which dissolves RTV silicones and silicone conformal coatings effectively at room temp.; for depotting, decapsulation; removal of RTV's from CRT tubes; removal of buffer coatings from optical fibers; paint stripping; may attack epoxies and urethanes
Properties: Amber liq.; chlorinated solv. odor; sol. 2.0 g/100g water; sp.gr. 1.272; vapor pressure 355 mm Hg (20 C); b.p. 104 F; pH 2.14; 100% volatile; 75-95% methylene chloride, 5-25% benzenesulfonic acid
Precaution: Corrosive liq.; keep away from contact with water, keep away from from heat, sparks, or open flame
Storage: Store in cool, dry, ventilated area

Dynasolve 220 [Dynaloy]
Chem. Descrip.: Hydrocarbon base
Uses: Solvent which dissolves RTV silicones and silicone conformal coatings effectively at room temp.; for removal of mold releases, parting agents, adhesives, polymers; electronic depotting, decapsulation, removal of RTV's from CRT tubes; may attack epoxies and urethanes
Properties: Amber liq.; faint petroleum hydrocarbon odor; sol. < 0.1% in water; sp.gr. 0.828; vapor pressure < 10 mm Hg: b.p. 311-348 F; flash pt. 110 F; pH 1.70; 100% volatile; 60-80% hydrotreated heavy naphtha, petroleum; 20-40% benzenesulfonic acid
Precaution: Corrosive liq.; keep away from contact with water, keep away from from heat, sparks, or open flame
Storage: Store in cool, dry, ventilated area

Dynasolve 225 [Dynaloy]
Chem. Descrip.: Hydrocarbon base
Uses: Solvent which dissolves RTV silicones and silicone conformal coatings effectively at room temp.; for removal of mold releases, parting agents, adhesives, polymers; electronic depotting, decapsulation, removal of RTV's from CRT tubes; may attack epoxies and urethanes
Properties: Amber liq.; faint petroleum hydrocarbon odor; sol. < 0.1% in water; sp.gr. 0.806; vapor pressure approx. 30 mm Hg; b.p. 235-290 F; flash pt. 52 F; pH 1.60; 100% volatile; 60-80% hydrotreated heavy naphtha, petroleum; 20-40% benzenesulfonic acid
Precaution: Corrosive, flamm. liq.; keep away from contact with water, keep away from from heat, sparks, or open flame
Storage: Store in cool, dry, ventilated area

Dynasolve 230 [Dynaloy]
Chem. Descrip.: Organic
Uses: Solvent which dissolves RTV silicones and silicone conformal coatings effectively at room temp.; for removal of mold releases, parting agents, adhesives, polymers; electronic depotting, decapsulation, removal of RTV's from CRT tubes; may attack epoxies and urethanes
Properties: Amber liq.; faint petroleum hydrocarbon odor; sol. < 0.1% in water; sp.gr. 0.707; vapor pressure approx. 60 mm Hg; b.p. 202-222 F; flash pt. 16 F; pH 1.30;

100% volatile; 60-80% hydrotreated heavy naphtha, petroleum; 20-40% benzenesulfonic acid

Precaution: Corrosive, flamm. liq.; keep away from contact with water, keep away from from heat, sparks, or open flame

Storage: Store in cool, dry, ventilated area

Dynasolve 611 [Dynaloy]

Chem. Descrip.: Ethylene glycol ether base

Uses: Solvent for surf. prep. and residual photoresist removal or removal of other org. contaminants; use hot (100-120 C)

Dynasolve 699 [Dynaloy]

Chem. Descrip.: Butyrolactone-based solv.

CAS 96-48-0; EINECS 202-509-5

Uses: Solvent for use at R.T. for thin acrylic coatings; used hot (120-150 C) for acrylic adhesives and epoxy powd. coatings; for removal of conformal coatings, fiber optics coatings, adhesives, and polymers, and for paint stripping; electronic depotting applics.

Dynasolve 700 [Dynaloy]

Chem. Descrip.: Propylene glycol ether base

Uses: Solvent for electronic depotting, decapsulation applics; for R.T. temp. use only; similar in str. to Uresolve HF

Precaution: Flamm.

Dynasolve 710 [Dynaloy]

Chem. Descrip.: Propylene glycol ether base

Uses: Solvent for electronic depotting, decapsulation, deinking, and surf. prep. applics.; suitable for cured epoxies, urethanes, and silicones; stronger than Dynasolve 700, but less selective; similar in str. to Uresolve Plus; slow at R.T

Dynasolve 711 [Dynaloy]

Chem. Descrip.: Propylene glycol ether base

Uses: Solvent for electronic depotting, decapsulation, deinking, surf. prep. and photo resist stripping applics.; suitable for cured epoxies, urethanes, and silicones; similar to Uresolve Plus SG; can be used in contact with aluminum

Dynasolve 750 [Dynaloy]

Chem. Descrip.: Propylene glycol ether base

Uses: Solvent for electronic depotting, decapsulation, deinking, and surf. prep. applics.; suitable for cured epoxies, urethanes, and silicones; stronger than Dynasolve 700, but less selective; similar in str. to Uresolve Plus; slow at R.T.; stronger than Dynasolve 710, but less selective; similar in str. to Uresolve Plus 500; slow at R.T., more effective at 65-110 C

Dynasolve CU-4 [Dynaloy]

Uses: Solvent similar to Dynaflush, but more effective for cleaning urethane residues and crystallized isocyanates; can be used hot (95 C) for faster results; cleans residues from mixing and metering equip., nozzles, spray guns, RIM molds

Dynasolve CU-5 [Dynaloy]

Uses: Solvent designed for use in cleaning urethane residues and crystallized isocyanates from various types of polyurethane processing equip.; nonchlorinated, non-ozone depleting; designed to replace methylene chloride, acetone, MEK, and 1,1,1-trichlororethane; can be used for extended periods, tolerates high resin loading; recyclable via vacuum distillation

Properties: Clear; sp.gr. 1.060; b.p. 202 C; flash pt. (CC) 191 F; pH 4-6; nonflamm.

Toxicology: Harmful if inhaled or swallowed; noncarcinogenic

Precaution: Incompat. with Viton, PVC, ABS, Buna-N, Kynar, Lexan, Noryl EN-265, PET, butyl rubber, neoprene

Storage: Contains powerful org. solvs., keep away from heat and flame

Dynasolve CU-6 [Dynaloy]

Uses: Nongelling solvent designed for use in cleaning urethane residues and crystallized isocyanates from various types of polyurethane processing equip.; nonchlorinated, non-ozone depleting; designed to replace methylene chloride, acetone, MEK, and 1,1,1-trichlororethane; can be used for extended periods, tolerates high resin loading; recyclable via vacuum distillation

Properties: Clear; sp.gr. 1.055; b.p. 202 C; flash pt. (CC) 210 F; pH 4-6; nonflamm.

Toxicology: Harmful if inhaled or swallowed; noncarcinogenic

Precaution: Incompat. with Viton, PVC, ABS, Buna-N, Kynar, Lexan, Noryl EN-265, PET, butyl rubber, neoprene

Storage: Contains powerful org. solvs., keep away from heat and flame

Dynasolve DF-1 [Dynaloy]

Uses: Solvent; replenisher conc. for Dynaflush; when added to vacuum distilled or spent Dynaflush, will restore its efficiency and loading capacity to original levels

Dynasolve M-10 [Dynaloy]

Chem. Descrip.: Hydrocarbon base

Uses: Solvent effective on silicone oils and parting agents at R.T.; also for general adhesive and polymer removal; for quicker removal, heat to 65 C

Dynasolve M-30 [Dynaloy]

Uses: Nonchlorinated, nonflamm. solvent for general cleaning and degreasing operations; esp. for curing of uncured polymers such as epoxies, urethanes and silicones

Dynasolve M-35 [Dynaloy]

Uses: Low-toxicity, low volatility solvent for general cleaning applics.; especially suited for use in cleaning of uncured polymers, such as epoxies, urethanes, and silicones; also effective for dissolving cured cyanoacrylate instant glues

Dynasolve MP-500 [Dynaloy]

Chem. Descrip.: Ethylene glycol ether base

Uses: Solvent for cleaning residues from mixing, metering equip., nozzles, RIM molds; removal of mold releases, adhesives and inks from rollers, general polymer removal; suitable for cured epoxies, silicones, and urethanes; stronger than Dynasolve MP-AL GR, but less selective; slow at R.T.; more effective at 65-110 C

Dynasolve MP-710 [Dynaloy]

Chem. Descrip.: Propylene glycol ether base

Uses: Solvent for cleaning residues from mixing, metering equip., nozzles, RIM molds; removal of mold releases, adhesives and inks from rollers, general polymer removal, paint stripping; suitable for cured epoxies, silicones, and urethanes; similar in str. to Dynasolve MP-AL GR, but less selective; slow at R.T.; more effective at 65-110 C

Dynasolve MP-750 [Dynaloy]

Chem. Descrip.: Propylene glycol ether base

Uses: Solvent for cleaning residues from mixing, metering equip., nozzles, RIM molds; removal of mold releases, adhesives and inks from rollers, general polymer removal, paint stripping; suitable for cured epoxies, silicones, and urethanes; stronger than Dynasolve MP-710, but less selective; similar in str. to Dynasolve MP-500; slow at R.T.; more effective at 65-110 C

Dynasolve MP-AL GR [Dynaloy]

Chem. Descrip.: Ethylene glycol ether base

Uses: Solvent for cleaning residues from mixing, metering equip., nozzles, RIM molds; removal of mold releases, adhesives and inks from rollers, general polymer removal; suitable for cured epoxies, silicones, and urethanes; performs slowly at R.T.; more effective at 65-110 C; can be used in contact with aluminum

Dynasolve XD 16-3 [Dynaloy]

Chem. Descrip.: Aq. sol'n.

Uses: Solvent for removal of silicone oils and uncured silicone polymers from molds, molded parts, and processing equip.; esp. effective for cleaning mold release residue from plastic molded parts

Properties: Lt. amber liq.; sp.gr. 1.010; b.p. 212 F; flash pt. none; pH 6-8
Toxicology: Avoid contact with eyes and skin, breathing of vapors

Dynasolve XD 16-4 [Dynaloy]
Uses: Solvent developed for removal of silicone oils and uncured silicone polymers from molds, molded parts, and processing equip.
Properties: Lt. amber; sp.gr. 1.032; b.p. > 300 F; flash pt. (CC) 262 F
Toxicology: Harmful if inhaled or swallowed
Storage: Contains powerful org. solvs., keep away from heat and flame

Dynasolve XD 22-1 [Dynaloy]
Uses: Solvent designed for use in cleaning and degreasing applics.; nonchlorinated, non-ozone depleting; low vapor pressure, minimizing VOC emissions; noncorrosive to metals, leaves no residue upon evaporation; designed to replace methylene chloride, acetone, MEK, min. spirits, kerosene, and 1, 1,1-trichloroethane; recyclable via vacuum distillation
Properties: Clear; sp.gr. 0.780; b.p. 360 C; flash pt. (CC) 143 F; nonflamm.
Toxicology: Harmful if inhaled or swallowed; noncarcinogenic
Precaution: Incompat. with silicone rubber, neoprene, butyl rubber, EPDM, ABS, Noryl, polypropylene
Storage: Contains powerful org. solvs., keep away from heat and flame

Dynasolve XD 27-2 [Dynaloy]
Uses: Safety solvent for use in cleaning, defluxing of printed circuit boards; usable in commercially available cleaning equip.; superior solvency thoroughly and quickly removes rosin and ionic contamination from circuit boards and other electronic devices

Dytek® EP [DuPont Nylon]
Chem. Descrip.: 1,3-Pentanediamine
CAS 589-37-7
UN No.: 2734
Uses: Intermediate; epoxy curative; solvent for gas treatment; surfactant (asphalt emulsifiers, textiles); scale/corrosion inhibitor; additive for fuel, oil, plastic, petrol. and chem. processing; extender, catalyst for PU; polyamide monomer for coatings, adhesives, sealants, elastomers
Regulatory: CERCLA reportable
Properties: Colorless liq., sl. odor; misc. with water; m.w. 102; dens. 0.855 g/ml; visc. 1.89 cp; vapor pressure 1 mm Hg (20 C); m.p. -121 C; b.p. 164 C; flash pt. (CC) 59 C; ref. index 1.4512 (20/20 C); pH 12.3 (10%); surf. tens. 32.2 dyn/cm; 99% act.
Toxicology: Causes eye burns; damage may be permanent and may incl. blindness; causes skin sensitization and burns; may causse nose, throat, and lung irritation
Environmental: Biodeg.
Precaution: Corrosive, flamm.; incompat. with strong oxidants; thermal decomp. produces toxic fumes of NO_x
HMIS: Health: 3; Flammability: 2; Reactivity: 1
Storage: Store in well ventilated area in tightly closed containers; keep away from heat, spark, flames

E

Eastman® C-7 Ketone [Eastman]
Chem. Descrip.: Ketone mixt.
Uses: Solvent
Properties: Pt-Co 25 max. color; sp.gr. 0.817; dens. 6.82 lb/gal; f.p. -57 C; b.p. 140-146 C; flash pt. (TCC) 33 C

Eastman® C-11 Ketone [Eastman]
Chem. Descrip.: Ketone mixt.
Uses: Very slow-evaporating retarder solvent for coating formulations, lacquers, screen inks
Properties: Sol. 0.2% in water (20 C); sp.gr. 0.84 (20/20 C); dens. 7.02 lb/gal (20 C); visc. 2.12 cP (neat); f.p. 11 F; b.p. 200-240 C; flash pt. (Seta) 84 C; ref. index 1.4355 (20 C); surf. tens. 27.5 dynes/cm (24 C)

Eastman® CDA-19 [Eastman]
Chem. Descrip.: Completely denatured alcohol, denatured with 4% MIBK, 1% heptane
CAS 64-17-5; EINECS 200-578-6
Uses: Solvent used in protective coatings, printing inks, paint and varnish removers, antifreeze sol'n., as fuel for light, heat, or power prod.; avail. 95% act. and anhyd.

Eastman® DB [Eastman]
Chem. Descrip.: Diethylene glycol monobutyl ether
CAS 112-34-5; EINECS 203-961-6
Uses: Solvent used in baking enamels to promoted increased flow-out and gloss; in screen printing inks; textile dyeing formulations; coalescing aid for architectural and industrial latex paints; organosol dispersant; diluent in hydraulic brake fluids; cleaners; cutting oils; waterborne coatings; detergents, soaps; deactivator in pesticides
Regulatory: SARA and CERCLA reportable
Properties: Colorless clear liq., sl. odor; completely sol. in water; m.w. 162.23; sp.gr. 0.955 (20/20 C); dens. 7.94 lb/gal; visc. 4.74 cP; vapor pressure 0.02 mm Hg (20 C); f.p. -76 C; i.b.p. 230 C; flash pt. (COC) 111 C; fire pt. 117 C; ref. index 1.4316 (20 C); 99.2% min. purity
Toxicology: LD50 (oral, rat) 7292 mg/kg, (dermal, rabbit) 2764 mg/kg; causes eye irritation; expected to be a low ingestion hazard
Environmental: Low biochem. oxygen demand; little potential to cause oxygen depletion in aq. systems; low potential to affect aquatic organisms; BOD-5 0.25 g oxygen/g; COD 2.08 g oxygen/g
Precaution: Can react with oxidizing materials; peroxide former; flamm. limits in air 0.85% vol. (135 C) to 24.6% vol. (199 C); hazardous combustion prods.: CO, CO_2
HMIS: Health: 2; Flammability: 1; Reactivity: 0
Storage: Store in closed container away from heat and light; after opening, purge container with nitrogen before reclosing; minimize exposure to air; do not allow to evaporate to near dryness

Eastman® DB Acetate [Eastman]
Chem. Descrip.: Diethylene glycol monobutyl ether acetate
CAS 124-17-4; EINECS 204-685-9
Uses: Solvent used in printing inks and high-bake enamels; coalescing aid in low-pH latex paints; esp. applic. in screen inks and coil coatings due to very slow evaporation

rate and limited water sol.; selective solvent in separation of alcohols and ketone

Regulatory: SARA and CERCLA reportable

Properties: Colorless clear liq., pleasant odor; mod. sol. in water; m.w. 204.27; sp.gr. 0.98 (20/20 C); dens. 8.16 lb/gal; visc. 3.20 cP; vapor pressure 0.04 mm Hg (20 C); f.p. -32 C; i.b.p. 235 C; flash pt. (TCC) 105 C; fire pt. 128 C; ref. index 1.43 (20 C)

Toxicology: LD50 (oral, rat) 7.1-12.2 ml/kg, (dermal, rabbit) 5.5-14.48 ml/kg; prolonged/repeated skin contact may cause drying, cracking, or irritation; expected to be low ingestion hazard

Precaution: Can react with strong oxidizing materials, strong bases, strong acids; peroxide former; flamm. limits in air 0.8% vol. (135 C) to 10.7% vol. (203 C); hazardous combustion prods.: CO, CO_2

HMIS: Health: 1; Flammability: 1; Reactivity: 0

Storage: Store in closed container away from heat and light; after opening, purge container with nitrogen before reclosing; minimize exposure to air; do not allow to evaporate to near dryness

Eastman® DBP [Eastman]

Chem. Descrip.: Dibutyl phthalate

CAS 84-74-2; EINECS 201-557-4

Uses: Primary plasticizer-solvent for nitrocellulose lacquers, architectural latex coatings, adhesives; plasticizer for rubbers and CAB, ethyl cellulose, PVAc, and syn. resins; solvent for oil-sol. dyes, insecticides, peroxides, and org. compds.; antifoamer and fiber lubricant in textile mfg.; cosmetics plasticizer

Regulatory: SARA reportable

Properties: Colorless liq., odorless; negligible sol. in water; m.w. 278.35; sp.gr. 1.05 (20 C); vapor pressure 0.0000142 mm Hg (20 C); m.p. -35 C; b.p. 340 C; flash pt. (COC) 191 C; 100% act.

Toxicology: ACGIH TLV/TWA 5 mg/m^3; OSHA PEL 5 mg/m^3; LD50 (oral, rat) 20-25 g/kg, (dermal, rabbit) > 20 mL/kg; low hazard by inh., skin and eye contact; expected to be a low ingestion hazard

Environmental: Mod. BOD; may cause oxygen depletion in aq. systems; high potential to affect some aquatic organisms; low potential to affect germination/early growth of some plants; expected to biodegrade; COD 1.71 g O_2/g; BOD-5 0.34 g O_2/g

Precaution: Incompat. with strong oxidizing materials; lower explosive limit 0.47% vol. % (236 C); hazardous combustion prods.: CO, CO_2

HMIS: Health: 1; Flammability: 1; Reactivity: 0

Storage: Store in closed containers

Eastman® DE [Eastman]

Chem. Descrip.: Diethylene glycol monoethyl ether

CAS 111-90-0; EINECS 203-919-7

Uses: Very polar, slow-evaporating coupling solvent for CAB, nitrocellulose, epoxy resins, polyamide resins, and other coating materials; coalescing aid for latex paints and some PVA emulsions; in non-grain-raising wood stains; stamp-pad and printing inks; improves surf. gloss in floor polishes based on modified acrylic polymers; component in heavy-duty brake fluids, textile dyeing applics.; deactivator in pesticides; diluent in brake fluids

Regulatory: SARA and CERCLA reportable

Properties: Colorless clear liq., mild odor; completely sol. in water; m.w. 134.17; sp.gr. 0.990 (20/20 C); dens. 8.25 lb/gal; visc. 4.5 cP; vapor pressure 0.12 mm Hg (20 C); f.p. -90 C; i.b.p. 198 C; flash pt. (TCC) 91 C; fire pt. 96 C; ref. index 1.4260 (20 C); 100% act.

Toxicology: LD50 (oral, rat) 10,502 mg/kg, (dermal, rabbit) 9143 mg/kg; low hazard

Environmental: May cause oxygen depletion in aq. systems; low potential to affect aquatic organisms, sec. waste treatment microbial metabolism; high potential to biodegrade; BOD-5 0.14 g oxygen/ml; COD 1.9 g O_2/g

Precaution: Combustible liq. and vapor; incompat. with strong oxidizing materials; peroxide former; flamm. limits in air 1.20% vol. (135 C) to 23.5% vol. (182 C); hazardous combustion prods.: CO, CO_2

HMIS: Health: 1; Flammability: 2; Reactivity: 0

Storage: Store in closed container away from heat and light; minimize exposure to air; do not allow to evaporate to near dryness

Eastman® DE Acetate [Eastman]

Chem. Descrip.: Diethylene glycol monoethyl ether acetate

CAS 112-15-2; EINECS 203-940-1

Uses: High-boiling solvent; coalescing aid in latex paints; also for textile and silk-screen printing inks, slow-drying lacquers

Regulatory: SARA and CERCLA reportable

Properties: Colorless clear liq., sl. odor; completely sol. in water; m.w. 176.21; sp.gr. 1.012 (20/20 C); dens. 8.42 lb/gal; visc. 4.5 cP; vapor pressure 0.04 mm Hg (20 C); f.p. -25 C; i.b.p. 214 C; flash pt. (COC) 107 C; fire pt. 110 C; ref. index 1.42 (20 C); 100% act.

Toxicology: LD50 (oral, rat) 11 g/kg, (dermal, rabbit) 15 g/kg; low hazard

Environmental: Mod. BOD; may cause oxygen depletion in aq. systems; high potential to biodegrade; BOD-5 0.42 g oxygen/g

Precaution: Incompat. with strong oxidizing materials; peroxide former; flamm. limits in air 0.98% vol. (135 C) to 19.4% vol. (185 C); hazardous combustion prods.: CO, CO_2

HMIS: Health: 1; Flammability: 1; Reactivity: 0

Storage: Store in closed container away from heat and light; after opening, purge container with nitrogen before reclosing; minimize exposure to air; do not allow to evaporate to near dryness

Eastman® DE-HG [Eastman]

Chem. Descrip.: Diethylene glycol monoethyl ether/ethylene glycol

Uses: Solvent

Properties: Pt-Co 10 max. color; sp.gr. 1.027; dens. 8.56 lb/gal; f.p. -70 C; b.p. 190-205 C; flash pt. (TCC) 96 C

Eastman® DM [Eastman]

Chem. Descrip.: Diethylene glycol methyl ether

CAS 111-77-3; EINECS 203-906-6

Uses: Solvent for coating formulations incl. nongrain-raising wood stains, printing inks, dye pastes for textiles; coalescing aid for PVAc latex paints; solvent in stamp pad and stencil inks; diluent for hydraulic brake fluids; deicer in jet fuel; deactivator in pesticides

Properties: Mild odor; sol. in water; m.w. 120.15; sp.gr. 1.023 (20/20 C); dens. 8.51 lb/gal; visc. 3.90 cP (neat); vapor pressure 0.2 torr (20 C); f.p. -121 F; b.p. 191-198 C; flash pt. (TCC) 191 F; ref. index 1.4268 (20 C); surf. tens. 34.8 dynes/cm

Toxicology: TLV 100 ppm

Precaution: DOT: combustible liq.; flamm. limits in air 1.38 vol. % (135 C) to 22.7 vol.% (168 C)

Eastman® DP [Eastman]

Chem. Descrip.: Diethylene glycol monopropyl ether

CAS 5881-94-3; EINECS 229-985-7

Uses: Very slow-evaporating coupling solvent for sol'n. and water-dilutable coatings; active solvent for NC, acrylic copolymers, epoxy resins, chlorinated rubber, alkyd resins; coalescing aid in architectural latex coatings, floor polishes; liq. cleaners; diluent in hydraulic brake fluids; household and general industrial cleaners

Regulatory: SARA and CERCLA reportable

Properties: Colorless clear liq., mild odor; completely sol. in water; m.w. 148.2; sp.gr. 0.963 (20/20 C); dens. 8.04 lb/gal; visc. 4.1 cP; vapor pressure 0.05 mm Hg (20 C); f.p. < -68 C; i.b.p. 210 C; flash pt.

(TCC) 93 C; fire pt. 103 C; ref. index 1.429 (20 C); 99% min. purity
Toxicology: LD50 (oral, rat) 6661 mg/kg, (dermal, guinea pig) 5048 mg/kg; causes eye irritation; low hazard by inh., ingestion or skin contact.
Environmental: May cause oxygen depletion in aq. systems; low potential to affect aquatic organisms or germination/early growth of some plants; mod. potential to biodegrade; COD 2 g O_2/g; BOD-5 0.14 g O_2/g
Precaution: Incompat. with strong oxidizing materials; potential peroxide former; lower flamm. limit in air 0.85% vol. (122 C); hazardous combustion prods.: CO, CO_2
HMIS: Health: 2; Flammability: 1; Reactivity: 0
Storage: Store in closed container away from heat and light; minimize exposure to air

Eastman® EB [Eastman]
Chem. Descrip.: Ethylene glycol monobutyl ether
CAS 111-76-2; EINECS 203-905-0
Uses: Solvent for alkyd, phenolic, maleic, epoxy, and NC resins; retarder solvent for lacquers; improves gloss and flow-out; reduces orange peel; in water-dilutable coatings, waterborne latex emulsions or disps., pesticides, inks, resin mfg.; household and general industrial cleaners, fuel additives; coupling agent for oil-sol. components with water; coalescing aid
Regulatory: SARA and CERCLA reportable
Properties: Colorless clear liq., sl. odor; completely sol. in water; m.w. 118.7; sp.gr. 0.902 (20/20 C); dens. 7.51 lb/gal; visc. 6.4 cP; vapor pressure 0.6 mm Hg (20 C); f.p. -75 C; i.b.p. 169 C; flash pt. (TCC) 62 C; fire pt. 70 C; ref. index 1.4193 (20 C); 99.4% min. purity
Toxicology: LD50 (oral, rat) 1746 mg/kg, (dermal, rabbit) 435 mg/kg; ACGIH TLV 25 ppm TWA, skin; harmful if inhaled, absorbed thru skin, or swallowed; may cause blood disorders based on animal data; causes skin and eye irritation
Environmental: High BOD; potential to cause oxygen depletion in aq. systems; low potential to affect aquatic organisms; high potential to biodegrade; COD 2.18 g O_2/g
Precaution: Combustible liq. and vapor; can react violently with strong oxidizing materials; peroxide former; flamm. limits in air 1.1% vol. (93 C) to 12.7% vol. (135 C); hazardous combustion prods.: CO, CO_2
HMIS: Health: 2; Flammability: 2; Reactivity: 0
Storage: Store in closed container away from heat and light; minimize exposure to air; after opening, purge container with nitrogen before reclosing; do not allow to evaporate to near dryness

Eastman® EB Acetate [Eastman]
Chem. Descrip.: Ethylene glycol monobutyl ether acetate
CAS 112-07-2; EINECS 203-934-9
Uses: High-boiling retarder solvent in automotive paints and coil coatings; flow and leveling agent in high-bake and air-dry coatings incl. epoxies; moisture-sensitive paints, e.g., PU coatings; screen and stamp pad inks; coalescing aid for latex paints
Regulatory: SARA and CERCLA reportable
Properties: Colorless liq.; mild odor; sol. 1.1% in water; m.w. 160.21; sp.gr. 0.941 (20/20 C); dens. 7.84 lb/gal; visc. 1.8 cP; vapor pressure 0.29 mm Hg (20 C); f.p. -64 C; i.b.p. 186 C; flash pt. (TCC) 71 C; fire pt. 82 C; ref. index 1.4142 (20 C); 99% min. assay
Toxicology: LD50 (oral, rat) 7.46 ml/kg, (dermal, rabbit) 1.5 g/kg; may cause blood disorders, kidney damage; harmful if inhaled or absorbed thru skin; expected to be low ingestion hazard
Precaution: Combustible liq. and vapor; can react with strong oxidizing agents; peroxide former; flamm. limits in air 0.88% vol. (93

C) to 8.54% vol. (135 C); hazardous combustion prods.: CO, CO_2

HMIS: Health: 2; Flammability: 2; Reactivity: 0

Storage: Store in closed container away from heat and light; minimize exposure to air; after opening, purge container with nitrogen before reclosing

Eastman® EEH [Eastman]

Chem. Descrip.: 85% Ethylene glycol 2-ethylhexyl ether, 15% diethylene glycol 2-ethylhexyl ether

Uses: Slow-evaporating retarder solvent and coalescing aid in coating systems incl. architectural and industrial emulsions/disps.; good solvency for polyesters, alkyds, polyamides, and other coating systems; low surf. tens., high elec. resist.; good for cathodic electrodeposition primers

Regulatory: SARA reportable

Properties: Colorless clear liq., ether odor; sol. 0.2% in water (20 C); sp.gr. 0.892; dens. 7.42 lb/gal; vapor pressure 0.08 mm Hg (20 C); f.p. < -45 C; i.b.p. 224 C; flash pt. (Seta CC) 98 C; ref. index 1.4361 (20 C)

Toxicology: LD50 (oral, male rat) 4674 mg/kg, (dermal, guinea pig) > 20 ml/kg; causes skin and eye irritation; ingestion may cause GI tract irritation

Environmental: Low BOD; little potential to cause oxygen depletion in aq. systems; mod. potential to affect some aquatic organisms and germination/early plant growth of some plants; some potential to biodegrade; COD 1.00-1.03 g O_2/g

Precaution: Can react with strong oxidizing materials; potential peroxide former; flamm. limits in air 0.61% vol. (121 C) to 8.02% vol. (174 C); hazardous combustion prods.: CO, CO_2

HMIS: Health: 2; Flammability: 1; Reactivity: 0

Storage: Store in closed container away from heat and light; minimize exposure to air

Eastman® EEP [Eastman]

Chem. Descrip.: Ethyl 3-epoxypropionate

Uses: Let-down solvent for thinning synthetic coating resins; provides easier handling, storage, shipping, resin stability; ideal for thinning moisture-sensitive resins for PU coatings; formulating air-dry and baking finishes; polymerization solvent for high-solids acrylic resins; retarder solvent for high-bake enamels

Properties: Pt-Co 15 ppm max. color; flash pt. 58 C; elec. resist. 20 megohms; 99% min. assay

Toxicology: Low toxicity

Eastman® EP [Eastman]

Chem. Descrip.: Ethylene glycol monopropyl ether

CAS 2807-30-9; EINECS 220-548-6

Uses: Slow-evaporating coupling solvent for resin/water systems incl. waterborne coatings; controls visc. of waterborne resins; retarder solvent for coatings systems based on NC, acrylic, epoxy, polyamide, and alkyd resins

Regulatory: SARA and CERCLA reportable

Properties: Colorless clear liq.; mild odor; completely misc. with water over wide temp. range; m.w. 104.15; sp.gr. 0.913 (20/20 C); dens. 7.59 lb/gal; visc. 2.42 cP; vapor pressure 1.3 mm Hg (20 C); f.p. < –90 C; i.b.p. 149.5 C; flash pt. (TCC) 49 C; fire pt.56 C; ref. index 1.4136 (20 C); 99.4% min. purity

Toxicology: LD50 (oral, rat) 3089 mg/kg, (dermal, rabbit) 1337 mg/kg; causes eye irritation; harmful if inhaled, absorbed thru skin, or swallowed; may cause blood disorders

Environmental: High BOD; potential to cause oxygen depletion in aq. systems; low potential to affect aquatic organisms or germination/early growth of some plants; high potential to biodegrade; COD 2.04 g O_2/g

Precaution: Combustible liq. and vapor; can react violently with strong oxidizing agents; peroxide former; flamm. limits in air 1.26% vol. (69

C) to 15.8% vol. (127 C); hazardous combustion prods.: CO, CO_2
HMIS: Health: 2; Flammability: 2; Reactivity: 0
Storage: Store in closed containers away from heat and light; minimize exposure to air; after opening, purge container with nitrogen before reclosing

Eastman® PM [Eastman]
Chem. Descrip.: Propylene glycol methyl ether
CAS 107-98-2; EINECS 203-539-1
Uses: Solvent for coating formulations
Properties: Sol. in water; m.w. 90.12; sp.gr. 0.923 (20/20 C); dens. 7.69 lb/gal; visc. 1.90 cP (neat); vapor pressure 8 torr (20 C); f.p. -139 F; b.p. 121 C; flash pt. (TCC) 91 F; ref. index 1.4036 (20 C); surf. tens. 28.3 dynes/cm
Toxicology: TLV 100 ppm

Eastman® PM Acetate [Eastman]
Chem. Descrip.: Propylene glycol monomethyl ether acetate
CAS 108-65-6; EINECS 203-603-9
Uses: Retarder solvent for gravure inks; glycol ether ester solvent for screen inks; process solvent for resin mfg., urethanes
Properties: Pt-Co 10 max. color; sp.gr. 0.970; dens. 8.06 lb/gal; f.p. -55 C; b.p. 140-150 C; flash pt. (Seta) 45 C; 99% min. assay

Eastman® TEP [Eastman]
Chem. Descrip.: Triethyl phosphate
CAS 78-40-0; EINECS 201-114-5
Uses: Intermediate for insecticides/pesticides, in floor polishes, as plasticizer for cellulose acetate, as solvent for textiles, dyeing assistant, in sizes, flame retardant synergist in unsat. polyesters, in lubricants, hydraulic fluids, as catalyst; visc. suppressant for general polymer applics.
Properties: Colorless clear liq., mild pleasant cider-like odor; misc. with water, ethanol, ethyl acetate, ethyl ether, acetone, chloroform, benzene, xylene, castor oil; partly sol. in min. oil, kerosene, CS_2; sp.gr. 1.067-1.072 (20/20 C); dens. 8.91 lb/gal; visc. 1.76 cP (20 C); vapor pressure 1 mm Hg (40 C); m.p. –56.4 C; b.p. 209 C; flash pt. (PMCC) 99 C; fire pt. (COC) 141 C; ref. index 1.4019
Toxicology: LD50 (oral, rat) 1311 mg/kg (sl. toxic), (dermal, guinea pig) > 20 ml/kg; causes eye irritation; sl. skin irritant; emits highly toxic fumes of phosphorus oxides when heated to decomp.
Precaution: Incompat. with oxidizing materials

Eastman® Triacetin [Eastman]
Chem. Descrip.: Glyceryl triacetate USP
CAS 102-76-1; EINECS 203-051-9
Uses: Plasticizer for coatings, film coatings on pharmaceutical tablets, beads, and gran.; also in topical antifungal creams and ointments; compat. with cellulosics, methacrylic acid copolymers, and PVAP; high solvent power
Use Level: 10-35% (tablet coatings)
Properties: Low color liq.; sol. 61.20 g/l in water (20 C); misc. with alcohol, ether, chloroform; m.w. 218.21; sp.gr. 1.1540; dens. 9.65 lb/gal (20 C); visc. 17.4 cP; f.p. 3.2 C; supercools to ≈ -70 C; m.p. -89 C; b.p. 258 C (760 mm); ref. index 1.4290

Eastone FSE [Eastman]
Chem. Descrip.: Glycol ether
Uses: Solvent
Properties: Pt-Co 10 max. color; sp.gr. 0.913 (20.20 C); dens. 7.60 lb/gal; f.p. -90 C; b.p. 150-159 C; flash pt. (TCC) 49 C; 99% min. assay

EB [Occidental]
Chem. Descrip.: Butoxyethanol
CAS 111-76-2; EINECS 203-905-0
Uses: Solvent

EB [ARCO]
Chem. Descrip.: Ethylene glycol butyl ether
CAS 111-76-2; EINECS 203-905-0
Uses: Solvent for coatings and cleaning applics.; HAPS listed; SARA reportable
Properties: 100% sol. in water @ 20 C; m.w. 118.17; sp.gr. 0.902 (20/20 C); dens. 7.51 lb/gal; visc. 6.4 cps; vapor pressure 0.6 mm Hg (20 C); b.p.

169 C; HLB 7.4; flash pt. (TCC) 143 F; ref. index 1.4193; surf. tens. 26.6 dynes/cm; sp. heat 0.56 cal/g/°C

EBA [ARCO]

Chem. Descrip.: Ethylene glycol butyl ether acetate

CAS 112-07-2

Uses: Solvent for coatings and cleaning applics.; HAPS listed; SARA reportable

Properties: 1% sol. in water @ 20 C; m.w. 160.21; sp.gr. 0.941 (20/20 C); dens. 7.84 lb/gal; visc. 1.8 cps; vapor pressure 0.29 mm Hg (20 C); b.p. 186C; HLB 7.4; flash pt. (TCC) 160 F; ref. index 1.4142; surf. tens. 30.3 dynes/cm

Eccoclean CR-46 [Eastern Color & Chem.]

Chem. Descrip.: Self-emulsifying solv. cleaner

Uses: Added solvency for extra cleaning power; useful on difficult cleaning problems

Properties: Water-disp

Eccoclean RPW [Eastern Color & Chem.]

Chem. Descrip.: Self-emulsifying solv. cleaner

Uses: Provides rapid penetration and solvency of difficult cleaning problems

Properties: Combustible

Eccoclean RPW-5/WA [Eastern Color & Chem.]

Uses: Sl. modified version of Eccoclean RPW solvent cleaner for stubborn cleaning problems

Ecco Cleaner 930 Conc. [Eastern Color & Chem.]

Chem. Descrip.: Emulsified sol'n. contg. ester-type solvs.

Uses: Solvent cleaner, detergent, and scouring agent for nat. and syn. fabrics and fibers; esp. for removal of spinning oils from woolen fabrics; contains ester-type solvs. most useful in general scouring needs for removing oils, greases, and soils

Use Level: 0.25-0.75% (wool scour), full str. (cleaner)

Regulatory: SARA nonreportable

Properties: Lt. amber clear liq.; mild odor; sol. in water; sp.gr. 1.04; dens. 8.2 lb/gal; b.p. 212 F; flash pt. none; pH 8.5 (2%); 60% act.; anionic

Toxicology: Nontoxic; will cause irritation and possible tearing on eye contact with undiluted form; may cause mild skin irritation, defatting of tissues, possible corneal damage; may be harmful if swallowed; may cause nausea, vomiting, diarrhea

Environmental: Biodeg.

Precaution: Incompat. with acids; hazardous dec. prods.: thermal dec. yields CO, CO_2, and other noxious vapors; spills may be slippery

Storage: Store indoors; keep from freezing

Ecco Cleaner HR-LO [Eastern Color & Chem.]

Chem. Descrip.: Emulsified ester sol'n. contg. ester type slvs., natural alcohols, and essential oils

Uses: Environmentally friendly solvent cleaner for removing oils, greases, adhesives, inks and coating from metal and nonmetallic surfaces

Regulatory: SARA nonreportable

Properties: Lt. amber clear liq.; mild odor; forms stable emulsions in water; sp.gr. 0.99; dens. 8.3 lb/gal; b.p. 212 F; flash pt. (COC) 167 F; pH 6.2-7.2; 100% act.; nonionic

Toxicology: Generally nontoxic; will cause irritation and possible tearing on eye contact with undiluted form; may cause mild skin irritation, defatting of tissues, possible corneal damage; may be harmful if swallowed; may cause nausea, vomiting, diarrhea

Environmental: Biodeg.

Precaution: Incompat. with acids; hazardous dec. prods.: thermal dec. yields CO, CO_2, and other noxious vapors; spills may be slippery

Storage: Store indoors; keep from freezing

Eccoflex 090 [Eastern Color & Chem.]

Chem. Descrip.: Mixed methyl esters

Uses: Environmentally friendly solvent cleaner used to replace common solvent cleaners, chlorinated solvs., aromatic, glycol ethers; spot cleaner for textiles; machinery and equip. cleaner; print screen cleaner; leather cleaner; wax stripper; degreaser for elec. components and motors; solvent for adhesives, inks, and coatings

Properties: Lt. amber clear liq., mild odor; m.w. 160; sp.gr. 1.09 (15.5 C); visc. 2.4-2.5 cps; vapor pressure 0.1 max. mm Hg (20 C); f.p. -25 C; i.b.p. 200 C; acid no. 0.5 max.; flash pt. (PMCC) 235 C; ref. index 1.423-1.425; 100% act.

Environmental: Biodeg.

Eccosolv C-14 [Eastern Color & Chem.]

Uses: Self-emulsifying solvent cleaner for cleaning and degreasing of machine parts

Discontinued prod.

Ecco Strip LS-W [Eastern Color & Chem.]

Chem. Descrip.: Emulsifiable ester sol'n. contg. < 30% methyl pyrrolidone and < 10% ethylene glycol butyl ether

Uses: Syn. detergent and stripping agent; solvent cleaner based on org. esters that can be used in place of common solvent cleaners, chlorinated solvs., aromatic and glycol esters; dries completely without leaving residue under controlled conditions; used as spot cleaner for textiles, machinery and equip. cleaner, print screen and leather cleaner, wax stripper, degreaser for elec. components and motors, solvent for adhesives, inks and coatings; environmentally friendly

Regulatory: Exempt from Rule 442; SARA reportable

Properties: Lt. amber clear liq.; mild odor; water-reducible; forms stable emulsions in water; sp.gr. 1.05 (15.5 C); visc. 2.0 cps; b.p. 212 F; acid no. 0.5 max.; flash pt. (COC) 198 F; VOC 76.5%; 100% act.

Toxicology: Generally nontoxic; eye contact will cause irritation and possible tearing in undiluted state; may cause mild skin irritation, defatting of tissues, possible corneal damage; may be hamful if swallowed; may cause nausea, vomiting, diarrhea

Environmental: Biodeg.

Precaution: Incompat. with acids; hazardous dec. prods.: thermal dec. yields CO, CO_2, and other noxious vapors

Storage: Store indoors; keep from freezing

Ecco Tipping Solution #3 [Eastern Color & Chem.]

Uses: Blend of solvs. providing exc. shoelace tipping with cellulose acetate and improved economy

EE [ARCO]

Chem. Descrip.: Ethylene glycol ethyl ether

CAS 110-80-5; EINECS 203-804-1

Uses: Solvent for coatings and cleaning applics.; HAPS listed; SARA reportable

Properties: 100% sol. in water @ 20 C; m.w. 90.12; sp.gr. 0.931 (20/20 C); dens. 7.75 lb/gal; visc. 2.1 cps; vapor pressure 3.8 mm Hg (20 C); b.p. 134 C; HLB 8.3; flash pt. (TCC) 110 F; ref. index 1.4076; surf. tens. 29.3 dynes/cm; sp. heat 0.56 cal/g/°C

EEA [ARCO]

Chem. Descrip.: Ethylene glycol ethyl ether acetate

CAS 111-15-9; EINECS 203-839-2

Uses: Solvent for coatings and cleaning applics.; HAPS listed; SARA reportable

Properties: 24% sol. in water @ 20 C; m.w. 132.16; sp.gr. 0.973 (20/20 C); dens. 8.11 lb/gal; visc. 1.3 cps; vapor pressure 1.7 mm Hg (20 C); b.p. 150 C; HLB 8.3; flash pt. (TCC) 130 F; ref. index 1.4030; surf. tens. 28.2 dynes/cm

EEH [ARCO]

Chem. Descrip.: Ethylene glycol ethyl hexyl ether

CAS 1559-35-9

Uses: Solvent for coatings and cleaning applics.; HAPS listed

Properties: 0.2% sol. in water @ 20 C; m.w. 174.29; sp.gr. 0.892 (20/20 C); dens. 7.42 lb/gal; visc. 7.0 cps; vapor pressure 0.08 mm Hg (20 C); b.p. 224 C; HLB 5.4; flash pt. (Seta) 208 F; ref. index 1.4361; surf. tens.

27.6 dynes/cm

EH [ARCO]

Chem. Descrip.: Ethylene glycol hexyl ether

CAS 112-25-4; EINECS 203-951-1

Uses: Solvent for coatings and cleaning applics.; HAPS listed; SARA reportable

Properties: 1% sol. in water @ 20 C; m.w. 146.23; sp.gr. 0.889(20/20 C); dens. 7.40lb/gal; visc. 5.2 cps; vapor pressure < 0.1 mm Hg (20 C); b.p. 208 C; HLB 6.4; flash pt. (TCC) 179 F; ref. index 1.4290; surf. tens. 30.3 dynes/cm

Ektapro® EEP Solvent [Eastman]

Chem. Descrip.: Ethyl 3-ethoxypropionate

CAS 763-69-9

Uses: High-performance retarder solvent for formulating enamels, lacquers, topcoats, and primers incl. auto OEM topcoats, coil, maintenance, appliance, aerospace, photoresist, marine, and plastic coatings; urethane-grade quality; polymerization solvent

Properties: Pt-Co 15 max. color; moderate odor; 2.9% sol. in water; m.w. 146.19; sp.gr. 0.95; dens. 7.91 lb/gal; visc. 1.0 cP; f.p. < -50 C; b.p. 165 C (760 mm); flash pt. (Seta) 58 C; surf. tens. 27.0 dynes/cm (23 C); elec. resist. 20 megohms

Ektasolve® DM [Eastman]

Chem. Descrip.: Diethylene glycol monomethyl ether

CAS 111-77-3; EINECS 203-906-6

Uses: Evaporating solvent used in brushing lacquers and dye stains; useful in wood stains, printing inks, and dye pastes for textiles; coalescing aid for PVAc latex paints; used in stamp pad and stencil inks; diluent for hydraulic brake fluids; also for cosmetics

Properties: Water-wh. liq.; agreeable odor; m.w. 120.15; f.p. -85 C; b.p. 191 C min.; water-sol.; sp.gr. 1.023; dens. 1.02 kg/l; ref. index 1.4263; flash pt. 96 C (COC); fire pt. 96 C; > 99% volatiles by vol.

Ektasolve® PM Acetate [Eastman]

Chem. Descrip.: Propylene glycol methyl ether acetate

CAS 108-65-6; EINECS 203-603-9

Uses: Retarder solvent for cosmetic and industrial applics.

Properties: Sol. 25.9% in water; dens. 7.91 lb/gal; b.p. 142-147 C; flash pt. 46 C; surf. tens. 26.4 dynes/cm; elec. resist. 6 megohms

Unverified

EM [Occidental]

Chem. Descrip.: Methoxyethanol

CAS 109-86-4; EINECS 203-713-7

Uses: Solvent

Unverified

EM [ARCO]

Chem. Descrip.: Ethylene glycol methyl ether

CAS 109-86-4; EINECS 203-713-7

Uses: Solvent for coatings and cleaning applics.; HAPS listed; SARA reportable

Properties: 100% sol. in water @ 20 C; m.w. 76.10; sp.gr. 0.966 (20/20 C); dens. 8.04 lb/gal; visc. 1.5 cps; vapor pressure 6.2 mm Hg (20 C); b.p. 124 C; HLB 8.8; flash pt. (TCC) 105 F; ref. index 1.4021; surf. tens. 30.8 dynes/cm; sp. heat 0.53 cal/g/°C

EMA [ARCO]

Chem. Descrip.: Ethylene glycol methyl ether acetate

CAS 110-49-6; EINECS 203-772-9

Uses: Solvent for coatings and cleaning applics.; HAPS listed; SARA reportable

Properties: 100% sol. in water @ 20 C; m.w. 118.14; sp.gr. 1.006 (20/20 C); dens. 8.37 lb/gal; visc. 1.1 cps; vapor pressure 2 mm Hg (20 C); b.p. 145 C; HLB 8.8; flash pt. (TCC) 120 F; ref. index 1.4025; surf. tens. 34.0 dynes/cm

Emcol® CC-42 [Witco/Oleo-Surf.; Witco SA]

Chem. Descrip.: PPG-40 diethylmonium chloride

CAS 9076-43-1

Uses: Detergent, pigment dispersant, particle suspension aid, emulsifier, solvent, conditioner, antistat, lubricant, corrosion inhibitor for toiletries, cosmetics, germicides, syn. fibers and plastics, textiles, industrial processes; ore flotation addi-

tive; dry-cleaning
Properties: Lt. amber oily liq.; sol. in IPA, acetone, MEK, min. spirits, ethanol; partly sol. in water; sp.gr. 1.01; flash pt. > 200 C (PMCC); pH 6.5; 100% conc.; cationic

Emcol® CC-55 [Witco/Oleo-Surf.]
Chem. Descrip.: Polypropoxy quat. ammonium acetate
Uses: Antistat; conditioner for hair rinse preparations; emulsifier for cosmetics and textile flame retardants; solvent for phenolic-type germicides for cosmetics and toiletries; antistat for syn. fibers and plastics; fabric conditioner; lubricant for textile and industrial formulations; solvent cleaning and scouring agent; corrosion inhibitor in protective coatings; pigment dispersant in nonaq. media; o/w emulsifier
Properties: Lt. amber oily liq.; sol. @ 25% in ethanol, IPA, acetone, MEK; water-disp.; sp.gr. 1.02; flash pt. > 93 C (PMCC); pH 6.5; 99% solids; cationic

Emcol® CC-57 [Witco/Oleo-Surf.]
Chem. Descrip.: Polypropoxy quat. ammonium phosphate
Uses: Antistat, conditioner, emulsifier, solvent, lubricant, solvent cleaning and scouring agent, corrosion inhibitor, and dispersant; used in syn. fibers and plastics, dry-cleaning; personal care prods., germicides, flame retardants, textile and industrial applic., protective coatings, and pigments
Properties: Lt. amber oily liq.; sol. in water, ethanol, IPA, acetone, MEK; sp.gr. 1.12; flash pt. > 93 C (PMCC); cationic

EmCon CO [Fanning]
Chem. Descrip.: Castor oil USP
CAS 8001-79-4; EINECS 232-293-8
Uses: Skin conditioning agent, occlusive solvent for use in lipsticks, eye shadow, blushes, makeup, nail polish, pharmaceuticals
Properties: Gardner 1 oil; sp.gr. 0.957-0.961; acid no. 2 max.; iodine no. 83-88; sapon. no. 176-182; hyd. no. 160-168

EmCon Olive [Fanning]
Chem. Descrip.: Olive oil USP
CAS 8001-25-0; EINECS 232-277-0
Uses: Skin and hair conditioning agent, occlusive solvent for shampoos, tonics, cleansing prods., face/body/hand lotions, suntan gels, creams, lotions, pharmaceutical ointments
Properties: Pale straw clear liq., bland odor; sp.gr. 0.910-0.915; acid no. 3.5 max.; iodine no. 79-88; sapon. no. 190-195

EmCon Rice Bran [Fanning]
Chem. Descrip.: Rice bran oil
CAS 68553-81-1; EINECS 271-397-8
Uses: Skin conditioning agent, occlusive solvent for face powds., cleansing prods., moisturizing creams and lotions, pharmaceutical ointments
Properties: Sp.gr. 0.971-0.973; acid no. 0.10 max.; iodine no. 102-110; sapon. no. 185-195

EmCon W [Fanning]
Chem. Descrip.: Wheat germ oil USP
CAS 8006-95-9
Uses: External lubricant, skin/hair conditioner, occlusive solvent; for hair conditioners, shampoos, lipsticks, cleansers, moisturizing creams and lotions, skin care prods., pharmaceuticals
Properties: Gardner 10 max. oil; acid no. 0.5 max.; iodine no. 120-135; sapon. no. 180-195

EmCon Walnut [Fanning]
Chem. Descrip.: Walnut oil
CAS 8024-09-7
Uses: Skin conditioning agent, occlusive solvent for suntan gels, creams, lotions, topical pharmaceuticals, ointments
Properties: Clear oil, sl. nutty odor; iodine no. 145-158; sapon. no. 180-200

Emerest® 2419 [Henkel/Cospha; Henkel Canada]
Chem. Descrip.: Glyceryl dioleate
CAS 25637-84-7; EINECS 247-144-2
Uses: Emulsifier, lubricant, rust preventive additive, mold release agent, solvent for dyes and pigments; used in leathers, lubricants

and softeners
Properties: Gardner 8 liq.; sol. in min. oil, toluene, IPA, xylene; dens. 7.7 lb/gal; visc. 55 cSt (100 F); HLB 1.6; pour pt. < 0 C; flash pt. 470 F; 100% conc.; nonionic
Discontinued prod.

Emerest® 2452 [Henkel/Cospha; Henkel Canada]
Chem. Descrip.: Polyglyceryl-3 diisostearate
CAS 66082-42-6; EINECS 291-548-1
Uses: Emulsifier, solubilizer, dye and pigment wetter; emollient; thickener; solvent; for creams, lotions, lip prods.
Properties: Amber visc. liq.; sol. @ 5% in min. oil, triolein, IPA, IPM; disp. in glycerin; dens. 8.2 lb/gal; visc. 990 cSt (100 F); HLB 6.7; acid no. 20 max.; iodine no. 0.5 max.; sapon. no. 165-175; pour pt. 4 C; flash pt. 455 F; 100% conc.; nonionic

Emerest® 2624 [Henkel/Emery; Henkel/Cospha; Henkel/Textiles; Henkel Canada]
Chem. Descrip.: PEG-4 oleate
CAS 9004-96-0; EINECS 233-293-0
Uses: Lubricant component in textile processing; softener/lubricant for leather during tanning; emulsifier for min. oils, fatty oils, and solvs. for cutting oils, solvs. in metal cleaners and degreasers, w/o emulsifier for consumer pesticide aerosols
Regulatory: EPA-exempt
Properties: Gardner 3 liq.; sol. 5% in glyceryl trioleate, xylene; disp. in water; dens. 8.1 lb/gal; visc. 34 cSt (100 F); HLB 8.3; pour pt. < -15 C; flash pt. 415 F; cloud pt. < 25 C; 100% conc.; nonionic

Emerest® 2648 [Henkel/Emery; Henkel/Cospha; Henkel/Textiles; Henkel Canada]
Chem. Descrip.: PEG-8 dioleate
CAS 9005-07-6
Uses: Lipophilic emulsifier and solubilizer for min. oils, fats, and solvs.; emulsifier for kerosene in agric. and pesticide sprays; emulsification of latex paints, metalworking fluids, solvs.; speciality and industrial lubricants; textiles
Regulatory: EPA-exempt
Properties: Gardner 3 liq.; sol. in min. oil, glyceryl trioleate, xylene; water-disp.; dens. 8.1 lb/gal; visc. 45 cSt (100 F); HLB 8.8; pour pt. 6 C; flash pt. 515 F; cloud pt. < 25; 100% act.; nonionic
Discontinued prod.

Emery® 912 [Henkel/Emery]
Chem. Descrip.: Glycerin CP/USP
CAS 56-81-5; EINECS 200-289-5
Uses: Skin softener, visc. modifier, flavor enhancer, moisturizer, solvent, humectant, thickener, and solubilizer in cosmetics, drug vehicles, food applics., glass, ceramics, agric., and adhesives
Regulatory: EPA-exempt; SARA nonreportable
Properties: Water-wh. visc. liq.; odorless; sol. in water; sp.gr. 1.2517; m.p. 18 C; b.p. 171 C; pour pt. 18 C; flash pt. (CC) 199 C; 96.0% min. glycerol
Toxicology: TLV:TWA 10 mg/m^3 (mist); LD50 (oral, rat) 12.6 g/kg; mild eye irritant, moderate skin irritant; inhalation of mist may cause respiratory irritation
Storage: Store in cool, dry, well-ventilated area; avoid excessive heat, open flames; incompat. with strong acids and oxidizers

Emery® 916 [Henkel/Emery]
Chem. Descrip.: Glycerin CP/USP
CAS 56-81-5; EINECS 200-289-5
Uses: Skin softener, visc. modifier, flavor enhancer, moisturizer, solvent, humectant, thickener and solubilizer in cosmetics, drug vehicles, food applics., glass, ceramics, agric., and adhesives
Regulatory: EPA-exempt
Properties: Water-wh. visc. liq.; odorless; sol. in water; sp.gr. 1.2607; m.p. 18 C; b.p. 171 C; pour pt. 18 C; flash pt. (CC) 199 C; 99.5% min. glycerol
Toxicology: TLV:TWA 10 mg/m^3 (mist); LD50 (oral, rat) 12.6 g/kg; mild eye irritant, moderate skin irritant; inhalation of mist may cause respiratory irritation

Storage: Store in cool, dry area; avoid excessive heat, open flames; incompat. with strong acids and oxidizers

Emery® 917 [Henkel/Emery]
Chem. Descrip.: Glycerin CP/USP kosher
CAS 56-81-5; EINECS 200-289-5
Uses: Skin softener, solubilizer, visc. modifier, flavor enhancer, moisturizer, solvent, humectant, thickener and solubilizer in cosmetics, drug vehicles, food applics., glass, ceramics, agric., and adhesives
Properties: Water-wh. liq.; sol. in water; sp.gr. 1.2607; m.p. 18 C; b.p. 171 C; flash pt. (CC) 199 C; 99.7% min. glycerol
Toxicology: TLV:TWA 10 mg/m^3 (mist); LD50 (oral, rat) 12.6 g/kg; mild eye irritant, moderate skin irritant; inhalation of mist may cause respiratory irritation
Storage: Store in cool, dry area; avoid excessive heat, open flames; incompat. with strong acids and oxidizers

Emery® 918 [Henkel/Emery]
Chem. Descrip.: Glycerin CP/USP
CAS 56-81-5; EINECS 200-289-5
Uses: Skin softener, visc. modifier, flavor enhancer, moisturizer, solvent, humectant, and solubilizer in cosmetics, drug vehicles, and food applics., glass, ceramics, and adhesives
Properties: Water-wh. liq.; sol. in water; sp.gr. 1.2612; m.p. 18 C; b.p. 171 C; flash pt. (CC) 199 C; 99.8% min. glycerol
Toxicology: TLV:TWA 10 mg/m^3 (mist); LD50 (oral, rat) 12.6 g/kg; mild eye irritant, moderate skin irritant; inhalation of mist may cause respiratory irritation
Storage: Store in cool, dry area; avoid excessive heat, open flames; incompat. with strong acids and oxidizers

Emery® 2203 [Henkel/Emery]
Chem. Descrip.: Methyl tallowate
CAS 61788-61-2; EINECS 262-989-7
Uses: Detergent intermediate; solvent for pesticides and herbicides
Regulatory: EPA-exempt
Properties: Gardner 1 color; dens. 7.2 lb/gal; visc. 6.7 cSt (100 C); m.p. 10 C; acid no. 1.0 max.; iodine no. 47-53; sapon. no. 190-200; flash pt. 355 F

Emery® 2209 [Henkel/Emery]
Chem. Descrip.: Methyl caprylate-caprate
CAS 68937-83-7
Uses: Detergent intermediate; solvent for pesticides and herbicides
Regulatory: EPA-exempt
Properties: Gardner 1 color; dens. 7.1 lb/gal; visc. 1.6 cSt (100 C); m.p. -30 C; acid no. 0.5 max.; iodine no. 0.4 max.; sapon. no. 330-336; flash pt. 180 F

Emery® 2214 [Henkel/Emery]
Chem. Descrip.: Methyl myristate
CAS 124-10-7; EINECS 204-680-1
Uses: Detergent intermediate; solvent for pesticides and herbicides
Regulatory: EPA-exempt
Properties: Gardner 1 color; dens. 7.2 lb/gal; visc. 3.4 cSt (100 C); m.p. 17 C; acid no. 1.0 max.; iodine no. 0.6 max.; sapon. no. 230-234; flash pt. 300 F; 95% C14 ester

Emery® 2216 [Henkel/Emery]
Chem. Descrip.: Methyl palmitate
CAS 112-39-0; EINECS 203-966-3
Uses: Detergent intermediate; solvent for pesticides and herbicides
Regulatory: EPA-exempt
Properties: Gardner 1 color; m.p. 27 C; acid no. 0.2 max.; iodine no. 0.2 max.; sapon. no. 206-210; flash pt. 330 F; 95% C14 ester

Emery® 2218 [Henkel/Emery]
Chem. Descrip.: Methyl stearate
CAS 112-61-8
Uses: Detergent intermediate; solvent for pesticides and herbicides
Regulatory: EPA-exempt
Properties: Gardner 1 color; m.p. 36 C; acid no. 0.5 max.; iodine no. 1.0 max.; sapon. no. 188-192; flash pt. 307 F; 95% C18 ester

Emery® 2219 [Henkel/Emery]
Chem. Descrip.: Methyl oleate
CAS 112-62-9
Uses: Detergent intermediate; solvent for pesticides and herbicides

Regulatory: EPA-exempt
Properties: Gardner 1 color; dens. 6.8 lb/gal; m.p. 18 C; acid no. 4.0 max.; iodine no. 66-74; sapon. no. 188-192; flash pt. 345 F; 24% C18 stearate, 58% C18 oleate (unsat.), 14% C18 linoleate (unsat.)

Emery® 2224 [Henkel/Emery]
Chem. Descrip.: Methyl soyate
Uses: Solvent for pesticides and herbicides
Regulatory: EPA-exempt
Properties: Gardner 11 color; dens. 7.3 lb/gal; visc. 4.3 cSt (100 C); m.p. 18 C; flash pt. 350 F
Unverified

Emery® 2230 [Henkel/Emery]
Chem. Descrip.: Methyl sunflowerate
Uses: Solvent for pesticides and herbicides
Regulatory: EPA-exempt
Properties: Gardner 1 color; dens. 7.3 lb/gal; visc. 4.6 cSt (100 C); m.p. -2 C; flash pt. 345 F

Emery® 2231 [Henkel/Emery]
Chem. Descrip.: Dist. methyl canolate
Uses: Solvent for pesticides and herbicides
Regulatory: EPA-exempt
Properties: Gardner 1 color; dens. 7.3 lb/gal; visc. 4.6 cSt (100 C); m.p. < -10 C; flash pt. 345 F

Emery® 2232 [Henkel/Emery]
Chem. Descrip.: Methyl canolate
Uses: Solvent for pesticides and herbicides
Regulatory: EPA-exempt
Properties: Gardner 10 color; dens. 7.3 lb/gal; visc. 5.0 cSt (100 C); m.p. < -10 C; flash pt. 345 F
Unverified

Emery® 2253 [Henkel/Emery]
Chem. Descrip.: Methyl coconate
CAS 61788-59-8; EINECS 262-988-1
Uses: Detergent intermediate; solvent for pesticides and herbicides
Regulatory: EPA-exempt
Properties: Gardner 1 color; dens. 7.5 lb/gal; visc. 2.8 cSt (100 C); m.p. 4 C; acid no. 0.5 max.; iodine no. 7-11; sapon. no. 250-260; flash pt. 250 F

Emery® 2254 [Henkel/Emery]
Chem. Descrip.: Stripped methyl coconate
CAS 61788-59-8; EINECS 262-988-1
Uses: Detergent intermediate; solvent for pesticides and herbicides
Regulatory: EPA-exempt
Properties: Gardner 1 color; acid no. 1.0 max.; iodine no. 5-10; sapon. no. 237-247; flash pt. 295 F

Emery® 2270 [Henkel/Emery]
Chem. Descrip.: Methyl laurate
CAS 111-82-0; EINECS 203-911-3
Uses: Detergent intermediate; solvent for pesticides and herbicides
Regulatory: EPA-exempt
Properties: Gardner 1 color; dens. 7.2 lb/gal; visc. 2.8 cSt (100 C); m.p. -1 C; acid no. 0.5 max.; iodine no. 0.5 max.; sapon. no. 251-255; flash pt. 270 F; 70% C12, 28% C14

Emery® 2290 [Henkel/Emery]
Chem. Descrip.: Methyl laurate
CAS 111-82-0; EINECS 203-911-3
Uses: Detergent intermediate; solvent for pesticides and herbicides
Regulatory: EPA-exempt
Properties: Gardner 1 color; dens. 7.2 lb/gal; visc. 2.5 cSt (100 C); m.p. 2 C; acid no. 0.5 max.; iodine no. 0.5 max.; sapon. no. 258-262; flash pt. 265 F; 90% C12 ester

Emery® 2296 [Henkel/Emery]
Chem. Descrip.: Methyl laurate
CAS 111-82-0; EINECS 203-911-3
Uses: Detergent intermediate; solvent for pesticides and herbicides
Regulatory: EPA-exempt
Properties: Gardner 1 color; dens. 7.2 lb/gal; visc. 2.3 cSt (100 C); m.p. 5 C; acid no. 0.5 max.; iodine no. 0.5 max.; sapon. no. 260-264; flash pt. 305 F; 96% C12 ester

Emery® 2301 [Henkel/Emery]
Chem. Descrip.: Methyl oleate
Uses: Solvent for pesticides and herbicides
Regulatory: EPA-exempt
Properties: Gardner 1 color; dens. 7.0 lb/gal; visc. 3.6 cSt (100 C); m.p. -16 C; flash pt. 350 F; 76% unsat. C18

Emery® 5794 [Henkel/Emery]
Chem. Descrip.: PEG-3 triethanolamine
CAS 36936-60-4
Uses: Solvent neutralizer for acid dyes

Properties: Dk. brn. liq.; m.w. 281; hyd. no. 590-630; amine no. 196-204; flash pt. (COC) 315 F
Discontinued prod.

Emery® 6686 [Henkel/Emery]
Chem. Descrip.: PEG 600
CAS 25322-68-3; EINECS 229-859-1
Uses: Chemical intermediate for coatings, adhesives, lubricants, metalworking, paper mfg., petrol. prod., ceramics, printing, electronics, solvs., cleaners, latex paints, mold release agent, rubber
Properties: Gardner < 1 liq.; sol. in water, xylene; disp. in Stod.; dens. 9.4 lb/gal; visc. 63 cSt (100 F); pour pt. 22 C; flash pt. 475 F
Discontinued prod.

Emery® 6687 [Henkel/Emery]
Chem. Descrip.: PEG-300
CAS 25322-68-3; EINECS 220-045-1
Uses: Chemical intermediate for coatings, adhesives, lubricants, metalworking, paper mfg., petrol. prod., ceramics, printing, electronics, solvs., cleaners, latex paints, mold release agent, rubber, textiles
Properties: Gardner < 1 liq.; sol. in water, xylene; disp. in Stod.; dens. 9.4 lb/gal; visc. 33 cSt (100 F); pour pt. -10 C; flash pt. > 400 F
Discontinued prod.

Emery® 6705 [Henkel/Emery]
Chem. Descrip.: Phenoxyethanol
CAS 122-99-6; EINECS 204-589-7
Uses: Solvent for NC, cellulose acetate, ethylcellulose, and vinyl, alkyl, and ester type resins; used in printing inks and special formulations for removing industrial finishes and coatings
Properties: Gardner 2 liq.; sol. 5% in Stod., xylene; water-insol.; dens. 9.1 lb/gal; visc. 10 cSt (100 F); pour pt. 13 C; flash pt. 250 F
Discontinued prod.

Emery® 6709 [Henkel/Cospha; Henkel Canada]
Chem. Descrip.: PEG 400
CAS 25322-68-3; EINECS 225-856-4
Uses: Humectant, binder, solvent, visc. modifier for personal care prods.; chemical intermediate for coatings, adhesives, lubricants, metalworking, paper mfg., petrol. prod., ceramics, printing, electronics, solvs., cleaners, mold release agent
Properties: Gardner < 1 clear, visc. liq.; sol. @ 5% in water, xylene; m.w. 400; dens. 9.4 lb/gal; visc. 45 cSt (100 F); pour pt. 6 C; flash pt. > 350 F; 100% act.; nonionic
Discontinued prod.

Emery® Methyl Lardate [Henkel/Emery]
Chem. Descrip.: Methyl lardate
Uses: Solvent-carrier for agric. spray prods.; defoaming component in metalworking, paper deinking, pharmaceutical fermentation
Discontinued prod.

Emery® Methyl Oleate [Henkel/Emery]
Chem. Descrip.: Methyl oleate
Uses: Solvent-carrier for agric. spray prods.; defoaming component in metalworking, paper deinking, pharmaceutical fermentation; also for prep. of lubricants for automotive, textile, metal rolling operations
Properties: Acid no. 4 max.; iodine no. 85-91; sapon. no. 190-200

Emkapon DAC-500 [Emkay]
Uses: Solvent scour for removing fatty soils, oils, and waxes used in sizing; for synthetics, nylon, Dacron, Kodel, Arnel, Spandex

Emkapon KW [Emkay]
Chem. Descrip.: Solvs. with anionic sulfonates
Uses: Emulsifier, solvent scour for removal of Dacron wax size and water repellent finishes; stable in caustic sol'n. up to 4% strength
Properties: Liq.; anionic

Emulbon LB-78 [Toho Chem. Ind.]
Chem. Descrip.: Six-membered borate with monoalkyl POE group
Uses: Lubricant, plasticizer, solvent
Properties: Liq.

Emulsifier Q [Consos]
Chem. Descrip.: Solvent/emulsifier blend
Uses: Solvent scour for heavily oiled and waxed textiles
Properties: Liq.; anionic

Environmental: Biodeg.

EP [ARCO]

Chem. Descrip.: Ethylene glycol propyl ether

CAS 2807-30-9; EINECS 220-548-6

Uses: Solvent for coatings and cleaning applics.; HAPS listed; SARA reportable

Properties: 100% sol. in water @ 20 C; m.w. 104.15; sp.gr. 0.913 (20/20 C); dens. 7.59 lb/gal; visc. 2.4 cps; vapor pressure 1.3 mm Hg (20 C); b.p. 149 C; HLB 7.8; flash pt. (TCC) 120 F; ref. index 1.4136; surf. tens. 27.9 dynes/cm

Eposolve® 299-R [Hardman]

Chem. Descrip.: Blend of chlorinated solvs.

Uses: Versatile clean-up solvent (non-red label) for cured and uncured epoxies; designed for attacking and disintegrating fully cured thermosetting plastics such as epoxies, urethanes, and polyesters; clean-up solvent for tools, brushes, automatic dispensing equip., spilled uncured resins; does not contain ODC

Properties: Clear liq.; dens. 11.1 lb/gal; visc. 10 cps; flash pt. 48 C

Toxicology: May be harmful if inhaled; can cause severe eye burns

Storage: Store in cool, dry place; keep containers sealed

Escaid 110 [Exxon]

Chem. Descrip.: Hydrocarbon

Uses: Solvent

Properties: M.w. 177; sp.gr. 0.80 (15.6 C); dens. 6.65 lb/gal; vapor pressure 0.2 mm Hg (20 C); flash pt. 78.9 C

Estalan 430 [Lanaetex Prods.]

Chem. Descrip.: Diethylene glycol dioctanoate/diisononanoate

Uses: Emollient, penetrant, spreading agent, perfume solvent in antiperspirants, deodorants, preshave lotions, and other liqs.

Estasan GT 8-40 3578 [Unichema Int'l.; Unichema N. Am.]

Chem. Descrip.: Caprylic/capric triglyceride (40% C8 + 60% C10)

CAS 65381-09-1; EINECS 265-724-3

Uses: Solvent and carrier for drugs, capsules, flavors, fragrances, food ingreds.; lubricant, mold release for food applics.; fat source in dietetic prods.; spreading/penetrating aid, solvent for cosmetics and toiletries, creams, lotions

Regulatory: BP, DAB compliance; Japan approval

Properties: Lovibond 5Y/0.5R oil, odorless, bland taste; visc. 29 mPa·s; acid no. 0.1 max.; iodine no. 2 max.; sapon. no. 300-315; hyd. no. 5 max.; cloud pt. 10 C

Environmental: Readily biodeg.

Storage: Store @ R.T. in dry atmosphere

Estasan GT 8-60 3575 [Unichema Int'l.; Unichema N. Am.]

Chem. Descrip.: Caprylic/capric triglyceride (60% C8 + 40% C10)

CAS 65381-09-1; EINECS 265-724-3

Uses: Solvent and carrier for drugs, capsules, flavors, fragrances, food ingreds.; lubricant, mold release for food applics.; fat source in dietetic prods.; spreading/penetrating aid, solvent for cosmetics and toiletries, creams, lotions

Regulatory: BP, DAB compliance; Japan approval

Properties: Lovibond 5Y/0.5R oil, odorless, bland taste; visc. 23 mPa·s; acid no. 0.1 max.; iodine no. 0.5 max.; sapon. no. 325-345; hyd. no. 5 max.; cloud pt. -8 C

Environmental: Readily biodeg.

Storage: Store @ R.T. in dry atmosphere

Estasan GT 8-60 3580 [Unichema Int'l.; Unichema N. Am.]

Chem. Descrip.: Caprylic/capric triglyceride (60% C8 + 40% C10)

CAS 65381-09-1; EINECS 265-724-3

Uses: Solvent and carrier for drugs, capsules, flavors, fragrances, food ingreds.; lubricant, mold release for food applics.; fat source in dietetic prods.; spreading/penetrating aid, solvent for cosmetics and toiletries, creams, lotions

Regulatory: BP, DAB compliance; Japan approval; kosher

Properties: Lovibond 5Y/0.5R oil, odorless, bland taste; visc. 23

mPa·s; acid no. 0.1 max.; iodine no. 0.5 max.; sapon. no. 325-345; hyd. no. 5 max.; cloud pt. -8 C
Environmental: Readily biodeg.
Storage: Store @ R.T. in dry atmosphere

Estasan GT 8-65 3577 [Unichema Int'l.; Unichema N. Am.]
Chem. Descrip.: Caprylic/capric triglyceride (65% C8 + 35% C10)
CAS 65381-09-1; EINECS 265-724-3
Uses: Solvent and carrier for drugs, capsules, flavors, fragrances, food ingreds.; lubricant, mold release for food applics.; fat source in dietetic prods.; spreading/penetrating aid, solvent for cosmetics and toiletries, creams, lotions
Regulatory: BP, DAB compliance; Japan approval
Properties: Lovibond 5Y/0.5R oil, odorless, bland taste; visc. 23 mPa·s; acid no. 0.1 max.; iodine no. 0.5 max.; sapon. no. 325-360; hyd. no. 5 max.; cloud pt. -5 C
Environmental: Readily biodeg.
Storage: Store @ R.T. in dry atmosphere

Estasan GT 8-65 3581 [Unichema Int'l.; Unichema N. Am.]
Chem. Descrip.: Caprylic/capric triglyceride (65% C8 + 35% C10)
CAS 65381-09-1; EINECS 265-724-3
Uses: Solvent and carrier for drugs, capsules, flavors, fragrances, food ingreds.; lubricant, mold release for food applics.; fat source in dietetic prods.; spreading/penetrating aid, solvent for cosmetics and toiletries, creams, lotions
Regulatory: BP, DAB compliance; Japan approval; kosher
Properties: Lovibond 5Y/0.5R oil, odorless, bland taste; visc. 23 mPa·s; acid no. 0.1 max.; iodine no. 0.5 max.; sapon. no. 335-360; hyd. no. 5 max.; cloud pt. -5 C
Environmental: Readily biodeg.
Storage: Store @ R.T. in dry atmosphere

Estasan GT 8-70 3579 [Unichema Int'l.; Unichema N. Am.]
Chem. Descrip.: Caprylic/capric triglyceride (70% C8 + 25-30% C10)
CAS 65381-09-1; EINECS 265-724-3
Uses: Solvent and carrier for drugs, capsules, flavors, fragrances, food ingreds.; lubricant, mold release for food applics.; fat source in dietetic prods.; spreading/penetrating aid, solvent for cosmetics and toiletries, creams, lotions
Regulatory: BP, DAB compliance; Japan approval; kosher
Properties: Lovibond 5Y/0.5R oil, odorless, bland taste; visc. 23 mPa·s; acid no. 0.1 max.; iodine no. 0.5 max.; sapon. no. 335-360; hyd. no. 5 max.; cloud pt. -12 C
Environmental: Readily biodeg.
Storage: Store @ R.T. in dry atmosphere

Estol IPM 1509 (BIO-IPM) [Unichema Int'l.]
Chem. Descrip.: Isopropyl myristate, veg. derived
CAS 110-27-0; EINECS 203-751-4
Uses: Penetrating emollient, solvent for cosmetics and topical medicinals with unique spreading props.; reduces heaviness and oiliness of prods. containing triglycerides, liq. waxes, and min. oils; provides gloss and conditioning in hair sprays; for skin creams, sun care preps., aerosols, bath oils, preshave lotions
Properties: Colorless low-visc. liq., odorless; sp.gr. 0.850-0.855; acid no. 0.1 max.; iodine no. 0.5 max.; sapon. no. 206-211; cloud pt. 0 C max.; 98% min. act.

Ethana® [Asahi Chem. Ind.]
Chem. Descrip.: 1,1,1-Trichloroethane
Uses: Solvent
Unverified

Ethylan® HB Series [Akcros UK]
Chem. Descrip.: Aromatic alkoxylates
Uses: Cosolvent with high flash pt. for use in iodophors, rinse aids, detergent sterilizers, glass cleaners, etc.
Properties: Liq.; nonionic
Unverified

Ethylan® HB1 [Akcros UK]
Chem. Descrip.: Short chain ethoxylate
Uses: Coalescing agent for surface coatings; solvent for acrylic and vinyl acetate copolymers; plasti-

cizer; preservative effective against a variety of microorganisms

Properties: Lt. amber clear liq., mild odor; sol. hazy in water; sp.gr. 1.098; visc. 27 cs; pour pt. -15 C; flash pt. (COC) > 100 C; pH 8.8 (1% aq.); 100% act.; nonionic

Toxicology: Sl. toxic; LD50 (rat, oral) 2-4 ml/kg; sl. skin irritant, severe eye irritant

Ethylan® HB4 [Akcros UK]

Chem. Descrip.: Aromatic ethoxylate

Uses: Penetrant, hydrotrope, cosolvent, solubilizer, coupling agent; enhances detergency emulsification and wetting; for iodophors, rinse aids, detergent sterilizers, glass cleaners

Properties: Water-wh. clear liq., faint odor; sol. in water; sp.gr. 1.121; visc. 64 cs; pour pt. < 0 C; cloud pt. > 100 C (1% aq.); flash pt. (COC) > 180 C; pH 6.4 (1% aq.); 100% conc.; nonionic

Ethylan® KELD [Akcros UK]

Chem. Descrip.: Modified fatty alkylolamide

Uses: Emulsifier for wax, kerosene and olein, solvent degreasing

Properties: Clear amber liq.; mild odor; disp. in water; sp.gr. 1.005; visc. 967 cs; flash pt. (COC) > 150 C; pour pt. 5 C; pH 9.0 (1% aq.); 100% act.; nonionic

Unverified

Ethyl Diglyme [Ferro/Grant]

Chem. Descrip.: Diethylene glycol diethyl ether

CAS 112-73-2; EINECS 204-001-9

Uses: Solvent which tends to solvate cations; used in electrochemistry, polymer and boron chemistry; physical processes such as gas absorption, extraction, stabilization; used in industrial prods. such as fuels, lubricants, textiles, pharmaceuticals, pesticides

Properties: Colorless; mild, nonresidual odor; water-sol.; misc. with ethanol, acetone, benzene, diethyl ether, octane; m.w. 162.23; sp.gr. 0.9082; dens. 7.56 lb/gal; visc. 1.4 cP; f.p. -44.3 C; b.p. 189 C; flash pt. (CC) 90 C; ref. index 1.4115; pH netural; surf. tens. 27.2 dynes/cm; 98.0% min. purity

Toxicology: LD50 5000 mg/kg; low to mod. acute toxicity; chronic exposure may cause reproductive effects

Environmental: Slowly biodeg.

Ethyl Di-Icinol [ICI Australia]

Chem. Descrip.: 2-(2-Ethoxyethoxy) ethanol

CAS 110-90-0; EINECS 203-919-7

Uses: Solvent for use in protective coatings, inks, cleaning prods., agric. chems.; aids wetting, penetration, and soil removal; coupling solvent

Unverified

Ethyl Glyme [Ferro/Grant]

Chem. Descrip.: Ethylene glycol diethyl ether

CAS 629-14-1; EINECS 211-076-1

Uses: Solvent which tends to solvate cations; used in electrochemistry, polymer and boron chemistry; physical processes such as gas absorption, extraction, stabilization; used in industrial prods. such as fuels, lubricants

Properties: Colorless; mild ethereal, nonresidual odor; water-sol.; misc. with ethanol, acetone, benzene, diethyl ether, octane; m.w. 118.18; sp.gr. 0.8417; dens. 7.0 lb/gal; visc. 0.7 cP; f.p. -74 C; b.p. 121 C; flash pt. (CC) 27 C; ref. index 1.3922; pH neutral; 97.0% min. purity

Toxicology: LD50 4400 mg/kg; low to mod. acute toxicity; chronic exposure may cause reproductive effects

Environmental: Slowly biodeg.

Ethyl Icinol [ICI Australia]

Chem. Descrip.: 2-Ethoxyethanol

CAS 110-80-5; EINECS 203-804-1

Uses: Solvent for use in protective coatings, inks, cleaning prods., agric. chems.; aids wetting, penetration, and soil removal; coupling solvent

Unverified

Eumulgin® PWM2 [Henkel/Pulcra]

Chem. Descrip.: Oleth-2

CAS 9004-98-2

Uses: W/o emulsifier, dispersant, lipophilic cosolvent; fragrance grade; broad pH and electrolyte tolerance
Properties: Liq.; HLB 5.0; hyd. no. 155-165; pH 6.0-7.5 (1%); 100% conc.; nonionic
Unverified

Exxal® 16 [Exxon]
Chem. Descrip.: Hexyldecanol
CAS 36311-34-9; EINECS 219-370-1
Uses: Extender and solvent for dyes and fragrances
Discontinued prod.

Exxate® 600 [Exxon]
Chem. Descrip.: Oxo-hexyl acetate
CAS 88230-35-7; DOT No.: UN 1993
Uses: Solvent for high solids, maintenance, marine, and industrial waterborne coatings, acrylic resin polymerization, metalworking fluids, urethane foam, industrial cleaners/degreasers, paint strippers, pigment dispersions, pesticide emulsifiable concs.; textile scouring aids, NC or screen printing inks, penetrating/lubricating oils, vinyl plastisols; good solvency, low surf. tens. and interfacial tens., good wetting, lubricity; visc. modifier; latex coalescer; no HAPS or SARA reportables
Properties: Pt-Co 10 color, sweet ester odor; sol. 0.02% in water; misc. with common petrol. solvs.; m.w. 144; sp.gr. 0.874 (20/20 C); dens. 7.27 lb/gal (20 C); visc. 1.04 cp; vapor pressure 1.4 mm Hg (20 C); f.p. < -60 C; i.b.p. 164 C; pour pt. < -70 F; cloud pt. < -70 F; flash pt. (TCC) 134 F; ref. index 1.4095 (20 C); surf. tens. 25 dynes/cm; 99% purity
Toxicology: LD50 (oral, rat) ≥ 5 g/kg, (dermal, rabbit) > 3 g/kg; low toxicity by ing. or skin contact; frequent/prolonged skin contact may cause irritation and dermatitis; sl. to mod. eye irritant; high concs. of vapor are irritating to mucous membranes
Environmental: Biodeg.
Precaution: DOT: combustible liq.
Storage: Keep containers closed; keep away from heat, sparks, and flames

Exxate® 700 [Exxon]
Chem. Descrip.: Oxo-heptyl acetate
CAS 90438-79-2; DOT No.: UN 1993
Uses: Solvent for high solids, maintenance, marine, and industrial waterborne coatings, acrylic resin polymerization, metalworking fluids, urethane foam, industrial cleaners/degreasers, paint strippers, pigment dispersions, pesticide emulsifiable concs.; textile scouring aids, NC or screen printing inks, penetrating/lubricating oils, vinyl plastisols; good solvency, low surf. tens. and interfacial tens., good wetting, lubricity; visc. modifier; latex coalescer; no HAPS or SARA reportables
Properties: Pt-Co 10 color, sweet ester odor; sol. 0.01% in water; misc. with common petrol. solvs.; m.w. 158; sp.gr. 0.874 (20/20 C); dens. 7.27 lb/gal (20 C); visc. 1.24 cp; vapor pressure 0.8 mm Hg (20 C); f.p. < -60 C; i.b.p. 176 C; pour pt. < -70 F; cloud pt. < -70 F; flash pt. (TCC) 150 F; ref. index 1.4158 (20 C); surf. tens. 26 dynes/cm
Toxicology: LD50 (oral, rat) ≥ 5 g/kg, (dermal, rabbit) > 3 g/kg; low toxicity by ing. or skin contact; frequent/prolonged skin contact may cause irritation and dermatitis; sl. to mod. eye irritant; high concs. of vapor are irritating to mucous membranes
Environmental: Biodeg.
Precaution: DOT: combustible liq.
Storage: Keep containers closed; keep away from heat, sparks, and flames

Exxate® 800 [Exxon]
Chem. Descrip.: C8 alkyl acetate
CAS 108419-32-5; DOT No.: UN 1993
Uses: Solvent for high solids, maintenance, marine, and industrial waterborne coatings, acrylic resin polymerization, metalworking fluids, urethane foam, industrial cleaners/degreasers, paint strippers, pigment dispersions, pesticide emulsifiable concs.; textile

scouring aids, NC or screen printing inks, penetrating/lubricating oils, vinyl plastisols; good solvency, low surf. tens. and interfacial tens., good wetting, lubricity; visc. modifier; latex coalescer; no HAPS or SARA reportables

Properties: Pt-Co 10 color, mild ester odor; sol. 0.02% in water; misc. with common petrol. solvs.; m.w. 172; sp.gr. 0.875 (20/20 C); dens. 7.28 lb/gal (20 C); visc. 1.74 cp; vapor pressure 0.75 mm Hg (20 C); f.p. < -60 C; i.b.p. 186 C; pour pt. < -70 F; cloud pt. < -70 F; flash pt. (TCC) 171 F; ref. index 1.420 (20 C); surf. tens. 26 dynes/cm

Toxicology: LD50 (oral, rat) ≥ 5 g/kg, (dermal, rabbit) > 3 g/kg; low toxicity by ing. or skin contact; frequent/prolonged skin contact may cause irritation and dermatitis; sl. to mod. eye irritant; high concs. of vapor are irritating to mucous membranes

Environmental: Biodeg.

Precaution: DOT: combustible liq.

Storage: Keep containers closed; keep away from heat, sparks, and flames

Exxate® 900 [Exxon]

Chem. Descrip.: C9 alkyl acetate

CAS 108419-33-6; DOT No.: UN 1993

Uses: Solvent for high solids, maintenance, marine, and industrial waterborne coatings, acrylic resin polymerization, metalworking fluids, urethane foam, industrial cleaners/degreasers, paint strippers, pigment dispersions, pesticide emulsifiable concs.; textile scouring aids, NC or screen printing inks, penetrating/lubricating oils, vinyl plastisols; good solvency, low surf. tens. and interfacial tens., good wetting, lubricity; visc. modifier; latex coalescer; no HAPS or SARA reportables

Properties: Pt-Co 10 color, mild ester odor; sol. 0.02% in water; misc. with common petrol. solvs.; m.w. 186 ; sp.gr. 0.873 (20/20 C); dens. 7.26 lb/gal (20 C); visc. 2.22 cp; vapor pressure 0.24 mm Hg (20 C); f.p. < -60 C; i.b.p. 205 C; pour pt. < -70 F; cloud pt. < -70 F; flash pt. (TCC) 194 F; ref. index 1.425 (20 C); surf. tens. 27 dynes/cm

Toxicology: LD50 (oral, rat) ≥ 5 g/kg, (dermal, rabbit) > 3 g/kg; low toxicity by ing. or skin contact; frequent/prolonged skin contact may cause irritation and dermatitis; sl. to mod. eye irritant; high concs. of vapor are irritating to mucous membranes

Environmental: Biodeg.

Precaution: DOT: combustible liq.

Storage: Keep containers closed; keep away from heat, sparks, and flames

Exxate® 1000 [Exxon]

Chem. Descrip.: C10 alkyl acetate

CAS 108419-34-7

Uses: Low vapor pressure solvent for high solids, maintenance, marine, and industrial waterborne coatings, acrylic resin polymerization, metalworking fluids, urethane foam, industrial cleaners and degreasers, paint strippers, pigment dispersions; pesticide emulsifiable concs., textile scouring aids, screen printing inks, penetrating/lubricating oils, vinyl plastisols; low surf. tens. and interfacial tens., good wetting, lubricity; visc. modifier; latex coalescer; no HAPS or SARA reportables

Properties: Pt-Co 10 color, mild ester odor; nil sol. in water; misc. with common petrol. solvs.; m.w. 200; sp.gr. 0.873 (20/20 C); dens. 7.26 lb/gal (20 C); visc. 2.60 cp; vapor pressure 0.1 mm Hg (20 C); f.p. < -60 C; i.b.p. 220 C; pour pt. < -70 F; cloud pt. < -70 F; flash pt. (TCC) 212 F; ref. index 1.429 (20 C); surf. tens. 27.5 dynes/cm; KB value > 100

Toxicology: LD50 (oral, rat) ≥ 5 g/kg, (dermal, rabbit) > 3 g/kg; low toxicity by ing. or skin contact; frequent/prolonged skin contact may cause irritation and dermatitis; sl. to mod. eye irritant; high concs. of vapor are irritating to mucous membranes

Environmental: Biodeg.

Storage: Keep containers closed; keep away from heat, sparks, and flames

Exxate® 1300 [Exxon]

Chem. Descrip.: C13 alkyl acetate

CAS 108419-35-8

Uses: Low vapor pressure solvent for high solids, maintenance, marine, and industrial waterborne coatings, acrylic resin polymerization, metalworking fluids, urethane foam, industrial cleaners and degreasers, paint strippers, pigment dispersions; pesticide emulsifiable concs., textile scouring aids, screen printing inks, penetrating/lubricating oils, vinyl plastisols; low surf. tens. and interfacial tens., good wetting, lubricity; visc. modifier; latex coalescer; no HAPS or SARA reportables

Properties: Pt-Co 10 color, nearly odorless; nil sol. in water; misc. with common petrol. solvs.; m.w. 242; sp.gr. 0.879 (20/20 C); dens. 7.28 lb/gal (20 C); visc. 4.63 cp; vapor pressure 0.03 mm Hg (20 C); f.p. < -60 C; i.b.p. 240 C; pour pt. < -70 F; cloud pt. < -70 F; flash pt. (TCC) 261 F; ref. index 1.438 (20 C); surf. tens. 28 dynes/cm; KB value 96

Toxicology: LD50 (oral, rat) ≥ 5 g/kg, (dermal, rabbit) > 3 g/kg; low toxicity by ing. or skin contact; frequent/prolonged skin contact may cause irritation and dermatitis; sl. to mod. eye irritant; high concs. of vapor are irritating to mucous membranes

Environmental: Biodeg.

Storage: Keep containers closed; keep away from heat, sparks, and flames

Exx-Print® 176 D [Exxon]

Chem. Descrip.: 53% Paraffin; 47% cycloparaffin; < 0.2% aromatic

Uses: Solvent for printing inks

Properties: Sp.gr. 0.80 (15.6 C); dens. 6.65 lb/gal (60 F); b.p. 207-234 C; flash pt. (TCC) 180 F; surf. tens. 27.6 dynes/cm

Exx-Print® 283 D [Exxon]

Chem. Descrip.: 59% Paraffin; 41% cycloparaffin; < 0.5% aromatic

Uses: Solvent for printing inks

Properties: Sp.gr. 0.81 (15.6 C); dens. 6.77 lb/gal (60 F); b.p. 251-269 C; flash pt. (TCC) 238 F; surf. tens. 28.6 dynes/cm

Exx-Print® 287 D [Exxon]

Chem. Descrip.: 100% Paraffin; < 0.01% aromatic

Uses: Solvent for printing inks

Properties: Sp.gr. 0.76 (15.6 C); dens. 6.35 lb/gal (60 F); b.p. 222-243 C; flash pt. (TCC) 203 F; surf. tens. 26.7 dynes/cm

Exx-Print® 314 A [Exxon]

Chem. Descrip.: 99.8% Aromatic; < 0.2% cycloparaffin

Uses: Solvent for printing inks

Properties: Sp.gr. 1.00 (15.6 C); dens. 8.34 lb/gal (60 F); b.p. 231-276 C; flash pt. (TCC) 219 F; surf. tens. 38.0 dynes/cm

Exx-Print® 392 D [Exxon]

Chem. Descrip.: 100% Paraffin; 0.01% aromatic

Uses: Solvent for printing inks

Properties: Sp.gr. 0.77 (15.6 C); dens. 6.44 lb/gal (60 F); b.p. 255-279 C; flash pt. (TCC) 248 F; surf. tens. 28.9 dynes/cm

Exx-Print® 493 D [Exxon]

Chem. Descrip.: 66% Paraffin; 34% cycloparaffin; < 0.5% aromatic

Uses: Solvent for printing inks

Properties: Sp.gr. 0.82 (15.6 C); dens. 6.82 lb/gal (60 F); b.p. 273-312 C; flash pt. (TCC) 265 F; surf. tens. 26.9 dynes/cm

Exx-Print® 588 D [Exxon]

Chem. Descrip.: 57% Paraffin; 43% cycloparaffin; < 0.5% aromatic

Uses: Solvent for printing inks

Properties: Sp.gr. 0.81 (15.6 C); dens. 6.99 lb/gal (60 F); b.p. 276-313 C; flash pt. (TCC) 279 F; surf. tens. 26.9 dynes/cm

Exxsol® D 40 [Exxon]

Chem. Descrip.: Dearomatized aliphatic hydrocarbon

CAS 64742-47-8

Uses: Solvent for coatings, printing inks, cleaning and degreasing formulations; no SARA reportables; < 100 ppm HAPS; improved sol-

vency; cost-effective diluents
Regulatory: FDA 21CFR §176.180, 176.210, 177.2600, 178.3570, 178.3620(b)
Properties: Mild pleasant odor; sp.gr. 0.78 (15.6/15.6 C); dens. 6.49 lb/gal (60 F); visc. 1.0 cP; vapor pressure 1.8 mm Hg (20 C); b.p. 159-204 C; flash pt. 38 C; surf. tens. 26 dynes/cm; KB value 34; 42% paraffin, 57% cycloparaffin, 0.4% aromatics
Toxicology: OEL 300 ppm

Exxsol® D 60 [Exxon]
Chem. Descrip.: Dearomatized aliphatic hydrocarbon
CAS 64742-47-8
Uses: Solvent for coatings, printing inks, agrochems., cleaning and degreasing formulations; no SARA reportables; < 100 ppm HAPS; improved solvency; cost-effective diluents
Regulatory: FDA 21CFR §176.180, 176.210, 177.2600, 178.3570, 178.3620(b), 40CFR §180.1001(c)(e)
Properties: Mild pleasant odor; sp.gr. 0.79 (15.6 C); dens. 6.59 lb/gal (60 F); visc. 1.2 cP; vapor pressure 1.1 mm Hg (20 C); b.p. 187-210 C; flash pt. 146 F; surf. tens. 27.3 dynes/cm; KB value 32; 49% paraffin, 51% cycloparaffin, 0.4% aromatics
Toxicology: OEL 300 ppm

Exxsol® D 75/100 Naphtha [Exxon; Imperial Oil]
Chem. Descrip.: Dearomatized aliphatic hydrocarbon solv.
CAS 64742-49-0
Uses: Solvent for architectural coatings, OEM and auto refinish coatings, industrial metal and wood coatings, maintenance and marine coatings
Properties: Sp.gr. 0.72 (15.6/15.6 C); dens. 5.99 lb/gal (60 F); vapor pressure 75 mm Hg (20 C); b.p. 76-96 C; flash pt. (TCC) < -18 C; surf. tens. 20.4 dynes/cm; KB value 36; 92% paraffin, 8% cycloparaffin, < 0.01% aromatics
Toxicology: OEL 300 ppm

Exxsol® D 80 [Exxon]
Chem. Descrip.: Dearomatized aliphatic hydrocarbon
CAS 64742-47-8
Uses: Solvent for coatings, cleaners, engine degreasers, agrochems., car dewaxing; no SARA reportables; < 100 ppm HAPS; improved solvency; good solvency for greases; cost-effective diluents; 50% compliance with criteria for LVP solvent
Regulatory: FDA 21CFR §176.180, 176.210, 177.2600, 178.3570, 178.3620(b), 40CFR §180.1001(c)(e)
Properties: Mild pleasant odor; sp.gr. 0.79 (15.6/15.6 C); dens. 6.65 lb/gal (60 F); visc. 1.7 cP; vapor pressure 0.2 mm Hg (20 C); b.p. 207-234 C; flash pt. 180 F; surf. tens. 27.6 dynes/cm; KB value 28; 53% paraffin, 47% cycloparaffin
Toxicology: OEL 300 ppm

Exxsol® D 110 [Exxon]
Chem. Descrip.: Dearomatized aliphatic hydrocarbon
CAS 64742-47-8
Uses: Solvent for coatings, carburetor cleaners, brake cleaners, degreasers, insecticides; no SARA reportables; < 100 ppm HAPS; improved solvency; cost-effective diluents; complies with criteria for LVP solvent
Regulatory: FDA 21CFR §176.180, 176.210, 177.2600, 178.3570, 178.3620(b), 40CFR §180.1001(c)(e)
Properties: Mild pleasant odor; m.w. 195; sp.gr. 0.81 (15.6/15.6 C); dens. 6.77 lb/gal (60 F); visc. 2.8 cP; vapor pressure < 0.1 mm Hg (20 C); b.p. 251-269 C; flash pt. 105 C; surf. tens. 28.6 dynes/cm; KB value 26; 59% paraffin, 41% cycloparaffin
Toxicology: OEL 300 ppm

Exxsol® D 115/145 Naphtha [Exxon; Imperial Oil]
Chem. Descrip.: Dearomatized aliphatic hydrocarbon
CAS 64742-49-0
Uses: Solvent for architectural coatings, OEM and auto refinish coatings, industrial metal and wood coatings, maintenance and marine coatings
Properties: Sp.gr. 0.75 (15.6/15.6 C);

dens. 6.22 lb/gal (60 F); vapor pressure 13 mm Hg (20 C); b.p. 116-144 C; flash pt. (TCC) 10 C; surf. tens. 22.1 dynes/cm; KB value 36; 59% paraffi, 41% cycloparaffin, 0.01% aromatics

Toxicology: OEL 300 ppm

Exxsol® D-130 [Exxon]

Chem. Descrip.: Dearomatized aliphatic hydrocarbon

CAS 64742-47-8

Uses: Solvent for agricultural pesticides, architectural coatings, OEM and auto refinish coatings, industrial metal and wood coatings, maintenance and marine coatings, cleaning and degreasing formulations

Regulatory: FDA 21CFR §176.180, 176.210, 177.2600, 178.3570, 178.3620(b), 40CFR §180.1001(c)(e)

Properties: M.w. 200; sp.gr. 0.84 (15.6/15.6 C); dens. 6.97 lb/gal (60 F); vapor pressure < 0.1 mm Hg (20 C); b.p. 276-316 C; flash pt. 129 C; surf. tens. 29.7 dynes/cm; KB value 24

Toxicology: OEL 300 ppm

Exxsol® D 3135 [Exxon; Imperial Oil]

Chem. Descrip.: Dearomatized aliphatic hydrocarbon

CAS 64742-47-8

Uses: Solvent for architectural coatings, OEM and auto refinish coatings, industrial metal and wood coatings, maintenance and marine coatings, cleaning and degreasing formulations

Regulatory: FDA 21CFR §176.180, 176.210, 177.2600, 178.3570, 178.3620(b)

Properties: Sp.gr. 0.77 (15.6/15.6 C); dens. 6.43 lb/gal (60 F); vapor pressure 1.1 mm Hg (20 C); b.p. 157-177 C; flash pt. (TCC) 41 C; surf. tens. 23.7 dynes/cm; KB value 34; 52% paraffin, 48% cycloparaffin, 0.02% aromatics

Toxicology: OEL 300 ppm

Exxsol® Heptane [Exxon]

Chem. Descrip.: Heptane

CAS 142-82-5; EINECS 205-563-8

Uses: Solvent for printing inks

Properties: M.w. 99; sp.gr. 0.70 (15.6/15.6 C); dens. 5.79 lb/gal (60 F); visc. 0.5 cP; vapor pressure 76 mm Hg (20 C); b.p. 94-98 C; flash pt. -7.2 C; surf. tens. 21.2 dynes/cm; KB value 29; 96% paraffin, 4% cycloparaffin, < 0.01% aromatic

Exxsol® Hexane [Exxon]

Chem. Descrip.: Hexane

CAS 64742-49-0; EINECS 203-777-6

Uses: Solvent

Properties: M.w. 86; sp.gr. 0.67 (15.6/15.6 C); dens. 5.59 lb/gal (60 F); vapor pressure 150 mm Hg (20 C); b.p. 66-69 C; flash pt. (TCC) < 0 F; surf. tens. 19.4 dynes/cm; KB value 29; 92% paraffin, 8% cycloparaffin, < 0.01% aromatic

Exxsol® Isopentane [Exxon]

Chem. Descrip.: Isopentane, dearomatized

CAS 78-78-4; EINECS 201-142-8

Uses: Solvent

Properties: M.w. 72; sp.gr. 0.62 (15.6/15.6 C); dens. 5.21 lb/gal (60 F); vapor pressure 620 mm Hg (20 C); flash pt. (TCC) < -18 C; surf. tens. 14.5 dynes/cm; 100% paraffin, < 0.01% aromatics

Toxicology: OEL 600 ppm

Exxsol® Methylpentane Naphtha [Exxon]

Chem. Descrip.: Dearomatized aliphatic hydrocarbon

CAS 64742-49-0

Uses: Solvent

Properties: M.w. 88; sp.gr. 0.66 (15.6/15.6 C); dens. 5.5 lb/gal (60 F); vapor pressure 200 mm Hg; flash pt. (TCC) < -18 C; surf. tens. 17.0 dynes/cm; KB value 28; 99% paraffin, 1% cycloparaffin, < 0.01% aromatics

Toxicology: OEL 400 ppm

EZA® [Albemarle]

Chem. Descrip.: Zeolite A, a hydrated sodium aluminosilicate

CAS 1344-00-9; EINECS 215-684-8

Uses: Cryststalline molecular sieve which selectively adsorbs molecules and exchange ions; detergent builder; solvent; anticaking agent for detergents and desiccants; gas separation; wh. pigment in paper, paint, and plastics; also in carpet cleaners, dentifrices, flame

retardants
Properties: Wh. cryst. powd.; 3 μ mean particle size; pH 10-12 (1% slurry); 100% act., 18.5% moisture

F

Fancol ALA [Fanning]
Chem. Descrip.: Cetyl acetate and acetylated lanolin alcohol
Uses: Cosmetic, toiletry conditioner, emollient, pigment binder, spreading agent, oil coupler, solvent for skin care prods., hand and body lotions, suntan and bath oils; glossing and plasticizing agent for hairsprays; pharmaceutical ointments
Properties: Clear lt. oily liq.; acid no. 1 max.; iodine no. 12 max.; sapon. no. 200 max.; hyd. no. 8 max.

Fancol CB [Fanning]
Chem. Descrip.: Cocoa butter obtained from roasted seeds of *Theobroma cacao*
CAS 8002-31-1
Uses: Skin conditioner, occlusive solvent, skin protectant for OTC drug prods., makeup, moisturizing creams and lotions, suntan preps.
Properties: Ylsh. wh. solid; iodine no. 35-40; sapon. no. 192-197; ref. index 1.4560-1.4580

Fancol CB Extra [Fanning]
Chem. Descrip.: Cocoa butter USP obtained from roasted seeds of *Theobroma cacao*
CAS 8002-31-1
Uses: Skin conditioner, occlusive solvent, skin protectant for OTC drug prods., makeup, moisturizing creams and lotions, suntan preps.
Properties: Ylsh. wh. solid; iodine no. 35-43; sapon. no. 188-195; ref. index 1.4560-1.4580

Fancol Karite Butter [Fanning]
Chem. Descrip.: Shea butter
CAS 68424-60-2
Uses: Ointment base, anti-irritant for skin, skin conditioner, occlusive agent, solvent for suntan preps., body lotions, winter sports prods., wrinkle creams, soaps, shave foams, shampoos, balsams
Properties: Gardner 4 max. color; m.p. 32-45 C; acid no. 0.5 max.; iodine no. 60-70; sapon. no. 178-190; hyd. no. 5

Fancol OA-95 [Fanning]
Chem. Descrip.: Oleyl alcohol
CAS 143-28-2; EINECS 205-597-3
Uses: Plasticizer, emulsion stabilizer, antifoam and coupling agent, aerosol lubricant, petrol. additive, pigment dispersant; rust preventive; detergent, release agent, cosolvent, softener, tackifier, spreading agent; used for metalworking, petrochemicals, pulp and paper, paints and coatings, plastics and polymers, inks, food applics., pharmaceuticals, cosmetics; chemical intermediate
Properties: Liq.; sol. in IPA, acetone, lt. min. oil, trichloroethylene, kerosene, VMP naphtha, benzene, turpentine; acid no. 0.05 max.; iodine no. 90-96; sapon no. 1 max.; hyd. no. 200-212; cloud pt. 5 C max.; nonionic

Fancorsil A [Fanning]
Chem. Descrip.: Cyclomethicone and dimethicone
Uses: Hair/skin conditioner, emollient, solvent; used in hair conditioners, makeup, moisturizing creams and lotions, topical pharmaceutical ointments; provides lubricity, long lasting gloss
Properties: Clear visc. liq.; sp.gr. 0.960; visc. 5000 cps min.; 100% solids

Filmer EHC [Union Carbide]
Uses: Solvent, coalescing aid for water-disp. and water-sol. coatings
Properties: Sol. 1.7% in water (20 C); m.w. 190.30; sp.gr. 0.933 (20/20 C); dens. 7.79 lb/gal (20 C); b.p. 259.0 C; flash pt. (CC) 271 F

Filmex® A-1 [Quantum]
Chem. Descrip.: 100 gal Ethyl alcohol, 5 gal methyl alcohol, 1.05 gal MIBK, 10.5 lb IPA (99%)
Uses: Solvent and thinner; general purpose industrial applics.
Properties: Sol. in conc. sulfuric acid; misc. in water, alcohol, chlorine, benzene

Filmex® A-2 [Quantum]
Chem. Descrip.: 100 gal Ethyl alcohol, 15.5 gal methyl alcohol, 1.05 gal MIBK
Uses: Industrial solvent for flexographic printing for cleaning equip., formulating and thinning inks, producing rotogravure etchings; in textile industry to promote adhesion, improve dye penetration, soften fibers; in chemical/pharmaceutical processing; in chemical specialties prod., such as latex coagulants and lacquers

Filmex® B [Quantum]
Chem. Descrip.: 100 gal Ethyl alcohol, 10.25 gal methyl alcohol, 1.05 gal MIBK, 5.25 lb isobutyl alcohol (99%)
Uses: Industrial solvent for flexographic printing for cleaning equip., formulating and thinning inks, producing rotogravure etchings; in textile industry to promote adhesion, improve dye penetration, soften fibers; in chemical/pharmaceutical processing; in chemical specialties prod., such as latex coagulants and lacquers

Filmex® C [Quantum]
Chem. Descrip.: 100 gal Ethyl alcohol, 5 gal methyl alcohol, 4.45 gal ethyl acetate (99%), 1.05 gal MIBK
Uses: Industrial solvent for flexographic printing for cleaning equip., formulating and thinning inks, producing rotogravure etchings; in textile industry to promote adhesion, improve dye penetration, soften fibers; in chemical/pharmaceutical processing; in chemical specialties prod., such as latex coagulants and lacquers

Filmex® D-1 [Quantum]
Chem. Descrip.: 100 gal Ethyl alcohol, 5 gal methyl alcohol, 1.05 gal MIBK, 15.75 lb isopropyl alcohol
Uses: Industrial solvent for flexographic printing for cleaning equip., formulating and thinning inks, producing rotogravure etchings; in textile industry to promote adhesion, improve dye penetration, soften fibers; in chemical/pharmaceutical processing; in chemical specialties prod., such as latex coagulants and lacquers

Filmex® D-2 [Quantum]
Chem. Descrip.: 100 gal Ethyl alcohol, 20.75 gal methyl alcohol, 1.05 gal MIBK
Uses: Industrial solvent for flexographic printing for cleaning equip., formulating and thinning inks, producing rotogravure etchings; in textile industry to promote adhesion, improve dye penetration, soften fibers; in chemical/pharmaceutical processing; in chemical specialties prod., such as latex coagulants and lacquers

Flexricin® 9 [CasChem]
Chem. Descrip.: Propylene glycol ricinoleate
CAS 26402-31-3; EINECS 247-669-7
Uses: Wetting agent, dye solvent, wax plasticizer, stabilizer for textile, household, and cosmetic applics., rewetting dried skins
Regulatory: FDA approval
Properties: Gardner 2+ liq.; sol. in toluene, butyl acetate, MEK, ethanol; sp.gr. 0.96; visc. 3 stokes; pour pt. –15 F; acid no. 2; iodine no. 77; sapon. no. 159; hyd. no. 288; 100% act.; nonionic

FoamFlush® Urethane Remover [ISP]
Chem. Descrip.: N-Methyl-2-pyrrolidone, butyrolactone, other ingreds.
Uses: Solvent cleaner for urethane removal from cured and uncured

polymers in the foam industry, as well as coatings and adhesives used in foam cutting and fabrication; safer alternative to methylene chloride, ketones, alcohols

Properties: Water-thin clear liq., sweet odor; misc. with water and most org. solvs.; sp.gr. 1.06; vapor pressure 0.29 mm Hg; f.p. -20 C; b.p. 202 C; flash pt. (Seta CC) 88 C; pH 4-6 (10% aq.)

Toxicology: TLV 100 ppm; low oral and inhalation toxicity; mod. dermal toxicity; mod. eye irritant; contains org. solvs.; avoid breathing vapors or mist

Precaution: Keep away from excessive heat and open flames

Foamid AME-70 [Alzo]

Chem. Descrip.: Acetamide MEA

CAS 142-26-7; EINECS 205-530-8

Uses: Wetting agent with broad solvent power for org. and inorg. materials and high b.p.; for cosmetic and health prods.

Properties: Gardner 2 clear liq., mild organoleptic odor; sol. in most alcohols, glycols, diols, polyols, glycol ethers, ketones, and water; sp.gr. 1.100; dens. 9.2 lb/gal; acid no. 15 max.; alkali no. 15 max.; flash pt. (OC) > 180 C; ref. index 1.4400; pH 6.0-8.5 (50% aq.); 70% min. act. in water; nonionic

Toxicology: Low toxicity

Foamid AME-75 [Alzo]

Chem. Descrip.: Acetamide MEA

CAS 142-26-7; EINECS 205-530-8

Uses: Wetting agent with broad solvent power for org. and inorg. materials and high b.p.; for cosmetic and health prods.

Properties: Gardner 2 clear liq., mild organoleptic odor; sol. in most alcohols, glycols, diols, polyols, glycol ethers, ketones, and water; sp.gr. 1.100; dens. 9.2 lb/gal; acid no. 15 max.; alkali no. 15 max.; flash pt. (OC) > 180 C; ref. index 1.4410; pH 6.0-8.5 (50% aq.); 75% min. act. in water; nonionic

Toxicology: Low toxicity

Foamid AME-100 [Alzo]

Chem. Descrip.: Acetamide MEA

CAS 142-26-7; EINECS 205-530-8

Uses: Wetting agent with broad solvent power for org. and inorg. materials and high b.p.; for cosmetic and health prods.

Properties: Gardner 2 clear liq., mild organoleptic odor; sol. in most alcohols, glycols, diols, polyols, glycol ethers, ketones, and water; sp.gr. 1.120; dens. 9.3 lb/gal; acid no. 15 max.; alkali no. 15 max.; flash pt. (OC) > 180 C; ref. index 1.4700; pH 6.0-8.5 (50% aq.); 95% min. act.; nonionic

Toxicology: Low toxicity

Foamid LM2E [Alzo]

Chem. Descrip.: Lactamide DGA

Uses: Humectant, emollient, o/w emulsifier, and cosolvent for skin and hair creams and lotions; imparts light, smooth, nongreasy feel to the skin, and sheen, conditioning, moisture balancing, and antistatic props. to overtreated hair; solvent, freeze pt. depressant, coupling and wetting agent in 'oil-free' pigmented prods.

Properties: Straw to lt. yel. clear liq., mild organoleptic odor; sol. in most low m.w. alcohols, diols, triols, polyols, glycol ethers, and water; sp.gr. 1.150; dens. 9.58 lb/gal; acid no. 30 max.; alkali no. 35 max.; flash pt. (OC) > 180 C; ref. index 1.4350; 80% min. act. in water; nonionic

Formula 111 [Chemtronics]

Uses: Heavy-duty solvent for strong cleaning/degreasing on interfaces, metal contacts, electromech. assemblies, PCB connectors

Unverified

Formula 500-C [Garratt Callahan]

Uses: Fuel oil tank and piping cleaner; strong solvents dissolve and disperse gunk on screens, fuel lines, pumps, burner guns, and in storage tanks; as a result, burning efficiency and handling ease is increased; high dispersing action which helps penetrate heaviest sludge masses; completely combustible when mixed with fuel oil

Properties: Lt. yel.-brn. clear liq.; dens.

8.6 lb/gal; flash pt. 185 F
Storage: Suggested shelf life of 6 mos. in cool, dry area

Frigen 113 TR-N [Hoechst AG]
Chem. Descrip.: Fluorinated hydrocarbon
Uses: Solvent for removing org. and inorg. pollution
Unverified

G

G-4972 [ICI Am.]
Chem. Descrip.: PEG-15 hydroxystearate
Uses: Solvent for injection sol'ns
Unverified

GAFsolver™ TMP [ISP]
Chem. Descrip.: N-Methyl-2-pyrrolidone
CAS 872-50-4; EINECS 212-828-1
Uses: Solvent for solv.-based cleaners, degreasers, and stripping agents, adhesives, PVC cements, PVC pipe and hose mfg., removal of carbon deposits (automotive and jet engine parts); aq. household and industrial cleaners, equip. (resin removal) cleaning, general maintenance solv.-based cleaners, base solvent for specialty cleaners
Properties: APHA < 200 water-thin liq.; completely misc. with water and most org. solvs. incl. alcohols, ethers, ketones, glycol ethers, aromatic and chlorinated hydrocarbons, and veg. oils; sp.gr. 1.027; dens. 8.55 lb/gal; visc. 165 cps; f.p. -24.4 C; b.p. 202 C; flash pt. (TCC) 93 C; fire pt. (COC) 99 C; 99% min. purity
Toxicology: Relatively low toxicity; severe eye irritant
Environmental: Biodeg.
Precaution: Combustible
Storage: Hygroscopic; avoid extended exposure to moist air

Gaftex® HI-8 [ISP]
Chem. Descrip.: N-Octyl pyrrolidone
Uses: Surf. act. solvent for use in hard surf. cleaners, heavy duty concs. and liq., glass cleaners, waxes and polishes; outstanding solvs. with additional surf. act., exc. wetting, and low foaming; low visc.; stable over wide pH range; hydophilic/lipophilic balance; NV, unaffected by VOC regulations
Properties: Liq.; sparingly sol. in water, but sol. in other polar and nonpolar solvs.; dens. 0.92 g/cm^3; f.p. –25 C; b.p. 100 C (0.3 mm Hg); HLB 6; flash pt. (TCC) 119 C; 97% act.; nonionic
Toxicology: LD50 (oral) > 2050 mg/kg; low toxicity, no skin/eye irritation
Environmental: Biodeg.

Gaftex® HI-12 [ISP]
Chem. Descrip.: N-Dodecyl pyrrolidone
CAS 2687-96-9
Uses: Surf. act. solvent for use in hard surf. cleaners, heavy duty concs. and liq., glass cleaners, waxes and polishes; outstanding solvs. with additional surf. act., exc. wetting, and low foaming; low visc.; stable over wide pH range; hydophilic/lipophilic balance; NV, unaffected by VOC regulations
Properties: Liq.; sparingly sol. in water, but sol. in other polar and nonpolar solvs.; dens. 0.90 g/cm^3; f.p. 10 C; b.p. 145 C (0.2 mm Hg); HLB 3; flash pt. (TCC) 116 C; 97% act.; nonionic
Toxicology: LD50 (oral) > 5000 mg/kg; low toxicity, no skin/eye irritation
Environmental: Biodeg.

Gaftex® HI-20H [ISP]
Chem. Descrip.: N-2-Hydroxyethyl-2-pyrrolidone
Uses: Used as replacement for hazardous solvs. in many types of

hard surf. and heavy duty cleaners; particularly useful in slowly settling or non-settling sol'ns. and suspensions; provides cleaning power to remove stubborn stains with reduced risk of damage to sensitive surfs.; reduces streaking effects and eliminates need for extra alkalinity; effective in removing ink stains; particularly useful for aerosol oven cleaners

Properties: Clear liq.; misc. with water, and most org. solvs.; immisc. with aliphatic hydrocarbons, ethers, vegetable and min. oils; sp.gr. 1.139; visc. 53 mPas; f.p. 20 C; b.p. 295 C; flash pt. (CC) > 100 C; 98% purity

Toxicology: LD50 (oral) 14,000 mg/kg; low toxicity, non-irritant

Gaftex® HI-B [ISP]

Chem. Descrip.: γ-Butyrolactone solv.

CAS 96-48-0; EINECS 202-509-5

Uses: Useful for formulations requiring non-irritating and non-sensitizing ingredients; replacement for hazardous solvs. in many hard surf. and heavy duty cleaners; incl. removing ink stains without damage to sensitive surfs.; offers extra stain cleaning power; no extra alkalinity is required and streaking effects reduced; for oven and automobile cleaners

Properties: Clear liq.; misc. with water, methanol, acetone, benzene, carbon tetrachloride, ethyl ether, monochlorobenzene; insol. in aliphatic hydrocarbons; sp.gr. 1.12; visc. 1.7 mPas; f.p. -44.4 C; b.p. 204 C; flash pt. (CC) 93 C; pH 4.5 (10% aq. sol'n.); 98% purity

Toxicology: LD50 (oral) 1600 mg/kg; low toxicity, non-irritant

Environmental: Biodeg.

GBL [ARCO]

Chem. Descrip.: γ-Butyrolactone

CAS 96-48-0; EINECS 202-509-5

Uses: Spinning and coagulating solvent for textiles, in nail polish removers, detergents, sunscreens, household cleaners, paint removers, agric. use, polymers, petrol. industry, mfg. of electronics, dyes; chem. process solvent; synthesis of vitamins; no ODP; HAPS-free, recyclable; replacement for regulated solvs.; exc. chem. and thermal stability

Properties: Completely misc. with water; sp.gr. 1.13 (20 C); vapor pressure < 1.5 mm Hg (20 C); m.p. 43 C; b.p. 204 C; flash pt. 98 C; nonflamm.

Genesolv® [AlliedSignal]

Uses: Solvents for specific cleaning, degreasing or dewatering tasks, flux removal

Gilugel MIG [Giulini]

Chem. Descrip.: Caprylic/capric triglyceride, 20% aluminum/magnesium hydroxide stearate

Uses: Occlusive solvent, stabilizer, and visc. control agent imparting smooth, nongreasy feel to eye shadows, foundations, lipsticks, and creams and lotions

Properties: Off-wh. soft gel

Unverified

Gilugel MIN [Giulini]

Chem. Descrip.: Min. oil, 20% aluminum/magnesium hydroxide stearate

Uses: Emollient, occlusive solvent, stabilizer, and visc. control agent imparting smooth, nongreasy feel to makeup, creams, lotions, and sunscreens

Properties: Moderately stiff translucent gel

Unverified

Gilugel SIL5 [Giulini]

Chem. Descrip.: Cyclomethicone and 20% aluminum/magnesium hydroxide stearate

Uses: Emollient, solvent, stabilizer, and visc. control agent imparting smooth, nongreasy feel to eyeliners, eye shadows, makeup bases, creams, lotions, sunscreens; hair conditioning agent

Unverified

Glidsafe® Avsol 7B [SCM Glidco Organics]

Chem. Descrip.: Terpene emulsion

Uses: Solvent cleaning agent

Properties: Flash pt. 142 F

Glidsol® 66-2 [SCM Glidco Organics]

Chem. Descrip.: Sat. terpene hydrocarbon
Uses: Solvent
Properties: Flash pt. 123 F

Glidsol® 100 [SCM Glidco Organics]
Uses: Industrial degreasing solvent
Properties: Flash pt. 106 F

Glidsol® 140 [SCM Glidco Organics]
Chem. Descrip.: Terpene alcohol
Uses: Solvent for ore flotation
Properties: Flash pt. 199 F

Glidsol® 175 [SCM Glidco Organics]
Chem. Descrip.: Terpene alcohol
Uses: Solvent for cleaning formulations
Properties: Flash pt. 183 F

Glidsol® 180 [SCM Glidco Organics]
Uses: High flash pt. limonene replacement solvent
Properties: Flash pt. 142 F

Glidsol® 5140 [SCM Glidco Organics]
Uses: Limonene extender
Properties: Low odor; flash pt. 108 F

Glidsol® Paint Purge 5125 [SCM Glidco Organics]
Uses: Paint line purging/cleaning solvent
Properties: Flash pt. 116 F

Glucam® E-10 [Amerchol]
Chem. Descrip.: Methyl gluceth-10
CAS 68239-42-9
Uses: Humectant for personal care prods.; freezing pt. depressant; emollient in aq. and hydroalcoholic prods.; moisturizer; foam modifier in detergent and shampoo systems; solvent and solubilizer for topical pharmaceuticals; film plasticizer; adds gloss, conditioning; used in emulsions, toilet articles, shampoos, shaving creams, shower gels, eye gels, facial cleansers
Properties: Pale yel. med. visc. syrup; pract. odorless; sol. in water, alcohol, hydroalcoholic systems; acid no. 1.5 max.; iodine no. 1 max.; sapon. no. 1.5 max; hyd. no. 350-370; 100% conc.; nonionic
Toxicology: LD50 (acute oral) > 5 ml/kg; nonirritating to eyes and skin

Glucam® E-20 [Amerchol]
Chem. Descrip.: Methyl gluceth-20
CAS 68239-43-0
Uses: Humectant for personal care prods.; freezing pt. depressant; emollient in aq. and hydroalcoholic prods.; moisturizer; foam modifier in detergent and shampoo systems; solvent and solubilizer for topical pharmaceuticals; for sunscreens, soap bars, dry skin creams, shampoos, liq. soaps
Properties: Pale yel. thin syrup; pract. odorless; sol. in water, alcohol, hydroalcoholic systems; acid no. 1.0 max.; iodine no. 1 max.; sapon. no. 1.0 max.; hyd. no. 205-225; 100% conc.
Toxicology: LD50 (acute oral) > 5 ml/kg; nonirritating to eyes and skin

Glucam® P-10 [Amerchol]
Chem. Descrip.: PPG-10 methyl glucose ether
CAS 61849-72-7
Uses: Humectant for personal care prods.; freezing pt. depressant; emollient in aq. and hydroalcoholic prods.; moisturizer; foam modifier in detergent and shampoo systems; solvent and solubilizer for topical pharmaceuticals; for eye makeup removers, hair conditioners, styling prods., makeups, moisturizing creams
Properties: Pale yel. heavy visc. syrups; pract. odorless; sol. in water, alcohol, and hydroalcoholic systems, castor oil, IPM, IPP; visc. 8500 cps; acid no. 1.0 max.; iodine no. 1 max.; sapon. no. 1.0 max.; hyd. no. 285-305; 100% conc.; nonionic
Toxicology: LD50 (acute oral) > 5 ml/kg; nonirritating to skin; mild transient irritation to eyes

Glucam® P-20 [Amerchol]
Chem. Descrip.: PPG-20 methyl glucose ether
CAS 61849-72-7
Uses: Humectant for personal care prods.; freezing pt. depressant; emollient in aq. and hydroalcoholic prods.; moisturizer; foam modifier in detergent and shampoo systems; solvent and solubilizer for topical pharmaceuticals; for face cleansing toners, nail con-

ditioners, fragrances, aftershaves
Properties: Pale yel. med. visc. syrup; pract. odorless; sol. in water, alcohol, and hydroalcoholic systems, castor oil, IPM, IPP; visc. 1700 cps; acid no. 1.0 max.; iodine no. 1 max.; sapon. no. 1.0 max.; hyd. no. 160-180; 100% conc.; nonionic
Toxicology: LD50 (acute oral) > 3 ml/kg; nonirritating to skin; mild transient irritation to eyes

Glyezin® [BASF AG]
Uses: Dye solvents and fixing assistants for textile finishing

Grindtek AML 60 [Grindsted Prods.]
Chem. Descrip.: Acetylated palm kernel glycerides
Uses: Lubricant and plasticizer for plastics and coatings; cosolvent for polar additives in low polarity systems
Properties: Lt. tan liq.; sol. in ethanol, peanut oil, toluene, wh. spirit, paraffin oil; HLB 2.1
Discontinued prod.

Grindtek AMOS 90 [Grindsted Prods.]
Chem. Descrip.: Acetylated lard glyceride
CAS 8029-92-3
Uses: Lubricant and plasticizer for plastics and coatings; cosolvent for polar additives in low polarity systems
Properties: Yel. liq.; sol. in ethanol, peanut oil, toluene, wh. spirit, paraffin oil; HLB 1.8
Discontinued prod.

H

Hallcomid® M-8-10 [C.P. Hall]
Chem. Descrip.: N,N-Dimethyl caprylamide-capramide
Uses: Mutual solvent for polar and nonpolar ingred. in cosmetics, cleaners, corrosion inhibitor; modifier for polymer coatings
Properties: Colorless to pale yel. liq.; m.w. 171-199; sol. 0.17 g/100 g water; sol. in most common organic solvs., hydrocarbons, plasticizers, fatty acids, and veg. oils; sp.gr. 0.88; visc. 10.5 cps; f.p. -21 C; b.p. 115-170 C (3 mm Hg); flash pt. 245 F; 95% conc.; nonionic

Hallcomid® M-18-OL [C.P. Hall]
Chem. Descrip.: N,N-Dimethyl oleamide
Uses: Solubilizer, solvent, dispersant, wetting agent; modifier for coatings
Properties: Slightly yel. liq.; m.w. 310; sol. 0.1 g/100 g water, most common org. solvs., hydrocarbons, plasticizers, fatty acids, veg. oils; sp.gr. 0.875; visc. 18.5 cps; f.p. -8 C; b.p. 196-211 C (3 mm Hg); HLB 7.0; flash pt. 410 F; 95% conc.; nonionic

Halso® 99 [Occidental]
Chem. Descrip.: Monochlorotoluene
CAS 25168-05-2
Uses: Extender, diluent or substitute for other org. solvents; for dye carrier, fuel oil additive, sludge solvent, in paint thinners, metal parts cleaners, adhesives
Properties: Dens. 9 lb/gal; visc. 0.747 cps (100 C); f.p. -32 C; b.p. 159-162 C; flash pt. (TCC) 123 F; surf. tens. 33 dynes/cm (20 C); KB value 110

HAN® 857 [Exxon]
Chem. Descrip.: Aromatic
CAS 64742-06-9
Uses: Solvent
Properties: Saybolt +22 color; sp.gr. 0.903 (60/60 F); flash pt. 60 C; KB value 78
Toxicology: TLV/OEL 1000 ppm
Discontinued prod.

Hartosolve OL [Hart Prods.]
Chem. Descrip.: Emulsified solvents
Uses: Solvent scour and cleaner
Properties: Liq.; 100% conc.; anionic

HEP® [ISP]
Chem. Descrip.: N-Hydroxyethyl pyrrolidone
CAS 3445-11-2
Uses: Solvent used in microelectronics, photoresists; to clean silk screens; prep. of polymer membranes for gas scrubbing; anticlogging jet printing ink; aq. veterinary tetracycline antibiotic; reaction solvent for prep. of retinoid esters for skin treatment; mfg. of haloalkyl lactams; alkaline paint stripper; solvent plant growth regulator; intermediate in lube oil additives; offers reactive hydroxyl group
Properties: Liq.; water-sol.; m.w. 129; sp.gr. 1.139; visc. 53 cps; f.p. 20 C; b.p. 295 C; flash pt. (CC) 320 F; ref. index 1.4951; surf. tens. 49 dynes/cm; 98% min. purity
Toxicology: LD50 (oral, rat) > 14.43 g/kg; low acute toxicity; minimally irritating to eyes; nonirritating to skin
Environmental: Biodeg.

Herco® Pine Oil [Hercules]
Chem. Descrip.: Pine oil
CAS 8002-09-3
Uses: Solvent; various grades for

household and industrial cleaners; disinfectant, antifoam agent, textile specialties
Properties: Disp. in water

Hetester PMA [Heterene; Bernel]
Chem. Descrip.: Propylene glycol myristyl ether acetate
Uses: Emollient, solvent, and plasticizer for anhyd. oil systems, emulsions
Regulatory: FDA, EU, and Japanese compliances
Properties: Colorless clear liq.; sol. @ 5% in most cosmetic oils; insol. in water, glycols, 70% ethanol; acid no. 2.0 max.; sapon. no. 140-160; hyd. no. 10.0 max.; pH 6.0-7.0 (5% in 50/50 IPA/water); cloud pt. ≈ 0 C

Hetoxide BN-13 [Heterene]
Chem. Descrip.: PEG-13 beta-naphthol ether
CAS 35545-57-4
Uses: Emollient, emulsifier, visc. control agent, lubricant, pigment dispersant, perfume solubilizer, used in cosmetics, household, textile industry, metal treating and plating; intermediate
Properties: Gardner 8 max. paste; sol. in water, IPA, min. oil; HLB 16.0; acid no. 1.0; hyd. no. 73-83; 99% conc.; nonionic

Hetoxide MPC [Heterene]
Chem. Descrip.: m,p-Cresol hydrophobe
Uses: Solvent for lacquers and coatings
Properties: Gardner 5 max. liq.; sol. in IPA, min. oil; disp. in water; hyd. no. 275-295

Hexcel FO 425A [Hexcel]
Chem. Descrip.: Petrol. solv.
Uses: General purpose cleaning and degreasing solvent for cleaning filled telecommunications cables and tools; noncorrosive
Properties: Clear; odorless; sp.gr. 0.753; flash pt. 200 F (COC)
Discontinued prod.

Hexyl Carbitol® [Union Carbide]
Chem. Descrip.: Diethylene glycol n-hexyl ether
CAS 112-59-4; EINECS 203-988-3
Uses: Solvent for waterborne coatings
Properties: Sol. 1.7% in water (20 C); m.w. 190.28; sp.gr. 0.935 (20/20 C); dens. 7.78 lb/gal (20 C); visc. 8.5 cP (20 C); vapor pressure < 0.01 mm Hg (20 C); b.p. 259.1 C; flash pt. (CC) 271 F

Hexyl Cellosolve® [Union Carbide]
Chem. Descrip.: Ethylene glycol n-hexyl ether
CAS 112-25-4; EINECS 203-951-1
Uses: Solvent, coalescing aid for water-disp. coatings
Properties: Sol. 0.99% in water (20 C); m.w. 146.23; sp.gr. 0.889 (20/20 C); dens. 7.40 lb/gal (20 C); visc. 5.2 cP (20 C); vapor pressure 0.05 mm Hg (20 C); b.p. 208.1 C; flash pt. (CC) 179 F

High Purity MTBE [ARCO]
Chem. Descrip.: Methyl t-butyl ether
CAS 1634-04-4; EINECS 216-653-1
Uses: Extraction solvent, reaction medium in pharmaceuticals, for polymerizations, Grignard reactions; octane enhancer for gasoline; highly resist. to peroxide formation
Properties: Clear liq., terpene-like odor; sl. sol. in water; misc. with most org. solvs.; sp.gr. 0.737; f.p. –108 C; b.p. 55 C (760 mm Hg); ref. index 1.3694 (20 C); surf. tension 18.3 dynes/cm (32 C); visc. 0.350 cps (20 C); flash pt. -30 C; > 99.9% purity

Hipochem MS-BW [High Point]
Chem. Descrip.: Emulsified aromatic and aliphatic solvs.
Uses: Low foaming detergent, solvent scour for removal of oils, grease, waxes, and fatty materials in textile operations
Properties: Colorless clear liq., mild aromatic odor; pH 6-8 (1%); 100% conc.; nonionic

Hiposcour® NFMS-2 [High Point]
Chem. Descrip.: Complex solv./surfactant blend
Uses: Solvent and detergent scour for removal of oil and grease; stable over broad pH range; can be used in sodium peroxide and sodium hypochlorite bleach baths

Properties: Amber clear liq., mild alcoholic odor; sp.gr. 0.93; dens. 7.75 lb/gal; pH 6.0-6.5; anionic
Unverified

Hi-Sol® 10 [Ashland]
Chem. Descrip.: Naphtha, light aromatic
CAS 64742-95-6
UN No.: 1255; DOT No.: UN 1255
Uses: Aromatic solvent
Properties: Sp.gr. 0.884-0.864 (60 F); dens. 7.18 lb/gal (15.6 C); i.b.p. 152 C; flash pt. (TCC) 40.6 C; KB value 87 min.

Hi-Sol® 15 [Ashland]
Chem. Descrip.: Naphtha
CAS 64742-94-5
UN No.: 1256; DOT No.: UN 1256
Uses: Aromatic solvent
Properties: Sp.gr. 0.880-0.910 (60 F); dens. 7.43 lb/gal (15.6 C); i.b.p. 176.7 C; flash pt. (TCC) 61.1 C; KB value 85 min.

Hi-Sol® 70 [Ashland]
Chem. Descrip.: Naphtha
Uses: Aromatic solvent

Hisolve DM [Toho Chem. Ind.]
Chem. Descrip.: Diethylene glycol monomethyl ether
CAS 111-77-3; EINECS 203-906-6
Uses: Solvent for lacquer thinners, inks, dyes, brake fluids, metal surf. finishing agents
Properties: Liq.

Hisolve MC [Toho Chem. Ind.]
Chem. Descrip.: Ethylene glycol monomethyl ether
CAS 109-86-4; EINECS 203-713-7
Uses: Solvent for paints, inks, adhesives, plastics and plasticizers, cleaners, paint removers
Properties: Liq.

Hisolve MDM [Toho Chem. Ind.]
Chem. Descrip.: Diethylene glycol dimethyl ether
CAS 111-96-6; EINECS 203-924-4
Uses: Reaction solvent in reduction, isomerization, substitution, alkylation condensation, polymerization, and metal including reaction; separation and extraction solvent
Properties: Liq.

Hisolve PM [Toho Chem. Ind.]
Chem. Descrip.: Polyethylene glycol monomethyl ether
Uses: Solvent for brake fluid
Properties: Liq.
Unverified

Hisolve TM [Toho Chem. Ind.]
Chem. Descrip.: Triethylene glycol monomethyl ether
CAS 112-35-6
Uses: Solvent for brake fluid and sens. liq. for wet process duplicator
Properties: Liq.
Unverified

Homogenol L-18 [Kao Corp. SA]
Chem. Descrip.: Polymeric surfactant
Uses: Dispersant in oil vehicles, nonpolar solvent
Properties: Liq.
Unverified

I

Icinol BE33 [ICI Australia]
Chem. Descrip.: Monobutyl ether of triethylene glycol and tetraethylene glycol
Uses: Solvent for lacquers, enamels, water-borne systems, inks, industrial cleaners, waterless hand cleaners; aids wetting, penetration and soil removal; coupling agent
Properties: Sol. in water, common organic chems.; sp.gr. 0.985-1.00; visc. 9.2 cps; b.p. 260 C; flash pt. (OC) > 250 C; 99% act.
Unverified

Icinol DPM [ICI Australia]
Chem. Descrip.: (2-Methoxy-methylethoxy) propanol
CAS 34590-94-8
Uses: Solvent for use in protective coatings, inks, cleaning prods., agric. chems.; aids wetting, penetration, and soil removal; coupling solvent
Unverified

Icinol EE22 [ICI Australia]
Chem. Descrip.: Monoethyl ether of diethylene glycol (80% min.), triethylene glycol, and tetraethylene glycol
Uses: Solvent for protective coatings, inks, cleaning prod., and agric. chems
Properties: Visc. 4.5 cps; flash pt. (OC) 96 C; 1.0% max.
Storage: Store and handle in stainless or mild steel; do not store in aluminum or aluminum alloys, black iron, or copper and some of its alloys

Icinol H260 [ICI Australia]
Chem. Descrip.: Polyoxyalkylene glycol ether
Uses: Syn. lubricant, defoamer; high visc. index solvent
Properties: Liq.; water-sol.; visc. 260 SUS (100 F)
Unverified

Icinol H280X [ICI Australia]
Chem. Descrip.: Polyoxyalkylene glycol ether
Uses: Inhibited version of Icinol H260; syn. lubricant, defoamer; high visc. index solvent
Unverified

Icinol H660 [ICI Australia]
Chem. Descrip.: Polyoxyalkylene glycol ether
Uses: Syn. lubricant, defoamer; high visc. index solvent
Properties: Liq.; water-sol.; visc. 660 SUS
Unverified

Icinol H660YA [ICI Australia]
Chem. Descrip.: Polyoxyalkylene glycol ether
Uses: Inhibited version of Icinol H660; syn. lubricant, defoamer; high visc. index solvent
Unverified

Icinol H5100 [ICI Australia]
Chem. Descrip.: Polyoxyalkylene glycol ether
Uses: Syn. lubricant, defoamer; high visc. index solvent
Properties: Liq.; water-sol.; visc. 5100 SUS
Unverified

Icinol L65 [ICI Australia]
Chem. Descrip.: Polyoxyalkylene glycol ether
Uses: Syn. lubricant, defoamer; high visc. index solvent
Properties: Liq.; water-insol.; visc. 65 SUS (100 F)

Unverified

Icinol L285 [ICI Australia]
Chem. Descrip.: Polyoxyalkylene glycol ether
Uses: Syn. lubricant, defoamer; high visc. index solvent
Properties: Liq.; water-insol.; visc. 285 SUS
Unverified

Icinol L300X [ICI Australia]
Chem. Descrip.: Polyoxyalkylene glycol ether
Uses: Inhibited version of Icinol L285
Unverified

Icinol L385 [ICI Australia]
Chem. Descrip.: Polyoxyalkylene glycol ether
Uses: Syn. lubricant, defoamer; high visc. index solvent
Properties: Liq.; water-insol.; visc. 385 SUS
Unverified

Icinol L625 [ICI Australia]
Chem. Descrip.: Polyoxyalkylene glycol ether
Uses: Syn. lubricant, defoamer; high visc. index solvent
Properties: Liq.; water-insol.; visc. 625 SUS
Unverified

Icinol L1715 [ICI Australia]
Chem. Descrip.: Polyoxyalkylene glycol ether
Uses: Syn. lubricant, defoamer; high visc. index solvent
Properties: Liq.; water-insol.; visc. 1715 SUS
Unverified

Icinol ME33 [ICI Australia]
Chem. Descrip.: Monomethyl ether of triethylene glycol and tetraethylene glycol
Uses: Solvent for protective coatings, inks, cleaning prod., and agric. chems.
Properties: Sp.gr. 1.058; visc. 7.5 cps; b.p. 258 C; flash pt. (OC) 118 C; 0.3% max.
Storage: Store and handle in stainless or mild steel; do not store in aluminum or aluminum alloys, black iron, or copper and some of its alloys

Icinol ME45 [ICI Australia]
Chem. Descrip.: Monomethyl ether of tetraethylene glycol and pentaethylene glycol
Uses: Solvent for protective coatings, inks, cleaning prod., and agric. chems.
Properties: Sp.gr. 1.065-1.075; b.p. 260 C; flash pt. (OC) > 118 C; 0.3% max.
Storage: Store and handle in stainless or mild steel; do not store in aluminum or aluminum alloys, black iron, or copper and some of its alloys

Icinol PM [ICI Australia]
Chem. Descrip.: 1-Methoxy 2-propanol
CAS 107-98-2; EINECS 203-539-1
Uses: Solvent for use in protective coatings, inks, cleaning prods., agric. chems.; aids wetting, penetration, and soil removal; coupling solvent
Unverified

Igepal® Cephene Lave [Rhône-Poulenc France]
Chem. Descrip.: Ethoxylated phenol
Uses: Surfactant; solvent for vinyl, phenolic, polyester, nitrocellulose, cellulose acetate, and other resins; in metal cleaners, paint strippers, and other cleaning compds. where solvs. and solv. boosters are required; vehicle for inks
Properties: Liq.; 99% conc.; nonionic

Igepal® LAVE [Rhône-Poulenc Surf. & Spec.]
Chem. Descrip.: Phenol ethoxylate
Uses: Solvent for vinyl, phenolic, polyester, nitrocellulose, cellulose acetate, and other resins; ingred. for metal cleaners, paint strippers, cleaning compds.; vehicle for inks
Properties: Liq.; 99% conc.; nonionic

Igepal® OD-410 [Rhône-Poulenc Surf. & Spec.; Rhône-Poulenc France]
Chem. Descrip.: Phenoxyethanol, phenoxyglycol, and phenoxydiglycol
Uses: Solvent for vinyl, phenolic, polyester, alkyd, NC, and cellulose acetate resins; ingred. of metal cleaners, paint strippers, and cleaning compds. where solvs. and solv. boosters are required; used as

ink vehicle
Properties: Liq.; 100% act.; nonionic

Imwitor® 408 [Hüls Am.]
Chem. Descrip.: Propylene glycol caprylate
CAS 31565-12-5
Uses: Coemulsifier, solubilizer, absorp. promoter for cosmetics, topical and rectal pharmaceuticals; solvent for lipophilic drugs
Properties: Sl. yel. liq.; acid no. 2 max.; iodine no. 1 max.; sapon. no. 250-275; 50% monoglycerides
Discontinued prod.

Imwitor® 412 [Hüls Am.]
Chem. Descrip.: Propylene glycol laurate
CAS 142-55-2; EINECS 205-542-3
Uses: Coemulsifier, solubilizer, absorp. promoter for cosmetics, topical and rectal pharmaceuticals; solvent for lipophilic drugs
Properties: Sl. yel. solid; acid no. 2 max.; iodine no. 1 max.; sapon. no. 220-235; 50% monoglycerides
Discontinued prod.

Imwitor® 742 [Hüls Am.; Hüls AG]
Chem. Descrip.: Caprylic/capric glycerides
CAS 26402-26-6
Uses: Plasticizer for hard fats; solvent for lipophilic ingreds., active substances; cosmetic emollient; coemulsifier, solubilizer, carrier for lipophilic drugs; dispersant, absorp. promoter; bacteriostatic effect
Regulatory: German pharmacopoeia compliance
Properties: Wh-ylsh. cryst. solid, sl. coconut odor, fatty bitter taste; sol. in ethanol, actone, ether, heptane, hexane; misc. with oils and fats; m.p. 25 C; HLB 3-4; acid no. 2 max.; iodine no. 1 max.; sapon. no. 250-280; 45% min. monoglycerides; nonionic
Toxicology: LD50 (oral, rat) > 5 g/kg; moderate skin irritant (undiluted)

Imwitor® 908 [Hüls Am.; Hüls AG]
Chem. Descrip.: Glyceryl caprylate
CAS 26402-26-6; EINECS 247-668-1
Uses: Coemulsifier for o/w emulsions, lipophilic materials; solvent, solubilizer for active substances, surfactant, absorp. promoter; carrier of drugs; bacteriostatic effect; cosmetics ingred.
Properties: Ylsh. oily liq., char. coconut odor, fatty sl. bitter taste; sol. in ether, n-hexane, alcohol; disp. in water; m.p. 20-25 C; HLB 3.0; acid no. 2 max.; iodine no. 1 max.; sapon. no. 250-275; 45% min. monoglycerides; nonionic
Discontinued prod.

Incrocas 35 [Croda Inc]
Chem. Descrip.: PEG-35 castor oil NF
CAS 61791-12-6
Uses: Surfactant, emulsifier, solvent, wetting agent, emollient, lubricant for topical and parenteral pharmaceuticals
Use Level: 0.5-5%
Properties: Pale yel. liq.; HLB 12.5; nonionic

Incromectant AMEA-70 [Croda Inc]
Chem. Descrip.: Acetamide MEA
CAS 142-26-7; EINECS 205-530-8
Uses: Solvent, humectant for cosmetics; clarifying agent for shampoos; conditioner for creme rinses
Use Level: 0.5-15%
Properties: Sol'n.; sp.gr. 1.07-1.17; dens. 9.3 lb/gal; pH 6.0-8.5; 70% min. act.

Intravon® SOL [Crompton & Knowles]
Chem. Descrip.: Aromatic solv./surfactant blend
Uses: Solvent scouring detergent for syn. and natural fibers and their blends, one-bath dyeing/scouring processes; aftersoaping agent, wetting agent, penetrant; alkali stable
Properties: Off-wh. to yel. gel; self-emulsifiable in water; flash pt. (PMCC) 52 C; pH 6.0 ± 0.5; anionic/nonionic
Toxicology: Defatting on prolonged skin contact.
Environmental: Biodeg.

Intravon® SOL-N [Crompton & Knowles]
Chem. Descrip.: Aliphatic solv./surfactant blend
Uses: Low odor solv.-based scouring detergent for all fibers, in one-bath dyeing/scouring processes; oil

solubilizer; good wetting and penetration, moderate foamer; stable to alkali
Properties: Off-wh. to yel. flowable gel, low odor; water sol./emulsifiable at all temps.; f.p. < -3 C; flash pt. (PMCC) > 95 C; pH 6.0 ± 0.5; anionic/nonionic
Toxicology: Defatting to skin on prolonged contact; eye irritant

Intravon® SOL-W [Crompton & Knowles]
Chem. Descrip.: Aromatic solv./surfactant blend
Uses: Fast wetting and penetrating solv.-based scouring detergent for syn. and natural fibers and their blends, in one-bath dyeing/scouring processes; oil solubilizer; alkali stable
Properties: Off-wh. to yel. gel; self-emulsifiable in water; flash pt. (PMCC) 52 C; pH 6.0 ± 0.5; anionic/nonionic
Toxicology: Defatting to skin on prolonged contact.
Environmental: Biodeg.

Isanol® [BASF AG]
Chem. Descrip.: 60% Iso- and 40% n-butanol
CAS 71-36-3; EINECS 200-751-6
Uses: Solvent for prod. of adhesives and surface coating resins
Discontinued prod.

Isoamyl Alcohol 95% [CPS]
Chem. Descrip.: Isoamyl alcohol
CAS 123-51-3; EINECS 204-633-5
Uses: Used for flavors, fragrances, fine chems., pharmaceuticals, extraction solvent, herbicides, plasticizers, mining chems., solvent for gums, resins, lacquers, oil additives, intermediates
Regulatory: FDA 21CFR §172.515
Properties: APHA 50 color; m.w. 88.15; sp.gr. 0.811 (20/20 C); distillation range 126-132 C (760 mm); flash pt. 47 C; 95% purity, 0.5% water

Isoamyl Alcohol 99% [CPS]
Chem. Descrip.: Isoamyl alcohol
CAS 123-51-3; EINECS 204-633-5
Uses: Used for flavors, fragrances, fine chems., pharmaceuticals, extraction solvent, herbicides, plasticizers, mining chems., solvent for gums, resins, lacquers, oil additives, intermediates
Regulatory: FDA 21CFR §172.515
Properties: APHA 20 color; m.w. 88.15; sp.gr. 0.811 (20/20 C); distillation range 128-132 C (760 mm); flash pt. 47 C; 99% purity, 0.2% water

Isopar® C [Exxon]
Chem. Descrip.: C7-8 isoparaffin
CAS 64742-48-9; EINECS 274-273-1
Uses: Solvent for coatings; minimizes VOC; virtually HAPS-free; extremely low surf. tens. to improve flow and wetting; effective thinner and cleaner; suitable for odorless alkyd and acrylic paints
Regulatory: FDA 21CFR §172.882, 178.3530; 40CFR §180.1001(c)(e)
Properties: Odorless; m.w. 114; sp.gr. 0.70 (15.6 C); dens. 5.83 lb/gal (60 F); visc. 0.5 cP; vapor pressure 63 mm Hg (20 C); b.p. 98-104 C; flash pt. 18 F; surf. tens. 20.3 dynes/cm; KB value 27; 100% paraffin
Toxicology: OEL 400 ppm; low toxicity

Isopar® E [Exxon]
Chem. Descrip.: C8-9 isoparaffin
CAS 64742-48-9
Uses: Solvent for coatings, thinners, cleaners; minimizes VOC; virtually HAPS-free; extremely low surf. tens. to improve flow and wetting; effective thinner and cleaner; suitable for odorless alkyd and acrylic paints
Regulatory: FDA 21CFR §172.882, 178.3530, 573.220; 40CFR §180.1001(c)(e)
Properties: Odorless; m.w. 123; sp.gr. 0.72 (15.6 C); dens. 6.02 lb/gal (60 F); visc. 0.6 cP; vapor pressure 36 mm Hg (20 C); b.p. 118-137 C; flash pt. 45 F; surf. tens. 22.1 dynes/cm; KB value 29; 99% paraffin, 1% cycloparaffin
Toxicology: OEL 400 ppm; low toxicity

Isopar® G [Exxon]
Chem. Descrip.: C10-11 isoparaffin
CAS 64742-48-9
Uses: Solvent, diluent, carrier for personal care sprays; solvent for coat-

ings, thinners, cleaners, degreasers; extremely low surf. tens. to improve flow and wetting; suitable for odorless alkyd/acrylic paints; no SARA reportables; minimizes VOC; virtually HAPS-free

Regulatory: FDA 21CFR §172.882, 172.884, 176.180, 176.210, 177.2600, 178.3530, 178.3570, 178.3620(b), 178.3650; 40CFR §180.1001(c)(e)

Properties: Odorless; m.w. 149; sp.gr. 0.75 (15.6 C); dens. 6.23 lb/gal (60 F); visc. 1.0 cP; vapor pressure 2.3 mm Hg (20 C); b.p. 160-176 C; flash pt. 106 F; surf. tens. 23.8 dynes/cm; KB value 28; 95% paraffin, 5% cycloparaffin

Toxicology: OEL 300 ppm; low toxicity

Isopar® G Naphtha [Exxon]

Chem. Descrip.: Isoparaffin hydrocarbon solv.

CAS 64742-48-9

Uses: Solvent

Properties: Sp.gr. 0.74 (15.6/15.6 C); dens. 3.18 lb/gal (60 F); b.p. 156-172 C; flash pt. (TCC) 39 C; surf. tens. 22.2 dynes/cm; KB value 28; 99% paraffin, 1% cycloparaffin, 0.03% aromatics

Toxicology: OEL 300 ppm

Isopar® H [Exxon]

Chem. Descrip.: C11-12 isoparaffin

CAS 64742-48-9

Uses: Solvent, diluent, carrier for personal care sprays; solvent for coatings, thinners, cleaners, degreasers; extremely low surf. tens. to improve flow and wetting; suitable for odorless alkyd/acrylic paints; no SARA reportables; minimizes VOC; virtually HAPS-free

Regulatory: FDA 21CFR §172.882, 172.884, 176.180, 176.210, 177.2600, 178.3530, 178.3570, 178.3620(b), 178.3650; 40CFR §180.1001(c)(e)

Properties: Odorless; m.w. 160; sp.gr. 0.76 (15.6 C); dens. 6.32 lb/gal (60 F); visc. 1.3 cP; vapor pressure 1.0 mm Hg (20 C); b.p. 178-188 C; flash pt. 129 F; surf. tens. 24.1 dynes/cm; KB value 27; 95% paraffin, 5% cycloparaffin

Toxicology: OEL 300 ppm; low toxicity

Isopar® K [Exxon]

Chem. Descrip.: C11-12 isoparaffin

CAS 64742-48-9

Uses: Solvent, diluent, carrier for personal care sprays; solvent for coatings, thinners, cleaners, degreasers; extremely low surf. tens. to improve flow and wetting; suitable for odorless alkyd/acrylic paints; no SARA reportables; minimizes VOC; virtually HAPS-free

Regulatory: FDA 21CFR §172.882, 172.884, 176.180, 176.210, 177.2600, 178.3530, 178.3570, 178.3620(b), 178.3650; 40CFR §180.1001(c)(e)

Properties: Odorless; m.w. 164; sp.gr. 0.76 (15.6 C); dens. 6.34 lb/gal (60 F); visc. 1.4 cP; vapor pressure 0.9 mm Hg (20 C); b.p. 177-197 C; flash pt. 135 F; surf. tens. 24.2 dynes/cm; KB value 27; 94% paraffin, 6% cycloparaffin

Toxicology: OEL 300 ppm; low toxicity

Isopar® K Naphtha [Exxon]

Chem. Descrip.: Isoparaffin hydrocarbon solv.

CAS 64742-48-9

Uses: Solvent

Properties: Sp.gr. 0.76 (15.6/15.6 C); dens. 6.35 lb/gal (60 F); b.p. 182-204 C; flash pt. (TCC) 60 C; surf. tens. 23.3 dynes/cm; KB value 28; 97% paraffin, 3% cycloparaffin, < 0.01% aromatics

Toxicology: OEL 300 ppm

Isopar® L [Exxon]

Chem. Descrip.: C11-13 isoparaffin

CAS 64742-48-9

Uses: Solvent for skin care formulations, coatings, agrochems., thinners, cleaners, and degreasers; suitable for odorless alkyd/acrylic paints; no SARA reportables; minimizes VOC; virtually HAPS-free; extremely low surf. tens. to improve flow and wetting

Regulatory: FDA 21CFR §172.882, 172.884, 176.180, 176.210, 177.2600, 178.3530, 178.3570, 178.3620(b), 178.3650, 573.680; 40CFR §180.1001(c)(e)

Properties: Odorless; m.w. 171; sp.gr. 0.77 (15.6 C); dens. 6.40 lb/gal (60 F); visc. 1.6 cP; vapor pressure 0.8 mm Hg (20 C); b.p. 189-207 C; flash pt. 147 F; surf. tens. 25.1 dynes/cm; KB value 27; 89% paraffin, 11% cycloparaffin

Toxicology: OEL 300 ppm; low toxicity

Isopar® M [Exxon]

Chem. Descrip.: C13-14 isoparaffin

CAS 64742-47-8

Uses: Solvent for hair spray, coatings, antiperspirants, laundry prewash, furniture/floor polish, waterless hand cleaner, agrochems, thinners, cleaners, degreasers; suitable for odorless alkyd/acrylic paints; carrier for insecticide actives; good dispersing props.; complies with criteria for LVP solvent; no SARA reportables; minimizes VOC; virtually HAPS-free; extremely low surf. tens. to improve flow and wetting

Regulatory: FDA 21CFR §172.882, 172.884, 176.180, 176.210, 177.2600, 178.3530, 178.3570, 178.3620(b), 573.680; 40CFR §180.1001(c)(e)

Properties: Nearly odorless; m.w. 191; sp.gr. 0.79 (15.6 C); dens. 6.57 lb/gal (60 F); visc. 2.7 cP; vapor pressure < 0.1 mm Hg (20 C); b.p. 223-254 C; flash pt. 199 F; surf. tens. 26.4 dynes/cm; KB value 27; 80% paraffin, 21% cycloparaffin

Toxicology: OEL 300 ppm; low acute toxicity

Isopar® V [Exxon]

Chem. Descrip.: Isoparaffin hydrocarbon solv.

CAS 64742-46-7

Uses: Solvent for household applics. incl. polishes, cleaning compds., degreasers, waterless hand cleaners, stain removers, air fresheners, and insecticides; also for use in inks and elastomer-based adhesives; complies with criteria for LVP solvent; no SARA reportables

Regulatory: FDA 21CFR §172.884, 176.180, 176.210, 178.3620(b), 178.3650, 573.220, 573.680; 40CFR §180.1001(c)(e)

Properties: Saybolt +30 color; low odor; misc. with other hydrocarbons; m.w. 197; sp.gr. 0.82 (15.6 C); dens. 6.82 lb/gal (60 F); visc 7.5 cps; vapor pressure 0.3 mm Hg (38 C); b.p. 273-311; flash pt. (ATSM D 93) 129 C; surf. tens. 26.9 dynes/cm; KB value 25

Toxicology: OEL 300 ppm

Precaution: Volatile; vapors can be ignited causing explosions or flash fires

Storage: Keep away from heat, sparks, and open flame

J

Jeffamine® D-230 [Huntsman]
Chem. Descrip.: Polyoxypropylenediamine
Uses: Epoxy curing agent used in surface coatings, heat-cured sol'n. coatings, laminates, potting, casting, reinforced composites, adhesives, polymer concrete, syntactic foam, reactive solvent
Use Level: 32 phr
Properties: Colorless to sl. yel sl. hazy liq.; sol. in most hydrocarbons and oxygenated solvs., misc. in water; m.w. 230; sp.gr. 0.9480; dens. 7.9 lb/gal; visc. 9 cps; vapor pressure 1 mm Hg (101 C); flash pt. (PMCC) 121 C; ref. index 1.4466; pH 11.7 (5% aq.); distort. temp 60-75 C (264 psi); 0.1% water
Toxicology: LD50 (oral, rat) 1.66 g/kg, (dermal, rabbit) 0.76 g/kg; can cause burns to skin and eyes, toxic effects if absorbed thru skin or swallowed
Storage: Nitrogen blanket suggested for all storage in case of accidental high temp.; pronounced discoloration is likely to occur above 140 F

Jeffox™ PPG-400 [Huntsman]
Chem. Descrip.: PPG-400
Uses: Intermediate yielding esters; useful as lubricants, defoaming agents in rubber and pharmaceuticals, solvs. and humectant modifiers for inks, plasticizers, and functional fluids
Properties: Water-wh. visc. liq., faint ether-like odor; m.w. 400; dens. 8.40 lb/gal; visc. 150-175 SUS (100 F); flash pt. (PMCC) 320 F; pour pt. -35 F; pH 5-7

Jeffox™ PPG-2000 [Huntsman]
Chem. Descrip.: PPG-2000
Uses: Intermediate yielding esters; useful as lubricants, defoaming agents in rubber and pharmaceuticals, solvs. and humectant modifiers for inks, plasticizers, and functional fluids
Properties: Water-wh. visc. liq., faint ether-like odor; m.w. 2000; dens. 8.37 lb/gal; flash pt. (PMCC) 370 F; pour pt. -25 F; pH 5-7

Jeffox™ WL-440 [Huntsman]
Chem. Descrip.: EO/PO derivs.
Uses: Functional fluids used as lubricants, plasticizers, solvs., coupling agents, frothing agents, heat transfer fluids, intermediates, defoaming agents; used in solder reflow applics., boiler defoaming, ore flotation, inks, dyes
Properties: Clear liq.; m.w. 488; water-sol.; sp.gr. 1.1515; dens. 9.55 lb/gal; visc. 100 cSt (100 F); flash pt. 580 F (COC); pour pt. -30 F; ref. index 1.4721

Jeffox™ WL-660 [Huntsman]
Chem. Descrip.: EO/PO deriv
Uses: Functional fluids used as lubricants, plasticizers, solvs., coupling agents, frothing agents, heat transfer fluids, intermediates, defoaming agents; used in solder reflow applics., boiler defoaming, ore flotation, inks, dyes
Properties: Clear liq.; m.w. 1800; water-sol.; sp.gr. 1.0502; dens. 8.75 lb/gal; visc. 143 cSt (100 F); cloud pt. 60 C (1% aq.); flash pt. 475 F (COC); pour pt. -45 F; ref. index 1.4586

Jeffox™ WL-1400 [Huntsman]
Chem. Descrip.: EO/PO deriv
Uses: Functional fluids used as lubri-

cants, plasticizers, solvs., coupling agents, frothing agents, heat transfer fluids, intermediates, defoaming agents; used in solder reflow applics., boiler defoaming, ore flotation, inks, dyes

Properties: Clear liq.; m.w. 2500; water-sol.; sp.gr. 1.095; dens. 9.15 lb/gal; visc. 306 cSt (100 F); cloud pt. > 99 C (1% aq.); flash pt. 475 F (COC); pour pt. 50 F; ref. index 1.4669

Jeffox™ WL-5000 [Huntsman]

Chem. Descrip.: EO/PO deriv

Uses: Functional fluids used as lubricants, plasticizers, solvs., coupling agents, frothing agents, heat transfer fluids, intermediates, defoaming agents; used in solder reflow applics., boiler defoaming, ore flotation, inks, dyes

Properties: Clear liq.; m.w. 4365; water-sol.; sp.gr. 1.0557; dens. 8.80 lb/gal; visc. 1080 cSt (100 F); cloud pt. 53 C (1% aq.); flash pt. 460 F (COC); pour pt. -10 F; ref. index 1.4600

Jeffsol™ EC [Huntsman]

Chem. Descrip.: Ethylene carbonate CAS 96-49-1; EINECS 202-510-0

Uses: Solvent, reactant, plasticizer for fibers, textiles, dyeing, plastics and resins, aromatic hydrocarbon extraction, electrolytes, photosensitive compositions for preparing lithographic plates; fuel for jet engines; hydraulic brake fluids

Properties: Colorless solid, pract. odorless; sol. in benzene, butyl acetate, chloroform, ethanol, ethyl acetate, methanol, nonylphenol, toluene, water; m.w. 88.06; sp.gr. 1.3218 (39/4 C); dens. 10.87 lb/gal (40 C); visc. 1.9 cs (40 C); m.p. 36.4 C; b.p. 238 C (760 mm); flash pt. 320 F; ref. index 1.4158 (50 C); dielec. const. 95.3; 99.5% min. act.

Toxicology: LD50 (oral, rat) > 5 g/kg, (dermal, rat) > 3 g/kg; pract. nontoxic; minimally irritating to skin

Precaution: May decompose with evolution of CO_2 in presence of an acid, base, metal oxide, or salt; protect from sources of ignition or static elec. discharge

Jeffsol™ PC [Huntsman]

Chem. Descrip.: Propylene carbonate CAS 108-32-7; EINECS 203-572-1

Uses: Solvent, reactant, plasticizer for fibers, textiles, dyeing, plastics and resins, gas treating, aromatic hydrocarbon extraction, metal extraction, surf. coatings (paints, varnishes, adhesives, plastics, epoxies, mastics); foundry sand binder, lubricants, electrolytes, cosmetics (polar additive for montmorillonite or bentonite clay gellants); fire extinguishing compds.; antifoam for antifreeze; hydraulic brake fluids; as plating medium

Properties: Clear mobile liq., pract. odorless; sol. in acetone, benzene, chloroform, glycols, ethanol, methanol, MEK, toluene, xylene; sol. 25 g/100 g water; m.w. 102.09; sp.gr. 1.2057 (20/4 C); dens. 10.05 lb/gal (20 C); visc. 1.9 cs (20 C); f.p. -49 C; b.p. 242 C (760 mm); pour pt. -100 F (supercools easily); flash pt. 275 F; fire pt. 280 F; ref. index 1.4210; dielec. const. 65; 99.8% min. act.

Toxicology: LD50 (dermal, rabbit) > 20 g/kg; pract. nontoxic; nonirritating to skin; ingestion may cause renal damage and death

Precaution: May decompose with evolution of CO_2 in presence of an acid, base, metal oxide, or salt; protect from sources of ignition or static elec. discharge

K

Kemester® 115 [Witco/Oleo-Surf.]
Chem. Descrip.: Methyl oleate
Uses: Intermediate in prod. of superamides, in metalworking lubricants, as solvent
Properties: Gardner 4 max.; acid no. 4.0 max.; iodine no. 90 max.; sapon. no. 185-205

Kemester® 143 [Witco/Oleo-Surf.]
Chem. Descrip.: Methyl tallowate
CAS 61788-61-2; EINECS 262-989-7
Uses: Intermediate in prod. of superamides, in metalworking lubricants, as solvent; lubricant, plasticizer for leather, rubber prods.
Properties: Gardner 8 max.; acid no. 4.0 max.; iodine no. 60 max.; sapon. no. 195 min.

Kemester® 205 [Witco/Oleo-Surf.]
Chem. Descrip.: Methyl oleate
Uses: Intermediate in prod. of superamides, in metalworking lubricants, as specialized solvent; opacifier, visc. control agent
Properties: Gardner 1 liq.; acid no. 4.0 max.; iodine no. 80-90; sapon. no. 194-203; 100% conc.

Kemester® 213 [Witco/Oleo-Surf.]
Chem. Descrip.: Methyl oleate/linoleate
Uses: Intermediate in prod. of superamides, in metalworking lubricants, as solvent
Properties: Gardner 3 max.; acid no. 10.0 max.; iodine no. 130-145; sapon. no. 194-204

Kemester® 226 [Witco/Oleo-Surf.]
Chem. Descrip.: Methyl soyate
Uses: Intermediate in prod. of superamides, in metalworking lubricants, as specialized solvent; opacifier, visc. control agent
Properties: Gardner 4 max. liq.; acid no. 7.0 max.; iodine no. 125-135; sapon. no. 195-200; 100% conc.

Kemester® 2050 [Witco/Oleo-Surf.]
Chem. Descrip.: Methyl eicosenate
CAS 1120-28-1; EINECS 214-304-8
Uses: Intermediate in prod. of alkanolamides, in metalworking lubricants, as specialized solvent; foam depressant and nutrient in fermentation
Properties: Gardner 14 max. color; acid no. 25 max.; iodine no. 90-110; sapon. no. 180-190

Kemester® 4516 [Witco/Oleo-Surf.]
Chem. Descrip.: Methyl stearate
Uses: Intermediate in prod. of alkanolamides, in metalworking lubricants, as specialized solvent; foam depressant and nutrient in fermentation
Properties: Gardner 1 max. color; acid no. 3 max.; iodine no. 2.5 max.; sapon. no. 192-202

Kemester® 9018 [Witco/Oleo-Surf.]
Chem. Descrip.: Methyl stearate
Uses: Intermediate in prod. of superamides, in metalworking lubricants, specialized solvent; opacifier, visc. control agent
Properties: Gardner 1 max.; acid no. 4.0 max.; sapon. no. 185-192

Kemester® 9022 [Witco/Oleo-Surf.]
Chem. Descrip.: Methyl behenate
CAS 929-77-1; EINECS 213-207-8
Uses: Intermediate in prod. of superamides, in metalworking lubricants, as solvent
Properties: Gardner 2 max.; acid no. 20 max.; iodine no. 1 max.; sapon. no. 150-160

Kemester® EGDS [Witco/Oleo-Surf.]

Chem. Descrip.: Glycol distearate
CAS 627-83-8; EINECS 211-014-3
Uses: Intermediate in prod. of superamides, in metalworking lubricants, specialized solvent; industrial lubricant; opacifier, pearling additive, thickener for cosmetics and pharmaceuticals; internal/external lubricant for PVC
Regulatory: FDA accepted
Properties: Gardner 2 max. solid; m.p. 60-63 C; acid no. 6 max.; iodine no. 1 max.; sapon. no. 190-200; 100% conc.

Kemester® EGMS [Witco/Oleo-Surf.]
Chem. Descrip.: Glycol stearate
CAS 111-60-4; EINECS 203-886-9
Uses: Intermediate in prod. of superamides, in metalworking lubricants, specialized solvent; industrial lubricant; opacifier, pearling additive, thickener for cosmetics and pharmaceuticals
Properties: Gardner 1 solid; m.p. 56-60 C; acid no. 1; iodine no. 1 max.; sapon. no. 179-195

Kemstrene® 96.0% USP [Witco/Oleo-Surf.]
Chem. Descrip.: Refined glycerin USP
CAS 56-81-5; EINECS 200-289-5
Uses: Humectant, solvent; used in cosmetics, pharmaceuticals, liq. soaps, confections, inks, and lubricants; intermediate used in polyester and PU formulations
Properties: Sp.gr. 1.25165 min; 96.0% purity

Kemstrene® 99.0% [Witco/Oleo-Surf.]
Chem. Descrip.: Refined glycerin USP
CAS 56-81-5; EINECS 200-289-5
Uses: Humectant, solvent; used in cosmetics, pharmaceuticals, liq. soaps, confections, inks, and lubricants; intermediate used in polyester and PU formulations
Properties: Sp.gr. 1.25945 min.; 99.0% purity

Kemstrene® 99.7% USP [Witco/Oleo-Surf.]
Chem. Descrip.: Refined glycerin USP
CAS 56-81-5; EINECS 200-289-5
Uses: Humectant, solvent; used in cosmetics, pharmaceuticals, liq. soaps, confections, inks, and lubricants; intermediate used in polyester and PU formulations
Properties: Sp.gr. 1.26124; 99.7% purity

Kemstrene® High Gravity [Witco/Oleo-Surf.]
Chem. Descrip.: Refined glycerin
CAS 56-81-5; EINECS 200-289-5
Uses: Humectant, solvent; used in cosmetics, liq. soaps, confections, inks, and lubricants; intermediate used in polyester and PU formulations
Properties: Sp.gr. 1.2587
Unverified

Kenflex® N1 [Kenrich Petrochems.]
Chem. Descrip.: Aromatic hydrocarbon polymeric plasticizer (50% alkylnaphthalene formalin, CAS 52613-22-6 and 50% petrol. naphthenic distillate, CAS 64742-53-6)
Uses: Discoloring plasticizer, softener, processing aid for chlorosulfonated PE providing superior processing, improved extrusion, exc. electricals, exc. mold flow and release, high temp. stability, permanence, nonextractability, nonmigrance, nonvolatility; coupling for poly blends; also for polysulfide, neoprene, butyl rubber and polar thermoplastics; solvent
Use Level: 5-20 phr (Hypalon)
Properties: Dk. brn. liq.; sp.gr. 1.01; dens. 8.41 lb/gal; visc. 12,000 cps; i.b.p. 176 C; flash pt. (COC) 190 ± 20 C; VOC none
Toxicology: Potential skin, eye, and respiratory tract irritant; may be harmful if swallowed; may contain trace amts. of formaldehyde, a suspected carcinogen; may emit toxic fumes on thermal decomp.
Precaution: Incompat. with oxidizers, acids; avoid excessive heat, ignition sources; thermal dec. may emit toxic fumes; hazardous dec. prods.: CO, SO_x, aldehydes
Storage: Store in well-ventilated area in tightly closed containers

Kenplast® A-450 [Kenrich Petrochems.]
Chem. Descrip.: 60-100% Petroleum

distillate (CAS 68477-31-6), 15-40% methylnaphthalenes, 3-7% naphthalene (CAS 91-20-3)
UN No.: 1993; 3082; DOT No.: NA 1993
Uses: Plasticizer, diluent, solvent for urethanes, epoxies, PVC, cellulosics
Properties: Yel. clear liq.; acetic odor; negligible sol. in water; sp.gr. 0.98; vapor pressure 0.15 mm Hg (20 C); b.p. 210-288 C; flash pt. > 180 F
Toxicology: LD50 (oral, rat) > 2622 mg/kg, (dermal, rabbit) > 2000 mg/kg; LC50 (inh., 1 h, rat) > 1.67 mg/l (no death); can be harmful if high concs. are inhaled; harmful or fatal if liq. is aspirated into lungs; causes mild skin irritation
Precaution: Combustible
Storage: Store away from heat, ignition sources, open flame

Kenplast® ES-2HP [Kenrich Petrochems.]
Chem. Descrip.: 95% Cumylphenyl acetate, 5% p-cumyl phenol
Uses: Plasticizer for urethanes; high flash pt. reactive diluent for epoxy, reducing odor and irritation; lowers visc. of highly filled systems; ideal for polyamide-cured epoxy floorings and coal tar pipe coatings; comonomer and impact modifier for phenolics; improves adhesion of vinyl plastisols coating metals and polar plastics; solvent
Properties: Lt. yel. clear liq.; sp.gr. 1.08; dens. 9.0 lb/gal; visc. 50-200 cps; i.b.p. 326 C; dist. range 326-343 C; flash pt. (COC) 370 ± 30 F; .pour pt. -13 C; acid no. < 1; sapon. no. 154; hyd. no. 13; pH 7 ± 1.0; VOC none
Toxicology: LD50 (rat) 6 g/kg; mild skin irritant; potential eye and respiratoyr tract irritant; may be harmful if swallowed
Precaution: Incompat. with oxidizers, acids; may emit toxic fumes on thermal decomp.

Kenplast® ESB [Kenrich Petrochems.]
Chem. Descrip.: Cumyl-phenyl benzoate
Uses: Primary plasticizer for PVC and PVC/nitrile; process aid for semi-rigid PVC extrudates; solvates conductive polyester and acrylic inks to improve impact.
Properties: Gardner > 18 liq.; sp.gr. 1.13 dens. 9.4 lb/gal; visc. 15,000 cps; dist. range 250-427 C; pour pt. 16 C; acid no. < 1; sapon. no. 124; flash pt. (COC) 171 C; 70% act. in α-methyl styrene oligomer
Discontinued prod.

Kenplast® ESN [Kenrich Petrochems.]
Chem. Descrip.: Cumylphenyl neodecanoate
Uses: Primary plasticizer for PVC, PVC/nitrile; process aid for semi-rigid PVC extrusions; solvates and impact modifies conductive polyester and acrylic inks
Properties: Gardner > 18 liq.; sp.gr. 1.05; dens. 8.75 lb/gal; visc. 5800 cps; pour pt. 1.66 C; acid no. < 1; sapon. no. 130; flash pt. (COC) 196 C; 85% act. in a-methyl styrene oligomer
Discontinued prod.

Kenplast® G [Kenrich Petrochems.]
Chem. Descrip.: Mixed alkylated phenanthrenes; 93-95% petrol. distillate (CAS 64741-81-7), 1-5% molten sulfur, < 3% 1,2-benzanthracene
Uses: Nonreactive discoloring primary plasticizer for dark PVC, nitrile, and urethane compds.; flexibilizer and diluent for epoxy resins; solvent
Properties: Dk. straw-colored liq.; sp.gr. 1.08 (60 F); visc. 25 cps (100 F); i.b.p. 420 F; flash pt. (COC) 196 C
Toxicology: LD50 (oral, rat) 2.5 g/kg; exposure may cause eye irritation, mild skin irritation, nose, throat, and respiratory tract irritation, CNS depression, death; ing. of lg. amts. may be harmful; nonmutagenic; 1,2-benzanthracene is carcinogenic
Precaution: Incompat. with strong oxidizing agents; hazardous decomp. prods.: CO_2, CO, various hydrocarbons, sulfur compds.
Storage: Will crystallize during storage below 50 F; warm before use to

eliminate crystals; do not store above 70 F

Kessco® IPM [Stepan; Stepan Canada]
Chem. Descrip.: Isopropyl myristate
CAS 110-27-0; EINECS 203-751-4
Uses: Replacement for min. oil; used as lubricants in textile spin finish, coning oils, carding, dye bath; emollient, solubilizer, vehicle, solvent for makeup, shaving preps., bath oils, hair preps., highly pigmented prods.
Properties: APHA 20 max. clear liq., pract. odorless; insol. in water; sol. in IPA, min. oil, peanut oil; f.p. -3 C; m.p. -3 C; b.p. 162.5 C; acid no. 1 max.; iodine no. 1 max.; sapon. no. 204-212; flash pt. (COC) 305 F; ref. index 1.433; 100% act.
Environmental: Biodeg.

Kessco® IPP [Stepan; Stepan Canada]
Chem. Descrip.: Isopropyl palmitate
CAS 142-91-6; EINECS 205-571-1
Uses: Replacement for min. oil; used as lubricants in textile spin finish, coning oils, carding, dye bath; emollient, solubilizer, vehicle, solvent for makeup, shaving preps., bath oils, hair preps., highly pigmented prods.
Properties: APHA 20 max. clear liq., pract. odorless; insol. in water; sol. in IPA, min. oil, peanut oil; sp.gr. 0.849-0.855; f.p. 13 C; b.p. 172 C (4 mm); acid no. 1 max.; iodine no. 1 max.; sapon. no. 185-191; flash pt. (COC) 323 F; ref. index 1.437; 100% act.
Environmental: Biodeg.

Kodaflex® DBP [Eastman]
Chem. Descrip.: Dibutyl phthalate
CAS 84-74-2; EINECS 201-557-4
Uses: Primary plasticizer-solvent for nitrocellulose lacquers; plasticizer for rubbers and CAB, ethyl cellulose, PVAc, and syn. resins; solvent for oil-sol. dyes, insecticides, peroxides, and org. compds.; antifoamer and fiber lubricant in textile mfg.; cosmetics plasticizer
Properties: Pt-Co 15 ppm liq.; sol. 0.46% in water; m.w. 278; sp.gr. 1.048; dens. 8.72 lb/gal (20 C); visc. 15 cP; f.p. -35 C; b.p. 340 C (760 mm); ref. index 1.4920; flash pt. (COC) 190 C; fire pt. (COC) 202 C; dissip. factor 0.08 x 10^{-2} (1 MHz); dielec. const. 5.8 (1 MHz); vol. resist. 3.0 x 10^{9} ohm cm; 99% min. assay
Precaution: Combustible

Kodaflex® DEP [Eastman]
Chem. Descrip.: Diethyl phthalate
CAS 84-66-2; EINECS 201-550-6
Uses: Plasticizer; wetting agent in grinding pigments; pigment-disp. medium in cellulose acetate sol'ns. and plastics, and solvent for natural resins and polymers; seldom used in PVC prods. due to relatively high volatility; also for cosmetics
Properties: Pt-Co 10 ppm liq.; sol. 0.12 g/l in water; m.w. 222; sp.gr. 1.120; dens. 9.32 lb/gal (20 C); visc. 9.5 cP; f.p. < -50 C; b.p. 298 C (760 mm); ref. index 1.4990; flash pt. (COC) 161 C; fire pt. (COC) 171 C; dissip. factor 0.10 x 10^{-2} (1 MHz); dielec const. 6.7 (1 MHz); vol. resist. 1.45 x 10^{9} ohm cm; 99% min. assay

Kodaflex® DMP [Eastman]
Chem. Descrip.: Dimethyl phthalate
CAS 131-11-3; EINECS 205-011-6
Uses: Plasticizer with high solvent power for cellulose acetate extrusion compds.; compatible with ethyl cellulose, CAB, PS, PVAc, polyvinyl butyral, and PVC; used in NC-based printing inks; also for cosmetics and personal care prods.
Properties: Pt-Co 5 ppm liq.; sol. 0.45 g/l in water; m.w. 194; sp.gr. 1.192 (20/20 C); dens. 9.93 lb/gal (20 C); visc. 11.0 cP; f.p. -1 C; b.p. 284 C; ref. index 1.513; flash pt. (COC) 157 C; fire pt. (COC) 168 C; dissip. factor 0.13 x 10^{-2} (1 mc); dielec. const. 7.5 (1 mc); vol. resist. 1.07 x 10^{9} ohm cm; 99% min. assay

Kodaflex® HS-3 [Eastman]
Chem. Descrip.: Butyl octyl phthalate
CAS 84-78-6; EINECS 201-562-1
Uses: High-solvating primary plasticizer for polyvinyl homopolymer and copolymer resins, vinyl plastisols, expanded vinyl foams, and rotational molding and dip coating

Properties: Pt-Co 35 ppm liq.; sol. < 0.01 g/l in water; sp.gr. 0.988-0.996; dens. 8.26 lb/gal (20 C); visc. 42 cP; f.p. -48 C; b.p. 350 C; ref. index 1.4848; flash pt. (COC) 208 C; fire pt. (COC) 224 C

Kodaflex® HS-4 [Eastman]

Chem. Descrip.: 60% Dioctyl phthalate, 5% dibutyl phthalate

Uses: High-solvating primary plasticizer; used in formulating vinyl plastisols having processing chars. required to be mechanically frothed; for PVC formulations, rotational molding, slush molding, dip coating, and filament coating applics.

Properties: Pt-Co 35 ppm liq.; sol. < 0.01 g/l in water; sp.gr. 0.993; dens. 8.26 lb/gal (20 C); visc. 31.8 cP; f.p. -47 C; b.p. 337 C; ref. index 1.4798; flash pt. (COC) 182 C; fire pt. (COC) 200 C; dissip. factor 7.9 x 10^{-2} (1 MHz); dielec. const. 3.4 (1 MHz); vol. resist. 2.0 x 10^{14} ohm-cm

Kronitex® TBP [FMC]

Chem. Descrip.: Tributyl phosphate CAS 126-73-8; EINECS 204-800-2

Uses: Primary plasticizer for NC, cellulose acetate, chlorinated rubber, vinyls; high boiling solvent for lithographic inks; antifoam for paints, paper coatings, aq. adhesives, inks, casein sol'ns., textile sizes, detergent sol'ns.; pigment dispersant; also for uranium extraction

Properties: APHA 50 max. color, char. odor; sol. in most org. liqs., min. oil, gasoline, 0.1% in water; m.w. 266; sp.gr. 0.977-0.983 (20/20 C); dens. 8.14 lb/gal (20 C); visc. 3.7 cp (20 C); vapor pressure 7.3 mm Hg (150 C); f.p. < -80 C; pour pt. < -80 C; b.p. 137-145 C; flash pt. (PMCC) 115 C; fire pt. 182 C; ref. index 1.423 ± 0.001; surf. tens. 29 dynes/cm (20 C)

Toxicology: Severe skin irritant

Kronitex® TOF [FMC]

Chem. Descrip.: Trioctyl phosphate CAS 78-42-2; EINECS 201-116-6

Uses: Low temp. plasticizer for use in PVC, rubber, paints, coatings; highly efficient solvent

Kwik Dri [Ashland]

Chem. Descrip.: Aliphatic solv.

Uses: Solvent

Unverified

L

Labrafac® CM10 [Gattefosse; Gattefosse SA]
Chem. Descrip.: PEG-8 caprylic/capric glycerides
CAS 85536-08-9
Uses: Surfactant for oral SE drug delivery systems; bioavailability enhancer, emulsifier, solvent, solubilizer for pharmaceutical orals, topicals, vaginals, rectals, and nasals, soft or hard gelatin capsules
Properties: Gardner < 5 oily liq., faint odor; sol. in veg. oil, alcohol, chloroform, methylene chloride; partly sol. in propylene glycol; partly disp. in water; sp.gr. 1.000-1.040 (20 C); HLB 10; acid no. < 2; iodine no. < 2; sapon. no. 160-200; hyd. no. 115-155; ref. index 1.430-1.485 (20 C); nonionic
Toxicology: LD0 (oral, rat) > 2 g/kg
Storage: Preserve in orig. container; prevent exposure to air, light, heat, and moisture

Labrafac® Hydro [Gattefosse; Gattefosse SA]
Chem. Descrip.: Caprylic/capric triglyceride PEG-4 complex
Uses: Hydrophilic rancidless oil, bioavailability enhancer, emulsifier, solvent, solubilizer for pharmaceutical orals, topicals, vaginals, rectals, and nasals, cosmetics (creams, lotions, shampoos, bath oils), veterinary prods.
Properties: Gardner < 5 oily liq., faint odor; sol. in veg. oil, alcohol; partly disp. in water; HLB 4-5; acid no. < 2; iodine no. < 3; sapon. no. 265-285; nonionic
Toxicology: LD0 (oral, rat) > 20 ml/kg; sl. skin and eye irritant

Labrafac® LIPO [Gattefosse; Gattefosse SA]
Chem. Descrip.: Caprylic/capric triglyceride
CAS 65381-09-1; EINECS 265-724-3
Uses: Nonrancidifying, temp. stable sat. lipophilic solvent, superfatting agent for cosmetic, pharmaceutical, or veterinary emulsions and microemulsions, aerosols, capsules, gels, creams, lotions, ointments, shampoos, massage and sun oils
Properties: Gardner < 1 oily liq., odorless to faint odor; acid no. < 0.2; iodine no. < 2; sapon. no. 320-340
Toxicology: LD50 (oral, rat) > 10 ml/kg; sl. eye irritant, nonirritating to skin

Labrafil® Isostearique [Gattefosse]
Chem. Descrip.: Triisostearin PEG-6 esters
Uses: Solvent, amphiphilic agent for improving drug delivery; nonrancidifying excipient for aerosols, nasal preps., creams, massage oils; for pharmaceuticals, cosmetics, veterinary preps.
Properties: Gardner < 5 oily liq., faint odor; HLB 3-4; acid no. < 2; iodine no. < 15; sapon. no. 150-170; nonionic
Toxicology: LD50 (oral, rat) > 20 ml/kg; nonirritating to skin and eyes

Labrasol [Gattefosse; Gattefosse SA]
Chem. Descrip.: PEG-8 caprylic/capric glycerides
CAS 85536-07-8
Uses: Hydrophilic oil, solvent, excipient, solubilizer for pharmaceutical, veterinary, and cosmetic formulations; surfactant for micro-

emulsions; wetting agent; penetration enhancer; improves topical, rectal, and oral absorption; superfatting agent for bath preps., shampoos

Properties: Gardner < 5 oily liq., faint odor; sol. in water, veg. oil, propylene glycol; very sol. in ethanol, chloroform, methylene chloride; sp.gr. 1.060-1.070; visc. 80-110 mPa·s; HLB 14.0; acid no. < 1; iodine no. < 2; sapon. no. 85-105; ref. index 1.450-1.470; 100% conc.; nonionic

Toxicology: LD50 (oral, rat) 22 g/kg; sl. ocular irritant at 0.1 ml; nonirritating to skin

Storage: Preserve in org. container; prevent exposure to air, light, heat, and moisture

Lacolene [Ashland]

Chem. Descrip.: Aliphatic

Uses: Solvent

Unverified

Lanpol 5 [Croda Chem. Ltd]

Chem. Descrip.: PEG-5 lanolate

CAS 68459-50-7

Uses: O/w emulsifier, solvent for bromo acid dyes in lipsticks, skin care and makeup prods., solubilizer, wetting agent, dispersant

Properties: Pale yel. pasty solid; sol. in oil, disp. in water; HLB 7.5; sapon. no. 105-120; pH 4-7 (3%); 97% conc.; nonionic

Lanpol 520 [Croda Chem. Ltd]

Chem. Descrip.: POE lanolin acid

Uses: O/w emulsifier, solvent for bromo acid dyes in lipsticks, skin care and makeup, solubilizer, wetting agent

Properties: Pale yel. pasty solid; HLB 14.1; sapon. no. 45-60; pH 4-7 (3%); 97% conc.; nonionic

Unverified

Lexol® 3975 [Inolex]

Chem. Descrip.: Isopropyl palmitate, isopropyl myristate, isopropyl stearate

Uses: Emollient and solvent for cosmetics and topical pharmaceuticals; replacement for IPM or IPP; oxidative resist.

Properties: Colorless liq., odorless; sol. in acetone, benzene, CCl_4, castor oil, chloroform, ethanol, heptane, IPA; insol. in water; sp.gr. 0.850-0.858; visc. 7 cps; f.p. 11 C; acid no. 1 max.; iodine no. 1 max.; sapon. no. 184-194; flash pt. (PMCC) 169 C; ref. index 1.4367; 100% conc.

Lexol® GT-855 [Inolex]

Chem. Descrip.: Caprylic/capric triglyceride

CAS 65381-09-1; EINECS 265-724-3

Uses: Emollient with nonoily skin feel; moisturizer; for creams, lotions, bath oils, lipstick, makeup; solvent for perfume and flavor ingreds.; vehicle for medicinals, antibiotics, vitamins; solubilizer; oxidative stability

Properties: APHA 100 max. clear liq., odorless; tasteless; alcohol-sol.; sp.gr. 0.943; visc. 27 cps; f.p. -19 C; acid no. 0.1 max.; iodine no. 1 max.; sapon. no. 325-355; flash pt. (PMCC) 224 C; ref. index 1.4479; 100% conc.

Discontinued prod.

Lexol® GT-865 [Inolex]

Chem. Descrip.: Caprylic/capric triglyceride

CAS 65381-09-1; EINECS 265-724-3

Uses: Emollient, emulsion stabilizer, vehicle for cosmetic creams, lotions, bath oils, preshave lotions, lipsticks, makeup; solvent for perfume; carrier for flavors; vehicle for medicinals, antibiotics, vitamins

Properties: APHA 100 max. clear liq., odorless, tasteless; sol. in alcohol; sp.gr. 0.947; visc. 25 cps; f.p. -19 C; acid no. 0.1 max.; iodine no. 1 max.; sapon. no. 335-355; flash pt. (PMCC) 233 C; ref. index 1.4471; 100% conc.

Lexol® IPM [Inolex]

Chem. Descrip.: Isopropyl myristate NF

CAS 110-27-0; EINECS 203-751-4

Uses: Emollient, solvent, penetrant, cloud pt. depressant; used in personal care prods.

Properties: Colorless clear liq.; odorless; sol. in acetone, benzene, CCl_4, castor oil, chloroform, ethanol, heptane, IPA; insol. in water; m.w. 270; sp.gr. 0.847-0.854; dens. 7.1 lb/gal; visc. 4.8 cp; f.p. 3 C; b.p. 170 C; acid no. 1 max.; iodine no. 1 max.; sapon. no. 202-212; flash pt. 305 F; ref. index 1.433

Lexol® IPM-NF [Inolex]

Chem. Descrip.: Isopropyl myristate NF

CAS 110-27-0; EINECS 203-751-4

Uses: Emollient, solvent, penetrant for cosmetic applics. (bath oils, elec. preshaves); low visc., low f.p., outstanding spreading, good sol., high oxidative stability

Properties: Colorless clear liq., odorless; sol. in acetone, benzene, CCl_4, castor oil, chloroform, ethanol, heptane, IPA; insol. in water, glycerol, propylene glycol; sp.gr. 0.852; visc. 0.5 cps; f.p. 2 C; acid no. 0.1; iodine no. 0.3; sapon. no. 208; flash pt. (COC) 155 C; ref. index 1.4327; 100% conc.

Lexol® PG-800 [Inolex]

Chem. Descrip.: Propylene glycol dioctanoate

CAS 56519-71-2

Uses: Emollient and moisturizer with nonoily feel, oxidation stability; for cosmetic/pharmaceutical creams, lotions, topicals, lipsticks, glossers, makeup bases, bath oils, aftershaves; solvent, carrier/vehicle for flavors, fragrance, vitamins, antibiotics, medicinals

Properties: Colorless clear liq., char. odor; sol. in alcohol, min. oil, acetone; sp.gr. 0.918; visc. 10 cps; f.p. -34 C; acid no. 1 max.; iodine no. 1 max.; sapon. no. 320-340; flash pt. (COC) 272 C; ref. index 1.4350; 100% conc.

Lexol® PG-865 [Inolex]

Chem. Descrip.: Propylene glycol dicaprylate/dicaprate

Uses: Emollient, moisturizer with lubricity and nonoily skin deposition for creams, lotions, makeup, bath oils, pre-electric shave lotions, aerosol systems; solubilizer, solvent, coupling agent, vehicle for flavors, fragrances, pigmented cosmetics, vitamins, antibiotics, medicinals

Properties: APHA 100 max. clear liq., odorless; sol. in alcohol, min. and veg. oil, acetone; sp.gr. 0.922; visc. 10 cps; f.p. -38 C; acid no. 0.1 max.; iodine no. 1 max.; sapon. no. 315-335; flash pt. (PMCC) 195 C; ref. index 1.4391; 100% conc.

Storage: Store in tightly closed original containers @ 13-38 C to avoid deterioration or contamination; store bulk liq. under dry inert atmosphere in SS or aluminum vessels

Lipolan R [Lipo]

Chem. Descrip.: Lanolin oil

Uses: Emollient, spreading agent, conditioner, cosolvent, plasticizer, and lubricant for personal care prods.; dispersant for pigments

Properties: Yel. to amber liq.; mild char. odor; sol. in min. and veg. oils, isopropyl esters, and anhyd. IPA; insol. in water; acid no. 2 max.; iodine no. 18-36; sapon. no. 85-110; cloud pt. 18 C max.

Liponate GC [Lipo]

Chem. Descrip.: Caprylic/capric triglyceride

CAS 65381-09-1; EINECS 265-724-3

Uses: Emollient ester for adjusting rub-in and afterfeel of personal care prods.; thickener and visc. control agent, pigment dispersant; clouding agent, diluent for vitamins, solvent for oil-disp. materials, useful in flavor systems and dietary formulations

Regulatory: FDA 21CFR §170.30; kosher grade avail.

Properties: Colorless to sl. yel. oily liq., bland odor and taste; sol. in anhyd. alcohol, min. and veg. oils; insol. in water; sp.gr. 0.935-0.955; acid no. 0.1 max.; iodine no. 1 max.; sapon. no. 335-360

Liponate PC [Lipo]

Chem. Descrip.: Propylene glycol dicaprylate/dicaprate

Uses: Emollient ester for adjusting rub-in and afterfeel of personal care prods.; thickener and visc. control agent, pigment dispersant; clouding agent, diluent for vita-

mins, solvent for oil-disp. materials, useful in flavor systems and dietary formulations
Regulatory: FDA 21CFR §172.860, 172.856; kosher grade avail.
Properties: Colorless clear liq., bland odor and taste; sol. in min. and veg. oils, ethanol; water-insol.; acid no. 0.1 max.; iodine no. 1 max.; sapon. no. 315-335

Lipovol SES [Lipo]
Chem. Descrip.: Sesame oil
CAS 8008-74-0; EINECS 232-370-6
Uses: Emollient, solvent, and vehicle used in cosmetics, toiletries, pharmaceuticals, and foods; offers lt., nontacky feel and enhances gloss and spread of pigmented sticks and pot prods.
Properties: Yel. clear liq.; bland char. odor; sol. in isopropyl esters and min. oil; insol. in water; sp.gr. 0.916-0.921; acid no. 0.2 max.; iodine no. 103-116; sapon. no. 188-195

Lonzest® 143-S [Lonza]
Chem. Descrip.: Myristyl propionate
CAS 6221-95-0; EINECS 226-300-9
Uses: Emollient, penetrant, and spreading agent in hair and skin cosmetics; perfume solvent in bath oils, antiperspirants, preshave lotions, sunburn lotions; solvent/emollient in lipsticks, waterless hand cleaners, makeup, facial cleaners; humectant
Properties: APHA 10 clear liq., odorless; sol. in acetone, CCl_4, castor, lanolin, peanut, silicone, and cottonseed oils, ethanol, ethyl acetate, heptane, IPA, toluene; sp.gr. 0.852; visc. 5 cps; f.p. -5 C; acid no. 0.5 max.; sapon. no. 195-210; ref. index 1.433; pH 6-8 (10% aq. disp.); 100% act.; nonionic
Toxicology: LD50 (oral) > 39 g/kg; not a primary skin or eye irritant; noncomedogenic

Lowenol 915 [Lowenstein]
Uses: Phenylether-based solvent for hair color systems

Lutrol® E 300 [BASF AG]
Chem. Descrip.: PEG-6
CAS 25322-68-3; EINECS 220-045-1
Uses: Emulsifier, emollient, lubricant, and solvent for liq. preps.
Properties: Liq.; 100% conc.

Lutrol® E 400 [BASF AG]
Chem. Descrip.: PEG-8
CAS 25322-68-3; EINECS 225-856-4
Uses: Emulsifier, emollient, lubricant, and solvent for liq. preps.
Properties: Liq.; 100% conc.

Lutrol® E 1500 [BASF AG]
Chem. Descrip.: PEG-32
CAS 25322-68-3; EINECS 203-989-9
Uses: Solvent, emulsifier, binder, solubilizer, resorption promoter for substances insol. or sparingly sol. in water
Properties: Microbeads; 100% conc.; nonionic

Lutrol® E 4000 [BASF AG]
Chem. Descrip.: PEG-75
CAS 25322-68-3; EINECS 203-989-9
Uses: Solvent, emulsifier, binder, solubilizer, resorption promoter for substances insol. or sparingly sol. in water
Properties: Microbeads; 100% conc.; nonionic

Lutrol® E 6000 [BASF AG]
Chem. Descrip.: PEG-150
CAS 25322-68-3; EINECS 203-989-9
Uses: Solvent, emulsifier, binder, solubilizer, resorption promoter for substances insol. or sparingly sol. in water
Properties: Microbeads; 100% conc.; nonionic

M

M-1421-7B, 12B Nylon Tipping Solutions [Eastern Color & Chem.]
Uses: Solvent system tipping prods. provide exc. nylon tips to nylon cord with proper heat/mold procedures

Macol® 57 [PPG/Spec. Chem.]
Chem. Descrip.: PPG-10 butanediol
Uses: Nongreasy emollient and nonwhitening aid for colognes, aftershaves, skin fresheners, antiperspirants, nail enamel removers; high ref. index, good spreading props.; glossing agent for hair; good solvent for many sunscreen actives; crystal growth inhibitor
Properties: Water-wh. clear liq., mild char. odor; sol. in water, SD alcohol 40B, butylene glycol, IPM, benzyl laurate, almond oil; disp. in cyclomethicone, min. and jojoba oils; insol. in glycerin; sp.gr. 1.005; visc. < 100 cps; m.p. < -25 C; b.p. > 150 C; cloud pt. 42 C; flash pt. (CC) > 95 C; ref. index 1.450; pH 6.0
Toxicology: LD50 (oral) < 5 g/kg; nontoxic inhalation hazard; safe for external use

Macol® 85 [PPG/Spec. Chem.]
Chem. Descrip.: m,p-Cresol propoxylate
CAS 9064-13-5
Uses: Wetting agent, dispersant, antistat, defoamer, gellant, solubilizer, lubricant base for cosmetic, medical, paper, pharmaceutical and textile industries; coalescing solvent for plasticizing acrylic emulsion polymers, latex caulks, epoxies, alkyds, polyesters, phenolics; increases flexibility and toughness, improves film formation
Regulatory: SARA nonreportable
Properties: Clear liq., bland odor; sol. in aromatic solvs.; disp. in water; m.w. 4600; sp.gr. 1.00; visc. 100 cps; vapor pressure < 1 mm Hg; b.p. > 300 F; HLB 16.0; acid no. 1 max.; hyd. no. 120-138; pour pt. < 10 C; cloud pt. 85 C (1% aq.); flash pt. (PMCC) > 200 F; ref. index 1.475; surf. tens. 42.5 dynes/cm (0.1%); Ross-Miles foam > 600 mm (0.1%); 100% conc.; nonionic
Toxicology: Contact with skin or eyes may cause temporary irritation
Precaution: Incompat. with strong oxidizing materials; thermal decomp. may produce carbon mono/dioxides
NFPA: Health: 1; Flammability: 1; Reactivity: 0
Storage: Store in well-ventilated area below 120 F

Marlican® [Hüls AG]
Chem. Descrip.: Straight-chain dodecylbenzene
CAS 67774-74-7
Uses: Detergent intermediate, solubilizer; sec. plasticizer; reduces visc. of PVC pastes; solvent for carbonless copy papers
Properties: Liq.; 100% act.; anionic
Environmental: Biodeg.

Marlophen® P 1 [Hüls Am.; Hüls AG]
Chem. Descrip.: Phenol ethoxylate (1 EO)
Uses: Solubilizer, solvent
Properties: Liq.; cloud pt. 76 C (5% in 25% BDG); 100% act.; nonionic

Marlophen® P 7 [Hüls Am.; Hüls AG]
Chem. Descrip.: Phenol ethoxylate (7 EO)
CAS 9004-78-8

Uses: Solubilizer, solvent
Properties: Liq.; cloud pt. 65 C (2% in 10% NaCl); 100% act.; nonionic
Unverified

Marvanol® Solvent Scour 34 [Marlowe-Van Loan]
Chem. Descrip.: Solvs. and detergents
Uses: Solvent scour for textiles
Properties: Amber liq.; 100% act.; anionic
Unverified

Mazon® EE-1 [PPG/Spec. Chem.]
Chem. Descrip.: Benzyl laurate
CAS 140-25-0; EINECS 204-405-8
Uses: Nonoily emollient, lubricant for skin care, bath, and hair care formulations; readily emulsified; spreads easily; solvent and degreaser in sunscreens
Properties: APHA < 150 clear liq., mild odor; sol. in min. oil, jojoba oil, almond oil, IPM, cyclomethicone, SD alcohol 40-B; misc. with sunscreens; insol. in water; sp.gr. 0.920; visc. < 50 cps; m.p. -5 C; b.p. 95 C; flash pt. (CC) 95 C
Toxicology: Nontoxic, nonirritating, nonsensitizing

Methyl Carbitol® [Union Carbide]
Chem. Descrip.: Methoxydiglycol
CAS 111-77-3; EINECS 203-906-6
Uses: Solvent, coalescing aid for water-disp. and water-sol. coatings
Properties: Sol. in water (20 C); m.w. 120.16; sp.gr. 1.021 (20/20 C); dens. 8.51 lb/gal (20 C); visc. 3.9 cP (20 C); vapor pressure 0.1 mm Hg (20 C); b.p. 194.0 C; flash pt. (CC) 188 F

Methyl Cellosolve® [Union Carbide]
Chem. Descrip.: Methoxyethanol
CAS 109-86-4; EINECS 203-713-7
Uses: Solvent for coatings (NC, cellulose acetate, ethyl cellulose, butyrate, acrylic)
Properties: Sol. in water (20 C); m.w. 76.10; sp.gr. 0.966 (20/20 C); dens. 8.04 lb/gal (20 C); b.p. 124.5 C; flash pt. (CC) 103 F

Methyl Cellosolve® Acetate [Union Carbide]
Chem. Descrip.: Methoxyethanol acetate
CAS 110-49-6; EINECS 203-772-9
Uses: Solvent for coatings (NC, cellulose acetate, ethyl cellulose, butyrate, acrylic)
Properties: Sol. in water (20 C); m.w. 118.14; sp.gr. 1.006 (20/20 C); dens. 8.37 lb/gal (20 C); b.p. 145.0 C; flash pt. (CC) 121 F

Methyl Di-Icinol [ICI Australia]
Chem. Descrip.: 2-(2-Methoxyethoxy) ethanol
CAS 111-77-3; EINECS 203-906-6
Uses: Solvent for use in protective coatings, inks, cleaning prods., agric. chems.; aids wetting penetration, and soil removal; coupling solvent
Unverified

Methyl Dipropasol [Union Carbide]
Chem. Descrip.: PPG-2 methyl ether
Uses: Solvent for acrylic coatings
Properties: Sol. in water (20 C); m.w. 148.20; sp.gr. 0.951 (20/20 C); dens. 7.91 lb/gal (20 C); b.p. 188.3 C; flash pt. (CC) 175 F

Methyl Icinol [ICI Australia]
Chem. Descrip.: 2-Methoxyethanol
CAS 109-86-4; EINECS 203-713-7
Uses: Solvent for use in protective coatings, inks, cleaning prods., agric. chems.; aids wetting, penetration, and soil removal; coupling solvent
Unverified

Methyl Propasol® [Union Carbide]
Chem. Descrip.: Methoxypropanol
CAS 107-98-2; EINECS 203-539-1
Uses: Solvent for coatings (NC, butyrate, ethyl cellulose, acrylic, water-disp. and -sol.)
Properties: Sol. in water; m.w. 90.12; sp.gr. 0.923 (20/20 C); dens. 7.69 lb/gal (20 C); b.p. 121 C; flash pt. (CC) 91 F

Methyl Propasol® Acetate [Union Carbide]
Chem. Descrip.: Propylene glycol methyl ether acetate
CAS 108-65-6; EINECS 203-603-9
Uses: Solvent for coatings (NC, cellulose acetate, ethyl cellulose, butyrate, acrylic, urethane), waterborne coatings
Properties: Sol. 18.5% in water (20 C); m.w. 132.16; sp.gr. 0.969 (20/20 C); dens. 8.09 lb/gal (20 C); b.p. 145.7

C; flash pt. (CC) 114 F

Miglyol® 810 [Hüls Am.; Hüls AG]
Chem. Descrip.: Caprylic/capric triglyceride
CAS 65381-09-1; EINECS 265-724-3
Uses: Emollient, dispersant, lubricant, suspending agent, solubilizer; act. ingred. for cosmetics and pharmaceuticals; carrier/vehicle and solvent for inj. prods., topical ointments, creams, lotions, suppositories; dietetic prods.; penetrating and spreading props. on skin
Properties: Colorless liq., neutral odor; sol. in diethyl ether, petrol. ether, chloroform, IPA, toluene, alcohol, min. oil, acetone; sp.gr. 0.94-0.95; visc. 25-35 mPa·s; acid no. 0.1 max.; iodine no. 0.5 max.; sapon. no. 335-355; pH neutral; cloud pt. 0 C.; nonionic

Miglyol® 812 [Hüls Am.; Hüls AG]
Chem. Descrip.: Caprylic/capric triglyceride
CAS 65381-09-1; EINECS 265-724-3
Uses: Dispersant, lubricant, anticaking agent, carrier, solvent, solubilizer, suspending agent for cosmetics, pharmaceuticals (oral, external, suppositories), dietetic prods.; spreading and penetrating props. on skin; stable to oxidation
Properties: Colorless liq.; sol. in alcohol, min. oil, acetone; visc. 25-35 mPa·s; acid no. 0.1 max.; iodine no. 0.5 max.; sapon. no. 325-345; cloud pt. 10 C; 100% conc.; nonionic

Miglyol® 840 [Hüls Am.; Hüls AG]
Chem. Descrip.: Propylene glycol dicaprylate/dicaprate
Uses: Emollient, dispersant, lubricant, suspending agent, solubilizer; act. ingred. for cosmetics and pharmaceuticals; carrier/vehicle and solvent for inj. prods., topical ointments, creams, lotions, suppositories; dietetic prods.; flavor fixative; oxidation-stable
Properties: Colorless liq.; sol. in alcohol, min. oil, acetone; visc. 8-14 mPa·s; acid no. 0.1 max.; iodine no. 0.5 max.; sapon. no. 320-340; cloud pt. -30 C

Miglyol® 8810 [Hüls Am.; Hüls AG]
Chem. Descrip.: Butylene glycol dicaprylate/caprate
Uses: Emollient, dispersant, lubricant, suspending agent; hypoallergenic; solubilizer; act. ingred. for cosmetic and pharmaceutical preps.; carrier/vehicle and solvent for inj. prods., topical ointments, creams, lotions, esp. baby skin care; penetrating and spreading props. on skin
Properties: Colorless liq., neutral odor; sol. in alcohol, diethyl ether, n-hexane, fatty oils; insol. in water; visc. 8-13 mPa·s; acid no. 0.1 max.; iodine no. 1.0 max.; sapon. no. 295-315; hyd. no. 5 max.

Mod Acid [Ruetgers-Nease]
Chem. Descrip.: Modified toluene sulfonic acid prod.
Uses: Hydrotrope, solvent, intermediate; catalyst; crosslinking catalyst for phenolic adhesives, melamine-formaldehyde and acrylic polymers; curing catalyst in polymerization of furan for foundry molds or cores; vulcanization of rubber; prep. of acetals, esters
Properties: Cryst. range 0-15 C; 92% min. assay
Toxicology: Corrosive; avoid contact with skin and eyes

Monoglyme [Ferro/Grant]
Chem. Descrip.: Ethylene glycol dimethyl ether
CAS 110-71-4; EINECS 203-794-9
Uses: Solvent which tends to solvate cations; used in electrochemistry, polymer and boron chemistry, physical processes such as gas absorp., extraction, stabilization, industrial prods. such as fuels, lubricants, textiles, pesticides; solvent vehicle for pharmaceuticals; reaction solvent for pharmaceutical mfg.
Properties: Colorless; ethereal, nonresidual odor; water-sol.; misc. with ethanol, acetone, benzene, diethyl ether, octane; m.w. 90.12; sp.gr. 0.8683 (20/20 C); dens. 7.24 lb/gal; visc. 1.1 cP (20 C); vapor pressure 54 mm Hg (20 C); f.p. -69 C; b.p.

85.2 C; flash pt. (CC) -6 C; ref. index 1.3792; pH neutral; surf. tens. 22.9 dynes/cm (20 C); sp. heat 0.438 cal/g/°C; 99.6% min. purity

Toxicology: LD50 5370 mg/kg; low to mod. acute toxicity; avoid exposure to vapors, skin contact; TLV/8h TWA: 5 ppm; STEL 25 ppm; LD50 5100 mg/kg); chronic exposure may cause reproductive effects

Environmental: Slowly biodeg.

Monolan® PPG440, PPG1100, PPG2200 [Akcros UK]

Chem. Descrip.: PPG

Uses: Lubricant, antistat, plasticizer, cosolvs. in dyestuff, ink, resin, rubber industries, cosmetic preparations, intermediates in surfactants and plastic prod.

Properties: Colorless liq.; faint odor; sol. in water (PPG440), insol. in water (PPG1100, 2200); m.w. 400, 1000, 2000 resp.; sp.gr. 1.010, 1.005, and 1.004; visc. 80, 180, and 450 cs; flash pt. > 450 F (COC); pour pt. < 0 C; 100% act.; nonionic

Unverified

MPDiol® Glycol [ARCO]

Chem. Descrip.: 2-Methyl-1,3-propanediol

CAS 2163-42-0

Uses: Intermediate used in prod. of urethanes, unsat. polyesters, gel coats, sat. polyester and alkyd coatings, polymeric plasticizers; produces unsat. polyester resins with light color, exc. styrene misc., exc. balance of tens. str., elong., flexibility; nonvolatile solvent, compatibilizer for personal care, adjuvants, fabric softener carriers, nonionic surfactants, pigment dispersions

Properties: Colorless clear low-visc. liq.; sol. in water; m.w. 90.1; sp.gr. 1.015; dens. < 8.47 lb/gal; visc. 174 cps; vapor pressure 4.3 mm Hg (100 C); m.p. -91 C; b.p. 213 C; hyd. no. 1200; flash pt. (OC) > 230 F; 98.5% assay

Precaution: Store away from strong min. acids, dehydrating agents, strong oxidizing agents

Storage: Avoid excessive exposure to moisture; keep away from heat, sparks, and flame

M-Pyrol® [ISP]

Chem. Descrip.: N-Methyl-2-pyrrolidone

CAS 872-50-4; EINECS 212-828-1

Uses: Solvent, stabilizer for vinyls, urethanes, acrylics, difficult sol. resins and waxes; spinning solvent for syn. fibers; used in commercial petrochemical processing, for agric. chemicals, in removal of strippable coatings, for lube oil processing; used as nonaq. syn. medium for polyol reactions; residue cleaning solvent for PVC and urethane prod. equipment; replaces toxic solvs.; reduces VOC emissions

Properties: Liq., mild amine-like odor; misc. with water; m.w. 99.1; sp.gr. 1.027; dens. 1.03 g/cc (20 C); visc. 1.7 cp (20 C); vapor pressure 1.0 torr (40 C); f.p. -24.4 C; b.p. 202 C; flash pt. (OC) 93 C; ref. index 1.4700; surf. tens. 41 dynes/cm; KB > 300; 99.8% min. purity

Toxicology: Low toxicity

Environmental: Rapidly biodeg.; low toxicity to aquatic organisms

Precaution: Combustible in bulk containers; when ignited, will sustain a fire; flamm. limits in air 1.3-9.5 vol. %

Storage: Stable but hygroscopic

MS-195/CO2 Safezone® Cleaning Solvent & Flux Remover [Miller-Stephenson]

Chem. Descrip.: 92% 1,1,1-Trichloroethane, 5% isopropyl alcohol, 3% carbon dioxide (as propellant)

UN No.: 1950

Uses: High purity solvent cleaner and flux remover; multipurpose cleaner for flushing away contaminants; removes most types of soil and flux; dissolves oil, grease, wax, tar, fat, gum, rosin; inert to most plastics; 100%-CFC free; rec. for printed circuit boards, hybrid microcircuits, silicon wafers, pressure transducers, molds, machinery/tooling, etc.

Regulatory: Complies with Mil-P-28809A
Properties: Clear liq.; ethereal odor; sl. sol. in water; dens. 1.05 g/cc; vapor pressure 79 mm Hg; b.p. 57 C; flash pt. none until 50% evaporation, then 120 F; pH neutral; nonflamm.
Toxicology: Inh. of conc. vapor can cause dizziness, headache, narcosis, anesthesia, suffocation, CNS damage, or death; prolonged/repeated skin contact can cause irritation and dermatitis
Precaution: Incompat. with finely divided reactive metals and strong caustics; trichloroethane may form explosive mixts. with liq. oxygen; can be dec. by high temps. forming halogen acids and phosgene in smaller amts.
Storage: 1 yr shelf life; rotate stock; store away from sources of heat, direct sunlight, or temps. above 49 C

Myritol® PC [Henkel/Cospha; Henkel KGaA/Cospha]
Chem. Descrip.: Propylene glycol dicaprylate/dicaprate
CAS 68583-51-7
Uses: Emollient oil and cosolvent for cosmetics, pharmaceuticals, bath oils, aerosols, and antiperspirants; exc. solvent for active ingreds.; stable vehicle for pigmented cosmetics
Regulatory: DAB, JCIC compliance
Properties: Colorless clear low visc. oil, faint char. odor; m.w. 340; sp.gr. 0.91-0.93 (20 C); visc. 9-12 mPa·s (20 C); acid no. 0.1 max.; iodine no. 0.5 max.; sapon. no. 325-335; hyd. no. 5 max.; cloud pt. < -20 C; ref. index 1.440-1.442 (20 C)
Storage: 1 yr min. storage stability in sealed original containers at temps. below 30 C in a dry environment

Myvacet® 9-45K [Eastman]
Chem. Descrip.: Acetylated hydrog. soybean oil glyceride
Uses: Emulsifier, lubricant, solvent for icings and shortenings
Properties: Liq.; sp.gr. 0.94 (80 C); m.p. 4-12 C; HLB 3.8-4.0; acid no. 3 max.; iodine no. 43-53; sapon. no. 370-382; hyd. no. 0-15; 96% min. acetylation; nonionic
Discontinued prod.

N

Nadone® [AlliedSignal]
Chem. Descrip.: Cyclohexanone
CAS 108-94-1; EINECS 203-631-1
Uses: General solvent; coating resins, printing inks, adhesives, fats, oils, waxes, agric.
Properties: Sol. in acetone, ethanol, benzene, ethylene glycol
Discontinued prod.

Naetex-L [Lanaetex Prods.]
Chem. Descrip.: PEG-2 lactamide
Uses: Coupling agent, lubricant, emollient, and humectant for personal care prods.; enhances wet combing, imparts softness and antistatic props. to hair; solvent in cosmetic creams and lotions; hygroscopic nature prevents dry skin
Discontinued prod.

140 Naphtha [Exxon]
Chem. Descrip.: Hydrocarbon solv
CAS 64742-47-8
Uses: Solvent
Properties: Sp.gr. 0.80 (15.6/15.6 C); dens. 6.67 lb/gal (60 F); b.p. 187-209 C; flash pt. (TCC) 65 C; surf. tens. 25.0 dynes/cm; KB value 36; 46% paraffin, 34% cycloparaffin, 20% aromatics
Toxicology: OEL 200 ppm

1520 Naphtha [Exxon]
Chem. Descrip.: Hydrocarbon solv.
CAS 64742-49-0
Uses: Solvent
Properties: Sp.gr. 0.67 (15.6/15.6 C); dens. 5.61 lb/gal (60 F); b.p. 64-94 C; flash pt. (TCC) < -18 C; surf. tens. 18.2 dynes/cm; KB value 28; 98% paraffin, 2% cycloparaffin, < 0.5% aromatics
Toxicology: OEL 100 ppm

2024 Naphtha [Exxon]
Chem. Descrip.: Hydrocarbon solv.
CAS 64742-48-9
Uses: Solvent
Properties: Sp.gr. 0.74 (15.6/15.6 C); dens. 6.15 lb/gal (60 F); b.p. 97-117 C; flash pt. (TCC) 3 C; surf. tens. 21.3 dynes/cm; KB value 39; 52% paraffin, 42% cycloparaffin, 6% aromatics
Toxicology: OEL 400 ppm

2429 Naphtha [Exxon]
Chem. Descrip.: Hydrocarbon solv.
CAS 64742-89-8
Uses: Solvent
Properties: Sp.gr. 0.76 (15.6/15.6 C); dens. 6.29 lb/gal (60 F); b.p. 115-143 C; flash pt. (TCC) 10 C; surf. tens. 22.2 dynes/cm; KB value 39; 58% paraffin, 30% cycloparaffin, 12% aromatics
Toxicology: OEL 400 ppm

3135 Naphtha [Exxon]
Chem. Descrip.: Stoddard solvent
CAS 8052-41-3
Uses: Solvent
Properties: Sp.gr. 0.79 (15.6/15.6 C); dens. 6.56 lb/gal (60 F); b.p. 159-176 C; flash pt. (TCC) 42 C; surf. tens. 23.7 dynes/cm; KB value 39; 45% paraffin, 35% cycloparaffin, 20% aromatics
Toxicology: OEL 200 ppm

Naturechem® GTR [CasChem]
Chem. Descrip.: Glyceryl triacetyl ricinoleate
CAS 101-34-8; EINECS 202-935-1
Uses: Mild emollient, pigment wetter, cosolvent; softener for waxes and resins
Properties: Gardner 2+ color; m.p. -40 C; acid no. < 1; iodine no. 76; hyd. no. 5

Naturechem® MAR [CasChem]
Chem. Descrip.: Methyl acetyl ricinoleate
CAS 140-03-4; EINECS 205-392-9
Uses: Light emollient; reduces greasiness of emollients such as min. oil; cosolvent props.; solubilizer for benzophenone-3; superior freeze/thaw props.
Properties: Gardner 1 color; m.p. -25 C; acid no. < 1; iodine no. 77; hyd. no. < 5

Naxol® [AlliedSignal]
Chem. Descrip.: Cyclohexanol
CAS 108-93-0; EINECS 203-630-6
Uses: Solvent for fats, oils, resins; coupling agent
Properties: Misc. with ether, ethanol, aromatic hydrocarbons

Naxonac™ 510 [Ruetgers-Nease]
Chem. Descrip.: Nonyl phenol ether phosphate
Uses: Detergent, wetting agent, emulsifier, lubricant, hydrotrope for heavy-duty and household detergents, waterless hand cleaners, solvent degreasers, emulsion polymerization, paint and wax strippers, electrolytic cleaners
Properties: Clear liq.; sol. and stable in aq. sol'ns. with high alkaline concs.; pH 2; 99% conc.; anionic

NE™ 101 [ANGUS]
Chem. Descrip.: Nitroethane
CAS 79-24-3; EINECS 201-188-9
Uses: Intermediate; specialty solvent for coating, adhesive, and ink formulations; low resistivity and med. evaporation rate; improves solvent release without blocking; not HAPs listed
Properties: Colorless liq., char. odor; sol. 4.5 vol. % in water (20 C); m.w. 75.068; sp.gr. 1.052 (20/20 C); visc. 0.677 cp (20 C); vapor pressure 15.6 mm Hg (20 C); f.p. -89.52 C; b.p. 114 C; flash pt. (TCC) 87 F; ref. index 1.39193 (20 C); pH 6.0 (0.01m aq.); surf. tens. 32.66 dynes/cm (20 C); dielec. const. 28.06 (30 C); 98% min. act.
Toxicology: LD50 (oral, rat) 1624 mg/kg; TLV/PEL 100 ppm; sl. toxic by ing.; harmful if swallowed; vapors may cause headache, nausea, vomiting, narcosis; chronic overexposure may cause liver/kidney damage; defatting skin irritant; nonirritating to eyes
Precaution: Flamm.; salts of nitroparaffins with inorg. bases or amines are thermally unstable; violent decomp. can occur; lower explosive limit 4%; incompat. with lead, copper and alloys, strong alkalies; produces CO_x, NO_x
Storage: Keep container closed; keep away from heat, sparks, flame

Neobee® 18 [Stepan Food Ingreds.; Stepan Europe]
Chem. Descrip.: Hybrid safflower oil, refined, deodorized
CAS 8001-23-8; EINECS 232-276-5
Uses: Highly stable oil for margarine, mayonnaise, salad dressings; emollient base for cosmetics and pharmaceuticals; solubilizer, solvent, lubricant, nutritional fluid
Regulatory: FDA 21CFR §170.30, GRAS; avail. in kosher grade
Properties: Sl. yel. liq., sl. fatty odor, bland taste; sol. in alcohol, min. oil, acetone; sp.gr. 0.915; visc. 70 cps; iodine no. 90; sapon. no. 190; surf. tens. 31.6 dynes/cm
Toxicology: Nonirritating
Storage: Avoid prolonged storage above 90 F

Neobee® 20 [Stepan Food Ingreds.]
Chem. Descrip.: Propylene glycol dicaprylate/dicaprate
Uses: Emollient oil for cosmetics and pharmaceuticals (bath oils, creams, lotions, lipsticks, makeup bases, shaving preps.), carrier/extender for flavors, fragrances, and colors in foods; vehicle for vitamins, antibiotics, and medicinals; solubilizer, cosolvent
Properties: APHA 50 max. liq.; sol. in alcohol, min. oil, acetone; sp.gr. 0.920; visc. 9 cps; acid no. 0.10 max.; sapon. no. 315-335; surf. tens. 31.0 dynes/cm
Discontinued prod.

Neobee® 1054 [Stepan Food Ingreds.]
Chem. Descrip.: Propylene glycol dicaprylate/dicaprate

Uses: Cosolvent for cosmetics and pharmaceuticals, carrier for flavors and colors; solubilizer for antibiotics, medicinals; diluent; emollient for creams, lotions, ointments
Regulatory: FDA 21CFR §170.30, GRAS; avail. in kosher grade
Properties: Sl. yel. liq., bland odor; sol. in alcohol, min. oil, acetone; sp.gr. 0.910-0.923; iodine no. 0.1; sapon. no. 325; hyd. no. 1; 0.05% moisture
Storage: Avoid prolonged storage above 90 F

Neobee® 1062 [Stepan Food Ingreds.]
Chem. Descrip.: Coconut oil-derived triglycerides
Uses: Cosolvent, carrier for fat-sol. vitamins, medicinals, and ointments in food, cosmetic and pharmaceutical industries
Regulatory: FDA 21CFR §170.30, GRAS; avail. in kosher grade
Properties: Liq.; sol. in min. oil, acetone; sp.gr. 0.925-0.945; sapon. no. 295-315
Discontinued prod.

Neobee® M-5 [Stepan Food Ingreds.; Stepan Europe]
Chem. Descrip.: Caprylic/capric triglyceride
CAS 65381-09-1; EINECS 265-724-3
Uses: Diluent vehicle/carrier for vitamins, antibiotics, nutritional fluids, and medicinals; solubilizer, cosolvent for fragrance/flavoring, medicinals, food colors; clouding agent for beverages; source of nutritional MCT; emollient oil for cosmetics and pharmaceuticals (bath oils, creams, lotions, lipsticks, makeup base, shaving preps.)
Regulatory: FDA 21CFR §170.30, GRAS; avail. in kosher grade
Properties: Sl. yel. liq., bland odor, tasteless; sol. in alcohol, min. oil, acetone; sp.gr. 0.930-0.960; visc. 23 cps; acid no. 0.10 max.; iodine no. 1; sapon. no. 340; hyd. no. 5; surf. tens. 32.3 dynes/cm; 0.1% moisture
Storage: Avoid prolonged storage above 90 F

Neobee® M-20 [Stepan Food Ingreds.; Stepan Europe]
Chem. Descrip.: Propylene glycol dicaprylate/dicaprate
CAS 68583-51-7
Uses: Low visc. emollient oil for creams, lotions, and ointments; solubilizer for antibiotics, medicinals; diluent for essential oils, injectables; solvent and lubricating agent; exc. deposition on skin; solubilizer, solvent, lubricant; diluent for essential oils in food applics.
Regulatory: FDA 21CFR§170.30, 172.856; avail. in kosher grade
Properties: Pale yel. liq., bland odor; sol. in alcohol containing up to 20% water; iodine no. 1 max.; sapon. no. 326; hyd. no. 0.6; 0.04% moisture
Storage: Avoid prolonged storage above 90 F

Neobee® O [Stepan Food Ingreds.]
Chem. Descrip.: Caprylic/capric triglyceride
CAS 65381-09-1; EINECS 265-724-3
Uses: Solubilizer, cosolvent, carrier for fat-sol. vitamins, flavors, medicinals, and ointments in food, cosmetic and pharmaceutical industries; diluent for food colors, essential oils, injectables; clouding agent for beverages; source of nutritional MCT; spreading agent, penetrant
Regulatory: FDA 21CFR §170.30, GRAS; avail. in kosher grade
Properties: Sl. yel. liq., very sl. fatty odor; sol. in min. oil, acetone; sp.gr. 0.938; visc. 30 cps; iodine no. 4.2; sapon. no. 298; hyd. no. 5; surf. tens. 32 dynes/cm; 0.03% moisture
Storage: Avoid prolonged storage above 90 F

Neosolv™ 2 [Shell]
Chem. Descrip.: C_{12} unsat. aliphatic hydrocarbon
Uses: Environmentally friendly solv. meeting VOC restrictions; used for adhesives, air fresheners, automotive brake, carburetor, and choke cleaners, bathroom and tile cleaners, dusting aids, engine degreasers, fabric protection sprays, floor polishes/waxes; furniture pol-

ishes/waxes, hair sprays/mousses/styling gels, insecticides, insect repellents, laundry aids, oven cleanes, shaving aids, waterless hand cleaners

Properties: Pt-Co 5 max. color; low odor; m.w. 168; sp.gr. 0.764 (60/60 F); dens. 6.36 lb/gal (60 F); visc. 1.42 cP (60 F); vapor pressure 0.0984 mm Hg (20 F); f.p. -33 F; b.p. 415-419 F; flash pt. (Seta) 168 F; KB value 27; VOC 0 lb/gal

Environmental: Readily biodeg.

Precaution: Reacts readily with air or oxygen to form trace peroxides and other oxygenates

Storage: Store under nitrogen blanketing

Neosolv™ 4 [Shell]

Chem. Descrip.: Unsat. aliphatic hydrocarbon

Uses: Environmentally friendly solv. meeting VOC restrictions; used for adhesives, air fresheners, automotive brake, carburetor, and choke cleaners, bathroom and tile cleaners, dusting aids, engine degreasers, fabric protection sprays, floor polishes/waxes; furniture polishes/waxes, hair sprays/mousses/styling gels, insecticides, insect repellents, laundry aids, oven cleanes, shaving aids, waterless hand cleaners

Properties: Pt-Co 10 max. color; low odor; sp.gr. 0.770-0.780 (60/60 F); dens. 6.46 lb/gal (60 F); visc. 1.42 cP (60 F); vapor pressure 0.0128 mm Hg (20 F); f.p. 6 F; b.p. 467-475 F; flash pt. (PMCC) 210 F min.; KB value 24-26; VOC 0 lb/gal

Environmental: Readily biodeg.

Precaution: Reacts readily with air or oxygen to form trace peroxides and other oxygenates

Storage: Store under nitrogen blanketing

Neosolv™ 6 [Shell]

Chem. Descrip.: Unsat. aliphatic hydrocarbon

Uses: Environmentally friendly solv. meeting VOC restrictions; used for adhesives, air fresheners, automotive brake, carburetor, and choke cleaners, bathroom and tile cleaners, dusting aids, engine degreasers, fabric protection sprays, floor polishes/waxes; furniture polishes/waxes, hair sprays/mousses/styling gels, insecticides, insect repellents, laundry aids, oven cleanes, shaving aids, waterless hand cleaners

Properties: Pt-Co 10 max. color; low odor; sp.gr. 0.780-0.790 (60/60 F); dens. 6.54 lb/gal (60 F); visc. 3.10-3.30 cP (60 F); vapor pressure 0.00129 mm Hg (20 F); f.p. 41 F; b.p. 514-536 F; flash pt. (PMCC) 240 F min.; KB value 22-24; VOC 0 lb/gal

Environmental: Readily biodeg.

Precaution: Reacts readily with air or oxygen to form trace peroxides and other oxygenates

Storage: To keep prod. fluid in cold weather, store at least 5 F above f.p., but < 100 F; store under nitrogen blanketing

Neosolv™ 8 [Shell]

Chem. Descrip.: Unsat. aliphatic hydrocarbon

Uses: Environmentally friendly solv. meeting VOC restrictions; used for adhesives, air fresheners, automotive brake, carburetor, and choke cleaners, bathroom and tile cleaners, dusting aids, engine degreasers, fabric protection sprays, floor polishes/waxes; furniture polishes/waxes, hair sprays/mousses/styling gels, insecticides, insect repellents, laundry aids, oven cleanes, shaving aids, waterless hand cleaners

Properties: Pt-Co 10 max. color; low odor; sp.gr. 0.787-0.797 (60/60 F); dens. 6.59 lb/gal (60 F); visc. 4.51-4.61 cP (60 F); vapor pressure < 0.001 mm Hg (20 F); f.p. 37 F; b.p. 517-602 F; flash pt. (PMCC) 250 F min.; KB value 20-22; VOC 0 lb/gal

Environmental: Readily biodeg.

Precaution: Reacts readily with air or oxygen to form trace peroxides and other oxygenates

Storage: To keep prod. fluid in cold weather, store at least 5 F above f.p., but < 100 F; store under nitrogen blanketing

Neosolv™ 24 [Shell]

Chem. Descrip.: Unsat. aliphatic hydrocarbon

Uses: Environmentally friendly solv. meeting VOC restrictions; used for adhesives, air fresheners, automotive brake, carburetor, and choke cleaners, bathroom and tile cleaners, dusting aids, engine degreasers, fabric protection sprays, floor polishes/waxes; furniture polishes/waxes, hair sprays/mousses/styling gels, insecticides, insect repellents, laundry aids, oven cleanes, shaving aids, waterless hand cleaners

Properties: Pt-Co 10 max. color; low odor; sp.gr. 0.763-0.773 (60/60 F); dens. 6.40 lb/gal (60 F); visc. 1.70-1.90 cP (60 F); vapor pressure 0.0589 mm Hg (20 F); f.p. -9 F; b.p. 430-473 F; flash pt. (PMCC) 190 F min.; KB value 26-28; VOC 0 lb/gal

Environmental: Readily biodeg.

Precaution: Reacts readily with air or oxygen to form trace peroxides and other oxygenates

Storage: Store under nitrogen blanketing

Neosolv™ 150 [Shell]

Chem. Descrip.: Unsat. aliphatic hydrocarbon

Uses: Environmentally friendly solv. meeting VOC restrictions; used for adhesives, air fresheners, automotive brake, carburetor, and choke cleaners, bathroom and tile cleaners, dusting aids, engine degreasers, fabric protection sprays, floor polishes/waxes; furniture polishes/waxes, hair sprays/mousses/styling gels, insecticides, insect repellents, laundry aids, oven cleaners, shaving aids, waterless hand cleaners

Properties: Pt-Co 10 max. color; low odor; sp.gr. 0.763-0.773 (60/60 F); visc. 1.66-1.91 cP (60 F); vapor pressure 0.097-0.161 mm Hg (20 F); f.p. 9 F max.; b.p. 389-471 F; flash pt. (PMCC) 150 F min.; KB value 21-23; VOC 2.5 lb/gal max.

Environmental: Readily biodeg.

Storage: Store below 100 F; nitrogen blanketing is rec. to prevent oxidation to peroxides and carbonyls

Neosolvent NSG [Toho Chem. Ind.]

Chem. Descrip.: Polypropylene polyethylene glycol monomethyl ether

Uses: Solvent for brake fluid

Properties: Liq.

NEP® [ISP]

Chem. Descrip.: N-Ethyl pyrrolidone

CAS 2687-91-4

Uses: Reaction solvent for chem. reactions down to -70 C; antibiotic prod.; elec. capacitors; mfg. of polyamide fibers and film; jet printing ink; photo resist stripper; polymerization (adipamide and hexamethylene diamine); synthesis of ethylene glycol; working fluid in advanced heat transformers

Properties: Liq.; water-sol.; m.w. 113; sp.gr. 0.993; visc. 3.5 cps; f.p. -70 C; b.p. 200 C; flash pt. (CC) 199 F; ref. index 1.4664; surf. tens. 36 dynes/cm; 98% min. purity

Toxicology: LD50 (oral, rat) > 14.43 g/kg; low acute toxicity; minimally irritating to eyes; nonirritating to skin

Environmental: Biodeg.

Nikkol BWA-5 [Nikko Chems. Co. Ltd]

Chem. Descrip.: PEG-5 lanolin alcohol

CAS 61791-20-6

Uses: Cosolvent, emulsifier, solubilizer, and softener for shampoos; provides hydrophilic props. to oil phase

Regulatory: JSCI listed

Properties: Ylsh. brn. paste; HLB 12.5; 100% conc.; nonionic

Nikkol BWA-10 [Nikko Chems. Co. Ltd]

Chem. Descrip.: PEG-10 lanolin alcohol

CAS 61791-20-6

Uses: Cosolvent; emollient for cream, milk lotion; provides hydrophilic props. to oil phase

Regulatory: JSCI listed

Properties: Ylsh. brn. solid; HLB 15.5; nonionic

Nikkol SO-30 [Nikko Chems. Co. Ltd]

Chem. Descrip.: Sorbitan trioleate

CAS 26266-58-0; EINECS 247-569-3
Uses: W/o emulsifier for cosmetics; exc. solvent; oil phase regulator; antirust
Regulatory: JSCI listed
Properties: Yel. liq.; HLB 1.7; 100% conc.; nonionic

Nikkol SO-30R [Nikko Chems. Co. Ltd]
Chem. Descrip.: Sorbitan trioleate
CAS 26266-58-0; EINECS 247-569-3
Uses: Cosmetics emulsifier; antirust; exc. solvent; oil phase regulator; refined grade (low color/odor)
Regulatory: JSCI listed
Properties: Pale yel. liq.; HLB 4.0

Nikkol Sefsol 218 [Nikko Chems. Co. Ltd]
Chem. Descrip.: Propylene glycol monocaprylate
CAS 31565-12-5
Uses: Cosmetic solvent, solubilizer of sl. sol. ingreds.; good compat. with oily materials
Properties: Colorless liq.

Nikkol Sefsol 228 [Nikko Chems. Co. Ltd]
Chem. Descrip.: Propylene glycol dicaprylate
CAS 7384-98-7; EINECS 230-962-9
Uses: Cosmetic solvent, solubilizer for sl. sol. ingreds.; good compat. with oily materials
Regulatory: JSCI listed
Properties: Colorless liq.

Niox KJ-10 [Henkel/Pulcra]
Chem. Descrip.: Ethoxylated oleyl alcohol
Uses: Emulsifier, dispersant, fragrance grade w/o emulsifier; lipophilic cosolvent; solubilizer
Properties: Liq.; oil-misc.
Unverified

Niox KS Series [Henkel/Pulcra]
Chem. Descrip.: PEGs
Uses: Humectant for food, cosmetic, pharmaceutical prods.; lubricant, solvent for dyestuffs and resins; solvent and ointment base for cosmetics and pharmaceuticals
Properties: Sol. in water; partly sol. in some org. solvs.
Discontinued prod.

NiPar S-10™ [ANGUS]
Chem. Descrip.: 1-Nitropropane
CAS 108-03-2; EINECS 203-544-9
Uses: Intermediate, solvs. for coatings; pharmaceutical intermediate
Properties: Liq.; sol. 1.5% in water (20 C); m.w. 89.1; dens. 1.0 g/ml (20 C), 8.35 lb/gal; visc. 0.844 cp (20 C); vapor pressure 10.23 mm Hg; f.p. –104 C; b.p. 131 C; flash pt. (TCC) 96 F; ref. index 1.40160 (20 C); pH 6.0 (0.01M aq.); surf. tens. 30.64 dynes/cm (20 C); dielec. const. 23.24 (30 C); 98.5% min. purity
Toxicology: Harmful if inhaled; overexposure can cause liver damage and death; OSHA PEL 25 ppm; avoid breathing vapors
Precaution: Flamm.; keep away from heat, sparks, flame
Storage: Keep container closed

Nirez® 9007 [Arizona]
Chem. Descrip.: Ester of selected tall oil fatty acids
Uses: Specialty ink solvent for use in high solids, low VOC ink systems and in soybean oil systems to reduce visc. and adjust tack

Nirez® 9011 [Arizona]
Chem. Descrip.: Ester of selected tall oil fatty acids
Uses: Specialty ink solvent for use in high solids, low VOC ink systems and in soybean oil systems to reduce visc. and adjust tack

NM-55® [ANGUS]
Chem. Descrip.: Blend of nitromethane
CAS 75-52-5; EINECS 200-876-6
Uses: Fuel for automotive racing and model engines; solvent
Discontinued prod.

NMP [ARCO]
Chem. Descrip.: N-Methyl pyrrolidone
CAS 872-50-4; EINECS 212-828-1; DOT No.: NA 1993
Uses: Environmentally friendly solvent, cosolvent; as coatings solv, in stripping and cleaning of paints/varnishes, industrial cleaning, mold cleaning, petrochem. processing, agric. solvent; solvent for adhesives, inks, dyes, soaps, metal complexes; polymer solvent for PVAc, PVDF, PS, vinyl copoly-

mers, nylon and aromatic polyamides and polyimides, polyesters, acrylics, PC, cellulose derivs., syn. elastomers, waxes; low flamm., noncorrosive, nonchlorine; low evaporation rate
Regulatory: SARA reportable
Properties: Colorless mobile liq., mild amine-like odor; completely sol. in water; m.w. 99.13; sp.gr. 1.028; dens. 8.57 lb/gal (20 C); vapor pressure 0.2 mm Hg (20 C); f.p. -25 C; b.p. 202-205 C; flash pt. 93 C; ref. index 1.4680-1.4695; surf. tens. 45.8 dynes/cm; VOC 99.7%; 99.9+% assay
Toxicology: LD50 (oral, rat) 3.9 g/kg; relatively nontoxic; low inhalation hazard; mod. eye irritant; not a skin absorption hazard; may cause delayed skin irritation and blistering; sl. heath hazard by ing. of large quantities
Environmental: Readily biodeg.; EC50 (rainbow trout, 96 h) ≥ 500 mg/l, (algae, 72 h) ≥ 500 mg/l
Precaution: DOT: combustible; avoid contact with strong acids or alkalies; incompat. with reducing agents, strong oxidizing agents
HMIS: Health: 2; Flammability: 1; Reactivity: 0
Storage: Store in dry area, out of direct sunlight; protect from weather; hygroscopic; air and moisture exclusion will prolong shelf life

NMP. [Dynaloy]
Chem. Descrip.: N-Methyl pyrrolidone-based
Uses: Solvent for removal of adhesives and inks from rollers, general polymer removal, paint stripping; suitable for cured epoxies; also for electronic depotting, deflashing, decapsulation, and deinking applics., surf. prep. and photoresist stripping; use hot (105-175 C); not selective; available filtered thru 0.2 μ filter

NMP-EL [ARCO]
Chem. Descrip.: N-Methyl-2-pyrrolidone
CAS 872-50-4; EINECS 212-828-1
Uses: A photoresist stripper, solvent in electronics applics. incl. semiaq. defluxing, degreasing and coatings; HAPS-free
Properties: Completely water misc.; sp.gr. 1.028; vapor pressure < 0.3 mm Hg (20 C); m.p. -25 C; b.p. 202 C; flash pt. 93 C
Toxicology: Causes irritation to skin and eyes
Storage: Air and moisture exclusion will prolong storage life; store in clean, mild steel or alloy drums

Noiox KS-10, -12, -13, -14, -16 [Henkel/Pulcra]
Chem. Descrip.: PEG
Uses: Lubricant, solvent for dyestuffs and resins; plasticizer for casein, gelatin compositions and printing inks; used in cosmetic and pharmaceutical items; intermediate prods.
Properties: Liq., solid, paste; water-sol.; 100% conc.; nonionic
Discontinued prod.

Nopalcol Series [Henkel/Functional Prods.]
Chem. Descrip.: PEG fatty esters
Uses: Dispersant, wetting agent, emulsifier, plasticizer, lubricant, binder, thickener for cosmetics, distillation defoamers, dry-cleaning agents, leather and paper industries; solvent, textiles, wall tile mastic
Properties: Liq. to solid; 100% conc.
Unverified

Norpar® 12 [Exxon]
Chem. Descrip.: Normal paraffin hydrocarbon solv.
CAS 64771-72-8
Uses: Solvent used in aluminum rolling oils, as diluent solvs. in carbonless copy paper, and in spark erosion machining; as clean fuels in lamp oil and barbecue lighter fluids; in pesticides; cleaning and degreasing formulations
Regulatory: FDA 21CFR §176.180, 176.210, 177.2600, 178.3570, 178.3620(b); 40CFR §180.1001(c)(e)
Properties: Little/no odor; m.w. 163; sp.gr. 0.749 (60/60 F); dens. 6.24 lb/gal (60 F); visc. 1.26 cps; b.p. 192-217 C; flash pt. (ATSM D 93) 73 C; surf. tens. 26.9 dynes/cm; KB

value 23; < 0.01% aromatics
Toxicology: OEL 300 ppm
Precaution: Combustible

Norpar® 13 [Exxon]
Chem. Descrip.: Normal paraffin hydrocarbon solvent
CAS 64771-72-8
Uses: Solvent used in aluminum rolling oils, as diluent solvs. in carbonless copy paper, and in spark erosion machining; as clean fuels in lamp oil and barbecue lighter fluids; in pesticides, furniture polishes, cleaning and degreasing formulations; complies with criteria for LVP solvent
Regulatory: FDA 21CFR §176.210, 177.2600, 178.3570, 178.3620(b); 40CFR §180.1001(c)(e)
Properties: Little/no odor; m.w. 189; sp.gr. 0.764 (60/60 F); dens. 6.35 lb/gal (60 F); visc. 1.84 cps; b.p. 225-243 C; flash pt. (ATSM D 93) 95 C; surf. tens. 26.7 dynes/cm; KB value 22; < 0.01% aromatics
Toxicology: OEL 300 ppm

Norpar® 15 [Exxon]
Chem. Descrip.: Normal paraffin hydrocarbon solvent
CAS 64771-72-8
Uses: Solvent used in aluminum rolling oils, as diluent solvs. in carbonless copy paper, and in spark erosion machining; as clean fuels in lamp oil and barbecue lighter fluids; in pesticides, floor polishes/cleaners, liq. candles, cleaners/degreasers; solvency for waxes; complies with criteria for LVP solvent
Regulatory: FDA 21CFR §176.180, 176.210, 177.2600, 178.3570, 178.3620(b); 40CFR §180.1001(c)(e)
Properties: Little/no odor; m.w. 212; sp.gr. 0.773 (60/60 F); dens. 6.44 lb/gal (60 F); visc. 2.53 cps; b.p. 251-275 C; flash pt. (ATSM D 93) 120 C; surf. tens. 28.9 dynes/cm; KB value 28; 0.01% aromatics
Toxicology: OEL 300 ppm; low acute toxicity

Novol NF [Croda Inc; Croda Chem. Ltd]
Chem. Descrip.: Super refined oleyl alcohol NF
CAS 143-28-2; EINECS 205-597-3
Uses: Emollient, emulsion stabilizer, solubilizer, superfatting agent, penetrant, pigment suspending aid used in cosmetics, personal care prods., pharmaceuticals; lipsticks, sunscreens, antiperspirants, bath oils; solvent and penetrant for topical and transdermal pharmaceuticals
Use Level: 2-30%
Properties: Gardner 1 max. liq.; mild odor; sol. in min. oil, IPA, propylene glycol; misc. with fat, oil and wax mixts.; sp.gr. 0.845-0.855 (15 C); visc. 24-32 cps; acid no. 0.1 max.; sapon. no. 0.3 max.; hyd. no. 195-210
Toxicology: Mild skin irritant, nonirritating to eyes

NusolvTM ABP-62 [Ridge Tech.]
Chem. Descrip.: 1,1´-Biphenyl (1-methylethyl)
CAS 25640-78-2
Uses: Solvent possessing exc. solvency, chem. and thermal stability, nonvolatility; for aerosols, adhesives, electronic parts mfg. and cleaning, industrial cleaning, inks and paper coatings, metal cleaning, paints, plastics, sealants, tile mfg.
Regulatory: FDA and EPA clearances
Properties: Colorless clear liq., fragrant aromatic odor; sp.gr. 0.985; visc. 6.2 cst (100 F); i.b.p. 480 F; pour pt. -55 F; flash pt. (COC) 300 F; fire pt. 310 F; ref. index 1.583; dielec. str. > 45 KV; dielec. const. 2.60 (100 C); vol. resist. 5×10^{12} ohm-cm (100 C)

NusolvTM ABP-74 [Ridge Tech.]
Chem. Descrip.: Alkyl biphenyl mixt.
CAS 81846-81-3
Uses: Solvent possessing exc. solvency, chem. and thermal stability, nonvolatility; for aerosols, adhesives, electronic parts mfg. and cleaning, industrial cleaning, inks and paper coatings, metal cleaning, paints, plastics, sealants, tile mfg.
Regulatory: FDA and EPA clearances
Properties: Colorless clear liq., sl. pleasant odor; sp.gr. 0.975; dens.

8.10 lb/gal (60 F); visc. 7.4 cst (100 F); i.b.p. 570 F; pour pt. -50 F; flash pt. (COC) > 300 F

Nusolv™ ABP-87 [Ridge Tech.]
Chem. Descrip.: Alkyl biphenyl mixt.
Uses: Solvent possessing exc. solvency, chem. and thermal stability, nonvolatility; for aerosols, adhesives, electronic parts mfg. and cleaning, industrial cleaning, inks and paper coatings, metal cleaning, paints, plastics, sealants, tile mfg.
Regulatory: FDA and EPA clearances
Properties: Pale yel. clear liq., sl. mild odor; sp.gr. 0.971; visc. 8.7 cst (100 F); i.b.p. 564 F; f.p. -27.2 C; flash pt. (COC) > 300 F; ref. index 1.571

Nusolv™ ABP-103 [Ridge Tech.]
Chem. Descrip.: Bis(methylethyl)-1,1´-biphenyl
CAS 69009-90-1
Uses: Solvent possessing exc. solvency, chem. and thermal stability, nonvolatility; for aerosols, adhesives, electronic parts mfg. and cleaning, industrial cleaning, inks and paper coatings, metal cleaning, paints, plastics, sealants, tile mfg.; non-V.O.C., high flash pt., sl. or no odor and color, chemically and thermally stable, nonconductive
Regulatory: FDA and EPA clearances
Properties: Sl. yel. clear liq., odorless; negligible sol. in water; m.w. 238; sp.gr. 0.965; dens. 8.049 lb/gal (60 F); visc. 10.3 cst (100 F); vapor pressure < 0.01 mm Hg; m.p. -40 F; b.p. 313 C; pour pt. -40 F; flash pt. (COC) 320 F; ref. index 1.567; pH neutral
Toxicology: LD50 (oral, rat) 4650 mg/kg (sl. toxic); sl./moderate eye irritant; prolonged/repeated exposure may cause dermatitis; may be irritating to respiratory tract.
Precaution: Combustible; incompat. with high heat, sparks, open flame, strong oxidizers; partial burning produces fumes, smoke, CO

Nusolv™ ABP-164 [Ridge Tech.]
Chem. Descrip.: Alkyl biphenyl mixt.
CAS 81846-81-3
Uses: Solvent possessing exc. solvency, chem. and thermal stability, nonvolatility; for aerosols, adhesives, electronic parts mfg. and cleaning, industrial cleaning, inks and paper coatings, metal cleaning, paints, plastics, sealants, tile mfg.; non-VOC
Regulatory: FDA and EPA clearances
Properties: Sl. yel. clear liq., sl. mild odor; sp.gr. 0.958; visc. 16.4 cst (100 F); i.b.p. 657 F; pour pt. -40 F; flash pt. (COC) 345 F

O

#15 Oil [CasChem]
Chem. Descrip.: Polymerized castor oil
Uses: Pigment wetting/dispersing agent; plasticizer for resins, gums, polymers; lubricant, penetrant; coupling solvent; adhesion promoter; for cellulose lacquers, inks, adhesives, industrial lubricants, polishes, caulks, leather dressing, hydraulic fluids; rubber compding.
Regulatory: FDA approval
Properties: Gardner 11 color; sp.gr. 1.013; visc. 250 stokes; pour pt. 35 F; acid no. 14; iodine no. 64; sapon. no. 220; hyd. no. 137

#30 Oil [CasChem]
Chem. Descrip.: Castor oil, polymerized
Uses: Pigment wetting/dispersing agent; plasticizer for resins, gums, polymers; lubricant, penetrant; coupling solvent; adhesion promoter; for cellulose lacquers, inks, adhesives, industrial lubricants, polishes, caulks, leather dressing, hydraulic fluids; rubber compding.
Regulatory: FDA approval
Properties: Gardner 13 visc. liq.; sp.gr. 1.019; visc. 500 stokes; pour pt. 45 F; acid no. 13; iodine no. 63; sapon. no. 220; hyd. no. 136

#40 Oil [CasChem]
Chem. Descrip.: Polymerized castor oil
Uses: Pigment wetting/dispersing agent; plasticizer for resins, gums, polymers; lubricant, penetrant; coupling solvent; adhesion promoter; for cellulose lacquers, inks, adhesives, industrial lubricants, polishes, caulks, leather dressing, hydraulic fluids; rubber compding.
Regulatory: FDA approval
Properties: Gardner 14 color; sp.gr. 1.020; visc. 800 stokes; pour pt. 50 F; acid no. 13; iodine no. 60; sapon. no. 225; hyd. no. 135

Olepal 1 [Gattefosse]
Chem. Descrip.: PEG-6 oleate
CAS 9004-96-0
Uses: Solvent for lipids and emulsifier for o/w disps
Properties: Liq.

Olepal ISO [Gattefosse SA]
Chem. Descrip.: PEG-6 isostearate
CAS 56002-14-3
Uses: Solvent and emulsifier for cosmetic/pharmaceutical emulsions and microemulsions
Properties: Liq.; HLB 12-13; acid no. < 6; iodine no. < 15; sapon. no. 95-125

Oxitol [Shell]
Chem. Descrip.: Ethylene glycol monoethyl ether
CAS 110-80-5; EINECS 203-804-1
Uses: Med. boiling solvent with toluene dilution ratio of 4.9; strong solvent action on NC and alkyd resins; also for phenolics, Epon resins, various oils and dyes; used in mfg. of surf. coatings, esp. lacquers where it improves gloss; in printing and dyeing textiles; used in leather finishing; coupling agent for soap-hydrocarbon systems; used in varnish removers and cleaning sol'ns
Properties: Colorless liq.; mild odor; water-misc.; sp.gr. 0.9256-0.9286; 0.10% max. water
Discontinued prod.

Oxsol® 73 [Occidental]

Chem. Descrip.: Monochlorotoluene (70%)/perchloroethylene (30%) blend
Uses: Non-ozone depleting solvent for industrial solvent cleaning, aerosols, adhesives, coatings, and inks; 1,1,1-trichloroethane alternative; no flash to boiling; aggressive solvency
Regulatory: Highly regulated
Properties: Odor; dens. 10.3 lb/gal; visc. 0.91 cps; vapor pressure 8 mm Hg; f.p. -33 C; b.p. 129-158 C; flash pt. none; surf. tens. 32 dynes/cm; KB value 114
Toxicology: May cause mild to severe eye irritation, sl. to severe skin irritation, respiratory tract irritation, CNS depression; contains suspected carcinogen

Oxsol® 100 [Occidental]
Chem. Descrip.: p-Chlorobenzotrifluoride
CAS 98-56-6
UN No.: 2234; DOT No.: UN2234
Uses: Non-ozone depleting solvent for industrial solvent cleaning, aerosols, adhesives, coatings, inks, electronic applics.; 1,1,1-trichloroethane alternative; VOC-exempt status
Regulatory: SARA reportable
Properties: Water-wh. clear liq., char. chlorinated aromatic odor; sp.gr. 1.33-1.35; dens. 11.2 lb/gal; visc. 0.79 cps; vapor pressure 8 mm Hg; f.p. -36 C; b.p. 139 C; flash pt. (TCC) 109 F; surf. tens. 25 dynes/cm; KB value 64; 97-100% act.
Toxicology: ACGIH TLV 20 ppm; LD50 (oral, rat) > 6.8 g/kg, (rabbit, dermal) > 2.7 g/kg; may cause mild to moderate eye, skin, and respiratory tract irritation; may cause CNS depression; not highly toxic if swallowed; prolonged ing. may cause liver/kidney damage
Precaution: Combustible; incompat. with oxidizing agents
HMIS: Health: 1; Flammability: 2; Reactivity: 1
Storage: Store containers in cool, dry, ventilated, fire-resist. area; keep away from heat, sparks, open flame; ground equip. to avoid static charges; vent indoor tanks to outdoors

Oxsol® 253 [Occidental]
Chem. Descrip.: p-Chlorobenzotrifluoride (20%), monochlorotoluene (50%), and perchloroethylene (30%) blend
Uses: Non-ozone depleting solvent for industrial solvent cleaning, aerosols, adhesives, coatings, and inks; 1,1,1-trichloroethane alternative
Regulatory: Highly regulated; partially VOC exempt
Properties: Odor; dens. 10.72 lb/gal; visc. 0.88 cps; vapor pressure 9 mm Hg; f.p. < -60 C; b.p. 127-159 C; flash pt. none; surf. tens. 32 dynes/cm; KB value 109
Toxicology: May cause mild to severe eye irritation, sl. to severe skin irritation, respiratory tract irritation, CNS depression; contains suspected carcinogen

Oxsol® 325 [Occidental]
Chem. Descrip.: p-Chlorobenzotrifluoride (30%), monochlorotoluen (20%), and perchloroethylene (50%) blend
Uses: Non-ozone depleting solvent for industrial solvent cleaning, aerosols, adhesives, coatings, and inks; 1,1,1-trichloroethane alternative
Regulatory: Highly regulated; partially VOC exempt
Properties: Odor; dens. 11.81 lb/gal; visc. 0.85 cps; vapor pressure 13 mm Hg; f.p. < -50 C; b.p. 121-155 C; flash pt. none; surf. tens. 32 dynes/cm; KB value 100
Toxicology: May cause mild to severe eye irritation, sl. to severe skin irritation, respiratory tract irritation, CNS depression; contains suspected carcinogen

Oxsol® 550 [Occidental]
Chem. Descrip.: p-Chlorobenzotrifluoride (50%), monochlorotoluene (50%) blend
Uses: Non-ozone depleting solvent for industrial solvent cleaning, aerosols, adhesives, coatings, and

inks; 1,1,1-trichloroethane alternative; aggressive solvency

Regulatory: Moderately regulated; partially VOC exempt

Properties: Odor; dens. 10.02 lb/gal; visc. 0.83 cps; vapor pressure 6 mm Hg; f.p. < -60 C; b.p. 142-159 C; flash pt. (TCC) 113 F; surf. tens. 32 dynes/cm; KB value 92

Toxicology: May cause mild to severe eye irritation, sl. to severe skin irritation, respiratory tract irritation, CNS depression

Precaution: Combustible

Oxsol® 1000 [Occidental]

Chem. Descrip.: 3,4-Dichlorobenzotrifluoride

CAS 328-84-7

Uses: Non-ozone depleting solvent for industrial solvent cleaning, aerosols, adhesives, coatings, and inks; 1,1,1-trichloroethane alternative; not hazardous waste

Properties: Odor; dens. 12.3 lb/gal; visc. 1.52 cps; vapor pressure 2 mm Hg; f.p. -12 C; b.p. 173.5 C; flash pt. (TCC) 170 F; surf. tens. 29 dynes/cm; KB value 69; 100% act.

Toxicology: May cause mild to severe eye irritation, sl. to severe skin irritation, respiratory tract irritation, CNS depression; ACGIH TLV 5 ppm

Precaution: Combustible

P

Palatinol® DBP [BASF]
Chem. Descrip.: Di-n-butylphthalate
CAS 84-74-2; EINECS 201-557-4
Uses: Plasticizer for PVAc-based coatings and adhesives; used to bond paper and in lamination of vinyl wall coverings to gypsum board; NC lacquers and cellophane coatings; mfg. of smokeless powds. for military use, processing aid for nitrile rubbers; carrier solvent for pigments and dyes, solvent for printing ink vehicles, fixative for perfumes, modifier of phenolic laminates, component of floor waxes
Properties: APHA 35 max.; mild, char. odor; m.w. 278.35; b.p. 182 C; sol. < 0.01% in water; sp.gr. 1.045-1.051; dens. 8.7 lb/gal; visc. 20.1 cps; pour pt. -40 C; acid no. 0.07 max.; ref. index 1.4926; flash pt. (COC) 171 C

Pale 4 [CasChem]
Chem. Descrip.: Polymerized castor oil
Uses: Pigment wetting/dispersing agent; plasticizer for resins, gums, polymers; lubricant, penetrant; coupling solvent; adhesion promoter; for cellulose lacquers, inks, adhesives, industrial lubricants, polishes, caulks, leather dressing, hydraulic fluids; rubber compounding, gasket cement
Regulatory: FDA approval
Properties: Gardner 4 color; sp.gr. 0.998; visc. 32 stokes; pour pt. 5 F; acid no. 16; iodine no. 70; sapon. no. 212; hyd. no. 158

Pale 16 [CasChem]
Chem. Descrip.: Polymerized castor oil
Uses: Pigment wetting/dispersing agent; plasticizer for resins, gums, polymers; lubricant, penetrant; coupling solvent; adhesion promoter; for cellulose lacquers, inks, adhesives, industrial lubricants, polishes, caulks, leather dressing, hydraulic fluids; rubber compounding, gasket cement
Regulatory: FDA approval
Properties: Gardner 9 color; sp.gr. 1.025; visc. 250 stokes; pour pt. 25 F; acid no. 24; iodine no. 56; sapon. no. 237; hyd.no. 136

Pale 170 [CasChem]
Chem. Descrip.: Polymerized castor oil
Uses: Pigment wetting/dispersing agent; plasticizer for resins, gums, polymers; lubricant, penetrant; coupling solvent; adhesion promoter; for cellulose lacquers, inks, adhesives, industrial lubricants, polishes, caulks, leather dressing, hydraulic fluids; rubber compounding, gasket cement
Regulatory: FDA approval
Properties: Gardner 2 color; sp.gr. 0.970; visc. 11 stokes; pour pt. -5 F; acid no. 4; iodine no. 80; sapon. no. 184; hyd. no. 160

Pale 1000 [CasChem]
Chem. Descrip.: Polymerized castor oil
Uses: Pigment wetting/dispersing agent; plasticizer for resins, gums, polymers; lubricant, penetrant; coupling solvent; adhesion promoter; for cellulose lacquers, inks, adhesives, industrial lubricants, polishes, caulks, leather dressing, hydraulic fluids; rubber compounding, gasket cement
Regulatory: FDA approval
Properties: Gardner 9 color; sp.gr. 1.018; visc. 120 stokes; pour pt. 25 F;

acid no. 20; iodine no. 59; sapon. no. 230; hyd. no. 139

Panasol™ Solvents [Amoco]

Uses: Solvents providing outstanding flow, control of evaporation rates, and high solvency for use in can and coil coatings

Partsprep™ Degreaser [ISP]

Chem. Descrip.: N-Methyl-2-pyrrolidone and surfactants

Uses: Solvent cleaner for removal of grease, cutting oils, buffing compds., and waxes in metal, photographic, aerospace industries, and precision mechanics; safer alternative to CFCs, trichloroethylene, semiaq. cleaners

Properties: Clear liq.; water-sol.; sp.gr. 1.02-1.04; ref. index 1.46-1.48

Toxicology: TLV 100 ppm; low oral and inhalation toxicity; mod. dermal toxicity; mod. eye irritant

Patogen PD-3 [Yorkshire Pat-Chem]

Uses: High flash pt., non 'red label' solvent scour for removal of stubborn sizes; high caustic stability

Unverified

PD-23 [Witco/Petrol. Spec.]

Chem. Descrip.: Min. oil

Uses: Non-VOC solvent used as defoamer, dust control agent, quench oil, pesticide diluent for fiber lubricants, metal foil and sheet rolling oils, in froth flotation of vegetables, mfg. of paper and board, in waterless hand cleaners; liq. candles, charcoal lighter fluid, polishes, cleaners, cosmetic/pharmaceutical creams and lotions

Regulatory: FDA 21CFR §172.884, 178.3650

Properties: Colorless clear liq., nearly odorless; sp.gr. 0.800 (60/60 C); visc. 2.6 cSt (40 C); vapor pressure < 0.01 mm Hg (20 C); b.p. 452 F; pour pt. 0 F; flash pt. (COC) 230 F; ref. index 1.442

PD-25 [Witco/Petrol. Spec.]

Chem. Descrip.: Petroleum distillate

Uses: Non-VOC solvent used as defoamer, dust control agent, quench oil, pesticide diluent for fiber lubricants, metal foil and sheet rolling oils, in froth flotation of vegetables, mfg. of paper and board, in waterless hand cleaners; liq. candles, charcoal lighter fluid, polishes, cleaners, pharmaceutical creams and lotions

Regulatory: FDA 21CFR §172.884, 178.3650

Properties: Colorless clear liq., nearly odorless; sp.gr. 0.810 (60/60 C); visc. 3.9 cSt (40 C); vapor pressure < 0.01 mm Hg (20 C); b.p. 510 F; pour pt. 20 F; flash pt. (COC) 260 F; ref. index 1.446

PD-28 [Witco/Petrol. Spec.]

Chem. Descrip.: Petroleum distillate

Uses: Non-VOC solvent used as defoamer, dust control agent, quench oil, pesticide diluent for fiber lubricants, metal foil and sheet rolling oils, in froth flotation of vegetables, mfg. of paper and board, in waterless hand cleaners; liq. candles, charcoal lighter fluid, polishes, cleaners, pharmaceutical creams and lotions

Regulatory: FDA 21CFR §172.878, 178.3620a

Properties: Colorless liq., nearly odorless; sp.gr. 0.820 (60/60 C); visc. 4.6 cSt (40 C); vapor pressure < 0.01 mm Hg (20 C); b.p. 526 F; pour pt. 30 F; flash pt. (COC) 285 F; ref. index 1.450

Pecosil® DCU [Phoenix]

Chem. Descrip.: Dimethicone copolyol undecylenate

CAS 159100-33-1

Uses: Emollient used in creams and lotions, cosmetics, sunscreens, perms, lip prods., aerosol systems, fragrances, bath prods., nail conditioning treatments, pharmaceutical topicals; coupling agent, cosolvent for water and oil systems

Properties: Off-wh. to yel. clear to hazy liq.; sol. in water and castor oil; sp.gr. 1.01 g/ml; acid no. 15 max.; flash pt. (TCC) > 200 F; pH 5-7 (10%); 100% act.

Toxicology: Nonirritating to eyes; nonprimary skin irritant; nontoxic orally; noncomedogenic; skin exposure can cause irritation; ingestion may cause nausea, vomiting,

abdominal pain
Precaution: Incompat. with strong oxidizers and reducers; hazardous decomp. prods.: thermal $(CO)_x$, $(SiO)_x$
HMIS: Health: 1; Flammability: 1; Reactivity: 0

Pegol® P-1000 [Rhône-Poulenc Surf. & Spec.]
Chem. Descrip.: PPG
Uses: Antifoam for industrial applics., additive for tire lubricants; carrier and cosolvent for personal care prods.
Properties: Liq.; pour pt. < -25 C; 100% conc.; nonionic
Discontinued prod.

Pegol® P-2000 [Rhône-Poulenc Surf. & Spec.]
Chem. Descrip.: PPG
Uses: Antifoam for industrial applics., additive for tire lubricants; carrier and cosolvent for personal care prods.
Properties: Liq.; pour pt. < -25 C; 100% conc.; nonionic
Discontinued prod.

Pelemol® G7A [Phoenix]
Chem. Descrip.: Glycereth-7 triacetate
CAS 57569-76-3
Uses: Humectant, emollient, coupling agent, solubilizer; solvent for sunscreen ingreds.; used in facial cleansers, creams and lotions, splash-on prods., shaving prods., hair prods., pharmaceutical topicals
Properties: Sl. yel. clear visc. liq., bland odor; sol. in water, alcohol, hydroalcoholic sol'ns.; sp.gr. ≈ 1.16; acid no. 7 max.; sapon. no. 305-330; hyd. no. 20 max.; flash pt. (PMCC) > 150 C; 100% act.
Toxicology: Nonirritating to eyes; nonprimary skin irritant; LD50 (oral) > 5 g/kg, nontoxic orally; noncomedogenic; skin exposure can cause irritation; ingestion may cause nausea, vomiting, abdominal pain
Precaution: Incompat. with oxidizing agents; hazardous decomp. prods.: may form oxides of carbon
HMIS: Health: 1; Flammability: 0; Reactivity: 0
Storage: Store in cool, dry place

Penreco 2251 Oil [Penreco]
Chem. Descrip.: Petroleum distillates
CAS 64742-14-9
Uses: Solvent for carbonless copy paper for office forms, industrial cleaning solvs., waterless hand cleaners, fragrance bases, automotive and furniture polishes, insecticide bases, lamp oils, food processing; defoamers in paper and coatings industries; froth flotation of vegetables, electrostatic discharge machine fluids, and magnetic particle inspection fluids
Regulatory: FDA 21CFR §172.884, 178.3650, 573.740
Properties: Sp.gr. 0.786 (60/60 F); dens. 6.56 lb/gal (60 F); visc. 1.66 cSt (100 F); i.b.p. 191 C min.; pour pt. -40 C; flash pt. 74 C; KB value 28; < 1% aromatics
Precaution: DOT: combustible liq.

Penreco 2257 Oil [Penreco]
Chem. Descrip.: Sat. hydrocarbon
CAS 64742-46-7
Uses: Hydrocarbon solvent for carbonless copy paper for office forms, industrial cleaning solvs., waterless handcleaners, fragrance bases, automotive and furniture polishes, insecticide bases, lamp oils, food processing; defoamers in paper and coatings industries, froth flotation of vegetables, electrostatic discharge machine fluids, and magnetic particle inspection fluids
Regulatory: FDA 21CFR §172.884, 178.3650, 573.740
Properties: Sp.gr. 0.795 (60/60 F); dens. 6.64 lb/gal (60 F); visc. 2.18 cSt (100 F); vapor pressure < 0.1 mm Hg; i.b.p. 221 C min.; pour pt. -26 C; flash pt. 99 C; KB value 26.5; < 1% aromatics

Penreco 2259 Oil [Penreco]
Chem. Descrip.: Sat. hydrocarbon
CAS 8042-47-5
Uses: Hydrocarbon solvent for carbonless copy paper for office forms, industrial cleaning solvs., waterless handcleaners, fragrance

bases, automotive and furniture polishes, insecticide bases, lamp oils, defoamers in paper and coatings industries; froth flotation of vegetables, electrostatic discharge machine fluids, and magnetic particle inspection fluids

Regulatory: FDA 21CFR §172.884, 178.3650, 573.740

Properties: Sp.gr. 0.804 (60/60 F); dens. 6.69 lb/gal (60 F); visc. 2.28 cSt (100 F); vapor pressure < 0.1 mm Hg; i.b.p. 235 C min.; pour pt. -18 C; flash pt. 102 C; KB value 24.5; < 1% aromatics

Penreco 2260 Oil [Penreco]

Chem. Descrip.: Sat. hydrocarbon
CAS 8042-47-5

Uses: Hydrocarbon solvent for carbonless copy paper for office forms, industrial cleaning solvs., waterless handcleaners, fragrance bases, automotive and furniture polishes, insecticide bases, lamp oils, defoamers in paper and coatings industries; froth flotation of vegetables, electrostatic discharge machine fluids, and magnetic particle inspection fluids

Regulatory: FDA 21CFR §172.884, 178.3620(b), 178.3650, 573.740

Properties: Sp.gr. 0.812 (60/60 F); dens. 6.76 lb/gal (60 F); visc. 4.3 cSt (100 F); vapor pressure < 0.1 mm Hg; i.b.p. 266 C min.; pour pt. -4 C; flash pt. 127 C; KB value 22.6; < 1% aromatics

Penreco 2263 Oil [Penreco]

Chem. Descrip.: Petroleum distillates
CAS 64742-47-8; EINECS 232-298-5

Uses: Solvent for carbonless copy paper for office forms, industrial cleaning solvs., waterless hand cleaners, fragrance bases, automotive and furniture polishes, insecticide bases, lamp oils, defoamers in paper and coatings industries; froth flotation of vegetables, electrostatic discharge machine fluids, and magnetic particle inspection fluids

Regulatory: FDA 21CFR §172.884, 178.3650, 573.740

Properties: Sp.gr. 0.786 (60/60 F); dens. 6.56 lb/gal (60 F); visc. 1.68 cSt (100 F); i.b.p. 191 C min.; pour pt. -40 C; flash pt. 74 C; KB value 29; < 1% aromatics

Precaution: DOT: combustible liq.

Penreco Snow [Penreco]

Chem. Descrip.: Wh. petrolatum USP
CAS 8009-03-8; EINECS 232-373-2

Uses: Emollient, base, solvent, and carrier for cosmetic and pharmaceutical preparations (creams, lotions, petrol. jellies, ointments, dental adhesives); foods (animal feed supplement, fruit/veg. coatings, food pkg.); sanitary lubricant in food prod. machinery; rust-preventive coating in food processing equipment; carrier in adhesive tapes and compds.; lubricant for textile, paper, PVC; shoe polishes

Regulatory: FDA 21CFR 172.880, 178.3700, 573.720

Properties: Wh. opaque semisolid, odorless; sol. in hydrocarbons; sp.gr. 0.86 (60/60 F); visc. 64-75 SUS (210 F); vapor pressure < 1 mm Hg (70 F); m.p. 122-135 F; b.p. 650 F; congeal pt. 123 F; solid. pt. 121 F; flash pt. (COC) 400 F

Toxicology: Pract. nontoxic by ingestion, but laxative props. may cause abdominal cramps, diarrhea; minimally irritating to eyes and skin on direct contact; STEL 10 mg/m^3 (as oil mist)

Precaution: Combustion may produce dense smoke, CO, CO_2, other oxides; may react with strong oxidizing agents

Storage: Store in closed containers away from heat, sparks, open flame, oxidizers

PentoXone [Shell]

Chem. Descrip.: 4-Methoxy-4-methylpentanone-2
CAS 107-70-0; EINECS 203-512-4

Uses: High boiling solvent used in preparation of NC, acrylic, vinyl, and cellulose acetate butyrate lacquers, alkyd and epoxy enamels, and urethane and thermosetting acrylic coatings; used in cleaning compds.

Properties: Colorless liq.; pleasant

odor; sp.gr. 0.899-0.909; 0.10% max. water

Permethyl® 99A [Presperse]
Chem. Descrip.: Isododecane
CAS 13475-82-6; EINECS 205-495-9
Uses: Cosolubilizer for nonhydrocarbon materials; for mascara, eyeliner, antiperspirant and where residual film is not desirable; solvent for debris on skin; clean feel, very little residual
Regulatory: JSCI/JCID approved
Properties: Colorless clear liq.; sol. in cyclomethicone, hydrocarbons, isoparaffins, min. spirits; m.w. 170; sp.gr. 0.747; visc. 1 cps (20 C); ref. index 1.424
Toxicology: Acute oral > 5 g/kg; pract. nonirritating to eyes; mild skin irritant; noncomedogenic

Petrosolv® A [Witco/Oleo-Surf.]
Chem. Descrip.: Aromatic solv.
Uses: Solvent additive for carburetor and soak cleaners
Unverified

PF-5030 [3M/Spec. Chems.]
Chem. Descrip.: Perfluorocarbon
Uses: Performance fluid offering low surf. tension, high chem./thermal stability, no VOC for improved coating flow and performance, exc. surf. wetting; dispersant for lubricants, mold release agents, protective coatings; reaction media for polymerizations, purification, separation processes; solvent for removal of halogenated lubricants, oils, greases; carrier for halogenated material; reduces aggressiveness of other solvs.; zero ozone depletion potential
Properties: Liq.; m.w. 188; dens. 1.35 g/ml (saturation pressure, 20 C); visc. 0.3 cp (saturation pressure, 0 C); vapor pressure 114.8 psia (21.1 C); b.p. -37 C; flash pt. none; surf. tens. 4.3 dynes/cm

PF-5040 [3M/Spec. Chems.]
Chem. Descrip.: Perfluorocarbon
Uses: Performance fluid offering low surf. tension, high chem./thermal stability, no VOC for improved coating flow and performance, exc. surf. wetting; dispersant for lubricants, mold release agents, protective coatings; reaction media for polymerizations, purification, separation processes; solvent for removal of halogenated lubricants, oils, greases; carrier for halogenated material; reduces aggressiveness of other solvs.; zero ozone depletion potential
Properties: Liq.; m.w. 238; dens. 1.52 g/ml (saturation pressure, 20 C); visc. 0.46 cp (saturation pressure, 20 C); vapor pressure 32.9 psia (21.1 C); b.p. -2 C; flash pt. none; surf. tens. 7.4 dynes/cm

PF-5050 [3M/Spec. Chems.]
Chem. Descrip.: Perfluoropentane
CAS 678-26-2
Uses: Performance fluid offering low surf. tension, high chem./thermal stability, no VOC for improved coating flow and performance, exc. surf. wetting; dispersant for lubricants, mold release agents, protective coatings; reaction media for polymerizations, purification, separation processes; solvent for removal of halogenated lubricants, oils, greases; carrier for halogenated material; reduces aggressiveness of other solvs.; zero ozone depletion potential
Properties: Liq.; m.w. 288; dens. 1.63 g/ml; visc. 0.45 cp; vapor pressure 11.80 psia; b.p. 30 C; flash pt. none; surf. tens. 9.5 dynes/cm

PF-5052 [3M/Spec. Chems.]
Chem. Descrip.: Perfluoro-N-methylmorpholine
Uses: Performance fluid offering low surf. tension, high chem./thermal stability, no VOC for improved coating flow and performance, exc. surf. wetting; dispersant for lubricants, mold release agents, protective coatings; reaction media for polymerizations, purification, separation processes; solvent for removal of halogenated lubricants, oils, greases; carrier for halogenated material; reduces aggressiveness of other solvs.; zero ozone depletion potential
Properties: Liq.; m.w. 299; dens. 1.70

g/ml; visc. 0.68 cp; vapor pressure 5.29 psia; b.p. 50 C; flash pt. none; surf. tens. 13.0 dynes/cm

PF-5060 [3M/Spec. Chems.]
Chem. Descrip.: Perfluorohexane
CAS 355-42-0; EINECS 206-585-0
Uses: Performance fluid offering low surf. tension, high chem./thermal stability, no VOC for improved coating flow and performance, exc. surf. wetting; dispersant for lubricants, mold release agents, protective coatings; reaction media for polymerizations, purification, separation processes; solvent for removal of halogenated lubricants, oils, greases; carrier for halogenated material; reduces aggressiveness of other solvs.; zero ozone depletion potential
Properties: Liq.; m.w. 338; dens. 1.68 g/ml; visc. 0.67 cp; vapor pressure 4.49 psia; b.p. 56 C; flash pt. none; surf. tens. 12.0 dynes/cm

PF-5070 [3M/Spec. Chems.]
Chem. Descrip.: Perfluoroheptane
CAS 335-57-9; EINECS 206-392-1
Uses: Performance fluid offering low surf. tension, high chem./thermal stability, no VOC for improved coating flow and performance, exc. surf. wetting; dispersant for lubricants, mold release agents, protective coatings; reaction media for polymerizations, purification, separation processes; solvent for removal of halogenated lubricants, oils, greases; carrier for halogenated material; reduces aggressiveness of other solvs.; zero ozone depletion potential
Properties: Liq.; m.w. 388; dens. 1.73 g/ml; visc. 0.95 cp; vapor pressure 1.53 psia; b.p. 80 C; flash pt. none; surf. tens. 13.0 dynes/cm

PF-5080 [3M/Spec. Chems.]
Chem. Descrip.: Perfluorocarbon
Uses: Performance fluid offering low surf. tension, high chem./thermal stability, no VOC for improved coating flow and performance, exc. surf. wetting; dispersant for lubricants, mold release agents, protective coatings; reaction media for polymerizations, purification, separation processes; solvent for removal of halogenated lubricants, oils, greases; carrier for halogenated material; reduces aggressiveness of other solvs.; zero ozone depletion potential
Properties: Liq.; m.w. 438; dens. 1.77 g/ml; visc. 1.4 cp; vapor pressure 0.85 psia; b.p. 101 C; flash pt. none; surf. tens. 15.0 dynes/cm

Pharmasolve™ [ISP]
Chem. Descrip.: N-Methyl-2-pyrrolidone, GMP grade
CAS 872-50-4; EINECS 212-828-1
Uses: Solvent for pharmaceuticals; increases water sol. of chloramphenicol, procaine, other therapeutic compds.; nonaq. chem. reaction medium; synthesis of vitamin E precursor; solubilizer for topicals for applic. to human or animal hair, skin, nails; skin penetration enhancer; extraction solvent; bioadhesive; spray bandages
Properties: Clear liq., amine-like odor; misc. with water, most org. solvs. incl. alcohols, ketones, aromatic and chlorinated hydrocarbons; sp.gr. 1.027 (25/4 C); visc. 1.65 cp; vapor pressure 0.29 mm Hg (20 C); m.p. -24.4 C; b.p. 202 C; flash pt.(OC) 95 C; ref. index 1.465-1.470; pH sl. alkaline; 99.85% min. purity
Toxicology: CEL 100 ppm; low toxicity; LD50 (oral, rat) 4200 mg/kg, (dermal, rabbit) 8000 mg/kg; causes eye irritation; not a primary skin irritant; nonsensitizing; prolonged or repeated exposure causes severe skin irritation
Environmental: Biodeg.
Precaution: Combustible; flamm. limits 1.3-9.5%; noncorrosive; incompat. with strong oxidizing or reducing agents; heated to decomp., emits CO and NO_x fumes
HMIS: Health: 2; Flammability: 2; Reactivity: 0
Storage: Keep container and vapors away from heat, sparks, flame; keep container closed

Pilot SXS-40 [Pilot]
Chem. Descrip.: Sodium xylene sul-

fonate
CAS 1300-72-7; EINECS 215-090-9
Uses: Coupling agent, hydrotrope, solubilizer, solvent, stabilizer; used in liq. cleaners, org. polymers and dyestuffs, petrol. industry, pulping, animal glues
Properties: Lt. yel. clear liq.; dens. 9.9 lb/gal; pH 7.5; 42% total solids; anionic

Pluracol® E200 [BASF]
Chem. Descrip.: PEG-4
CAS 25322-68-3; EINECS 203-989-9
Uses: Intermediate for preparation of nonionic surfactants; binder, base, coating, stabilizer, solvent, vehicle, extender, and coupling agent for cosmetics and toiletries; lubricant for metal applics., rubber industry; wood treatment; textile conditioning, antistat, and sizing agent, softener
Properties: Colorless clear liq.; m.w. 200; sol. in water and org. solvs. except aliphatic hydrocarbons; dens. 9.4 lb/gal; sp.gr. 1.12; visc. 4.36 cs (210 F); flash pt. 360 F; surf. tens. 57.2 dynes/cm (1%); pH 6.5 (5% aq.)

Pluracol® E300 [BASF]
Chem. Descrip.: PEG-6
CAS 25322-68-3; EINECS 220-045-1
Uses: Intermediate for preparation of nonionic surfactants; binder, base, coating, stabilizer, solvent, vehicle, extender, and coupling agent for pharmaceutical, cosmetic, and toiletries; lubricant for metal applics., rubber industry; wood treatment; textile conditioning, antistat, and sizing agent, softener; dispersant in food tablets and preparations; plasticizer
Properties: Colorless clear liq.; m.w. 300; sol. in water and org. solvs. except aliphatic hydrocarbons; dens. 9.4 lb/gal; sp.gr. 1.12; visc. 5.9 cs (99 C); flash pt. (COC) 210 C; pour pt. -13 C; surf. tens. 62.9 dynes/cm (1%); pH 5.7 (5% aq.)

Pluracol® E400 [BASF]
Chem. Descrip.: PEG-8
CAS 25322-68-3; EINECS 225-856-4
Uses: Intermediate for preparation of nonionic surfactants; binder, base, coating, stabilizer, solvent, vehicle, extender, and coupling agent for pharmaceutical, cosmetic, and toiletries; lubricant for metal applics., rubber industry; wood treatment; textile conditioning, antistat, and sizing agent, softener; dispersant in food tablets and preparations; plasticizer
Properties: Colorless clear liq.; sol. in water and org. solvs. except aliphatic hydrocarbons; dens. 9.4 lb/gal; sp.gr. 1.12; visc. 7.39 cs (210 F); flash pt. 460 F; surf. tens. 66.6 dynes/cm (1%); pH 6.2 (5% aq.)

Pluracol® E400 NF [BASF]
Chem. Descrip.: PEG-8
CAS 25322-68-3; EINECS 225-856-4
Uses: Chemical intermediate, base, coupler, thickener, lubricant, mold release agent, defoamer, softener, conditioner, antistat, sizing agent, dispersant for cosmetic and oral care preps.; excipient, carrier, solvent in pharmaceuticals; in metal polishing and cleaning formulations, rubber prods., paper and wood prods., textile processing, ink formulations
Properties: Liq.; m.w. 400; visc. 7.4 cs (99 C); pour pt. 5 C; flash pt. (COC) 182 C

Pluracol® E600 [BASF]
Chem. Descrip.: PEG-12
CAS 25322-68-3; EINECS 229-859-1
Uses: Intermediate for preparation of nonionic surfactants; binder, base, coating, stabilizer, solvent, vehicle, extender, and coupling agent for pharmaceutical, cosmetic, and toiletries; lubricant for metal applics., rubber industry; wood treatment; textile conditioning, antistat, and sizing agent, softener; dispersant in food tablets and preparations; plasticizer
Properties: Colorless clear liq.; sol. in water and org. solvs. except aliphatic hydrocarbons; dens. 9.4 lb/gal; sp.gr. 1.12; visc. 10.83 cs (210 F); flash pt. 480 F; surf. tens. 65.2 dynes/cm (1%); pH 5.3 (5% aq.)

Pluracol® E600 NF [BASF]

Chem. Descrip.: PEG-12
CAS 25322-68-3; EINECS 229-859-1
Uses: Chemical intermediate, base, coupler, thickener, lubricant, mold release agent, defoamer, softener, conditioner, antistat, sizing agent, dispersant, excipient, carrier, solvent for pharmaceutical, cosmetic, and oral care preps.; in metal polishing and cleaning formulations, rubber prods., paper and wood prods., textile processing, ink formulations
Properties: Liq.; m.w. 600; visc. 10.8 cs (99 C); pour pt. 20 C; flash pt. (COC) 249 C

Pluracol® E1450 NF [BASF]
Chem. Descrip.: PEG-32
CAS 25322-68-3; EINECS 203-989-9
Uses: Chemical intermediate, base, coupler, thickener, lubricant, mold release agent, defoamer, softener, conditioner, antistat, sizing agent, dispersant, excipient, carrier, solvent for pharmaceutical, cosmetic, and oral care preps.; in metal polishing and cleaning formulations, rubber prods., paper and wood prods., textile processing, ink formulations
Properties: Solid; m.w. 600; visc. 28.5 cs (99 C); m.p. 45 C; flash pt. (COC) 255 C

Pluracol® E1500 [BASF]
Chem. Descrip.: PEG-6-32
Uses: Intermediate for preparation of nonionic surfactants; binder, base, coating, stabilizer, solvent, vehicle, extender, and coupling agent for pharmaceutical, cosmetic, and toiletries; lubricant for metal applics., rubber industry; wood treatment; textile conditioning, antistat, and sizing agent, softener; dispersant in food tablets and preparations; plasticizer
Properties: Wh. waxy solid; sol. in water and org. solvs. except aliphatic hydrocarbons; dens. 10.0 lb/gal; sp.gr. 1.20; m.p. 46.0-47.5 C; flash pt. > 490 F; surf. tens. 62.8 dynes/cm (1%); pH 6.7 (5% aq.)
Discontinued prod.

Pluracol® E4000 [BASF]
Chem. Descrip.: PEG-75
CAS 25322-68-3; EINECS 203-989-9
Uses: Intermediate for preparation of nonionic surfactants; binder, base, coating, stabilizer, solvent, vehicle, extender, and coupling agent for pharmaceutical, cosmetic, and toiletries; lubricant for metal applics., rubber industry; wood treatment; textile conditioning, antistat, and sizing agent, softener; dispersant in food tablets and preparations; plasticizer
Properties: Wh. waxy solid; sol. in water and org. solvs. except aliphatic hydrocarbons; dens. 10.0 lb/gal; sp.gr. 1.20; m.p. 59.5 C; flash pt. > 490 F; surf. tens. 61.9 dynes/cm (1%); pH 6.7 (5% aq.)

Pluracol® E6000 [BASF]
Chem. Descrip.: PEG-150
CAS 25322-68-3; EINECS 203-989-9
Uses: Intermediate for preparation of nonionic surfactants; binder, base, coating, stabilizer, solvent, vehicle, extender, and coupling agent for pharmaceutical, cosmetic, and toiletries; lubricant for metal applics., rubber industry; wood treatment; textile conditioning, antistat, and sizing agent, softener
Properties: Wh. waxy solid; sol. in water and org. solvs. except aliphatic hydrocarbons; dens. 10.0 lb/gal; sp.gr. 1.21; m.p. 61.0 C; flash pt. > 490 F; surf. tens. 62.1 dynes/cm (1%); pH 6.7 (5% aq.)
Discontinued prod.

Pogol 400 NF [Huntsman]
Chem. Descrip.: PEG, NF grade
Uses: Solubilizer, solvent for pharmaceuticals; mold release agent and lubricant for both natural and syn. prods.; dye carrier
Properties: Clear liq.; m.w. 380-420; sp.gr. 1.12; visc; 92 cps; 100% act.; nonionic
Discontinued prod.

Polarin® Range [Grünau GmbH]
Chem. Descrip.: Glycerol and derivs.
Uses: Solvent, humectant for food industry

Polychol 5 [Croda Inc; Croda Chem. Ltd]

Chem. Descrip.: Laneth-5
CAS 61791-20-6
Uses: O/w emulsifier, dispersant, gellant, emollient, solubilizer for personal care prods., topical pharmaceuticals; bromo dye solvent
Use Level: 1-10%
Properties: Soft yel. wax; sol. in IPA, disp. in water, propylene glycol, min. oil; m.p. 35-48 C; HLB 7.5; acid no. 5 max.; hyd. no. 120-135; pH 4.5-6.0 (10% aq.); 100% conc.; nonionic
Toxicology: LD50 (oral, rat) 7.7 g/kg; mild skin irritant, nonirritating to eyes

Polychol 15 [Croda Inc; Croda Chem. Ltd]
Chem. Descrip.: Laneth-15
CAS 61791-20-6
Uses: O/w emulsifier, dispersant, solubilizer, emollient, thickener, and gelling agent for cosmetics and topical pharmaceuticals; cosolvent for nonmiscible oils; reduces greasy feeling of lanolin
Use Level: 1-10%
Properties: Yel. soft wax; sol. in water, IPA, propylene glycol; disp. in min. oil; m.p. 34-42 C; HLB 12.7; acid no. 5 max.; hyd. no. 82-92; pH 3.5-5.5 (10% aq.); 100% conc.; nonionic
Toxicology: Nonirritating to skin and eyes

Poly-G® WI 285 [Olin]
Chem. Descrip.: Polyalkylene glycol derivs.
Uses: Noncorrosive fluids used in metalworking fluids (cutting, drawing, stamping, rolling, etc.); as hydraulic fluids for boundary lubrication and low-temp. applics.; for corrosion and wear protection; as chemical intermediates, defoamers, plasticizers, and solvs. for both inks and dyes
Properties: Liq.; water-insol.; sp.gr. 0.991; dens. 8.25 lb/gal; visc. 285 SUS (38 C); pour pt. -48 C; flash pt. (COC) 218 C

Poly-G® WI 625 [Olin]
Chem. Descrip.: Polyalkylene glycol derivs.
Uses: Noncorrosive fluids used in metalworking fluids (cutting, drawing, stamping, rolling, etc.); as hydraulic fluids for boundary lubrication and low-temp. applics.; for corrosion and wear protection; as chemical intermediates, defoamers, plasticizers, and solvs. for both inks and dyes
Properties: Liq.; water-insol.; sp.gr. 0.996; dens. 8.29 lb/gal; visc. 616 SUS (38 C); pour pt. -39 C; flash pt. (COC) 227 C
Unverified

Poly-G® WI 1715 [Olin]
Chem. Descrip.: Polyalkylene glycol derivs.
Uses: Noncorrosive fluids used in metalworking fluids (cutting, drawing, stamping, rolling, etc.); as hydraulic fluids for boundary lubrication and low-temp. applics.; for corrosion and wear protection; as chemical intermediates, defoamers, plasticizers, and solvs. for both inks and dyes
Properties: Liq.; water-insol.; sp.gr. 1.003; dens. 8.35 lb/gal; visc. 1715 SUS (38 C); pour pt. -23 C; flash pt. (COC) 221 C
Unverified

Poly-G® WS 100 [Olin]
Chem. Descrip.: Polyalkylene glycol derivs.
Uses: Noncorrosive fluids used in metalworking fluids (cutting, drawing, stamping, rolling, etc.); hydraulic fluids for boundary lubrication and low-temp. applics.; for corrosion and wear protection; as chemical intermediates, defoamers, plasticizers, and solvs. for both inks and dyes
Properties: Liq.; water-sol.; sp.gr. 1.016; dens. 8.46 lb/gal; visc. 100 SUS (38 C); pour pt. -65 C; flash pt. (COC) 166 C

Poly-G® WS 170 [Olin]
Chem. Descrip.: Polyalkylene glycol derivs.
Uses: Noncorrosive fluids used in metalworking fluids (cutting, drawing, stamping, rolling, etc.); hydraulic fluids for boundary lu-

brication and low-temp. applics.; for corrosion and wear protection; as chemical intermediates, defoamers, plasticizers, and solvs. for both inks and dyes

Properties: Liq.; water-sol.; sp.gr. 1.035; dens. 8.62 lb/gal; visc. 170 SUS (38 C); pour pt. -61 C; flash pt. (COC) 210 C

Unverified

Poly-G® WS 260 [Olin]

Chem. Descrip.: Polyalkylene glycol derivs.

Uses: Noncorrosive fluids used in metalworking fluids (cutting, drawing, stamping, rolling, etc.); hydraulic fluids for boundary lubrication and low-temp. applics.; for corrosion and wear protection; as chemical intermediates, defoamers, plasticizers, and solvs. for both inks and dyes

Properties: Liq.; water-sol.; sp.gr. 1.038; dens. 8.64 lb/gal; visc. 260 SUS (38 C); pour pt. -62 C; flash pt. (COC) 227 C

Unverified

Poly-G® WS 660 [Olin]

Chem. Descrip.: Polyalkylene glycol derivs.

Uses: Noncorrosive fluids used in metalworking fluids (cutting, drawing, stamping, rolling, etc.); hydraulic fluids for boundary lubrication and low-temp. applics.; for corrosion and wear protection; as chemical intermediates, defoamers, plasticizers, and solvs. for both inks and dyes

Properties: Liq.; water-sol.; sp.gr. 1.052; dens. 8.75 lb/gal; visc. 660 SUS (38 C); pour pt. -42 C; flash pt. (COC) 227 C

Unverified

Poly-G® WS 2000 [Olin]

Chem. Descrip.: Polyalkylene glycol derivs.

Uses: Noncorrosive fluids used in metalworking fluids (cutting, drawing, stamping, rolling, etc.); hydraulic fluids for boundary lubrication and low-temp. applics.; for corrosion and wear protection; as chemical intermediates, defoamers, plasticizers, and solvs. for both inks and dyes

Properties: Liq.; water-sol.; sp.gr. 1.058; dens. 8.80 lb/gal; visc. 2000 SUS (38 C); pour pt. -33 C; flash pt. (COC) 243 C

Unverified

Poly-G® WS 3520 [Olin]

Chem. Descrip.: Polyalkylene glycol derivs.

Uses: Noncorrosive fluids used in metalworking fluids (cutting, drawing, stamping, rolling, etc.); hydraulic fluids for boundary lubrication and low-temp. applics.; for corrosion and wear protection; as chemical intermediates, defoamers, plasticizers, and solvs. for both inks and dyes

Properties: Liq.; water-sol.; sp.gr. 1.061; dens. 8.83 lb/gal; visc. 3520 SUS (38 C); pour pt. -28 C; flash pt. (COC) 227 C

Unverified

Poly-G® WS 5100 [Olin]

Chem. Descrip.: Polyalkylene glycol derivs.

Uses: Noncorrosive fluids used in metalworking fluids (cutting, drawing, stamping, rolling, etc.); hydraulic fluids for boundary lubrication and low-temp. applics.; for corrosion and wear protection; as chemical intermediates, defoamers, plasticizers, and solvs. for both inks and dyes

Properties: Liq.; water-sol.; sp.gr. 1.063; dens. 8.85 lb/gal; visc. 5100 SUS (38 C); pour pt. -31 C; flash pt. (COC) 227 C

Unverified

Poly-G® WT 9150 [Olin]

Chem. Descrip.: Polyalkylene glycol derivs.

Uses: Noncorrosive fluids used in metalworking fluids (cutting, drawing, stamping, rolling, etc.); hydraulic fluids for boundary lubrication and low-temp. applics.; for corrosion and wear protection; as chemical intermediates, defoamers, plasticizers, and solvs. for both inks and dyes

Properties: Liq.; water-thickening;

sp.gr. 1.089; dens. 9.06 lb/gal; visc. 9150 SUS (38 C); pour pt. -2 C; flash pt. (COC) 207 C

Unverified

Poly-G® WT 90,000 [Olin]

Chem. Descrip.: Polyalkylene glycol derivs.

Uses: Noncorrosive fluids used in metalworking fluids (cutting, drawing, stamping, rolling, etc.); hydraulic fluids for boundary lubrication and low-temp. applics.; for corrosion and wear protection; as chemical intermediates, defoamers, plasticizers, and solvs. for both inks and dyes

Properties: Liq.; water-thickening; sp.gr. 1.095; dens. 9.11 lb/gal; visc. 90,000 SUS (38 C); pour pt. 3 C; flash pt. (COC) 252 C

Unverified

Polyglyme [Ferro/Grant]

Chem. Descrip.: Poly(ethylene glycol) dimethyl ether

CAS 24991-55-7

Uses: Solvent for removing CO, CO_2, SO_2, H_2S and other sour-gas components from ammonia gas streams, for absorp. of fluorocarbons and other halogenated liqs./gases; stack gas scrubbing; brake fluid component; solubilizing agent for plastic, textile, and paper

Properties: Sl. yel.; very mild nonresidual odor; water-sol.; misc. with ethanol, acetone, benzene, diethyl ether, octane; m.w. 275; sp.gr. 1.04 (20/20 C); dens. 8.6 lb/gal; visc. 12 cP (20 C); vapor pressure < 0.01 mm Hg (20 C); flash pt. (CC) > 130 C

Environmental: Slowly biodeg.

Polylan® [Amerchol; Amerchol Europe]

Chem. Descrip.: Oleyl linoleate, lanolin linoleate

Uses: Emollient, dispersant, conditioner, penetrant, lubricant, cosolvent, dye solubilizer used in cosmetics and pharmaceuticals

Properties: Amber visc. oily liq.; sol. in min. and castor oil, absolute ethyl alcohol, IPA, alkylolamides, ethyl acetate, and D.C. Silicone Fluid #555; insol. in water; sp.gr. 0.890-0.915; visc. 100-120 cps; HLB 8.0; acid no. 5.0 max.; iodine no. 110-130; sapon. no. 90-105; 100% act.; nonionic

Discontinued prod.

Poly-Solv® DE (High Gravity) [Olin]

Chem. Descrip.: Diethylene glycol monoethyl ether (75%) and ethylene glycol (25%)

Uses: Solvent for brake fluids, hard-surf. cleaners, leather dyeing, paints, coatings, printing inks, textile vat dyeing and printing, adhesives, antifreeze, floor waxes/polishes, insect repellents; solubilizer for dyes; plasticizer; extraction and crystallization solvent; vinyl chloride dispersant; coupler in cosmetic preparations

Properties: APHA 10 max. liq.; mild, char. odor; f.p. -75 C; b.p. 195 C (760 mm); sol. in water; misc. with many org. solvs.; sp.gr. 1.0253; dens. 8.53 lb/gal; visc. 6.9 cP; ref. index 1.4297; flash pt. (COC) 96 C; sp. heat 2.308 joules/g/°C

Discontinued prod.

Poly-Solv® DE (Low Gravity) [Olin]

Chem. Descrip.: Diethylene glycol monoethyl ether

CAS 110-90-0; EINECS 203-919-7

Uses: Solvent for brake fluids, hard-surf. cleaners, leather dyeing, paints, coatings, printing inks, textile vat dyeing and printing, adhesives, antifreeze, floor waxes/polishes, insect repellents; solubilizer for dyes; plasticizer

Properties: APHA 10 max. liq.; mild, char. odor; m.w. 134.17; f.p. -76 C; b.p. 202 C (760 mm); sol. in water; misc. with many org. solvs.; sp.gr. 0.989; dens. 8.24 lb/gal; visc. 4.59 cP; ref. index 1.4273; flash pt. (TCC) 85 C; sp. heat 2.308 joules/g/°C; 100% act.

Discontinued prod.

Poly-Solv® DM [Olin]

Chem. Descrip.: Diethylene glycol monomethyl ether

CAS 111-77-3; EINECS 203-906-6

Uses: Solvent for brake fluids, hard-surf. cleaners, leather dyeing,

paints, coatings, printing inks, textile vat dyeing and printing, adhesives, antifreeze, floor waxes/polishes, insect repellents; solubilizer for dyes; plasticizer
Properties: APHA 15 max. liq.; mild, char. odor; m.w. 120.15; f.p. -85 C; b.p. 194 C (760 mm); sol. in water; misc. with many org. solvs.; sp.gr. 1.021; dens. 8.51 lb/gal; visc. 3.9 cP; ref. index 1.4263; flash pt. (TCC) 87 C; sp. heat 2.149 joules/g/°C

Poly-Solv® DPM [Olin]
Chem. Descrip.: Dipropylene glycol monomethyl ether
Uses: Solvent for brake fluids, hard-surf. cleaners, leather dyeing, paints, coatings, printing inks, textile vat dyeing and printing, adhesives, antifreeze, floor waxes/polishes, insect repellents; solubilizer for dyes; plasticizer
Properties: APHA 15 max. liq.; mild, char. odor; m.w. 148.20; f.p. -83 C; b.p. 187 C (760 mm); misc. with water and many org. solvs.; sp.gr. 0.954; dens. 7.94 lb/gal; visc. 3.9 cP; ref. index 1.419; flash pt. (TCC) 169 C

Poly-Solv® EE [Olin]
Chem. Descrip.: Ethylene glycol monoethyl ether
CAS 110-80-5; EINECS 203-804-1
Uses: Solvent for brake fluids, hard-surf. cleaners, leather dyeing, paints, coatings, printing inks, textile vat dyeing and printing, adhesives, antifreeze, floor waxes/polishes, insect repellents; solubilizer for dyes; plasticizer
Properties: APHA 10 max. liq.; mild char. odor; m.w. 90.12; f.p. -70 C; b.p. 135 C (760 mm); sol. in water; misc. with many org. solvs.; sp.gr. 0.931; dens. 7.76 lb/gal; visc. 2.1 cP; ref. index 1.4076; flash pt. (TCC) 45 C; sp. heat 2.321 joules/g/°C
Discontinued prod.

Poly-Solv® EM [Olin]
Chem. Descrip.: Ethylene glycol monomethyl ether
CAS 109-86-4; EINECS 203-713-7
Uses: Solvent for brake fluids, hard-surf. cleaners, leather dyeing, paints, coatings, printing inks, textile vat dyeing and printing, adhesives, antifreeze, floor waxes/polishes, insect repellents; solubilizer for dyes; plasticizer
Properties: APHA 10 max. liq.; mild char. odor; m.w. 76.09; f.p. -85 C; b.p. 124 C (760 mm); sol. in water; misc. with many org. solvs.; sp.gr. 0.966; dens. 8.05 lb/gal; visc. 1.7 cP; ref. index 1.4021; flash pt. (TCC) 41 C; sp. heat 2.233 joules/g/°C

Poly-Solv® MPM [Olin]
Chem. Descrip.: Monopropylene glycol monomethyl ether
CAS 107-98-2; EINECS 203-539-1
Uses: Solvent for brake fluids, hard-surf. cleaners, leather dyeing, paints, coatings, printing inks, textile vat dyeing and printing, adhesives, antifreeze, floor waxes/polishes, insect repellents; solubilizer for dyes; plasticizer
Properties: APHA 10 max. liq.; mild char. odor; m.w. 90.12; f.p. -96 C; b.p. 121 C (760 mm); misc. with water and many org. solvs.; sp.gr. 0.923; dens. 7.68 lb/gal; visc. 1.9 cP; ref. index 1.4036; flash pt. (TCC) 36 C

Poly-Solv® TM [Olin]
Chem. Descrip.: Triethylene glycol monomethyl ether
CAS 112-35-6
Uses: Solvent for brake fluids, hard-surf. cleaners, leather dyeing, paints, coatings, printing inks, textile vat dyeing and printing, adhesives, antifreeze, floor waxes/polishes, insect repellents; solubilizer for dyes; plasticizer
Properties: APHA 50 max. liq.; mild, char. odor; m.w. 164.20; f.p. -55 C; b.p. 249 C (760 mm); misc. with water and many org. solvs.; sp.gr. 1.048; dens. 8.74 lb/gal; visc. 7.5 cP; ref. index 1.4381; flash pt. (COC) 118 C

Poly-Solv® TPM [Olin]
Chem. Descrip.: PPG-3 methyl ether
Uses: Solvent for brake fluids, hard-surf. cleaners, leather dyeing, paints, coatings, printing inks, textile vat dyeing and printing, adhe-

sives, antifreeze, floor waxes/polishes, insect repellents; solubilizer for dyes; plasticizer
Properties: APHA 15 max. liq.; mild char. odor; m.w. 206.28; f.p. -78 C; b.p. 242 C (760 mm); misc. with water and many org. solvs.; sp.gr. 0.969; dens. 8.06 lb/gal; visc. 6.1 cP; ref. index 1.428; flash pt. (COC) 121 C

Polysolvan® O [Hoechst AG]
Chem. Descrip.: Glycolic acid-n-butyl ester
CAS 7397-62-8; EINECS 230-991-7
UN No.: 8027
Uses: Paint additive, flow agent for nitrocellulose lacquers, brush-applied paints, alkyd resin and oil paints; high-boiling solvent in stoving finishes; mfg. of printing inks
Properties: Colorless liq.; very faint ester-like odor; misc. with common org. solvs.; only partly misc. with water; dens. 1.015-1.023 g/cc (20 C); visc. 4.4 mPa·s (20 C); vapor pressure 1.33 mbar (20 C); f.p. ≈ -26 C; b.p. 82-97 C; flash pt. ≈ 75 C; ref. index 1.423-1.426 (20 C); sp. heat ≈ 2.1 kJ/kg·K (20 C); dielec. const. 13 (20 C); 97% min.act.
Toxicology: LD50 (oral, rat) 4595 mg/kg; irritant to mucous membranes; risk of serious eye damage
Environmental: Readily biodeg.

Polystab RB1483 [Akcros UK]
Chem. Descrip.: Aromatic alkoxyates blend
Uses: Solvent; reduces visc. of polyether polyols; gives improved in-mold flowout for PU mfg.
Properties: Liq.

PPG Perchlor [PPG Ind.]
Chem. Descrip.: Perchlorethylene
CAS 127-18-4; EINECS 204-825-9
Uses: Solvent avail. in dry-cleaning grade with unique stabilizer components that are completely removed from garments and leave no residual odor and which are recoverable with solvent after distilling; degreasing and general solvent grade for vapor degreasing, cold cleaning, or dry-cleaning applics.; heavy duty grade which is more heavily stabilized and intended for use in more rigorous vapor degreasing applics.
Properties: Water-wh. clear liq.; m.w. 165.85; f.p. -22.3 C; b.p. 121.1 C; sol. 0.015 g sol./100 g water; sp.gr. 1.623-1.628; dens. 13.57 lb/gal; visc. 0.88 cps; ref. index 1.5053

PPG Trichlor [PPG Ind.]
Chem. Descrip.: Trichlorethylene
CAS 79-01-6; EINECS 201-167-4
Uses: Solvent avail. in standard grade, degreasing and general solvent grade, stabilized for general, heavy-duty vapor degreasing and cold cleaning applics.; special grades are avail. for LOX flushing, vapor degreasing, and cleaning electronic components and synthesizing chemicals
Properties: Water-wh. clear liq.; sweet odor; m.w. 131.40; f.p. -86.4 C; b.p. 86.9 C; sol. 0.11 g sol./100 g water; sp.gr. 1.460-1.470; dens. 12.2 lb/gal; visc. 0.58 cps; ref. index 1.4782; pH 6.7-7.5

PrepRite™ Coatings Remover [ISP]
Chem. Descrip.: N-Methyl-2-pyrrolidone, butyrolactone, and other ingreds.
Uses: Solvent for one-step removal of paints, coatings, and residue from plant and maintenance equip., tools, etc.; also removes adhesives and resins (polyester, acrylic, urethane, vinyl ester, epoxy); low VOC emissions; replaces strippers contg. methylene chloride, phenol, alkali
Properties: Thick liq., sweet odor; water-sol.; sp.gr. 1.06; dens. 8.8 lb/gal; visc. 6000 cps (20 C); flash pt. (PMCC) 197 F; pH 4-5 (10 vol.%, 20 C)
Toxicology: TLV 100 ppm; low oral and inhalation toxicity; mod. dermal toxicity; mod. eye irritant; contains org. solvs.; avoid breathing vapors or mist
Precaution: Keep away from excessive heat and open flames

Primarol 1006 [Henkel Canada]
Chem. Descrip.: 2-Hexyl decanol-1;

EINECS 219-370-1
Uses: Extender and solvent for dyes and fragrance oils; intermediate for prod. of comps. for applics. in lubricants, emulsifiers, metal processing, and textiles
Properties: Water-wh. clear liq.; sol. in solvs.; flash pt. 154 C (COC); cloud pt. < -20 C; sapon. no. < 5
Unverified

Primarol 1107 [Henkel Canada]
Chem. Descrip.: C16-C20 2-alkyl alcohols
Uses: Extender and solvent for dyes and fragrance oils; intermediate for prod. of comps. for applics. in lubricants, emulsifiers, metal processing, and textiles
Properties: Water-wh. clear liq.; cloud pt. < 0 C
Unverified

Printsolve™ Ink Remover [ISP]
Chem. Descrip.: N-Methyl-2-pyrrolidone, dipropylene glycol methyl ether
Uses: Solvent cleaner for removal of water-and solv.-based inks, uv-cured inks, and heat-set inks; used in pulp/paper, flexographic/rotogravure, silk-screen printing, and graphics arts industries; safer alternative to xylene, acetates, aromatic hydrocarbons, naphthas, MEK, and alkaline cleaners
Properties: Water-sol.
Toxicology: TLV 100 ppm; low oral and inhalation toxicity; mod. dermal toxicity; mod. eye irritant

Procetyl 10 [Croda Inc; Croda Chem. Ltd]
Chem. Descrip.: PPG-10 cetyl ether
CAS 9035-85-2
Uses: Emollient, coupler, cosolvent, plasticizer, superfatting, wetting and spreading agent, penetrant; lubricant in cosmetics and personal care prods.; alcoholic and aq. alcoholic compositions
Properties: APHA 150 max. clear liq.; faint, char. sweet odor; sol. in min. oil, acetone, IPM, lanolin oil, alcohol; acid no. 1 max.; hyd. no. 80-100; pH 6.0-7.5 (3% disp.); 100% act.; nonionic
Toxicology: Nontoxic

Procetyl 30 [Croda Inc; Croda Chem. Ltd]
Chem. Descrip.: PPG-30 cetyl ether
CAS 9035-85-2
Uses: Emollient, coupler, cosolvent, plasticizer, superfatting, wetting and spreading agent, penetrant; lubricant in cosmetics and personal care prods.; alcoholic and aq. alcoholic compositions
Properties: APHA 150 max. clear liq.; faint, char. sweet odor; pH 6.0-7.5 (3% disp.); 100% act.

Procetyl 50 [Croda Inc; Croda Chem. Ltd]
Chem. Descrip.: PPG-50 cetyl ether
CAS 9035-85-2
Uses: Emollient, coupler, cosolvent, plasticizer, superfatting, wetting and spreading agent, penetrant; lubricant in cosmetics and personal care prods.; alcoholic and aq. alcoholic compositions
Properties: APHA 150 max. clear liq.; faint, char. sweet odor; sol. in PEG-200, lanolin oil, alcohol; acid no. 1 max.; hyd. no. 30-50; pH 6-7.5 (3% disp); 100% act.
Toxicology: Nontoxic

Progasol® SN [Rhône-Poulenc/Textiles]
Chem. Descrip.: Solv./surfactant blend
Uses: Solvent scouring agent, detergent
Properties: Liq.; 84% conc.
Discontinued prod.

Proglyde DMM [Dow]
Chem. Descrip.: Dipropylene glycol dimethyl ether
CAS 111109-77-4; EINECS 404-640-5
Uses: Solvent, coupling agent, coalescent, water remover for coatings (PU coatings, latexes, solv.-based), agric., resins, cleaning prods., electronics cleaners, paint removers; diluent for thixotropic systems; in extraction processes; formulation stabilizer; aprotic reaction medium for polymerization reactions
Properties: Colorless liq.; low odor; sol. 53 g/100 g water; m.w. 162; sp.gr. 0.898; visc. 1.1 cps; vapor

pressure 0.55 mm Hg (20 C); f.p. < -71 C; b.p. 175 C; flash pt. (PMCC) 65 C; ref. index 1.407; surf. tens. 26.8 dynes/cm; sp. heat 0.43 cal/g
Toxicology: Low acute toxicity by oral, dermal, or inh. routes; nonmutagenic; may cause sedation at very high dose levels and sl. to mod. but reversible skin or eye irritation
Environmental: Biodeg.

Promyr [Amerchol]
Chem. Descrip.: Isopropyl myristate
CAS 110-27-0; EINECS 203-751-4
Uses: Emollient and solvent for cosmetics, toiletries, makeups; nongreasy rub in
Properties: Water-wh. low visc. liq.; odorless; oil-sol.; acid no. 1.0 max.; sapon. no. 202-211

Propal [Amerchol]
Chem. Descrip.: Isopropyl palmitate
CAS 142-91-6; EINECS 205-571-1
Uses: Emollient and solvent for cosmetics, toiletries, makeups; nongreasy rub in
Properties: Water-wh. liq.; odorless; acid no. 1.0 max.; sapon. no. 182-191

Propsolv I-1 [Quantum]
Chem. Descrip.: 100 gal SDA 1-1, 1.0 gal rubber hydrocarbon solv., 4.25 gal ethyl acetate
Uses: Solvent for shellac, chemical specialties, latex coagulants

Propsolv I-2 [Quantum]
Chem. Descrip.: 100 gal SDA 1-2, 1.0 gal rubber hydrocarbon solv., 4.25 gal ethyl acetate
Uses: Solvent for shellac, chemical specialties, latex coagulants

Propsolv III-1 [Quantum]
Chem. Descrip.: 100 gal SDA 1-1, 1.0 gal MIBK, 1.0 gal rubber hydrocarbon solv., 0.87 gal ethyl acetate
Uses: Solvent for shellac, chemical specialties, latex coagulants

Propsolv III-2 [Quantum]
Chem. Descrip.: 100 gal SDA 1-2, 1.0 gal MIBK, 1.0 gal rubber hydrocarbon solv., 0.87 gal ethyl acetate
Uses: Solvent for shellac, chemical specialties, latex coagulants

Propyl Carbitol® [Union Carbide]
Uses: Solvent for water-disp. and water-sol. coatings
Properties: Sol. in water (20 C); m.w. 148.20; sp.gr. 0.969 (20/20 C); dens. 8.04 lb/gal (20 C); b.p. 215.8 C; flash pt. (CC) 200 F

Propyl Cellosolve® [Union Carbide]
Chem. Descrip.: Ethylene glycol propyl ether
CAS 2807-30-9; EINECS 220-548-6
Uses: Solvent for high-solids coatings (NC, vinyls, ethyl cellulose, butyrate, acrylic, water-sol.)
Properties: Sol. in water (20 C); m.w. 104.15; sp.gr. 0.912 (20/20 C); dens. 7.59 lb/gal (20 C); visc. 3.8 cP (20 C); vapor pressure 1.6 mm Hg (20 C); b.p. 151.5 C; flash pt. (CC) 120 F

Propyl Dipropasol [Union Carbide]
Uses: Solvent, coalescing aid for water-disp. and water-sol. coatings
Properties: Sol. 17% in water (20 C); m.w. 176.26; sp.gr. 0.930 (20/20 C); dens. 7.74 lb/gal (20 C); b.p. 211.3 C; flash pt. (CC) 190 F

Propylene Glycol USP/FCC Ultra Grade [Eastman]
Chem. Descrip.: Propylene glycol USP/FCC
CAS 57-55-6; EINECS 200-338-0
Uses: Solvent, humectant, preservative for food industry; plasticizer for cork seals for food pkg., in cellulose coatings for sausage and cheese; solvent for inks for food pkg.; lubricant for food pkg. equip.; antifreeze; solvent for org. chems. for medicine, water-sol. vitamins, dyes, perfumes; softening agent, spreader, emollient, humectant and preservative for cosmetics and pharmaceuticals; vehicle for drugs, ointments; dispensing aid in laundry detergents
Regulatory: FDA 21CFR §184.1666, GRAS; EPA exempt from pesticide tolerances; USDA accepted; kosher
Properties: Colorless clear visc. liq., essentially odorless, sl. char. taste; misc. with water, acetone, chloroform; sp.gr. 1.0351-1.0364; i.b.p. 186 C min.; 99.5% min. purity
Storage: Hygroscopic; storage in lined steel tank recommended; becomes

quite visc. at low temps

Propyl Propasol [Union Carbide]
Uses: Solvent for water-disp. and water-sol. coatings
Properties: Sol. in water (20 C); m.w. 116.18; sp.gr. 0.886 (20/20 C); dens. 7.36 lb/gal (20 C); b.p. 149.8 C; flash pt. (CC) 119 F

Prox-onic MG-020 P [Protex SA]
Chem. Descrip.: POP (20) methyl glucoside
CAS 61849-72-7
Uses: Cosolvent, perfume fixative
Properties: Liq.
Unverified

Punctilious® Ethyl Alcohol [Quantum]
Chem. Descrip.: Ethanol
CAS 64-17-5; EINECS 200-578-6
Uses: Solvent and extraction medium
Properties: Sol. in methanol; misc. in ether, water, chloroform

Punctilious® SDA 1-1 [Quantum]
Chem. Descrip.: Specially denatured ethanol
CAS 64-17-5; EINECS 200-578-6
Uses: Solvent or thinner; toilet prods. used externally, cosmetics, pharmaceuticals, detergents
Properties: Sol. in methanol; misc. in ether, water, chloroform

Purasolv® BL [Purac Am.]
Chem. Descrip.: Butyl lactate
CAS 138-22-7
Uses: Non-ozone depleting solvent for electronics, paints and coatings esp. waterborne and high solids systems, printing inks, cleaning agents, mold release agents; coalescing agent for aq. formulations; pharmaceutical intermediate; food additive
Regulatory: FDA approved GRAS flavor
Properties: Colorless clear liq., char. odor; partly misc. with water and most org. solvs.; m.w. 146; sp.gr. 0.975-0.985 (20 C); b.p. 180-195 C; flash pt. (CC) 79 C; ref. index 1.418-1.422; 97% min. assay
Environmental: Biodeg.

Purasolv® EL [Purac Am.]
Chem. Descrip.: Ethyl lactate
CAS 97-64-3; EINECS 202-598-0
Uses: Non-ozone depleting solvent for electronics, paints and coatings (solv.-based, high-solids systems, stripping formulations, as coalescing agent), printed circuit boards, metal industry; food additive
Regulatory: FDA approved
Properties: Completely misc. with water; m.w. 118; sp.gr. 1.03 (20 C); visc. 2.7 cP (20 C); vapor pressure 1.6 mbar (20 C); b.p. 154 C; flash pt. (CC) 55 C; surf. tens. 34 dynes/cm; KB value > 1000
Environmental: Biodeg.

Purasolv® ELECT [Purac Am.]
Chem. Descrip.: Ethyl lactate
CAS 97-64-3; EINECS 202-598-0
Uses: Photoresist grade precision cleaning solvent for wide variety of contaminants; non-ozone depleting solvent for electronics, paints/coatings esp. waterborne and high solids systems, printing inks, cleaning agents, mold release agents; food additive
Regulatory: FDA GRAS; FAO/WHO, JECFA approved
Properties: Colorless clear liq., char. mild pleasant odor; misc. with water and most org. solvs.; m.w. 118; sp.gr. 1.030-1.038; b.p. 309 F; flash pt. (CC) 52 C; ref. index 1.410-1.420; 99% min. assay
Toxicology: LD50 (oral) > 2000 mg/kg; immediately degrades into lactic acid and ethanol; sl. erythema on percutaneous applic. of 2068 mg/kg/day; possible sl. temporary eye irritation
Environmental: Readily biodeg.
Storage: Store in well-closed containers

Purasolv® ELS [Purac Am.]
Chem. Descrip.: Ethyl lactate
CAS 97-64-3; EINECS 202-598-0
Uses: Precision cleaning grade; non-ozone depleting solvent for electronics, paints and coatings esp. waterborne and high solids systems, printing inks, cleaning agents, mold release agents; food additive
Regulatory: FDA GRAS; FAO/WHO, JECFA approved

Properties: Colorless clear liq., char. mild pleasant odor; misc. with water and most org. solvs.; m.w. 118; sp.gr. 1.030-1.038; b.p. 309 F; flash pt. (CC) 52 C; ref. index 1.410-1.420; 98% min. assay
Toxicology: LD50 (oral) > 2000 mg/kg; immediately degrades into lactic acid and ethanol; sl. erythema on percutaneous applic. of 2068 mg/kg/day; possible sl. temporary eye irritation
Environmental: Readily biodeg.
Storage: Store in well-closed containers

Purasolv® IPL [Purac Am.]
Chem. Descrip.: Isopropyl lactate
CAS 617-51-6
Uses: Non-ozone depleting solvent for electronics, paints and coatings (solv.-based, high-solids systems, stripping formulations, as coalescing agent), printed circuit boards, metal industry
Properties: Colorless clear liq., char. odor; completely misc. with water and most org. solvs.; m.w. 132; sp.gr. 0.984-0.987 (20 C); visc. 2.9 cP (20 C); vapor pressure 1.0 mbar (20 C); b.p. 157 C; flash pt. (CC) 55 C; ref. index 1.407-1.409; surf. tens. 33 dynes/cm; KB value > 1000; 97% min. assay
Environmental: Biodeg.

Purasolv® ML [Purac Am.]
Chem. Descrip.: Methyl lactate
CAS 547-64-8; EINECS 208-930-0
Uses: Non-ozone depleting solvent for electronics, paints and coatings (solv.-based, high-solids systems, stripping formulations, as coalescing agent), printed circuit boards, metal industry
Properties: Colorless clear liq., char. odor; completely misc. with water and most org. solvs.; m.w. 104; sp.gr. 1.085-1.098 (20 C); visc. 2.9 cP (20 C); vapor pressure 3 mbar (20 C); b.p. 145 C; flash pt. (CC) 55 C; ref. index 1.410-1.415; surf. tens. 39 dynes/cm; KB value 62; 97% min. assay
Environmental: Biodeg.

2-Pyrol® [ISP]
Chem. Descrip.: 2-Pyrrolidone
CAS 616-45-5; EINECS 204-648-7
Uses: Monomer for polypyrrolidone; solvent for polymers, insecticides, polyhydroxylic alcohols, sugars, iodine, specialty inks; used in petrol. processing and acrylonitrile mfg.; solvent, solubilizer, intermediate for pharmaceuticals; plasticizer and coalescing agent for acrylic latices and acrylic-styrene copolymers in emulsion-type floor coatings; intermediate for N-methylol derivs., alkaloids related to peganine, amino-butyric acid derivs.
Properties: Liq.; solidfies below 25 C; misc. with water, ethanol, ethyl ether, chloroform, benzene, ethyl acetate, CS_2; insol. in aliphatic hydrocarbons; sp.gr. 0.11 (25/4 C); visc. 13.3 cp; m.p. 25 C; b.p. 245 C; flash pt. (OC) 130 C
Precaution: Noncorrosive; low fire hazard

Q

QO® Furfural [Great Lakes]
Chem. Descrip.: 2-Furaldehyde
CAS 98-01-1; EINECS 202-627-7
Uses: Chemical intermediate for mfg. of derivatives (furan and THF); solvent for separating sat. from unsat. compds. in petrol. lubricating oil, gas oil, and diesel fuel; extractive distillation of C4 and C5 hydrocarbons for the mfg. of syn. rubber; decolorizing agent for wood rosin; solvent and processing aid for anthracene; ingred. in resins; reactive solvent and wetting agent in abrasive wheels and brake linings
Properties: Colorless when freshly distilled; industrial furfural is lt. yel. to brn. liq.; pungent almond-like odor; m.w. 96.08; dens. 1.1545; visc. 1.49 cps; b.p. 161.7 C (760 mm); f.p. -36.5 C; TCC flash pt. 61.7 C; surf. tens. 40.7 dynes/cm (29.9 C); ref. index 1.5235

QO® Furfuryl Alcohol (FA®) [Great Lakes]
Chem. Descrip.: Furfuryl alcohol
CAS 98-00-0; EINECS 202-626-1
Uses: Used in the prod. of foundry sand binders and corrosion-resistant resins; intermediate for esterification and etherification; impregnating sol'n. and carbon binder; wood adhesive component; solvent and temporary plasticizer for phenolic resins in the mfg. of cold-molded abrasive wheels; visc. reducer, cure promoter, and carrier in amine-cured epoxy resins
Properties: Pale yel. liq.; sol. in water and many common org. solvs.; m.w. 98.10; sp.gr. 1.14 (20 C); dens. 9.45 lb/gal (20 C); visc. 5 cps; f.p. –14.6 C; b.p. 170 C; cloud pt. 10 C max.; flash pt. (TCC) 77 C; ref. index 1.4869; 98% min. assay

QO® Tetrahydrofuran (THF) [Great Lakes]
Chem. Descrip.: Tetrahydrofuran
CAS 109-99-9; EINECS 203-726-8
Uses: Industrial solvent with high solvency for broad range of materials and low boiling pt.; also used in the prod. of heterocyclic and open-chain compds.; chemical intermediate
Properties: Colorless liq.; misc. with water; b.p. 66 C

QO® Tetrahydrofurfuryl Alcohol (THFA®) [Great Lakes]
Chem. Descrip.: Tetrahydrofurfuryl alcohol
CAS 97-99-4; EINECS 202-625-6
Uses: High boiling solvent and carrier for pesticides; FDA approved for use in paper processing; chemical intermediate; also in industrial and consumer cleaners, leather and textile dyeing, epoxies, coatings, inks, paints, and adhesives; plasticizer and vinyl stabilizer carrier
Regulatory: FDA compliance
Properties: Colorless liq.; sol. in water; m.w. 102.13; sp.gr. 1.05 (20 C); dens. 8.79 lb/gal (20 C); visc. 6.24 cps (20 C); f.p. < -80 C; b.p. 178 C; flash pt. (TCC) 74 C; ref. index 1.4520 (20 C)
Environmental: Biodeg.

R

Renex® 648 [ICI Am.; ICI Surf. Belgium]
Chem. Descrip.: Nonoxynol-5
CAS 9016-45-9; EINECS 247-555-7
Uses: Detergent, solvent and emulsion cleaning applications; metal cleaning; personal care applics.
Properties: Yel. clear liq.; sol. in propylene glycol, perchlorethylene, IPA, cottonseed oil; sp.gr. 1.03; visc. 223 cps; HLB 10.5; flash pt. > 200 F; pour pt. -20 F; 100% act.; nonionic
Unverified

Renex® PEG 300 [ICI Surf. Am.; ICI Surf. UK]
Chem. Descrip.: PEG-6 USP/NF
CAS 25322-68-3; EINECS 220-045-1
Uses: Surfactant, solvent for personal care, pharmaceutical, industrial, and agrochem. applics.
Regulatory: USP/NF and EP compliance
Properties: Colorless liq.; m.w. 300; 100% conc.

Renex® PEG 400 [ICI Surf. Am.; ICI Surf. UK]
Chem. Descrip.: PEG-8 USP/NF
CAS 25322-68-3; EINECS 225-856-4
Uses: Surfactant, solvent for personal care, pharmaceutical, industrial, and agrochem. applics.
Regulatory: USP/NF and EP compliance
Properties: Colorless liq.; m.w. 400; 100% conc.

Renex® PEG 600 [ICI Surf. Am.; ICI Surf. UK]
Chem. Descrip.: PEG-12 USP/NF
CAS 25322-68-3; EINECS 229-859-1
Uses: Surfactant, solvent for personal care, pharmaceutical, industrial, and agrochem. applics.
Regulatory: USP/NF and EP compliance
Properties: Wh. wax; m.w. 600; 100% conc.

Renex® PEG 1000 [ICI Surf. Am.; ICI Surf. UK]
Chem. Descrip.: PEG-20 USP/NF
CAS 25322-68-3; EINECS 203-989-9
Uses: Surfactant, solvent for personal care, pharmaceutical, industrial, and agrochem. applics.
Regulatory: USP/NF and EP compliance
Properties: Wh. solid; m.w. 1000; 100% conc.

Renex® PEG 1500 [ICI Surf. Am.; ICI Surf. UK]
Chem. Descrip.: PEG USP/NF
Uses: Surfactant, solvent for cosmetics and pharmaceutical prods.
Regulatory: USP/NF and EP compliance
Properties: Wh. solid; m.w. 1500; 100% conc.

Renex® PEG 4000 [ICI Surf. Am.; ICI Surf. UK]
Chem. Descrip.: PEG
Uses: Surfactant, solvent for cosmetics and pharmaceuticals
Regulatory: USP/NF and EP compliance
Properties: Wh. solid; m.w. 4000; 100% conc.; nonionic

Renex® PEG 6000 [ICI Surf. Am.; ICI Surf. UK]
Chem. Descrip.: PEG
Uses: Surfactant, solvent for cosmetics and pharmaceuticals
Regulatory: USP/NF and EP compliance
Properties: Wh. solid; m.w. 6000; 100% conc.; nonionic

Renex® PEG 8000 [ICI Surf. Am.; ICI

Surf. UK]
Chem. Descrip.: PEG
Uses: Surfactant, solvent for cosmetics and pharmaceuticals
Regulatory: USP/NF and EP compliance
Properties: Wh. solid; m.w. 8000; 100% conc.; nonionic

Reomol® TBP [FMC]
Chem. Descrip.: Tributyl phosphate
CAS 126-73-8; EINECS 204-800-2
Uses: Defoaming agent for drilling muds, cements, fracturing fluids, plaster compositions, paper coatings, pulp bleaching, mold coating, glass luminescent coatings, aq. emulsion paints, adhesives, mercerizing liquors/dye baths; antifreeze sol'ns., fermentation, detergents; primary plasticizer and solvent; pigment dispersant
Properties: Liq.; sol. in min. oil, gasoline, most org. liqs.; sol. 0.1% in water; insol. in glycerine, glycols; sp.gr. 0.97-0.983 (20/20 C); dens. 8.14 lb/gal (20 C); visc. 3.7 cp (20 C); vapor pressure 7.3 mm Hg (150 C); f.p. < -80 C; b.p. 137-145 C (4 mm Hg); pour pt. < 80 C; flash pt. 166 C; fire pt. 182 C; surf. tens. 29 dynes/cm (20 C); < 0.1% moisture; 7.4% P
Toxicology: Severe skin irritant
Precaution: Low degree of fire hazard

Reomol® TOP [FMC]
Chem. Descrip.: Trioctyl phosphate
CAS 78-42-2; EINECS 201-116-6
Uses: Low temp. plasticizer for PVC, rubber, paints, coatings; highly efficient solvent

Rewopal® MPG 10 [Witco Surf. GmbH]
Chem. Descrip.: Phenoxyethanol
CAS 122-99-6; EINECS 204-589-7
Uses: Solvent, solubilizer, preservative
Properties: Liq.; 100% conc.; nonionic

Rewopal® MPG 40 [Witco Surf. GmbH]
Chem. Descrip.: Phenol polyglycol ether
CAS 122-99-6; EINECS 204-589-7
Uses: Solvent
Properties: Liq.; 100% conc.; nonionic

Rexoclean [Emkay]
Chem. Descrip.: Syn. detergents, terpenes, and solvs.
Uses: Complete scouring agent for stubborn oil and grease stains and for removal of nylon warp size; 85% act.
Unverified

Rexoclean 200N [Emkay]
Chem. Descrip.: Emulsified blend of solvs. and detergents
Uses: Multipurpose cleaner for heavy duty cleaning incl. rubber marks and stains, in soap dispensers
Unverified

Rexoclean GA [Emkay]
Uses: Solvent scour, detergent for low or high pH applics, bleaching, scouring, and cleansing obstinate fatty soils

Rexoclean JA [Emkay]
Chem. Descrip.: Emulsified solvs./detergent blend
Uses: Scouring agent for oil and grease soils and fatty sizing
Properties: Liq.; 35% conc.; anionic

Rexoclean NB Conc. [Emkay]
Chem. Descrip.: Mixt. of solvs., emulsifiers, and detergents
Uses: Can be cut 50-50 with water to make an efficient blanket cleaner
Unverified

Rexosolve [Emkay]
Chem. Descrip.: Blend of petrol. solvs. and terpenes
Uses: Cleaner for oil or grease soils, dirty knots, etc. in textile industry
Properties: Sol. in water

Rexosolve 150 [Emkay]
Chem. Descrip.: Sulfonated oils/emulsifier blend
Uses: Detergent, scouring assistant; oil and grease remover
Properties: Liq.; easily disp. in water; 100% act.; anionic

Rexosolve BCT [Emkay]
Uses: Dewaxing agent and solvent scour; used with caustic for cleaning rubber print blankets and to remove metallic pigment dispersions; leveling agent for overdyed Dacron; stripper on many colors

Rexosolve CR [Emkay]
Chem. Descrip.: Emulsified blend of solvs. and cresylic acid
Uses: Wool scouring agent; for re-

moving min. oil

Rexosolve EP, EPN [Emkay]
Chem. Descrip.: Blend of aromatic solvs.
Uses: Cleaners for oily and greasy soils; recommended for Burlington machines
Properties: Misc. with hot or cold water to produce highly stable emulsions

Rexosolve GA, GAX [Emkay]
Uses: Solvent scours for bleach, scour and dye baths to remove fatty soils; stable in acid and alkaline baths

Rexosolve HP [Emkay]
Chem. Descrip.: Conc. cleaner containing high flash solvs.
Uses: Cleaner for removing stubborn tars and greases

Rexosolve MS [Emkay]
Uses: Solvent scour used as general cleaner and washing compd. for back grays; softens and removes dried pigment; usually used with 2.5% caustic at 140 F; 100% act.

Rexosolve OA, OAC [Emkay]
Uses: Solvent scours for cleaning stains, rust and grease in one bath; compat. with oxalic acid

Rexosolve RF-2 [Emkay]
Chem. Descrip.: Blend of org. solvs. with syn. emulsifiers and detergents
Uses: Cleaner for removal of stubborn tars, greases and stains
Properties: Lt. free-flowing liq.; disp. readily in water to give stable emulsion

Rexosolve SL [Emkay]
Uses: Stain and soil remover for use in neutral or alkaline baths; may be used in the dye liquor

Rhodacal® 330 [Rhône-Poulenc/ Deterg. & Personal Care]
Chem. Descrip.: Branched isopropylamine dodecylbenzene sulfonate
CAS 26264-05-1; EINECS 247-556-2
Uses: Emulsifier, wetting agent, grease/pigment dispersant, lubricant, solubilizer, solvent, penetrant, high foaming base for shampoos and cleaners, metalworking fluids, agric. pesticides, emulsion polymerization, dry-cleaning, latex paints, metal cleaning; acid aluminum cleaners; oil slick emulsifier
Properties: Amber liq.; oil-sol.; sp.gr. 1.03; dens. 8.497 lb/gal; visc. 6500 cps; HLB 11.7; flash pt. > 200 C; pH 5-6.8 (5%); 90% act.; anionic

Rhodasurf® PEG 300 [Rhône-Poulenc Surf. & Spec.]
Chem. Descrip.: PEG-6
CAS 25322-68-3; EINECS 220-045-1
Uses: Intermediate for surfactants; binder/lubricant in pharmaceuticals; plasticizer; paper softener; humectant; solvent; antistat; for cosmetics, textile, plastics processing, dyes and inks
Properties: Liq.; m.w. 285-315; water-sol.; dens. 1.13 g/ml; pH 5-8 (5% DW); 100% conc.

Rhodasurf® PEG 400 [Rhône-Poulenc Surf. & Spec.]
Chem. Descrip.: PEG-9
CAS 25322-68-3; EINECS 222-206-1
Uses: Intermediate, plasticizer, solvent, lubricant, coupler in cosmetic lotions
Properties: Wh. liq.; water-sol.; m.w. 380-420; sp.gr. 1.13; f.p. 4-8 C; pH 5-8; 100% conc.
Discontinued prod.

Rhodasurf® PEG 600 [Rhône-Poulenc Surf. & Spec.]
Chem. Descrip.: PEG-14
CAS 25322-68-3
Uses: Plasticizer, solvent; lubricant, binder for pharmaceuticals; color stabilizer for fuel oils; intermediate for processing surfactants
Properties: Wh. liq.; m.w. 570-630; water sol.; sp.gr. 1.13 (30/15.5 C); f.p. 20-25 C; pH 5-8; 100% conc.

Rhodasurf® PEG 4000 [Rhône-Poulenc Surf. & Spec.]
Chem. Descrip.: PEG 4000
Uses: Solvent or cosolvent for industrial applics. where lubricity, water sol., low toxicity, and chemical stability are important
Properties: Solid; 100% conc.
Discontinued prod.

Ricinion [Gattefosse]
Chem. Descrip.: PEG-33 castor oil
CAS 61791-12-6

Uses: Solvent, emulsifier for cosmetic/pharmaceutical emulsions
Properties: Liq.; HLB 12-13; acid no. 30-40; sapon. no. 60-80
Toxicology: LD0 (oral, rat) > 10 ml/kg; nonirritating to skin and eyes

Ritalafa® 5 [R.I.T.A.]
Chem. Descrip.: PEG-5 lanolate
CAS 68459-50-7
Uses: Emulsifier, solvent
Discontinued prod.

Ritalafa® 10 [R.I.T.A.]
Chem. Descrip.: PEG-10 lanolate
CAS 68459-50-7
Uses: Emulsifier, solvent
Discontinued prod.

Ritawax AEO [R.I.T.A.]
Chem. Descrip.: Polysorbate 80, acetylated lanolin alcohol, cetyl acetate
Uses: Emollient, lubricant, moisturizer, penetrant, solubilizer, dispersant, plasticizer, cosolvent, emulsifier for personal care prods.; hair and skin softening and conditioning; vehicle for medicaments; degreases and detackifies emulsion systems
Properties: Lemon yel. to straw-colored clear oily liq., char. bland odor; sol. in water, IPA, castor oil, min. oil, veg. oil; acid no. 2 max.; sapon. no. 65-80; hyd. no. 60-70; 100% conc.; nonionic

Ritawax ALA [R.I.T.A.]
Chem. Descrip.: Cetyl acetate and acetylated lanolin alcohol
Uses: Emollient, lubricant, moisturizer, penetrant, plasticizer, cosolvent, solubilizer, emulsifier; hair and skin softening and conditioning; vehicle for medicaments; degreases and detackifies emulsion systems
Properties: Lemon yel. to straw-colored oily liq., char. bland odor; sol. in min. oil, castor oil, veg. oil, alcohol; acid no. 1 max.; iodine no. 10 max.; sapon. no. 180-200; hyd. no. 8 max.; 100% conc.; nonionic

RS Naphtha [Exxon]
Chem. Descrip.: Hydrocarbon solv.
CAS 64742-49-0
Uses: Solvent
Properties: Sp.gr. 0.69 (15.6/15.6 C); dens. 5.75 lb/gal (60 F); b.p. 66-102 C; flash pt. (TCC) < -18 C; surf. tens. 19.1 dynes/cm; KB value 34; 89% paraffin, 2% cycloparaffin, 9% aromatics
Toxicology: OEL 100 ppm

S

Santosol™ DMA [Monsanto]
Chem. Descrip.: Dimethyl adipate
CAS 627-93-0; EINECS 211-020-6
Uses: Solvent used in coil, container, and automotive coatings, printing inks, acrylic lacquers, enamels, surf. primers for epoxies and polyesters, paint strippers, foundry core binders, resin cleanup, intermediates, industrial cleaners, wire enamel resins; offers VOC compliance, noncorrosive
Properties: APHA 50 max.; sol. in alcohols, esters, glycol ethers, halogenated solvs., ketones, hydrocarbons; sol. 2.1% in water; m.w. 174; sp.gr. 1.057; dens. 8.82 lb/gal; visc. 4.33 cps; f.p. 9.4 C; b.p. 216-230 C; acid no. 0.1; flash pt. 124 C; surf. tens. 35.1 dynes/cm; ≥ 99.0% act.; 0.1% max. water content
Toxicology: May cause eye and respiratory tract irritation
Environmental: Biodeg.
Storage: May be stored/used with mild steel containers and equip.; no special precautions

Santosol™ DME [Monsanto]
Chem. Descrip.: 55-75% Dimethyl glutarate, 15-25% dimethyl succinate, 10-25% dimethyl adipate
Uses: Solvent used in coil, container, and automotive coatings, printing inks, acrylic lacquers, enamels, surf. primers for epoxies and polyesters, paint strippers, foundry core binders, resin cleanup, intermediates, industrial cleaners, wire enamel resins; offers VOC compliance, noncorrosive
Properties: APHA 50 max.; sol. in alcohols, esters, glycol ethers, halogenated solvs., ketones, hydrocarbons; sol. 5.5% in water; m.w. 160; sp.gr. 1.086; dens. 9.06 lb/gal; visc. 3.88 cps; f.p. -40.2 C; b.p. 200-220 C; acid no. 0.1; flash pt. 102 C; surf. tens. 35.4 dynes/cm; 0.1% max. water content
Toxicology: May cause eye and respiratory tract irritation
Environmental: Biodeg.
Storage: May be stored/used with mild steel containers and equip., no special precautions

Santosol™ DME-2 [Monsanto]
Chem. Descrip.: 72-76% Dimethyl glutarate, 23-27% dimethyl adipate, ≤ 1.0% dimethyl succinate
Uses: Used in coil, container, and automotive coatings, printing inks, acrylic lacquers, enamels, surf. primers for epoxies and polyesters, paint strippers, foundry core binders, resin cleanup, intermediates, industrial cleaners, and wire enamel resins; offers VOC compliance, noncorrosive
Properties: APHA 50 max.; sol. in alcohols, esters, glycol ethers, halogenated solvs., ketones, hydrocarbons; sol. 4.3% in water; m.w. 163; sp.gr. 1.076; dens. 8.98 lb/gal; visc. 3.31 cps; f.p. -42.4 C; b.p. 203-220 C; acid no. 0.1; flash pt. 108 C; surf. tens. 35 dynes/cm; 0.1% max. water content
Toxicology: May cause eye and respiratory tract irritation
Environmental: Biodeg.
Storage: May be stored/used with mild steel containers and equip., no special precautions

Santosol™ DME-3 [Monsanto]

Chem. Descrip.: 87-91% Dimethyl adipate, 8-12% dimethyl glutarate, ≤ 1.0% dimethyl succinate
Uses: Used in coil, container, and automotive coatings, printing inks, acrylic lacquers, enamels, surf. primers for epoxies and polyesters, paint strippers, foundry core binders, resin cleanup, intermediates, industrial cleaners, and wire enamel resins; offers VOC compliance, noncorrosive
Properties: APHA 50 max.; sol. in alcohols, esters, glycol ethers, halogenated solvs., ketones, hydrocarbons; sol. 2.6% in water; m.w. 172; sp.gr. 1.058; dens. 8.83 lb/gal; visc. 3.71 cps; f.p. 2.8 C; b.p. 211-229 C; acid no. 0.1; flash pt. 113 C; surf. tens. 32.5 dynes/cm; 0.1% max. water content
Toxicology: May cause eye and respiratory tract irritation
Environmental: Biodeg.
Storage: May be stored/used with mild steel containers and equip., no special precautions

Santosol™ DMG [Monsanto]
Chem. Descrip.: Dimethyl glutarate
CAS 1119-40-0; EINECS 214-277-2
Uses: Used in coil, container, and automotive coatings, printing inks, acrylic lacquers, enamels, surf. primers for epoxies and polyesters, paint strippers, foundry core binders, resin cleanup, intermediates, industrial cleaners, and wire enamel resins; offers VOC compliance, noncorrosive
Properties: APHA 50 max.; sol. in alcohols, esters, glycol ethers, halogenated solvs., ketones, hydrocarbons; sol. 5.1% in water; m.w. 160; sp.gr. 1.082; dens. 9.03 lb/gal; visc. 3.66 cps; f.p. -37.5 C; b.p. 203-214 C; acid no. 0.1; flash pt. 107 C; surf. tens. 35.6 dynes/cm; ≥ 99.0% act.; 0.1% max. water content
Toxicology: May cause eye and respiratory tract irritation
Environmental: Biodeg.
Storage: May be stored/used with mild steel containers and equip., no special precautions

Santosol™ DMS [Monsanto]
Chem. Descrip.: Dimethyl succinate
CAS 106-65-0; EINECS 203-419-9
Uses: Used in coil, container, and automotive coatings, printing inks, acrylic lacquers, enamels, surf. primers for epoxies and polyesters, paint strippers, foundry core binders, resin cleanup, intermediates, industrial cleaners, and wire enamel resins; offers VOC compliance, noncorrosive
Properties: APHA 50 max.; sol. in alcohols, esters, glycol ethers, halogenated solvs., ketones, hydrocarbons; sol. 10.3% in water; m.w. 146; sp.gr. 1.112; dens. 9.28 lb/gal; visc. 3.91 cps; f.p. 16.8 C; b.p. 192-201 C; acid no. 0.1; flash pt. 94 C; surf. tens. 34.6 dynes/cm; ≥ 99.0% act.; 0.1% max. water content
Toxicology: May cause eye and respiratory tract irritation
Environmental: Biodeg.
Storage: May be stored/used with mild steel containers and equip., no special precautions

Schercemol 1818 [Scher]
Chem. Descrip.: Isostearyl isostearate
CAS 41669-30-1; EINECS 255-485-3
Uses: Substantive emollient imparting luxurious softness to skin; used in cosmetics imparting slip and lubricity, luster, and sheen; cosolvent and solubilizer in perfumes
Properties: Yel. clear liq., slight typ. odor; sol. in org. solvs., insol. in water; m.w. 536; sp.gr. 0.865-0.875; f.p. -5 C; acid no. 2.0 max.; iodine no. 13 max.; sapon. no. 95-110; cloud pt. 17 C; flash pt. (OC) > 180 C; ref. index 1.4610

Schercemol CO [Scher]
Chem. Descrip.: Cetyl octanoate
CAS 59130-69-7; EINECS 261-619-1
Uses: Solvency properties for use in make-up removers
Properties: Clear liq.; f.p. 10 C; acid no. 3 max.; iodine no. nil; sapon. no. 140-155

Schercemol DIA [Scher]
Chem. Descrip.: Diisopropyl adipate
CAS 6938-94-9; EINECS 248-299-9
Uses: Nonoily penetrating emollient,

lubricant, and solvent with mild drying effects used in hydro-alcoholic cosmetic formulations

Properties: Water-wh. clear liq., faint ester odor; sol. in alcohols, higher glycols, ketones, ester, aromatic, chlorinated, and aliphatic hydrocarbons; insol. in water; m.w. 230; sp.gr. 0.960 ± 0.01; dens. 8.0 lb/gal; f.p. -1 C; b.p. 87-89 C; acid no. 2 max.; iodine no. nil; sapon. no. 480-500; flash pt. > 170 C; ref. index 1.423 ± 0.001

Schercemol DICA [Scher]

Chem. Descrip.: Diisocetyl adipate

CAS 57533-90-1; EINECS 260-799-9

Uses: Nonoily emollient and spreading agent for cosmetics; release and anticlogging additive in aerosols due to strong plasticizing action; film spreader for bath oils; solvent for lipstick colors

Properties: Gardner 2.0 max. clear liq.; slight odor; m.w. 594; f.p. < -10.0 C; sol. see Schercemol CM; sp.gr. 0.885 ± 0.005; dens. 7.4 lb/gal; acid no. 2.0 max.; sapon. no. 170-185; ref. index 1.4538 ± 0.001; flash pt. > 180 C (OC)

Discontinued prod.

Schercemol DIS [Scher]

Chem. Descrip.: Diisopropyl sebacate

CAS 7491-02-3; EINECS 231-306-4

Uses: Nonoily emollient, lubricant, solubilizer with mild drying effects used in hydro-alcoholic personal care prods.; solvent and coupling properties; fast spreading action

Properties: Colorless clear liq., bland odor; sol. in alcohols, higher glycols, ketones, esters, aromatic, aliphatic, and chlorinated hydrocarbons, min. and natural oils; insol. in water; m.w. 286; sp.gr. 0.932 ± 0.01; dens. 7.8 lb/gal; f.p. 0 C; acid no. 1.0 max.; iodine no. nil; sapon. no. 380-400; flash pt. > 170 C; ref. index 1.4320; 99% min. ester conc.

Schercemol IPM [Scher]

Chem. Descrip.: IPM

CAS 110-27-0; EINECS 203-751-4

Uses: Emollient and film spreader used in personal care prods.; solvent and solubilizer for additives in cosmetic preparations; nonoily

Properties: Water wh. clear liq.; odorless; m.w. 270; f.p. -4.0 C; sol. in org. solvs., insol. in water; sp.gr. 0.850; acid no. 1.0 max.; sapon. no. 202-212; ref. index 1.4330-1.4350; flash pt. > 170 C (OC); 99% min. ester conc.

Discontinued prod.

Schercemol NGDC [Scher]

Chem. Descrip.: Neopentyl glycol dicaprate

CAS 27841-06-1

Uses: Solvency properties for use in make-up removers

Properties: Clear liq.; f.p. 2 C; acid no. 3 max.; iodine no. nil; sapon. no. 255-270

Schercemol NGDO [Scher]

Chem. Descrip.: Neopentyl glycol dioctanoate

CAS 28510-23-8

Uses: Low freeze pt. emollient; solvent for makeup remover

Properties: Clear liq.; f.p. < -12 C; acid no. 3 max.; iodine no. nil; sapon. no. 290-310

Unverified

Schercemol OLO [Scher]

Chem. Descrip.: Oleyl oleate

CAS 3687-45-4; EINECS 222-980-4

Uses: Nonoily emollient for cosmetic formulations contributing luster, softness, and high degree of lubricity in skin and hair preparations; cosolvent and solubilizer in perfumes; lubricant for metalworking and wire drawing

Properties: Amber clear liq., mild oleic odor; sol. in org. solvs., insol. in water; m.w. 332; sp.gr. 0.860 ± 0.01; dens. 7.2 lb/gal; acid no. 2 max.; iodine no. 95 max.; sapon. no. 95-110; cloud pt. 13 C; flash pt. (OC) > 180 C; ref. index 1.4630 ± 0.001

Schercemol PGDP [Scher]

Chem. Descrip.: Propylene glycol dipelargonate

CAS 41395-83-9; EINECS 255-350-9

Uses: Emollient offering low f.p.; cosolvent for perfumed bath oils, creams, and lotions

Properties: Straw-colored clear liq.,

slight typ. odor; sol. in org. solvs., insol. in water; m.w. 360; sp.gr. 0.917; dens. 7.6 lb/gal; f.p. -25 C; acid no. 5 max.; iodine no. nil; sapon. no. 300-320; flash pt. (OC) > 170 C; ref. index 1.440; nonionic

Schercemol PGML [Scher]
Chem. Descrip.: Propylene glycol laurate
CAS 27194-74-7; EINECS 205-542-3
Uses: Emollient and solvent; stable base for cosmetics; emulsion stabilizer; solubilizes and couples ingred. such as perfumes, coloring and flavoring agents, sunscreen compds. into natural fatty veg. or min. oils; solvent for sparyed org. pesticides; produces sprayable oils which spread well, have good adherence, and are not readily removed by rainfall; plasticizer and stabilizer in vinyl copolymers made from PVAc and PVC; defoaming agent in PVAc emulsions
Properties: Yel. clear liq., mild odor; sol. in most org. solvs. such as alcohols, ketones, esters, glycol ethers, veg. oil, min. oil, aliphatic, aromatic, chlorinated hydrocarbons; disp. in glycols, triols, polyols; m.w. 258 (theoret.); sp.gr. 0.905 ± 0.01; dens. 7.45 lb/gal; f.p. 10 C; acid no. 5 max.; iodine no. 1 max.; sapon no. 225-240; flash pt. (OC) 160 C min.; pH 7.0 (10% disp.); nonionic

Schercophos NP-9 [Scher]
Chem. Descrip.: Nonoxynol-9 phosphate
EINECS 266-231-6
Uses: Low to moderate foam liq. detergent and emulsifier for heavy-duty all-purpose liq. detergents; solvent cleaner for dry-cleaning
Properties: Liq.; 100% conc.; anionic
Discontinued prod.

Scour KSV, KSV Special [Catawba-Charlab]
Uses: Emulsified solvent containing stain remover and wetting and scouring agent
Discontinued prod.

Scour NFP [Sybron]
Uses: High flash pt. solvent scour for textile roller cleaner, screen clear applics.

Secoster® A [Stepan Europe]
Chem. Descrip.: Fatty acid ethoxylate
Uses: Dispersant, solvent and oil emulsifier for household/industrial cleaners, oilfield prod.
Properties: Yel. to brn. liq.; 100% act.; nonionic

Secoster® DO 600 [Stepan Europe]
Chem. Descrip.: PEG 600 dioleate
CAS 9005-07-6; EINECS 288-459-5
Uses: Additive for cutting oils; solvent, emulsifier for solvs. and oils, creams, cleansing milks, pesticides, textile lubricants, oilfield prod.; dispersant; for household/industrial cleaners
Properties: Pale yel. to amber liq.; 100% act.; nonionic

Secoster® MA 300 [Stepan Europe]
Chem. Descrip.: PEG 300 abietate
Uses: Additive for cutting oils; solvent, emulsifier for solvs. and oils, creams, cleansing milks and pesticides; dispersant; for household/industrial cleaners
Properties: Brn. liq.; 100% act.; nonionic

Secoster® MO 400 [Stepan Europe]
Chem. Descrip.: PEG 400 oleate
CAS 9004-96-0
Uses: Lubricant, antistat, solvent and oil emulsifier; for household/industrial cleaners, oilfield prod.
Properties: Pale yel. to amber liq.; 100% act.; nonionic

Sepasolv® MPE [BASF AG]
Uses: For purification of natural gas
Discontinued prod.

Sesolvan® L [BASF AG]
Uses: Dye solvent for textile dyeing

Shell TS-28 [Shell]
Chem. Descrip.: Aromatic hydrocarbon
Uses: Solvent for dry-cleaning, coatings, automotive, and chemical specialties
Properties: Sp.gr. 0.852 (60 F); dens. 7.09 lb/gal (60 F); flash pt. (CC) 122 F

Shell VM&P Naphtha EC [Shell]
Uses: Solvent for dry-cleaning, coatings, automotive, and chemical

specialties; low boiling, fast evaporating
Properties: Sp.gr. 0.750 (60 F); dens. 6.24 lb/gal (60 F); flash pt. (CC) 55 F
Unverified

Shell VM&P Naphtha HT [Shell]
Chem. Descrip.: VM&P naphtha
Uses: Hydrotreated solvent; hydrotreating removes unwanted compds. (e.g., sulfur, nitrogen, oxygen), converts unsat. hydrocarbons (olefins, acetylenes, aromatics), results in relatively low odor; no more than trace amts. of HAPS; fast/med.-evaporating, high solvency solvent for agric. oil/fat extraction, polishes, industrial cleaners, adhesives, inks, paints, lighter fluids, pharmaceuticals, paper/textile coatings, petrol. gas absorp., asphalt, resins, plastics, rubber
Regulatory: SARA nonreportable
Properties: Sp.gr. 0.752 (60/60 F); vapor pressure 10 mm Hg (68 F); b.p. 247-282 F; flash pt. 55 F; KB value 35; VOC 6.19 lb/gal; 54 vol. % paraffins, 48 vol. % cycloparaffins

Shell Cyclo Sol 42 [Shell]
Chem. Descrip.: Aromatic
Uses: Solvent
Properties: Sp.gr. 0.8532 (60 F); dens. 7.103 lb/gal (60 F); b.p. 321-373 F; flash pt. (TCC) 113 F; KB value 76

Shell Cyclo Sol 53 [Shell]
Chem. Descrip.: Predominantly C_9 aromatic hydrocarbon solv.
Uses: Narrow boiling range, very high solvency, slow-evaporating solvent used in dry-cleaning, paints, coatings, automotive and chemical specialties, insecticides, herbicides, additives, industrial cleaners, inks, asphalts, wood preservatives; solvent for catalysts; low sulfur content; exc. for toluene/xylene replacement
Properties: Sp.gr. 0.874 (60 F); dens. 7.27 lb/gal (60 F); b.p. 30 F; flash pt. (CC) 111 F; KB value 92; VOC 7.19 lb/gal; 99.4% aromatics

Shell Cyclo Sol 63 [Shell]
Chem. Descrip.: C_9 -C_{11} aromatic hydrocarbon solv.
Uses: Solvent used in dry-cleaning, coatings, automotive and chemical specialties, insecticides, wood preservatives, industrial cleaning, inks, petrol. drilling, catalytic and reaction medium, textile printing, asphalt prods.
Properties: Sp.gr. 0.887 (60 F); dens. 7.39 lb/gal (60 F); flash pt. (CC) 128 F; KB value 89; VOC 7.35 lb/gal; 97.5% aromatics

Shell Mineral Spirits 135 [Shell]
Chem. Descrip.: Min. spirits
EINECS 232-453-7
Uses: Hydrocarbon solvent for dry-cleaning, coatings, adhesives, agric., automotive, and chemical specialties, cleaning and polishing compds., pharmaceuticals, inks, textile coatings, asphalt, concrete curing
Properties: Sp.gr. 0.783 (60 F); dens. 6.52 lb/gal (60 F); flash pt. (CC) 110 F

Shell Mineral Spirits 145-EC [Shell]
Chem. Descrip.: Min. spirits
EINECS 232-453-7
Uses: Hydrocarbon solvent for dry-cleaning, coatings, automotive, and chemical specialties
Properties: Sp.gr. 0.793 (60 F); dens. 6.60 lb/gal (60 F); flash pt. (CC) 103 F

Shell Mineral Spirits 146 HT [Shell]
Chem. Descrip.: Min. spirits
EINECS 232-453-7
Uses: Hydrotreated solvent; hydrotreating removes unwanted compds. (e.g., sulfur, nitrogen, oxygen), converts unsat. hydrocarbons (olefins, acetylenes, aromatics), results in relatively low odor; no more than trace amts. of HAPS; relatively high solvency; slow-evaporating solvent for paints, coatings, dry-cleaning, commercial and industrial cleaners, insecticides, lighter fluids, resins, plastics, adhesives, polishes, pharmaceuticals, inks, textile coating, asphalt, concrete
Regulatory: SARA nonreportable
Properties: Flash pt. 108 F; KB value 32; VOC 6.36 lb/gal

Shell Mineral Spirits 200 HT [Shell]
Chem. Descrip.: Min. spirits
EINECS 232-453-7
Uses: Hydrotreated solvent; hydrotreating removes unwanted compds. (e.g., sulfur, nitrogen, oxygen), converts unsat. hydrocarbons (olefins, acetylenes, aromatics), results in relatively low odor; no more than trace amts. of HAPS; relatively high solvency; slow-evaporating solvent for paints, coatings, dry-cleaning, commercial and industrial cleaners, insecticides, lighter fluids, resins, plastics, adhesives, polishes, pharmaceuticals, inks, textile coating, asphalt, concrete
Regulatory: SARA nonreportable
Properties: Low odor; flash pt. 111 F; KB value 32; VOC 6.40 lb/gal

Shell Sol 70 [Shell]
Chem. Descrip.: Hydrocarbon solv.
Uses: Solvent for dry-cleaning, coatings, automotive, and chemical specialties
Properties: Sp.gr. 0.747 (60 F); dens. 6.22 lb/gal (60 F); flash pt. (CC) 104 F

Shell Sol 71 [Shell]
Chem. Descrip.: Min. spirits
EINECS 232-453-7
Uses: Slow-evaporating solvent for dry-cleaning, paints, coatings, automotive, chemical specialties, commercial and industrial cleaners, insecticides, lighter fluids, resins, plastics
Regulatory: FDA approvals
Properties: Virtually odorless; flash pt. 125 F; KB value 26; VOC 6.24 lb/gal

Shell Sol 71 (H), 71 (WR) [Shell]
Uses: Solvent for dry-cleaning, coatings, automotive, and chemical specialties
Properties: Sp.gr. 0.759 (60 F); dens. 6.32 lb/gal (60 F); flash pt. (CC) 125 F

Shell Sol 72 [Shell]
Uses: Solvent for dry-cleaning, coatings, automotive, and chemical specialties
Properties: Sp.gr. 0.768 (60 F); dens. 6.4 lb/gal (60 F); flash pt. (CC) 137 F

Shell Sol 140 [Shell]
Uses: Solvent for dry-cleaning, coatings, automotive, and chemical specialties
Properties: Sp.gr. 0.787 (60 F); dens. 6.56 lb/gal (60 F); flash pt. (CC) 143 F

Shell Sol 142 HT [Shell]
Chem. Descrip.: Min. spirits
EINECS 232-453-7
Uses: Hydrotreated solvent; hydrotreating removes unwanted compds. (e.g., sulfur, nitrogen, oxygen), converts unsat. hydrocarbons (olefins, acetylenes, aromatics), results in relatively low odor; no more than trace amts. of HAPS; slow-evaporating solvent for paints, coatings, dry-cleaning, commercial and industrial cleaners, insecticides, lighter fluids, resins, plastics, polishes, textile coatings, minerals/mining
Regulatory: SARA nonreportable
Properties: Low odor; flash pt. 145 F; KB value 30; VOC 6.49 lb/gal
Precaution: Combustible

Shell Sol 340 [Shell]
Uses: Solvent for dry-cleaning, coatings, automotive, and chemical specialties
Properties: Sp.gr. 0.765 (60 F); dens. 6.37 lb/gal (60 F); flash pt. (CC) 104 F

Shell Sol 340 HT [Shell]
Chem. Descrip.: Min. spirits
EINECS 232-453-7
Uses: Hydrotreated solvent; hydrotreating removes unwanted compds. (e.g., sulfur, nitrogen, oxygen), converts unsat. hydrocarbons (olefins, acetylenes, aromatics), results in relatively low odor; no more than trace amts. of HAPS; narrow boiling range, slow-evaporating solvent for paints, coatings, dry-cleaning, commercial and industrial cleaners, insecticides, lighter fluids, resins, plastics, wood preservatives, polishes, petrol. gas absorp.; catalytic/reaction medium
Regulatory: SARA nonreportable
Properties: Low odor; flash pt. 103 F;

KB value 32; VOC 6.37 lb/gal

Shell Sol B [Shell]

Chem. Descrip.: Aliphatic naphtha solv.

Uses: Solvent for dry-cleaning, coatings, automotive, and chemical specialties; low boiling, fast evaporating

Properties: Sp.gr. 0.670 (60 F); dens. 5.58 lb/gal (60 F); flash pt. (CC) < 0 F

Shell Sol B HT [Shell]

Chem. Descrip.: Mixt. of predominantly C_6 (hexanes) and about 10% C_7 (heptanes) aliphatic hydrocarbons

Uses: Fast evaporating hydrotreated solvent; hydrotreating removes unwanted compds. (e.g., sulfur, nitrogen, oxygen), converts unsat. hydrocarbons (olefins, acetylenes, aromatics), results in relatively low odor; solvent for agric. fat/oil extraction, automotive undercoating, pharmaceuticals, adhesives, inks, paints, paper coating, lighter fluids, resins, plastics, rubber; lower hexane content than typ. commercial hexanes, thus reducing worker exposure

Properties: Sp.gr. 0.674 (60/60 F); vapor pressure 140 mm Hg (68 F); distillation ragne 140-183 F; flash pt. < 0 F; KB value 29; VOC 5.53 lb/gal; 27-31% n-hexane

Shell Super VM&P Naphtha EC [Shell]

Chem. Descrip.: Aliphatic naphtha solv.

Uses: Solvent for dry-cleaning, coatings, automotive, and chemical specialties; low boiling, fast evaporating

Properties: Sp.gr. 0.754 (60 F); dens. 6.27 lb/gal (60 F); flash pt. (CC) 55 F

Shell Toluene [Shell]

Chem. Descrip.: Toluene

CAS 108-88-3; EINECS 203-625-9

Uses: Med.-evaporating aromatic solvent for dry-cleaning, coatings, automotive, and chemical specialties; used primarily in blends; starting component for rubber solvs. and lacquer diluents

Properties: Sp.gr. 0.871 (60 F); dens. 7.25 lb/gal (60 F); flash pt. (CC) 41 F; KB value 105; VOC 7.16 lb/gal

Shell Tolu-Sol® 5 [Shell]

Chem. Descrip.: Aliphatic naphtha solv.

Uses: Solvent used in dry-cleaning, coatings, automotive, and chemical specialties; low boiling, fast evaporating

Properties: Sp.gr. 0.702 (60 F); dens. 5.85 lb/gal (60 F); flash pt. (CC) 20 F

Shell Tolu-Sol® 6-EC [Shell]

Uses: Solvent used in dry-cleaning, coatings, automotive, and chemical specialties; low boiling, fast evaporating

Properties: Sp.gr. 0.752 (60 F); dens. 6.26 lb/gal (60 F); flash pt. (CC) 21 F

Shell Tolu-Sol® 19-EC [Shell]

Uses: Solvent used in dry-cleaning, coatings, automotive, and chemical specialties; low boiling, fast evaporating

Properties: Sp.gr. 0.730 (60 F); dens. 6.08 lb/gal (60 F); flash pt. (CC) 24 F

Shell Tolu-Sol® A HT Solv. [Shell]

Chem. Descrip.: Heptane (C_7)

CAS 64742-89-8; EINECS 205-563-8

Uses: Hydrotreated solvent; hydrotreating removes unwanted compds. (e.g., sulfur, nitrogen, oxygen), converts unsat. hydrocarbons (olefins, acetylenes, aromatics), results in relatively low odor; exc. for recycling; no more than trace amts. of HAPS; solvent for agric. fat/oil extraction, automotive undercoating, polishes, adhesives, inks, paints, lighter fluids, resins, plastics, rubber, petrol. gas absorption, textile coatings, minerals; starting component for rubber solvs. and lacquer diluents

Regulatory: FDA 21CFR §172.250; SARA nonreportable

Properties: Saybot 30 min color; sp.gr. 0.695 (60/60 F); vapor pressure 41 mm Hg (68 F); i.b.p. 189 F min.; distillation range 194-208 F; flash pt. 7 F; KB value 32 min.; VOC 5.71

lb/gal; 91.2 vol. % paraffins, 8.8 vol % cycloparaffins

Shell Tolu-Sol® W HT Solv. [Shell]

Chem. Descrip.: Mixt. of predominantly C_7 and about 20% C_8 aliphatic hydrocarbons

CAS 64742-89-8

Uses: Hydrotreated solvent; hydrotreating removes unwanted compds. (e.g., sulfur, nitrogen, oxygen), converts unsat. hydrocarbons (olefins, acetylenes, aromatics), results in relatively low odor; no more than trace amts. of HAPS; high cycloparaffin content; high solvency solvent for agric. fat/oil extraction, automotive undercoating, polishes, adhesives, inks, paints, paper coatings, petrol. gas absorption, lighter fluids, resins, plastics, rubber; starting component for rubber solvs., lacquer diluents

Regulatory: FDA 21CFR §172.250; SARA nonreportable

Properties: Saybolt 30 min color; sp.gr. 0.748 (60/60 F); vapor pressure 41 mm Hg (68 F); i.b.p. 205 F min.; distillation range 209-230 F; flash pt. 20 F; KB value 32 min.; VOC 6.17 lb/gal; 26.6 vol. % paraffins, 74.4 vol. % cycloparaffins

Shell Xylene [Shell]

Chem. Descrip.: Xylene

CAS 1330-20-7; EINECS 215-535-7

Uses: Med.-evaporating aromatic solvent for dry-cleaning, coatings, automotive, and chemical specialties; used primarily in blends; starting component for rubber solvs. and lacquer diluents

Properties: Sp.gr. 0.871 (60 F); dens. 7.25 lb/gal (60 F); flash pt. (CC) 81 F; KB value 95; VOC 7.17 lb/gal

Ship Shape® Resin Cleaner [ISP]

Chem. Descrip.: N-Methyl-2-pyrrolidone, butyrolactone, deceth-6, fragrance

Uses: Solvent cleaner for removal of adhesives, paints, varnishes, and resins (polyester, acrylic, urethane, vinyl ester, epoxy) in fiberglass fabrication, furniture refinishing, jewelry, paint and resin mfg., rubber, aircraft, and automotive industries; safer alternative to acetone, methylene chloride, MEK, alkalines

Properties: Water-sol.

Toxicology: TLV 100 ppm; low oral and inhalation toxicity; mod. dermal toxicity; mod. eye irritant

Skellite® [Texaco]

Uses: Solvent used as stove and lamp fuel

Properties: Flash pt. –14 F

Discontinued prod.

Solimate [United-Guardian]

Uses: Mutual solvent and solubilizing agent esp. suited for solubilizing vitamins A, D, E; physiologically inert to tissue; can solubilize many org. substances to make clear aq. sol'ns.; for pharmaceutical, medical, health-care, household, cosmetic fields

Custom prod.

Solulan® 5 [Amerchol; Amerchol Europe]

Chem. Descrip.: Laneth-5, ceteth-5, oleth-5 and steareth-5

Uses: Emulsifier, wetting agent, dispersant, lubricant, solvent, conditioner, plasticizer, emollient used in personal care and dermatology prods.; foam stabilizer for detergent systems

Properties: Amber semisolid; faint pleasant odor; sol. @ 5% in anhyd. ethanol; disp. in water; HLB 8.0; acid no. 3 max.; sapon. no. 14 max.; pH 4.5-7.0 (10% aq.); 100% conc.; nonionic

Solulan® 16 [Amerchol; Amerchol Europe]

Chem. Descrip.: Laneth-16, ceteth-16, oleth-16 and steareth-16

Uses: Emulsifier, wetting agent, dispersant, lubricant, solvent, conditioner, plasticizer, emollient used in personal care and dermatology prods.; foam stabilizer for detergent systems

Properties: Lt. tan waxy solid; sol. @ 5% in water, 30% ethanol; HLB 15; cloud pt. 64-70 C; acid no. 3 max.; sapon. no. 8 max.; pH 4.5-7.5 (10% aq.); 100% conc.; nonionic

Solulan® 75 [Amerchol; Amerchol Europe]
Chem. Descrip.: PEG-75 lanolin
CAS 61790-81-6
Uses: Emulsifier, wetting agent, dispersant, lubricant, solvent, conditioner, plasticizer, emollient used in personal care and dermatology prods.; foam stabilizer for detergent systems
Properties: Lt. yel.-amber waxy solid; faint pleasant odor; sol. @ 5% in water, 30% ethanol; HLB 15; cloud pt. 80-87 C; acid no. 3 max.; sapon. no. 10-20; pH 4.5-7.5 (10% aq.); 100% conc.; nonionic

Solulan® C-24 [Amerchol; Amerchol Europe]
Chem. Descrip.: Choleth-24 and ceteth-24
Uses: Emulsifier, wetting agent, dispersant, lubricant, solvent, conditioner, plasticizer, emollient used in personal care and dermatology prods.; foam stabilizer for detergent systems
Properties: Off-wh. to pale yel. waxy solid; faint pleasant odor; sol. @ 5% in water; HLB 14; cloud pt. 88-95 C; acid no. 1; sapon. no. 3 max.; 100% conc.; nonionic

Solulan® L-575 [Amerchol; Amerchol Europe]
Chem. Descrip.: PEG-75 lanolin
CAS 61790-81-6
Uses: Emulsifier, wetting agent, dispersant, lubricant, solvent, conditioner, plasticizer, emollient used in personal care and dermatology prods.; foam stabilizer for detergent systems
Properties: Lt. yel.-amber visc. liq.; faint pleasant odor; sol. @ 5% in water, 30% ethanol; HLB 15; cloud pt. 80-87 C; acid no. 1 max.; sapon. no. 10 max.; 50% conc.; nonionic

Soluphor® P [BASF; BASF AG]
Chem. Descrip.: Pyrrolidone-2
CAS 616-45-5; EINECS 204-648-7
Uses: Solvent for pharmaceutical injectables, veterinary medicine

Solutol® HS 15 [BASF AG]
Chem. Descrip.: PEG-660 hydroxystearate
Uses: Solvent for injection sol'ns.; nonionic

Solvenol® No. 2 [Hercules Ltd]
Chem. Descrip.: Mixed terpenes
Uses: Hydrocarbon solvent; constituent of industrial cleaners, deodorants
Properties: Liq.

Solvenon® BB [BASF AG]
Uses: Difficultly volatile solvent for surface coatings and paints; additive in cleaning and pickling agents
Discontinued prod.

Solvenon® DIP [BASF AG]
Chem. Descrip.: Mixt. of isomers of dipropylene glycol monoisopropyl ethers
Uses: Solvent for resins and dyes, surface coatings industry; leveling agent for water-dilutable baking finishes; film-binding assistant for dispersions; component in cleaning agents and printing inks

Solvenon® DPM [BASF AG]
Chem. Descrip.: Mixt. of isomeric dipropylene glycol monomethyl ethers
Uses: Solvent for domestic cleaners, acid rust remover, prod. of printing inks; film-binding assistant for aq. polymer dispersions in paint and varnish industry

Solvenon® I [BASF AG]
Chem. Descrip.: 85% Isobutyl formate, 15% isobutanol
Uses: Solvent and diluent for adhesives, paints, varnishes
Discontinued prod.

Solvenon® IPP [BASF AG]
Chem. Descrip.: 1-Isopropoxy-2-propanol, 2-isopropoxy-1-propanol
Uses: Solvent for resins and dyes, surface coatings, cleaning agent for printing plates

Solvenon® PC [BASF AG]
Chem. Descrip.: Propylene carbonate
CAS 108-32-7; EINECS 203-572-1
Uses: Solvent for pigments and dyes, in screen printing dyes; extracting agent; washing liq. for natural and syn. gases; intermediate for org. syntheses

Solvenon® PM [BASF AG]
Chem. Descrip.: 1-Methoxypropa-

nol-2
CAS 107-98-2; EINECS 203-539-1
Uses: Solvent for cleaning agents, printing inks, surface coatings

Solvenon® PP [BASF AG]
Chem. Descrip.: 1-Phenoxypropanol-2 and 2-Phenoxypropanol-1
Uses: Solvent in surface coatings, e.g., electrodipping varnishes by cathodic deposition; component of cleaning agents

Solvent APV Spec. [Hüls AG]
Chem. Descrip.: Ethoxydiglycol
CAS 111-90-0; EINECS 203-919-7
Uses: High boiling solvent for fragrance materials, essential oils, terpene oils for cosmetics use
Properties: Colorless liq.; dens. 0.986-0.990 kg/l; b.p. 198-203 C; acid no. ≤ 0.05; hyd. no. 420-490
Unverified

Solvent B-100 [Eastern Color & Chem.]
Uses: Universally used shoelace tipping solvent exhibiting exc. drying props. and quick solubilization of acetate films normally used in shoelace tipping; does not cause cloudiness or whiteness in finished tip

Solvent B-200 (B-2) [Eastern Color & Chem.]
Uses: Shoelace tipping solvent modified to alter drying props. without altering other desirable characteristics

Solvent GC [BASF AG]
Chem. Descrip.: Mixt. of ethylene glycol acetates
Uses: Solvent for printing inks and rubber stamp inks
Discontinued prod.

Solvent PM [BASF AG]
Chem. Descrip.: 1-Methoxypropan-2-ol
CAS 107-98-2; EINECS 203-539-1
Uses: Solvent for cosmetic industry
Discontinued prod.

Solvent Scour 25/27 [Huntsman]
Chem. Descrip.: Alkanolamides and alkyl phenol ethoxylate
Uses: Scouring agent and prescour for greasy and oil soiled fabrics
Properties: Yel. liq.; dens. 0.899; 100% act.; nonionic
Unverified

Solvent-Scour 263 [ICI Surf. UK]
Chem. Descrip.: Chlorinated solv. and emulsifier
Uses: General purpose scouring agent for all fibers
Properties: Liq.; anionic/nonionic
Unverified

Solvent-Scour 880 [ICI Surf. UK]
Chem. Descrip.: Chlorinated solv., detergent, and emulsifier
Uses: General purpose scouring agent for all fibers
Properties: Liq.; nonionic
Unverified

Sotex CW [Morton Int'l./Spec. Chem.]
Chem. Descrip.: Fatty acid ester, long chain
Uses: Visc. depressant and stabilizer in vinyl plastisols; dispersant for paints and coatings; pigments in plasticizers; cosolvent
Properties: Visc. liq.; sol. in hydrocarbons, ketones; sp.gr. 0.970-1.010; b.p. > 200 C; 100% act.; nonionic

Sovatex C Series [Standard Chem. UK]
Chem. Descrip.: Dodecylbenzene sulfonate blends
Uses: Detergent, solvent scour, wetting agent, emulsifier for textiles, laundry, dry-cleaning, leather, and chemical industries
Properties: Liq., paste, solid; water-disp.; anionic
Unverified

SR-256 [Sartomer]
Chem. Descrip.: 2-(2-Ethoxyethoxy)-ethyl acrylate; with 1000 ppm MEHQ inhibitor
Uses: Monofunctional monomer which can be polymerized by the use of heat, catalysts, and/or irradiation; esp. sensitive to uv lt., and is a good solvent, making it very suitable for radiation-curable ink and coating formulations
Properties: APHA 70 clear liq.; mild odor; m.w. 188; sp.gr. 1.013; dens. 9.4 lb/gal; flash pt. 216 F (PMCC); ref. index 1.4355
Storage: Store away from direct sunlight, oxidizing agents and materials which may generate free radicals; storage temps. should not be >

90 F

Standamul® G [Henkel/Cospha]
Chem. Descrip.: Octyl dodecanol
CAS 5333-42-6; EINECS 226-242-9
Uses: Carrier for oil-sol. act. ingred.; dissolves dyes; pigment dispersant; coupler for waxes and other materials in personal care prods.
Properties: Water-wh. clear liq.; odorless; cloud pt. -50 C; acid no. 0.5 max.; sapon. no. 5.0 max.; 100% conc.
Unverified

Standamul® G-16 [Henkel/Cospha]
Chem. Descrip.: Isocetyl alcohol
CAS 36311-34-9; EINECS 252-964-9
Uses: Emollient; extender and solvent for pigments and personal care prods.; binder in cosmetics
Properties: Water-wh. clear liq.; visc. 100 cps; cloud pt. -50 C; acid no. 0.5 max.; sapon. no. 5.0 max.; 100% conc.
Unverified

Stearate 400 WL 817 [Gattefosse SA]
Chem. Descrip.: PEG-8 palmitostearate
Uses: Solvent and emulsifier for cosmetic/pharmaceutical emulsions
Properties: Drop pt. 29-34 C; HLB 11-12; acid no. < 6; iodine no. < 3; sapon. no. 70-95

Stearate 1500 [Gattefosse SA]
Chem. Descrip.: PEG (1500) palmitostearate
Uses: Solvent and emulsifier for cosmetic/pharmaceutical emulsions
Properties: Drop pt. 46-50 C; HLB 16; acid no. < 6; iodine no. < 3; sapon. no. 30-45
Toxicology: Nonirritating to skin and eyes

Stearate 6000 WL 1644 [Gattefosse SA]
Chem. Descrip.: PEG-150 palmitostearate
Uses: Solvent and emulsifier for cosmetic/pharmaceutical emulsions
Properties: Drop pt. 55-60 C; HLB 18; acid no. ≤ 2; iodine no. < 3; sapon. no. 15-25
Toxicology: Nonirritating to skin; sl. irritating to eyes

Stepanate® AXS [Stepan; Stepan Canada]
Chem. Descrip.: Ammonium xylene sulfonate
CAS 26447-10-9; EINECS 247-710-9
Uses: Hydrotrope, solubilizer, coupler in detergent field; heavy duty cleaners, wax strippers, dishwashing detergents; solvent or fluidizer; freeze-thaw stabilizer; additive for high electrolyte or brine systems
Properties: Colorless liq.; water-sol.; dens. 9.05 lb/gal; pH 6.5-7.2 (1% aq.); 40% act.; anionic

Striptron Stripper [Dow]
Chem. Descrip.: Inhibited methylene chloride with additives
CAS 75-09-2; EINECS 200-838-9
Uses: Solvent stripper for dry film photoresist and screen inks
Properties: Sol. in most org. solvs.; low water sol.
Discontinued prod.

Sucro Ester 15 [Gattefosse SA]
Chem. Descrip.: Saccharose palmitate
CAS 26446-38-8; EINECS 247-706-7
Uses: Food emulsifier; o/w emulsifier wetting agent, crystallization inhibitor for veg. oils and fats; inhibitor of thermal denaturation of proteins; solvent and emulsifier for pharmaceutical/cosmetic emulsions
Properties: Fine powd., faint odor; sol. in water and ethanol @ 60 C; insol. in min. and veg. oils; HLB 15.0; acid no. < 5; sapon. no. 95-135; 100% conc.; nonionic
Toxicology: LD0 (oral, rat) > 8 g/kg

Sulfa-Hitech® 0382 [Elf Atochem N. Am.]
Chem. Descrip.: DMDS-based
Uses: Solvent used to dissolve sulfur in the prod. of sour gas wells, in sour-gas pipelines, and in refinery and chemical plant flowlines
Properties: Sp.gr. 1.060; b.p. 107-111 C

Sulfolane W [Shell]
Chem. Descrip.: Tetramethylene sulfone
CAS 126-33-0; EINECS 204-783-1
Uses: Highly polar compd. with outstanding solvent properties and chemical and thermal stability; for extraction of benzene, toluene, other aromatic hydrocarbons from

oil refinery streams; used in Sulfinol process to remove acid gases; used for separation of low boiling alcohols, min. oils, tars; plasticizer; polymerization solvent; dielec. in elec. equipment; solvent in surf. coatings; component in hydraulic fluids
Properties: Sp.gr. 1.2600-1.2615; 2.6-3.0% water

Super Lacolene [Ashland]
Chem. Descrip.: Aliphatic
Uses: Solvent
Unverified

Super Refined® Almond Oil [Croda Inc; Croda Surf. Ltd]
Chem. Descrip.: Almond oil NF
CAS 8007-69-0
Uses: Emollient; provides elegant skin feel and promotes spreading in creams, lotions, bath oils; solvent, vehicle for pharmaceuticals (parenteral formulations for intramuscular administration, nutritional supplements, IV emulsions, liniments, ointments)
Properties: APHA 40 oil, odorless, tasteless; sol. in min. oil, IPA
Toxicology: LD50 (oral, rat) > 5 g/kg; nonirritating to skin

Super Refined® Corn Oil [Croda Inc]
Chem. Descrip.: Corn oil NF
CAS 8001-30-7; EINECS 232-281-2
Uses: Solvent and vehicle for pharmaceuticals (parenteral formulations for intramuscular administration, nutritional supplements, IV emulsions, liniments, ointments)
Properties: APHA 40 oil, odorless, tasteless; sol. in min. oil, IPA

Super Refined® Cottonseed Oil [Croda Inc]
Chem. Descrip.: Cottonseed oil NF/USP
CAS 8001-29-4; EINECS 232-280-7
Uses: Solvent and vehicle for pharmaceuticals (parenteral formulations for intramuscular administration, nutritional supplements, IV emulsions, liniments, ointments)
Properties: APHA 50 oil, odorless, tasteless; sol. in min. oil, IPA

Super Refined® Olive Oil [Croda Inc; Croda Surf. Ltd]
Chem. Descrip.: Olive oil
CAS 8001-25-0; EINECS 232-277-0
Uses: Cosmetic emollient; lubricant for hair care prods.; solvent and vehicle for pharmaceuticals (parenteral formulations for intramuscular administration, nutritional supplements, IV emulsions, liniments, ointments)
Properties: APHA 30 oil, odorless, tasteless; sol. in min. oil, IPA

Super Refined® Peanut Oil [Croda Inc]
Chem. Descrip.: Peanut oil NF/USP
CAS 8002-03-7; EINECS 232-296-4
Uses: Emollient in skin care prods.; solvent and vehicle for pharmaceuticals (parenteral formulations for intramuscular administration, nutritional supplements, IV emulsions, liniments, ointments)
Properties: APHA 30 oil, odorless, tasteless; sol. in min. oil, IPA

Super Refined® Safflower Oil [Croda Inc; Croda Surf. Ltd]
Chem. Descrip.: Safflower oil USP
CAS 8001-23-8; EINECS 232-276-5
Uses: Emollient; solvent and vehicle for pharmaceuticals (parenteral formulations for intramuscular administration, nutritional supplements, IV emulsions, liniments, ointments)
Properties: APHA 30 mobile oil, odorless, tasteless; sol. in min. oil, IPA
Toxicology: LD50 (oral, rat) > 5 g/kg; minimal eye and skin irritant

Super Refined® Sesame Oil [Croda Inc; Croda Surf. Ltd]
Chem. Descrip.: Sesame oil NF/USP
CAS 8008-74-0; EINECS 232-370-6
Uses: Emollient oil used in skin care preps.; solvent and vehicle for pharmaceuticals (parenteral formulations for intramuscular administration, nutritional supplements, IV emulsions, liniments, ointments)
Regulatory: NF, BP compliance
Properties: APHA 30 clear oily liq., odorless, tasteless; sol. in min. oil, IPA; sp.gr. 0.916-0.921; iodine no. 103-116; sapon. no. 188-195; hyd. no. 5 max.; ref. index 1.472-1.476; 0.05% max. moisture

Toxicology: LD50 (oral, rat) > 5 g/kg; mild skin irritant, minimal eye irritant

Super Refined® Soybean Oil USP [Croda Inc; Croda Surf. Ltd]
Chem. Descrip.: Soybean oil USP
CAS 8001-22-7; EINECS 232-274-4
Uses: Emollient oil used in skin care preps.; solvent and vehicle for pharmaceuticals (parenteral formulations for intramuscular administration, nutritional supplements, IV emulsions, liniments, ointments)
Properties: APHA 60 max. clear oily liq., odorless, tasteless; sol. in min. oil, IPA; sp.gr. 0.916-0.922; iodine no. 120-141; sapon. no. 180-200; hyd. no. 5 max.; ref. index 1.465-1.475
Toxicology: LD50 (oral, rat) > 5 g/kg; mild skin irritant, minimal eye irritant

Supersol ICS [Sybron]
Uses: Textile solvent scour to remove wax and oil; caustic stable; anionic/nonionic
Discontinued prod.

Surfadone® LP-100 [ISP]
Chem. Descrip.: Caprylyl pyrrolidone
Uses: Surfactant for personal care use; specialty solvent for commercial cleaning, textile processing, lithographic printing prods., water-borne coatings, inks; leveling agent, dispersant, emulsification aid, visc. builder, pigment dispersant; VOC replacement
Properties: Pale yel. low visc., clear to sl. hazy liq.; sol. in polar and nonpolar solvs. incl. ethanol, acetone, xylene, heptane, paraffin oil, Stod., perchloroethylene; sparingly sol. in water; m.w. 192; sp.gr. 0.92; b.p. 100 C (0.3 mm Hg); solid. pt. -25 C; HLB 6; flash pt. (TCC) 119 C; surf. tens. 28 dynes/cm; Draves wetting 3.5 s (0.1%); Ross-Miles foam 25 mm (initial, 0.1%); 97% min. act.
Toxicology: LD50 (oral) 2.05 g/kg; not a primary skin or eye irritant; noncomedogenic
Environmental: Biodeg.

Surfadone® LP-300 [ISP]
Chem. Descrip.: Lauryl pyrrolidone
CAS 2687-96-9
Uses: Surfactant for personal care use; provides enhanced body, improved combability, higher sheen, foam boosting/stabilization, visc., and clarity to shampoos; leveling agent, visc. builder, dispersant, emulsification aid, pigment dispersant; specialty solvent for commercial cleaning, textile processing, lithographic printing prods., water-borne coatings, inks; VOC replacement
Properties: Pale yel. low visc., clear to sl. hazy liq.; sol. in polar and nonpolar solvs. incl. ethanol, acetone, xylene, heptane, paraffin oil, Stod., perchloroethylene; sparingly sol. in water; m.w. 253; sp.gr. 0.90; b.p. 145 C (0.2 mm Hg); solid. pt. 10 C; HLB 3; flash pt. (TCC) 116 C; surf. tens. 26 dynes/cm; Draves wetting > 300 s (0.1%); Ross-Miles foam 13 mm (initial, 0.1%); 97% min. act.
Toxicology: LD50 (oral) > 5 g/kg; not a primary skin or eye irritant; noncomedogenic
Environmental: Biodeg.

Sylfat® DX [Arizona]
Chem. Descrip.: Tall oil fatty acid ester
Uses: For plasticizers, extenders, surface-act. agents in grinding and cutting oils, specialty lubricant additives, corrosion inhibitors, specialty solvs. for printing inks, metalworking, and oil well servicing
Properties: Gardner 5+ color; sp.gr. 0.94; visc. 700 cps; acid no. 12; flash pt. (OC) 190 C
Custom prod.

Sylfat® MM [Arizona]
Chem. Descrip.: Tall oil fatty acid ester
Uses: For plasticizers, extenders, surface-act. agents in grinding and cutting oils, specialty lubricant additives, corrosion inhibitors, specialty solvs. for printing inks, metalworking, and oil well servicing
Properties: Gardner 9 color; sp.gr. 0.91; visc. 20 cps; acid no. 5; iodine no. 105; flash pt. (OC) 190 C

Syn Fac® TEA-97 [Milliken]
Chem. Descrip.: Ethoxylated amine

Uses: Solvent and dispersant for acidic materials; used in the dye industry for solubilizing acid dyes, pigments, and optical brighteners; disperses soils in metal cleaning

Properties: Gardner < 10 liq.; sp.gr. 1.10; visc. 400 cps; acid no. 22; hyd. no. 610; cloud pt. > 100 C (1%); pH 8.4 (10% sol'n.); 100% conc.; nonionic

Toxicology: Strongly alkaline; avoid contact with skin and eyes

T

T4250 [Hüls Am.]
Chem. Descrip.: Phenyltris(trimethylsiloxy) silane
Uses: Low visc. silicone fluid with high control over volatility, visc., dens. and ref. index, and unique solvent props.; as lubricant
Discontinued prod.

TBP [FMC]
Chem. Descrip.: Tributyl phosphate
CAS 126-73-8; EINECS 204-800-2
Uses: Antifoam agent used in paper coating compds., water adhesives, inks, casein sol'ns., textile sizes, and detergent sol'ns.; plasticizer and solvent for NC, cellulose acetate, chlorinated rubber, and vinyls; solvent for natural gums and syn. resins; component in extraction of rare earth metal salts
Properties: Pt-Co 50 max. liq.; char. odor; m.w. 266; f.p. < -80 C; b.p. 137-145 C; sol. in org. liqs., min. oil, and gasoline; sp.gr. 0.978 ± 0.003; dens. 975 kg/m^3; visc. 3.7 cp; pour pt. < -80 C; ref. index 1.423 ± 0.001; flash pt. 166 C; fire pt. 182 C

T-Det® N-4 [Harcros USA]
Chem. Descrip.: Nonoxynol-4
CAS 9016-45-9; EINECS 230-770-5
Uses: Detergent, wetting agent; agric. toxicant formulations; intermediate for household detergents; drycleaning soap additive, sludge dispersant additive in fuel oils; gasoline additive; solvent cleaner for metal processing; color development aid, stabilizer and emulsifier for latex paints
Regulatory: FDA 21CFR §175.105, 178.3400, 180.1001c,d,e
Properties: Pale yel. oily liq.; mild aromatic odor; sol. in butyl Cellosolve, CCl_4, min. and corn oil, ethanol, kerosene, min. spirits, perchloroethylene, Stod., toluene; dens. 8.5 lb/gal; sp.gr. 1.02; visc. 52 SUS (210 F); HLB 8.9; flash pt. > 400 F (OC); pour pt. < O F; pH 7; 99.5% min. act.; nonionic

T-Det® N-6 [Harcros USA]
Chem. Descrip.: Nonoxynol-6
CAS 9016-45-9
Uses: Emulsifier for silicones, household waxes and polishes; wetting agent; detergent and dispersant for petrol. and fuel oils; agric. toxicant formulations; dry-cleaning soap additive; leather and metal processing; solvent cleaner; deinking agent for paper
Regulatory: FDA 21CFR §175.105, 178.3400, 180.1001c,d,e
Properties: Pale yel. liq.; mild aromatic odor; sol. in butyl Cellosolve, CCl_4, corn oil, diesel fuel, ethanol, kerosene, Stod., xylene; dens. 8.7 lb/gal; sp.gr. 1.04; visc. 58 SUS (210 F); HLB 10.9; flash pt. > 400 F (OC); pour pt. < 0 F; pH 7; 99.5% min. act.; nonionic

Tebol™ [ARCO]
Chem. Descrip.: Tert. butyl alcohol
CAS 75-65-0; EINECS 200-889-7
Uses: Solvent, cosolvent, compatibilizer, coupling agent, processing aid for pharmaceuticals, personal care prods., aq. coatings and adhesives, agric. formulations, polymer processing, cleaners/disinfectants; chlorinated hydrocarbon stabilizer
Properties: Sol. 100% in water; m.w. 74.12; sp.gr. 0.80 (20/20 C); dens.

6.7 lb/gal (20 C); vapor pressure 30 mm Hg (20 C); b.p. 175 F; flash pt. 52 F

Tebol™ 99 [ARCO]

Chem. Descrip.: High purity tert. butyl alcohol

CAS 75-65-0; EINECS 200-889-7

Uses: Solvent, cosolvent, compatibilizer, coupling agent, processing aid for pharmaceuticals, personal care prods., aq. coatings and adhesives, agric. formulations, polymer processing, cleaners/disinfectants; chlorinated hydrocarbon stabilizer

Properties: Pt-Co ≤ 10 color; sol. in water; misc. with most org. solvs.; sp.gr. 0.78; dens. 6.5 lb/gal; visc. 3.3 cps (30 C); i.b.p. 81.5 C (760 mm Hg); f.p. 24.5 C; flash pt. (CC) 11 C; 99.3% act.

Tecsol® 1 [Eastman]

Chem. Descrip.: SD alcohol 1 (100 parts) with ethyl acetate (5) and heptane (1)

Uses: Proprietary solvent for use in shellac, lacquer, and varnish markets; latent solvent in NC lacquers to reduce sol'n. visc.

Properties: Completely sol. in water; m.w. 46.07; sp.gr. 0.789-0.820 (20/20C); dens. 6.576.83 lb/gal; visc. 1.201.5 cP (20 C); f.p. -173 F; b.p. 74-82; flash pt. (COC) 49-72 F; ref. index 1.3614 (20 C); surf. tens. 22.4 dynes/cm (20 C)

Tecsol® 3 [Eastman]

Chem. Descrip.: SD alcohol 1 (100 parts) with MIBK (1), ethyl acetate (1), and heptane (1)

Uses: Proprietary solvent for use in shellac, lacquer, and varnish markets; latent solvent in NC lacquers to reduce sol'n. visc.

Properties: Completely sol. in water; m.w. 46.07; sp.gr. 0.789-0.820 (20/20C); dens. 6.576.83 lb/gal; visc. 1.201.5 cP (20 C); f.p. -173 F; b.p. 74-82; flash pt. (COC) 49-72 F; ref. index 1.3614 (20 C); surf. tens. 22.4 dynes/cm (20 C)

Tecsol® A [Eastman]

Chem. Descrip.: SD alcohol 3A (100 parts) with IPA (10), MIBK (1)

Uses: Industrial solvent for manufacturing processes, prod. of printing inks, as latent solvent in cellulosic coatings

Properties: Completely sol. in water; m.w. 46.07; sp.gr. 0.789-0.820 (20/20C); dens. 6.576.83 lb/gal; visc. 1.201.5 cP (20 C); f.p. -173 F; b.p. 74-82; flash pt. (COC) 49-72 F; ref. index 1.3614 (20 C); surf. tens. 22.4 dynes/cm (20 C)

Tecsol® A-2 [Eastman]

Chem. Descrip.: SD alcohol 3A (100 parts) with methyl alcohol (10), MIBK (1)

Uses: Industrial solvent for manufacturing processes, prod. of printing inks, as latent solvent in cellulosic coatings

Properties: Completely sol. in water; m.w. 46.07; sp.gr. 0.789-0.820 (20/20C); dens. 6.576.83 lb/gal; visc. 1.201.5 cP (20 C); f.p. -173 F; b.p. 74-82; flash pt. (COC) 49-72 F; ref. index 1.3614 (20 C); surf. tens. 22.4 dynes/cm (20 C)

Tecsol® B [Eastman]

Chem. Descrip.: SD alcohol 3A (100 parts) with isopropyl alcohol(5), MIBK (1), methyl alcohol (5)

Uses: Industrial solvent for manufacturing processes, prod. of printing inks, as latent solvent in cellulosic coatings

Properties: Completely sol. in water; m.w. 46.07; sp.gr. 0.789-0.820 (20/20C); dens. 6.576.83 lb/gal; visc. 1.201.5 cP (20 C); f.p. -173 F; b.p. 74-82; flash pt. (COC) 49-72 F; ref. index 1.3614 (20 C); surf. tens. 22.4 dynes/cm (20 C)

Tecsol® C [Eastman]

Chem. Descrip.: SD alcohol 3A (100 parts) with ethyl acetate (5), MIBK (1)

Uses: Industrial solvent for manufacturing processes, prod. of printing inks, as latent solvent in cellulosic coatings; cosolvent; reduces surf. tens., controls evaporation rate for waterborne, industrial, and trade sales coatings

Properties: Completely sol. in water; m.w. 46.07; sp.gr. 0.789-0.820 (20/

20C); dens. 6.576.83 lb/gal; visc. 1.201.5 cP (20 C); f.p. -173 F; b.p. 74-82; flash pt. (COC) 49-72 F; ref. index 1.3614 (20 C); surf. tens. 22.4 dynes/cm (20 C)

Tecsol® D [Eastman]
Chem. Descrip.: SD alcohol 3A (100 parts) with isopropyl alcohol (15), MIBK (1)
Uses: Industrial solvent for manufacturing processes, prod. of printing inks, as latent solvent in cellulosic coatings
Properties: Completely sol. in water; m.w. 46.07; sp.gr. 0.789-0.820 (20/20C); dens. 6.576.83 lb/gal; visc. 1.201.5 cP (20 C); f.p. -173 F; b.p. 74-82; flash pt. (COC) 49-72 F; ref. index 1.3614 (20 C); surf. tens. 22.4 dynes/cm (20 C)

Tecsol® D-1 [Eastman]
Chem. Descrip.: Ethyl alcohol, anhyd.
CAS 64-17-5; EINECS 200-578-6
Uses: Special industrial solvent
Properties: Pt-Co 20 max. color; sp.gr. 0.794-0.797; dens. 6.64 lb/gal; f.p. –114 C; b.p. 74-80 C; flash pt. (TCC) 4-15 C; also avail. as 95%

Tecsol® D-2 [Eastman]
Chem. Descrip.: SD alcohol 3A (100 parts) with methyl alcohol (15), MIBK (1)
Uses: Industrial solvent for manufacturing processes, prod. of printing inks, as latent solvent in cellulosic coatings
Properties: Completely sol. in water; m.w. 46.07; sp.gr. 0.789-0.820 (20/20C); dens. 6.576.83 lb/gal; visc. 1.201.5 cP (20 C); f.p. -173 F; b.p. 74-82; flash pt. (COC) 49-72 F; ref. index 1.3614 (20 C); surf. tens. 22.4 dynes/cm (20 C)

Tecsol® H [Eastman]
Chem. Descrip.: SD alcohol 3A (100 parts) with isopropyl alcohol (100)
Uses: Industrial solvent for manufacturing processes, prod. of printing inks, as latent solvent in cellulosic coatings
Properties: Completely sol. in water; m.w. 46.07; sp.gr. 0.789-0.820 (20/20C); dens. 6.576.83 lb/gal; visc. 1.201.5 cP (20 C); f.p. -173 F; b.p. 74-82; flash pt. (COC) 49-72 F; ref. index 1.3614 (20 C); surf. tens. 22.4 dynes/cm (20 C)

Tegosoft® 189 [Goldschmidt]
Chem. Descrip.: Isostearyl isononanoate
Uses: Emollient, superfatting agent, solvent for lipophilic components
Properties: Liq.
Discontinued prod.

Tegosoft® CI [Goldschmidt; Goldschmidt AG]
Chem. Descrip.: Cetearyl isononanoate
Uses: Cosmetic esters for emolliency, skin softening, and moisture retention in hair and skin care prods., creams and lotions, hair conditioners/glossing systems, sunscreen preps., soaps; solvent for lipophilic additives, e.g., uv-filters

Tegosoft® CT [Goldschmidt; Goldschmidt AG]
Chem. Descrip.: Caprylic/capric triglycerides
CAS 65381-09-1
Uses: Cosmetic esters for emolliency, skin softening, and moisture retention in hair and skin care prods., creams and lotions, hair conditioners/glossing systems, sunscreen preps., soaps; solvent for lipophilic additives

Tegosoft® M [Goldschmidt; Goldschmidt AG]
Chem. Descrip.: Isopropyl myristate
CAS 110-27-0; EINECS 203-751-4
Uses: Cosmetic esters for emolliency, skin softening, and moisture retention in hair and skin care prods., creams and lotions, hair conditioners/glossing systems, sunscreen preps., soaps; solvent for lipophilic additives

Tegosoft® P [Goldschmidt; Goldschmidt AG]
Chem. Descrip.: Isopropyl palmitate
CAS 142-91-6; EINECS 205-571-1
Uses: Cosmetic esters for emolliency, skin softening, and moisture retention in hair and skin care prods., creams and lotions, hair conditioners/glossing systems, sunscreen preps., soaps; solvent for lipophilic

additives

Teric PEG 200 [ICI Australia]
Chem. Descrip.: PEG 200
CAS 25322-68-3; EINECS 203-989-9
Uses: Binder for glazing; intermediate for PEG esters, methacrylate resins, PU foams; plasticizer/solvent for cork; toiletries; metalworking lubricants; paints/resins; paper/film; printing inks; textile emulsifier
Properties: APHA 25 color; sol. in water and most polar org. solvs.; m.w. 190-210; sp.gr. 1.127; visc. 4.2 cst (99 C); pour pt. -50 C; flash pt. (OC) > 170 C; ref. index 1.459; pH 4.5-7.5; surf. tens. 44.6 dynes/cm; 99% act.
Environmental: Biodeg.
Unverified

Teric PEG 300 [ICI Australia]
Chem. Descrip.: PEG 300 USP
CAS 25322-68-3; EINECS 220-045-1
Uses: Pesticide solubilizer/carrier; visc. modifier for brake fluids; intermediate for PEG esters, PU foams; plasticizer/solvent; cosmetics; metalworking lubricants; paints/resins; paper/film; printing inks; rubber release; textile aux.; liq. pharmaceutical preps., ointment bases; binder for capsules/pills
Properties: APHA 25 color; sol. in water and most polar org. solvs.; m.w. 285-315; sp.gr. 1.128; visc. 5.8 cst (99 C); pour pt. -12 C; flash pt. (OC) > 190 C; ref. index 1.463; pH 4.5-7.5; surf. tens. 44.6 dynes/cm; 99% act.
Environmental: Biodeg.

Teric PEG 400 [ICI Australia]
Chem. Descrip.: PEG 400 USP
CAS 25322-68-3; EINECS 225-856-4
Uses: Emulsifier, antistat for textiles; rubber; inks; cosmetics; paper/film; pesticide solubilizer/carrier; intermediate for PEG esters, PU foams; plasticizer/solvent for cork; metalworking lubricants; paints/resins; pharmaceutical liq. preps., ointment bases, suppository bases; binder for capsules/pills; preservation of pathological specimens
Properties: APHA 25 color; sol. in water and most polar org. solvs.; m.w. 380-420; sp.gr. 1.130; visc. 7.3 cst (99 C); pour pt. 6 C; flash pt. (OC) > 215 C; ref. index 1.465; pH 4.5-7.5; surf. tens. 44.6 dynes/cm; 99% act.
Environmental: Biodeg.

Teric PPG 400 [ICI Australia]
Chem. Descrip.: PPG 400
Uses: Intermediate for surfactants, ethers, esters; metalworking lubricants; solvent for paints/varnishes, printing inks, veg. oils; textile lubricants
Properties: APHA 100 color; sol. in polar org. solvs.; sp.gr. 1.009; visc. 82 cps; hyd. no. 220-250; pour pt. –26 C; flash pt. (OC) 204 C; ref. index 1.445; pH 5-8 (1% aq. disp.); surf. tens. 31.1 dynes/cm; 99% act.
Environmental: Biodeg.
Unverified

Teric PPG 1650 [ICI Australia]
Chem. Descrip.: PPG 1650
Uses: Biodeg.; intermediate for surfactants, ethers, esters; antifoam for ceramics, rubber; lubricant/softener for leather; metalworking lubricant; demulsifier for petrol. industry; plasticizer for plastics; solvent for cosmetics
Properties: Sol. in polar org. solvs.; sp.gr. 1.006; visc. 330 cps; hyd. no. 67-71; pour pt. -32 C; flash pt. (OC) 230 C; ref. index 1.449; pH 5-8 (1% aq. disp.); surf. tens. 31.8 dynes/cm; 99% act.
Unverified

Teric PPG 2250 [ICI Australia]
Chem. Descrip.: PPG 2250
Uses: Biodeg.; intermediate for surfactants, ethers, esters; hydraulic brake fluids; cosmetics; dyeing; lubricant/softener for leather; metalworking lubricant; demulsifier for petrol.; latex coagulant; rubber release agent; solvent for veg. oils
Properties: APHA 50 color; sol. in polar org. solvs.; sp.gr. 1.005; visc. 482 cps; hyd. no. 47-54; pour pt. -29 C; flash pt. (OC) 240 C; ref. index 1.449; pH 4.0-7.5 (10% in aq.

methanol); surf. tens. 32.1 dynes/cm; 99% act.
Unverified

Tetraglyme [Ferro/Grant]
Chem. Descrip.: Tetraethylene glycol dimethyl ether
CAS 24991-55-7
Uses: Solvent which tends to solvate cations; used in electrochemistry, polymer and boron chemistry; processes such as gas absorp., extraction, stabilization; used in industrial prods. such as fuels, lubricants, textiles, pesticides; solvent for pharmaceuticals; drug carrier, stabilizer
Properties: Colorless clear; mild, nonresidual odor; water-sol.; misc. with ethanol, acetone, benzene, diethyl ether, octane; m.w. 222.28; sp.gr. 1.0132; dens. 8.45 lb/gal; visc. 4.1 cP; f.p. -29.7 C; b.p. 275 C; flash pt. (CC) 141 C; ref. index 1.4330 (20 C); pH neutral; surf. tens. 33.8 dynes/cm (20 C); 98.0% min. purity
Toxicology: LD50 5100 mg/kg; low to mod. acute toxicity; chronic exposure may cause reproductive effects
Environmental: Slowly biodeg.

Tetralin [DuPont]
Chem. Descrip.: 1,2,3,4-Tetrahydronaphthalene
CAS 119-64-2; EINECS 204-340-2
Uses: Solvent for oils, resins, waxes, rubber, and asphalt
Properties: Colorless to pale yel. clear liq.; m.w. 132.2; f.p. -31 C; b.p. 207 C (760 mm Hg); negligible sol. in water; sp.gr. 0.970; dens. 8.1 lb/gal; flash pt. 82 C (TCC); 97.0% min. tetrahydronaphthalene

Texacar® EC [Huntsman]
Chem. Descrip.: Ethylene carbonate
CAS 96-49-1; EINECS 202-510-0
Uses: Solvent for org. and inorg. materials; Rule 66 exempt; also used as reactant and plasticizer in fibers and textiles, plastics and resins, aromatic hydrocarbon extraction, electrolytes, hydraulic brake fluids
Properties: Wh. cake solid, mild odor; melts to clear liq. > 95 F; m.w. 88.06; sp.gr. 1.3218 (39/4 C); dens. 10.87 lb/gal (40 C); visc. 1.9 cs (40 C); f.p. 35.5 C min.; b.p. 248.2 C (760 mm); flash pt. 305 F; ref. index 1.4158; pH 6.5-7.5 (10% aq.); 99% min. assay

Texacar® PC [Huntsman]
Chem. Descrip.: Propylene carbonate
CAS 108-32-7; EINECS 203-572-1
Uses: Solvent for org. and inorg. materials; Rule 66 exempt; also used as reactant and plasticizer in fibers and textiles, hydraulic fluids, plastics and resins, gas treating, aromatic hydrocarbon extraction, metal extraction, surf. coatings; foundry sand binders, lubricants, electrolytes, personal care prods.; gellant for clays in greases and cosmetics
Properties: Clear mobile liq., pract. odorless; m.w. 102.09; sp.gr. 1.2057 (20/4 C); dens. 10.05 lb/gal; visc. 2.8 cs; b.p. 241.9 C (760 mm); f.p. –49.2 C; flash pt. (PMCC) 275 F; pour pt. -100 F; ref. index 1.4210; pH 6.5-7.5 (10% aq.); 99.4% min. assay

Texanol® Ester-Alcohol [Eastman]
Chem. Descrip.: 2,2,4-Trimethyl-1,3-pentanediol monoisobutyrate
CAS 25265-77-4; EINECS 246-771-9
Uses: Slow-evaporating solvent used as coalescing agent in latex finishes and water-base inks, PVAc latices, PVAc-acrylic latices, and acrylic, EVA, and B/S latices; used as solvent in electrodeposition coatings, lacquer coatings; defoamer in waterborne systems and drilling muds for oil industry; chemical intermediate for prod. of esters. used as stain-resistance plasticizers, lubricant base stocks, solvents, syn. detergents, and herbicides; solvent for high-bake enamels
Regulatory: SARA nonreportable
Properties: Colorless clear liq.; mild char. odor; water-insol.; m.w. 216.3; sp.gr. 0.950; dens. 0.95 kg/l; visc. 13.5 cP (20 C); vapor pressure 0.013 mm Hg; f.p. -50 C; b.p. 244 C min. (initial, 760 mm); pour pt. -57 C; ref. index 1.4423; flash pt. 120 C (COC); fire pt. 132 C; > 99%

volatiles by vol.
Toxicology: LD50 (oral, rat) 6.86 ml/kg, (dermal, rabbit) > 15 ml/kg; may cause transient eye irritation; expected to be low ingestion hazard
Environmental: Mod. potential to affect some aquatic organisms, algal growth; low potential to affect sec. waste treatment microbial metabolism, germination or early growth of some plants; expected to biodegrade
Precaution: Incompat. with oxidizing materials; hazardous combustion prods.: CO, CO_2
HMIS: Health: 1; Flammability: 1; Reactivity: 0
Storage: Store in closed containers

Texsolve B [Texaco]
Chem. Descrip.: Highly refined commercial hexane
CAS 110-54-3; EINECS 203-777-6
Uses: Solvent used for veg. oil and pharmaceutical extraction, compounding rubber cements and sealants, and polyolefin prod.
Properties: Liq.; sp.gr. 0.679 (60 F); dens. 5.65 lb/gal (60 F); i.b.p. 152 F; flash pt. -25 F; 56% act.
Toxicology: TLV 500 ppm (OSHA), 50 ppm (ACGIH)
Precaution: Flamm.
Discontinued prod.

Texsolve C [Texaco]
Chem. Descrip.: Mixed heptane fraction
CAS 142-82-5; EINECS 205-563-8
Uses: Solvent used for compounding rubber cements and sealants, extraction of oils and fats, and recrystallization of org. chemicals
Properties: Liq.; sp.gr. 0.700 (60 F); dens. 5.85 lb/gal (60 F); i.b.p. 185 F; flash pt. (TCC) 13 F
Toxicology: TLV 500 ppm (OSHA), 400 ppm (ACGIH)
Precaution: Flamm.
Discontinued prod.

Texsolve E [Texaco]
Chem. Descrip.: Heptane, hexane, octane, and toluene
Uses: Solvent used in rubber tire mfg., rubber cements, sealants, inks, lacquers, and adhesives, extraction of oils and fats, recrystallization of org. chems.
Properties: Liq.; sp.gr. 0.694 (60/60); i.b.p. 190 F min.; flash pt. -20 F
Precaution: Flamm.
Discontinued prod.

Texsolve H [Texaco]
Chem. Descrip.: Hexane-heptane fraction
Uses: Solvent used in rubber tire mfg., rubber cements, sealants, inks, lacquers, and adhesives; offers fast evaporation and extremely low evaporation residue
Properties: Liq.; i.b.p. 158 F; flash pt. –20 F
Precaution: Flamm.
Discontinued prod.

Texsolve S [Texaco]
Chem. Descrip.: Min. spirits
EINECS 232-453-7
Uses: Solvent used in paint and protective coatings, dry-cleaning, degreasing, wood treating, and charcoal lighter fluid; meets or exceeds standards for min. spirits and Stod
Properties: Flash pt. 107 F
Discontinued prod.

Texsolve S-66 [Texaco]
Chem. Descrip.: Exempt min. spirits with max. aromatic content
EINECS 232-453-7
Uses: Solvent used in paint and protective coatings, dry-cleaning, degreasing, wood treating, charcoal lighter fluid; meets or exceeds standards for min. spirits and Stod
Properties: Flash pt. 107 F
Discontinued prod.

Texsolve S-LO [Texaco]
Chem. Descrip.: Low aromatic min. spirits
EINECS 232-453-7
Uses: Solvent used in charcoal lighter fluid, floor finishes, paints, coatings, hand cleaners
Properties: Low odor; flash pt. 108 F
Discontinued prod.

Texsolve V [Texaco]
Chem. Descrip.: VM&P naphtha
CAS 68920-06-9
Uses: Rule 66 solvent used in paints,

coatings, rubber compounding, sealants, and chemical absorption
Properties: Liq., pleasant odor; sp.gr. 0.751 (60 F); dens. 6.25 lb/gal (60 F); i.b.p. 256 F; flash pt. 58 F
Toxicology: TLV 300 ppm (ACGIH)
Precaution: Flamm.
Discontinued prod.

TF Solvent [Chemtronics]
Uses: Mild cleaning agent for removal of flux residue, oil, grease and contaminants from magnetic tape heads, thermal print heads, optical equip., contacts, relays, and PCB assemblies
Unverified

THF [ARCO]
Chem. Descrip.: Tetrahydrofuran
CAS 109-99-9; EINECS 203-726-8
Uses: Solvent for vinyls; reaction solvent; for PTMEG, magnetic tape paint, adhesives, PU coatings, PVC cements, vinyl films and cellophane
Properties: Colorless clear low-visc. liq.; m.w. 72.1; sol. in water; sp.gr. 0.89 (20/20 C); dens. 7.4 lb/gal (20 C); vapor pressure 155 mm Hg (20 C); b.p. 149 F; flash pt. 1 F

THF [ISP]
Chem. Descrip.: Tetrahydrofuran
CAS 109-99-9; EINECS 203-726-8
Uses: Solvent in PVC pipe cements, top coatings, other resin applics.; chem. intermediate and reaction medium; extraction solvent for natural and syn. prods.; high solvent capacity for resins (high m.w. vinyls, thermoplastic PU, polyester, etc.)
Properties: APHA 20 max. liq.; misc. with water, alcohols, diethyl ether, esters, ketones, aliphatic, aromatic, and chlorinated hydrocarbons; sp.gr. 0.886-0.889 (20/4 C); dens. 7.41 lb/gal; visc. 0.53 cps (20 C); f.p. -108.5 C; b.p. 66 C; flash pt. (TCC) -14.4 C; ref. index 1.4073 (20 C); surf. tens. 26.4 dynes/cm; sp. heat 0.469 cal/g/°C (liq.); dielec. const. 7.58 (20 C); 99.8% min. assay
Toxicology: LD50 (oral, rat) 1.65 g/kg; mod. eye irritant; mild skin irritant
Precaution: DOT: flamm. liq.; peroxides may form and accumulate under certain storage conditions

T-Mulz® 426 [Harcros USA]
Chem. Descrip.: Phosphate ester, free acid
Uses: Dry-cleaning detergent, industrial formulations, solvent degreasers, textile lubricant, agric. applics.
Regulatory: FDA compliance
Properties: Amber visc. liq.; sol. in ethanol, ethylene, xyelne; disp. in min. oil, kerosene; sp.gr. 1.00; dens. 8.3 lb/gal; acid no. 97 (pH 5.5); pour pt. 25 F; flash pt. (PMCC) > 350 F; pH 2-3 (1% aq.); surf. tens. 34.8 dynes/cm (0.05% aq.); 99.5% min. act.; anionic

TP-35 Solvent [Chemtronics]
Uses: Precision cleaner/defluxer; removes residual moisture from electronic equip.; for critical cleaning of PCBs, components, magnetic tape heads, optical lenses, precision instruments; safe for most plastics
Unverified

Transcutol [Gattefosse SA]
Chem. Descrip.: Ethoxydiglycol USP
CAS 111-90-0; EINECS 203-919-7
Uses: Solvent for active ingreds. in pharmaceutical preps.; solubilizer for difficult-to-formulate compds. for oral, topical, sublingual, and transdermal applics.; cosurfactant for microemulsions; useful in veterinary injections; in cosmetics for skin care, shave, hair, and nail prods., self-tanning prods.
Regulatory: Japanese compliance
Properties: Transparent limpid liq., faint odor; sol. in water, ethanol 96°; partly sol. in veg. oils; insol. in min. oil; sp.gr. 0.985-0.989 (20 C); f.p. -76 C; b.p. 195-202 C; acid no. < 0.10; flash pt. 96 C; ref. index 1.425-1.429 (20 C); 100% conc.
Toxicology: LD50 (oral, rat) 5.9 ml/kg, nontoxic; nonirritating to skin and eyes
Storage: Preserve in orig. hermetically closed container; stored under nitrogen; use rapidly after opening

Tri-Ethane [PPG Ind.]

Chem. Descrip.: Trichloroethane
Uses: All-purpose solvent for cold cleaning, vapor degreasing used in resin applics.; for dry-film photoresist processing; adhesive solvent; used in aerosols as solvent and vapor pressure depressant; formulating solvent for printing inks and lubricants; textile scouring, desizing, and dyeing to minimize pollution; contains effective stabilizers; for non-emissive use only
Properties: Water-wh. clear liq.; m.w. 133.42; f.p. -45 C; b.p. 74.1 C; sol. 0.07 g sol./100 g water; sp.gr. 1.304; dens. 10.84 lb/gal; visc. 0.735 cps; ref. index 1.4379; pH 7.0; nonflamm.

Triglyme [Ferro/Grant]
Chem. Descrip.: Triethylene glycol dimethyl ether
CAS 112-49-2
Uses: Solvent which tends to solvate cations; used in electrochemistry, polymer/boron chemistry; gas absorp., extraction, stabilization; used in industrial prods. such as fuels, lubricants, textiles, pesticides; solvent for pharmaceuticals
Properties: Colorless clear; mild, nonresidual odor; water-sol.; misc. with ethanol, acetone, benzene, diethyl ether, octane; m.w. 178.22; sp.gr. 0.9862; dens. 8.23 lb/gal; visc. 3.8 cP; f.p. -45 C; b.p. 216 C; flash pt. (CC) 111 C; ref. index 1.4224; pH neutral; surf. tens. 29.4 dynes/cm (20 C); 98.0% min. purity
Toxicology: LD50 5000 mg/kg; low to mod. acute toxicity; chronic exposure may cause reproductive effects
Environmental: Slowly biodeg.

Trivent DIA [Trivent]
Chem. Descrip.: Diisopropyl adipate
CAS 6938-94-9; EINECS 248-299-9
Uses: Lubricating dry emollient with strong solvent props.; hydro/alcoholic compat. for aftershave lotions and afterbath splashes; lubricant and detackifier in antiperspirants; drying agent in heavy oily systems
Properties: Colorless clear liq., faint typ. odor; sol. in min. oils, lower alcohols, ketones, esters, chlorinated hydrocarbons; insol. in water, glycerin; m.w. 230; sp.gr. 0.960; f.p. -2 C; acid no. 1 max.; iodine no. nil; sapon. no. 480-500; flash pt. (COC) > 170 C

Trivent NP-13 [Trivent]
Chem. Descrip.: Tridecyl neopentanoate
CAS 106436-39-9
Uses: Very dry emollient for nonoily creams and lotions; SPF booster for sunscreens; binder and pigment dispersant; plasticizer for hair setting polymers; cosolvent for oils and waxes in anhyd. sticks; wetting agent for Carbopol
Properties: Colorless to pale yel. liq., sl. mild odor; sol. @ 5% in 95% ethanol, min. oil, cyclomethicone, IPM, dimethicone, isocetyl alcohol, hexyelne glycol; insol. in water, glycerin, propylene glycol; m.w. 284; sp.gr. 0.850-0.860; acid no. 2 max.; sapon. no. 185-205; ref. index 1.434-1.437
Toxicology: LD50 (acute oral) > 5 g/kg; may cause eye and skin irritation; noncomedogenic

U

Ucar® n-Butyl Propionate [Union Carbide]
Chem. Descrip.: n-Butyl propionate
CAS 590-01-2
Uses: Solvent for high-solids and waterborne coatings; provides strong solvency, high elec. resistivity, good odor chars.
Properties: Sol. 0.2% in water (20 C); m.w. 130.19; sp.grr. 0.875 (20/20 C); dens. 7.30 lb/gal (20 C); visc. 0.9 cP (20 C); vapor pressure 3.4 mm Hg (20 C); b.p. 144.7 C; flash pt. (CC) 100 F; surf. tens. 25.3 dynes/cm (20 C); elec. resist. > 100 megohm

Ucar® Ester EEP [Union Carbide]
Chem. Descrip.: Ethyl 3-ethoxypropionate
CAS 763-69-9
Uses: Solvent for waterborne and high-solids coatings (NC, cellulose acetate, ethyl cellulose, butyrate, acrylic, urethane)
Properties: Sol. 1.6% in water (20 C); m.w. 146.19; sp.gr. 0.950 (20/20 C); dens. 7.91 lb/gal (20 C); visc. 1.3 cP (20 C); vapor pressure 0.8 mm Hg (20 C); b.p. 169.7 C; flash pt. (CC) 136 F

Ucar® Filmer IBT [Union Carbide]
Chem. Descrip.: 2,2,4-Trimethyl-1,3-pentanediol monoisobutyrate
CAS 25265-77-4; EINECS 246-771-9
Uses: Solvent, coalescing aid for water-disp. and water-sol. coatings
Properties: Sol. < 0.1% in water (20 C); m.w. 216.30; sp.gr. 0.950 (20/20 C); dens. 7.91 lb/gal (20 C); visc. 17.8 cP (20 C); vapor pressure < 0.01 mm Hg (20 C); b.p. 244.0 C; flash pt. (CC) 248 F

Ucar® n-Pentyl Propionate [Union Carbide]
Chem. Descrip.: n-Pentyl propionate
CAS 624-54-4
Uses: Solvent for high-solids and waterborne coatings; provides strong solvency, high elec. resistivity, good odor chars.
Properties: Sol. < 0.02% in water (20 C); m.w. 144.12; sp.grr. 0.874 (20/20 C); dens. 7.28 lb/gal (20 C); visc. 1.0 cP (20 C); vapor pressure 1.5 mm Hg (20 C); b.p. 164.9 C; flash pt. (CC) 135 F

Ucon® 50-HB-55 [Union Carbide]
Chem. Descrip.: PPG-2-buteth-3
Uses: Emollient for cosmetics; lubricant for compressor, rubber, greases, textiles; intermediate for prep. of resins, plasticizers, modifiers, and surfactants; solvent for inks and dyes; demulsifier; solvent, softener, and plasticizer for textile and paper coatings, NC coatings; mold release agents and dye solvs. for vinyl resins; antifoam in boiler water, latex processing, washing sol'ns., paints, adhesives, salt water evaporators
Properties: Sol. in water, alcohol; m.w. 270 (avg.); sp.gr. 0.971 (20/20 C); dens. 8.08 lb/gal (20 C); visc. 55 SUS (38 C); pour pt. -65 C; flash pt. (TOC) 90 C; ref. index 1.4402 (20 C)

Ucon® 50-HB-100 [Union Carbide]
Chem. Descrip.: PPG-3-buteth-5
Uses: Emollient for cosmetics; lubricant for compressor, rubber, greases, textiles; intermediate for prep. of resins, plasticizers, modifiers, and surfactants; solvent for inks and dyes; demulsifier; sol-

vent, softener, and plasticizer for textile and paper coatings, NC coatings; mold release agents and dye solvs. for vinyl resins; antifoam in boiler water, latex processing, washing sol'ns., paints, adhesives, salt water evaporators
Properties: M.w. 520 (avg.); sp.gr. 1.023 (20/20 C); dens. 8.51 lb/gal (20 C); visc. 100 SUS (38 C); pour pt. -51 C; flash pt. (PMCC) 141 C; ref. index 1.4491 (20 C)

Ucon® 50-HB-170 [Union Carbide]
Chem. Descrip.: PPG-5-buteth-7
Uses: Emollient for cosmetics; lubricant for compressor, rubber, greases, textiles; intermediate for prep. of resins, plasticizers, modifiers, and surfactants; solvent for inks and dyes; demulsifier; solvent, softener, and plasticizer for textile and paper coatings, NC coatings; mold release agents and dye solvs. for vinyl resins; antifoam in boiler water, latex processing, washing sol'ns., paints, adhesives, salt water evaporators
Properties: M.w. 750 (avg.); sp.gr. 1.031 (20/20 C); dens. 8.58 lb/gal (20 C); visc. 170 SUS (38 C); pour pt. -43 C; flash pt. (PMCC) 185 C; ref. index 1.4531 (20 C)

Ucon® 50-HB-260 [Union Carbide]
Chem. Descrip.: PPG-7-buteth-10
Uses: Emollient for cosmetics; lubricant for compressor, rubber, greases, textiles; intermediate for prep. of resins, plasticizers, modifiers, and surfactants; solvent for inks and dyes; demulsifier; solvent, softener, and plasticizer for textile and paper coatings, NC coatings; mold release agents and dye solvs. for vinyl resins; antifoam in boiler water, latex processing, washing sol'ns., paints, adhesives, salt water evaporators
Regulatory: FDA 21CFR §175.105, 176.210(d)(3), 178.3120
Properties: M.w. 970 (avg.); sp.gr. 1.035 (20/20 C); dens. 8.62 lb/gal (20 C); visc. 260 SUS (38 C); pour pt. -40 C; flash pt. (PMCC) 194 C; ref. index 1.4542 (20 C)

Ucon® 50-HB-400 [Union Carbide; Amerchol]
Chem. Descrip.: PPG-9-buteth-12
Uses: Emollient for cosmetics; lubricant for compressor, rubber, greases, textiles; intermediate for prep. of resins, plasticizers, modifiers, and surfactants; solvent for inks and dyes; demulsifier; solvent, softener, and plasticizer for textile and paper coatings, NC coatings; mold release agents and dye solvs. for vinyl resins; antifoam in boiler water, latex processing, washing sol'ns., paints, adhesives, salt water evaporators
Regulatory: FDA 21CFR §175.105, 176.210(d)(3), 178.3120, 178.3910
Properties: Clear liq., char. odor; m.w. 1230 (avg.); sp.gr. 1.041 (20/20 C); dens. 8.67 lb/gal (20 C); visc. 400 SUS (38 C); pour pt.-38 C; flash pt. (PMCC) 204 C; ref. index. 1.4576 (20 C); pH 5.5-7.5 (10% aq.)

Ucon® 50-HB-660 [Union Carbide; Amerchol]
Chem. Descrip.: PPG-12-buteth-16
Uses: Emollient for cosmetics; lubricant for compressor, rubber, greases, textiles; intermediate for prep. of resins, plasticizers, modifiers, and surfactants; solvent for inks and dyes; demulsifier; food processing; solvent, softener, and plasticizer for textile and paper coatings, NC coatings; mold release agents and dye solvs. for vinyl resins; antifoam in boiler water, latex processing, washing sol'ns., paints, adhesives, salt water evaporators
Regulatory: FDA 21CFR §173.310, 175.105, 176.210(d)(3), 178.3120, 178.3570, 178.3910
Properties: Clear liq., low char. odor; m.w. 1590 (avg.); sp.gr. 1.051 (20/20 C); dens. 8.75 lb/gal (20 C); visc. 660 SUS (38 C); pour pt. -34 C; flash pt. (PMCC) 202 C; ref. index 1.4576 (20 C); pH 5.5-7.5 (10% aq.)

Ucon® 50-HB-2000 [Union Carbide]
Chem. Descrip.: PPG-20-buteth-30
Uses: Emollient for cosmetics; lubricant for compressor, rubber,

greases, textiles; intermediate for prep. of resins, plasticizers, modifiers, and surfactants; solvent for inks and dyes; demulsifier; food processing; solvent, softener, and plasticizer for textile and paper coatings, NC coatings; mold release agents and dye solvs. for vinyl resins; antifoam in boiler water, latex processing, washing sol'ns., paints, adhesives, salt water evaporators

Regulatory: FDA 21CFR §173.310, 175.105, 176.210(d)(3), 178.3120, 178.3570, 178.3910

Properties: M.w. 2660 (avg.); sp.gr. 1.062 (20/20 C); dens. 8.84 lb/gal (20 C); visc. 2000 SUS (38 C); pour pt. -32 C; flash pt. (PMCC) 172 C; ref. index 1.4590 (20 C)

Ucon® 50-HB-3520 [Union Carbide; Amerchol]

Chem. Descrip.: PPG-28-buteth-35

Uses: Emollient for cosmetics; lubricant for compressor, rubber, greases, textiles; intermediate for prep. of resins, plasticizers, modifiers, and surfactants; solvent for inks and dyes; demulsifier; food processing; solvent, softener, and plasticizer for textile and paper coatings, NC coatings; mold release agents and dye solvs. for vinyl resins; antifoam in boiler water, latex processing, washing sol'ns., paints, adhesives, salt water evaporators

Regulatory: FDA 21CFR §173.310, 175.105, 176.210(d)(3), 178.3120, 178.3570, 178.3910

Properties: Clear liq., char. odor; m.w. 3380 (avg.); sp.gr. 1.062 (20/20 C); dens. 8.81 lb/gal (20 C); visc. 3520 SUS (38 C); pour pt. -29 C; flash pt. (PMCC) 172 C; ref. index 1.4598 (20 C); pH 4.5-7.5 (10% aq.)

Ucon® 50-HB-5100 [Union Carbide; Amerchol]

Chem. Descrip.: PPG-33-buteth-45

Uses: Emollient for cosmetics; lubricant for compressor, rubber, greases, textiles; intermediate for prep. of resins, plasticizers, modifiers, and surfactants; solvent for inks and dyes; demulsifier; food processing; solvent, softener, and plasticizer for textile and paper coatings, NC coatings; mold release agents and dye solvs. for vinyl resins; antifoam in boiler water, latex processing, washing sol'ns., paints, adhesives, salt water evaporators

Regulatory: FDA 21CFR §173.310, 173.340, 175.105, 1176.200, 176.210(d)(3), 177.1200, 178.3120, 178.3570, 178.3910

Properties: Clear visc. liq., low char. odor; m.w. 3930 (avg.); sp.gr. 1.063 (20/20 C); dens. 8.85 lb/gal (20 C); visc. 5100 SUS (38 C); pour pt. -29 C; flash pt. (PMCC) 180 C; ref. index 1.4596 (20 C); pH 5.5-7.5 (10% aq.)

Ucon® 75-H-450 [Union Carbide; Amerchol]

Chem. Descrip.: PEG/PPG-17/6 copolymer

CAS 9003-11-6

Uses: Lubricant for antiperspirants; emollient, solvent, bodying agent; lubricant for textile spin finishes; intermediate for prep. of resins, plasticizers, modifiers, and surfactants; demulsifier; solvent for inks and dyes; solvent, softener, and plasticizer for textile and paper coatings, NC coatings; mold release agents and dye solvents for vinyl resins

Regulatory: FDA 21CFR §175.105, 176.210(d)(3), 178.3120

Properties: Clear to cloudy liq., mild char. odor; m.w. 980 (avg.); sp.gr. 1.097 (20/20 C); dens. 9.13 lb/gal (20 C); visc. 450 SUS (38 C); pour pt.-15 C; flash pt. (PMCC) 188 C; ref. index 1.4655 (20 C); pH 5.5-7.5 (10% aq.)

Ucon® Fluid AP [Amerchol]

Chem. Descrip.: PPG-14 butyl ether

CAS 9003-13-8

Uses: Lubricant for antiperspirants; emollient, solvent, bodying agent

Properties: Clear liq., char. odor; visc. 85-105 cps; acid no. 0.5 max.; flash pt. (COC) 400 F min.

Ucon® LB-65 [Union Carbide]

Chem. Descrip.: PPG-5 butyl ether
CAS 9003-13-8
Uses: Lubricant for two-stroke cycle engines, rubber, greases, textiles; reactive intermediate for resins, plasticizers, modifiers, and surfactant mfg.; antifoam in boiler water, latex processing, washing sol'ns., paints, adhesives, salt water evaporators; solvent for inks and dyes; solvent, softener, and plasticizer for textile and paper coatings, NC coatings; mold release agents and dye solvents for vinyl resins; demulsifier; cosmetic emollient
Properties: M.w. 340 (avg.); sp.gr. 0.960 (20/20 C); dens. 7.99 lb/gal (20 C); visc. 65 SUS (38 C); pour pt. -57 C; flash pt. (PMCC) 119 C; ref. index 1.3281 (20 C)

Ucon® LB-135 [Union Carbide]
Chem. Descrip.: PPG-9 butyl ether
CAS 9003-13-8
Uses: Lubricant for two-stroke cycle engines, rubber, greases, textiles; reactive intermediate for resins, plasticizers, modifiers, and surfactant mfg.; antifoam in boiler water, latex processing, washing sol'ns., paints, adhesives, salt water evaporators; solvent for inks and dyes; solvent, softener, and plasticizer for textile and paper coatings, NC coatings; mold release agents and dye solvents for vinyl resins; demulsifier; cosmetic emollient
Properties: M.w. 640 (avg.); sp.gr. 0.981 (20/20 C); dens. 8.16 lb/gal (20 C); visc. 135 SUS (38 C); pour pt. -48 C; flash pt. (PMCC) 177 C; ref. index 1.4436 (20 C)

Ucon® LB-165 [Union Carbide]
Uses: Lubricant for two-stroke cycle engines, rubber, greases, textiles; reactive intermediate for resins, plasticizers, modifiers, and surfactant mfg.; antifoam in boiler water, latex processing, washing sol'ns., paints, adhesives, salt water evaporators; solvent for inks and dyes; solvent, softener, and plasticizer for textile and paper coatings, NC coatings; mold release agents and dye solvents for vinyl resins; demulsifier; cosmetic emollient
Properties: M.w. 740 (avg.); sp.gr. 0.983 (20/20 C); dens. 8.18 lb/gal (20 C); visc. 165 SUS (38 C); pour pt. -46 C; flash pt. (PMCC) 190 C; ref. index 1.4453 (20 C)

Ucon® LB-285 [Union Carbide]
Chem. Descrip.: PPG-15 butyl ether
CAS 9003-13-8
Uses: Lubricant for two-stroke cycle engines, rubber, greases, textiles; reactive intermediate for resins, plasticizers, modifiers, and surfactant mfg.; antifoam in boiler water, latex processing, washing sol'ns., paints, adhesives, salt water evaporators; solvent for inks and dyes; solvent, softener, and plasticizer for textile and paper coatings, NC coatings; mold release agents and dye solvents for vinyl resins; demulsifier; cosmetic emollient
Regulatory: FDA 21CFR §175.105, 176.210(d)(2) and (3), 178.3120, 178.3910
Properties: M.w. 1020 (avg.); sp.gr. 0.989 (20/20 C); dens. 8.23 lb/gal (20 C); visc. 285 SUS (38 C); pour pt. -40 C; flash pt. (PMCC) 179 C; ref. index 1.4482 (20 C)

Ucon® LB-385 [Union Carbide]
Chem. Descrip.: PPG-18 butyl ether
CAS 9003-13-8
Uses: Lubricant for two-stroke cycle engines, rubber, greases, textiles; reactive intermediate for resins, plasticizers, modifiers, and surfactant mfg.; antifoam in boiler water, latex processing, washing sol'ns., paints, adhesives, salt water evaporators; solvent for inks and dyes; solvent, softener, and plasticizer for textile and paper coatings, NC coatings; mold release agents and dye solvents for vinyl resins; demulsifier; cosmetic emollient
Regulatory: FDA 21CFR §175.105, 176.200, 176.210(d)(2) and (3), 177.1200, 178.3120, 178.3910
Properties: Low color and odor; sol. in alcohol; insol. in water; m.w. 1200 avg.; sp.gr. 0.994 (20/20 C); dens. 8.27 lb/gal (20 C); visc. 385 SUS (100 F); pour pt. -37 C; flash pt.

(PMCC) 177 C; ref. index 1.4489 (20 C)

Ucon® LB-525 [Union Carbide]
Chem. Descrip.: PPG-22 butyl ether
CAS 9003-13-8
Uses: Lubricant for two-stroke cycle engines, rubber, greases, textiles; reactive intermediate for resins, plasticizers, modifiers, and surfactant mfg.; antifoam in boiler water, latex processing, washing sol'ns., paints, adhesives, salt water evaporators; solvent for inks and dyes; solvent, softener, and plasticizer for textile and paper coatings, NC coatings; mold release agents and dye solvents for vinyl resins; demulsifier; cosmetic emollient
Regulatory: FDA 21CFR §175.105, 176.200, 176.210(d)(2) and (3), 177.1200, 178.3120, 178.3910
Properties: M.w. 1420 (avg.); sp.gr. 0.997 (20/20 C); dens. 8.30 lb/gal (20 C); visc. 525 SUS (38 C); pour pt. -34 C; flash pt. (PMCC) 185 C; ref. index 1.4487 (20 C)

Ucon® LB-625 [Union Carbide]
Chem. Descrip.: PPG-24 butyl ether
CAS 9003-13-8
Uses: Lubricant for two-stroke cycle engines, rubber, greases, textiles; reactive intermediate for resins, plasticizers, modifiers, and surfactant mfg.; antifoam in boiler water, latex processing, washing sol'ns., paints, adhesives, salt water evaporators; solvent for inks and dyes; solvent, softener, and plasticizer for textile and paper coatings, NC coatings; mold release agents and dye solvents for vinyl resins; demulsifier; cosmetic emollient
Regulatory: FDA 21CFR §175.105, 176.200, 176.210(d)(2) and (3), 177.1200, 178.3120, 175.3570, 178.3910
Properties: Low color and odor; sol. in alcohol; m.w. 1550; sp.gr. 1.00 (20/20 C); dens. 8.32 lb/gal (20 C); visc. 625 SUS (38 C); pour pt. -32 C; flash pt. (PMCC) 168 C; ref. index 1.4491 (20 C)

Ucon® LB-1145 [Union Carbide; Amerchol]
Chem. Descrip.: PPG-33 butyl ether
CAS 9003-13-8
Uses: Lubricant for two-stroke cycle engines, rubber, greases, textiles; reactive intermediate for resins, plasticizers, modifiers, and surfactant mfg.; antifoam in boiler water, latex processing, washing sol'ns., paints, adhesives, salt water evaporators; solvent for inks and dyes; solvent, softener, and plasticizer for textile and paper coatings, NC coatings; mold release agents and dye solvents for vinyl resins; demulsifier; cosmetic emollient
Regulatory: FDA 21CFR §175.105, 176.200, 176.210(d)(2) and (3), 177.1200, 178.3120, 175.3570, 178.3910
Properties: Low color and odor; sol. in alcohol; m.w. 2080 (avg.); sp.gr. 1.002 (20/20 C); dens. 8.34 lb/gal (20 C); visc. 1145 SUS (38 C); pour pt. -29 C; flash pt. (PMCC) 190 C; ref. index 1.4499 (20 C)

Ucon® LB-1715 [Union Carbide; Amerchol]
Chem. Descrip.: PPG-40 butyl ether
CAS 9003-13-8
Uses: Lubricant for two-stroke cycle engines, rubber, greases, textiles; reactive intermediate for resins, plasticizers, modifiers, and surfactant mfg.; antifoam in boiler water, latex processing, washing sol'ns., paints, adhesives, salt water evaporators; solvent for inks and dyes; solvent, softener, and plasticizer for textile and paper coatings, NC coatings; mold release agents and dye solvents for vinyl resins; demulsifier; cosmetic emollient
Regulatory: FDA 21CFR §175.105, 176.200, 176.210(d)(2) and (3), 177.1200, 178.3120, 175.3570, 178.3910
Properties: Clear liq., low char. odor; sol. in alcohol; m.w. 2490 (avg.); sp.gr. 1.002 (20/20 C); dens. 8.34 lb/gal (20 C); visc. 1715 SUS (38 C); pour pt. -23 C; flash pt. (PMCC) 188 C; ref. index 1.4503 (20 C); pH 5.0-8.5 (10% w/v)

Ucon® LB-3000 [Union Carbide]

Chem. Descrip.: PPG-53 butyl ether
CAS 9003-13-8
Uses: Lubricant for two-stroke cycle engines, rubber, greases, textiles; reactive intermediate for resins, plasticizers, modifiers, and surfactant mfg.; antifoam in boiler water, latex processing, washing sol'ns., paints, adhesives, salt water evaporators; solvent for inks and dyes; solvent, softener, and plasticizer for textile and paper coatings, NC coatings; mold release agents and dye solvents for vinyl resins; demulsifier; cosmetic emollient
Regulatory: FDA 21CFR §176.200, 177.1200, 178.3570, 178.3910
Properties: Low color and odor; sol. in alcohol; visc. 3000 SUS (38 C)

Unichem BZBN [Universal Preserv-A-Chem]
Chem. Descrip.: Benzyl benzoate
CAS 120-51-4; EINECS 204-402-9
Uses: Fixative and solvent for musk in perfumes and flavors; external medicine

Unimate® BYS [Union Camp]
Chem. Descrip.: n-Butyl stearate
CAS 123-95-5; EINECS 204-666-5
Uses: Skin emollient and solvent used in lipsticks, creams, lotions
Properties: Sol. in min. oil

Uniplex 110 [Unitex]
Chem. Descrip.: Dimethyl phthalate
CAS 131-11-3; EINECS 205-011-6
Uses: Solvent and plasticizer for cellulose acetate butyrate compositions; solvent for org. catalysts; plasticizer and solvent for aerosol hair sprays
Properties: APHA 10 max. clear liq., essentially odorless; sp.gr. 1.188-1.192; visc. 14-15 cps; ref. index 1.512-1.514; 99% min. assay

Uniplex 150 [Unitex]
Chem. Descrip.: Dibutyl phthalate
CAS 84-74-2; EINECS 201-557-4
Uses: Solvent/plasticizer in fingernail polish, nail polish remover, hair sprays, org. peroxide catalysts, adhesives, coatings; compat. with cellulosics, methacrylate, PS, PVB, vinyl chloride, urea-formaldehyde, melamine-formaldehye, phenolics; suitable for food contact use
Regulatory: FDA 21CFR §175.105, 175.300, 175.320, 176.170, 177.1200, 177.2430
Properties: APHA 25 max. clear liq., mild odor; pract. insol. in water; m.w. 278; sp.gr. 1.042; dens. 8.72 lb/gal; visc. 59 cps (32 F); f.p. -35 C; b.p. 340 C (760 mm Hg); ref. index 1.4905; 99% min. assay

Uniplex 155 [Unitex]
Chem. Descrip.: Diisobutyl phthalate
CAS 84-69-5
Uses: Plasticizer for thermoplastic and thermoset resins; solvent/plasticizer in cellophane, resin coated sand for foundry casting, org. peroxides; suitable for food contact use
Regulatory: FDA 21CFR §175.105, 175.300, 175.320, 176.170, 177.1200, 177.2510, 177.2550, 177.2600, 177.2800, 177.2910
Properties: APHA 25 max. clear liq., mild odor; pract. insol. in water; m.w. 278; sp.gr. 1.038-1.040; dens. 8.65 lb/gal; b.p. 327 C (760 mm Hg); 99% min. assay

Upamide ACMEA [Universal Preserv-A-Chem]
Chem. Descrip.: Acetamide MEA
CAS 142-26-7; EINECS 205-530-8
Uses: Solvent, humectant for cosmetics

Upiwax 350 [Universal Preserv-A-Chem]
Chem. Descrip.: PEG-6 methyl ether
CAS 9004-74-4
Uses: Intermediate, solvent

Ureresolve [Dynaloy]
Chem. Descrip.: Glycol-based solv.
Uses: Solvent used hot (175-205 C) for removing heavily encrusted urethanes; cleans residues from mixing and metering equip., nozzles, spray guns, etc.

Uresolve 411 [Dynaloy]
Chem. Descrip.: Methanol base
Uses: Flamm. solvent for urethane coatings, RTV and silicone oils; for electronic use (conformal coating removal, cable assembly repair, depotting)

Uresolve Blue [Dynaloy]
Chem. Descrip.: Methanol base
Uses: Solvent for electronic conformal coating removal, cable assembly repair, connector repair, depotting; for R.T. use only; most selective; suitable for cured urethanes
Precaution: Flamm.

Uresolve HF [Dynaloy]
Chem. Descrip.: Ethylene glycol ether base
Uses: Solvent for electronic depotting, decapsulation; stronger than Uresolve 411, but less selective; can be heated (to 37 C) for faster action

Uresolve Plus [Dynaloy]
Chem. Descrip.: Ethylene glycol ether base
Uses: Solvent for electronic depotting, decapsulation, and deinking applics.; surf. prep.; suitable for cured epoxies, urethanes, and silicones; stronger than Uresolve HF, but less selective; slow at R.T.; more effective @ 65-110 C

Uresolve Plus 500 [Dynaloy]
Chem. Descrip.: Ethylene glycol ether base
Uses: Solvent for electronic depotting, decapsulation, deinking, and surf. prep. applics.; stronger than Uresolve Plus, but less selective; slow @ R.T., more effective @ 65-100 C

Uresolve Plus SG [Dynaloy]
Chem. Descrip.: Ethylene glycol ether base
Uses: Semiconductor grade solvent; for depotting, deflashing, decapsulation, surf. prep., photo resist stripping applics.; can be used in contact with aluminum

V

Varonic® DM55 [Witco/Oleo-Surf.]
Chem. Descrip.: Methyl capped glycol ether
Uses: Solvent with low toxicity and exc. grease cutting properties for surf. cleaners; textile scouring; skin and hair care prods.
Properties: Liq.; water-emulsifiable; 100% solids
Unverified

Varsol® 1 [Exxon]
Chem. Descrip.: Aliphatic hydrocarbon solv. (Stoddard)
CAS 8052-41-3; EINECS 232-489-3
Uses: Solvent for coatings, printing inks
Properties: M.w. 140; sp.gr. 0.80 (15.6 C); dens. 6.65 lb/gal (60 F); visc. 1.0 cP; vapor pressure 1.6 mm Hg (20 C); b.p. 160-205 C; flash pt. 111 F; surf. tens. 26.6 dynes/cm; KB value 39; 38% paraffin, 47% cycloparaffin, 15% aromatic
Toxicology: OEL 100 ppm

Varsol® 18 [Exxon]
Chem. Descrip.: Aliphatic hydrocarbon solv. (Stoddard)
CAS 8052-41-3; EINECS 232-489-3
Uses: Solvent for coatings, printing inks
Properties: M.w. 145; sp.gr. 0.80 (15.6 C); dens. 6.62 lb/gal (60 F); visc. 0.9 cP; vapor pressure 5.6 mm Hg (20 C); b.p. 160-205 C; flash pt. 110 F; surf. tens. 26.3 dynes/cm; KB value 36; 40% paraffin, 53% cycloparaffin, 7% aromatic

Vista 47 [Condea Vista]
Chem. Descrip.: Hydrotreated mixts. of mainly isoparaffins and naphthenes with very low levels of aromatics
Uses: Special blend of hydrocarbon solvents designed to meet needs of printing ink formulations; esp. suited as an ink oil in heatset printing formulations
Regulatory: FDA 21CFR §172.884, 178.3650
Properties: Saybolt Universal +20 color liq.; m.w. 197; sp.gr. 0.812 (60/60 F); dens. 6.78 lb/gal (60 F); visc. 3.8 cSt (70 F); vapor pressure 0.10 mm Hg (100 F); f.p. 0 F; i.b.p. 464 F; pour pt. 0 F; flash pt. (PM) 228 F; KB value 27
Storage: Store in carbon steel tanks

Vista LPA [Condea Vista]
Chem. Descrip.: Hydrotreated mixts. of mainly isoparaffins and naphthenes with very low levels of aromatics
Uses: High purity, med. flash pt. solvent with broad boiling range; chemical process solvent; carrier in pesticide formulations, waterless hand cleaners; also for dry-cleaning, household/industrial cleaners, metal rolling oils, paints, coatings, thinners; paper and mining chems., water treatment chems., food applics.; as charcoal lighter fluids, degreasers, freeze pt. depressants, lamp oils, lubricating oils, printing ink oils
Regulatory: FDA 21CFR §172.884, 173.340, 175.105, 176.180, 176.200, 176.210, 177.1200, 177.1400, 177.2260, 177.2600, 177.2800, 178.3120, 178.3570, 178.3620(b), 178.3650, 178.3910(b)(2)
Properties: Clear liq., mild odor; m.w. 167; sp.gr. 0.809 (60/60 F); dens. 6.75 lb/gal (60 F); visc. 2.2 cSt (70

F); vapor pressure 1 mm Hg (100 F); f.p. -90 F; pour pt. -95 F; i.b.p. 362 F; flash pt. (TCC) 148 F; KB value 32

Precaution: DOT: combustible liq.

Vista LPA-110 [Condea Vista]

Chem. Descrip.: Hydrotreated mixts. of mainly isoparaffins and naphthenes with very low levels of aromatics

Uses: Solvent for use in dry-cleaning prods., household/industrial cleaning prods., metal rolling oils, paints, coatings, thinners, stains, paper and mining chems., pesticide formulations, water treatment chems., waterless hand cleaners; also offer exceptional performance as charcoal lighter fluids, chem. process solvs., degreasers, freeze pt. depressants, lamp oils, lubricating and printing ink oils

Regulatory: FDA 21CFR §172.884, 173.340, 175.105, 176.180, 176.200, 176.210, 177.1200, 177.1400, 177.2260, 177.2600, 177.2800, 178.3120, 178.3570, 178.3620(b), 178.3650, 178.3910(b)(2)

Properties: Saybolt Universal +30 color liq.; m.w. 141; sp.gr. 0.794 (60/60 F); dens. 6.63 lb/gal (60 F); visc. 1.4 cSt (70 F); vapor pressure 2 mm Hg (100 F); f.p. < -103 F; pour pt. < -112 F; i.b.p. 335 F; flash pt. (TCC) 112 F; KB value 36

Precaution: DOT: flamm. liq.

Vista LPA-142 [Condea Vista]

Chem. Descrip.: Hydrotreated mixts. of mainly isoparaffins and naphthenes with very low levels of aromatics

Uses: Solvent for industrial cleaning prods., as diluent in polymer processing, in charcoal lighter fluid; also for pesticides, coatings, household prods., chemical process solvs., ink oils, water treatment, paper, and mining chems., and food applics.

Regulatory: FDA 21CFR §172.884, 173.340, 175.105, 176.180, 176.200, 176.210, 177.1200, 177.1400, 177.2260, 177.2600, 177.2800, 178.3120, 178.3570, 178.3620(b), 178.3650, 178.3910(b)(2)

Properties: Clear liq., mild odor; m.w. 152; sp.gr. 0.809 (60/60 F); dens. 6.75 lb/gal (60 F); visc. 1.8 cSt (70 F); vapor pressure 1.1 mm Hg (100 F); f.p. < -103 F; i.b.p. 368 F; pour pt. < -30 F; flash pt. (TCC) 146 F; KB value 35

Precaution: DOT: combustible liq.

Vista LPA-170 [Condea Vista]

Chem. Descrip.: Hydrotreated mixts. of mainly isoparaffins and naphthenes with very low levels of aromatics

Uses: Solvent for prod. of polymer emulsions, in mining solvent extraction processes, metal lubricants, specialty coatings; also for pesticides, household prods., chemical process solvs., ink oils, water treatment, paper, and mining chems.

Regulatory: FDA 21CFR §172.884, 173.340, 175.105, 176.180, 176.200, 176.210, 177.1200, 177.1400, 177.2260, 177.2600, 177.2800, 178.3120, 178.3570, 178.3620(b), 178.3650, 178.3910(b)(2)

Properties: Clear liq., mild odor; m.w. 171; sp.gr. 0.811 (60/60 F); dens. 6.77 lb/gal (60 F); visc. 2.4 cSt (70 F); vapor pressure 0.37 mm Hg (100 F); f.p. -81 F; i.b.p. 413 F; pour pt. < -30 F; flash pt. (TCC) 178 F; KB value 32

Precaution: DOT: combustible liq.

Vista LPA-210 [Condea Vista]

Chem. Descrip.: Hydrotreated mixts. of mainly isoparaffins and naphthenes with very low levels of aromatics

Uses: Solvent for applics. where odor is critical, esp. chem. processing, printing ink oil, metal lubricant, pesticide carrier; also for coatings, household prods., water treatment, paper, and mining chems.

Regulatory: FDA 21CFR §178.3620(c), 178.3910(a)(3)

Properties: Clear liq., mild odor; m.w. 194; sp.gr. 0.823 (60/60 F); dens. 6.87 lb/gal (60 F); visc. 3.8 cSt (70 F); vapor pressure 0.10 mm Hg (100 F); f.p. -43 F; i.b.p. 465 F; pour pt. < -30 F; flash pt. (PM) 226 F; KB value 29

Vorite 105 [CasChem]

Chem. Descrip.: Polymerized castor oil

Uses: Pigment wetting/dispersing agent; plasticizer for resins, gums, polymers; lubricant, penetrant; coupling solvent; adhesion promoter; for cellulose lacquers, inks, adhesives, industrial lubricants, polishes, caulks, leather dressing, hydraulic fluids; rubber compounding, gasket cement

Properties: Gardner 4 color; sp.gr. 0.975; visc. 26 stokes; pour pt. -5 F; acid no. 2; iodine no. 85; sapon. no. 170; hyd. no. 130

Vorite 110 [CasChem]

Chem. Descrip.: Polymerized castor oil

Uses: Pigment wetting/dispersing agent; plasticizer for resins, gums, polymers; lubricant, penetrant; coupling solvent; adhesion promoter; for cellulose lacquers, inks, adhesives, industrial lubricants, polishes, caulks, leather dressing, hydraulic fluids; rubber compounding, gasket cement

Properties: Gardner 3 oil; sp.gr. 0.990; visc. 115 stokes; pour pt. 15 F; acid no. 2; iodine no. 82; sapon. no. 166; hyd. no. 102

Vorite 115 [CasChem]

Chem. Descrip.: Polymerized castor oil

Uses: Pigment wetting/dispersing agent; plasticizer for resins, gums, polymers; lubricant, penetrant; coupling solvent; adhesion promoter; for cellulose lacquers, inks, adhesives, industrial lubricants, polishes, caulks, leather dressing, hydraulic fluids; rubber compounding, gasket cement

Properties: Gardner 3 oil; sp.gr. 0.995; visc. 192 stokes; pour pt. 20 F; acid no. 2; iodine no. 85; sapon. no. 165; hyd. no. 93

Vorite 120 [CasChem]

Chem. Descrip.: Polymerized castor oil

Uses: Pigment wetting/dispersing agent; plasticizer for resins, gums, polymers; lubricant, penetrant; coupling solvent; adhesion promoter; for cellulose lacquers, inks, adhesives, industrial lubricants, polishes, caulks, leather dressing, hydraulic fluids; rubber compounding, gasket cement

Properties: Gardner 4 oil; sp.gr. 1.001; visc. 700 stokes; pour pt. 50 F; acid no. 2; iodine no. 82; sapon. no. 160; hyd. no. 78

Vorite 125 [CasChem]

Chem. Descrip.: Polymerized castor oil

Uses: Pigment wetting/dispersing agent; plasticizer for resins, gums, polymers; lubricant, penetrant; coupling solvent; adhesion promoter; for cellulose lacquers, inks, adhesives, industrial lubricants, polishes, caulks, leather dressing, hydraulic fluids; rubber compounding, gasket cement

Properties: Gardner 4 oil; sp.gr. 1.007; visc. 900 stokes; pour pt. 55 F; acid no. 2; iodine no. 84; sapon. no. 157; hyd. no. 72

W

Wickenol® 127 [Alzo]
Chem. Descrip.: Isopropyl stearate
CAS 112-10-7; EINECS 203-934-9
Uses: Emollient, cosolvent, and lubricant
Properties: Gardner -1 color; sol. in acetone, min. and veg. oils, ethyl acetate, alcohol; insol. in water; pour pt. 18 C; acid no. 0.1; ref. index 1.438

Wickenol® 136 [Alzo]
Chem. Descrip.: IPP, IPM, and isopropyl stearate
Uses: Emollient, conditioner, moisturizer, solubilizer, vehicle, and solvent for cosmetic and toilet preparations
Properties: Gardner -1 liq.; sol. in alcohols, ketones, aromatic and aliphatic hydrocarbons; insol. in water; pour pt. 7 C; acidno. 0.1; ref. index 1.435
Unverified

Wickenol® 151 [Alzo]
Chem. Descrip.: Isononyl isononanoate
CAS 42131-25-9
Uses: Emollient, moisturizer, pigment wetter; silky emolliency and solvent chars.; for hair care prods.
Properties: Gardner -1 color; sol. in alcohol, animal, veg., and min. oil; insol. in water; pour pt. < -15 C; acid no. 0.10; ref. index 1.435
Unverified

Wickenol® 152 [Alzo]
Chem. Descrip.: Isodecyl isononanoate
CAS 41395-89-5
Uses: Silky emollience and solvent char. for skin and hair care prods.; pigment wetter
Properties: Gardner -1 color; sol. in alcohol, animal, veg., and min. oils; pour pt. < -15 C; acid no. 0.10; ref. index 1.438
Unverified

Wickenol® 153 [Alzo]
Chem. Descrip.: Isotridecyl isononanoate
Uses: Silky emollience and solvent char. for skin and hair care prods.; pigment wetter, moisturizer
Properties: Gardner -1 color; sol. in alcohol, animal, veg., and min. oils; pour pt. < -18 C; acid no. 0.10; ref. index 1.444
Unverified

Wickenol® 158 [Alzo]
Chem. Descrip.: Dioctyl adipate
CAS 103-23-1; EINECS 203-090-1
Uses: Emollient, moisturizer, pigment wetter/dispersant, cosolvent; increases water vapor porosity of fatty components used in cosmetic and topical pharmaceutical preparations
Properties: Gardner -1 color; sol. in alcohol, animal, veg., and min. oils; pour pt. -20 C; acid no. 0.10; ref. index 1.445
Unverified

Wickenol® 171 [CasChem]
Chem. Descrip.: Octyl hydroxystearate
CAS 29383-26-4
Uses: Emollient, moisturizer, pigment wetter/dispersant; increases water vapor porosity of fatty components used in cosmetic and topical pharmaceutical preparations; refatting agent, counterirritant, cosolvent, solubilizer
Properties: Gardner 2+ color; sol. in

acetone, castor, corn, and min. oil, chloroform, ethyl acetate, ethanol; pour pt. 5 C; acid no. 0.30; ref. index 1.456

Unverified

Witcamide® CMEA [Witco/Oleo-Surf.]

Chem. Descrip.: Acetamide MEA

CAS 142-26-7; EINECS 205-530-8

Uses: Solvent, humectant for cosmetics

Properties: Liq.; 70-100% conc.; nonionic

Witcohol™ 42 [Witco/Oleo-Surf.]

Chem. Descrip.: Cetyl-stearyl-oleyl alcohol

Uses: Solvent; raw material for detergents, emulsifiers, lube oil additives, textile auxiliaries, germicides, chem. intermediates, org. synthesis, resins, and lubricants

Properties: Wh. soft waxy solid; mild char. fatty odor; sol. in IPA, benzene, trichloroethylene, ethyl ether, acetone, turpentine, VM&P naptha, kerosene, lt. min. oil; m.w. 250-267; sp.gr. 0.814 (60/25 C); visc. 40 SSU (210 F); b.p. 315-345 C; acid no. 1.5 max.; iodine no. 45-55; sapon. no. 3.5 max.; hyd. no. 210-225; cloud pt. 36 C max.

Witcohol™ 66 [Witco/Oleo-Surf.]

Chem. Descrip.: Isostearyl alcohol

CAS 27458-93-1; EINECS 248-470-8

Uses: Emulsion thickener and stabilizer, cosolvent, emollient for use in cosmetic applics.; chem. intermediates, high-temp. lubricant; offers low pour pt., reduced greasiness compared to oleyl alcohol and other emollient agents, low aftertack, oxidation stability; imparts luxurious feel to skin

Properties: APHA 100 max. liq.; sol. in IPA, ethyl ether, acetone, VM&P naptha, lt. min. oil; m.w. 268; sp.gr. 0.881; visc. 58 cps; acid no. 1.0 max.; sapon. no. 1.5 max.; hyd. no. 200 min.; pour pt. 18 F; cloud pt. 5 C max.

Witcohol™ 320 [Witco/Oleo-Surf.]

Chem. Descrip.: Oleyl, cetyl and stearyl alcohols; straight-chain, monohydric, mono-unsat.

Uses: Commercial grade alcohol for use in antifoam agents, coupling agents, plasticizers, lubricants, emulsion stabilizers, petrol. additives, chem. intermediates

Properties: APHA 100 max. liq.; sol. in warm ethanol, IPA, trichloroethylene, benzene, ethyl ether, acetone, turpentine, VM&P naptha, kerosene, lt. min. oil; acid no. 1.0 max.; iodine no. 75-85; sapon. no. 3.0 max.; hyd. no. 205-225; cloud pt. 66 F

Witcohol™ 330 [Witco/Oleo-Surf.]

Chem. Descrip.: Oleyl, cetyl and stearyl alcohols; straight-chain, monohydric, mono-unsat.

Uses: Commercial grade alcohol for use in antifoam agents, coupling agents, plasticizers, lubricants, emulsion stabilizers, petrol. additives, chem. intermediates

Properties: APHA 100 max. liq.; sol. in warm ethanol, IPA, trichloroethylene, benzene, ethyl ether, acetone, turpentine, VM & P naptha, kerosene, lt. min. oil; acid no. 1.0 max.; iodine no. 72-82; sapon. no. 3.0 max.; hyd. no. 205-225; cloud pt. 72 F

Witcohol™ 340 [Witco/Oleo-Surf.]

Chem. Descrip.: Oleyl, cetyl and stearyl alcohols; straight-chain, monohydric, mono-unsat.

Uses: Commercial grade alcohol for use in antifoam agents, coupling agents, plasticizers, lubricants, emulsion stabilizers, petrol. additives, chem. intermediates

Properties: APHA 100 max. liq.; sol. in warm ethanol, IPA, trichloroethylene, benzene, ethyl ether, acetone, turpentine, VM & P naptha, kerosene, lt. min. oil; acid no. 1.0 max.; iodine no. 60-70; sapon. no. 3.0 max.; hyd. no. 205-225; cloud pt. 82 F

Witcohol™ 350 [Witco/Oleo-Surf.]

Chem. Descrip.: Oleyl alcohol

CAS 143-28-2; EINECS 205-597-3

Uses: Chem. intermediate for antifoam agents, coupling agents, plasticizers, lubricants, emulsion stabilizers, petrol. additives

Properties: APHA 100 max. liq.; acid no. 0.2 max.; iodine no. 70-75; sapon. no. 1.0 max.; hyd. no. 210-220; cloud pt. 18-22 C

Witconol™ APEM [Witco/Oleo-Surf.]
Chem. Descrip.: PPG-3-myreth-3
CAS 37311-04-9
Uses: Lubricant, emulsifier, wetting agent, penetrant, dye dispersant, scouring aid, antistat, solvent coupler; syn. oils for personal care prods.; metal processing; textile surfactant
Properties: Liq.; sol. in 3A ethanol, min. oil.; nonionic

Witconol™ APM [Witco/Oleo-Surf.; Witco Surf. GmbH]
Chem. Descrip.: PPG-3 myristyl ether
CAS 63793-60-2
Uses: Lubricant, emulsifier, wetting agent, penetrant, dye dispersant, scouring aid, antistat, solvent coupler; metal processing; textile surfactant; emollient oil for cosmetics and toiletries, esp. stick prods.; solubilizer for alcohol in oils
Properties: Gardner 1 liq.; sol. in oil; sp.gr. 0.90; pH 7.0; 100% act.; nonionic

Witconol™ APS [Witco/Oleo-Surf.]
Chem. Descrip.: PPG-11 stearyl ether
CAS 25231-21-4
Uses: Lubricant, emulsifier, wetting agent, penetrant, dye dispersant, scouring aid, antistat, solvent coupler; syn. oils for personal care prods.; metal processing; textile surfactant; emollient oil for cosmetics and toiletries, solubilizer
Properties: Gardner 1 liq.; sol. in oil; sp.gr. 0.94; pH 7.0; nonionic

Y

York Krystal Kleer Castor Oil [United Catalysts]
Chem. Descrip.: Castor oil USP contg. an antioxidant
CAS 8001-79-4; EINECS 232-293-8
Uses: High quality oil imparting gloss and emollience and as a dye solvent in personal care prods., e.g., lipsticks, pharmaceuticals, and urethane applics.
Properties: Gardner 1 color; sp.gr. 0.959; dens. 7.98 lb/gal; visc. 7.5 stokes; acid no. 1; iodine no. 86; sapon. no. 180; hyd. no. 164

York USP Castor Oil [United Catalysts]
Chem. Descrip.: Castor oil USP
CAS 8001-79-4; EINECS 232-293-8
Uses: High quality oil imparting gloss and emollience and as a dye solvent in personal care prods., e.g., lipsticks, pharmaceuticals, urethane applics.
Properties: Gardner 1 color; sp.gr. 0.959; dens. 7.98 lb/gal; visc. 7.5 stokes; acid no. 1; iodine no. 86; sapon. no. 180; hyd. no. 164

Part II
Chemical Dictionary/ Cross-Reference

Chemical Dictionary/Cross-Reference

2-AB. *See* s-Butylamine
Absolute alcohol. *See* Alcohol
Aceite de Algodon. *See* Cottonseed oil
Aceite de ricino. *See* Castor oil

Acetal
CAS 105-57-7; EINECS 203-310-6
UN No. 1088 (DOT)
Synonyms: Polyacetal; Acetaldehyde diethyl acetal; 1,1-Diethoxyethane; Diethyl acetal; Ethylidene diethyl ether
Empirical: $C_6H_{14}O_2$
Formula: $CH_3CH(OC_2H_5)_2$
Properties: Colorless volatile liq.; agreeable odor; fruity green flavor; sol. in heptane, ethyl acetate; misc. with alcohol, ether; sl. sol. in water; m.w. 118.18; dens. 0.831; vapor pressure 10 mm Hg (8 C); b.p. 103-104 C; flash pt. (CC) 36 C
Toxicology: LD50 (oral, rat) 4.57 g/kg, (IP, rat) 900 mg/kg; moderately toxic by ingestion and IP route; skin and eye irritant; narcotic
Precaution: Highly flamm.; dangerous fire hazard exposed to heat or flames; tends to polymerize on standing; can react vigorously with oxidizers; forms heat-sensitive explosive peroxides on contact with air; Hazardous decomp. prods.: Heated to decomp., emits acrid smoke and fumes
Uses: Solvent, cosmetics, perfumes, synthetic flavoring for nonalcoholic beverages, ice cream, pharmaceuticals
Regulatory: FDA §172.515; FEMA GRAS
Manuf./Distrib.: Aldrich; Ashland; Berje; Coz; Fluka; Hoechst Celanese; Hüls AG; LNP; Penta Mfg.; Westlake Plastics

Acetaldehyde diethyl acetal. *See* Acetal

Acetamide MEA
CAS 142-26-7; EINECS 205-530-8
Synonyms: N-Acetyl ethanolamine; N-(2-Hydroxyethyl)acetamide
Classification: Aliphatic amide
Empirical: $C_4H_9NO_2$
Uses: Solvent, humectant, skin and hair conditioner, intermediate, coupling agent, pigment dispersant, clarifying agent for shampoos, moisturizer
Regulatory: FDA 21CFR §175.105
Manuf./Distrib.: Aldrich; Heterene; McIntyre; Scher; Witco/Oleo-Surf.
Trade names: Acetamide MEA; Foamid AME-70; Foamid AME-75; Foamid AME-100; Incromectant AMEA-70; Upamide ACMEA; Witcamide® CMEA

Acetate C-8. *See* Octyl acetate
1-Acetate-1,2,3-propanetriol. *See* Glyceryl acetate

Acetdimethylamide. *See* Dimethyl acetamide

Acetic acid
CAS 64-19-7; EINECS 200-580-7
UN No. 2789 (DOT), 2790 (DOT)
Synonyms: Ethanoic acid; Vinegar acid; Methanecarboxylic acid; Ethyllic acid; Pyroligneus acid
Classification: Aliphatic organic acid
Empirical: $C_2H_4O_2$
Formula: CH_3COOH
Properties: Clear colorless liq., pungent odor, sharply acid taste; misc. with water, alcohol, glycerol, ether; insol. in carbon disulfide; m.w. 60.03; dens. 1.0492 (20/4 C); m.p. 16.63 C; b.p. 118 C (765 mm); visc. 1.22 cps (20 C); vapor pressure 11.4 mm Hg (20 C); flash pt. (OC) 43 C; ref. index 1.3715 (20 C)
Toxicology: LD50 (oral, rat) 3310 mg/kg; ACGIH TLV 10 ppm; STEL 15 ppm; mod. toxic by ingestion, inhalation; corrosive; strong irritant to eyes, skin, and tissue; caustic; can cause burns, lachrymation; human systemic effects by ing.
Precaution: Flamm.; moderate fire and explosion hazard exposed to heat or flame; can react vigorously with oxidizers; explosive or violent reactions possible; incompat. with many chems.; Hazardous decomp. prods.: Heated to decomp., emits irritating fumes
NFPA: Health 2, Flammability 3, Reactivity 0
Uses: Mfg. of acetic anhydride, cellulose acetate, vinyl acetate monomer; acetic esters; prod. of plastics, pharmaceuticals, cosmetics, aroma chems., dyes, insecticides, photographic chemicals, food additives; solvent reagent; acidifier; flavoring agent; boiler water additive; coagulant for latex; solvent/vehicle in foods, pharmaceuticals
Regulatory: FDA 21CFR §73.85, 133, 172.814, 178.1010, 184.1005, GRAS; USDA 9CFR §318.7; BP; Europe listed; UK approved; USP/NF, BP compliance
Manuf./Distrib.: Air Prods.; Aldrich; Ashland; BASF; BP Chems. Inc; Brown; Chemical; Coyne; Fluka; General Chem.; Hoechst Celanese; Janssen Chimica; Olin; PMC Spec.; Quantum; Rhône-Poulenc N. Am.; Ruger; Seeler Ind.; Sigma; Spectrum Chem. Mfg.; Veckridge; Wacker-Chemie GmbH

Acetic acid amyl ester. *See* Amyl acetate
Acetic acid benzyl ester. *See* Benzyl acetate
Acetic acid-2-butoxy ester. *See* s-Butyl acetate
Acetic acid 2-(2-butoxyethoxy) ethyl ester. *See* Diethylene glycol butyl ether acetate
Acetic acid, butyl ester. *See* n-Butyl acetate
Acetic acid 2-butyl ester. *See* s-Butyl acetate
Acetic acid n-butyl ester. *See* n-Butyl acetate
Acetic acid, sec-butyl ester. *See* s-Butyl acetate
Acetic acid t-butyl ester. *See* t-Butyl acetate
Acetic acid, C7-9-branched alkyl esters, C8-rich. *See* C8 alkyl acetate
Acetic acid cyclohexyl ester. *See* Cyclohexyl acetate
Acetic acid dimethylamide. *See* Dimethyl acetamide
Acetic acid-1,3-dimethylbutyl ester. *See* Methyl amyl acetate
Acetic acid 1,1-dimethylethyl ester. *See* t-Butyl acetate
Acetic acid esters of mono- and diglycerides of fatty acids. *See* Acetylated mono- and diglycerides of fatty acids
Acetic acid 2-ethoxyethyl ester. *See* Ethoxyethanol acetate
Acetic acid, ethyl ester. *See* Ethyl acetate
Acetic acid 2-ethylhexyl ester. *See* 2-Ethylhexyl acetate

Acetic acid, glacial
CAS 64-19-7; EINECS 200-580-7

UN No. 2789, 2790
Synonyms: Ethanoic acid; Ethanolic acid; Concentrated acetic acid; Vinegar acid; Glacial acetic acid
Classification: Aliphatic organic acid
Empirical: $C_2H_4O_2$
Formula: CH_3COOH
Properties: Clear colorless liq., pungent char. odor, acid taste when dil. with water; misc. with water, alcohol, acetone, glycerol, ether; insol. in chloroform; m.w. 60.05; dens. 1.049 (20/4 C); m.p. 16.2 C; b.p. 118 C; flash pt. 40 C; ref. index 1.3720
Toxicology: TLV:TWA 10 ppm; human poison by unspecified routes; mod. toxic by various routes; corrosive; severe eye and skin irritant; can produce lung obstruction
Precaution: DOT: Corrosive material
NFPA: Health 3, Flammability 2, Reactivity 0
Uses: Acidifying agent, buffering agent, solvent for pharmaceuticals
Regulatory: USP/NF, BP, Ph.Eur. compliance
Manuf./Distrib.: Aldrich; Am. Biorganics; Ashland; Chemical; Eastman; Fluka; Houghton Chem.; Penta Mfg.; Quantum; Ruger; Seeler Ind.; Sigma; Spectrum Chem. Mfg.

Acetic acid hexyl ester. *See* Hexyl acetate
Acetic acid, 2-hydroxyethyl ester. *See* Ethylene glycol acetate
Acetic acid isobutyl ester. *See* Isobutyl acetate
Acetic acid isopentyl ester. *See* Isoamyl acetate
Acetic acid isopropyl ester. *See* Isopropyl acetate
Acetic acid-3-methoxybutyl ester. *See* Methoxybutyl acetate
Acetic acid (3-methoxy-n-butyl) ester. *See* Methoxybutyl acetate
Acetic acid 2-methoxyethyl ester. *See* Methoxyethanol acetate
Acetic acid, 2-methoxy-1-methyl ethyl ester. *See* Propylene glycol methyl ether acetate
Acetic acid methyl ester. *See* Methyl acetate
Acetic acid 1-methylethyl ester. *See* Isopropyl acetate
Acetic acid 1-methylpropyl ester. *See* s-Butyl acetate
Acetic acid 2-methylpropyl ester. *See* Isobutyl acetate
Acetic acid, monoglyceride. *See* Glyceryl acetate
Acetic acid octyl ester. *See* Octyl acetate
Acetic acid pentyl ester. *See* Amyl acetate
Acetic acid phenylmethyl ester. *See* Benzyl acetate
Acetic acid propyl ester. *See* Propyl acetate
Acetic acid n-propyl ester. *See* Propyl acetate
Acetic acid, trifluoro-. *See* Trifluoroacetic acid
Acetic ester. *See* Ethyl acetate
Acetic ether. *See* Ethyl acetate
Acetin. *See* Glyceryl acetate
Acetin. *See* Triacetin
Acetoacetic acid ethyl ester. *See* Ethylacetoacetate
Acetoacetic acid methyl ester. *See* Methyl acetoacetate
Acetoacetic ester. *See* Ethylacetoacetate
Acetoacetone. *See* Acetylacetone
Aceto-caustin. *See* Trichloroacetic acid

Acetone
CAS 67-64-1; EINECS 200-662-2
UN No. 1090 (DOT)

Acetonic acid

Synonyms: Dimethylketone; Dimethylketal; Dimethyl formaldehyde; Methyl ketone; Propanone; 2-Propanone; Ketone propane; β-Ketopropane; Pyroacetic acid; Pyroacetic ether
Classification: Aliphatic ketone
Empirical: C_3H_6O
Formula: CH_3COCH_3
Properties: Colorless volatile transparent liq., sweetish odor, pungent sweetish taste; sol. in water, alcohol, chloroform, dimethylformamide, ether, most volatile oils; m.w. 58.09; sp.gr. 0.792 (20/20 C); vapor pressure 400 mm Hg (39.5 C); m.p. -94.3 C; b.p. 56.2 C; flash pt. (CC) -18 C
Toxicology: LD50 (oral, mouse) 3000 mg/kg; ACGIH TLV/TWA 750 ppm; STEL 1000 ppm; narcotic in high conc.; moderately toxic by ingestion and inhalation; lg. doses may cause narcosis; extreme concs. may cause collapse, coma, death; inh. irritates lungs
Precaution: DOT: flamm. liq.; dangerous fire risk; explosive limit in air 2.6-12.8%; reacts violently with oxidizing agents, chlorinated solv./alkali mixts.; reacts vigorously with sulfur dichloride, potassium t-butoxide, hexachloromelamine
NFPA: Health 1, Flammability 3, Reactivity 0
Uses: Solvent for paints, varnishes, lacquers, resins, fats, waxes, adhesives, printing inks, plastics, pharmaceuticals; degreasing agent; for cleaning and drying precision equip.; delustrant for cellulose acetate fibers; extraction solvent for food use; mfg. of acetate rayon, photographic film, paint removers, glues, cements, plastics, rubber, dyes, explosives
Regulatory: FDA 21CFR §73.1, 73.30, 73.345, 73.615, 173.210, 175.105, 175.320, 176.300, 182.99; 40CFR §180.1001; 30 ppm tolerance in spice oleoresins; FEMA GRAS; Japan approved with restrictions; USP/NF, BP compliance; VOC-exempt
Manuf./Distrib.: Aldrich; Allchem Ind.; AlliedSignal; Aristech; Ashland; BASF; J.T. Baker; Baychem; BP Chems. Ltd; R.E. Carroll; Chemcentral; Chemical; Dow; Eastman; Elf Atochem N. Am.; Exxon; Fluka; General Chem.; Georgia Gulf; C.P. Hall; Harcros USA; Houghton Chem.; Mitsui Petrochem. Ind.; Montedipe SpA; Olin; Primachem; Samson; Shell; Sigma; Spectrum Chem. Mfg.; Sunnyside; Texaco; Union Carbide; Van Waters & Rogers; Veckridge

Acetonic acid. *See* Lactic acid

Acetonitrile

CAS 75-05-8; EINECS 200-835-2
UN No. 1648 (DOT)
Synonyms: ACN; Methyl cyanide; Methanecarbonitrile; Cyanomethane; Ethanenitrile; Ethyl nitrile
Classification: Nitrile/cyano compd.
Empirical: C_2H_3N
Formula: CH_3CN
Properties: Colorless liq., ether-like odor; sol. in water, methanol, ether, acetone, chloroform, CCl_4, ethylene chloride, methyl acetate; m.w. 41.05; sp.gr. 0.783; m.p. -45 C; b.p. 81.6 C; flash pt. (CC) 12.8 C; ref. index 1.3440
Toxicology: LD50 (oral, rat) 3800 mg/kg; ACGIH TLV/TWA 40 ppm; STEL 60 ppm (skin); poisonous; toxic by skin absorp. and inh.; skin irritant; may cause severe eye irritation; lachrymator; inh. of vapors can cause nose/throat irritation
Precaution: DOT: Flamm. liq.; dangerous fire risk; incompat. with oxidizing agents, acids, reducing agents (violent/explosive reactions possible); reacts slowly with water or steam to release toxic and flamm. vapors (ammonia, hydrogen cyanide); Hazardous decomp. prods.: Acetic acid, ammonia, hydro-

gen cyanide; heated to decomp., emits highly toxic fumes of CN^-, NO_x
NFPA: Health 3, Flammability 3, Reactivity 0
Uses: Solvent for hydrocarbon extraction processes, esp. for butadiene; intermediate; catalyst; for separation of fatty acids from vegetable oils; mfg. of synthetic pharmaceuticals, Vitamin B, perfumes, syn. fibers, other chems.; refining of hydrocarbons, waxes, oils, greases; chem. purification; rubber antioxidant; alcohol denaturant
Regulatory: BP compliance; SARA reportable
Manuf./Distrib.: Aldrich; Allchem Ind.; BP Chems. Ltd; Burdick & Jackson; DuPont; EniChem Am.; Fluka; R.W. Greeff; ICC Ind.; Janssen Chimica; Mitsui Petrochem. Ind.; Mitsui Toatsu; Montedipe SpA; Penta Mfg.; Sigma; Spectrum Chem. Mfg.

Acetophenone
CAS 98-86-2; EINECS 202-708-7
UN No. 1993
Synonyms: Ketone methyl phenyl; Methyl phenyl ketone; Phenyl methyl ketone; Acetylbenzene; 1-Phenylethanone; Hypnone; Benzoyl methide
Classification: Aromatic organic compd.
Empirical: C_8H_8O
Formula: $C_6H_5COCH_3$
Properties: Colorless liq., sweet pungent odor and taste; sol. in alcohol, chloroform, ether, fatty oils, glycerol; sl. sol. in water; m.w. 120.15; dens. 1.030 (20/20 C); vapor pressure 1 mm Hg (15 C); m.p. 20.5 C; b.p. 201.7 C; flash pt. 82.2 C; ref. index 1.5339 (20 C)
Toxicology: LD50 (oral, rat) 815 mg/kg; poison by IP, subcut. routes; moderate toxicity by ingestion; skin and eye irritant; narcotic in high concs.; a hypnotic; may cause allergic reaction
Precaution: DOT: Combustible liq.; flamm. exposed to heat, flames, or oxidizers; Hazardous decomp. prods.: Heated to decomp., emits acrid smoke and fumes
NFPA: Health 1, Flammability 2, Reactivity 0
Uses: Perfumery, solvent, intermediate for pharmaceuticals, resins, paints, etc.; flavoring agent for foods, pharmaceuticals; polymerization catalyst, organic synthesis
Regulatory: FDA 21CFR §172.515; FEMA GRAS; Japan approved for flavoring
Manuf./Distrib.: Aldrich; Alemark; Allchem Ind.; Amber Syn.; ARCO; BP Chems. Ltd; EniChem Am.; Fluka; R.W. Greeff; Janssen Chimica; Mitsui Petrochem. Ind.; Mitsui Toatsu; Montedipe SpA; Penta Mfg.; Rhône-Poulenc N. Am.; Sigma; Spectrum Chem. Mfg.

1-Acetoxy-2-butoxyethane. *See* Butoxyethanol acetate
Acetoxyethane. *See* Ethyl acetate
1-Acetoxy-2-methoxyethane. *See* Methoxyethanol acetate
Acetoxymethylbenzene. *See* Benzyl acetate
2-Acetoxypentane. *See* s-Amyl acetate
1-Acetoxypropane. *See* Propyl acetate
2-Acetoxypropane. *See* Isopropyl acetate
α-Acetoxytoluene. *See* Benzyl acetate

Acetylacetone
CAS 123-54-6; EINECS 204-634-0
UN No. 2310 (DOT)
Synonyms: 2,4-Pentanedione; Pentanedione-2,4; Diacetylmethane; Acetoacetone
Empirical: $C_5H_8O_2$
Formula: $CH_3COCH_2COCH_3$
Properties: Colorless to sl. yel. liq., pleasant odor; misc. with alcohol, benzene,

ether, chloroform, acetone, glac. acetic acid, propylene glycol; insol. in water; m.w. 100.13; dens. 0.952-0.962; m.p. -23.2 C; b.p. 139 C (746 mm); flash pt. (OC) 105 F

Toxicology: LD50 (oral, rat) 1000 mg/kg, (IP, mouse) 750 mg/kg; mod. toxic by ingestion, IP, inh. routes; skin and severe eye irritant; narcotic in high doses

Precaution: Flamm. exposed to heat or flame; incompat. with oxidizing materials; light-sensitive; refrigerate

NFPA: Health 2, Flammability 2, Reactivity 0

Uses: Solvent for cellulose acetate; intermediate; chelating agent for metals, paint driers, lubricant additives, pesticides; synthesis of vitamin B_6 and K; intermediate for drugs, pesticides; prod. of corrosion inhibitors; synthetic flavoring agent

Regulatory: FDA 21CFR §172.515; BP compliance

Manuf./Distrib.: Aceto; AC Ind.; Aldrich; Fluka; Monomer-Polymer & Dajac Labs; Penta Mfg.; Sigma; Spectrum Chem. Mfg.; Union Carbide; Wacker Chems.; Wacker-Chemie GmbH

Acetylated hydrogenated lard glyceride

CAS 8029-91-2

Synonyms: Glycerides, lard mono-, hydrogenated, acetates

Definition: Acetyl ester of the monoglyceride derived from hydrog. lard

Uses: Emulsifier, emollient, plasticizer for cosmetic and technical prods.; food additive; lubricant, solvent, plasticizer for foodstuffs; forms films with good moisture vapor barrier properties

Regulatory: FDA 21CFR §172.828

Trade names: Axol® E 61

Acetylated hydrogenated soybean oil glyceride

Uses: Emulsifier; food-grade lubricant and emollient; deaerator in some systems; food additive; lubricant, solvent for foods

Regulatory: FDA 21CFR §172.828

Trade names: Myvacet® 9-45K

Acetylated lanolin alcohol

CAS 61788-49-6; EINECS 262-980-8

Synonyms: Lanolin, alcohols, acetates

Classification: Acetyl ester

Uses: Emollient for cosmetic and pharmaceutical creams and lotions

Trade names containing: Acetol® 1706; Acetulan®; Fancol ALA; Ritawax AEO; Ritawax ALA

Acetylated lard glyceride

CAS 8029-92-3

Synonyms: Glycerides, lard mono-, acetates

Classification: Acetyl ester

Uses: Emulsifier, lubricant, emollient, deaerator; food emulsifier; cosolvent for polar additives in low polarity systems

Regulatory: FDA 21CFR §172.828, 175.230

Trade names: Grindtek AMOS 90

Acetylated mono- and diglycerides of fatty acids

CAS 68990-55-6; 68990-58-9

Synonyms: Acetic acid esters of mono- and diglycerides of fatty acids

Definition: Partial or complete esters of glycerin with a mixture of acetic acid and edible fat-forming fatty acids

Properties: White to pale yel. liq., bland taste; sol. in alcohol, acetone; insol. in water; HLB 2-3

Toxicology: Heated to decomp., emits acrid smoke and fumes
Uses: Food emulsifier, coating agent, stabilizer, lubricant, solvent, texture modifying agent; also for pharmaceutical use
Regulatory: FDA 21CFR §172.828; Europe listed; UK approved
Trade names: Axol® E 66

Acetylated palm kernel glycerides
Synonyms: Glycerides, palm kernel oil mono-, di-, and tri-, acetates
Classification: Acetyl ester
Uses: Lubricant and plasticizer for plastics and coatings; cosolvent for polar additives in low polarity systems
Trade names: Grindtek AML 60

Acetylbenzene. *See* Acetophenone
Acetylene trichloride. *See* Trichloroethylene
N-Acetyl ethanolamine. *See* Acetamide MEA
2-(Acetyloxy)-1,2,3-propanetricarboxylic acid, tributyl ester. *See* Acetyl tributyl citrate
2-(Acetyloxy)-1,2,3-propanetricarboxylic acid, triethyl ester. *See* Acetyl triethyl citrate
2-Acetyl propane. *See* Methyl isopropyl ketone

Acetyl tributyl citrate
CAS 77-90-7; EINECS 201-067-0
Synonyms: ATBC; 2-(Acetyloxy)-1,2,3-propanetricarboxylic acid, tributyl ester; Tributyl acetyl citrate
Classification: Aliphatic ester
Definition: Ester of citric acid
Empirical: $C_{20}H_{34}O_8$
Formula: $CH_3COOC_3H_4(COOC_4H_9)_3$
Properties: Colorless sl. visc. liq., sweet herbaceous odor; sol. in alcohol; insol. in water; m.w. 402.49; dens. 1.14; b.p. > 300 C; flash pt. 204 C; ref. index 1.4408
Toxicology: Heated to decomp., emits acrid smoke and irritating fumes
Uses: Plasticizer for vinyl resins, food pkg.; synthetic flavoring for foods and pharmaceuticals; solvent in household, I&I cleaners; solvent, film-former, and plasticizer for inks, adhesives, coatings
Regulatory: FDA 21CFR §172.515, 175.105, 175.300, 175.320, 178.3910, 181.22, 181.27; FEMA GRAS
Manuf./Distrib.: AC Ind.; Aldrich; Allchem Ind.; Morflex; Pfizer Int'l.; Reilly Chems. SA; Unitex
Trade names: Citroflex® A-4

Acetyl triethyl citrate
CAS 77-89-4; EINECS 201-066-5
Synonyms: ATEC; 2-(Acetyloxy)-1,2,3-propanetricarboxylic acid, triethyl ester; Tricarballylic acid-β-acetoxytributyl ester; Triethyl acetylcitrate
Classification: Aliphatic ester
Empirical: $C_{14}H_{22}O_8$
Formula: $CH_3COOC_3H_4(COOC_2H_5)_3$
Properties: Colorless, odorless liq.; sl. sol. in water; m.w. 318.36; dens. 1.135 (25 C); flash pt. 187 C
Toxicology: Moderate toxicity by intraperitoneal route; mild toxicity by ingestion
Precaution: Combustible
Uses: Plasticizer for cellulosics, primarily ethyl cellulose, food pkg.; solvent for household, I&I cleaners, pharmaceuticals; plasticizer for pharmaceutical coatings; solvent, film-former, plasticizer for inks, adhesives, coatings

Acetyl trimethyl citrate

Regulatory: FDA 21CFR §175.105, 175.300, 175.320, 178.3910, 181.22, 181.27
Manuf./Distrib.: Aldrich; Morflex; Unitex
Trade names: Citroflex® A-2

Acetyl trimethyl citrate
CAS 146340-22-9
Empirical: $C_{11}H_{16}O_8$
Properties: Sol. in toluene, sol. 5.5 g/100 ml water; insol. in heptane; m.w. 276; sp.gr. 1.2324; vapor pressure 0.16 mm Hg (20 C); flash pt. (COC) 311 F; ref. index 1.4440
Uses: Solvent in household, I&I cleaners; solvent, film-former, plasticizer for inks, adhesives, coatings
Trade names: Citroflex® A-1

ACN. *See* Acetonitrile
Acrylic acid, polymer with sucrose-polyallyl ether. *See* Carbomer
Active primary amyl alcohol. *See* 2-Methyl-1-butanol
Adeps lanae. *See* Lanolin
Adipic acid bis (2-ethylhexyl) ester. *See* Dioctyl adipate
Adipic acid, dibutyl ester. *See* Dibutyl adipate
Adipic acid diisopropyl ester. *See* Diisopropyl adipate
Adipic acid dimethyl ester. *See* Dimethyl adipate
Adipic acid dinitrile. *See* Adiponitrile
Adipic acid nitrile. *See* Adiponitrile
Adipodinitrile. *See* Adiponitrile

Adiponitrile
CAS 111-69-3
UN No. 2205 (DOT)
Synonyms: Adipic acid dinitrile; Adipic acid nitrile; Adipodinitrile; Adipylnitrile; 1,4-Dicyanobutane; Hexanedinitrile; Hexanedioic acid dinitrile; Tetramethylene cyanide
Classification: Alkyl nitrile; cyano compd.
Empirical: $C_6H_8N_2$
Formula: $NC(CH_2)_4CN$
Properties: Water-wh. oily liq.; pract. odorless; sol. in methanol, ethanol, chloroalkanes; mod. sol. in water; m.w. 108.1; sp.gr. 0.965 (20/4 C); m.p. 1 C; b.p. 295 C; flash pt. (CC) 159 C
Toxicology: LD50 (oral, rat) 155 mg/kg, (IP, mouse) 40 mg/kg; ACGIH TLV/TWA 2 ppm (skin); poison by inh., ing., subcut., IP routes; may cause skin and eye irritation; inh. of mist can probably cause nose/throat irritation at high concs.
Precaution: Flamm. exposed to heat or flame; incompat. with oxidizing agents (increases fire/explosion risk), strong acids (violent reactions possible releasing toxic and flamm. hydrogen cyanide); Hazardous decomp. prods.: Hydrogen cyanide; heated to decomp., emits toxic fumes of CN^-
NFPA: Health 4, Flammability 2, Reactivity 1
Uses: Intermed. in mfg. of nylon; mfg. of other chems.; as polar solvent
Manuf./Distrib.: Aldrich; Fluka

Adipylnitrile. *See* Adiponitrile
Adronol acetate. *See* Cyclohexyl acetate
Albacol. *See* n-Propyl alcohol

Alcohol
CAS 64-17-5; EINECS 200-578-6
UN No. 1170 (DOT), 1986

Synonyms: EtOH; Ethyl alcohol; Ethyl alcohol, undenatured; Ethanol; Ethanol, undenatured; Distilled spirits; Absolute alcohol
Definition: Undenatured ethyl alcohol
Empirical: C_2H_6O
Formula: CH_3CH_2OH
Properties: Colorless limpid, volatile liq., vinous odor, pungent taste; misc. with water, methanol, ether, chloroform, 95% acetone; m.w. 46.08; dens. 0.816 (15.5 C); b.p. 78.3 C; f.p. -117.3 C; flash pt. (CC) 12.7 C; ref. index 1.365 (15 C)
Toxicology: LD50 (oral, rat) 7060 mg/kg, (IP, rat) 3750 mg/kg; ACGIH TLV/TWA 1000 ppm; CNS depressant; moderately toxic to humans by ingestion; eye and mucous membrane irritant; severe skin irritant; ing. of lg. doses can cause alcohol poisoning
Precaution: DOT: Flamm. liq.; dangerous fire and explosion hazard; can react vigorously with oxidizers; reacts violently with many chemicals
NFPA: Health 0, Flammability 3, Reactivity 0
Uses: Solvent; extraction medium; mfg. of acetaldehyde, denatured alcohol, paints, waterborne coatings, pharmaceuticals (tonics, colognes), perfumery, organic synthesis; octane booster in gasoline; topical antiseptic; antimicrobial, extraction solvent, vehicle for foods; active ingred. in OTC drug prods.
Regulatory: FDA 21CFR §169.3, 169.175, 169.176, 169.177, 169.178, 169.180, 169.181, 172.340, 172.560, 175.105, 176.200, 176.210, 177.1440, 184.1293, GRAS; 27CFR §2.5, 2.12; FEMA GRAS; USP/NF, BP, Ph.Eur. compliance; Japan restricted
Manuf./Distrib.: Aldrich; Ashland; J.T. Baker; Baychem; BP Chems. Inc; Chemcentral; Condea Vista; Coyne; Eastman; Fluka; Georgia-Pacific Resins; Gist-brocades SpA; Grain Processing; Great Western; C.P. Hall; Harcros USA; Houghton Chem.; Quantum; Samson; Sigma; Spectrum Chem. Mfg.; Sunnyside; Union Carbide; Van Waters & Rogers
Trade names: Punctilious® Ethyl Alcohol; Punctilious® SDA 1-1; Tecsol® D-1
Trade names containing: Filmex® A-1; Filmex® A-2; Filmex® B; Filmex® C; Filmex® D-1; Filmex® D-2; Propsolv I-1; Propsolv I-2; Propsolv III-1; Propsolv III-2

Alcohol C-3. *See* n-Propyl alcohol
Alcohol C-5. *See* n-Amyl alcohol
Alcohol C-6. *See* Hexyl alcohol
Alcohol C-7. *See* Heptyl alcohol
Alcohol C-8. *See* Caprylic alcohol
Alcohol C_8. *See* 2-Ethylhexanol; Isooctyl alcohol
Alcohol C-10. *See* n-Decyl alcohol
Alcohol C16. *See* Cetyl alcohol
Alcohol C_{22}. *See* Behenyl alcohol

Alcohol denatured
CAS 64-17-5; EINECS 200-578-6
UN No. 1170 (DOT), 1986 (DOT), 1987 (DOT)
Synonyms: Denatured alcohol; Denatured ethyl alcohol; Denatured ethanol; Denatured spirits; Potato alcohol
Definition: Alcohol denatured with one or more denaturing agents
Empirical: C_2H_6O
Formula: CH_3CH_2OH
Properties: Colorless clear volatile liq.; sweet alcohol odor; hygroscopic; sol. in water, methanol, diethyl ether, chloroform, acetone; m.w. 46.08; sp.gr. 0.789 (20 C); m.p. -114 C; b.p. 78 C; flash pt. (CC) 13 C
Toxicology: TLV: 1000 ppm in air; potentially poisonous by ingestion

Precaution: Flamm. liq.; fire hazard; mod. explosion hazard
NFPA: Flammability:; 3, Reactivity 0
Uses: Mfg. of acetaldehyde and other chemicals, solvents, antifreeze, brake fluids, fuels; solvent used in protective coatings, printing inks, paint and varnish removers
Manuf./Distrib.: Aldrich; Ashland; BP Chems. Ltd; Fluka; Midwest Grain Prods.; Sigma; Spectrum Chem. Mfg.
Trade names: Eastman® CDA-19

Alcohols, lanolin. *See* Lanolin alcohol
Alcohols, tallow. *See* Tallow alcohol
Alginic acid, ester with 1,2-propanediol. *See* Propylene glycol alginate
Alkane C-4. *See* Butane
Alkane C_5. *See* n-Pentane
Alkane C-6. *See* Hexane
Alkane C_7. *See* Heptane
Alkanes, C7-8-iso-. *See* C7-8 isoparaffin
Alkanes, C8-9-iso-. *See* C8-9 isoparaffin
Alkanes, C10-11-iso-. *See* C9-14 isoparaffin
Alkanes, C10-11-iso-. *See* C10-11 isoparaffin
Alkanes, C10-13-iso-. *See* C10-13 isoparaffin
Alkanes, C11-12-iso-. *See* C11-12 isoparaffin
Alkanes, C11-13-iso-. *See* C11-13 isoparaffin
Alkanes, C13-14-iso-. *See* C13-14 isoparaffin
Almond artificial essential oil. *See* Benzaldehyde
Almond oil. *See* Sweet almond oil
Almond oil, sweet. *See* Sweet almond oil

Aluminum/magnesium hydroxide stearate
Definition: Complexation prod. of stearic acid with aluminum/magnesium hydroxide
Trade names containing: Gilugel MIG; Gilugel MIN; Gilugel SIL5

Aluminum sodium silicate. *See* Sodium silicoaluminate
Amide C_1. *See* Formamide
Amine C_5. *See* Pentylamine
Aminic acid. *See* Formic acid
2-Aminobutane. *See* s-Butylamine
γ-Aminobutyric acid lactam. *See* 2-Pyrrolidone
Aminoethane. *See* Ethylamine
1-Aminoethane. *See* Ethylamine
2-Aminoethoxyethanol. *See* 2-(2-Aminoethoxy) ethanol

2-(2-Aminoethoxy) ethanol
CAS 929-06-6; EINECS 213-195-4
UN No. 3055 (DOT)
Synonyms: 2-Aminoethoxyethanol; Diethylene glycol amine; Diglycolamine; Ethanol, 2-(2-aminoethoxy)-; β-(β-Hydroxyethoxy) ethylamine
Empirical: $C_4H_{11}NO_2$
Formula: $NH_2CH_2CH_2OCH_2CH_2OH$
Properties: Colorless, slightly viscous liq., amine odor; misc. in water, alcohols; m.w. 105.16; dens. 1.0572 (20/20 C); b.p. 221 C; f.p. -12.5 C; flash pt. 126.6 C
Toxicology: LD50 (oral, rat) 5660 mg/kg, (skin, rabbit) 1190 mg/kg; mod. toxic by skin contact; mildly toxic by ing.; primary irritant; strong irritant to tissue; severe eye, skin irritant; anesthetic, somnolent effect
Precaution: Corrosive; Hazardous decomp. prods.: Heated to decomp., emits

toxic fumes of NO_x
Uses: Solvent for removal of acid components from gases, recovery of aromatics from refinery streams; intermediate; prep. of foam stabilizers, wetting agents, emulsifiers, condensation polymers
Manuf./Distrib.: Aldrich; BASF AG; Fluka; Sigma
Trade names: Diglycolamine® Agent (DGA®)

Aminoethylethandiamine. *See* Diethylenetriamine
N-(2-Aminoethyl)ethylenediamine. *See* Diethylenetriamine
1-Aminopentane. *See* Pentylamine
2-Aminopentane. *See* Diethylamine
2-Aminopropane. *See* Isopropylamine
Ammonia aqua. *See* Ammonium hydroxide
Ammonia aqueous. *See* Ammonium hydroxide
Ammonia solution. *See* Ammonium hydroxide
Ammonia solution, strong. *See* Ammonium hydroxide
Ammonia water. *See* Ammonium hydroxide

Ammonium hydroxide
CAS 1336-21-6; EINECS 215-647-6
UN No. 2672 (DOT)
Synonyms: Ammonia solution; Ammonia solution, strong; Strong ammonia solution; Ammonia water; Aqueous ammonia; Ammonia aqueous; Ammonia aqua; Aqua ammonia; Aqua ammonium; Spirit of Hartshorn
Classification: Inorganic base
Empirical: H_5NO
Formula: NH_4OH
Properties: Clear colorless liq., very pungent odor, acrid taste; sol. in water; m.w. 35.06; dens. 0.90; m.p. -77 C; flash pt. none; pH 13.6
Toxicology: LD50 (oral, rat) 350 mg/kg; human poison by ingestion; experimental poison by inh. and ing.; inhalation irritant; severe eye irritant; liq. can inflict burns
Precaution: DOT: Corrosive material; vapor will ignite, but not readily; incompat. with acrolein, nitromethane, acrylic acid, chlorosulfonic acid, dimethyl sulfate, halogens, HCl, HF, HNO_3, propylene oxide, etc.; Hazardous decomp. prods.: Heated to decomp., emits NH_3 and NO_x
NFPA: Health 3, Flammability 1, Reactivity 0
Uses: Food additive, alkali, leavening agent, pH control agent, processing aid; boiler water additive; food coloring diluent/solvent; taste and odor control in water treatment; alkalizing agent for pharmaceuticals
Regulatory: FDA 21CFR §163.110, 182.90, 184.1139, GRAS; Europe listed; UK approved; USP/NF compliance
Manuf./Distrib.: Aldrich; Alexander; AlliedSignal; Ashland; Brown; Chemical; Coyne; Cytec Ind.; LaRoche Ind.; Ruger; Sigma; Spectrum Chem. Mfg.; Vigoro Ind.

Ammonium xylenesulfonate
CAS 26447-10-9; EINECS 247-710-9
Synonyms: Benzenesulfonic acid, dimethyl-, ammonium salt
Definition: Ammonium salt of ring sulfonated mixed xylene isomers
Formula: $C_8H_{10}O_3S \cdot H_3N$
Properties: Sol. in water; m.w. 203.26
Toxicology: Chronic toxicity or skin effects are not known; narcotic in high doses
Precaution: Flamm.
Uses: Hydrotrope, stabilizer, solubilizer used in formulating detergents, inks, electroplating baths, dyestuffs, polymers; lacquer solvent used in nail polishes

Manuf./Distrib.: Witco/Oleo-Surf.
Trade names: Stepanate® AXS

Amyl acetate
CAS 628-63-7; EINECS 211-047-3
UN No. 1104 (DOT)
Synonyms: n-Amyl acetate; Amylacetic ester; Amylacetic ether; Primary amyl acetate; Pear oil; Pent-acetate; Acetic acid amyl ester; Acetic acid pentyl ester; Pentyl acetate; 1-Pentanol acetate; n-Pentyl acetate; 1-Pentyl acetate
Classification: Carboxylic acid ester
Definition: Ester of amyl alcohol and acetic acid
Empirical: $C_7H_{14}O_2$
Formula: $CH_3COOC_5H_{11}$
Properties: Colorless liq. (pure), yel. liq. (commercial); pear or banana-like odor; completely sol. in alcohol, ether; sl. sol. in water; m.w. 130.19; sp.gr. 0.876 (20 C); m.p. -70.8 C; b.p. 149 C; flash pt. (CC) 25 C; ref. index 1.4013
Toxicology: LD50 (oral, rat) 6500 mg/kg; ACGIH TLV/TWA 100 ppm; mod. toxic by IP route; skin and eye irritant; inhalation of vapor can irritate nose/throat; higher exposure may cause breathing difficulty, increased heart rate, CNS depression
Precaution: Flamm.; dangerous fire hazard exposed to heat or flame; incompat. with reducing agents, strong acids, strong bases (decomp. can occur), oxidizing agents (increased risk of fire and explosion); mod. explosive as vapor exposed to flame; Hazardous decomp. prods.: Heated to decomp., emits acrid smoke and irritating fumes
NFPA: Health 1, Flammability 3, Reactivity 0
Uses: Solvent for lacquers and paints; extraction of penicillin; mfg. of photographic film, artificial leather, nail polishes; warning odor; flavoring agent for foods and pharmaceuticals; printing and finishing fabrics; solvent for phosphors in fluorescent; anti-inflammatory agent
Regulatory: FDA 21CFR §175.105; FEMA GRAS; BP compliance
Manuf./Distrib.: Aldrich; Ashland; BP Chems. Ltd; Chemcentral; Coyne; Fluka; Houghton Chem.; Penta Mfg.; Sigma; Spectrum Chem. Mfg.; Union Carbide; Van Waters & Rogers

n-Amyl acetate. *See* Amyl acetate

s-Amyl acetate
CAS 626-38-0
UN No. 1104 (DOT)
Synonyms: 2-Acetoxypentane; 2-Pentyl acetate; 1-Methylbutyl acetate; 2-Pentanol acetate
Classification: Carboxylic acid ester
Empirical: $C_7H_{14}O_2$
Formula: $CH_3COOCH(CH_3)CH_2CH_2CH_3$
Properties: Colorless clear liq.; fruity odor; sol. in alcohol, ether; sl. sol. in water; m.w. 130.19; sp.gr. 0.862-0.866 (20/20 C); m.p. -78.5 C; b.p. 134 C; flash pt. (CC) 32 C
Toxicology: LCLo (inh., guinea pig, 5 h) 10,000 ppm; ACGIH TLV/TWA 125 ppm; mild skin and eye irritant; inh. of vapor may cause nose/throat irritation; higher exposure may cause difficulty breathing, increased heart rate, CNS depression symptoms
Precaution: Flamm.; mod. explosive as vapor exposed to heat or flame; incompat. with oxidizing materials (increase fire/explosion hazard), strong acids, strong bases, reducing materials (decomp. can occur); Hazardous decomp. prods.: Heated to decomp., emits acrid smoke and irritating fumes

NFPA: Health 1, Flammability 3, Reactivity 0
Uses: Solvent for nitrocellulose, ethylcellulose; mfg. of lacquers, varnishes, cements, artificial leather, nail enamels, coated paper, printing compds.

Amylacetic ester. *See* Amyl acetate
Amylacetic ether. *See* Amyl acetate

n-Amyl alcohol
CAS 71-41-0; EINECS 200-752-1
UN No. 1105 (DOT)
Synonyms: 1-Pentanol; Pentan-1-ol; n-Pentanol; n-Pentyl alcohol; Alcohol C-5; n-Butyl carbinol; Pentyl alcohol
Classification: Primary aliphatic alcohol
Empirical: $C_5H_{12}O$
Formula: $CH_3(CH_2)_4OH$
Properties: Colorless clear liq., somewhat sweet balsamic odor; sl. sol. in water; misc. with alcohol, ether; m.w. 88.15; sp.gr. 0.812 (20/4 C); m.p. -79 C; b.p. 137-139 C; flash pt. (CC) 38 C; ref. index 1.409
Toxicology: LD50 (oral, rat) 3030 mg/kg; highly toxic and narcotic; ingestion of 30 mg can cause death in humans; irritating to respiratory tract; severe eye and skin irritant; can be absorbed thru skin in toxic amts.
Precaution: DOT: Flamm. liq.; incompat. with oxidizing agents (increases fire/explosion risk); NPFA; Health 1, Flammability 3, Reactivity 0
Uses: Solvent in organic synthesis; solvent in nail lacquers; raw material for pharmaceutical preparations; plastics processing; synthetic flavoring agent for foods and pharmaceuticals
Regulatory: FDA 21CFR §172.515; FEMA GRAS; BP compliance
Manuf./Distrib.: Aldrich; Allchem Ind.; Ashland; Condea Vista; Fluka; Hoechst Celanese; Houghton Chem.; Sigma; Spectrum Chem. Mfg.; Union Carbide

s-Amyl alcohol
CAS 6032-29-7
UN No. 1105 (DOT)
Synonyms: 2-Pentanol; Pentanol-2; sec-Amyl alcohol; 2-Hydroxypentane; 1-Methyl-1-butanol; Methyl propyl carbinol; Pentan-2-ol; sec-Pentyl alcohol
Classification: Sec. aliphatic alcohol
Empirical: $C_5H_{12}O$
Formula: $CH_3CH_2CH_2CH(OH)CH_3$
Properties: Colorless liq.; mild odor; sol. in alcohol, ether; sl. sol. in water; m.w. 88.15; sp.gr. 0.8098 (20 C); f.p. -50 C; m.p. -75 C; b.p. 119 C; flash pt. (CC) 33 C
Toxicology: LD50 (oral, rat) 1470 mg/kg; mod. toxic by ing., IP route; narcotic; skin and severe eye irritant; can be absorbed thru skin; inh. may cause nose/throat irritation, headache, dizziness, coughing, nausea, vomiting
Precaution: Flamm. exposed to heat or flame; severe explosion hazard exposed to heat or flame; incompat. with oxidizing agents (increases fire/explosion hazard), hydrogen trisulfide (explosive reaction possible); Hazardous decomp. prods.: Heated to decomp., emits acrid smoke and irritating fumes
NFPA: Health 0, Flammability 2, Reactivity 0
Uses: Solvent for paints and lacquers; intermediate for pharmaceuticals
Manuf./Distrib.: Aldrich; Air Prods.; Fluka; Penta Mfg.

sec-Amyl alcohol. *See* s-Amyl alcohol

t-Amyl alcohol
CAS 75-85-4; EINECS 200-908-9
UN No. 1105 (DOT)
Synonyms: 2-Methyl-2-butanol; 2-Methyl butanol-2; 3-Methylbutan-3-ol; tert-Amyl alcohol; t-Pentyl alcohol; t-Pentanol; Dimethyl ethyl carbinol; Amylene

hydrate
Classification: Tert. aliphatic alcohol
Empirical: $C_5H_{12}O$
Formula: $CH_3CH_2C(CH_3)_2OH$
Properties: Colorless clear volatile liq., char. camphoraceous odor, burning taste; sol. in 8 parts water; misc. with alcohol, ether, benzene, chloroform, glycerin, oils; m.w. 88.15; sp.gr. 0.808 (20/4 C); vapor pressure 10 mm Hg (17.2 C); m.p. -9 C; b.p. 100-103 C; flash pt. (CC) 19 C; ref. index 1.405
Toxicology: LD50 (oral, rat) 1 g/kg; toxic; mod. irritating to human mucous membranes; drying to skin; can be absorbed thru skin; inh. of vapors may cause nose/throat irritation, headache, dypsnea, coughing, nausea, vomiting
Precaution: Flamm. exposed to heat, flame, or oxidizing materials; mod. explosive as vapor exposed to heat or flame; Hazardous decomp. prods.: Heated to decomp., emits acrid smoke and irritating fumes
NFPA: Health 1, Flammability 3, Reactivity 0
Uses: Solvent for epoxy and PU coatings; pharmaceutical solvent; flotation agent; stabilizer and intermediate for organic synthesis; medicine (sedative)
Regulatory: USP/NF compliance
Manuf./Distrib.: AC Ind.; Aldrich; Allchem Ind.; BASF; BASF AG; Fluka; Sigma; Spectrum Chem. Mfg.

tert-Amyl alcohol. *See* t-Amyl alcohol
Amylamine. *See* Pentylamine
n-Amylamine. *See* Pentylamine
Amylcarbinol. *See* Hexyl alcohol
Amylene hydrate. *See* t-Amyl alcohol
Amyl hydride. *See* n-Pentane
Amyl methyl alcohol. *See* 2-Methyl-1-pentanol
Amyl methyl carbinol. *See* 2-Heptanol
Amyl methyl ketone. *See* Methyl n-amyl ketone
n-Amyl methyl ketone. *See* Methyl n-amyl ketone
Amyl propanoate. *See* n-Pentyl propionate
Anesthetic ether. *See* Ethyl ether
1,4-Anhydro-D-glucitol, 6-hexadecanoate. *See* Sorbitan palmitate
Anhydrosorbitol monolaurate. *See* Sorbitan laurate
Anhydrosorbitol trioleate. *See* Sorbitan trioleate
Anhydrous hydrobromic acid. *See* Hydrobromic acid
Anhydrous lanolin. *See* Lanolin
Aniline, N,N-dimethyl. *See* n,n-Dimethylaniline
Annulene. *See* Benzene

Apricot kernel oil PEG-6 esters
CAS 97488-91-0
Definition: Complex mixture formed from the transesterification of apricot kernel oil and PEG-6
Uses: Hydrophilic oil, solvent, emulsifier for pharmaceuticals

Aqua ammonia. *See* Ammonium hydroxide
Aqua ammonium. *See* Ammonium hydroxide
Aqueous ammonia. *See* Ammonium hydroxide
Arachis oil. *See* Peanut oil
Aromatic castor oil. *See* Castor oil
Artificial almond oil. *See* Benzaldehyde
Artificial ant oil. *See* Furfural
Artificial oil of ants. *See* Furfural
ATBC. *See* Acetyl tributyl citrate

ATEC. *See* Acetyl triethyl citrate
Automotive gasoline. *See* Gasoline
Azabenzene. *See* Pyridine
Azacyclohexane. *See* Piperidine
3-Azapentane-1,5-diamine. *See* Diethylenetriamine
Azeite. *See* Olive oil
Azine. *See* Pyridine

2-BA. *See* s-Butylamine
Banana oil. *See* Isoamyl acetate

Behenyl alcohol
CAS 661-19-8; EINECS 211-546-6
Synonyms: 1-Docosanol; Alcohol C_{22}
Definition: Mixture of fatty alcohols chiefly of n-docosanol
Empirical: $C_{22}H_{46}O$
Formula: $CH_3(CH_2)_{20}CH_2OH$
Properties: Colorless waxy solid; insol. in water, sol. in ethanol, chloroform; m.w. 326.61; m.p. 71 C; b.p. 180 C (0.22 mm)
Toxicology: Low toxicity
Uses: Synthetic fibers, lubricants, evaporation retardant on water surfaces; surfactant for polymerization; pharmaceutical raw material; foam control agent, cosolvent, plasticizer, emollient
Regulatory: FDA 21CFR §178.3910
Manuf./Distrib.: AC Ind.; Aldrich; Brown; Condea Vista; Fluka; M. Michel; Schweizerhall; Sigma
Trade names: Adol® 60

Benne oil. *See* Sesame oil
Bentanol. *See* Benzyl alcohol
Benylate. *See* Benzyl benzoate
Benyl phnylformate. *See* Benzyl benzoate

Benzaldehyde
CAS 100-52-7; EINECS 202-860-4
UN No. 1989 (DOT), 1990
Synonyms: Benzoic aldehyde; Benzic aldehyde; Benzene methylal; Benzene carbonal; Almond artificial essential oil; Artificial almond oil; Synthetic oil of bitter almond; Benzenecarbaldehyde; Benzene carboxaldehyde; Phenylmethanal
Classification: Aromatic aldehyde
Empirical: C_7H_6O
Formula: C_6H_5CHO
Properties: Colorless to yel. oily liq., bitter almond odor, burning taste; very sol. in alcohol, ether, benzene, acetone, petrol. ether, conc. sulfuric acid, liq. CO_2, liq. ammonia; sl. sol. in water; m.w. 106.13; sp.gr. 1.041; vapor pressure 1 mm Hg (26.2 C); m.p. -26 C; b.p 179 C; flash pt. (CC) 62 C; ref. index 1.544
Toxicology: LD50 (oral, rat) 1300 mg/kg; highly toxic; poison by ingestion and intraperitoneal routes; allergen; feeble local anesthetic; skin and eye irritant; vapors can irritate nose, throat, lungs
Precaution: DOT: Combustible liq.; a strong reducing agent; light-sensitive; acts

violently with oxidizers; incompat. with strong bases, strong reducing agents, alkaline cyanide sol'n., heavy metal salts
NFPA: Health 2, Flammability 2, Reactivity 0
Uses: Chemical intermediate for dyes, odorants, pharmaceuticals, pesticides; aromatic alcohols; paints; solvent for oils, resins, cellulose acetate and nitrate; mfg. of cinnamic acid, benzoic acid; photographic chemicals; flavoring agent and odorant in foods, drinks, cosmetics, pharmaceuticals, perfumes, cleaning prods.
Regulatory: FDA 21CFR §182.60, 27CFR §21.65, 21.151, GRAS; FEMA GRAS; Japan approved as flavoring; USP/NF, BP compliance
Manuf./Distrib.: Aceto; Aldrich; Allchem Ind.; J.T. Baker; Bayer; Chemical; Chemisphere; DSM; Elf Atochem SA; Fluka; R.W. Greeff; Haarmann & Reimer; Hoechst Celanese; Janssen Chimica; Kalama; Mitsubishi Chem.; Penta Mfg.; Punda Mercantile; Rit-Chem; Schweizerhall; Sigma; SNIA UK; Spectrum Chem. Mfg.; Van Waters & Rogers

Benzamide, N,N-diethyl-3-methyl. *See* Diethyl toluamide

Benzene
CAS 71-43-2; EINECS 200-753-7
UN No. 1114 (DOT)
Synonyms: Benzol; Benzole; Benzolene; Bicarburet of hydrogen; Carbon oil; Coal naphtha; Cyclohexatriene; Mineral naphtha; Motor benzol; Nitration benzene; Phene; Annulene; Phenyl hydride; Pyrobenzol; Pyrobenzole
Classification: Aromatic hydrocarbon
Empirical: C_6H_6
Properties: Colorless to light-yel. mobile nonpolar liq., aromatic hydrocarbon odor; misc. with alcohol, ether, chloroform, CCl_4, acetone, oils; sl. sol. in water; m.w. 78.11; sp.gr. 0.8790 (20/4 C); vapor pressure 100 mm Hg (26 C); f.p. 5.5 C; b.p. 80.1 C; flash pt. (CC) -11 C; ref. index 1.50110 (20 C)
Toxicology: LD50 (oral, rat) 3.8 ml/kg, (skin, mouse) 48 mg/kg; ACGIH TLV/TWA 10 ppm; very toxic; narcotic; poison by inh. causing CNS depression, dizziness, headache, nausea, unconsciousness; absorbed thru the skin causing systemic effects
Precaution: Extremely flamm.; fire risk; incompat. with oxidizers, peroxides, chlorine, nitric acid, ozone, interhalogens, metal perchlorates, etc.; explosion risk; potential violent reactions; Hazardous decomp. prods.: None known
NFPA: Health 2, Flammability 3, Reactivity 0
Uses: Mfg. of ethylbenzene, dodecylbenzene, cyclohexane, phenol, nitrobenzene, chlorobenzene, cumene, maleic anhydride; dyes, medicinals, artificial leather, airplane dopes, lacquers; solvent for waxes, resins, oils, paints; very minor component of gasoline
Regulatory: SARA reportable
Manuf./Distrib.: Aldrich; ARCO; Ashland; J.T. Baker; BP Chems. Ltd; Exxon; Fluka; Janssen Chimica; Marathon Oil; Mitsubishi Petrochem.; Mitsui Petrochem. Ind.; Mobil; Occidental; Shell; Sigma; Spectrum Chem. Mfg.

Benzenecarbaldehyde. *See* Benzaldehyde
Benzenecarbinol. *See* Benzyl alcohol
Benzene carbonal. *See* Benzaldehyde
Benzene carboxaldehyde. *See* Benzaldehyde
Benzene, 1-chloro-4 (trifluoromethyl). *See* p-Chlorobenzotrifluoride
1,2-Benzenedicarboxylic acid bi (2-methoxyethyl) ester. *See* Dimethoxyethyl phthalate
1,2-Benzenedicarboxylic acid, butyl octyl ester. *See* Butyl octyl phthalate
1,2-Benzenedicarboxylic acid, dibutyl ester. *See* Dibutyl phthalate

o-Benzenedicarboxylic acid dibutyl ester. *See* Dibutyl phthalate
Benzene-o-dicarboxylic acid di-n-butyl ester. *See* Dibutyl phthalate
1,2-Benzenedicarboxylic acid diethyl ester. *See* Diethyl phthalate
1,2-Benzenedicarboxylic acid dimethyl ester. *See* Dimethyl phthalate
1,2-Benzenedicarboxylic acid dioctyl ester. *See* Dioctyl phthalate
Benzene, o-dichloro-. *See* o-Dichlorobenzene
Benzene, 1,2-dichloro-4-(trifluoromethyl)-. *See* 3,4-Dichlorobenzotrifluoride
Benzene, 1,2-dimethyl. *See* o-Xylene
Benzeneethanol. *See* Phenethyl alcohol
Benzene hexahydride. *See* Cyclohexane
Benzene isopropyl. *See* Cumene
Benzenemethanaminium, N,N,N-trimethyl-, chloride. *See* Benzyl trimethyl ammonium chloride
Benzenemethanol. *See* Benzyl alcohol
Benzene methylal. *See* Benzaldehyde
Benzenenitrile. *See* Benzonitrile
Benzenesulfonic acid, dimethyl-, ammonium salt. *See* Ammonium xylenesulfonate
Benzenesulfonic acid, 4-methyl-, sodium salt. *See* Sodium toluenesulfonate
Benzene, 1,2,4-trichloro-. *See* 1,2,4-Trichlorobenzene
Benzic aldehyde. *See* Benzaldehyde
Benzin. *See* Gasoline; Naphtha
Benzine. *See* Naphtha
Benzoic acid, benzyl ester. *See* Benzyl benzoate
Benzoic acid, butyl ester. *See* Butyl benzoate
Benzoic acid n-butyl ester. *See* Butyl benzoate
Benzoic acid methyl ester. *See* Methyl benzoate
Benzoic acid nitrile. *See* Benzonitrile
Benzoic acid phenylmethyl ester. *See* Benzyl benzoate
Benzoic aldehyde. *See* Benzaldehyde
Benzol. *See* Benzene
Benzole. *See* Benzene
Benzolene. *See* Benzene

Benzonitrile
CAS 100-47-0; EINECS 202-855-7
UN No. 2224 (DOT)
Synonyms: Phenyl cyanide; Cyanobenzene; Benzenenitrile; Benzoic acid nitrile
Empirical: C_7H_5N
Formula: C_6H_5CN
Properties: Colorless liq., almond-like odor; sol. in alcohol; sl. sol. in cold water; m.w. 103.13; dens. 1.246 (20/4 C); m.p. -13 C; b.p. 190.7 C
Toxicology: LD50 (oral, mouse) 971 mg/kg; high toxicity; absorbed by skin; poison by IP and subcut. routes; skin irritant; heated to decomp., emits toxic fumes of CN^- and NO_x
Precaution: Combustible; DOT: poisonous material
Uses: Manufacture of benzoquanamine; intermediate for rubber chemicals; solvent for nitrile rubber, specialty lacquers, resins and polymers, anhydrous metallic salts
Manuf./Distrib.: Aldrich; Alemark; Amber Syn.; BASF; J.T. Baker; Fluka; Penta Mfg.; PMC Spec.; Sigma; SKW Chems.; Spectrum Chem. Mfg.; Triple Crown Am.

Benzoyl alcohol. *See* Benzyl alcohol
Benzoyl methide. *See* Acetophenone

Benztrimonium chloride. *See* Benzyl trimethyl ammonium chloride

Benzyl acetate

CAS 140-11-4; EINECS 205-399-7

Synonyms: Acetic acid benzyl ester; Acetic acid phenylmethyl ester; Acetoxymethylbenzene; Phenylmethyl acetate; α-Acetoxytoluene; Benzyl ethanoate

Classification: Carboxylic acid ester

Definition: Ester of benzyl alcohol and acetic acid

Empirical: $C_9H_{10}O_2$

Formula: $CH_3COOCH_2C_6H_5$

Properties: Colorless liq., sweet floral fruity odor; sol. in alcohol, chloroform, benzene, most fixed oils, propylene glycol; pract. insol. in water, glycerin; m.w. 150.19; sp.gr. 1.06; m.p. -51.5 C; b.p. 213.5 C; flash pt. (CC) 90 C; ref. index 1.501

Toxicology: LD50 (oral, rat) 2490 mg/kg; ACGIH TLV/TWA 10 ppm; mod. toxic by ingestion and subcutaneous routes; poison by inhalation; may be sl. absorbed thru skin; mild eye irritant; antipsychotic

Precaution: Combustible liq.; incompat. with strong oxidizing agents (increases fire/explosion risk), acids, bases, reducing agents (decomp. can occur); Hazardous decomp. prods.: Benzyl alcohol, benzoic acid, acetic acid; heated to decomp., emits irritating fumes

NFPA: Health 1, Flammability 1, Reactivity 0

Uses: Artificial jasmine and other perfumes; soap perfume; syn. flavoring agent for foods and pharmaceuticals; solvent and high boiler for cellulose acetate and nitrate, natural and synthetic resins; oils; lacquers; polishes; printing inks; varnish removers

Regulatory: FDA 21CFR §172.515; FEMA GRAS; Japan approved as flavoring

Manuf./Distrib.: Aldrich; Allan; Berje; Chemisphere; Fluka; Haarmann & Reimer; Janssen Chimica; Kalama; Koyo; Penta Mfg.; Quest Int'l.; Rhône-Poulenc N. Am.

Benzyl alcohol

CAS 100-51-6; EINECS 202-859-9

Synonyms: Bentanol; Benzenecarbinol; Benzoyl alcohol; α-Hydroxytoluene; Phenolcarbinol; Phenylmethanol; Phenylmethyl alcohol; Phenylcarbinol; Benzenemethanol; α-Toluenol

Classification: Aromatic alcohol

Empirical: C_7H_8O

Formula: $C_6H_5CH_2OH$

Properties: Water-wh. liq., faint aromatic odor, sharp burning taste; misc. with alcohol, chloroform, ether, water @ 206 C (dec.); m.w. 108.15; sp.gr. 1.042; vapor pressure 1 mm Hg (58 C); m.p. -15.3 C; b.p. 205.7 C; flash pt. (CC) 93 C; ref. index 1.540

Toxicology: LD50 (oral, rat) 1230 mg/kg, (IP, rat) 400 mg/kg, (skin, rabbit) 2000 mg/kg; poison by ing., IP, IV routes; mod. toxic by inh., skin contact; mod. skin and severe eye irritant; may cross-react with balsam Peru in sensitive persons

Precaution: Combustible liq.; light-sensitive; incompat. with oxidizers (increased fire and explosion risk), acids (exothermic polymerization), sulfuric acid; Hazardous decomp. prods.: Dec. explosively at 180 C with sulfuric acid; heated to decomp., emits acrid smoke and fumes

NFPA: Health 2, Flammability 1, Reactivity 0

Uses: Perfumes; flavoring; photographic developer; dyeing nylon, textiles, sheet plastics; solvent for dyestuffs, cellulose esters, casein, waxes, lacquers, inks; heat-sealing polyethylene films; intermediate for benzyl esters and ethers; bacteriostat; cosmetics; food and pharmaceutical antimicrobial preservative,

solubilizer, flavoring; astringent
Use level: 10% (disinfectant, local anesthetic), 1% (additive to injectables)
Regulatory: FDA 21CFR §172.515; FEMA GRAS; USA EPA registered; Japan approved as flavoring; Europe listed; FDA approved for use in injectables, parenterals, orals, topicals; USP/NF, BP, Ph.Eur. compliance
Manuf./Distrib.: Aldrich; Allchem Ind.; Ashland; Elf Atochem SA; Bayer/Fibers, Orgs., Rubbers; Charkit; Elf Atochem N. Am.; Fluka; Givaudan Iberica SA; R.W. Greeff; Haarmann & Reimer; Janssen Chimica; Kalama; E. Merck; Osaka Org. Chem. Ind.; Penta Mfg.; Quest Int'l.; Sigma; Spectrum Chem. Mfg.; Takasago Int'l.; Tosoh; Velsicol

Benzyl benzene carboxylate. *See* Benzyl benzoate

Benzyl benzoate
CAS 120-51-4; EINECS 204-402-9
UN No. 2810
Synonyms: Benylate; Benzyl benzene carboxylate; Benyl phnylformate; Phenylmethyl benzoate; Benzoic acid phenylmethyl ester; Benzoic acid, benzyl ester
Definition: Ester of benzyl alcohol and benzoic acid
Empirical: $C_{14}H_{12}O_2$
Formula: $C_6H_5COOCH_2C_6H_5$
Properties: Colorless oily liq., sl. aromatic odor; misc. with alcohol, chloroform, ether; insol. in water, glycerin; m.w. 212.26; sp.gr. 1.116-1.120; m.p. 21 C; b.p. 324 C; flash pt. (CC) 298 F; ref. index 1.568
Toxicology: LD50 (oral, rat) 500 mg/kg, (skin, rabbit) 4000 mg/kg; mod. toxic by ingestion, skin contact
Precaution: Combustible liq.; light-sensitive; reactive with oxidizing materials; Hazardous decomp. prods.: Heated to decomp., emits acrid and irritating fumes and smoke
NFPA: Health 1, Flammability 1, Reactivity 0
Uses: Fixative and solvent for musk in perfumes and flavors; syn. flavoring agent for foods and pharmaceuticals; external medicine; plasticizer for nitrocellulose and cellulose acetate; miticide, scabicide; insect repellent; dye carrier; solvent, preservative for pharmaceuticals
Regulatory: FDA 21CFR §172.515, 175.105; FEMA GRAS; FDA approved for injectables, orals; USP/NF, BP, Ph.Eur. compliance
Manuf./Distrib.: Aldrich; Allchem Ind.; Berje; Fluka; Haarmann & Reimer; Janssen Chimica; Kalama; Morflex; Novachem Corp.; Penta Mfg.; Pentagon Chems. Ltd; Rhône-Poulenc N. Am.; Schweizerhall; Sigma; Spectrum Chem. Mfg.
Trade names: Unichem BZBN

Benzyl carbinol. *See* Phenethyl alcohol
Benzyl dodecanoate. *See* Benzyl laurate
Benzyl ethanoate. *See* Benzyl acetate

Benzyl laurate
CAS 140-25-0; EINECS 204-405-8
Synonyms: Dodecanoic acid benzyl ester; Dodecanoic acid, phenylmethyl ester; Benzyl dodecanoate; Lauric acid, benzyl ester; Phenylmethyl dodecanoate
Definition: Ester of lauric acid and benzyl alcohol
Empirical: $C_{19}H_{30}O_2$
Properties: M.w. 290.49
Toxicology: Skin irritant; Hazardous decomp. prods.: Heated to decomp., emits acrid smoke and irritating fumes

Benzyltriethyl ammonium chloride

Uses: Emollient, lubricant for skin and hair care formulations; solvent and degreaser in sunscreens
Manuf./Distrib.: Penta Mfg.
Trade names: Mazon® EE-1

Benzyltriethyl ammonium chloride
CAS 56-37-1; EINECS 200-270-1
Synonyms: TEBAC; N,N,N-Triethylbenzenemethanaminium chloride
Classification: Quaternary ammonium salt
Empirical: $C_{13}H_{22}N \cdot Cl$
Formula: $C_6H_5CH_2N(Cl)(C_2H_5)_3$
Properties: M.w. 227.8
Uses: Solvent for cellulose; gelling inhibitor in polyester resins; intermediate
Manuf./Distrib.: Aldrich; Fluka; Janssen Chimica; Lindau Chems; Raschig; R S A; Spectrum Chem. Mfg.; Triple Crown Am.; Zeeland

Benzyl trimethyl ammonium chloride
CAS 56-93-9; EINECS 200-300-3
Synonyms: TMBAC; Trimethylbenzylammonium chloride; Benztrimonium chloride; Benzenemethanaminium, N,N,N-trimethyl-, chloride
Classification: Quaternary ammonium salt
Empirical: $C_{10}H_{16}ClN$
Formula: $C_6H_5CH_2N(CH_3)_3Cl$
Properties: Colorless cryst.; stable to 135 C; readily sol. in water, ethanol, butanol; sl. sol. in butyl phthalate, tributyl phosphate; m.w. 185.7; den. 1.07 (20/20 C, 61% sol'n.); m.p. 234-236 C; b.p. 135 C (some decomp.)
Toxicology: LDLo (oral, rat) 250 mg/kg; poison by ingestion; tumorigen
Precaution: Combustible; Hazardous decomp. prods.: Heated to decomp., emits very toxic fumes of NH_3, NO_x, Cl^-
Uses: Dispersant, dye leveler and retarder, emulsifier used in textile industry; solvent for cellulose; gelling inhibitor in polyester resins; intermediate; agric. chem.
Manuf./Distrib.: Aldrich; Ashland; Fluka; Janssen Chimica; Lindau Chems; Piedmont Chem. Ind.; R S A; Spectrum Chem. Mfg.; Sybron; Witco/Oleo-Surf.

Bicarburet of hydrogen. *See* Benzene
1,2-Bichloroethane. *See* Ethylene dichloride
Bicyclo (2.2.1) heptane, 2,2-dimethyl-3-methylene-. *See* Camphene
Bihexyl. *See* n-Dodecane
Biphenyl, isopropyl. *See* 1,1´-Biphenyl (1-methylethyl)

1,1´-Biphenyl (1-methylethyl)
CAS 25640-78-2
Synonyms: Biphenyl, isopropyl; Isopropyl diphenyl; Isopropylbiphenyl; Monoisopropyl biphenyl; (1-Methylethyl) 1,1´-biphenyl
Empirical: $C_{15}H_{16}$
Properties: M.w. 196.31
Toxicology: LD50 (oral, rat) 4570 mg/kg; LDLo (oral, rabbit) 4 g/kg; mod. toxic by ingestion; mutagen; may cause altered sleep time, tremors, ataxia; Hazardous decomp. prods.: Heated to decomp., emits acrid smoke and fumes
Trade names: Nusolv™ ABP-62

Bis(2-aminoethyl) amine. *See* Diethylenetriamine
Bis(β-aminoethyl)amine. *See* Diethylenetriamine
Bis (butoxyethyl)ether. *See* Diethylene glycol dibutyl ether
Bis (2-butoxyethyl) ether. *See* Diethylene glycol dibutyl ether
Bis (2-ethoxyethyl) ether. *See* Diethylene glycol diethyl ether

Bis (2-ethylhexyl) decanedioate. *See* Dioctyl sebacate
Bis(2-ethylhexyl) hexanedioate. *See* Dioctyl adipate
Bis (2-ethylhexyl) maleate. *See* Dioctyl maleate
Bis (2-ethylhexyl) phthalate. *See* Dioctyl phthalate
N,N-Bis (2-hydroxyethyl) coco amides. *See* Cocamide DEA
Bis(2-hydroxyethyl) ether. *See* Diethylene glycol
N,N-Bis(2-hydroxyethyl)ethylamine. *See* N-Ethyldiethanolamine
Bis(2-hydroxyethyl) sulfide. *See* Thiodiglycol
Bis(β-hydroxyethyl)sulfide. *See* Thiodiglycol
Bis (hydroxypropyl) ether. *See* Dipropylene glycol
Bis (2-methoxyethyl) ether. *See* Diethylene glycol dimethyl ether
Bis (methoxyethyl) phthalate. *See* Dimethoxyethyl phthalate
Bis (2-methoxyethyl) phthalate. *See* Dimethoxyethyl phthalate
Bis(methoxypropyl) ether. *See* Dipropylene glycol dimethyl ether
1,3-Bis(1-methylethyl) benzene. *See* 1,3-Diisopropyl benzene
1,4-Bis(1-methylethyl) benzene. *See* 1,4-Diisopropyl benzene

Bis(methylethyl)-1,1´-biphenyl
CAS 69009-90-1
Synonyms: Diisopropyl biphenyl
Empirical: $C_{18}H_{22}$
Properties: M.w. 238.37
Uses: Solvent possessing exc. solvency, chem. and thermal stability, nonvolatility; for aerosols, adhesives, electronic parts mfg. and cleaning, industrial cleaning, inks and paper coatings, metal cleaning, paints, plastics, sealants, tile mfg.
Trade names: Nusolv™ ABP-103

Bis(1-methylethyl) decanedioate. *See* Diisopropyl sebacate
Bis(1-methylethyl)hexanedioate. *See* Diisopropyl adipate
Bis(2-methylpropyl) hexanedioate. *See* Diisobutyl adipate
Bisulfate. *See* Sulfur dioxide
Bisulfite. *See* Sulfur dioxide
BLO. *See* Butyrolactone
BOP. *See* Butyl octyl phthalate
Borneol acetate. *See* Bornyl acetate

Bornyl acetate
CAS 76-49-3
Synonyms: levo-Bornyl acetate; Borneol acetate; 1,7,7-Trimethylbicyclo [2.2.1] heptan-2-ol acetate; Bornyl acetic ether; Bornyl ethanoate
Empirical: $C_{12}H_{20}O_2$
Properties: Colorless liq., white crystalline solid; piney odor; fresh, burning taste; sol. in alcohol, most fixed oils; sl. sol. in water; insol. in propyl glycol; m.w. 196.29; m.p. 27.5 C; b.p. 225-226 C; flash pt. 89 C; ref. index 1.462-1.466
Toxicology: Heated to decomp., emits acrid smoke and fumes
Precaution: Combustible liq.
Uses: Synthetic flavoring agent, solvent for foods and pharmaceuticals
Regulatory: FDA 21CFR §172.515; FEMA GRAS

levo-Bornyl acetate. *See* Bornyl acetate
Bornyl acetic ether. *See* Bornyl acetate
Bornyl ethanoate. *See* Bornyl acetate

Bromoform
CAS 75-25-2
UN No. 2515 (DOT)
Synonyms: Tribromomethane; Methenyl tribromide

Bromomethane

Classification: Halogenated aliphatic hydrocarbon
Empirical: $CHBr_3$
Properties: Colorless to yel. liq. or hexagonal cryst.; sweet chloroform-like odor; sol. in many org. solvs.; sl. sol. in water; m.w. 252.75; sp.gr. 2.9 (20 C); m.p. 7-9 C; b.p. 148 C; flash pt. none
Toxicology: LD50 (oral, rat) 1147 mg/kg; ACGIH TLV/TWA 0.5 ppm (skin); very toxic; human poison by ing.; may cause skin irritation; can be absorbed thru skin; lachrymator; mod. eye irritant; vapor may cause nose/throat irritation; anesthetic props.
Precaution: Incompat. with strong bases, metals, acetone/potassium hydroxide, cyclic polyethylene oxides/potassium hydroxide; possible violent reactions, explosive hazards; Hazardous decomp. prods.: Heated to decomp., emits highly toxic fumes of Br^-
Uses: Chem. intermediate; solvent; prod. of high-dens. liqs. for geological assays (petrographic analysis) and gauges
Manuf./Distrib.: Aldrich; Fluka; Geoliquids; Sigma; Spectrum Chem. Mfg.

Bromomethane. *See* Methyl bromide
2-Butanamine. *See* s-Butylamine

Butane
CAS 106-97-8; EINECS 203-448-7
UN No. 1011 (DOT)
Synonyms: n-Butane; Diethyl; Alkane C-4; Butyl hydride; Methylethylmethane
Classification: Alkane; hydrocarbon
Empirical: C_4H_{10}
Formula: $CH_3CH_2CH_2CH_3$
Properties: Colorless gas; faint disagreeable odor; easily liquefied under pressure @ R.T.; sol. in ethanol, ether, chloroform; sl. sol. in water; m.w. 58.12; sp.gr. 0.599; vapor pressure 1620 mm Hg (17 psig, 21 C); f.p. -138 C; b.p. -0.5 C; flash pt. (CC) -60 C
Toxicology: LC50 (inh., rat, 4 h) 658 g/m^3; ACGIH TLV/TWA 800 ppm; mildly toxic by inh.; causes drowsiness; anesthetic and asphyxiant props.; narcotic in high concs.; liq. may cause burns or frostbite to skin and eyes
Precaution: Flamm. gas; very dangerous fire hazard exposed to heat, flame, oxidizers; highly explosive; explosive limits 1.9-8.5%; incompat. with strong oxidizing agents (increases fire/explosion risk); Hazardous decomp. prods.: Heated to decomp., emits acrid smoke and fumes
NFPA: Health 1, Flammability 4, Reactivity 0
Uses: Hydrocarbon propellant; paint aerosols; org. synthesis; fuel for household and industry; raw material for syn. rubber; high octane liq. fuels; mfg. of ethylene; solvent; refrigerant; pure grades in calibrating instruments; food additive; propellant in food and pharmaceutical applics.
Regulatory: FDA 21CFR §173.350, 184.1165, GRAS; approved for topicals; USP/NF compliance
Manuf./Distrib.: Air Prods.; Aldrich; Phillips; Stanchem

n-Butane. *See* Butane
1,3-Butanediol. *See* Butylene glycol
Butane-1,3-diol. *See* Butylene glycol
Butane, 2-methyl-. *See* Isopentane
Butanethiol. *See* n-Butyl mercaptan
1-Butanethiol. *See* n-Butyl mercaptan
n-Butanethiol. *See* n-Butyl mercaptan
1-Butanol. *See* Butyl alcohol
Butan-1-ol. *See* Butyl alcohol

2-Butanol

CAS 78-92-2; EINECS 201-158-5

UN No. 1120 (DOT)

Synonyms: SBA; 2-Butyl alcohol; s-Butyl alcohol; sec-Butyl alcohol; Secondary butyl alcohol; Methyl ethyl carbinol; Ethylmethyl carbinol; sec-Butanol; Butan-2-ol; Butylene hydrate; 1-Methyl propanol; 2-Hydroxybutane

Classification: Sec. aliphatic alcohol

Empirical: $C_4H_{10}O$

Formula: $CH_3CH_2CH(OH)CH_3$

Properties: Colorless liq.; sweet odor; sol. in water, ethanol, ether, acetone, benzene; m.w. 74.12; sp.gr. 0.807 (20/4 C); m.p. -89 C; b.p. 99-100 C; flash pt. 23 C; ref. index 1.3972 (20 C)

Toxicology: ACGIH TLV/TWA 100 ppm; LD50 (oral, rat) 6480 mg/kg; poison by intravenous and intraperitoneal routes; mildly toxic by ingestion; skin and eye irritant; excessive inh. may cause nose/throat irritation, headache, nausea, fatigue, dizziness

Precaution: Flamm.; incompat. with alkali metals (increases fire/explosion risk, forms flamm. hydrogen gas), acids, acid chlorides, acid anhydrides (vigorous reactions), chlorine, isocyanates, ethylene oxide; Hazardous decomp. prods.: Unstable peroxides may form after prolonged storage

NFPA: Health 1, Flammability 3, Reactivity 0

Uses: Solvent for coatings, paint removers, adhesives, resins, lacquers, industrial cleaners; prep. of methyl ethyl ketone; extraction of fish protein; ingred. in hydraulic brake fluids; perfumes; syn. flavoring agent in foods

Regulatory: FDA 21CFR §172.515

Manuf./Distrib.: Ashland; Chemcentral; Exxon; Fluka; Houghton Chem.; Shell; Spectrum Chem. Mfg.

Butan-2-ol. *See* 2-Butanol
n-Butanol. *See* Butyl alcohol
sec-Butanol. *See* 2-Butanol
t-Butanol. *See* t-Butyl alcohol
2-Butanol acetate. *See* s-Butyl acetate
4-Butanolide. *See* Butyrolactone
Butanone. *See* Methyl ethyl ketone
2-Butanone. *See* Methyl ethyl ketone
3-Butanone. *See* Methyl ethyl ketone
2-Butanone, 3-methyl-. *See* Methyl isopropyl ketone
1-Butoxybutane. *See* Dibutyl ether
Butoxydiethylene glycol. *See* Butoxydiglycol

Butoxydiglycol

CAS 112-34-5; EINECS 203-961-6

Synonyms: DEGBE; PEG-2 butyl ether; Butyl 'Carbitol'; Butyldiglycol; Diglycol butyl ether; Diglycol monobutyl ether; Butoxydiethylene glycol; Diethylene glycol butyl ether; Diethylene glycol n-butyl ether; Diethylene glycol monobutyl ether; 2-(2-Butoxyethoxy) ethanol; o-Butyl diethylene glycol

Classification: Aliphatic ether alcohol

Empirical: $C_8H_{18}O_3$

Formula: $C_4H_9OCH_2CH_2OCH_2CH_2OH$

Properties: Colorless liq.; faint butyl odor; very sol. in ether, alcohol, acetone; sol. in oils and water; m.w. 162.26; sp.gr. 0.9553 (20/4 C); vapor pressure 0.02 mm Hg (20 C); m.p. -68.1 C; b.p. 230.6 C; flash pt. (CC) 78 C; ref. index 1.4316 (20 C)

Toxicology: LD50 (oral, rat) 6.56 g/kg, (skin, rabbit) 4120 mg/kg; mod. toxic by ingestion, intraperitoneal routes; mildly toxic by skin contact; severe eye

irritant
Precaution: Flamm. exposed to heat or flame; incompat. with strong oxidizers (increases risk of fire and explosion); Hazardous decomp. prods.: Heated to decomp., emits acrid smoke and irritating fumes
NFPA: Health 1, Flammability 2, Reactivity 0
Uses: Solvent for nitrocellulose, coatings, oils, dyes, gums, soaps, polymers, stamp pad inks, high-based enamels, household cleaners; plasticizer intermediate; diluent for hydraulic brake fluids; coalescing agent for latex paints; dispersant for vinyl chloride resins in organosols; wetting-out sol'n. in textile industry
Regulatory: FDA 21CFR §175.105, 176.180
Manuf./Distrib.: Aldrich; Allchem Ind.; ARCO; Ashland; Brown; Eastman; Fluka; Occidental; Shell; Union Carbide
Trade names: Butyl Carbitol®; Butyl Di-Icinol; Butyl Dioxitol®; DB; DB; Dowanol® DB; Eastman® DB

Butoxyethanol
CAS 111-76-2; EINECS 203-905-0
UN No. 2369 (DOT)
Synonyms: 2-Butoxyethanol; 2-Butoxy-1-ethanol; n-Butoxyethanol; o-Butyl ethylene glycol; Ethylene glycol monobutyl ether; Glycol butyl ether; Monobutyl glycol ether; 3-Oxa-1-heptanol; Ethylene glycol butyl ether; Ethylene glycol n-butyl ether; Butyl glycol
Classification: Ether alcohol
Empirical: $C_6H_{14}O_2$
Formula: $HOCH_2CH_2OC_4H_9$
Properties: Colorless clear mobile liq.; mild pleasant odor; sol. in water, alcohol, ether; m.w. 118.20; sp.gr. 0.9019 (20/20 C); vapor pressure 300 mm Hg (140 C); m.p. -74.8 C; b.p. 171.2 C; flash pt. 62 C
Toxicology: LD50 (oral, rat) 1480 mg/kg, (IP, rat) 220 mg/kg, (skin, rabbit) 490 mg/kg; ACGIH TLV/TWA 25 ppm (skin); poison by ing., skin contact, IP, IV routes; mod. toxic by inh., subcut.; human systemic effects; skin/eye irritant
Precaution: Flamm. exposed to heat or flame; incompat. with oxidizers, strong caustics; attacks some types of rubber, plastics, and coatings; Hazardous decomp. prods.: Heated to decomp., emits acrid smoke and irritating fumes
NFPA: Health 2, Flammability 2, Reactivity 0
Uses: Solvent for nitrocellulose, natural and syn. resins, sol. oils, lacquers, varnishes, enamels; cosolvent; gas chromatography; flume wash water additive for food processing; in textile dyeing/printing, leather treatment, prod. of plasticizers; stabilizer in metal and household cleaners, hydraulic fluids, insecticides, herbicides, rust removers
Regulatory: FDA 21CFR §173.315, 175.105, 176.210, 177.1650, 178.1010, 178.3297, 178.3570
Manuf./Distrib.: Aldrich; Allchem Ind.; ARCO; Ashland; Brown; Eastman; Fluka; Occidental; Shell; Sigma; Spectrum Chem. Mfg.; Union Carbide
Trade names: Butyl Cellosolve®; Butyl Icinol; Butyl Oxitol®; Dowanol® EB; Eastman® EB; EB; EB
Trade names containing: Ecco Strip LS-W

2-Butoxyethanol. *See* Butoxyethanol
2-Butoxy-1-ethanol. *See* Butoxyethanol
n-Butoxyethanol. *See* Butoxyethanol

Butoxyethanol acetate
CAS 112-07-2; EINECS 203-933-3; 203-934-9
Synonyms: Ethylene glycol butyl ether acetate; Ethylene glycol monobutyl ether

acetate; 1-Acetoxy-2-butoxyethane; 2-Butoxyethanol acetate; 2-Butoxyethyl acetate; 2-Butoxyethyl ester acetic acid; Ethanol, 2-butoxy-, acetate; Glycol monobutyl ether acetate

Empirical: $C_8H_{16}O_3$

Formula: $CH_3CH_2CH_2CH_2OCH_2CH_2OCOCH_3$

Properties: Colorless liq., mild fruity odor; sol. 1.1% in water; sol. in hydrocarbons and org. solvs.; m.w. 160.2; dens. 0.941 (20/20 C); b.p. -64 C; b.p. 186 C (760 mm); flash pt. 190 F; ref. index 1.4142 (20 C)

Toxicology: LD50 (oral, rat) 2400 mg/kg, (skin, rabbit) 1500 mg/kg; mod. toxic by ing., skin contact; mild skin irritant; may cause blood disorders, kidney damage; harmful if inhaled or absorbed thru skin

Precaution: Combustible liq. and vapor; peroxide former; Hazardous decomp. prods.: Heated to decomp., emits acrid smoke and irritating fumes

Uses: Solvent for high-solids coatings (NC, butyrate, ethyl cellulose acrylic); coalescing aid for water-disp. coatings

Manuf./Distrib.: Aldrich; Allchem Ind.; ARCO; Ashland; Eastman; Fluka; Occidental; Oxiteno

Trade names: Butyl Cellosolve® Acetate; Eastman® EB Acetate; EBA

2-Butoxyethanol acetate. *See* Butoxyethanol acetate

2-(2-Butoxyethoxy) ethanol. *See* Butoxydiglycol

2-(2-Butoxyethoxy)ethanol acetate. *See* Diethylene glycol butyl ether acetate

2-(2-(2-Butoxyethoxy)ethoxy)ethanol. *See* Butoxytriglycol

2-(2-Butoxyethoxy) ethyl acetate. *See* Diethylene glycol butyl ether acetate

1-Butoxyethoxy-2-propanol

CAS 124-16-3

Synonyms: 1-(2-Butoxyethoxy)-2-propanol; Propylene glycol butoxyethyl ether; 2-Butoxy-1-(2´-hydroxypropoxy) ethane; 4,7-Dioxaundecan-2-ol

Classification: Aliphatic ether alcohol

Empirical: $C_9H_{20}O_3$

Formula: $CH_3CH(OH)CH_2OCH_2CH_2OC_4H_9$

Properties: Colorless clear liq.; mild odor; sol. in water, variety of org. solvs.; m.w. 176.29; sp.gr. 0.931 (20/20 C); f.p. -90 C; b.p. 230 C; flash pt. (OC) 121 C

Toxicology: LD50 (oral, rat) 4000 mg/kg, (skin, rabbit) 2830 mg/kg; mod. toxic by ing. and skin contact; low toxicity by inh.; may cause mild skin irritation; absorbed by skin, but health effects unlikely; may cause mod. to severe eye irritation

Precaution: Combustible exposed to heat or flame; incompat. with oxidizing agents (increases fire risk), strong acids (may cause decomp. of ethers); Hazardous decomp. prods.: Heated to decomp., emits acrid and irritating fumes

NFPA: Health 2, Flammability 1, Reactivity 0

Uses: Solvent for paints, lacquers, resins, inks, dyes, oils, greases, cleaners

Manuf./Distrib.: Ashland

1-(2-Butoxyethoxy)-2-propanol. *See* 1-Butoxyethoxy-2-propanol

2-Butoxyethyl acetate. *See* Butoxyethanol acetate

2-Butoxyethyl ester acetic acid. *See* Butoxyethanol acetate

2-Butoxyethyl ether. *See* Diethylene glycol dibutyl ether

2-Butoxy-1-(2´-hydroxypropoxy) ethane. *See* 1-Butoxyethoxy-2-propanol

'Butoxyl'. *See* Methoxybutyl acetate

Butoxypropanol

CAS 5131-66-8; EINECS 225-878-4

Synonyms: 1-Butoxy-2-propanol; n-Butoxypropanol; Propylene glycol n-butyl

ether; Propylene glycol mono-n-butyl ether; Propylene glycol butyl ether
Classification: Aliphatic ether alcohol
Empirical: $C_7H_{16}O$
Formula: $CH_3CH_2CH_2CH_2OCH_2CHOHCH_3$
Properties: Colorless clear liq.; mild char. odor; mod. sol. in water; sol. in variety of org. solvs.; m.w. 132.23; sp.gr. 0.884 (20/20 C); f.p. -148 F; b.p. 170.2 C; flash pt. (OC) 65.6 C; ref.index 1.4074 (20 C)
Toxicology: LD50 (oral, rat) 2200 mg/kg; mod. toxic by ingestion and by skin contact; skin absorption may produce CNS depression; low toxicity by inh.; may cause eye irritation, possible corneal injury
Precaution: Combustible; incompat. with oxidizing agents (increases risk of fire and explosion); Hazardous decomp. prods.: Heated to decomp., emits acrid smoke and irritating fumes
Uses: Solvent for coatings, cleaners, electronic, and ink applics.; strong coupling ability; used in lt.-duty and hard-surf. cleaners, water-reducible polyester and alkyd resin prod.
Trade names: Arcosolv® PNB; Dowanol® PnB

1-Butoxy-2-propanol. *See* Butoxypropanol
1-t-Butoxy-2-propanol. *See* Propylene glycol t-butyl ether
n-Butoxypropanol. *See* Butoxypropanol
Butoxytriethylene glycol. *See* Butoxytriglycol

Butoxytriglycol
CAS 143-22-6; EINECS 205-592-6
Synonyms: Triethylene glycol butyl ether; Triethylene glycol monobutyl ether; Triglycol monobutyl ether; PEG-3 butyl ether; Butoxytriethylene glycol; Triethylene glycol n-butyl ether; Ethanol, 2-(2-(2-butoxyethoxy)ethoxy)-; 2-(2-(2-Butoxyethoxy)ethoxy)ethanol
Empirical: $C_{10}H_{22}O_4$
Formula: $C_4H_9O(C_2H_4O)_3H$
Properties: Liq.; sol. in water; m.w. 206.32; dens. 1.0021 (20/20 C); f.p. -47.6 C; b.p. dec.; flash pt. 143 C
Toxicology: LD50 (oral, rat) 6730 mg/kg, (skin, rabbit) 3540 mg/kg; moderately toxic by skin contact; mildly toxic by ingestion; skin and severe eye irritant; neuromuscular blockage; potential human reproductive effects
Precaution: Combustible exposed to heat or flames; Hazardous decomp. prods.: Heated to decomp., emits acrid smoke and irritating fumes
Uses: Plasticizer, intermediate, solvent
Manuf./Distrib.: Ashland; Fluka; Oxiteno
Trade names: Dowanol® TBH
Trade names containing: Icinol BE33

Butyl acetate. *See* n-Butyl acetate
1-Butyl acetate. *See* n-Butyl acetate
2-Butyl acetate. *See* s-Butyl acetate

n-Butyl acetate
CAS 123-86-4; EINECS 204-658-1
UN No. 1123 (DOT)
Synonyms: Acetic acid, butyl ester; Acetic acid n-butyl ester; Butyl acetate; 1-Butyl acetate; Butyl ethanoate
Classification: Carboxylic acid ester
Definition: Ester of butyl alcohol and acetic acid
Empirical: $C_6H_{12}O_2$
Formula: $CH_3COOC_4H_9$

Properties: Colorless liq., fruity odor; sol. in alcohol, ether, benzene, hydrocarbons; sl. sol. in water; m.w. 116.18; sp.gr. 0.8826 (20/20 C); vapor pressure 15 mm Hg; f.p. -75 C; b.p. 126.3 C; flash pt. (TOC) 36.6 C; ref. index 1.2951 (20 C)
Toxicology: LD50 (oral, rat) 14.13 g/kg; ACTIH TLV/TWA 150 ppm; STEL 200 ppm; mildly toxic by inhalation, ingestion; moderately toxic by intraperitoneal route; skin irritant; severe eye irritant; mild allergen
Precaution: Flamm.; dangerous fire hazard exposed to flame; mod. explosive when exposed to flame; incompat. with strong oxidizers (increases fire risk), strong acids or strong bases (decomp. can occur), potassium t-butylate; Hazardous decomp. prods.: Heated to decomp., emits acrid and irritating fumes
NFPA: Health 1, Flammability 3, Reactivity 0
Uses: Solvent for paints, lacquers, thinners; mfg. of artificial leather, photographic films, plastics, safety glass; syn. flavoring agent for foods and pharmaceuticals
Regulatory: FDA 21CFR §172.515, 175.105, 175.320, 177.1200; FEMA GRAS; Japan approved as flavoring
Manuf./Distrib.: Aldrich; Allchem Ind.; Ashland; J.T. Baker; BASF; Baychem; BP Chems. Ltd; Chemcentral; Chisso Am.; Coyne; Eastman; Fluka; General Chem.; Great Western; Harcros USA; Hoechst AG; Houghton Chem.; Hüls AG; Janssen Chimica; Penta Mfg.; Samson; Sigma; Spectrum Chem. Mfg.; Sunnyside; Union Carbide; Van Waters & Rogers

s-Butyl acetate
CAS 105-46-4; EINECS 203-300-1
UN No. 1123 (DOT)
Synonyms: 2-Butyl acetate; Acetic acid, sec-butyl ester; Acetic acid 2-butyl ester; Acetic acid-2-butoxy ester; Acetic acid 1-methylpropyl ester; 1-Methylpropyl acetate; 2-Butanol acetate; sec-Butyl alcohol acetate
Classification: Carboxylic acid ester
Definition: Ester of butyl alcohol and acetic acid
Empirical: $C_6H_{12}O_2$
Formula: $CH_3COOCH(CH_3)(C_2H_5)$
Properties: Colorless liq.; mild fruity odor; sol. in alcohol, ether, acetone; insol. in water; m.w. 116.16; sp.gr. 0.870 (20/4 C); m.p. -146 C; b.p. 112.2 C; flash pt. (OC) 31 C; ref. index 1.389 (20 C)
Toxicology: ACGIH TLV/TWA 200 ppm; STEL 150 ppm; mild skin and eye irritant; inh. of vapors can probably irritate nose and throat; ing. may irritate mouth and throat; high concs. may cause CNS depression, headache, dizziness, nausea, unconsciousness
Precaution: Flamm.; incompat. with strong oxidizers (increases fire/explosion risk), strong acids, strong bases (decomp. can occur); Hazardous decomp. prods.: s-Butyl alcohol, acetic acid; heated to decomp., emits acrid and irritating fumes
NFPA: Health 1, Flammability 3, Reactivity 0
Uses: Solvent for nitrocellulose lacquers, thinners, nail enamels, leather finishes
Manuf./Distrib.: Aldrich; Schweizerhall

t-Butyl acetate
CAS 540-88-5; EINECS 208-760-7
UN No. 1123 (DOT)
Synonyms: Acetic acid 1,1-dimethylethyl ester; Acetic acid t-butyl ester; tert-Butyl acetate
Classification: Carboxylic acid ester
Definition: Ester of t-butyl alcohol and acetic acid
Empirical: $C_6H_{12}O_2$
Formula: $CH_3COOC(CH_3)_3$

tert-Butyl acetate

Properties: Colorless clear liq.; fruity odor; pract. insol. in water; sol. in alcohol, ether, acetic acid; m.w. 116.16; sp.gr. 0.896 (20 C); b.p. 96 C; flash pt. 15 C; ref. index 1.3870; flamm.

Toxicology: ACGIH TLV/TWA 200 ppm; poison by inh. and ing.; vapors can probably irritate nose and throat; high concs. may cause CNS depression; mild skin and eye irritant

Precaution: Incompat. with strong oxidizers (increases risk of fire and explosion), strong bases, strong acids (decomp. can occur); Hazardous decomp. prods.: t-Butyl alcohol, acetic acid; heated to decomp., emits acrid smoke and irritating fumes

Uses: As gasoline additive; solvent; intermediate for cholesterol-lowering drugs

Manuf./Distrib.: Aldrich; Fluka; Schweizerhall; Sigma; Spectrum Chem. Mfg.; Wacker-Chemie GmbH

tert-Butyl acetate. *See* t-Butyl acetate
Butylacetone. *See* Methyl n-amyl ketone
Butyl adipate. *See* Dibutyl adipate

Butyl alcohol

CAS 71-36-3; EINECS 200-751-6

UN No. 1120 (DOT)

Synonyms: n-Butyl alcohol; 1-Hydroxybutane; Butyric alcohol; 1-Butanol; n-Butanol; Butan-1-ol; Butyl hydroxide; Methylolpropane; Propyl carbinol; Propylmethanol

Classification: Primary aliphatic alcohol

Empirical: $C_4H_{10}O$

Formula: $CH_3(CH_2)_2CH_2OH$

Properties: Colorless clear mobile liq., char. penetrating vinous odor; sol. in water; misc. with alcohol, ether, many org. solvs.; m.w. 74.14; sp.gr. 0.8109 (20/20 C); vapor pressure 5.5 mm Hg (20 C); m.p. -90 C; b.p. 117.7 C; flash pt. 35 C; ref. index 1.3993 (20 C)

Toxicology: LD50 (oral, rat) 790 mg/kg, (IV, mouse) 377 mg/kg, (skin, rabbit) 3400 mg/kg; ACGIH TLV:CL 50 ppm (skin); poison by IV route; mod. toxic by skin contact, ing., subcut., IP routes; skin and severe eye irritant; CNS depressant

Precaution: Flamm.; dangerous fire hazard exposed to heat, flame, oxidizers; mod. explosive exposed to flame; incompat. with oxidizing materials, aluminum, chromium trioxides; increased fire and explosion hazard; reacts at elevated temps. with Al

NFPA: Health 1, Flammability 3, Reactivity 0

Uses: Preparation of esters; solvent for fats, waxes, resins, shellac, varnish, plasticizers, paints; mfg. of rayon, detergents, other butyl compds.; hydraulic fluids; dehydration agents; syn. flavoring agent for foods and pharmaceuticals; solvent in pharmaceutical orals; color diluent for confectionery

Regulatory: FDA 21CFR §73.1, 172.515, 172.560, 175.105, 175.320, 176.200, 177.1200, 177.1440, 177.1650; 27CFR §21.99; FEMA GRAS; FDA approved for orals; USP/NF, BP compliance

Manuf./Distrib.: Aldrich; Allchem Ind.; Ashland; J.T. Baker; BASF; Baychem; BP Chems. Ltd; R.E. Carroll; Chemcentral; Chisso Am.; Condea Vista; Coyne; Eastman; Fluka; Harcros USA; Hoechst Celanese; Houghton Chem.; Hüls Am.; Penta Mfg.; Punda Mercantile; Shell; Sigma; Spectrum Chem. Mfg.; Sunnyside; Union Carbide; Van Waters & Rogers

Trade names: Isanol®

2-Butyl alcohol. *See* 2-Butanol
n-Butyl alcohol. *See* Butyl alcohol
s-Butyl alcohol. *See* 2-Butanol

sec-Butyl alcohol. *See* 2-Butanol

t-Butyl alcohol

CAS 75-65-0; EINECS 200-889-7

UN No. 1120 (DOT)

Synonyms: TBA; 2-Methyl-2-propanol; 1,1-Dimethylethanol; t-Butanol; Trimethyl carbinol; t-Butyl hydroxide; 2-Propanol, 2-methyl-

Classification: Tertiary aliphatic alcohol

Empirical: $C_4H_{10}O$

Formula: $(CH_3)_3COH$

Properties: Colorless liq. or rhombic prisms; unpleasant camphor-like odor; sol. in ethanol, ether, water; m.w. 74.12; sp.gr. 0.786 (20 C); vapor pressure 40 mm Hg (24.5 C); m.p. 25-26 C; b.p. 83 C; flash pt. (CC) 11 C; ref. index 1.3870 (20 C)

Toxicology: LD50 (oral, rat) 3500 mg/kg; ACGIH TLV/TWA 100 ppm; STEL 150 ppm; toxic; irritant to mucous membranes; inh. of 25 ppm causes pulmonary problems in man; CNS depressant; inh. of high concs. can cause headache, drowsiness, unconsciousness

Precaution: Flamm.; dangerous fire hazard exposed to heat, flame; mod. explosive as vapor; incompat. with strong oxidizers, potassium-sodium alloy, hydrogen peroxide and sulfuric acid (fire/explosion risk), strong min. acids

NFPA: Health 1, Flammability 3, Reactivity 0

Uses: Solvent, coupling agent, processing aid for pharmaceuticals, personal care prods., aq. coatings and adhesives, agric. formulations, polymer processing, cleaners/disinfectants; alcohol denaturant; dehydration agent; perfumery; chem. intermediate; paint removers; prep. of methyl methacrylate; octane improver in unleaded gasoline; denaturant, solvent, and processing aid for pharmaceuticals

Regulatory: FDA 21CFR §176.200, 178.3910, 27CFR §21.100

Manuf./Distrib.: AC Ind.; Aldrich; Allchem Ind.; ARCO; Ashland; Fluka; Hüls Am.; Sigma; Spectrum Chem. Mfg.

Trade names: Tebol™; Tebol™ 99

sec-Butyl alcohol acetate. *See* s-Butyl acetate

s-Butylamine

CAS 13952-84-6

UN No. 1993; DOT No. 1125

Synonyms: 2-AB; 2-BA; 2-Aminobutane; 2-Butanamine; sec-Butylamine; 1-Methylpropylamine

Classification: Sat. aliphatic amine

Empirical: $C_4H_{11}N$

Formula: $CH_3CH_2CH(CH_3)NH_2$

Properties: Colorless liq.; ammonia-like or fish odor; sol. in water, many org. solvs. incl. ethanol, acetone, ether, chloroform; m.w. 73.14; sp.gr. 0.724 (20 C); m.p. –104 C; b.p. 63 C; flash pt. (CC) –9 C

Toxicology: LD50 (oral, rat) 152 mg/kg, (dermal, rabbit) 2500 mg/kg; toxic, corrosive; poison by ingestion; powerful irritant; mod. toxic by skin contact; severe eye irritant; inh. may cause nose/throat/lung irritation, coughing, chest pain, difficulty breathing

Precaution: Extremely flamm.; incompat. with oxidizing materials, acids, acid chlorides, acid anhydrides, calcium or sodium hypochlorite, nitromethane, nitrosyl perchlorate; violent/explosive reactions possible; Hazardous decomp. prods.: Heated to decomp., emits toxic fumes of NO_x

NFPA: Health 3, Flammability 3

Uses: Fungistat for fruits; solvent; used to produce other org. chems.

Manuf./Distrib.: Aldrich; BASF; BASF AG; Elf Atochem N. Am.

sec-Butylamine. *See* s-Butylamine

Butyl benzoate
CAS 136-60-7; EINECS 205-252-7
Synonyms: Benzoic acid, butyl ester; Benzoic acid n-butyl ester; n-Butyl benzoate
Empirical: $C_{11}H_{14}O_2$
Formula: $C_6H_5COOC_4H_9$
Properties: Colorless oily liq.; insol. in water; misc. with alcohol or ether; m.w. 178.23; dens. 1.00 (20 C); vapor pressure < 0.01 mm Hg (20 C); f.p. -22 C; b.p. 247.3 C; flash pt. 107 C
Toxicology: LD50 (oral, rat) 5140 mg/kg, (skin, rabbit) 4 g/kg; moderately toxic by skin contact; mildly toxic by ingestion; severe skin, moderate eye irritant
Precaution: Combustible exposed to heat or flame; can react with oxidizing materials; Hazardous decomp. prods.: Heated to decomp., emits acrid and irritating fumes
Uses: Solvent for cellulose ether, plasticizer, perfume ingredient, dyeing of textiles
Manuf./Distrib.: Aldrich; Allchem Ind.; Fluka; Novachem Corp.; Pentagon Chems. Ltd; Penta Mfg.; Piedmont Chem. Ind.; Stockhausen; Sybron; Unitex
Trade names containing: Arosolve 9D5R

n-Butyl benzoate. *See* Butyl benzoate
N-Butyl-1-butanamine. *See* Dibutylamine
dl-sec-Butylcarbinol. *See* 2-Methyl-1-butanol
n-Butyl carbinol. *See* n-Amyl alcohol
s-Butyl carbinol. *See* 2-Methyl-1-butanol
t-Butyl carbinol. *See* 2,2-Dimethylpropanol
Butyl 'Carbitol'. *See* Butoxydiglycol
Butyl Carbitol acetate. *See* Diethylene glycol butyl ether acetate
Butyl citrate. *See* Tributyl citrate

Butylcyclohexane
CAS 1678-93-9
Synonyms: 1-Cyclohexylbutane
Classification: Sat. alicyclic hydrocarbon
Empirical: $C_{10}H_{10}$
Formula: $C_4H_9C_6H_{11}$
Properties: Colorless liq.; sol. in most org. solvs.; insol. in water; m.w. 140.27; sp.gr. 0.818 (20 C); m.p. -75 C; b.p. 178-180 C; flash pt. 41 C
Toxicology: May cause mild skin irritation; conc. vapor may cause eye irritation, nose/throat irritation, headache, lightheadedness, nausea, drowsiness; probably no significant CNS effects
Precaution: Combustible; incompat. with strong oxidizing agents (increased fire and explosion hazard); Hazardous decomp. prods.: None
NFPA: Health 0, Reactivity 0
Uses: Solvent; org. synthesis
Manuf./Distrib.: Aldrich; Fluka

s-Butylcyclohexane
CAS 52993-54-1
Classification: Sat. alicyclic hydrocarbon
Empirical: $C_{10}H_{20}$
Formula: $CH_3CH_2CH(CH_3)C_6H_{11}$
Properties: Colorless liq.; sol. in most org. solvs.; insol. in water; m.w. 140.27; sp.gr. 0.813 (20 C); b.p. 177 C
Toxicology: May cause mild skin irritation; conc. vapor may cause eye irritation,

nose/throat irritation, headache, lightheadedness, nausea, drowsiness; probably no significant CNS effects
Precaution: Combustible; incompat. with strong oxidizing agents (increased risk of fire and explosion); Hazardous decomp. prods.: None
NFPA: Health 0, Reactivity 0
Uses: Solvent; org. synthesis

t-Butylcyclohexane
CAS 3178-22-1
Synonyms: (1,1-Dimethylethyl) cyclohexane
Classification: Sat. alicyclic hydrocarbon
Empirical: $C_{10}H_{20}$
Formula: $C_6H_{11}C(CH_3)_3$
Properties: Colorless liq.; sol. in most org. solvs.; insol. in water; m.w. 140.27; sp.gr. 0.831 (20 C); m.p.-41 C; b.p. 167-169 C; flash pt. 42 C
Toxicology: May cause mild skin irritation; conc. vapor may cause eye irritation, nose/throat irritation, headache, lightheadedness, nausea, drowsiness; probably no significant CNS effects
Precaution: Combustible; incompat. with strong oxidizing agents (increased risk of fire and explosion); Hazardous decomp. prods.: None
NFPA: Health 0, Reactivity 0
Uses: Solvent; org. synthesis
Manuf./Distrib.: Aldrich; Fluka

Butylcyclopentane
CAS 2040-95-1
Classification: Sat. alicyclic hydrocarbon
Empirical: C_9H_8
Formula: $C_4H_9C_5H_9$
Properties: Colorless liq.; sol. in most org. solvs.; insol. in water; sp.gr. 0.785 (20 C); m.w. 126.23; m.p. -108 C; b.p. 157 C
Toxicology: May cause mild skin irritation; conc. vapor may be irritating to eyes; ing. may cause sore throat, nausea, diarrhea; probably a CNS depressant on inh.; very high concs. may cause unconsciousness and death
Precaution: Combustible; incompat. with strong oxidizing agents (increased fire and explosion hazard); Hazardous decomp. prods.: None
NFPA: Health 0, Reactivity 0
Uses: Solvent; org. synthesis
Manuf./Distrib.: Sigma

s-Butylcyclopentane
CAS 4850-32-2
Classification: Sat. alicyclic hydrocarbon
Empirical: C_9H_{18}
Formula: $CH_3CH_2CH(CH_3)C_5H_9$
Properties: Colorless liq.; sol. in most org. solvs.; insol. in water; m.w. 125.23
Toxicology: May cause mild skin irritation; conc. vapor may be irritating to eyes; probably a CNS depressant; inh. of high vapor concs. may cause nose/throat irritation, nausea, drowsiness; very high concs. may cause unconsciousness and death
Precaution: Combustible; incompat. with strong oxidizing agents (increased fire and explosion hazard); Hazardous decomp. prods.: None
Uses: Solvent; org. synthesis

t-Butylcyclopentane
CAS 3875-52-3

o-Butyl diethylene glycol

Classification: Sat. alicyclic hydrocarbon
Empirical: C_9H_{18}
Formula: $C_5H_9C(CH_3)_3$
Properties: Colorless liq.; sol. in most org. solvs.; insol. in water; m.w. 126.23
Toxicology: May cause mild skin irritation; conc. vapor may be irritating to eyes; probably a CNS depressant; inh. of high vapor concs. may cause nose/throat irritation, nausea, drowsiness; very high concs. may cause unconsciousness and death
Precaution: Combustible; incompat. with strong oxidizing agents (increased fire and explosion hazard); Hazardous decomp. prods.: None
Uses: Solvent; org. synthesis

o-Butyl diethylene glycol. *See* Butoxydiglycol
Butyldiglycol. *See* Butoxydiglycol
Butyldiglycol acetate. *See* Diethylene glycol butyl ether acetate
Butyl diglyme. *See* Diethylene glycol dibutyl ether
Butyl dipropasol solvent. *See* Dipropylene glycol butyl ether

Butylene glycol
CAS 107-88-0; EINECS 203-529-7
Synonyms: 1,3-Butanediol; Butane-1,3-diol; 1,3-Butylene glycol; β-Butylene glycol; 1,3-Dihydroxybutane; Methyltrimethylene glycol
Classification: Aliphatic diol
Empirical: $C_4H_{10}O_2$
Formula: $HOCH_2CH_2CHOHCH_3$
Properties: Visc. liq., sweet flavor with bitter aftertaste; m.w. 90.12; dens. 1.004-1.006 (20/20 C); vapor pressure 0.06 mm Hg (20 C); f.p. $<$ -50 C; b.p. 207.5 C; flash pt. 250 F
Toxicology: LD50 (oral, rat) 23 g/kg, (subcut., rat) 20 g/kg; mildly toxic by ingestion and subcutaneous routes; eye irritant
Precaution: Combustible when exposed to heat or flame; incompat. with oxidizing materials; Hazardous decomp. prods.: Heated to decomp., emits acrid smoke and irritating fumes
Uses: Solvent for flavoring agents, for foods and pharmaceuticals; pharmaceutical carrier, vehicle, solvent
Regulatory: FDA 21CFR §173.220, 184.1278 GRAS
Manuf./Distrib.: Aldrich; Allchem Ind.; Fluka; Hoechst Celanese; Hüls AG; Kyowa Hakko USA

1,3-Butylene glycol. *See* Butylene glycol
β-Butylene glycol. *See* Butylene glycol

Butylene glycol dicaprylate/caprate
Uses: Emollient, dispersant, lubricant, suspending agent, solubilizer, carrier/vehicle, solvent, penetrant, spreading agent for pharmaceuticals
Trade names: Miglyol® 8810

Butylene hydrate. *See* 2-Butanol
Butylene oxide. *See* Tetrahydrofuran
Butyl ethanoate. *See* n-Butyl acetate
Butyl ether. *See* Dibutyl ether
Butyl ethyl carbinol. *See* 3-Heptanol
o-Butyl ethylene glycol. *See* Butoxyethanol
Butyl ethylhexyl phthalate. *See* Butyl octyl phthalate
Butyl glycol. *See* Butoxyethanol
Butyl hydride. *See* Butane
Butyl hydroxide. *See* Butyl alcohol

t-Butyl hydroxide. *See* t-Butyl alcohol
Butyl 2-hydroxypropanoate. *See* Butyl lactate
Butyl α-hydroxypropionate. *See* Butyl lactate
n-Butyl-S(-)-2-hydroxypropionate. *See* Butyl lactate

Butyl lactate
CAS 138-22-7
Synonyms: n-Butyl lactate; n-Butyl-S(-)-2-hydroxypropionate; Butyl 2-hydroxypropanoate; 2-Hydroxypropanoic acid butyl ester; Butyl α-hydroxypropionate; Lactic acid butyl ester
Classification: Ester
Empirical: $C_7H_{14}O_3$
Formula: $CH_3CHOHCOOC_4H_9$
Properties: Water-wh. liq., mild odor; misc. with many lacquer solvs., diluents, oils, alcohol, ether; sl. sol. in water; hydrolyzed in acids and alkalies; m.w. 146.19; dens. 0.974-0.984 (20/20 C); vapor pressure 0.4 mm Hg (20 C); m.p. -43 C; b.p. 188 C; flash pt. (TOC) 75.5 C
Toxicology: LD50 (subcut., rat) 12 g/kg; LDLo (IP, mouse) 200 mg/kg; ACGIH TLV/TWA 5 ppm; poison by IP route; skin irritant
Precaution: Flamm. exposed to heat or flame; can react with oxidizing materials; Hazardous decomp. prods.: Heated to decomp., emits acrid smoke and irritating fumes
Uses: Dipped latex prods., coatings, specialty finishes, inks, diluents, adhesives, intermediates, photoresist solvents, screen printing of electronic parts; solvent for nitrocellulose; antiskinning agent, dry-cleaning fluids; non-ozone depleting solvent for electronics, coatings, inks, cleaning agents, mold release agents; syn. flavor and fragrance for foods and pharmaceuticals; pharmaceutical intermediate
Regulatory: FDA 21CFR §172.515; FEMA GRAS
Manuf./Distrib.: Aldrich; CPS; Penta Mfg.; Purac Biochem; Van Waters & Rogers
Trade names: Purasolv® BL

n-Butyl lactate. *See* Butyl lactate
Butyl mercaptan. *See* n-Butyl mercaptan
1-Butyl mercaptan. *See* n-Butyl mercaptan

n-Butyl mercaptan
CAS 109-79-5; EINECS 203-705-3
UN No. 2347 (DOT)
Synonyms: Butyl mercaptan; 1-Butyl mercaptan; Butanethiol; 1-Butanethiol; n-Butanethiol; n-Butyl thioalcohol; 1-Mercaptobutane; Thiobutylalcohol
Classification: Alkyl mercaptan
Empirical: $C_4H_{10}S$
Formula: n-C_4H_9SH
Properties: Colorless liq.; strong offensive skunk-like odor; very sol. in alcohol, ether, liq. hydrogen sulfide; pract. insol. in water; m.w. 90.2; sp.gr. 0.8365 (25/4 C); m.p. -116 C; b.p. 98 C; flash pt. (TCC) -7 C
Toxicology: LD50 (oral, rat) 1500 mg/kg, (IP, rat) 399 mg/kg; ACGIH TLV/TWA 0.5 ppm; poison by intraperitoneal route; moderately toxic by inhalation and ingestion; eye irritant; sl. skin irritant
Precaution: Flamm.; dangerous fire hazard exposed to heat, flame, sparks, oxidizers; incompat. with strong bases, strong oxidizing agents, reducing agents, strong acids, alkali metals, nitric acid; violent reactions possible; Hazardous decomp. prods.: Reaction with strong acids can produce toxic fumes of hydrogen sulfide; heated to decomp., emits toxic SO_x
NFPA: Health 2, Flammability 3, Reactivity 0

1-t-Butyl-4-methylbenzene

Uses: Intermediate, solvent for pesticide prod.; additive in oils, rubber; polymerization regulator
Manuf./Distrib.: Aldrich; Elf Atochem N. Am.; Fluka; Phillips

1-t-Butyl-4-methylbenzene. *See* p-t-Butyl toluene
t-Butyl methyl ether. *See* Methyl t-butyl ether
Butyl octadecanoate. *See* Butyl stearate
n-Butyl octadecanoate. *See* Butyl stearate

Butyl octyl phthalate
CAS 84-78-6; EINECS 201-562-1
Synonyms: BOP; 1,2-Benzenedicarboxylic acid, butyl octyl ester; Butyl ethylhexyl phthalate; Phthalic acid, butyl octyl ester
Empirical: $C_{20}H_{30}O_4$
Formula: $C_4H_9OOCC_6H_4COOC_8H_{17}$
Properties: Water-wh. liq., mild char. odor; m.w. 334.46; dens. 0.988-0.996 (20/20 C); f.p. -48 C; b.p. 350 C (760 mm); sapon. no. 298-308; flash pt. (COC) 208 C; ref. index 1.4868
Precaution: Combustible
Uses: High solvating plasticizer for PVC, PS, PVB, PVAc, cellulosics, and rotational molding and dip coating
Manuf./Distrib.: Aristech; Ashland; BASF; BP Oil; Eastman; Harwick; Unitex
Trade names: Kodaflex® HS-3

Butyl phosphate. *See* Tributyl phosphate
Butyl phthalate. *See* Dibutyl phthalate
n-Butyl phthalate. *See* Dibutyl phthalate
Butyl propionate. *See* n-Butyl propionate

n-Butyl propionate
CAS 590-01-2
Synonyms: Propanoic acid butyl ester; Butyl propionate
Empirical: $C_7H_{14}O_2$
Formula: $CH_3CH_2COOC_4H_9$
Properties: Colorless liq., earthy faintly sweet odor, apricot-like taste; very sol. in alcohol, ether; very sl. sol. in water; m.w. 130.19; dens. 0.8754 (20/4 C); f.p. 32 C; m.p. -89 C; b.p. 145-146 C; ref. index 1.401 (20 C)
Toxicology: LD50 (oral, rat) 5000 mg/kg; mildly toxic by ingestion; skin and eye irritant
Precaution: Flamm.; dangerously flamm. exposed to heat or flame; incompat. with oxidizers
Uses: Solvent for nitrocellulose, paints, retarder in lacquer thinner, ingredient of perfumes; syn. flavor for foods and pharmaceuticals
Regulatory: FDA 21CFR §172.515; FEMA GRAS
Manuf./Distrib.: Aldrich; Penta Mfg.; Shell; Union Carbide
Trade names: Ucar® n-Butyl Propionate

Butyl stearate
CAS 123-95-5; EINECS 204-666-5
Synonyms: Butyl octadecanoate; n-Butyl octadecanoate; Octadecanoic acid butyl ester
Definition: Ester of butyl alcohol and stearic acid
Empirical: $C_{22}H_{44}O_2$
Formula: $CH_3(CH_2)_{16}COO(CH_2)_3CH_3$
Properties: Crystals; sl. sol. in water; sol. in alcohol, ether; m.w. 340.60; dens. 0.86 (20/4 C); m.p. 17-22 C; b.p. 343 C; flash pt. (CC) 160 C; ref. index 1.4430 (20 C)
Toxicology: No known toxicity; heated to decomp., emits acrid smoke and

irritating fumes

Uses: Solvent, spreading and softening agent in plastics, textiles, cosmetics, rubbers; syn. flavoring agent for foods and pharmaceuticals; plasticizer for food pkg. materials; lubricant and slip agent for coatings and inks

Regulatory: FDA 21CFR §172.515, 173.340, 175.105, 175.300, 175.320, 176.200, 176.210, 177.2600, 177.2800, 178.3910, 181.22, 181.27; FDA approved for topicals; FEMA GRAS

Manuf./Distrib.: Aldrich; Amerchol; Anar; Aquatec Quimica SA; Ashland; Fluka; C.P. Hall; Henkel/Organic Prods.; Inolex; Kenrich Petrochems.; Mosselman NV; Penta Mfg.; Sea-Land; Sigma; Spectrum Chem. Mfg.; Stepan; Union Camp; Unitex; Witco/Oleo-Surf.

Trade names: Dermol EB; Unimate® BYS

n-Butyl thioalcohol. *See* n-Butyl mercaptan
4-t-Butyl toluene. *See* p-t-Butyl toluene

p-t-Butyl toluene

CAS 98-51-1

Synonyms: TBT; PTBT; 4-t-Butyl toluene; Toluene, p-tert-butyl-; 1-t-Butyl-4-methylbenzene; p-Methyl-t-butylbenzene; 1-Methyl-4-t-butylbenzene

Classification: Aromatic hydrocarbon

Empirical: $C_{11}H_{16}$

Formula: $(CH_3)_3CC_6H_4CH_3$

Properties: Colorless liq.; insol. in water; m.w. 148.25; dens. 0.857-0.863 (20/20 C); b.p. 192.8 C; flash pt. 54 C

Toxicology: LD50 (oral, mouse) 900 mg/kg; ACTIH TLV/TWA 10 ppm; STEL 20 ppm; mod. toxic by inhalation, ingestion, skin absorption; skin irritant; human eye irritant (conjunctive irritation); causes nausea and vomiting

Precaution: Flamm. exposed to heat or flame; incompat. with oxidizers; Hazardous decomp. prods.: Heated to decomp., emits acrid smoke and fumes

Uses: Intermediate for perfumes and fragrances; solvent

Manuf./Distrib.: Aldrich; Fluka; Hüls AG

Butyric acid lactone. *See* Butyrolactone
Butyric alcohol. *See* Butyl alcohol
Butyrolactam. *See* 2-Pyrrolidone

Butyrolactone

CAS 96-48-0; EINECS 202-509-5

Synonyms: BLO; 4-Deoxytetronic acid; Dihydro-2(3H)-furanone; 4-Hydroxybutanoic acid lactone; γ-Hydroxybutyrolactone; γ-Butyrolactone; Butyric acid lactone; Butyryl lactone; γ-Hydroxybutyric acid cyclic ester; 4-Butanolide; 4-Hydroxybutyric acid γ-lactone; Tetrahydro-2-furanone

Classification: Lactone

Empirical: $C_4H_6O_2$

Properties: Colorless oily liq., mild caramel odor; misc. with water; sol. in methanol, ethanol, acetone, ether, benzene; m.w. 86.09; dens. 1.120; m.p. -45 C; b.p. 204-205 C; flash pt. 98 C; ref. index 1.4348

Toxicology: LD50 (oral, rat) 1800 mg/kg, (IP, rat) 1000 mg/kg; mod. toxic by ingestion, IV, IP routes; suspected tumorigen; experimental reproductive effects; mutagenic data

Precaution: Combustible when exposed to heat or flame; reactive with oxidizing materials; Hazardous decomp. prods.: Heated to decomp., emits acrid and irritating fumes

NFPA: Health 0, Flammability 1, Reactivity 0

Uses: Intermediate for synthesis of polyvinylpyrrolidone, piperidine,

phenylbutyric acid; solvent for polyacrylonitrile, cellulose acetate, methyl methacrylate polymers, polystyrene, paints; constituent of paint removers, textile aids, drilling oils; flavoring agent for candy, soy milk; solvent for PAN, PS, fluorinated hydrocarbons, cellulose triacetate, shellac; in paint removers, petrol. processing, inks; solvent, stabilizer for pharmaceuticals; pharmaceutical intermediate
Regulatory: FEMA GRAS
Manuf./Distrib.: Aldrich; Allchem Ind.; BASF; Fabrichem; Fluka; Gelest; Great Western; Houghton Chem.; ISP; Janssen Chimica; Schweizerhall; Sigma; Spectrum Chem. Mfg.; UCB SA
Trade names: Agrisynth® BLO; Agsol Ex BLO; BLO®; Dynasolve 699; Gaftex® HI-B; GBL
Trade names containing: FoamFlush® Urethane Remover; PrepRite™ Coatings Remover; Ship Shape® Resin Cleaner

γ-Butyrolactone. *See* Butyrolactone
Butyryl lactone. *See* Butyrolactone

Cacao butter. *See* Cocoa butter
Cajeputene. *See* d-Limonene; dl-Limonene; l-Limonene
Cajeputol. *See* Eucalyptol
C-8 alcohols. *See* Caprylic alcohol
C16-18 alcohols. *See* Cetearyl alcohol
C6 alkyl acetate. *See* Oxo-hexyl acetate
C7 alkyl acetate. *See* Oxo-heptyl acetate

C8 alkyl acetate
CAS 108419-32-5
Synonyms: Acetic acid, C7-9-branched alkyl esters, C8-rich
Properties: Sp.gr. 0.87; flash pt. (TCC) 171 F
Toxicology: LD50 (oral, rat) 5 g/kg; reproductive effects
Uses: Solvent for high solids, maintenance, marine, and industrial waterborne coatings, acrylic resin polymerization media, metalworking fluids, urethane foam and industrial cleaners and degreasers, paint strippers, pigment dispersions; pesticide emulsifiable concs., textile scouring aids, NC or screen printing inks, penetrating and lubricating oils, vinyl plastisols; good solvency, low surf. tens. and interfacial tens., good wetting, lubricity; visc. modifier; latex coalescer
Trade names: Exxate® 800

C9 alkyl acetate
CAS 108419-33-6
Properties: Sp.gr. 0.87; flash pt. (TCC) 194 F
Uses: Solvent for high solids, maintenance, marine, and industrial waterborne coatings, acrylic resin polymerization media, metalworking fluids, urethane foam and industrial cleaners and degreasers, paint strippers, pigment dispersions; pesticide emulsifiable concs., textile scouring aids, NC or screen printing inks, penetrating and lubricating oils, vinyl plastisols; good solvency, low surf. tens. and interfacial tens., good wetting, lubricity; visc. modifier; latex coalescer
Trade names: Exxate® 900

C10 alkyl acetate
CAS 108419-34-7
Properties: Sp.gr. 0.87; flash pt. (TCC) 212 F
Uses: Solvent for high solids, maintenance, marine, and industrial waterborne coatings, acrylic resin polymerization media, metalworking fluids, urethane foam and industrial cleaners and degreasers, paint strippers, pigment dispersions; pesticide emulsifiable concs., textile scouring aids, NC or screen printing inks, penetrating and lubricating oils, vinyl plastisols; good solvency, low surf. tens. and interfacial tens., good wetting, lubricity; visc. modifier; latex coalescer
Trade names: Exxate® 1000

C13 alkyl acetate
CAS 108419-35-8
Properties: Sp.gr. 0.87; flash pt. (TCC) 261 F
Uses: Solvent for high solids, maintenance, marine, and industrial waterborne coatings, acrylic resin polymerization media, metalworking fluids, urethane foam and industrial cleaners and degreasers, paint strippers, pigment dispersions; pesticide emulsifiable concs., textile scouring aids, NC or screen printing inks, penetrating and lubricating oils, vinyl plastisols; good solvency, low surf. tens. and interfacial tens., good wetting, lubricity; visc. modifier; latex coalescer
Trade names: Exxate® 1300

C12-15 alkyl benzoate
CAS 68411-27-8; EINECS 270-112-4
Definition: Ester of benzoic acid and C12-15 alcohols
Uses: Emollient, solubilizer, sunscreen vehicle, deoiler for creams and oils; binder in pressed powds.; solvent and perfume fixative
Trade names: Crodamol AB

Camphene
CAS 79-92-5
UN No. 9011 (DOT)
Synonyms: 2,2-Dimethyl-3-methylenenorbornane; 3,3-Dimethyl-2-methylene norcamphane; Bicyclo (2.2.1) heptane, 2,2-dimethyl-3-methylene-
Empirical: $C_{10}H_{16}$
Properties: Colorless cubic cryst., oily odor, terpene camphoraceous taste; sol. in alcohol; misc. with fixed oils; insol. in water; m.w. 136.26; dens. 0.842 (54/4 C); m.p. 50-51 C; b.p. 159 C; ref. index 1.452 (55 C)
Toxicology: Mutagenic data
Precaution: Combustible; yields flamm. vapors when heated; reacts with oxidizing materials; Hazardous decomp. prods.: Heated to decomp., emits acrid smoke and fumes
Uses: Syn. flavoring agent for foods and pharmaceuticals
Regulatory: FDA 21CFR §172.515; FEMA GRAS
Manuf./Distrib.: Berje; Penta Mfg.; SCM Glidco Organics; Spectrum Chem. Mfg.
Trade names containing: Acintene® AC-P; Acintene® DP 738; Acintene® P-AC Alpha-Pinene

Camphocean. *See* 1,1,2-Trimethylcyclopentane
Capric alcohol. *See* n-Decyl alcohol

Capric triglyceride
Uses: Food emulsifier; solvent for flavors and oil-sol. food additives
Trade names: Delios® C

Caprinic alcohol. *See* n-Decyl alcohol

Caproyl alcohol. *See* Hexyl alcohol
Capryl alcohol. *See* Caprylic alcohol; 2-Octanol
Caprylic acid, monoester with 1,2-propanediol. *See* Propylene glycol caprylate

Caprylic alcohol
CAS 111-87-5; EINECS 203-917-6
Synonyms: n-Octyl alcohol; pri-Octyl alcohol; Heptyl carbinol; Octanol; 1-Octanol; n-Octanol; 1-Hydroxyoctane; C-8 alcohols; Alcohol C-8; Capryl alcohol
Classification: Primary aliphatic alcohol
Empirical: $C_8H_{18}O$
Formula: $CH_3(CH_2)_6CH_2OH$
Properties: Colorless oily liq., fresh orange-rose odor, sl. herbaceous taste; pract. insol. in water; misc. in alcohol, ether, and chloroform; m.w. 130.26; sp.gr. 0.827; m.p. -16.7 C; b.p. 194.5 C; flash pt. 81 C; ref. index 1.429 (20 C)
Toxicology: LD50 (oral, mouse) 1790 mg/kg; poison by intravenous route; moderately toxic by ingestion; skin and eye irritant; does not readily form a vapor; vapors may cause CNS depression; can probably cause 'alcohol' intoxication effects
Precaution: Flamm. exposed to heat or flame; can react with oxidizers
NFPA: Health 1, Flammability 2, Reactivity 0
Uses: Wetting agent, emulsifier, surfactant intermediate, raw material for plasticizers; emulsion polymerization; tobacco-offshoots controller; detergent intermediate; solvent for paints, foods; defoaming agent; syn. flavoring agent for foods and pharmaceuticals; mfg. of perfumes
Regulatory: FDA 21CFR §172.230, 172.515 (only for encapsulating lemon, dist. lime, orange, peppermint, and spearmint oils), 172.864, 173.280, 175.105, 175.300, 176.210, 177.1010, 177.1200, 177.2800, 178.3480; FEMA GRAS
Manuf./Distrib.: Albemarle; Aldrich; Condea Vista; Fluka; Henkel/Emery; M. Michel; Penta Mfg.; Schweizerhall; Sigma; Spectrum Chem. Mfg.; Union Camp; Van Waters & Rogers

Caprylic/capric glycerides
CAS 26402-26-6
Definition: Mixture of mono-, di-, and triglycerides of caprylic and capric acids
Uses: Plasticizer for hard fats; solvent for lipophilic ingreds.; emollient; solvent, coemulsifier, solubilizer, carrier for lipophilic drugs; antifoam for foods
Trade names: Imwitor® 742

Caprylic/capric/lauric triglyceride
CAS 68991-68-4
Definition: Mixed triester of glycerin with caprylic, capric and lauric acids
Uses: Emollient, solvent, carrier, fixative, and extender for pharmaceutical, nutritional, and cosmetic applics.
Trade names: Captex® 350

Caprylic/capric/linoleic triglyceride
CAS 67701-28-4
Definition: Mixed triester of glycerin with caprylic, capric, and linoleic acids
Uses: Emollient, solvent, carrier, fixative, and extender in pharmaceuticals, nutritional, and cosmetic applics.
Trade names: Captex® 810A; Captex® 810B; Captex® 810C; Captex® 810D

Caprylic/capric/oleic triglyceride
CAS 67701-28-4
Uses: Emollient, solvent, and extender in nutritional applics.; carrier for flavors

and fragrances
Trade names: Captex® 910A; Captex® 910B; Captex® 910C; Captex® 910D

Caprylic/capric triglyceride
CAS 65381-09-1; 85409-09-2; 73398-61-5; EINECS 265-724-3
Synonyms: Octanoic/decanoic acid triglyceride
Definition: Mixed triester of glycerin and caprylic and capric acids
Uses: Solvent for colors and perfumes; emollient, moisturizer, dispersant, solubilizer, suspending agent in cosmetics, toiletries, pharmaceuticals; plasticizer; lubricity vehicle; food emulsifier
Regulatory: Approved for topicals
Manuf./Distrib.: Calgene; Croda Inc; Inolex; Lonza; Stepan
Trade names: Aldo® MCT KFG; Captex® 300; Delios® S; Delios® V; Estasan GT 8-40 3578; Estasan GT 8-60 3575; Estasan GT 8-60 3580; Estasan GT 8-65 3577; Estasan GT 8-65 3581; Estasan GT 8-70 3579; Labrafac® LIPO; Lexol® GT-855; Lexol® GT-865; Liponate GC; Miglyol® 810; Miglyol® 812; Neobee® M-5; Neobee® O; Tegosoft® CT
Trade names containing: Gilugel MIG

Caprylic/capric triglyceride PEG-4 esters
Definition: Complex mixture formed from transesterification of caprylic/capric triglyceride and PEG-4
Uses: Hydrophilic oil, bioavailability enhancer, emulsifier, solvent, solubilizer for cosmetic, pharmaceutical, and veterinary creams, lotions, ointments, shampoos, bath oils
Trade names: Labrafac® Hydro

Caprylyl acetate. *See* Octyl acetate

Caprylyl pyrrolidone
Synonyms: N-Octyl-2-pyrrolidone
Classification: Substituted heterocyclic compd.
Uses: Surfactant for personal care; specialty solvent for commercial cleaning, textile processing, lithographic printing prods., coatings, inks
Trade names: Surfadone® LP-100

Carbamaldehyde. *See* Formamide
Carbinol. *See* Methyl alcohol
'Carbitol'. *See* Ethoxydiglycol

Carbomer
CAS 9007-16-3; 9003-01-4; 9007-17-4; 9007-20-9; 76050-42-5
Synonyms: Carboxypolymethylene; Acrylic acid, polymer with sucrose-polyallyl ether
Definition: Homopolymer of acrylic acid crosslinked with an allyl ether of pentaerythritol or an allyl ether of sucrose
Properties: Wh. fluffy powd., odorless; hygroscopic; dens. 1.41
Toxicology: LD50 (oral, rat) 4100 mg/kg; mod. toxic by ingestion; Hazardous decomp. prods.: Heated to decomp., emits acrid smoke and fumes
Uses: Emulsifier, thickener, stabilizer, suspending agent; for lubricating, quenching and silicone emulsions; graphite, polyethylene, fiber and paper suspensions, textile back coatings, pharmaceuticals, cosmetics; solvent thickener for cosmetics, pharmaceuticals
Use level: 0.4% (suspending agent), 0.5-5% (aq. ointment base)
Regulatory: FDA approved for orals, rectals, topicals; BP compliance
Manuf./Distrib.: Aldrich; Sigma; Spectrum Chem. Mfg.
Trade names: Carbopol® 940; Carbopol® 980

Carbon bisulfide. *See* Carbon disulfide

Carbon dioxide
CAS 124-38-9; EINECS 204-696-9
UN No. 1013 (DOT), 1015 (DOT), 1041 (DOT), 1845 (DOT), 1952 (DOT), 2187 (DOT)
Synonyms: Carbonic acid gas; Carbonic anhydride
Classification: Gas
Empirical: CO_2
Properties: Colorless gas; odorless; sol. 1 vol. dissolves in about 1 vol. of water; m.w. 44.01; m.p. subl. @ -78.5 C; nonflamm.
Toxicology: LCLo, (inh., human, 1 min) 10 pph; ACGIH TLV/TWA 5000 ppm; STEL 30,000 ppm; asphyxiant at > 10%; experimental teratogen; skin contact can cause burns
Precaution: Various dusts explode in CO_2 atmospheres; several bulk metals will burn in CO_2; reacts vigorously with various chems.; incompat. with acrylaldehyde, aziridine, metal acetylides, sodium peroxide
Uses: Refrigeration, aerosol propellant, chemical intermediate, fire extinguishing, inert atmospheres, medicine, oil wells, mining, blowing agent; municipal water treatment, precipitant for waste stream treatment, pH control, silica activation; recarbonation in water softening; food additive, preservative, aerating agent, carbonation, cooling agent, leavening agent, pH control agent, processing aid, propellant, carbonated beverages; air displacement in pharmaceutical inhalants
Regulatory: FDA 21CFR §169.115, 169.140, 169.150, 184.120, GRAS, 193.45 (modified atm. for pest control); USDA 9CFR §318.7, 381.147; BATF 27CFR 240.1051; Japan approved; Europe listed; UK approved; FDA approved for inhalants; USP/NF, BP, Ph.Eur. compliance
Manuf./Distrib.: Air Liquide Am.; Air Prods.; Aldrich; BOC Gases; Carbonic Ind.; Great Western; Liquid Carbonic; Nissan Chem. Ind.; Norsk Hydro AS; Scott Spec. Gases; Showa Denko
Trade names containing: MS-195/CO2 Safezone® Cleaning Solvent & Flux Remover

Carbon disulfide
CAS 75-15-0; EINECS 200-843-6
UN No. 1131 (DOT)
Synonyms: Carbon bisulfide; Carbon sulfide; Dithiocarbonic anhydride
Classification: Inorg. compd. of carbon and sulfur
Empirical: CS_2
Formula: S=C=S
Properties: Colorless clear liq.; mild sweet odor (pure), strong disagreeable odor (impure); m.w. 76.13; sol. 2 g/l in water (20 C); dissolves in ethanol, methanol, benzene, chloroform, CCl_4, ether, oils; vapor pressure 400 mm Hg (28 C); m.p. -112 C; b.p. 46.3 C; flash pt. (CC) -30 C
Toxicology: LD50 (oral, rat) > 505 mg/kg; LCLo (inh., human, 30 min) 4000 ppm; ACGIH TLV/TWA 10 ppm (skin); readily absorbed thru skin; dissolves oils from skin; splashes in eyes cause immediate severe irritation; tumorigen
Precaution: Highly flamm.; incompat. with air, chemically active metals, metal oxides, metal azides, amines, bases, imines, halogens, nitrogen oxides, permanganates and sulfuric acid; violent and explosive reactions possible; Hazardous decomp. prods.: Heated to decomp., emits highly toxic fumes of SO_x
NFPA: Health 3, Flammability 4, Reactivity 0
Uses: Mfg. of viscose rayon, cellophane, carbon tetrachloride, fluorocarbon chems., pharmaceutical, agric., mining, and rubber chems.; flotation agents;

solvent for rubbers, waxes, plastics, sulfur, selenium, bromine, and iodine
Regulatory: FDA 21CFR 193.225, 193.230 (fumigant); SARA reportable
Manuf./Distrib.: Akzo Nobel; Aldrich; Elf Atochem N. Am.; Fluka; PPG Ind.; Rhône-Poulenc; Sigma; Spectrum Chem. Mfg.

Carbonic anhydride. *See* Carbon dioxide
Carbon oil. *See* Benzene
Carbon sulfide. *See* Carbon disulfide

Carbon tetrachloride
CAS 56-23-5; EINECS 200-262-8; UN No. 1846
Synonyms: Tetrachloromethane; Perchloromethane; Carbon tet
Empirical: CCl_4
Properties: Colorless liq.; sweetish distinct odor; misc. with alcohol, ether, chloroform, benzene; insol. in water; m.w. 153.84; dens. 1.585 (25/4 C); m.p. -23 C
Toxicology: TLV 5 ppm in air; LD50 (oral, rat) 2800-2920 mg/kg; toxic by ing., inh., skin absorp.; carcinogen; dec. to phosgene at high temps.
Uses: Refrigerants, metal degreasing, agric. fumigant, chlorinating org. compds., prod. of semiconductors, solvents
Manuf./Distrib.: Aldrich; Ashland; Fluka; Janssen Chimica; Mitsui Toatsu; Sigma

(+)-Carvene. *See* d-Limonene
Carvene. *See* l-Limonene

Castor oil
CAS 1323-38-2; 8001-79-4; EINECS 232-293-8
Synonyms: Aromatic castor oil; Castor oil aromatic; Ricinus oil; Aceite de ricino; Huile de ricini; Oleum ricini; Ricini oleum; Oil of Palma Christi; Tangantangan oil
Definition: Fixed oil obtained from seeds of *Ricinus communis*
Properties: Colorless to pale yel. viscous liq., char. odor; sol. in alcohol; misc. with glac. acetic acid, chloroform, ether; dens. 0.961; m.p. -12 C; b.p. 313 C; flash pt. (CC) 445 F; iodine no. 83-88; sapon. no. 176-187; hyd. no. 160-168; ref. index 1.478
Toxicology: Moderately toxic by ingestion; allergen; eye irritant; purgative, laxative in large doses
Precaution: Combustible when exposed to heat; spontaneous heating may occur
NFPA: Health 0, Flammability 1, Reactivity 0
Uses: Plasticizer in lacquers and nitrocellulose; polyurethane coatings; drying oil; hydraulic fluids; industrial lubricants; used in mfg. of Turkey red oil, cosmetics, leather treatment; antisticking agent, release agent, drying oil, flavoring for foods; component of protective coatings; pharmaceutical vehicle, plasticizer, solvent, cosolvent
Use level: 0.08-23 mg (solid oral dosage forms); 5-12.5% (topicals)
Regulatory: FDA 21CFR §73.1, 172.510, 172.876, 175.300, 176.210, 177.2600, 177.2800, 178.3120, 178.3570, 178.3910, 181.22, 181.26, 181.28; FEMA GRAS; FDA approved for injectables, orals, topicals; USP/NF, BP, Ph.Eur., JP compliance
Manuf./Distrib.: Air Prods.; Aldrich; Amber Syn.; Am. Oil & Supply; Arista Ind.; Ashland; CasChem; Climax Perf. Materials; Degen; Fanning; Fluka; Harcros USA; ICI Surf. Am.; Lipo; Norman, Fox; Penta Mfg.; Sigma; Spectrum Chem. Mfg.; Süd-Chemie Rheologicals; Van Waters & Rogers; Welch, Holme & Clark; Zeochem
Trade names: Cosmetol® X; Crystal® O; Crystal® Crown; Diamond Quality®; EmCon CO; York Krystal Kleer Castor Oil; York USP Castor Oil

Castor oil, acetylated. *See* Glyceryl triacetyl ricinoleate

Castor oil, polymerized
Definition: Rubber-like polymer resulting from combination of castor oil with sulfur or diisocyanates
Uses: Pigment wetting/dispersing; plasticizer, processing aid for resins, gums, polymers, rubber; lubricant, penetrant, coupling solvent; adhesion promoter; for lacquers, inks, adhesives, lubricants, polishes, hydraulic fluids
Trade names: #15 Oil; #30 Oil; #40 Oil; Pale 4; Pale 16; Pale 170; Pale 1000; Vorite 105; Vorite 110; Vorite 115; Vorite 120; Vorite 125

Castorwax. *See* Hydrogenated castor oil
Caustic potash. *See* Potassium hydroxide
Caustic soda. *See* Sodium hydroxide
'Cellosolve'. *See* Ethoxyethanol

Cetearyl alcohol
CAS 8005-44-5; 67762-27-0; EINECS 267-008-6
Synonyms: Cetostearyl alcohol; Cetyl/stearyl alcohol; C16-18 alcohols
Definition: Mixture of fatty alcohols, predominantly cetyl and stearyl alcohols
Properties: Wh. unctuous flakes or gran., faint char. odor, bland mild taste; sol. in alcohol, ether; insol. in water; m.p. 48-55 C; acid no. ≤ 2; iodine no. ≤ 4; hyd. no. 208-228
Uses: Surfactant intermediate; raw material for ethoxylation, sulfation, etc.; stabilizer in emulsion polymerization; lubricant in rigid PVC; cosolvent, emollient, consistency factor for pharmaceutical creams, hand lotions, bath oils, shaving creams; antiseptic in topical anti-infective prods.
Regulatory: FDA 21CFR §175.105; FDA approved for orals, otics, topicals; USP/NF, BP, Ph.Eur. compliance
Trade names: Adol® 640

Cetearyl isononanoate
Synonyms: Isononanoic acid, cetyl/stearyl ether
Definition: Ester of cetearyl alcohol and a branched chain of nanoic acid
Uses: Emollient for applic. in skin care, pharmaceutical skin care, massage, and sun protection preps.; solvent for lipophilic ingreds.
Trade names: Tegosoft® CI

Cetearyl octanoate
CAS 59130-70-7; EINECS 261-619-1
Synonyms: Cetyl/stearyl 2-ethyl hexanoate; 2-Ethylhexanoic acid, cetyl/stearyl ester
Definition: Ester of cetearyl alcohol and octanoic acid
Uses: Emollient, waterproofing agent in skin care preps., pharmaceuticals
Trade names containing: Crodamol CAP

Ceteth-5
CAS 9004-95-9 (generic); 4478-97-1
Synonyms: PEG-5 cetyl ether; POE (5) cetyl ether; 3,6,9,12,15-Pentaoxa-hentriacontan-1-ol
Definition: PEG ether of cetyl alcohol
Formula: $CH_3(CH_2)_{14}CH_2(OCH_2CH_2)_nOH$, avg. n = 5
Uses: Emulsifier, cleaner, dispersant for cosmetics and pharmaceuticals, esp. creams and lotions
Manuf./Distrib.: Aldrich; Fluka; Sigma
Trade names containing: Solulan® 5

Ceteth-16
CAS 9004-95-9 (generic)
Synonyms: PEG-16 cetyl ether; POE (16) cetyl ether

Definition: PEG ether of cetyl alcohol
Formula: $CH_3(CH_2)_{14}CH_2(OCH_2CH_2)_nOH$, avg. n = 16
Uses: Emulsifier, lubricant in cosmetics and pharmaceuticals
Regulatory: FDA 21CFR §176.200
Manuf./Distrib.: Aldrich; Fluka; Sigma
Trade names containing: Solulan® 16

Ceteth-24
CAS 9004-95-9 (generic)
Synonyms: PEG-24 cetyl ether; POE (24) cetyl ether
Definition: PEG ether of cetyl alcohol
Formula: $CH_3(CH_2)_{14}CH_2(OCH_2CH_2)_nOH$, avg. n = 24
Uses: Emulsifier, cleaner, dispersant for cosmetics and pharmaceuticals, esp. creams and lotions
Regulatory: FDA 21CFR §177.2800
Manuf./Distrib.: Aldrich; Fluka; Sigma
Trade names containing: Solulan® C-24

Cetoleth-2
Synonyms: PEG-2 cetyl/oleyl ether; POE (2) cetyl/oleyl ether
Definition: Polyethylene glycol ether of cetyl alcohol and oleyl alcohol
Uses: Emulsifier, mfg. of textile lubricants, solvent and waterless hand cleaners; coemulsifier; detergent additive in petrol. oils; intermediate for anionic surfactants

Cetoleth-3
Synonyms: PEG-3 cetyl/oleyl ether; POE (3) cetyl/oleyl ether
Definition: Polyethylene glycol ether of cetyl alcohol and oleyl alcohol
Uses: Emulsifier, mfg. of textile lubricants, solvent and waterless hand cleaners; coemulsifier; detergent additive in petrol. oils; intermediate for anionic surfactants

Cetoleth-6
Synonyms: PEG-6 cetyl/oleyl ether; POE (6) cetyl/oleyl ether
Definition: Polyethylene glycol ether of cetyl alcohol and oleyl alcohol
Uses: Emulsifier, mfg. of textile lubricants, solvent and waterless hand cleaners; coemulsifier; detergent additive in petrol. oils; intermediate for anionic surfactants

Cetostearyl alcohol. *See* Cetearyl alcohol

Cetyl acetate
CAS 629-70-9; EINECS 211-103-7
Synonyms: Hexadecyl acetate; 1-Hexadecanol, acetate; Palmityl acetate
Definition: Ester of cetyl alcohol and acetic acid
Empirical: $C_{18}H_{36}O_2$
Formula: $CH_3COOCH_2(CH_2)_{14}CH_3$
Properties: M.w. 284.54
Toxicology: LD50 (oral, rat) > 5 g/kg; primary irritant
Uses: Used in pharmaceutical topicals
Manuf./Distrib.: Sigma
Trade names containing: Acetol® 1706; Acetulan®; Fancol ALA; Ritawax AEO; Ritawax ALA

Cetyl alcohol
CAS 124-29-8; 36653-82-4; EINECS 253-149-0
Synonyms: Palmityl alcohol; C16 linear primary alcohol; Alcohol C16; 1-Hexadecanol; Hexadecyl alcohol

Cetyl 2-ethylhexanoate

Classification: Fatty alcohol
Empirical: $C_{16}H_{34}O$
Formula: $CH_3(CH_2)_{14}CH_2OH$
Properties: Wh. waxy solid; partially sol. in alcohol and ether; insol. in water; m.w. 242.27; dens. 0.8176 (49.5 C); m.p. 49.3 C; b.p. 344 C; acid no. ≤ 2; iodine no. ≤ 5; hyd. no. 218-238; flash pt. > 110 C; ref. index 1.4283; surface-active dispersant, wetting agent, and compatibility aid for enhancing color acceptance in high solids coatings (alkyds, epoxies, uv and eb-cured coatings); emulsifier for emulsion polymerization
Toxicology: LD50 (oral, rat) 6400 mg/kg; mod. toxic by ingestion, intraperitoneal routes; eye and human skin irritant; can cause hives; heated to decomp., emits acrid smoke and fumes
Precaution: Flamm. when exposed to heat or flame; can react with oxidizing materials
Uses: Perfumery; emulsifier, emollient, coupling agent, surface-active dispersant, wetting agent; foam stabilizer in detergents; opacifier; thickener; chemical intermediate; cosmetics, pharmaceuticals; external mold release for acrylics; emulsion polymerization; color enhancer in high solids coatings; flavoring agent, color diluent, intermediate for confectionery, food supplements in tablet form, gum; cosolvent
Regulatory: FDA 21CFR §73.1, 73.1001, 172.515, 172.864, 175.105, 175.300, 176.200, 177.1010, 177.1200, 177.2800, 178.3480, 178.3910; FEMA GRAS; FDA approved for ophthalmics, orals, otics, rectals, topicals; USP/NF, BP, Ph.Eur. compliance
Manuf./Distrib.: Aarhus Oliefabrik A/S; AC Ind.; Albemarle; Aldrich; Allchem Ind.; Amerchol; Brown; Chemron; Condea Vista; Croda Chem. Ltd; Fluka; R.W. Greeff; Henkel/Emery; Lipo; Lonza; M. Michel; Norman, Fox; Penta Mfg.; Procter & Gamble; Rhône-Poulenc Surf. & Spec.; Sea-Land; Sigma; Spectrum Chem. Mfg.; Stepan
Trade names: Adol® 52 NF; Adol® 520 NF
Trade names containing: Witcohol™ 42; Witcohol™ 320; Witcohol™ 330; Witcohol™ 340

Cetyl 2-ethylhexanoate. *See* Cetyl octanoate
Cetyl isooctanoate. *See* Cetyl octanoate

Cetyl lactate
CAS 35274-05-6; EINECS 252-478-7
Synonyms: n-Hexadecyl lactate; 2-Hydroxypropanoic acid hexadecyl ester; 1-Hexadecanol lactate
Definition: Ester of cetyl alcohol and lactic acid
Empirical: $C_{19}H_{38}O_3$
Formula: $CH_3CHOHCOO(CH_2)_{15}CH_3$
Properties: Waxy solid; sol. in oil, ethanol, IPM, oleyl alcohol; partly sol. in water, propylene glycol; m.w. 314.49; m.p. 41 C; b.p. 132 C (0.1 mm), 170 C (1 mm); ref. index 1.4410
Toxicology: LD50 (rat, oral) > 20 ml/kg; no known toxicity; nonirritating to eyes and skin
Uses: Nonionic emollient, lubricant for cosmetics and pharmaceuticals; binder for pressed powds.; solvent for lipstick dyes
Manuf./Distrib.: Am. Biorganics
Trade names: Ceraphyl® 28

Cetyl octanoate
CAS 59130-69-7; EINECS 261-619-1
Synonyms: Cetyl 2-ethylhexanoate; Cetyl isooctanoate; n-Hexadecyl 2-ethylhexanoate; Hexanoic acid, 2-ethyl-, hexadecyl ester

Definition: Ester of cetyl alcohol and 2-ethylhexanoic acid
Uses: Emollient for cosmetics; solvency properties for makeup removers
Trade names: Schercemol CO

Cetyl/stearyl alcohol. *See* Cetearyl alcohol
Cetyl/stearyl 2-ethyl hexanoate. *See* Cetearyl octanoate
CFC 12. *See* Dichlorodifluoromethane
CFC 22. *See* Chlorodifluoromethane
CFC 112. *See* 1,1,2,2-Tetrachloro-1,2-difluoroethane
CFC 113. *See* Trichlorotrifluoroethane
CFC 114. *See* Dichlorotetrafluoroethane
Chinese bean oil. *See* Soybean oil
Chlorcosane. *See* Paraffin, chlorinated
Chlorethyl. *See* Ethyl chloride
Chlorinated paraffin. *See* Paraffin, chlorinated
Chloroben. *See* o-Dichlorobenzene
4-Chlorobenzotrifluoride. *See* p-Chlorobenzotrifluoride

p-Chlorobenzotrifluoride
CAS 98-56-6
UN No. 2234 (DOT)
Synonyms: PCBTF; Parachlorobenzotrifluoride; Toluene, p-chloro-α,α,α,-trifluoro; α,α,α-Trifluoro-4-chlorotoluene; Benzene, 1-chloro-4 (trifluoromethyl); 1-Chloro-4-(trimethyl)benzene; p-Trifluoromethylphenyl chloride; p-(Trifluoromethyl)chlorobenzene; 4-Chlorobenzotrifluoride; (p-Chlorophenyl)trifluoromethane; p-Chlorotrifluoromethylbenzene; 4-Chlorotrifluoromethylbenzene
Empirical: $C_7H_4F_3Cl$
Properties: Water-wh. liq.; aromatic odor; m.w. 180.56; f.p. -36 C; m.p. -33.2 C; b.p. 138.7 C; flash pt. (CC) 116 F
Toxicology: LD50 (oral, rat) 13 g/kg, (inh., rat) 22 g/m^3, (oral, mouse) 11,500 mg/kg; mildly toxic by inh. or ingestion; human mutagenic data
Precaution: Flamm.; strong exothermic reactions with sodium dimethylsulfinate; Hazardous decomp. prods.: Heated to decomp., emits toxic fumes of F^- and Cl^-
Uses: Non-ozone depleting solvent for industrial solvent cleaning, aerosols, adhesives, coatings, inks, electronic applics.; 1,1,1-trichloroethane alternative
Manuf./Distrib.: Aldrich; Fluka; Occidental; Sigma
Trade names: Oxsol® 100
Trade names containing: Oxsol® 253; Oxsol® 325; Oxsol® 550

Chlorocosane. *See* Paraffin, chlorinated
Chloroden. *See* o-Dichlorobenzene
1-Chloro-2,2-dichloroethylene. *See* Trichloroethylene
Chlorodifluoroethane. *See* 1,1,1-Chlorodifluoroethane

1,1,1-Chlorodifluoroethane
CAS 75-68-3
UN No. 2517 (DOT)
Synonyms: FC142b; Chlorodifluoroethane; α-Chloroethylidene fluoride; 1,1,1-Difluorochloroethane; 1,1-Difluoro-1-chloroethane; Difluoromonochloroethane; Fluorocarbon FC142b; Freon 142; Freon 142b; Ethane, 1-chloro-1,1-difluoro-
Empirical: $C_2H_3ClF_2$
Formula: CH_3CClF_2
Properties: Colorless gas, nearly odorless; insol. in water; m.w. 100.50; dens. 1.194 (-9 C); m.p. -131 C; b.p. -9.2 C

Toxicology: LC50 (inh., rat, 4 h) 2050 g/m^3; very mildly toxic by inh.; mutagenic data
Precaution: Flamm. gas; very dangerous fire hazard exposed to heat, flame, oxidizers; can react vigorously with oxidizing materials; Hazardous decomp. prods.: Heated to decomp., emits toxic fumes of F^- and Cl^-
Uses: Refrigerant, solvent, intermediate
Manuf./Distrib.: Aldrich; Elf Atochem N. Am.

Chlorodifluoromethane
CAS 75-45-6
UN No. 1018 (DOT)
Synonyms: F 22; R 22; CFC 22; Monochlorodifluoromethane; Difluorochloromethane; Difluoromonochloromethane; Fluorocarbon 22; Freon 22; Propellant 22; Refrigerant 22
Classification: Halogenated aliphatic hydrocarbon
Empirical: $CHClF_2$
Formula: $ClCHF_2$
Properties: Colorless gas or liquefied compressed gas; nearly odorless to sl. sweetish odor; sol. in ether, acetone, chloroform; partly sol. in water; m.w. 86.47; sp.gr. 1.209 (liq., 21 C), 4.82 g/l (gas at b.p.); f.p. -160 C; b.p. -40.8 C; noncombustible
Toxicology: LC50 (inh., mouse, 20 min) 28 pph; LCLo (inh., dog) 70 pph; ACGIH TLV/TWA 1000 ppm; STEL 1250 ppm; asphyxiant; low toxicity by inh.; high concs. may cause CNS effects
Precaution: Compressed gas; 50% mixts. with air are combustible @ elevated pressures, although ignition is difficult; incompat. with reactive metals, e.g., sodium, potassium, powd. aluminum (can react violently or explosively); Hazardous decomp. prods.: Heated to decomp., emits toxic fumes of F^- and Cl^-
Uses: Refrigerant; low-temp. solvent; aerosol propellant; intermediate for mfg. of fluorocarbon resins
Manuf./Distrib.: Air Prods.; Aldrich; Elf Atochem N. Am.; Fluka; Harris Spec. Chems.; LaRoche Chem.; LaRoche Ind.; Sigma

Chloroethane. *See* Ethyl chloride
Chloroethene. *See* 1,1,1-Trichloroethane
α-Chloroethylidene fluoride. *See* 1,1,1-Chlorodifluoroethane
Chlorofluorocarbon 113. *See* Trichlorotrifluoroethane

Chloroform
CAS 67-66-3; EINECS 200-663-8
UN No. 1888 (DOT)
Synonyms: Trichloromethane; Methenyl chloride; Trichloroform; Methane, trichloro-; Methane trichloride; Formyl trichloride
Classification: Halogenated aliphatic hydrocarbon
Empirical: $CHCl_3$
Properties: Colorless clear mobile liq., ethereal odor, sweet burning taste; misc. with alcohol, benzene, hexane, acetone, diethyl ether, naphtha, fixed and volatile oils; sl. sol. in water; m.w. 119.38; sp.gr. 1.474-1.478; m.p. -63 C; b.p. 61 C; flash pt. none; ref. index 1.4460
Toxicology: LDLo (oral, human) 2514 mg/kg; ACGIH TLV/TWA 10 ppm; prolonged skin contact may produce irritation, burning; may be absorbed thru skin; vapor may cause stinging of eyes; respiratory and skin allergies; inh. affects CNS, heart, liver, kidneys
Precaution: DOT: Poisonous material; incompat. with strong bases, reactive metals, alkali metals, strong oxidizers, nitromethane, potassium t-butoxide; vigorous reactions, ignition, explosions possible; Hazardous decomp. prods.:

Slowly decomp. to produce hydrogen chloride; reacts with oxidizers to form highly toxic phosgene gas and chlorine
NFPA: Health 2, Flammability 0, Reactivity 0
Uses: Prod. of chlorodifluormethane (a refrigerant and propellant), fluorocarbon plastics; solvent; analytical chemistry; intermediate in prep. of dyes, drugs, pesticides; fumigant; insecticides; ingred. in drugs and cosmetics; preservative, solvent, flavoring in pharmaceuticals
Regulatory: SARA reportable; BP compliance
Manuf./Distrib.: Aldrich; Allchem Ind.; Burdick & Jackson; Cross World Sales; Elf Atochem N. Am.; Fluka; Houghton Chem.; Hüls AG; Mallinckrodt; Mitsui Toatsu; Occidental; Sigma; Spectrum Chem. Mfg.; Vulcan

Chloroform, methyl-. *See* 1,1,1-Trichloroethane
Chloromethane. *See* Methyl chloride
Chloromethylbenzene. *See* Monochlorotoluene
Chloronitropropane. *See* 1-Chloro-1-nitropropane

1-Chloro-1-nitropropane
CAS 600-25-9
Synonyms: Chloronitropropane
Classification: Halogenated aliphatic nitro compd.; nitroparaffin
Empirical: $C_3H_6ClNO_2$
Formula: $CH_3CH_2CHClNO_2$
Properties: Colorless liq.; unpleasant odor; sol. in gycols, oils, chloroform, ethanol, diethyl ether; sl. sol. in water; m.w. 123.5; sp.gr. 1.21 (20 C); b.p. 139-143 C; flash pt. (OC) 62 C
Toxicology: LD50 (oral, mouse) 510 mg/kg, (subcut., mouse) 165 mg/kg; ACGIH TLV/TWA 2 ppm; toxic; poison by ing., subcut. routes; mod. toxic by inh.; lachrymatory (vapor irritates eyes, causes tearing)
Precaution: Flamm. exposed to heat, flame, oxidizers; mod. explosive exposed to heat; incompat. with strong acids (may form highly toxic chloride fumes) and strong oxidizers (increased risk of fire and explosion); Hazardous decomp. prods.: Highly toxic chloride fumes on contact with acids; heated to decomp., emits toxic fumes of Cl^- and NO_x
NFPA: Flammability 2, Reactivity 3
Uses: Solvent and intermediate in chem. and syn. rubber industries; previously used as fungicide

(p-Chlorophenyl)trifluoromethane. *See* p-Chlorobenzotrifluoride
Chlorothene. *See* 1,1,1-Trichloroethane
4-Chlorotrifluoromethylbenzene. *See* p-Chlorobenzotrifluoride
p-Chlorotrifluoromethylbenzene. *See* p-Chlorobenzotrifluoride
1-Chloro-4-(trimethyl)benzene. *See* p-Chlorobenzotrifluoride
Chloryl. *See* Ethyl chloride

Choleth-24
CAS 27321-96-6 (generic)
Synonyms: PEG-24 cholesteryl ether; POE (24) cholesteryl ether
Definition: Polyethylene glycol ether of cholesterol with avg. ethoxylation value 9f 24
Uses: Emulsifier, solubilizer, dispersant, thickener for cosmetics, hair care prods., creams and lotions, pharmaceutical topicals
Regulatory: Approved for topicals
Trade names containing: Solulan® C-24

Cinene. *See* d-Limonene; dl-Limonene
Cineol. *See* Eucalyptol

1,8-Cineol. *See* Eucalyptol

Cineole. *See* Eucalyptol

1,4-Cineole

CAS 470-67-7

Synonyms: 1,4-Epoxy-p-menthane; p-Menthane, 1,4-epoxy-; Isocineole; 7-Oxabicyclo (2.2.1) heptane, 1-methyl-4-(1-methylethyl)-

Empirical: $C_{10}H_{18}O$

Properties: Colorless liq., camphoraceous odor, spicy flavor; sol. in alcohol; insol. in water; m.w. 154.25; dens. 0.887; m.p. -46 C; b.p. 65 C (16 mm); flash pt. 118 F

Toxicology: LD50 (oral, rat) 3100 mg/kg, (skin, rabbit) > 5 g/kg

Uses: Synthetic flavoring agent for foods and pharmaceuticals

Regulatory: FDA 21CFR §172.515; FEMA GRAS

Manuf./Distrib.: Aldrich; Fluka; SCM Glidco Organics

Trade names containing: Acintene® AC-P; Acintene® DP 738

1,8-Cineole. *See* Eucalyptol

C7-8 isoparaffin

CAS 64742-48-9; EINECS 274-273-1

Synonyms: Alkanes, C7-8-iso-

Definition: Mixt. of branched chain aliphatic hydrocarbons with 7 or 8 carbons in the alkyl chain

Uses: Solvent for coatings; minimizes VOC; virtually HAPS free; extremely low surf. tens. to improve flow and wetting; effective thinner and cleaner; suitable for odorless alkyd and acrylic paints

Regulatory: FDA 21CFR §172.882, 173.340

Trade names: Isopar® C

C8-9 isoparaffin

CAS 64742-48-9

Synonyms: Alkanes, C8-9-iso-

Definition: Mixt. of branched chain aliphatic hydrocarbons with 8 or 9 carbons in alkyl chain

Uses: Solvent for coatings

Regulatory: FDA 21CFR §172.882, 173.340

Trade names: Isopar® E

C9-14 isoparaffin

Synonyms: Alkanes, C10-11-iso-

Definition: Mixture of branched chain aliphatic hydrocarbons with 10 or 11 carbons in alkyl chain

Uses: Solvent

C10-11 isoparaffin

CAS 64742-48-9

Synonyms: Alkanes, C10-11-iso-

Definition: Mixt. of branched chain aliphatic hydrocarbons with 9-11 carbons in the alkyl chain

Uses: Solvent, diluent, carrier for personal care prods. for sprays; solvent for coatings; minimizes VOC; virtually HAPS free; extremely low surf. tens. to improve flow and wetting; effective thinner and cleaner; suitable for odorless alkyd/acrylic paints

Regulatory: FDA 21CFR §172.882, 173.340

Trade names: Isopar® G

C10-13 isoparaffin
CAS 68551-17-7; EINECS 271-366-9
Synonyms: Alkanes, C10-13-iso-
Definition: Mixture of branched chain aliphatic hydrocarbons with 10 to 13 carbons in the alkyl chain
Uses: Solvent

C11-12 isoparaffin
CAS 64742-48-9
Synonyms: Alkanes, C11-12-iso-
Definition: Mixt. of branched chain aliphatic hydrocarbons with 11 or 12 carbons in the alkyl chain
Uses: Solvent, diluent, carrier for personal care prods. for sprays; solvent for coatings; minimizes VOC; virtually HAPS free; extremely low surf. tens. to improve flow and wetting; effective thinner and cleaner; suitable for odorless alkyd/acrylic paints
Regulatory: FDA 21CFR §172.882, 173.340
Trade names: Isopar® H; Isopar® K

C11-13 isoparaffin
CAS 64742-48-9
Synonyms: Alkanes, C11-13-iso-
Definition: Mixture of branched chain aliphatic hydrocarbons with 11 to 13 carbons in the alkyl chain
Uses: Solvent for skin care formulations; solvent for coatings; minimizes VOC; virtually HAPS free; extremely low surf. tens. to improve flow and wetting; effective thinner and cleaner; suitable for odorless alkyd/acrylic paints
Regulatory: FDA 21CFR §172.882, 173.340
Trade names: Isopar® L

C13-14 isoparaffin
CAS 64742-48-9
Synonyms: Alkanes, C13-14-iso-
Definition: Mixture of branched chain aliphatic hydrocarbons with 13 or 14 carbons in the alkyl chain
Uses: Solvent for hair prods., coatings; thinner and cleaner; suitable for odorless alkyd/acrylic paints; in antibiotic gels; minimizes VOC; virtually HAPS free
Regulatory: FDA 21CFR §172.882, 173.340
Trade names: Isopar® M

Citric acid, tributyl ester. *See* Tributyl citrate
C18 linear alcohol. *See* Stearyl alcohol
C13 linear primary alcohol. *See* Tridecyl alcohol
C16 linear primary alcohol. *See* Cetyl alcohol
Cloroben. *See* o-Dichlorobenzene
Coal naphtha. *See* Benzene
Coal oil. *See* Kerosene
Coal tar naphtha. *See* Naphtha

Cocamide DEA
CAS 8051-30-7; 61791-31-9; 68603-42-9; EINECS 263-163-9; 271-657-0
Synonyms: Coconut diethanolamide; Cocoyl diethanolamide; N,N-Bis (2-hydroxyethyl) coco amides
Definition: Ethanolamides of coconut acid
Formula: $RCO-N(CH_2CH_2OH)_2$, RCO represents the coconut acid radical
Toxicology: May produce contact sensitivity
Uses: Detergent, thickener, emulsifier, foam booster/stabilizer, wetting agent,

solubilizer for cosmetics, industrial and household cleaners, textiles, etc.; plastics antistat
Regulatory: FDA 21CFR §172.710, 173.322 (0.2% max.), 175.105, 176.210, 177.1200, 177.2260, 177.2800; FDA approved for topicals
Manuf./Distrib.: Aquatec Quimica SA
Trade names: Comperlan® COD

Cocoa butter
CAS 8002-31-1
Synonyms: Theobroma oil; Cacao butter
Definition: Ylsh. wh. solid obtained from roasted seeds of *Theobroma cacao*
Properties: Ylsh.-wh. solid, chocolate-like odor and taste; sol. in ether, chloroform; sl. sol. in alcohol; insol. in water; dens. 0.858-0.864 (100/25 C); m.p. 30-35 C; iodine no. 33-42; sapon. no. 188-198; ref. index 1.4537-1.4585 (40 C)
Toxicology: May cause allergic skin reactions
Precaution: Combustible
Uses: Chewing gum base; pharmaceutical emollient, carrier, filler, suppository base; skin conditioner, occlusive solvent, skin protectant for OTC drug prods., makeup, sun preps.
Regulatory: FDA 21CFR §182.20, GRAS; FDA approved for orals, rectals, topicals; USP/NF, BP compliance
Manuf./Distrib.: ABITEC; R.W. Greeff; Kaufholz & Co.; Penta Mfg.; Spectrum Chem. Mfg.; Vrymeer Commodities
Trade names: Fancol CB; Fancol CB Extra

Coconut diethanolamide. *See* Cocamide DEA
Cocoyl diethanolamide. *See* Cocamide DEA
Columbian spirits. *See* Methyl alcohol
Concentrated acetic acid. *See* Acetic acid, glacial
Copper (II) oxide. *See* Copper oxide (ic)

Copper oxide (ic)
CAS 1317-38-0; EINECS 215-269-1
Synonyms: Cupric oxide; Copper (II) oxide
Empirical: CuO
Formula: CuO
Properties: Blk. fine powd.; m.w. 79.54; dens. 6.4; b.p. dec. @ 1026 C
Precaution: Incompat. with metals and reductants; explodes when heated with powd. aluminum, hydrogen, magnesium, etc.; ignites on contact with hydrogen trisulfide; incandescent reactions
Uses: Ceramic colorant, reagent in analytical chemistry, fungicide, insecticide, catalyst, purification of hydrogen, batteries and electrodes, electroplating, solvent, desulfurizing oils, rayon, metallurgical and welding fluxes, antifouling paints; phosphors; trace mineral in animal feeds
Manuf./Distrib.: Aldrich; Am. Chemet; Carbochem; Cerac; Chemisphere Ltd; Fluka; Goldschmidt Ind. Chems.; Nihon Kagaku Sangyo; Noah; SCM Metal Prods.; Shance; Sigma; Spectrum Chem. Mfg.

Corn oil
CAS 8001-30-7; EINECS 232-281-2
Synonyms: Maize oil; Zea mays oil
Definition: Refined fixed oil obtained from wet milling of corn, *Zea mays*
Properties: Pale yel. oily liq., faint char. odor and taste; insol. in water; sol. in ether, chloroform, amyl acetate, benzene, CS_2; sl. sol. in alcohol; dens. 0.914-0.921; m.p. -10 C; acid no. 2-6; iodine no. 109-133; sapon. no. 187-193; flash pt. 321 C; ref. index 1.470-1.474

Toxicology: Nontoxic; human skin irritant; experimental teratogen; may be an allergen
Precaution: Combustible exposed to heat or flame; dangerous spontaneous heating may occur; light-sensitive
NFPA: Health 0, Flammability 1, Reactivity 0
Uses: Used in preparing foodstuffs, coating agent, emulsifier, formulation aid, texturizer, in bakery prods., margarine, salad oil, dietary supplement; soap, lubricants, leather finishing, hair dressing; pharmaceutical solvent, vehicle; paints
Regulatory: FDA 21CFR §175.105, 175.300, 176.200, 176.210, GRAS; FDA approved for injectables, orals, topicals; USP/NF, BP, JP compliance
Manuf./Distrib.: ABITEC; Alba Int'l.; Arista Ind.; Ashland; Fluka; Grain Processing; Penta Mfg.; Sigma; Spectrum Chem. Mfg.; A.E. Staley Mfg.; Tri-K Ind.; Welch, Holme & Clark
Trade names: Super Refined® Corn Oil

Cotton oil. *See* Cottonseed oil

Cottonseed oil
CAS 8001-29-4; EINECS 232-280-7
Synonyms: Deodorized winterized cottonseed oil; Cotton oil; Oleum Gossypii seminis; Aceite de Algodon
Definition: Refined fixed oil from seeds of *Gossypium hirsutum*
Properties: Pale yel. oily liq., nearly odorless; sol. in ether, benzene, chloroform; sl. sol. in alcohol; dens. 0.915-0.921; f.p. 0-5 C; solid. pt. 31-33 C; iodine no. 109-120; sapon. no. 190-198; flash pt. (CC) 486 F
Toxicology: Experimental tumorigen and teratogen; allergen
Precaution: Combustible exposed to heat or flame; may be dangerous hazard due to spontaneous heating; light-sensitive
NFPA: Health 0, Flammability 1, Reactivity 0
Uses: Coating agent, emulsifier, formulation aid, texturizer for use in cooking oil, margarine, salad oil, shortening; drying oil for paints; pharmaceutical vehicle, solvent
Use level: 56-92% (intramuscular preps.), 0.002-402 mg (solid oral dosage forms)
Regulatory: FDA 21CFR §175.105, 175.300, 176.200, 176.210, 177.2800, GRAS; FDA approved for intramuscular injectables, orals; USP/NF compliance
Manuf./Distrib.: ABITEC; Aldrich; Alnor Oil; Amber Syn.; Arista Ind.; Ruger; Sigma; Spectrum Chem. Mfg.; Tri-K Ind.; Welch, Holme & Clark
Trade names: Super Refined® Cottonseed Oil

Cryofluoran. *See* Dichlorotetrafluoroethane
Cryofluorane. *See* Dichlorotetrafluoroethane

Cumene
CAS 98-82-8
UN No. 1918 (DOT)
Synonyms: Isopropylbenzene; Isopropyl benzol; Benzene isopropyl; (1-Methylethyl) benzene; 2-Phenylpropane
Empirical: C_9H_{12}
Formula: $C_6H_5CH(CH_3)_2$
Properties: Colorless liq.; m.w. 120.21; dens. 0.864 (20/4 C); vapor pressure 10 mm Hg (38.3 C); m.p. -96 C; b.p. 152 C; flash pt. 111 F
Toxicology: LD50 (oral, rat) 1400 mg/kg, (skin, rabbit) 12,300 mg/kg; TCLo (inh., human) 200 ppm; ACGIH TLV/TWA 50 ppm (skin); mod. toxic by ing.; mildly toxic by inh., skin contact; primary eye and skin irritant; human systemic effects

Precaution: Flamm. exposed to heat or flame; can react with oxidizing materials; violent reaction with HNO_3, chlorosulfonic acid, oleum
NFPA: Health 2, Flammability 3, Reactivity 1
Uses: Production of phenol, acetone and α-methylstyrene; org. solvent
Manuf./Distrib.: Aldrich; Ashland; Chevron; Fluka; Georgia Gulf; Hüls AG; JLM Marketing; Mitsubishi Petrochem.; Mitsui Petrochem. Ind.; Sigma; Spectrum Chem. Mfg.; Texaco

Cumyl phenol
CAS 599-64-4
Synonyms: 4-Cumylphenol; 4-(Dimethylphenylmethyl) phenol; 4-(2-Phenylisopropyl)phenol; 4-Hydroxydiphenyldimethylmethane; p-Cumylphenol
Empirical: $C_{15}H_{16}$
Formula: $C_6H_5C(CH_3)_2C_6H_4OH$
Properties: Wh. to tan crystals, phenol odor; m.w. 212.29; dens. 1.115 g/ml (25 C); m.p. 74-76 C; b.p. 335 C; f.p. 72 C; flash pt. 320 F
Toxicology: Irritant
Precaution: Combustible
Uses: Intermediate for resins, insecticides, lubricants, surfactants; polycarbonate chain terminator; modifier for epoxy, furan, and phenolic resins
Manuf./Distrib.: AC Ind.; Aldrich; Howard Hall; Hüls AG; ICI Spec. Chems.; PMC Spec.; Schenectady
Trade names containing: Kenplast® ES-2HP

4-Cumylphenol. *See* Cumyl phenol
p-Cumylphenol. *See* Cumyl phenol

Cumylphenyl acetate
CAS 24133-73-1
Properties: Sp.gr. 1.03; m.p. < 0 C; flash pt. (COC) 250 C
Uses: Plasticizer for urethanes; high flash pt. reactive diluent for epoxy, for polyamide-cured epoxy floorings and coal tar pipe coatings; comonomer and impact modifier for phenolics
Manuf./Distrib.: Kenrich Petrochems.
Trade names containing: Kenplast® ES-2HP

Cumylphenyl benzoate
Properties: Sp.gr. 1.10; m.p. 48-52 C; flash pt. (COC) 350 C
Uses: Primary plasticizer for PVC and PVC/nitrile; process aid for semirigid PVC extrudates; solvates conductive polyester and acrylic inks to improve impact
Manuf./Distrib.: Kenrich Petrochems.
Trade names: Kenplast® ESB

Cumylphenyl neodecanoate
Uses: Primary plasticizer for PVC, PVC/nitrile; process aid for semirigid PVC extrusions; solvates and impact modifies conductive polyester and acrylic inks
Trade names: Kenplast® ESN

Cupric oxide. *See* Copper oxide (ic)
Cyanobenzene. *See* Benzonitrile
Cyanoethane. *See* Propionitrile
Cyanomethane. *See* Acetonitrile
Cyclic dimethyl polysiloxane with n= 3-6. *See* Cyclomethicone
Cyclic dimethylsiloxane. *See* Cyclomethicone
Cyclic ethylene carbonate. *See* Ethylene carbonate

Cyclic methylethylene carbonate. *See* Propylene carbonate
Cyclic propylene carbonate. *See* Propylene carbonate
Cyclic 1,2-propylene carbonate. *See* Propylene carbonate
1,3-Cyclohexadiene, 1-methyl-4-isopropyl-. *See* α-Terpinene

Cyclohexane
CAS 110-82-7; EINECS 203-806-2
UN No. 1145 (DOT)
Synonyms: Hexahydrobenzene; Benzene hexahydride; Hexamethylene; Hexanaphthene
Classification: Sat. aliphatic hydrocarbon
Empirical: C_6H_{12}
Properties: Colorless mobile liq., pungent gasoline-like odor; insol. in water; sol. in alcohol, acetone, benzene; m.w. 84.16; sp.gr. 0.779 (20/4 C); f.p. 4.6 C; m.p. 6.5 C; b.p. 80.7 C; flash pt. (CC) -18.3 C; ref. index 1.4264
Toxicology: LD50 (oral, rat) 29,820 mg/kg; ACGIH TLV/TWA 300 ppm; poison by IV; mod. toxic by ingestion, inhalation, skin contact; systemic and skin irritant; high concs. may act as narcotic; mutagenic data
Precaution: DOT: Flamm. liq.; dangerous fire hazard exposed to heat or flame; reactive with oxidizers; mod. explosion hazard as vapor exposed to flame; explosive mixed hot with liq. dinitrogen tetraoxide; Hazardous decomp. prods.: Heated to decomp., emits acrid smoke, irritating fumes
NFPA: Health 1, Flammability 3, Reactivity 0
Uses: Mfg. of nylon; solvent for cellulose ethers, fats, oils, waxes, coatings; paint and varnish remover; extraction solvent for essential oils; glass substitutes; laboratory reagent; chemical intermediate; in fungicidal formulations; color diluent for foods; in pharmaceutical orals
Regulatory: FDA approved for orals; BP compliance
Manuf./Distrib.: Aldrich; Ashland; Exxon Europe; Fluka; Huntsman; Phillips; Sigma; Spectrum Chem. Mfg.; Texaco; Triple Crown Am.

Cyclohexane, methyl-. *See* Methyl cyclohexane

Cyclohexanol
CAS 108-93-0; EINECS 203-630-6
Synonyms: 1-Cyclohexanol; Cyclohexyl alcohol; Hexahydrophenol; Hexalin; Hydroxycyclohexane
Classification: Alicyclic alcohol
Empirical: $C_6H_{12}O$
Formula: $CH_2(CH_2)_4CHOH$
Properties: Colorless oily liq. or cryst. (below 25 C); camphor-like odor; sol. in methanol, ethanol, acetone, ether, benzene, most org. solvs.; mod. sol. in water; m.w. 100.09; sp.gr. 0.937 (37.4 C); m.p. 23 C; b.p. 160.9 C; flash pt. 67.7 C; ref. index 1.4656 (20 C)
Toxicology: ACGIH TLV/TWA 50 ppm (skin); LD50 (oral, rat) 2.06 g/kg; toxic; mild skin irritant; lachrymator; mod. to severe eye irritant on direct contact; may cause nose/throat irritation; high concs. may cause headache, nausea, vomiting, tremors
Precaution: Combustible; incompat. with strong oxidizing agents (reacts vigorously or explosively generating heat and pressure)
NFPA: Health 1, Flammability 2, Reactivity 0
Uses: Stabilizer, homogenizer in soap making and detergents to incorporate solvents and phenolic insecticides; solvent for alkyd and phenolic resins, paints, lacquers, shellacs, varnishes, rubber, fats, oils; source of adipic acid for nylon, textile finishing; plasticizer prod.; dye solvent and carrier
Manuf./Distrib.: Aldrich; AlliedSignal; Ashland; J.T. Baker; BASF; BASF AG;

Chemcentral; Fluka; Harcros USA; Hüls Am.; Spectrum Chem. Mfg.; Sunnyside; UCB SA; Van Waters & Rogers
Trade names: Naxol®

1-Cyclohexanol. *See* Cyclohexanol
Cyclohexanol acetate. *See* Cyclohexyl acetate

Cyclohexanone
CAS 108-94-1; EINECS 203-631-1
UN No. 1915 (DOT)
Synonyms: Ketohexamethylene; Pimelic ketone; Pimelin ketone
Classification: Cyclic ketone
Empirical: $C_6H_{10}O$
Formula: $CO(CH_2)_4CH_2$
Properties: Water-wh. to pale yel. oily liq., peppermint-acetone odor; sl. sol. in water; sol. in alcohol, ether, common org. solvs.; m.w. 98.08; sp.gr. 0.947 (20/4 C); vapor pressure 10 mm Hg (38.7 C); m.p. -32.1 C; b.p. 156.7 C; flash pt. (CC) 44 C; ref. index 1.4507
Toxicology: LD50 (oral, rat) 1535 mg/kg, (skin, rabbit) 1000 mg/kg; ACGIH TLV/TWA 25 ppm (skin); mod. toxic by ing., inh., subcut, IV, IP; skin and severe eye irritant; inh. may cause throat irritation
Precaution: Flamm. exposed to heat or flame; incompat. with strong oxidizing agents (increases fire and explosion hazard); sl. explosion hazard as vapor exposed to flame; explosive reaction with nitric acid @ 75 C; can form explosive peroxides
NFPA: Health 1, Flammability 2, Reactivity 0
Uses: As paint and varnish remover; solvent for cellulose acetate, nitrocellulose, natural resins, vinyl resins, paints, rubber, waxes, fats; in prod. of adipic acid for nylon, cyclohexanone resins, caprolactam, polymers, resins; wood stains; metal degreasing; polisher; textile dyeing and processing; lubricating oil additives
Regulatory: SARA reportable
Manuf./Distrib.: Aldrich; Allchem Ind.; AlliedSignal; Ashland; J.T. Baker; BASF; Chemcentral; Coyne; Fluka; Harcros USA; Houghton Chem.; Sigma; Spectrum Chem. Mfg.; Stanchem; Sunnyside; Union Carbide; Van Waters & Rogers
Trade names: Nadone®

Cyclohexanyl acetate. *See* Cyclohexyl acetate
Cyclohexatriene. *See* Benzene
Cyclohexene, 1-methyl-4-(1-methylethenyl)-, (S)-. *See* l-Limonene

Cyclohexyl acetate
CAS 622-45-7; EINECS 210-736-6
UN No. 2243 (DOT)
Synonyms: Acetic acid cyclohexyl ester; Adronol acetate; Cyclohexanol acetate; Cyclohexanyl acetate; Hexalin acetate; Hexaline acetate
Classification: Acetic acid ester
Empirical: $C_8H_{14}O_2$
Formula: $CH_3COOC_6H_{11}$
Properties: Colorless to pale yel. oily liq.; fruity odor; sol. in alcohol, ether; pract. insol. in water; m.w. 142.20; sp.gr. 0.966; b.p. 172-173 C; flash pt. (CC) 58 C; ref. index 1.4400-1.4410
Toxicology: LD50 (oral, rat) 6730 mg/kg, (skin, rabbit) 10 g/kg; mod. toxic by subcutaneous route; mildly toxic by ingestion, skin contact; human systemic effects by inh.: conjunctiva irritation and unspecified respiratory changes
Precaution: Flamm. exposed to heat or flame; incompat. with strong oxidizing

agents (increases risk of fire and explosion), strong bases (decomp. can occur); Hazardous decomp. prods.: Cyclohexanol, acetic acid; heated to decomp., emits acrid smoke and irritating fumes
NFPA: Health 1, Flammability 2, Reactivity 0
Uses: Solvent for nitrocellulose, cellulose ether, bitumens, metallic soaps, basic dyes, brown oils, crude rubber, resins, lacquers; syn. flavor and fragrance for foods and pharmaceuticals
Regulatory: FDA 21CFR §172.515; FEMA GRAS; Japan approved as flavoring
Manuf./Distrib.: Aldrich

Cyclohexyl alcohol. *See* Cyclohexanol
1-Cyclohexylbutane. *See* Butylcyclohexane
Cyclohexylmethane. *See* Methyl cyclohexane
1-Cyclohexyl-2-pyrrolidone. *See* N-Cyclohexyl pyrrolidone

N-Cyclohexyl pyrrolidone
CAS 6837-24-7; EINECS 229-919-7
Synonyms: 1-Cyclohexyl-2-pyrrolidone; 2-Pyrrolidinone, 1-cyclohexyl-
Empirical: $C_{10}H_{17}NO$
Properties: Liq.; m.w. 167.25; sp.gr. 1.026; f.p. 12 C; b.p. 284 C; flash pt. (CC) 293 F; ref. index 1.4950
Toxicology: LD50 (oral, rat) 370 mg/kg, (skin, rabbit) 1600 mg/kg; primary skin irritant; may cause permanent eye damage
Uses: Solvent for plating engineering plastics, esp. EMS applics.; dyeing of aromatic amide fibers; photo resists; gas separation; elec. insulating coatings; chem. synthesis; absorp. of sulfur dioxide and nitric acid; aromatic extraction; high temp. polyimide processing aid
Manuf./Distrib.: Aldrich; Fluka
Trade names: CHP®

Cyclomethicone
CAS 69430-24-6
Synonyms: Cyclic dimethylsiloxane; Polydimethyl siloxy cyclics; Cyclic dimethyl polysiloxane with n= 3-6; Cyclopolydimethylsiloxane
Definition: Fully methylated cyclic dimethyl polysiloxane
Empirical: $(C_2H_6OSi)_n$
Formula: $[—(CH_3)_2SiO—]_n$, avg. n = 3-6
Toxicology: LD50 (skin, rabbit) > 16 ml/kg
Uses: Conditioner, emollient for hair care prods., aerosols, sticks, shaving preparations, deodorants, antiperspirants; lubricant, spreading agent, detackifier for skin cleansers; water repellent agent in pharmaceutical preps.
Regulatory: FDA approved for orals, injectables, topicals; USP/NF compliance
Manuf./Distrib.: Dow Corning
Trade names containing: Fancorsil A; Gilugel SIL5

Cyclopentane
CAS 287-92-3; EINECS 206-016-6
UN No. 1146 (DOT)
Synonyms: Pentamethylene
Classification: Sat. alicyclic hydrocarbon
Empirical: C_5H_{10}
Properties: Colorless liq.; misc. with most org. solvs.; insol. in water; m.w. 70.13; sp.gr. 0.75 (20 C); vapor pressure 400 mm Hg (31 C); m.p. -93 C; b.p. 49.5 C; flash pt. 19.4 F
Toxicology: ACGIH TLV/TWA 600 ppm; mildly toxic by ing. and inh.; eye and skin irritant; CNS depressant; ing. may cause sore throat, nausea, diarrhea; inh.

of high vapor concs. can cause nose and throat irritation, light-headedness
Precaution: Flamm.; very dangerous fire hazard exposed to heat or flame; incompat. with strong oxidizing agents (increased fire and explosion hazard); Hazardous decomp. prods.: Heated to decomp., emits acrid smoke and fumes
NFPA: Health 1, Flammability 3, Reactivity 0
Uses: Solvent for cellulose acetate; fat and oil extraction; hydrocarbon propellant; org. synthesis; laboratory reagent
Manuf./Distrib.: Aldrich; Fluka; Phillips; Sigma; Spectrum Chem. Mfg.; Triple Crown Am.

Cyclopentimine. *See* Piperidine
2-Cyclopentylpropane. *See* Isopropylcyclopentane
Cyclopentyl-2 propane. *See* Isopropylcyclopentane
Cyclopolydimethylsiloxane. *See* Cyclomethicone
Cyclotetramethylene oxide. *See* Tetrahydrofuran

DAMP. *See* 1,3-Pentanediamine
Dapsone. *See* Sulfolane
DBA. *See* Dibutylamine
DBP. *See* Dibutyl phthalate
DCB. *See* o-Dichlorobenzene
DCBTF. *See* 3,4-Dichlorobenzotrifluoride
1,2-DCE. *See* Ethylene dichloride
DCM. *See* Methylene chloride

Decahydronaphthalene
CAS 91-17-8; EINECS 202-046-9
UN No. 1147 (DOT)
Synonyms: Perhydronaphthalene; Dekalin; Naphthalane
Definition: Mixture of cis and trans
Empirical: $C_{10}H_{18}$
Properties: Liq.; insol. in water; sol. in alcohol; m.w. 138.14; dens. 0.8963; m.p. –43.26 C; b.p. 185.5 C; flash pt. 58 C; ref. index 1.474
Toxicology: TCLo (inh., human) 100 ppm; primary irritant; conjunctive irritant; human tumorigen
Precaution: Flamm.
NFPA: Health 2, Flammability 2, Reactivity 0
Uses: Solvent and stabilizer for shoe creams and floor waxes; solvent in paint and lacquers, oils, resins, rubber, and asphalt
Manuf./Distrib.: Aldrich; Fluka; Sigma; Spectrum Chem. Mfg.
Trade names: Decalin

n-Decane
CAS 124-18-5
UN No. 2247
Classification: Sat. aliphatic hydrocarbon
Empirical: $C_{10}H_{22}$
Formula: $CH_3(CH_2)_8CH_3$
Properties: Colorless liq.; gasoline-like odor; sol. in most org. solvs.; insol. in water; m.w. 142.29; sp.gr. 0.734; m.p. -29.6 C; b.p. 174 C; flash pt. 46 C
Toxicology: mod. to severe skin irritant; probably mod. to severe eye irritant; conc.

vapor may cause nose/throat irritation, headache, drowsiness, nausea, difficulty breathing

Precaution: Combustible; incompat. with strong oxidizing agents (increases fire and explosion hazard); Hazardous decomp. prods.: None

NFPA: Health 0, Flammability 2, Reactivity 0

Uses: Component of gasoline, jet fuel, kerosene, petrol. solvents such as wh. spirit; solvent; org. synthesis; paper and rubber industries; constituent of polyolefin mfg. wastes

Manuf./Distrib.: Aldrich; Fluka; Phillips; Sigma; Triple Crown Am.

Decanedioic acid, bis(2-ethylhexyl) ester. *See* Dioctyl sebacate
Decanedioic acid, bis(1-methylethyl) ester. *See* Diisopropyl sebacate
Decanedioic acid, dimethyl ester. *See* Dimethyl sebacate
Decanoic acid, 2,2-dimethyl-1,3-propanediol diester. *See* Neopentyl glycol dicaprate
Decanoic acid, 1-methyl-1,2-ethanediyl ester mixed with 1-methyl-1,2-ethanediyl dioctanoate. *See* Propylene glycol dicaprylate/dicaprate
1-Decanol. *See* n-Decyl alcohol
n-Decanol. *See* n-Decyl alcohol

Deceth-6

CAS 26183-52-8 (generic)

Synonyms: PEG-6 decyl ether; PEG 300 decyl ether; POE (6) decyl ether

Definition: PEG ether of decyl alcohol

Formula: $CH_3(CH_2)_8CH_2(OCH_2CH_2)_nOH$, avg. n = 6

Uses: Wetting agent, detergent , penetrant, emulsifier for textiles, clay soils, fire fighting; surfactant, detergent raw material and additive used in dishwashing agents, in mfg. of tablet soaps

Trade names containing: Ship Shape® Resin Cleaner

n-Decyl alcohol

CAS 112-30-1; 68526-85-2; EINECS 203-956-9

Synonyms: Alcohol C-10; Capric alcohol; Caprinic alcohol; Noncarbinol; Nonylcarbinol; 1-Hydroxydecane; 1-Decanol; n-Decanol; Decylic alcohol

Classification: Primary aliphatic alcohol

Empirical: $C_{10}H_{22}O$

Formula: $CH_3(CH_2)_8CH_2OH$

Properties: Colorless to lt. yel. mod. visc. liq., sweet odor; sol. in alcohol, ether, acetone, benzene, chloroform, acetic acid; pract. insol. in water; m.w. 158.32; sp.gr. 0.8297 (20/4 C); m.p. 7 C; b.p. 232.9 C; flash pt. (OC) 82 C; ref. index 1.43587

Toxicology: LD50 (oral, rat) 4720 mg/kg; mod. toxic by skin contact; irritating to eyes, skin, respiratory system; inh. can probably cause nose/throat irritation; high concs. may cause CNS depression, headache, nausea, drowsiness, but is unlikely

Precaution: Combustible; flamm. when exposed to heat or flame; incompat. with oxidizing agents (increases fire/explosion risk); Hazardous decomp. prods.: Heated to decomp., emits acrid smoke and irritating fumes

NFPA: Health 0, Flammability 2, Reactivity 0

Uses: In mfg. of plasticizers, perfumes, syn. lubricants, petrol. additives, herbicides, surfactants, solvents, coatings; emulsion polymerization; antifoaming agent in flavorings; as solvent; syn. flavoring agent in foods and pharmaceuticals

Regulatory: FDA 21CFR §172.515, 172.864, 175.300, 176.170, 178.3480, 178.3910; FEMA GRAS; Japan approved as flavoring

Manuf./Distrib.: Albemarle; Aldrich; Brown; Condea Vista; Fluka; Henkel/Emery; M. Michel; Penta Mfg.; Schweizerhall; Sigma

Decylic alcohol. *See* n-Decyl alcohol
Decyl 9-octadecenoate. *See* Decyl oleate
Decyl octyl alcohol. *See* Stearyl alcohol

Decyl oleate
CAS 3687-46-5; 59231-34-4; EINECS 222-981-6
Synonyms: Decyl 9-octadecenoate; 9-Octadecenoic acid decyl ester
Definition: Ester of decyl alcohol and oleic acid
Empirical: $C_{28}H_{54}O_2$
Formula: $CH_3(CH_2)_7CH{=}CH(CH_2)_7COOCH_2(CH_2)_8CH_3$
Properties: Wh. to straw liq.; sol. in min. oil, 95% ethanol, IPM, oleyl alcohol; insol. in water; sp.gr. 0.855-0.865; ref. index 1.4530-1.4555
Uses: Emollient, cosolvent, lubricant for creams, lotions, moisturizers, bath oils, pressed powd., pharmaceuticals
Trade names: Ceraphyl® 140

DEET. *See* Diethyl toluamide
DEG. *See* Diethylene glycol
DEGBE. *See* Butoxydiglycol
DEGEE. *See* Ethoxydiglycol
DEGHE. *See* Diethylene glycol n-hexyl ether
DEGiBE. *See* Diethylene glycol isobutyl ether
DEGMPE. *See* Diethylene glycol methylpentyl ether
DEHP. *See* Dioctyl phthalate
DEK. *See* Diethyl ketone
Dekalin. *See* Decahydronaphthalene
Delphene. *See* Diethyl toluamide
m-Delphene. *See* Diethyl toluamide
Denatured alcohol. *See* Alcohol denatured
Denatured ethanol. *See* Alcohol denatured
Denatured ethyl alcohol. *See* Alcohol denatured
Denatured spirits. *See* Alcohol denatured
Deodorized winterized cottonseed oil. *See* Cottonseed oil
4-Deoxytetronic acid. *See* Butyrolactone
DEP. *See* Diethyl phthalate
DET. *See* Diethyl toluamide
DETA. *See* Diethylenetriamine
Dewaxed lanolin. *See* Lanolin oil
Dextro-limonene. *See* d-Limonene
DFBM. *See* Dibromodifluoromethane
Diacetone. *See* Diacetone alcohol

Diacetone alcohol
CAS 123-42-2; EINECS 204-626-7
UN No. 1148 (DOT)
Synonyms: Diacetone; Diketone alcohol; 4-Hydroxy-1-keto-4-methylpentane; 4-Hydroxy-4-methyl-2-pentanone; 4-Hydroxy-4-methylpentanone-2; 4-Hydroxy-4-methyl pentan-2-one; 2-Pentanone, 4-hydroxy-4-methyl-; 2-Methyl-2-pentanol-4-one
Classification: Ketone
Empirical: $C_6H_{12}O_2$
Formula: $(CH_3)_2C(OH)CH_2COCH_3$
Properties: Liq.; faint pleasant odor; sol. in water; m.w. 116.16; sp.gr. 0.940 (20/20

C); vapor pressure 1.1 mm Hg (20 C); f.p. -47 F; b.p. 145.2-172 C; flash pt. (TCC) 126 F; ref. index 1.4234

Toxicology: LD50 (oral, rat) 4 g/kg, (IP, mouse) 933 mg/kg, (skin, rabbit) 13,500 mg/kg; TCLo (inh., human) 100 ppm; ACGIH TLV/TWA 50 ppm; mod. toxic by ing., IP routes; mildly toxic by skin contact; primary irritant; may cause headache, nausea, vomiting

Precaution: Flamm. exposed to heat or flame; can react with oxidizing materials; explosive as vapor exposed to heat or flame; Hazardous decomp. prods.: Heated to decomp., emits acrid smoke and irritating fumes

NFPA: Health 1, Flammability 2, Reactivity 0

Uses: Org. solvent for nitrocellulose, cellulose acetate, oils, resins, waxes, fats, dyes, tars, lacquers, dopes, coatings, wood preservatives, rayon, artificial leather, metal cleaning; laboratory reagent; hydraulic fluids; textile stripping agent

Regulatory: FDA 21CFR §175.105

Manuf./Distrib.: Aldrich; Allchem Ind.; Ashland; Baychem; BP Chems. Ltd; Browning; R.E. Carroll; Chemcentral; Coyne; Elf Atochem N. Am.; Fabrichem; Fluka; Great Western; Harcros USA; Hoechst Celanese; Houghton Chem.; Rhône-Poulenc Surf. & Spec.; Shell; Spectrum Chem. Mfg.; Sunnyside; Union Carbide; Van Waters & Rogers

1,2-Diacetoxyethane. *See* Ethylene glycol diacetate

Diacetylated monoglycerides

CAS 8029-92-3

Definition: Glycerin esterified with edible fatty acids and acetic acid

Properties: Clear liq.; sol. in 80% aq. alcohol, veg. and min. oils; acid no. 3 max.; sapon. no. 365-395; hyd. no. 15 max.

Uses: Excipient, solvent, plasticizer for pharmaceuticals

Regulatory: FDA approved for orals; USP/NF compliance

Diacetyl ether. *See* Ethylacetoacetate
Diacetylmethane. *See* Acetylacetone
2,2´-Diaminodiethylamine. *See* Diethylenetriamine
1,2-Diaminoethane. *See* Ethylenediamine
1,3-Diaminopentane. *See* 1,3-Pentanediamine
Diamylamine. *See* Di-n-amylamine

Di-n-amylamine

CAS 2050-92-2

Synonyms: Diamylamine; Di-n-amylamine; Dipentylamine; Di-n-pentylamine; N-Pentyl-1-pentanamine

Classification: Sat. aliphatic amine

Empirical: $C_{10}H_{23}N$

Formula: $(CH_3CH_2CH_2CH_2CH_2)_2NH$

Properties: Colorless liq.; ammoniacal odor; sol. in dimethyl sulfoxide, ethanol, ether, acetone; pract. insol. in water; m.w. 157.3; sp.gr. 0.777 (20 C); m.p. -44 C; b.p. 210-211 C; flash pt. (CC) 52 C; strong base

Toxicology: Corrosive; very toxic; can cause skin irritation, burns; may be absorbed thru skin in toxic amts.; can cause eye irritation, 'halo vision,' possible permanent injury

Precaution: Incompat. with oxidizing agents (fire/explosion risk), strong acids, acid chlorides, acid anhydrides, mercury (violent reactions possible)

NFPA: Health 3, Flammability 2, Reactivity 0

Uses: Anticorrosion agent for ferrous metals; solvent

Manuf./Distrib.: Aldrich; Elf Atochem N. Am.; Fluka

DIBA. *See* Diisobutyl adipate
DIBK. *See* Diisobutyl ketone
DIBP. *See* Diisobutyl phthalate

Dibromodifluoromethane
CAS 75-61-6
UN No. 1941
Synonyms: DFBM; Difluorodibromomethane
Classification: Halogenated aliphatic hydrocarbon
Empirical: CBr_2F_2
Formula: Br_2CF_2
Properties: Colorless heavy liq. or gas; odorless or char. odor; sol. in methanol, ether, acetone, benzene; insol. in water; m.w. 209.83; sp.gr. 2.288 (15 C); m.p. –141 C; b.p. 24.5 C; noncombustible
Toxicology: ACGIH TLV/TWA 100 ppm; inh. may cause nose/throat irritation, CNS depression; liq. may cause cooling of skin and eyes due to rapid evaporation; liq. may cause defatting of skin
Precaution: Reacts with chemically active metals (sodium, potassium, calcium, powd. aluminum, zinc, magnesium)
Uses: Fire-extinguishing agent; solvent for prep. of explosive mixtures; polymer intermediate; prod. of pharmaceuticals, dyes, quat. ammonium compds.; some use as tracer gas in environmental studies
Manuf./Distrib.: Aldrich; Fluka

Dibromomethane. *See* Methylene bromide
2,2´-Dibutoxyethyl ether. *See* Diethylene glycol dibutyl ether

Dibutyl adipate
CAS 105-99-7; EINECS 203-350-4
Synonyms: Hexanedioic acid, dibutyl ester; Adipic acid, dibutyl ester; Butyl adipate; Di-n-butyl adipate; Dibutyl adipinate; Dibutyl hexanedioate
Definition: Diester of butyl alcohol and adipic acid
Empirical: $C_{14}H_{26}O_4$
Formula: $[-CH_2CH_2CO_2(CH_2)_3CH_3]_2$
Properties: M.w. 258.36; dens. 0.962; b.p. 305 C; flash pt. > 110 C
Toxicology: LD50 (oral, rat) 12,900 mg/kg, (skin, rabbit) 20,000 mg/kg; mildly toxic; primary irritant to skin and eyes; experimental teratogen, reproductive effects; Hazardous decomp. prods.: Heated to decomp., emits acrid smoke and irritating fumes
Uses: Emollient; oily component for day creams, pharmaceutical sun and skin protective oils; plasticizer for cellulosics, PVAc, PVC, VCA; agric. chem.; solvent for lipoid-sol. ingreds.
Manuf./Distrib.: Aldrich; Aquatec Quimica SA; Novachem Corp.
Trade names: Cetiol® B

Di-n-butyl adipate. *See* Dibutyl adipate
Dibutyl adipinate. *See* Dibutyl adipate

Dibutylamine
CAS 111-92-2
UN No. 2248 (DOT)
Synonyms: DBA; DNBA; Di-n-butylamine; N-Butyl-1-butanamine; n-Dibutylamine
Classification: Sat. aliphatic amine
Empirical: $C_8H_{19}N$
Formula: $CH_3CH_2CH_2CH_2NHCH_2CH_2CH_2CH_3$
Properties: Colorless liq.; ammonia-like or fishy odor; sol. in org. solvs., ethanol,

acetone, ether, most hydrocarbons incl. benzene; sl. sol. in water; m.w. 129.28; sp.gr. 0.760; vapor pressure 2 mm Hg (20 C); m.p. -60 to -59 C; b.p. 159 C; flash pt. (CC) 41 C

Toxicology: LD50 (oral, rat) 220 mg/kg, (skin rabbit) 1010 mg/kg; corrosive; very toxic; poison by ing. and subcut. routes; may cause serious skin and severe eye irritation and 'halo vision'; inh. of mist can irritate nose/throat/lungs

Precaution: Corrosive; flamm. exposed to heat, flames; incompat. with oxidizing agents (risk of fire/explosion), strong acids, acid chlorides, acid anhydrides (violent reactions possible); exothermic reaction with cellulose nitrate does not proceed to ignition; Hazardous decomp. prods.: May react with nitrosating agents to form carcinogenic nitrosamines; heated to decomp., emits toxic fumes of NO_x

NFPA: Health 3, Flammability 2, Reactivity 0

Uses: Catalyst for formaldehyde polymerization; corrosion inhibitor; solvent; mfg. of pesticides, polymers, rubber, dyes

Manuf./Distrib.: Aldrich; Allchem Ind.; Ashland; BASF; BASF AG; Elf Atochem N. Am.; Fluka; Sigma; Spectrum Chem. Mfg.

Di-n-butylamine. *See* Dibutylamine

n-Dibutylamine. *See* Dibutylamine

Dibutyl-1,2-benzene dicarboxylate. *See* Dibutyl phthalate

N,N-Dibutyl-1-butanamine. *See* Tributylamine

Dibutyl 'Carbitol'. *See* Diethylene glycol dibutyl ether

Dibutyldiglycol. *See* Diethylene glycol dibutyl ether

2,2-Dibutyldihydro-6H-1,3,2-oxathiastannin-6-one. *See* Dibutyltin β-mercaptopropionate

Dibutyl ether

CAS 142-96-1

UN No. 1149 (DOT)

Synonyms: Butyl ether; 1-Butoxybutane; n-Dibutyl ether; Di-n-butyl ether; Dibutyloxide; 1,1′-Oxybis (butane)

Empirical: $C_8H_{18}O$

Formula: $CH_3(CH_2)_3O(CH_2)_3CH_3$

Properties: Colorless clear liq.; mild ether-like odor; stable; sol. in water, alcohol; m.w. 130.26; sp.gr. 0.768 (20 C); m.p. -95.2 C; b.p. 142 C; flash pt. 77 F

Toxicology: LD50 (oral, rat) 7400 mg/kg, (skin, rabbit) 10 g/kg; TCLo (inh., human) 200 ppm; mildly toxic by inhalation, ingestion, and skin contact; eye irritant; human systemic effects by inh.; skin irritant

Precaution: Flamm.; dangerous fire hazard exposed to heat, flame, oxidizers; incompat. with NCl_3, oxidizing materials; Hazardous decomp. prods.: Heated to decomp., emits acrid smoke and fumes

Uses: Solvent for paints, Grignard synthesis, fats, oils, org. acids, alkaloids, nat. and syn. resins; extractant; constituent of catalysts for copolymerization; mfg. of pesticides

Manuf./Distrib.: Focus; Penta Mfg.; Spectrum Chem. Mfg.; Wacker-Chemie GmbH; Wacker Chems.

n-Dibutyl ether. *See* Dibutyl ether

Di-n-butyl ether. *See* Dibutyl ether

Dibutyl hexanedioate. *See* Dibutyl adipate

Dibutyl ketone. *See* Diisobutyl ketone

Dibutyl(3-mercaptopropionato(2-)tin. *See* Dibutyltin β-mercaptopropionate

2,2-Dibutyl-1-oxa-2-stanna-3-thiacyclohexan-6-one. *See* Dibutyltin β-mercaptopropionate

Dibutyloxide. *See* Dibutyl ether

Dibutyl phthalate
CAS 84-74-2; EINECS 201-557-4
UN No. 9095 (DOT)
Synonyms: DBP; 1,2-Benzenedicarboxylic acid, dibutyl ester; o-Benzenedicarboxylic acid dibutyl ester; Benzene-o-dicarboxylic acid di-n-butyl ester; Dibutyl-1,2-benzene dicarboxylate; n-Butyl phthalate; Butyl phthalate; Di-n-butyl phthalate
Definition: Aromatic diester of butyl alcohol and phthalic acid
Empirical: $C_{16}H_{22}O_4$
Formula: $C_6H_4(COOC_4H_9)_2$
Properties: Colorless stable oily liq., odorless; misc. with common org. solvs.; insol. in water; m.w. 278.17; dens. 1.0484 (20/20 C); f.p. -35 C; b.p. 340 C; flash pt. (COC) 340 F; ref. index 1.4920
Toxicology: LD50 (oral, rat) 8000 mg/kg, (IP, rat) 3050 mg/kg; ACGIH TLV/TWA 5 mg/m^3; mod. toxic by IP, IV routes; mildly toxic by ing.; irritant; if ingested can cause gastrointestinal upset, human systemic effects (hallucinations, nausea, vomiting)
Precaution: Combustible exposed to heat or flame; can react with oxidizers; reacts explosively with chlorine; Hazardous decomp. prods.: Heated to decomp., emits acrid smoke and fumes
NFPA: Health 0, Flammability 1, Reactivity 0
Uses: Plasticizer-solvent in nitrocellulose, lacquers, elastomers, explosives, nail polishes; solvent for perfumes, oils; perfume fixative; textile lubricating agent; as insect repellent; in oral pharmaceuticals
Regulatory: FDA 21CFR §175.105, 175.300, 176.170, 176.300, 177.1200, 177-2420, 177.2600; FDA approved for orals; SARA reportable; BP, Ph.Eur. compliance
Manuf./Distrib.: Aldrich; Allchem Ind.; Aristech; Ashland; BP Chems. Ltd; Chisso Am.; Coyne; Daihachi Chem. Ind.; Eastman; Fluka; Great Western; C.P. Hall; Houghton Chem.; Mitsubishi Gas; Sigma; Spectrum Chem. Mfg.; Unitex
Trade names: Eastman® DBP; Kodaflex® DBP; Palatinol® DBP; Uniplex 150
Trade names containing: Kodaflex® HS-4

Di-n-butyl phthalate. *See* Dibutyl phthalate

Dibutyltin β-mercaptopropionate
CAS 78-06-8
Synonyms: 2,2-Dibutyl-1-oxa-2-stanna-3-thiacyclohexan-6-one; 2,2-Dibutyl-dihydro-6H-1,3,2-oxathiastannin-6-one; Mercaptopriopionic acid, dibutyltin salt; Dibutyl(3-mercaptopropionato(2-)tin; Dibutyltin-S,O-3-mercaptopropionate; Dibutyltin-S,O-β-mercaptopropionate; Dibutyltin-O,S-mercaptopropionate
Empirical: $C_{11}H_{22}O_2SSn$
Properties: Wh. crystals; m.w. 337.08
Toxicology: LD50 (IV, mouse) 100 mg/kg; poison by IV route; Hazardous decomp. prods.: Heated to decomp., emits toxic fumes of SO_x
Uses: Solvent; stabilizer for PVC

Dibutyltin-O,S-mercaptopropionate. *See* Dibutyltin β-mercaptopropionate
Dibutyltin-S,O-3-mercaptopropionate. *See* Dibutyltin β-mercaptopropionate
Dibutyltin-S,O-β-mercaptopropionate. *See* Dibutyltin β-mercaptopropionate
o-Dichlorbenzene. *See* o-Dichlorobenzene
o-Dichlorbenzol. *See* o-Dichlorobenzene
1,2-Dichlorobenzene. *See* o-Dichlorobenzene

o-Dichlorobenzene
CAS 95-50-1; EINECS 202-425-9

UN No. 1591 (DOT)
Synonyms: DCB; ODB; ODCB; Chloroben; Chloroden; Cloroben; o-Dichlorbenzene; o-Dichlorbenzol; 1,2-Dichlorobenzene; Dizene; Orthodichlorobenzene; Orthodichlorobenzol; Benzene, o-dichloro-
Empirical: $C_6H_4Cl_2$
Properties: Colorless clear liq., pleasant odor; sol. in alcohol, ether; misc. with most org. solvs.; insol. in water; m.w. 146.95; dens. 1.2884 (20/4 C); f.p. -17 C; m.p. -17.5 C; b.p. 172-179 C; flash pt. 150 F; ref. index 1.5459
Toxicology: LD50 (oral, rat) 500 mg/kg, (IP, rat) 840 mg/kg; ACGIH TLV:CL 50 ppm (skin); toxic by inh. and ing.; primary irritant to eyes, skin, mucous membranes; chronic exposure can cause CNS effects; causes liver/kidney injury
Precaution: Flamm. exposed to heat or flame; can react vigorously with oxidizing materials; slow reaction with Al may cause explosion during storage in sealed aluminum container; Hazardous decomp. prods.: Heated to decomp., emits toxic fumes of Cl^-
NFPA: Health 2, Flammability 2, Reactivity 0
Uses: Mfg. of 3,4-dichloroaniline; solvent for wide range of org. materials and for oxides of nonferrous metals; degreasing hides and wool; industrial odor control; pesticide, herbicide, insecticide, soil fumigant
Regulatory: SARA reportable
Manuf./Distrib.: Aldrich; Allchem Ind.; Ashland; Bayer/Fibers, Orgs., Rubbers; Fluka; Mitsubishi Gas; Monsanto; PPG Ind.; Sigma; Spectrum Chem. Mfg.
Trade names: Dizene

3,4-Dichlorobenzotrifluoride
CAS 328-84-7
Synonyms: DCBTF; Benzene, 1,2-dichloro-4-(trifluoromethyl)-
Empirical: $C_7H_3F_3Cl_2$
Properties: M.w. 215.00; dens. 1.478; m.p. -13 to -12 C; b.p. 173-174 C; flash pt. 65 C
Toxicology: Irritant
Uses: Non-ozone depleting solvent for industrial solvent cleaning, aerosols, adhesives, coatings, and inks
Manuf./Distrib.: Aldrich; Occidental
Trade names: Oxsol® 1000

1,1-Dichloro-2-chloroethylene. *See* Trichloroethylene
1,1-Dichloro-2,2-dichloroethane. *See* 1,1,2,2-Tetrachloroethane

Dichlorodifluoromethane
CAS 75-71-8; EINECS 200-893-9
UN No. 1028 (DOT)
Synonyms: CFC 12; F 12; FC 12; R 12; Difluorodichloromethane; Halon; Fluorocarbon-12; Refrigerant 12; Propellant 12; Food freezant 12
Classification: Halogenated aliphatic hydrocarbon
Empirical: CCl_2F_2
Formula: $ClCF_2Cl$
Properties: Colorless gas; may be liquefied under pressure; odorless; noncorrosive; insol. in water; sol. in most org. solvs., alcohol, ether, acetic acid; m.w. 120.9; sp.gr. 1.325 (21 C, liq.); vapor pressure 5 atm (16 C); f.p. -158 C; b.p. -29.8 C; noncombustible
Toxicology: LC50 (inh., rat, 30 min) 80 pph; ACGIH TLV/TWA 1000 ppm; human systemic effects by inh.; narcotic in high concs.; may cause nose/throat/respiratory irritation, lightheadedness, drowsiness, incoordination, liver changes

Precaution: Compressed gas; incompat. with reactive metals, e.g., sodium, potassium, aluminum, magnesium, zinc (can react violently or explosively), water (can cause violent spattering or explosion); Hazardous decomp. prods.: Heated to decomp., emits highly toxic fumes of phosgene, Cl^-, and F^-
Uses: Refrigerant; blowing agent; aerosol propellant; low-temp. solvent; degreasing agent; leak detecting agent; monomer for resins; freezing of foods by direct contact; prep. of frozen tissue sections
Regulatory: FDA 21CFR §173.355; FDA approved for implants, inhalers, rectals, topicals; USP/NF, BP compliance; SARA reportable
Manuf./Distrib.: Air Prods.; Allchem Ind.; Elf Atochem N. Am.; ICI Spec. Chems.; LaRoche Chem.; LaRoche Ind.; PCR; Sigma

1,2-Dichloroethane. *See* Ethylene dichloride
Dichloro-1,2-ethane. *See* Ethylene dichloride
α,β-Dichloroethane. *See* Ethylene dichloride
sym-Dichloroethane. *See* Ethylene dichloride
Dichloroethylene. *See* Ethylene dichloride
Dichloromethane. *See* Methylene chloride
1,2-Dichloropropane. *See* Propylene dichloride
α,β-Dichloropropane. *See* Propylene dichloride

Dichlorotetrafluoroethane
CAS 76-14-2
UN No. 1958
Synonyms: CFC 114; F 114; FC 114; R 114; sym-Dichlorotetrafluoroethane; 1,2-Dichlorotetrafluoroethane; 1,2-Dichloro-1,1,2,2-tetrafluoroethane; Fluorane 114; Fluorocarbon-114; Freon 114; Halocarbon 114; Propellant 114; Refrigerant 114; Cryofluoran; Cryofluorane; Tetrafluorodichloroethane; Ethane, 1,2-dichloro-1,1,2,2-tetrafluoro-
Classification: Halogenated aliphatic hydrocarbon
Empirical: $C_2Cl_2F_4$
Formula: $CClF_2CClF_2$
Properties: Colorless clear gas or liquefied compressed gas; nearly odorless to faint ethereal odor; pract. insol. in water; sol. in alcohol, ether; m.w. 170.92; sp.gr. 1.5312 (liq., 0 C); vapor pressure 1620 mm Hg; f.p. -94 C; b.p. 3.55 C; ref. index 1.3092; noncombustible
Toxicology: LC50 (inh., rat, 30 min) 72 pph; ACGIH TLV/TWA 1000 ppm; toxic by inhalation; an asphyxiant; may cause mild irritation of nose, throat, upper airways; high concs. may cause lightheadedness, dizziness, drowsiness
Precaution: Compressed gas; incompat. with reactive metals (sodium, potassium, magnesium, zinc, powd. aluminum); can react violently or explosively
Uses: Refrigerant; aerosol propellant for industrial and pharmaceutical applics.; solvent; degreasing agent; fire extinguishing agent; foam blowing agent; dielectric fluid
Regulatory: FDA approved for inhalants, nasals, rectals, topicals; USP/NF, BP compliance
Manuf./Distrib.: Aldrich; Allchem Ind.; Harris Spec. Chems.; LaRoche Ind.

1,2-Dichlorotetrafluoroethane. *See* Dichlorotetrafluoroethane
1,2-Dichloro-1,1,2,2-tetrafluoroethane. *See* Dichlorotetrafluoroethane
sym-Dichlorotetrafluoroethane. *See* Dichlorotetrafluoroethane
1,4-Dicyanobutane. *See* Adiponitrile
Diethanolethylamine. *See* N-Ethyldiethanolamine
1,1-Diethoxyethane. *See* Acetal
1,2-Diethoxyethane. *See* Ethylene glycol diethyl ether
Diethyl. *See* Butane

Diethyl acetal. *See* Acetal

Diethylamine

CAS 109-89-7; EINECS 203-716-3
UN No. 1154 (DOT)
Synonyms: 2-Aminopentane; N,N-Diethylamine; N-Ethylethanamine
Classification: Aliphatic organic compd.
Empirical: $C_4H_{11}N$
Formula: $(C_2H_5)_2NH$
Properties: Colorless liq.; ammoniacal odor; misc. with water, alcohol; m.w. 73.14; dens. 0.707; vapor pressure 400 mm Hg (38 C); m.p. -50 C; b.p. 55 C; flash pt. -28 C; ref. index 1.3850 (20 C)
Toxicology: LD50 (oral, rat) 540 mg/kg, (skin, rabbit) 820 mg/kg; LC50 (inh., mammal) 5000 mg/m^3; mod. toxic by ing., inh., skin contact; corrosive; skin and severe eye irritant; inh. of vapor can cause severe cough and chest pains
Precaution: Flamm.; dangerous fire hazard exposed to heat, flame, oxidizes; explodes on contact with dicyanofurazan; violent reaction with sulfuric acid; ignites on contact with cellulose nitrate; Hazardous decomp. prods.: Heated to decomp., emits toxic fumes of NO_x
NFPA: Health 2, Flammability 3, Reactivity 0
Uses: Rubber chemicals, textile specialties, selective solvent, dyes, flotation agents, resins, pesticides, polymerization inhibitor, pharmaceuticals, petroleum chemicals, electroplating, corrosion inhibitors; in pharmaceutical injectables
Regulatory: FDA approved for injectables (IM, IV); BP compliance
Manuf./Distrib.: AC Ind.; Air Prods.; Aldrich; Allchem Ind.; Ashland; BASF; BASF AG; Coyne; Elf Atochem N. Am.; Fluka; Sigma; Spectrum Chem. Mfg.; Union Carbide

N,N-Diethylamine. *See* Diethylamine
(Diethylamino) ethane. *See* Triethylamine
Diethyl 1,2-benzenedicarboxylate. *See* Diethyl phthalate
Diethyl carbinol. *See* 3-Pentanol
Diethyl 'Carbitol'. *See* Diethylene glycol diethyl ether
Diethyldiglycol. *See* Diethylene glycol diethyl ether
Diethylene dioxide. *See* 1,4-Dioxane
1,4-Diethylene dioxide. *See* 1,4-Dioxane
Diethylene ether. *See* 1,4-Dioxane

Diethylene glycol

CAS 111-46-6; EINECS 203-872-2
Synonyms: DEG; Dihydroxydiethyl ether; β,β´-Dihydroxydiethyl ether; Bis(2-hydroxyethyl) ether; Diglycol; 2,2´-Dihydroxyethyl ether; Ethylene diglycol; Glycol ether; Glycol ethyl ether; 3-Oxapentane-1,5-diol; 3-Oxa-1,5-pentanediol; 2,2´-Oxybisethanol; 2,2´-Oxydiethanol
Classification: Aliphatic diol
Empirical: $C_4H_{10}O_3$
Formula: $CH_2OHCH_2OCH_2CH_2OH$
Properties: Colorless clear syrupy liq., pract. odorless, sweetish taste; extremely hygroscopic; lowers f.p. of water; misc. with water, ethanol, acetone, ether, ethylene glycol; m.w. 106.12; dens. 1.1184 (20/20 C); vapor pressure 1 mm Hg (91.8 C); f.p. -80 C; b.p. 245 C; flash pt. 255 F
Toxicology: LD50 (oral, human) 1000 mg/kg, (oral, rat) 12,565 mg/kg, (skin, rabbit) 11,890 mg/kg; poison by inh.; mod. toxic to humans by ing., IV routes; hazardous for household use in conc. of 10% or more; may cause acidosis and renal failure

Precaution: Combustible exposed to heat or flame; reactive with oxidizing materials; dec. exothermically with sodium hydroxide @ 230 C and releases explosive hydrogen gas; Hazardous decomp. prods.: Heated to decomp., emits acrid smoke and irritating fumes
NFPA: Health 1, Flammability 1, Reactivity 0
Uses: Prod. of polyurethane and unsat. polyester resins, triethylene glycol, paints; textile softener; solvent for nitrocellulose, dyes and oils; dehydration of natural gas, elasticizers, and surfactants; humectant for tobacco, casein, and synthetic sponge
Regulatory: FDA 21CFR §175.105, 177.2420
Manuf./Distrib.: Aldrich; Allchem Ind.; Ashland; BASF; Baychem; BP Chems. Ltd; Chemcentral; Coyne; DuPont; Eastman; Ferro/Grant; Fluka; C.P. Hall; Harcros USA; Hoechst Celanese; Houghton Chem.; Itochu Spec.; Mitsui Petrochem. Ind.; Mobil; Occidental; Olin; Samson; Shell; Sigma; Spectrum Chem. Mfg.; Sunnyside; Texaco; Union Carbide; Van Waters & Rogers

Diethylene glycol amine. *See* 2-(2-Aminoethoxy) ethanol
Diethylene glycol butyl ether. *See* Butoxydiglycol
Diethylene glycol n-butyl ether. *See* Butoxydiglycol

Diethylene glycol butyl ether acetate
CAS 124-17-4; EINECS 204-685-9
Synonyms: 2-(2-Butoxyethoxy) ethyl acetate; 2-(2-Butoxyethoxy)ethanol acetate; Acetic acid 2-(2-butoxyethoxy) ethyl ester; Diethylene glycol monobutyl ether acetate; Diglycol monobutyl ether acetate; Butyl Carbitol acetate; Butyldiglycol acetate
Empirical: $C_{10}H_{20}O_4$
Formula: $CH_3COOCH_2CH_2OCH_2CH_2O(CH_2)_3CH_3$
Properties: Colorless liq.; mod. water sol.; m.w. 204.3; dens. 0.977 (20/4 C); vapor pressure 0.01 mm Hg (20 C); f.p. -32.2 C; i.b.p. 235 C; flash pt. 102 C; ref. index 1.43 (20 C)
Toxicology: LD50 (oral, rat) 6500 mg/kg, (skin, rabbit) 14,500 mg/kg; mod. toxic by ing.; mild skin/eye irritant; prolonged/repeated skin contact may cause drying, cracking, or irritation
Precaution: Combustible exposed to heat or flame; incompat. with oxidizing materials; peroxide former; Hazardous decomp. prods.: Heated to decomp., emits acrid and irritating fumes
NFPA: Health 1, Flammability 1, Reactivity 0
Uses: Solvent for printing inks, coatings, separation processes; coalescing aid
Manuf./Distrib.: Aldrich; Allchem Ind.; ARCO; Ashland; Eastman; Fluka; Occidental; Spectrum Chem. Mfg.; Union Carbide
Trade names: Butyl Carbitol® Acetate; Dba; DB Acetate; Eastman® DB Acetate

Diethylene glycol dibutyl ether
CAS 112-73-2; EINECS 204-001-9
Synonyms: 2,2´-Dibutoxyethyl ether; Bis (butoxyethyl)ether; Bis (2-butoxyethyl) ether; 2-Butoxyethyl ether; Ether, bis (2-butoxyethyl); Dibutyldiglycol; Dibutyl 'Carbitol'; Butyl diglyme; Diethylene glycol di-n-butyl ether; 1,1´-(Oxybis(2,1-ethanediyloxy))bisbutane; 5,8,11-Trioxapentadecane
Empirical: $C_{12}H_{26}O_3$
Formula: $C_4H_9O(C_2H_4O)_2C_4H_9$
Properties: Almost colorless liq.; char. odor; sl. sol. in water; m.w. 218.38; dens. 0.8853 (20/20 C); f.p. -60.2 C; b.p. 256 C; flash pt. 118 C; ref. index 1.423 (20 C)
Toxicology: LD50 (oral, rat) 3900 mg/kg, (skin, rabbit) 4040 mg/kg; moderately toxic by ingestion; mildly toxic by skin contact; skin and eye irritant; experimental reproductive effects

Precaution: Combustible exposed to heat or flame; Hazardous decomp. prods.: Heated to decomp., emits acrid smoke and irritating fumes
Uses: Solvent used in electrochemistry, polymer and boron chemistry; physical processes such as gas absorption, extraction, stabilization; coatings and inks; diluent in VC dispersions; solvent, drug carrier, stabilizer in pharmaceuticals
Manuf./Distrib.: Aldrich; Brand-Nu Labs; Ferro/Grant; Fluka; Great Western; Hoechst AG; Sigma
Trade names: Butyl Diglyme; Ethyl Diglyme

Diethylene glycol di-n-butyl ether. *See* Diethylene glycol dibutyl ether

Diethylene glycol diethyl ether
CAS 112-36-7; EINECS 203-963-7
Synonyms: Diglycol diethyl ether; Bis (2-ethoxyethyl) ether; Diethyldiglycol; Diethyl 'Carbitol'; 1-Ethoxy-2-(β-ethoxyethoxy) ethane; Ethyl diglyme; Ethoxyethyl ether; 2-Ethoxyethyl ether; 3,6,9-Trioxaundecane
Empirical: $C_8H_{18}O_3$
Formula: $C_2H_5O(C_2H_4O)_2C_2H_5$
Properties: Colorless liq.; sol. in hydrocarbons and water; m.w. 162.93; dens. 0.9082 (20/20 C); f.p. -44.3 C; b.p. 189 C; flash pt. 180 F; ref. index 1.412 (20 C); very stable
Toxicology: LD50 (oral, rat) 4970 mg/kg; mod. toxic by ing.; primary irritant; eye irritant; experimental reproductive effects
Precaution: Flamm. exposed to heat or flame; Hazardous decomp. prods.: Heated to decomp., emits acrid smoke and irritating fumes
Uses: Solvent for nitrocellulose, resins, lacquers; organic synthesis
Manuf./Distrib.: Aldrich; Ferro/Grant; Fluka; Sigma

Diethylene glycol dimethyl ether
CAS 111-96-6; EINECS 203-924-4
Synonyms: Diethyl glycol dimethyl ether; Diglycol methyl ether; Dimethyldiglycol; Diglyme; Dimethoxydiglycol; 2-Methoxyethyl ether; Methyl diglyme; Trioxanonane; Bis (2-methoxyethyl) ether
Empirical: $C_6H_{14}O_3$
Formula: $CH_3(OCH_2CH_2)_2OCH_3$
Properties: Colorless liq., mild odor; misc. with water, alcohol, ether, and hydrocarbons; m.w. 134.18; dens. 0.9451 (20/20 C); f.p. -58 C; b.p. 162 C; flash pt. (OC) 70 C; ref. index 1.408 (20 C)
Toxicology: TCLo (inh., rat, 6 h) 1100 ppm, (oral, mouse) 5 g/kg; mutagen; reproductive effects
Precaution: Combustible; readily forms explosive peroxides on exposure to air, light, or heat; other explosive reactions possible; Hazardous decomp. prods.: Heated to decomp., emits toxic fumes of NO_x
Uses: Solvent, anhydrous reaction medium for organo-metallic synthesis; reaction solvent in reduction, isomerization, alkylation condensation, polymerization; extract and separation solvent; pharmaceutical solvent; paints
Manuf./Distrib.: Aldrich; Brand-Nu Labs; Ferro/Grant; Fluka; Great Western; Hoechst AG; Sigma
Trade names: Diglyme; Diglyme; Hisolve MDM

Diethylene glycol dioctanoate/diisononanoate
Definition: Diester of a mixt. of 2-ethylhexanoic and isononanoic acids and diethylene glycol
Uses: Emollient, penetrant, spreading agent, perfume solvent in antiperspirants, deodorants, preshave lotions
Trade names: Estalan 430

Diethylene glycol ethyl ether. *See* Ethoxydiglycol
Diethylene glycol ethyl ether acetate. *See* Ethoxydiglycol acetate

Diethylene glycol 2-ethylhexyl ether
CAS 1559-36-0
Synonyms: Diethylene glycol mono-2-ethylhexyl ether
Trade names containing: Eastman® EEH

Diethylene glycol ethylvinyl ether
CAS 10143-53-0
Synonyms: 2-[2-(2-Ethoxyethoxy)ethoxy]ethene; 1-(2-Ethoxyethoxy)-2-vinyloxy ethane; 2-Ethoxyethyl 2-(vinyloxy)ethyl ether; Vinyl ether 'Carbitol'
Classification: Aliphatic glycol diether
Empirical: $C_8H_{16}O_3$
Formula: $CH_2=CHOCH_2CH_2OCH_2CH_2OCH_2CH_3$
Properties: Colorless liq.; sol. in water, ethanol, many solvs.; m.w. 160.24; sp.gr. 0.941 (20 C); m.p. -80 C (sets to a glass); b.p. 191.2 C; flash pt. (OC) 32.8 C
Toxicology: LD50 (oral, rat) 11 g/kg, (skin, rabbit) 8410 mg/kg; mildly toxic by ing., skin contact; skin irritant; ing. may cause vomiting, cyanosis, headache, rapid respiration and heart rate, low blood pressure, muscle tenderness, unconsciousness
Precaution: Flamm.; incompat. with oxidizing materials (possible fire risk); Hazardous decomp. prods.: Heated to decomp., emits acrid smoke and fumes
Uses: Solvent

Diethylene glycol hexyl ether. *See* Diethylene glycol n-hexyl ether

Diethylene glycol n-hexyl ether
CAS 112-59-4; EINECS 203-988-3
Synonyms: DEGHE; 3,6-Dioxadodecanol-1; Diethylene glycol monohexyl ether; Diethylene glycol hexyl ether; n-Hexoxyethoxyethanol; 2-(2-Hexyloxyethoxy)ethanol; 3,6-Dioxa-1-dodecanol; Hexyl 'Carbitol'; Hexyldiglycol; Ethanol, 2-((2-hexyloxy)ethoxy)-
Classification: Aliphatic ether alcohol
Empirical: $C_{10}H_{22}O_3$
Formula: $CH_3(CH_2)_4CH_2OCH_2CH_2OCH_2CH_2OH$
Properties: Water-wh. liq.; sol. 3% in water; sl. to mod. sol. in water; m.w. 190.28; sp.gr. 0.9346 (20/20 C); vapor pressure 0.01 mm Hg (20 C); f.p. -33 C; b.p. 259.1 C; flash pt. 140.5 C; ref. index 1.438 (20 C)
Toxicology: LD50 (oral, rat) 4920 mg/kg, (skin, male rabbit) 2000 mg/kg; mod. toxic by skin contact; mildly toxic by ing.; eye and severe skin irritant; can be absorbed thru skin; possible allergic reactions; mod. to severe eye irritant
Precaution: Combustible exposed to heat or flame; incompat. with oxidizng materials (increased fire and explosion hazard); Hazardous decomp. prods.: Heated to decomp., emits acrid smoke and fumes
Uses: High-boiling solvent for coatings
Manuf./Distrib.: Aldrich; Fluka; Union Carbide
Trade names: Hexyl Carbitol®

Diethylene glycol isobutyl ether
CAS 18912-80-6
Synonyms: DEGiBE; Diethylene glycol monoisobutyl ether; Isobutoxy-2-ethoxy-2-ethanol; 2-(2-Isobutoxyethoxy) ethanol; 2-(2-(2-Methoxypropoxy)ethoxy) ethanol
Classification: Aliphatic ether alcohol
Empirical: $C_8H_{18}O_3$
Formula: $CH_3CH(CH_3)CH_2OCH_2CH_2OCH_2CH_2OH$

Properties: Colorless liq.; sol. in water; m.w. 162.26; sp.gr. 0.95-1.0; b.p. 217-225 C; flash pt. (CC) 106 C
Toxicology: LD50 (oral, rat) 4920 mg/kg, (skin, rabbit) 3560 mg/kg; mod. toxic by skin contact; mildly toxic by ing., inh.; nonirritating to skin; can be absorbed thru skin, but health effects unlikely; mod. to severe eye irritant
Precaution: Incompat. with strong oxidizing agent (increases fire and explosion hazard); Hazardous decomp. prods.: Heated to decomp., emits acrid smoke and irritating fumes
NFPA: Health 1, Flammability 1, Reactivity 0
Uses: Solvent
Manuf./Distrib.: Oxiteno

Diethylene glycol lactamide. *See* PEG-2 lactamide
Diethylene glycol methyl ether. *See* Methoxydiglycol

Diethylene glycol methylpentyl ether
CAS 10143-56-3
Synonyms: DEGMPE; Diethylene glycol monomethylpentyl ether; 2-Methylpentyl 'Carbitol'; 2-[2-((2-Methylpentyl)oxy)ethoxy]ethanol; Diethylene glycol mono-2-methylpentyl ether; 8-Methyl-3,6-dioxa-1-undecanol
Classification: Aliphatic ether alcohol
Empirical: $C_{10}H_{22}O_3$
Formula: $CH_3CH_2CH_2CH(CH_3)CH_2OCH_2CH_2OCH_2CH_2OH$
Properties: Colorless liq.; m.w. 190.32
Toxicology: LD50 (oral, rat) 5660 mg/kg, (skin, rabbit) 1580 mg/kg; mod. toxic by skin contact; mildly toxic by inj.; mod. to severe eye irritant; skin irritant
Precaution: Incompat. with oxidizing materials; possible fire risk; Hazardous decomp. prods.: Heated to decomp., emits acrid smoke and irritating fumes
Uses: Solvent

Diethylene glycol monobutyl ether. *See* Butoxydiglycol
Diethylene glycol monobutyl ether acetate. *See* Diethylene glycol butyl ether acetate
Diethylene glycol monoethyl ether. *See* Ethoxydiglycol
Diethylene glycol monoethyl ether acetate. *See* Ethoxydiglycol acetate
Diethylene glycol mono-2-ethylhexyl ether. *See* Diethylene glycol 2-ethylhexyl ether
Diethylene glycol monohexyl ether. *See* Diethylene glycol n-hexyl ether
Diethylene glycol monoisobutyl ether. *See* Diethylene glycol isobutyl ether
Diethylene glycol monomethyl ether. *See* Methoxydiglycol
Diethylene glycol monomethylpentyl ether. *See* Diethylene glycol methylpentyl ether
Diethylene glycol mono-2-methylpentyl ether. *See* Diethylene glycol methylpentyl ether
Diethylene glycol monopropyl ether. *See* Diethylene glycol propyl ether

Diethylene glycol propyl ether
CAS 6881-94-3; EINECS 229-985-7
Synonyms: Diethylene glycol monopropyl ether; 2-(2-Propoxyethoxy) ethanol
Empirical: $C_7H_{16}O_3$
Formula: $CH_2OHCH_2OCH_2CH_2OCH_2CH_2CH_3$
Properties: Liq.; sol. in water; m.w. 148.2; dens. 0.963 (20 C); b.p. 202 C (760 mm); f.p. < -90 C; flash pt. (TCC) 93 C
Toxicology: Causes eye irritation
Precaution: Potential peroxide former
Uses: Coupling solvent for sol'n. and water-dilutable coatings; active solvent for

NC, acrylic copolymers, epoxy resins, chlorinated rubber, alkyd resins; coalescing aid in architectural latex coatings, floor polishes; hydraulic brake fluids; household and general industrial cleaners
Manuf./Distrib.: Ashland; Eastman; Great Western
Trade names: DP; Eastman® DP

Diethyleneimide oxide. *See* Morpholine
Diethylene imidoxide. *See* Morpholine
Diethylene oxide. *See* 1,4-Dioxane; Tetrahydrofuran
Diethylene oximide. *See* Morpholine

Diethylenetriamine
CAS 111-40-0; EINECS 203-865-4
UN No. 2079 (DOT)
Synonyms: DETA; Aminoethylethandiamine; N-(2-Aminoethyl)ethylenediamine; 3-Azapentane-1,5-diamine; 2,2´-Diaminodiethylamine; 2,2´-Iminobisethylamine; 2,2´-Iminodiethylamine; Bis(2-aminoethyl) amine; Bis(β-aminoethyl)amine
Empirical: $C_4H_{13}N_3$
Formula: $NH_2C_2H_4NHC_2H_4NH_2$
Properties: Yel. visc. liq.; mild ammoniacal odor; sol. in water and hydrocarbons; strongly alkaline; hygroscopic; m.w. 103.17; dens. 0.9542 (20/20 C); vapor pressure 0.22 mm Hg (20 C); f.p. -39 C; b.p. 206.7 C; flash pt. 215 F
Toxicology: LD50 (oral, rat) 1080 mg/kg, (IP, rat) 74 mg/kg, (skin, rabbit) 1090 mg/kg; ACGIH TLV/TWA 1 ppm (skin); toxic by IP, ingestion, inhalation and skin absorption; strong irritant to eyes and skin
Precaution: Combustible exposed to heat or flame; can react with oxidizing materials; mixt. with nitromethane is a shock-sensitive explosive; ignites on contact with cellulose nitrate; corrosive to copper and its alloys; Hazardous decomp. prods.: Heated to decomp., emits toxic fumes of NO_x
Uses: Solvent for sulfur, acid gases, various resins, dyes; saponification agent for acidic materials; fuel component; epoxy curing agent; corrosion inhibitors; surfactants; intermediate for textile finishes; curing agent for two-pkg. protective coatings
Manuf./Distrib.: Akzo Nobel; Aldrich; Allchem Ind.; Ashland; BASF AG; Coyne; Diamine & Chems. Ltd; Fabrichem; Fluka; Houghton Chem.; Janssen Chimica; Nayler Chem. Ltd; Rhône-Poulenc Surf. & Spec.; Sigma; Tosoh; Union Carbide

Diethylenimide oxide. *See* Morpholine
N,N-Diethylethanamine. *See* Triethylamine
Diethyl ether. *See* Ethyl ether
Diethyl glycol dimethyl ether. *See* Diethylene glycol dimethyl ether
Di(2-ethylhexyl) adipate. *See* Dioctyl adipate
Di-(2-ethylhexyl) maleate. *See* Dioctyl maleate
Di(2-ethylhexyl) phthalate. *See* Dioctyl phthalate
Di-2-ethylhexyl sebacate. *See* Dioctyl sebacate

Diethyl ketone
CAS 96-22-0; EINECS 202-490-3
UN No. 1156 (DOT)
Synonyms: DEK; 3-Pentanone; Pentanone-3; Dimethylacetone; Propione; Metacetone; Methacetone
Empirical: $C_5H_{10}O$
Formula: $CH_3CH_2COCH_2CH_3$
Properties: Colorless mobile liq.; acetone odor; misc. with alcohol, ether; sol. in 25 parts water; m.w. 86.14; dens. 0.814; m.p. -42 C; b.p. 99-103 C; flash pt. 55 F; ref.

index 1.392
Toxicology: LD50 (oral, rat) 2.1 g/kg, (skin, rabbit) 20 g/kg; mod. toxic by ing., IP, IV routes; skin and eye irritant; mutagenic data
Precaution: Flamm.; dangerous fire hazard exposed to heat or flame; can react vigorously with oxidizing materials; reacts with hydrogen peroxide + nitric acid to form a shock- and heat-sensitive explosive peroxide; Hazardous decomp. prods.: Heated to decomp., emits acrid smoke and irritating fumes
Uses: Flavorings, medicine, organic synthesis; solvent for paints
Manuf./Distrib.: Aldrich; BASF; BASF AG; Esprit; Fluka; Hüls AG; Hüls Am.; Janssen Chimica; Penta Mfg.; Sigma; Union Carbide

N,N-Diethyl-3-methylbenzamide. *See* Diethyl toluamide
Diethyl oxide. *See* Ethyl ether

Diethyl phthalate
CAS 84-66-2; EINECS 201-550-6
Synonyms: DEP; Diethyl-o-phthalate; Ethyl phthalate; 1,2-Benzenedicarboxylic acid diethyl ester; Phthalic acid, diethyl ester; Phthalol; Diethyl 1,2-benzenedicarboxylate
Definition: Aromatic diester of ethyl alcohol and phthalic acid
Empirical: $C_{12}H_{14}O_4$
Formula: $C_6H_4(CO_2C_2H_5)_2$
Properties: Water-wh. oily liq., odorless, bitter taste; misc. with alcohol, ketones, esters, aromatic HC; insol. in water; m.w. 222.24; dens. 1.120 (25/25 C); f.p. -40.5 C; b.p. 298 C; flash pt. (OC) 325 F; visc. 31.3 cs (0 C); ref. index 1.500-1.505; stable
Toxicology: LD50 (oral, rat) 8600 mg/kg, (IP, rat) 5058 mg/kg; ACGIH TLV/TWA 5 mg/m^3; poison by IV route; mod. toxic by ingestion, subcut., IP routes; human systemic effects; strong irritant to eyes and mucous membranes
Precaution: Combustible when exposed to heat or flame; Hazardous decomp. prods.: Heated to decomp., emits acrid smoke and irritating fumes
NFPA: Health 0, Flammability 1, Reactivity 0
Uses: Solvent for nitrocellulose and cellulose acetate; plasticizer for food pkg., pharmaceuticals; wetting agent; pesticide, insecticidal preps.; in perfumery as solvent and fixative; plasticizer in solid rocket propellants; pharmaceutical fragrance
Regulatory: FDA 21CFR §175.105, 175.300, 175.320, 178.3910, 181.22, 181.27, 212.177; 27CFR §21.105; FDA approved for orals; USP/NF, BP compliance; SARA reportable
Manuf./Distrib.: Aldrich; Allan; Allchem Ind.; Berje; BP Chems. Ltd; Daihachi Chem. Ind.; Eastman; Fluka; Houghton Chem.; Hüls Am.; Morflex; Penta Mfg.; Sigma; Spectrum Chem. Mfg.; Unitex
Trade names: Kodaflex® DEP

Diethyl-o-phthalate. *See* Diethyl phthalate

Diethyl toluamide
CAS 134-62-3; EINECS 205-149-7
Synonyms: DEET; DET; Diethyl-m-toluamide; m-Toluic acid diethylamide; Benzamide, N,N-diethyl-3-methyl; N,N-Diethyl-m-toluamide; N,N-Diethyl-3-methylbenzamide; 3-Methyl-N,N-diethylbenzamide; m-Delphene; Delphene
Classification: Aromatic amide
Empirical: $C_{12}H_{17}NO$
Formula: $CH_3C_6H_4CON(C_2H_5)_2$
Properties: Colorless liq., mild bland odor; sol. in water, alcohol, ether, benzene;

Diethyl-m-toluamide

m.w. 191.30; dens. 0.996-1.002 (25/25 C); b.p. 160 C (19 mm); flash pt. > 230 F; ref. index 1.5230 (20 C)

Toxicology: LD50 (oral, rat) 1950 mg/kg, (skin, rabbit) 3180 mg/kg; LDLo (oral, human female) 950 mg/kg (coma), (oral, human male) 679 mg/kg; poison by IV route; mod. toxic by ingestion, skin contact; irritant to eyes, mucous membranes

Precaution: Combustible; Hazardous decomp. prods.: Heated to decomp., emits toxic fumes of NO_x

Uses: Pesticide, insect repellents, resin solvent, film formers

Manuf./Distrib.: Aldrich; DuPont; Fluka; Hoechst Celanese; Honeywill & Stein; Morflex; Sigma

Diethyl-m-toluamide. *See* Diethyl toluamide
N,N-Diethyl-m-toluamide. *See* Diethyl toluamide
1,1,1-Difluorochloroethane. *See* 1,1,1-Chlorodifluoroethane
1,1-Difluoro-1-chloroethane. *See* 1,1,1-Chlorodifluoroethane
Difluorochloromethane. *See* Chlorodifluoromethane
Difluorodibromomethane. *See* Dibromodifluoromethane
Difluorodichloromethane. *See* Dichlorodifluoromethane
Difluoromonochloroethane. *See* 1,1,1-Chlorodifluoroethane
Difluoromonochloromethane. *See* Chlorodifluoromethane
1,2-Difluoro-1,1,2,2-tetrachloroethane. *See* 1,1,2,2-Tetrachloro-1,2-difluoroethane
Diglycol. *See* Diethylene glycol
Diglycolamine. *See* 2-(2-Aminoethoxy) ethanol
Diglycol butyl ether. *See* Butoxydiglycol
Diglycol diethyl ether. *See* Diethylene glycol diethyl ether
Diglycol methyl ether. *See* Diethylene glycol dimethyl ether
Diglycol monobutyl ether. *See* Butoxydiglycol
Diglycol monobutyl ether acetate. *See* Diethylene glycol butyl ether acetate
Diglycol monoethyl ether. *See* Ethoxydiglycol
Diglyme. *See* Diethylene glycol dimethyl ether
2,3-Dihdyroxypropyl acetate. *See* Glyceryl acetate
Dihexyl. *See* n-Dodecane
Dihydro-1,3-dioxole. *See* 1,3-Dioxolane
Dihydro-2(3H)-furanone. *See* Butyrolactone
1,3-Dihydroxybutane. *See* Butylene glycol
Dihydroxydiethyl ether. *See* Diethylene glycol
β,β´-Dihydroxydiethyl ether. *See* Diethylene glycol
β,β´-Dihydroxydiethyl sulfide. *See* Thiodiglycol
2,2´-Dihydroxydipropyl ether. *See* Dipropylene glycol
1,2-Dihydroxyethane. *See* Glycol
Di-β-hydroxyethoxyethane. *See* Triethylene glycol
2,2´-Dihydroxyethyl ether. *See* Diethylene glycol
Dihydroxyethyl sulfide. *See* Thiodiglycol
β,β´-Dihydroxyethyl sulfide. *See* Thiodiglycol
2,2´-Dihydroxyisopropyl ether. *See* Dipropylene glycol
2,4-Dihydroxy-2-methylpentane. *See* Hexylene glycol
1,2-Dihydroxypropane. *See* Propylene glycol
2,3-Dihydroxypropyl octadecanoate. *See* Glyceryl stearate

Diisoamyl ketone

Uses: Solvent for paints

Manuf./Distrib.: Eastman

Diisobutyl adipate

CAS 141-04-8; EINECS 205-450-3

Synonyms: DIBA; Bis(2-methylpropyl) hexanedioate; Diisobutyl hexanedioate; Isobutyl adipate; Hexanedioic acid diisobutyl ester
Definition: Diester of isobutyl alcohol and adipic acid
Empirical: $C_{14}H_{26}O_4$
Formula: $[C_2H_4COOCH_2CH(CH_3)_2]_2$
Properties: Colorless liq., odorless; sol. in most org. solvs.; insol. in water; m.w. 258.40; dens. 0.950 (25 C); b.p. 278-280 C; f.p. -20 C
Toxicology: LD50 (IP, rat) 5950 mg/kg; mod. toxic by IP route; mildly toxic by ingestion; experimental teratogen, reproductive effects
Precaution: Combustible; Hazardous decomp. prods.: Heated to decomp., emits acrid smoke and irritating fumes
Uses: Plasticizer for some cellulosics, PVAc, PVC, VCA, in food pkg.
Regulatory: FDA 21CFR §175.105, 181.22, 181.27
Manuf./Distrib.: Aceto; Aldrich
Trade names containing: DBE-IB

Diisobutyl glutarate
Trade names containing: DBE-IB

Diisobutyl hexanedioate. *See* Diisobutyl adipate

Diisobutyl ketone
CAS 108-83-8; EINECS 203-620-1
UN No. 1157 (DOT)
Synonyms: DIBK; 2,6-Dimethyl-4-heptanone; 2,6-Dimethylheptan-4-one; s-Diisopropylacetone; Isovalerone; Dibutyl ketone; Isobutyl ketone; Valerone
Classification: Ketone
Empirical: $C_9H_{18}O$
Formula: $(CH_3)_2CHCH_2COCH_2CH(CH_3)_2$
Properties: Colorless liq.; mild peppermint odor; misc. with most org. liqs.; pract. insol. in water; m.w. 142.24; sp.gr. 0.8076; m.p. -46 C; b.p. 165 C; flash pt. (CC) 49 C; ref. index 1.4130 (20 C); pH neutral
Toxicology: LD50 (oral, rat) 4300 mg/kg, (skin, rabbit) 16 g/kg; ACGIH TLV/TWA 25 ppm; mod. toxic by ing. and inh.; irritant to respiratory system; extremely high vapor concs. may cause dizziness, nausea, incoordination, possibly death
Precaution: Flamm. exposed to heat or flame; incompat. with strong oxidizing agents (increases fire and explosion hazard); Hazardous decomp. prods.: Heated to decomp., emits acrid smoke and fumes
NFPA: Health 1, Flammability 2, Reactivity 0
Uses: Solvent for nitrocellulose, rubber, synthetic resins, lacquers, coatings; organic synthesis; roll-coating inks; stains; flavoring agent for foods and pharmaceuticals
Regulatory: FEMA GRAS
Manuf./Distrib.: Aldrich; Allchem Ind.; Ashland; Chemcentral; Coyne; Eastman; Fluka; Great Western; Harcros USA; Houghton Chem.; Hüls AG; Hüls Am.; Samson; Stanchem; Sunnyside; Union Carbide; Van Waters & Rogers

Diisobutyl phthalate
CAS 84-69-5
Synonyms: DIBP; Phthalic acid diisobutyl ester
Empirical: $C_{16}H_{22}O_4$
Properties: Sp.gr. 1.040 (20 C); f.p. -50 C; m.p. -64 C; flash pt. (PMCC) 185 C; ref. index 1.490
Toxicology: LD50 (oral, rat) 15 g/kg, (oral, mouse) 13 g/kg; mod. toxic by intraperitoneal route; mildly toxic by ingestion and skin contact

Diisobutyl succinate

Uses: Plasticizer-solvent for PVC, PVB, PS, cellulosics, paints, as vehicles for pigment dispersions
Manuf./Distrib.: Aldrich; Allchem Ind.; Hüls; Unitex
Trade names: Uniplex 155

Diisobutyl succinate
Trade names containing: DBE-IB

Diisocetyl adipate
CAS 57533-90-1; 58262-41-2; EINECS 260-799-9
Synonyms: Hexanedioic acid, bis (2-hexyl decyl) ester; Hexanedioic acid, bis (2-methylpentadecyl) ester
Definition: Diester of hexadecyl alcohol and adipic acid
Empirical: $C_{38}H_{74}O_4$
Formula: $C_{16}H_{33}OCO(CH_2)_4COOC_{16}H_{33}$
Uses: Emollient and spreading agent for cosmetics; film spreader for bath oils; solvent for lipstick dyes
Trade names: Schercemol DICA

s-Diisopropylacetone. *See* Diisobutyl ketone

Diisopropyl adipate
CAS 6938-94-9; EINECS 230-072-0
Synonyms: Adipic acid diisopropyl ester; Bis(1-methylethyl)hexanedioate; Hexanedioic acid, bis (1-methylethyl) ester; Isopropyl adipate
Definition: Diester of isopropyl alcohol and adipic acid
Empirical: $C_{12}H_{22}O_4$
Formula: $(CH_3)_2CHOCO(CH_2)_4COOCH(CH_3)_2$
Properties: Clear liq.; sol. in min. oil, ethanol, propylene glycol, IPM, oleyl alcohol; insol. in water, glycerin; sp.gr. 0.950-0.962; ref. index 1.4216-1.4245
Toxicology: LD50 (oral, rat) > 20 ± 3 ml/kg, (IV, rat) 640 mg/kg; mod. toxic by IV route; nonirritating to eyes and skin; Hazardous decomp. prods.: Heated to decomp., emits acrid smoke and fumes
Uses: Emollient, coupling agent in aq. alcoholic systems, shave lotions, hair tonics, topical pharmaceuticals; imparts spreadability to bath oils; plasticizer in hair sprays; solvent for aromatic oils and perfumes, lipstick dyes
Regulatory: FDA approved for topicals
Manuf./Distrib.: Inolex; ISP Van Dyk; Union Camp
Trade names: Crodamol DA; Schercemol DIA; Trivent DIA

1,3-Diisopropyl benzene
CAS 99-62-7; EINECS 202-773-1
Synonyms: 3-Isopropylcumene; 1,3-Bis(1-methylethyl) benzene; m-Diisopropylbenzene
Empirical: $C_{12}H_{18}$
Properties: M.w. 162.30
Toxicology: LD50 (oral, rat) 7400 mg/kg, (oral, mouse) 1650 mg/kg, (IP, mouse) 1650 mg/kg; mod. toxic by ingestion and IP routes; reproductive effects; Hazardous decomp. prods.: Heated to decomp., emits acrid smoke and fumes
Uses: Solvent; intermediate
Manuf./Distrib.: Aldrich; Fluka; Koch Chems. Ltd

1,4-Diisopropyl benzene
CAS 100-18-5; EINECS 202-826-9
Synonyms: 4-Isopropylcumene; 1,4-Bis(1-methylethyl) benzene; p-Diisopropyl benzene
Empirical: $C_{12}H_{18}$
Properties: M.w. 162.30

Toxicology: LD50 (oral, mouse) 3400 mg/kg
Uses: Solvent; intermediate
Manuf./Distrib.: Aldrich; Eastman; Fluka; Koch Chems. Ltd

m-Diisopropylbenzene. *See* 1,3-Diisopropyl benzene
p-Diisopropyl benzene. *See* 1,4-Diisopropyl benzene
Diisopropyl biphenyl. *See* Bis(methylethyl)-1,1′-biphenyl
Diisopropyl oxide. *See* Isopropyl ether

Diisopropyl sebacate
CAS 7491-02-3; EINECS 231-306-4
Synonyms: Bis(1-methylethyl) decanedioate; Decanedioic acid, bis(1-methylethyl) ester
Definition: Diester of isopropyl alcohol and sebacic acid
Empirical: $C_{16}H_{30}O_4$
Formula: $(CH_3)_2CHOCO(CH_2)_8COOCH(CH_3)_2$
Properties: Sp.gr. 0.936; flash pt. (COC) 190 C; ref. index 1.4310
Uses: Emollient, solubilizer, coupling agent in creams, lotions, bath oils, men's grooming aids, topical pharmaceuticals; plasticizer; solvent props.
Manuf./Distrib.: Union Camp
Trade names: Schercemol DIS

Diketone alcohol. *See* Diacetone alcohol

Dimethicone
CAS 9006-65-9; 9016-00-6; 63148-62-9; 68037-74-1 (branched)
Synonyms: PDMS; Dimethylpolysiloxane; Silicone rubber; Dimethyl silicone; Polydimethylsiloxane; Poly(oxy(dimethylsilylene))
Classification: Organosilicon polymer
Definition: Silicone oil consisting of dimethylsiloxane polymers
Empirical: $(C_2H_6OSi)_xC_4H_{12}Si$
Formula: $(CH_3)_3SiO[SI(CH_3)_2O]_nSi(CH_3)_3$
Properties: Colorless clear visc. oily liq., odorless; sol. in hydrocarbon solvs., benzene, toluene, xylene, hexane, petrol. spirits, amyl acetate; misc. in chloroform, ether; insol. in water, methanol, acetone, alcohol; m.w. 340-250,000; dens. 0.96-0.97; visc. 20-12,500 cst; flash pt. > 300 C; ref. index 1.400-1.404
Toxicology: Very low toxicity; ingestion of large amts. may cause stomach upset; suspected neoplastigen; heated to dec., emits acrid smoke and irritating fumes
Uses: Ointment and topical drug vehicle, skin protectant, foam preventative, surface active agent, release material, antifoaming agent; used in adhesive, ink, latex, soap, starch, and paint mfg. and other aq. industrial systems; heat transfer agents; defoamer for food applics.; chewing gum base
Regulatory: FDA 21CFR §145.180, 145.181, 146.185, 173.340, 175.105, 175.300, 176.170, 176.200, 176.210, 177.2260, 177.2600, 177.2800, 178.3570, 178.3910, 181.22, 181.28; USDA 9CFR §318.7, 381.147; Europe listed; UK approved; FDA approved for topicals; USP/NF, BP, Ph.Eur. compliance
Manuf./Distrib.: Aldrich; Dow Corning; Sigma
Trade names containing: Fancorsil A

Dimethicone copolyol undecylenate
CAS 159100-33-1
Synonyms: Dimethicone copolyol undecylinate
Uses: Emollient, coupling agent, cosolvent in cosmetics, pharmaceuticals
Trade names: Pecosil® DCU

Dimethicone copolyol undecylinate. *See* Dimethicone copolyol undecylenate
Dimethoxydiglycol. *See* Diethylene glycol dimethyl ether
1,2-Dimethoxyethane. *See* Ethylene glycol dimethyl ether

Dimethoxyethyl phthalate
CAS 117-82-8
Synonyms: DMEP; 1,2-Benzenedicarboxylic acid bi (2-methoxyethyl) ester; Phthalic acid bis(2-methoxyethyl) ester; Phthalic acid, di(methoxyethyl) ester; 2-Methoxyethyl phthalate; Bis (methoxyethyl) phthalate; Bis (2-methoxyethyl) phthalate; Di (2-methoxyethyl)phthalate
Empirical: $C_{14}H_{18}O_6$
Formula: $C_6H_4(COOCH_2CH_2OCH_3)_2$
Properties: Light-colored clear oily liq.; mild aromatic odor; m.w. 282.32; dens. 1.172 (20/20 C); vapor pressure 0.3 mm Hg (150 C); f.p. -45 C; m.p. -40 C; b.p. 340 C; flash pt. 194 C
Toxicology: LD50 (oral, rat) 2750 mg/kg, (IP, rat) 3735 mg/kg; LCLo (inh., rat, 16 h) 1595 ppm; mod. toxic by ing., inh., and IP routes; primary irritant; mutagen; reproductive effects
Precaution: Combustible when exposed to heat or flame; can react with oxidizing materials; Hazardous decomp. prods.: Heated to decomp., emits acrid smoke and irritating fumes
Uses: Plasticizer, solvent; pesticide
Manuf./Distrib.: Ashland; CTC Organics; Fluka

Di (2-methoxyethyl)phthalate. *See* Dimethoxyethyl phthalate

Dimethyl acetamide
CAS 127-19-5; EINECS 204-826-4
Synonyms: DMA; DMAC; N,N-Dimethylacetamide; Dimethylacetone amide; Acetdimethylamide; Acetic acid dimethylamide; Dimethylamide acetate
Empirical: C_4H_9NO
Formula: $CH_3CON(CH_3)_2$
Properties: Liq.; m.w. 87.12; dens. 0.942; vapor pressure 1.3 mm Hg; m.p. -20 C; b.p. 165-167 C; flash pt. 70 C; ref. index 1.437
Toxicology: LD50 (oral, rat) 4300 mg/kg, (IP, rat) 2750 mg/kg, (skin, rabbit) 2240 mg/kg; mod. toxic by skin contact, inh., IV, IP routes; mildly toxic by ing.; primary skin irritant; mutagen; reproductive effects
Precaution: Flamm. exposed to heat or flame; mod. explosion hazard; violent reaction with halogenated compds. above 90 C; Hazardous decomp. prods.: Heated to decomp., emits toxic fumes of NO_x
NFPA: Health 2, Flammability 2, Reactivity 0
Uses: Solvent; reagent
Manuf./Distrib.: Aldrich; Allchem Ind.; BASF; BASF AG; Cross World Sales; Fluka; Int'l. Chem. Inc; Monomer-Polymer & Dajac Labs; Primachem; Sigma; Spectrum Chem. Mfg.

N,N-Dimethylacetamide. *See* Dimethyl acetamide
Dimethylacetone. *See* Diethyl ketone
Dimethylacetone amide. *See* Dimethyl acetamide
Dimethyl acetylene carbinol. *See* Methyl butynol

Dimethyl adipate
CAS 627-93-0; EINECS 211-020-6
Synonyms: Dimethyl hexanedioate; Methyl adipate; Adipic acid dimethyl ester
Definition: Diester of methyl alcohol and adipic acid
Empirical: $C_8H_{14}O_4$
Formula: $CH_3O_2C(CH_2)_4CO_2CH_3$
Properties: Colorless liq.; m.w. 174.22; dens. 1.062 (20/4 C); m.p. 9-11 C; b.p. 115C (13 mm); flash pt. 107 C
Toxicology: LD50 (IP, rat) 1809 mg/kg; moderately toxic by intraperitoneal route;

suspected teratogen and reproductive effects; Hazardous decomp. prods.: Heated to decomp., emits acrid smoke and irritating fumes

Uses: Solvent for coatings, cleaners, inks, textile lubricants, urethane prod.; plasticizer for flexible thermoset polyester; polymer intermediate for polyester polyols for urethanes, wet-str. paper resins, polyester resins; specialty chemical intermediate

Manuf./Distrib.: Aldrich; Ashland; DuPont; Eastern Chem.; Fluka; Morflex; Sigma; UCB SA

Trade names: DBE-6; Santosol™ DMA

Trade names containing: DBE; DBE-2, -2SPG; DBE-3; DBE-9; Santosol™ DME; Santosol™ DME-2; Santosol™ DME-3

Dimethylamide acetate. *See* Dimethyl acetamide

Dimethylamine

CAS 124-40-3

UN No. 1032 (DOT), 1160 (DOT)

Synonyms: DMA; Dimethylamine anhydrous; N-Methylmethanamine

Classification: Sat. aliphatic amine

Empirical: C_2H_7N

Formula: CH_3NHCH_3

Properties: Colorless gas; strong fishy odor (below 100 ppm), ammonia-like odor (above 100 ppm); very sol. in water; sol. in ethanol, ether; m.w. 45.08; sp.gr. 0.68 (0 C); m.p. -92.2 C; b.p. 6.9 C; pH 11.9 (0.1 M aq, sol'n.)

Toxicology: ACGIH TLV/TWA 5 ppm; STEL 15 ppm; very toxic; poison on ing.; may cause skin irritation, burns on high exposure; lachrymator; can cause severe eye irritation, 'halo vision'; inh. of gas can cause nose/throat/lung irritation

Precaution: Flamm., corrosive, very toxic, compressed gas; incompat. with mercury, acrylaldehyde, acids, acid chlorides, acid anhydrides, nitrosating agents, strong oxidizing agents, fluorine, maleic anhydride; violent/explosive reactions possible; Hazardous decomp. prods.: Ammonia; reaction with nitrosating agents can form nitrosodimethylamine, a carcinogen; heated to decomp., emits toxic fumes of NO_x

NFPA: Health 3, Flammability 4, Reactivity 0

Uses: Solvent; fuel additive; antioxidant, stabilizer in natural rubber latex and some resins; prep. of solvents, pesticides, surfactants, soaps, detergents, rubber mfg. chems., dyes, pharmaceuticals, textile chems.

Manuf./Distrib.: Air Prods.; Aldrich; Allchem Ind.; BASF AG; Fluka; Sigma; Spectrum Chem. Mfg.

Dimethylamine anhydrous. *See* Dimethylamine
(Dimethylamino) benzene. *See* n,n-Dimethylaniline

n,n-Dimethylaniline

CAS 121-69-7; EINECS 204-493-5

UN No. 2253 (DOT)

Synonyms: Aniline, N,N-dimethyl; (Dimethylamino) benzene; N,N-Dimethylphenylamine; N,N-Dimethylbenzeneamine; Dimethylphenylamine; Xylidine

Empirical: $C_8H_{11}N$

Formula: $C_6H_5N(CH_3)_2$

Properties: Ylsh. to brnsh. oily liq.; sol. in alcohol, ether; insol. in water; m.w. 121.18; dens. 0.956 (20/4 C); m.p. 2-4 C; b.p. 76-78 C (10 mm); flash pt. ≈ 48 C; ref. index 1.558

Toxicology: LD50 (oral, rat) 1410 mg/kg; toxic by inhalation, skin contact, ingestion; CNS depressant; suspected carcinogen; TLV: 5 ppm in air; heated to decomp., emits highly toxic fumes of aniline and NO_x

Precaution: Photosensitive; combustible; explodes on contact with benzoyl peroxide

NFPA: Health 3, Flammability 1, Reactivity 0

Uses: Dyes, intermediates, solvent, manufacture of vanillin, stabilizer (acid acceptor), reagent; intermediate for paints

Manuf./Distrib.: Aceto; Aldrich; Allchem Ind.; BASF; BASF AG; Buffalo Color; DuPont; Fabrichem; Fluka; Hoechst Celanese; ICI Am.; Monomer-Polymer & Dajac Labs; Sigma; Spectrum Chem. Mfg.; Zeneca

Dimethylbenzene. *See* Xylene
1,2-Dimethylbenzene. *See* o-Xylene
1,3-Dimethylbenzene. *See* m-Xylene
1,4-Dimethylbenzene. *See* p-Xylene
m-Dimethylbenzene. *See* m-Xylene
o-Dimethylbenzene. *See* o-Xylene
p-Dimethylbenzene. *See* p-Xylene
N,N-Dimethylbenzeneamine. *See* n,n-Dimethylaniline
Dimethyl 1,2-benzenedicarboxylate. *See* Dimethyl phthalate
Dimethylbenzene sulfonic acid, sodium salt. *See* Sodium xylenesulfonate
Dimethyl butanedioate. *See* Dimethyl succinate
1,3-Dimethyl butanol. *See* 2-Methyl-1-pentanol
1,3-Dimethylbutyl acetate. *See* Methyl amyl acetate

Dimethyl caprylamide-capramide

Uses: Mutual solvent for polar and nonpolar ingred. in cosmetics, cleaners, corrosion inhibitor; modifier for polymer coatings

Trade names: Hallcomid® M-8-10

Dimethyl carbinol. *See* Isopropyl alcohol

1,1-Dimethylcyclohexane

CAS 590-66-9

Synonyms: Dimethyl-1,1-cyclohexane

Classification: Sat. alicyclic hydrocarbon

Empirical: C_8H_{16}

Properties: Colorless liq.; sol. in most org. solvs.; insol. in water; m.w. 122.22; sp.gr. 0.7809 (20 C); m.p. -34 C; b.p. 120 C; flash pt. 7 C

Toxicology: May cause mild skin irritation; vapor may be irritating to eyes; probably a CNS depressant; sl. nose/throat irritant; high vapor concs. may cause headache, lightheadedness, nausea, drowsiness; very high concs. may cause unconsciousness, death

Precaution: Flamm.; incompat. with strong oxidizing agents (increased fire and explosion hazard); Hazardous decomp. prods.: None

Uses: Solvent; org. synthesis

Manuf./Distrib.: Aldrich; Fluka

Dimethyl-1,1-cyclohexane. *See* 1,1-Dimethylcyclohexane

1,2-Dimethylcyclohexane

CAS 583-57-3

UN No. 2263 (DOT)

Synonyms: Hexahydro-o-xylene; o-Dimethylcyclohexane; cis-1,2-Dimethylcyclohexane

Classification: Sat. alicyclic hydrocarbon

Empirical: C_8H_{16}

Formula: $CH_3C_6H_{10}CH_3$

Properties: Colorless liq.; sol. in most org. solvs.; insol. in water; m.w. 122.22; sp.gr. 0.778 (20 C); b.p. 124 C; flash pt. 15 C
Toxicology: May cause mild skin irritation; conc. vapor may be irritating to eyes; probably a CNS depressant; sl. nose/throat irritation; high vapor concs. may cause headache, lightheadedness, nausea, drowsiness; very high concs. may cause unconsciousness, death
Precaution: Flamm.; very dangerous fire hazard exposed to heat, flame, oxidizers; incompat. with strong oxidizing agents (increased fire and explosion hazard); Hazardous decomp. prods.: Heated to decomp., emits acrid smoke and fumes
NFPA: Health 0, Reactivity 0
Uses: Solvent; org. synthesis
Manuf./Distrib.: Aldrich; Fluka

1,3-Dimethylcyclohexane
CAS 591-21-9
UN No. 2263 (DOT)
Synonyms: Hexahydroxylene; Hexahydro-m-xylene; m-Dimethylcyclohexane
Classification: Sat. alicyclic hydrocarbon
Empirical: C_8H_{16}
Formula: $CH_3C_6H_{10}CH_3$
Properties: Colorless liq.; sol. in most org. solvs.; insol. in water; m.w. 122.22; sp.gr. 0.767 (20 C); b.p. 124 C; flash pt. 10 C
Toxicology: May cause mild skin irritation; conc. vapor may be irritating to eyes; probably a CNS depressant; high vapor concs. may cause sl. nose/throat irritation, headache, lightheadedness, nausea, drowsiness; very high concs. may cause unconsciousness, death
Precaution: Flamm.; very dangerous fire hazard exposed to heat, flame, or oxidizers; incompat. with strong oxidizing agents (increased fire and explosion hazard); Hazardous decomp. prods.: Heated to decomp., emits acrid smoke and fumes
NFPA: Health 0, Flammability 3, Reactivity 0
Uses: Solvent; org. synthesis
Manuf./Distrib.: Aldrich

1,4-Dimethylcyclohexane
CAS 589-90-2
UN No. 2263 (DOT)
Synonyms: Hexahydro-p-xylene
Classification: Sat. alicyclic hydrocarbon
Empirical: C_8H_{16}
Formula: $CH_3C_6H_{10}CH_3$
Properties: Colorless liq.; sol. in most org. solvs.; insol. in water; m.w. 122.22; sp.gr. 0.783 (20 C); vapor pressure 10 mm Hg (10.2 C); m.p. -88 C; b.p. 120 C; flash pt. 11 C
Toxicology: May cause mild skin irritation; conc. vapor may be irritating to eyes; probably a CNS depressant; high vapor concs. may cause sl. nose/throat irritation, headache, nausea, drowsiness; very high concs. may cause unconsciousness, death
Precaution: Flamm.; dangerous fire hazard exposed to heat or flame; incompat. with strong oxidizing agents (increased fire and explosion hazard); Hazardous decomp. prods.: None
NFPA: Health 1, Flammability 3, Reactivity 0
Uses: Solvent; org. synthesis
Manuf./Distrib.: Aldrich

cis-1,2-Dimethylcyclohexane
CAS 2207-01-4
UN No. 2263
Classification: Sat. alicyclic hydrocarbon
Empirical: C_8H_{16}
Formula: $CH_3C_6H_{10}CH_3$
Properties: Colorless liq.; sol. in most org. solvs.; insol. in water; m.w. 122.22; sp.gr. 0.796 (20 C); m.p. -50 C; b.p. 130 C; flash pt. 12 C
Toxicology: May cause mild skin irritation; conc. vapor may be irritating to eyes; probably a CNS depressant; sl. nose/throat irritation; high vapor concs. may cause headache, lightheadedness, nausea, drowsiness; very high concs. may cause unconsciousness, death
Precaution: Flamm.; incompat. with strong oxidizing agents (increased fire and explosion hazard); Hazardous decomp. prods.: None
NFPA: Health 0, Reactivity 0
Uses: Solvent; org. synthesis
Manuf./Distrib.: Aldrich

cis-1,2-Dimethylcyclohexane. *See* 1,2-Dimethylcyclohexane

cis-1,3-Dimethylcyclohexane
CAS 638-04-0
UN No. 2263
Classification: Sat. alicyclic hydrocarbon
Empirical: C_8H_{16}
Formula: $CH_3C_6H_{10}CH_3$
Properties: Colorless liq.; sol. in most org. solvs.; insol. in water; m.w. 122.22; sp.gr. 0.784 (20 C); m.p. -76 C; b.p. 120 C; flash pt. 5 C
Toxicology: May cause mild skin irritation; conc. vapor may be irritating to eyes; probably a CNS depressant; sl. nose/throat irritant; high vapor concs. may cause headache, lightheadedness, nausea, drowsiness; very high concs. may cause unconsciousness, death
Precaution: Flamm.; incompat. with strong oxidizing agents (increased fire and explosion hazard); Hazardous decomp. prods.: None
NFPA: Health 0, Flammability 3, Reactivity 0
Uses: Solvent; org. synthesis
Manuf./Distrib.: Aldrich

cis-1,4-Dimethylcyclohexane
CAS 624-29-3; EINECS 210-840-1
UN No. 2263
Classification: Sat. alicyclic hydrocarbon
Empirical: C_8H_{16}
Formula: $CH_3C_6H_{10}CH_3$
Properties: Colorless liq.; sol. in most org. solvs.; insol. in water; m.w. 122.22; sp.gr. 0.783 (20 C); m.p. -87 C; b.p. 124 C; flash pt. 16 C
Toxicology: May cause mild skin irritation; conc. vapor can irritate eyes; probably a CNS depressant; sl. nose/throat irritant; high vapor concs. may cause headache, lightheadedness, nausea, drowsiness; very high concs. may cause unconsciousness, death
Precaution: Flamm.; incompat. with strong oxidizing agents (increased fire and explosion hazard); Hazardous decomp. prods.: None
NFPA: Health 0, Flammability 3, Reactivity 0
Uses: Solvent; org. synthesis
Manuf./Distrib.: Aldrich; Fluka

m-Dimethylcyclohexane. *See* 1,3-Dimethylcyclohexane
o-Dimethylcyclohexane. *See* 1,2-Dimethylcyclohexane

trans-1,2-Dimethylcyclohexane
CAS 6876-23-9
UN No. 2263
Classification: Sat. alicyclic hydrocarbon
Empirical: C_8H_{16}
Formula: $CH_3C_6H_{10}CH_3$
Properties: Colorless liq.; sol. in most org. solvs.; insol. in water; m.w. 122.22; sp.gr. 0.776 (20 C); m.p. -89 C; b.p. 123 C; flash pt. 6 C
Toxicology: May cause mild skin irritation; conc. vapor may be irritating to eyes; probably a CNS depressant; high vapor concs. may cause sl. nose/throat irritation, headache, lightheadedness, nausea, drowsiness; very high concs. may cause unconsciousness, death
Precaution: Flamm.; incompat. with strong oxidizing agents (increased fire and explosion hazard); Hazardous decomp. prods.: None
NFPA: Health 0, Reactivity 0
Uses: Solvent; org. synthesis
Manuf./Distrib.: Aldrich; Fluka

trans-1,3-Dimethylcyclohexane
CAS 2207-03-6
UN No. 2263
Classification: Sat. alicyclic hydrocarbon
Empirical: C_8H_{16}
Formula: $CH_3C_6H_{10}CH_3$
Properties: Colorless liq.; sol. in most org. solvs.; insol. in water; m.w. 122.22; sp.gr. 0.785 (20 C); m.p. -90 C; b.p. 124 C
Toxicology: May cause mild skin irritation; conc. vapor may be irritating to eyes; probably a CNS depressant; high vapor concs. may cause sl. nose/throat irritation, headache, lightheadedness, nausea, drowsiness; very high concs. may cause unconsciousness, death
Precaution: Flamm.; incompat. with strong oxidizing agents (increased fire and explosion hazard); Hazardous decomp. prods.: None
NFPA: Health 0, Flammability 3, Reactivity 0
Uses: Solvent; org. synthesis
Manuf./Distrib.: Aldrich

trans-1,4-Dimethylcyclohexane
CAS 2207-04-7
UN No. 2263
Classification: Sat. alicyclic hydrocarbon
Empirical: C_8H_{16}
Formula: $CH_3C_6H_{10}CH_3$
Properties: Colorless liq.; sol. in most org. solvs.; insol. in water; m.w. 122.22; sp.gr. 0.766 (20 C); m.p. -37 C; b.p. 119 C; flash pt. 11 C
Toxicology: May cause mild skin irritation; conc. vapor can irritate eyes; probably a CNS depressant; high vapor concs. may cause sl. nose/throat irritation, headache, lightheadedness, nausea, drowsiness; very high concs. may cause unconsciousness, death
Precaution: Flamm.; incompat. with strong oxidizing agents (increased fire and explosion hazard); Hazardous decomp. prods.: None
NFPA: Health 0, Flammability 3, Reactivity 0
Uses: Solvent; org. synthesis
Manuf./Distrib.: Fluka

1,1-Dimethylcyclopentane
CAS 1638-26-2
Classification: Sat. alicyclic hydrocarbon
Empirical: C_7H_{14}
Formula: $(CH_3)_2C_5H_8$
Properties: Colorless liq.; sol. in most org. solvs.; insol. in water; m.w. 98.19; sp.gr. 0.749 (20 C); m.p. -70 C; b.p. 88 C; flash pt. < 21 C
Toxicology: May cause mild skin irritation; conc. vapor may be irritating to eyes; probably a CNS depressant; high vapor concs. may cause nose/throat irritation, headache, lightheadedness, nausea, dizziness; very high concs. may cause unconsciousness, death
Precaution: Flamm.; incompat. with strong oxidizing agents (increased fire and explosion hazard); Hazardous decomp. prods.: None
Uses: Solvent; org. synthesis

1,2-Dimethylcyclopentane
CAS 822-50-4; 1192-18-3; 2452-99-5
Classification: Sat. alicyclic hydrocarbon
Empirical: C_7H_{14}
Formula: $CH_3C_5H_8CH_3$
Properties: Colorless liq.; sol. in most org. solvs.; insol. in water; m.w. 98.19; sp.gr. 0.744 (20 C); m.p. -62 C (cis), -120 C (trans); b.p. 99 C (cis), 92 C (trans); flash pt. < 21 C
Toxicology: May cause mild skin irritation; conc. vapor may be irritating to eyes; probably a CNS depressant; high vapor concs. may cause nose/throat irritation, headache, lightheadedness, nausea, dizziness; very high concs. may cause unconsciousness, death
Precaution: Flamm.; incompat. with strong oxidizing agents (increased fire and explosion hazard); Hazardous decomp. prods.: None
Uses: Solvent; org. synthesis

1,3-Dimethylcyclopentane
CAS 2453-00-1; 2532-58-3
Synonyms: Dimethyl-1,3 cyclopentane
Classification: Sat. alicyclic hydrocarbon
Empirical: C_7H_{14}
Formula: $(CH_3)_2C_5H_8$
Properties: Colorless liq.; sol. in most org. solvs.; insol. in water; m.w. 98.19; sp.gr. 0.744 (20 C); m.p. -134 C; b.p. 91-92 C; flash pt. < 21 C
Toxicology: May cause mild skin irritation; conc. vapor may be irritating to eyes; probably a CNS depressant; high vapor concs. may cause nose/throat irritation, headache, lightheadedness, nausea, dizziness; very high concs. may cause unconsciousness, death
Precaution: Flamm.; incompat. with strong oxidizing agents (increased fire and explosion hazard); Hazardous decomp. prods.: None
Uses: Solvent; org. synthesis

Dimethyl-1,3 cyclopentane. *See* 1,3-Dimethylcyclopentane
2,5-Di-o-methyl-1,4:3,6-dianhydro-D-glucitol. *See* Dimethyl isosorbide
Dimethyldiglycol. *See* Diethylene glycol dimethyl ether
1,4-Dimethyl-3,6-dioxa-1-heptanol. *See* PPG-2 methyl ether
Dimethyl disulfide. *See* Methyl disulfide
Dimethylenediamine. *See* Ethylenediamine
1,1-Dimethylethanol. *See* t-Butyl alcohol
Dimethyl ethyl carbinol. *See* t-Amyl alcohol
(1,1-Dimethylethyl) cyclohexane. *See* t-Butylcyclohexane

Dimethyl ethynyl carbinol. *See* Methyl butynol
Dimethyl formaldehyde. *See* Acetone

Dimethyl formamide
CAS 68-12-2; EINECS 200-679-5
UN No. 2265 (DOT)
Synonyms: DMF; DMFA; N,N-Dimethylformamide; N-Formyldimethylamine
Classification: Amide
Empirical: C_3H_7NO
Formula: $CHCON(CH_3)_2$
Properties: Water-wh. liq.; fishy unpleasant odor; a dipolar aprotic solv.; sol. in water; misc. with alcohol, ether, acetone, benzene, most org. solvs.; m.w. 73.10; sp.gr. 0.953-0.954 (15.6/15.6 C); f.p. -61 C; b.p. 152.8 C; flash pt. 57.7 C
Toxicology: ACGIH TLV/TWA 10 ppm (skin); LD50 (oral, rat) 2800 mg/kg; toxic by skin absorption; strong irritant to skin and tissue; vapor can cause mild eye irritation, liq. may cause temporary pain and reddening
Precaution: Combustible; moderate fire risk; incompat. with strong oxidizers, halogen gases, halogenated compds., 2,4,6-trichloro-1,3,5-triazine; vigorous reactions possible; polymerizes violently on contact with methylene bisphenyl diisocyanate; Hazardous decomp. prods.: Emits toxic fumes of NO_x
NFPA: Health 1, Flammability 2, Reactivity 0
Uses: Solvent for liqs. and gases, in vinyl resins and acetylene, butadiene, acid gases, paints; mfg. of polyacrylic fibers, butadiene, pharmaceuticals, dyes, petrol. prods.; catalyst in carboxylation reactions; organic synthesis; carrier for gases; reagent
Manuf./Distrib.: Aceto; Air Prods.; Aldrich; Allchem Ind.; Ashland; J.T. Baker; BASF; Baychem; Brown; Browning; Chemcentral; Coyne; DuPont; Fluka; Great Western; Harcros USA; Houghton Chem.; ICI Spec. Chems.; Mallinckrodt; Mitsubishi Gas; Monomer-Polymer & Dajac Labs; Nissan Chem. Ind.; Sigma; Spectrum Chem. Mfg.; UCB SA; Van Waters & Rogers
Trade names: Dynasolve 100

N,N-Dimethylformamide. *See* Dimethyl formamide

Dimethyl glutarate
CAS 1119-40-0; EINECS 214-277-2
Synonyms: Dimethyl pentanedioate
Empirical: $C_7H_{12}O_4$
Formula: $CH_3OCO(CH_2)_3COOCH_3$
Properties: M.w. 160.17; dens. 1.087 (20/4 C); b.p. 96-103 C; flash pt. 97 C; ref. index 1.424 (20 C)
Uses: Solvent for coatings, cleaners, inks, textile lubricants, urethane prod.; plasticizer for flexible thermoset polyester; polymer intermediate for polyester polyols for urethanes, wet-str. paper resins, polyester resins; specialty chemical intermediate
Manuf./Distrib.: Aldrich; Ashland; Fluka; Sigma
Trade names: DBE-5; Santosol™ DMG
Trade names containing: DBE; DBE-2, -2SPG; DBE-3; DBE-9; Santosol™ DME; Santosol™ DME-2; Santosol™ DME-3

Dimethylglycol. *See* Ethylene glycol dimethyl ether
2,6-Dimethyl-4-heptanone. *See* Diisobutyl ketone
2,6-Dimethylheptan-4-one. *See* Diisobutyl ketone
Dimethyl hexanedioate. *See* Dimethyl adipate
Dimethylhydroxybenzene. *See* Xylenol

Dimethyl isosorbide
CAS 5306-85-4; EINECS 226-159-8
Synonyms: DMI; 2,5-Di-o-methyl-1,4:3,6-dianhydro-D-glucitol; D-Glucitol, 1,4:3,6-dianhydro-2,5-di-o-methyl; Isosorbide dimethyl ether
Definition: Dimethyl ether of an anhydride of an isomer of sorbitol
Empirical: $C_8H_{14}O_4$
Properties: Wh. to ylsh. visc. liq.; m.w. 174.20; dens. 1.150; b.p. 93-95 C (0.1 mm); flash pt. 108 C; ref. index 1.4610
Uses: Carrier, solvent for cosmetic and pharmaceutical sol'ns.
Manuf./Distrib.: Aldrich; Fluka; Sigma
Trade names: Arlasolve® DMI

Dimethylketal. *See* Acetone
Dimethylketone. *See* Acetone
Dimethylmethane. *See* Propane
6,6-Dimethyl-2-methylenebicyclo[3.1.1]heptane. *See* β-Pinene
2,2-Dimethyl-3-methylenenorbornane. *See* Camphene
3,3-Dimethyl-2-methylene norcamphane. *See* Camphene

Dimethyl methylphosphonate
CAS 756-79-6; EINECS 212-052-3
Synonyms: DMMP; Methylphosphonic acid, dimethyl ester; Phosphonic acid, methyl-, dimethyl ester
Empirical: $C_3H_9O_3P$
Formula: $CH_3P(O)(OCH_3)_2$
Properties: M.w. 124.09; dens. 1.145; f.p. 68 C; b.p. 181 C
Toxicology: LD50 (oral, mouse) > 6810 mg/kg, (IV, rat) 1050 mg/kg; mutagenic data; suspected carcinogen; experimental reproductive effects; experimental nerve gas stimulant; Hazardous decomp. prods.: Heated to decomp., emits toxic fumes of PO_x
Uses: Flame retardant for rigid urethane foams, unsat. polyester resins, water-borne latexes; as solvent
Manuf./Distrib.: Aceto; Akzo Nobel; Albright & Wilson Am.; Aldrich; Fluka; Focus
Trade names: Antiblaze® DMMP

Dimethylnitromethane. *See* 2-Nitropropane

Dimethyl oleamide
Uses: Solubilizer, solvent, dispersant, wetting agent; modifier for coatings
Trade names: Hallcomid® M-18-OL

2,2-Dimethylpentane. *See* Neoheptane
Dimethyl-2,2-pentane. *See* Neoheptane

2,3-Dimethylpentane
CAS 565-59-3
UN No. 1206
Classification: Sat. aliphatic hydrocarbon
Empirical: C_7H_{16}
Formula: $CH_3CH(CH_3)CH(CH_3)CH_2CH_3$
Properties: Colorless liq.; gasoline-like odor; sol. in most org. solvs.; insol. in water; m.w. 100.2; sp.gr. 0.699 (20 C); m.p. -135 C; b.p. 90 C; flash pt. (CC) -6 C
Toxicology: ACGIH TLV/TWA 400 ppm; STEL 500 ppm; may cause mild skin irritation; conc. vapor may cause sl. eye irritation; conc. vapor may cause sl. nose/throat irritation, lightheadedness, dizziness, nausea, difficulty breathing
Precaution: Flamm.; very dangerous fire hazard epxoed to heat, flame, oxidizers;

incompat. with strong oxidizing agents (increases fire and explosion hazard); Hazardous decomp. prods.: Heated to decomp., emits acrid smoke and fumes
NFPA: Health 0, Flammability 3, Reactivity 0
Uses: Ingred. in gasoline, aviation fuels; solvent; org. synthesis
Manuf./Distrib.: Aldrich

2,4-Dimethylpentane
CAS 108-08-7
UN No. 1206
Classification: Sat. aliphatic hydrocarbon
Empirical: C_7H_{16}
Formula: $(CH_3)_2CHCH_2CH(CH_3)_2$
Properties: Colorless liq.; gasoline-like odor; sol. in most org. solvs.; insol. in water; m.w. 100.2; sp.gr. 0.677 (20 C); m.p. -119 C; b.p. 81 C; flash pt. (CC) -12 C
Toxicology: ACGIH TLV/TWA 400 ppm; STEL 500 ppm; may cause mild skin irritation; conc. vapor may cause sl. eye irritation; conc. vapor may cause sl. nose/throat irritation, lightheadedness, dizziness, nausea, difficulty breathing
Precaution: Flamm.; incompat. with strong oxidizing agents (increases fire and explosion hazard); Hazardous decomp. prods.: None
NFPA: Health 0, Flammability 3, Reactivity 0
Uses: Ingred. in gasoline, aviation fuels; solvent; org. synthesis
Manuf./Distrib.: Aldrich; Fluka

3,3-Dimethylpentane
CAS 562-49-2
UN No. 1206
Classification: Sat. aliphatic hydrocarbon
Empirical: C_7H_{16}
Formula: $CH_3CH_2C(CH_3)_2CH_2CH_3$
Properties: Colorless liq.; gasoline-like odor; sol. in most org. solvs.; insol. in water; m.w. 100.2; sp.gr. 0.693 (20 C); m.p. -135 C; b.p. 86 C; flash pt. (CC) -6 C
Toxicology: ACGIH TLV/TWA 400 ppm; STEL 500 ppm; may cause mild skin irritation; conc. vapor may cause sl. eye irritation; conc. vapor may cause sl. nose/throat irritation, lightheadedness, dizziness, nausea, difficulty breathing
Precaution: Flamm.; incompat. with strong oxidizing agents (increases fire and explosion hazard); Hazardous decomp. prods.: None
NFPA: Health 0, Flammability 2, Reactivity 0
Uses: Ingred. in gasoline, aviation fuels; solvent; org. synthesis
Manuf./Distrib.: Aldrich; Fluka

Dimethyl pentanedioate. *See* Dimethyl glutarate
Dimethylphenol. *See* Xylenol
Dimethylphenylamine. *See* n,n-Dimethylaniline
N,N-Dimethylphenylamine. *See* n,n-Dimethylaniline
4-(Dimethylphenylmethyl) phenol. *See* Cumyl phenol

Dimethyl phthalate
CAS 131-11-3; EINECS 205-011-6
Synonyms: DMP; Dimethyl 1,2-benzenedicarboxylate; Methyl phthalate; 1,2-Benzenedicarboxylic acid dimethyl ester; Phthalic acid dimethyl ester
Definition: Diester of methyl alcohol and phthalic acid
Empirical: $C_{10}H_{10}O_4$
Formula: $C_6H_4(CO_2CH_3)_2$
Properties: Colorless oily liq., sl. aromatic odor; misc. with alcohol, ether; insol. in

water, paraffinic hydrocarbons; sl. sol. in min. oil; m.w. 194.08; dens. 1.189 (25/25 C); visc. 17.2 cps; m.p. 5.5 C; b.p. 282.0 C; flash pt. 149 C; ref. index 1.5138 (25 C)

Toxicology: LD50 (oral, rat) 6.9 ml/kg; TLV 5 mg/m^3 of air (TWA); moderately toxic by ingestion and intraperitoneal routes; mildly toxic by inhalation; heated to decomp., emits acrid smoke and irritating fumes

Precaution: Combustible

NFPA: Health 0, Flammability 1, Reactivity 0

Uses: Solvent for resin, plasticizer for cellulose acetate and nitrocellulose lacquers; plastics, rubber; coating agents; safety glass

Regulatory: FDA 21CFR §175.105, 177.1010, 177.2420; SARA reportable

Manuf./Distrib.: Aldrich; Allan; Allchem Ind.; Ashland; BASF; BP Chems. Inc; Daihachi Chem. Ind.; Eastman; Fluka; Great Western; Houghton Chem.; Hüls Am.; Morflex; Spectrum Chem. Mfg.; UCB SA; Unitex

Trade names: Kodaflex® DMP; Uniplex 110

Dimethylpolysiloxane. *See* Dimethicone

2,2-Dimethylpropanol

CAS 75-84-3

Synonyms: Neopentyl alcohol; t-Butyl carbinol; Neoamyl alcohol

Classification: Primary aliphatic alcohol

Empirical: $C_5H_{12}O$

Formula: $CH_3C(CH_3)_2CH_2OH$

Properties: Volatile cryst.; peppermint odor; sol. in ethanol, ether; sl. sol. in water; m.w. 88.15; sp.gr. 0.812 (20 C); m.p. 53 C; b.p. 114 C; flash pt. (CC) 37 C

Toxicology: toxic; may cause severe eye irritation, sl. skin irritation; can be absorbed thru skin; inh. may cause nose/throat irritation, headache, dizziness, dyspnea, nausea, vomiting

Precaution: Flamm. solid; incompat. with oxidizing agents (increases risk of fire/explosion), hydrogen trisulfide (explosive reaction possible)

NFPA: Health 2, Flammability 3, Reactivity 0

Uses: Mfg. of flotation agents, pharmaceuticals, lacquers, rubber, plastics, fruit essences, explosives; component of paint strippers, hydraulic fluids; solvent

Manuf./Distrib.: Aldrich; Fluka

1,1-Dimethylpropargyl alcohol. *See* Methyl butynol

α,α-Dimethylpropargyl alcohol. *See* Methyl butynol

Dimethylpropylmethane. *See* Isohexane

Dimethyl sebacate

CAS 106-79-6; EINECS 203-431-4

Synonyms: Decanedioic acid, dimethyl ester

Formula: $[(CH_2)_4COOCH_3]_2$

Properties: Water-white liq.; m.w. 230.29; dens. 0.9896 (20/20 C); m.p. 24.5 C; b.p. ≈ 294 C; flash pt. 145 C

Precaution: Combustible

Uses: Solvent or plasticizer for nitrocellulose, vinyl resins; intermediate

Manuf./Distrib.: Aceto; Aldrich; Fluka; Henkel; Hüls AG; Penta Mfg.; Richman; Sigma; Union Camp; Unitex

Dimethyl silicone. *See* Dimethicone

Dimethyl succinate

CAS 106-65-0; EINECS 203-419-9

Synonyms: Dimethyl butanedioate; Methyl succinate

Empirical: $C_6H_{10}O_4$

Formula: $CH_3OCOCH_2CH_2COOCH_3$

Properties: Colorless liq., ethereal winey odor; m.w. 146.14; dens. 1.119 (20/4 C);

m.p. 16-19 C; b.p. 190-193 C; flash pt. 90 C; ref. index 1.419

Uses: Light and heat stabilizer for polyolefins, ABS polymer systems, flexible PVC, food pkg.; syn. flavoring agent for foods and pharmaceuticals; solvent for coatings, cleaners, inks, textile lubricants, urethane prod.; plasticizer for flexible thermoset polyester; polymer intermediate for polyester polyols for urethanes, wet-str. paper resins, polyester resins; specialty chemical intermediate

Regulatory: FDA 21CFR §172.515; FEMA GRAS

Manuf./Distrib.: Aldrich; Ashland; Chemie Linz N. Am.; DuPont; Fluka; Penta Mfg.; Schweizerhall; Sigma

Trade names: DBE-4; Santosol™ DMS

Trade names containing: DBE; DBE-2, -2SPG; DBE-3; DBE-9; Santosol™ DME; Santosol™ DME-2; Santosol™ DME-3

Dimethyl sulfoxide

CAS 67-68-5; EINECS 200-664-3

Synonyms: DMSO; Methyl sulfoxide; Sulfinylbis (methane); Methyl sulfinylmethane

Classification: Aliphatic organic compd.

Empirical: C_2H_6OS

Formula: $(CH_3)_2SO$

Properties: Colorless clear liq., strong char. odor; hygroscopic; sol. in water; pract. insol. in alcohol, benzene, acetone, chloroform, ether; m.w. 78.13; dens. 1.101; vapor pressure 0.37 mm Hg (20 C); m.p. 18.4 C; b.p. 189 C; flash pt. 95 C; ref. index 1.4790 (20 C)

Toxicology: LD50 (oral, rat) 17,500 mg/kg, (IP, rat) 8200 mg/kg; poison by ing.; mod. toxic by IV, IP routes; human mutagenic data

Precaution: Combustible exposed to heat or flame; can react with oxidizing materials; violent or explosive reaction with many acyl, aryl, and nonmetal halides, boron compds., etc.; incompat. with metal oxosalts, sulfur trioxide, magnesium perchlorate; Hazardous decomp. prods.: Heated to decomp., emits toxic fumes of SO_x

NFPA: Health 1, Flammability 1, Reactivity 0

Uses: Solvent for polymerization; analytical reagent; industrial cleaners, pesticides, paint stripping, hydraulic fluids, medicine (anti-inflammatory), veterinary medicine, plant pathology and nutrition, spinning syn. fibers, acrylic fiber; organic synthesis and solvents; carrier, solubilizer, solvent for pharmaceuticals

Regulatory: BP, Ph.Eur. compliance

Manuf./Distrib.: Aldrich; Allchem Ind.; Elf Atochem N. Am.; Fluka; Gaylord; Howard Hall; Itochu Spec.; Monomer-Polymer & Dajac Labs; Research Organics; Sigma; Spectrum Chem. Mfg.; Toray Fine Chems.

Trade names: Decap; DMSO

Dimethyl thiodipropionate

CAS 4131-74-2

Formula: $S(CH_2CH_2COOCH_3)_2$

Properties: m.w. 206; b.p. 130 C (1-2 mm)

Uses: Reaction intermediate in synthesis of thiodipropionic acid and esters; specialty solvent

Manuf./Distrib.: Aldrich; Hampshire; Witco/Polymer Addit.

Dinonyl phenol

CAS 1323-65-5; EINECS 215-356-4

Definition: Mixture of dialkyl substituted phenols

Formula: $(C_9H_{19})_2C_6H_3OH$

Properties: Colorless liq.; insol. in water; sol. in org. solvs.

Dioctyl adipate

Uses: Solvent; intermediate for mfg. nonionic surfactants, lube oil additives, petrol. demulsifiers, phenolic resins, plasticizers, and insecticides
Regulatory: FDA 21CFR §175.105
Manuf./Distrib.: Allchem Ind.; Huntsman; Texaco

Dioctyl adipate
CAS 103-23-1; EINECS 203-090-1
Synonyms: DOA; Bis(2-ethylhexyl) hexanedioate; Adipic acid bis (2-ethylhexyl) ester; Di(2-ethylhexyl) adipate; Hexanedioic acid, bis (2-ethylhexyl) ester
Classification: Aliphatic organic compd.
Definition: Diester of 2-ethylhexyl alcohol and adipic acid
Empirical: $C_{22}H_{42}O_4$
Formula: $[CH_2CH_2COOCH_2CH(C_2H_5)C_4H_9]_2$
Properties: Lt.-colored oily liq.; insol. in water; m.w. 370.64; dens. 0.9268 (20/20 C); b.p. 417 C; flash pt. 196 C; ref. index 1.4472
Toxicology: LD50 (oral, rat) 9110 mg/kg); suspected carcinogen and teratogen; moderately toxic by IV route; mildly toxic by ingestion and skin contact; mutagenic data; eye and skin irritant; on decomp., emits acrid smoke and irritating fumes
Precaution: Combustible
Uses: Plasticizer, commonly blended with general purpose plasticizers (DOP, DIOP); solvent, aircraft lubricants; paints, as vehicles for pigment dispersions; emollient, moisturizer, pigment dispersant, cosolvent in pharmaceuticals
Regulatory: FDA 21CFR §175.105, 175.300, 177.1200, 177.1210, 177.1400, 177.2600, 178.3740; FDA approved for injectables
Manuf./Distrib.: Aldrich; Allchem Ind.; Ashland; BASF; Chisso Am.; Coyne; Eastman; Esprit; Fluka; Houghton Chem.; Hüls AG; Inolex; Monsanto; Sigma
Trade names: Wickenol® 158

Dioctyl maleate
CAS 142-16-5; 2915-53-9; EINECS 205-524-5
Synonyms: DOM; Bis (2-ethylhexyl) maleate; Di-N-octyl maleate; Di-(2-ethylhexyl) maleate
Definition: Diester of 2-ethylhexyl alcohol and maleic acid
Empirical: $C_{20}H_{38}O_5$
Formula: $C_8H_{17}OCOCH{=}CHCOOC_8H_{17}$
Properties: Liq., char. mild odor; m.w. 358.52; sp.gr. 0.960-0.970; m.p. -50 to -85 C; flash pt. (COC) 182 C; ref. index 1.4480-1.4500
Toxicology: LD50 (oral, rat) 14,200 mg/kg; mildly toxic by ingestion; heated to decomp., emits acrid smoke and fumes
Uses: Emollient, fragrance coupler, solubilizer for benzophenone-3, antitackifier in antiperspirants, conditioner for hair prods.; in copolymerization of PVC and vinyl acetates; plasticizer for vinyl resins; also in latex paints; cleanser and wax solvent
Manuf./Distrib.: Allchem Ind.; Aristech; BASF; Finetex; Hoechst AG; Hüls; Monomer-Polymer & Dajac Labs; Unitex
Trade names: Bernel® Ester DOM

Di-N-octyl maleate. *See* Dioctyl maleate

Dioctyl phthalate
CAS 117-81-7; 117-84-0; EINECS 204-211-0
Synonyms: DOP; DEHP; Di(2-ethylhexyl) phthalate; Bis (2-ethylhexyl) phthalate; Di-s-octyl phthalate; 1,2-Benzenedicarboxylic acid dioctyl ester
Definition: Diester of 2-ethylhexyl alcohol and phthalic acid
Empirical: $C_{24}H_{38}O_4$

Formula: $C_6H_4[COOCH_2CH(C_2H_5)C_4H_9]_2$
Properties: Lt.-colored liq., odorless, bitter taste; insol. in water; misc. with min. oil; m.w. 390.62; dens. 0.9861 (20/20 C); b.p. 231 C (5 mm); flash pt. 218 C; ref. index 1.4836
Toxicology: TLV 5 mg/m^3; STEL 10 mg/m^3; LD50 (oral, rat) 30,600 mg/kg; poison by IV; mildly toxic by ingestion; skin and severe eye irritant; suspected human carcinogen; experimental teratogen; affects human GI tract; heated to decomp., emits acrid smoke
Precaution: Combustible
NFPA: Health 0, Flammability 1, Reactivity 0
Uses: Plasticizer for many resins and elastomers, in food pkg.; solvent, fixative in pharmaceuticals
Regulatory: FDA 21CFR §175.105, 175.300, 175.310, 715.380, 175.390, 176.170, 176.210, 176.1210, 177.1010, 177.1200, 177.1210, 177.1400, 177.2600, 178.3120, 178.3910, 181.22, 181.27; FDA approved for ophthalmics, injectables; BP compliance
Manuf./Distrib.: Aldrich; Allchem Ind.; Aristech; BASF; Chemisphere Ltd; Chisso Am.; Coyne; Daihachi Chem. Ind.; Eastman; Fluka; Great Western; C.P. Hall; Hoechst AG; Houghton Chem.; Hüls AG; Mitsubishi Gas; Sigma; Spectrum Chem. Mfg.; UCB SA
Trade names containing: Kodaflex® HS-4

Di-s-octyl phthalate. *See* Dioctyl phthalate

Dioctyl sebacate
CAS 122-62-3; 2432-87-3; EINECS 204-558-8
Synonyms: DOS; Di-2-ethylhexyl sebacate; Decanedioic acid, bis(2-ethylhexyl) ester; Bis (2-ethylhexyl) decanedioate; 2-Ethylhexyl sebacate; Octyl sebacate
Empirical: $C_{26}H_{50}O_4$
Formula: $(CH_2)_6(CH_2COOC_8H_{17})_2$
Properties: Pale straw-colored oily liq.; mild odor; insol. in water; m.w. 426.68; dens. 0.91 (25 C); f.p. -55 C; b.p. 248 C (4 mm); flash pt. (COC) 210 C; ref. index 1.447 (28 C)
Toxicology: LD50 (oral, rat) 12,800 mg/kg, (IV, rat) 900 mg/kg; mod. toxic by ing. and IV routes
Precaution: Combustible exposed to heat or flame; can react with oxidizers; Hazardous decomp. prods.: Heated to decomp., emits acrid and irritating fumes
Uses: Plasticizer for PVC, PS, cellulosics; lubricating oil; solvent; grease
Regulatory: FDA 21CFR §175.105, 175.300, 177.1210, 177.2600, 178.3910
Manuf./Distrib.: Aldrich; Allchem Ind.; Ashland; Fluka; C.P. Hall; Harwick; Hatco; New Japan Chem.; Novachem Corp.; Penta Mfg.; Sigma; Union Camp; Unitex; Velsicol

1,4-Dioxacyclohexane. *See* 1,4-Dioxane
1,3-Dioxacyclopentane. *See* 1,3-Dioxolane
3,6-Dioxadodecanol-1. *See* Diethylene glycol n-hexyl ether
3,6-Dioxa-1-dodecanol. *See* Diethylene glycol n-hexyl ether
Dioxan. *See* 1,4-Dioxane
1,4-Dioxan. *See* 1,4-Dioxane

1,4-Dioxane
CAS 123-91-1; EINECS 204-661-8
UN No. 1165
Synonyms: Diethylene ether; p-Dioxane; Dioxan; 1,4-Dioxan; Dioxyethylene ether; 1,4-Diethylene dioxide; Diethylene dioxide; Tetrahydro-1,4-dioxin;

p-Dioxane

Tetrahydro-p-dioxin; Diethylene oxide; 1,4-Dioxacyclohexane; Glycol ethylene ether

Classification: Sat. cyclic aliphatic ether

Empirical: $C_4H_8O_2$

Formula: $OCH_2CH_2OCH_2CH_2$

Properties: Colorless liq., ethereal odor; stable; sol. in water, most org. solvs.; m.w. 88.11; sp.gr. 1.0356 (20/20 C); f.p. 10-12 C; b.p. 101.3 C; flash pt. 18.3 C; ref. index 1.4220

Toxicology: ACGIH TLV/TWA 25 ppm (skin); OSHA 100 ppm min. air; LD50 (oral, mouse) 5700 mg/kg, (skin, rabbit) 7600 mg/kg; toxic by inhalation, absorbed by skin; probable eye irritant; inh. may cause nose/throat irritation, tearing, and burning

Precaution: DOT: Flamm. liq.; dangerous fire risk; stable if properly inhibited; may form explosive peroxides on exposure to air/moisture; incompat. with strong oxidizers, catalytic hydrogen, sulfur trioxide, triethynylaluminum; Hazardous decomp. prods.: May form peroxides on prolonged contact with air in the absence of inhibitors

NFPA: Health 2, Flammability 3, Reactivity 1

Uses: Stabilizer for chlorinated hydrocarbons; solvent for adhesives, dyes, cellulose, lacquer, paints, wax, pharmaceuticals, coatings, natural and syn. rubbers, dry-cleaning, metal surf. finishes, chem. reaction and extraction, transistors; dyeing, staining, and printing adjuvant; cleaning and detergent preps.; scintillation fluids

Regulatory: SARA reportable; BP compliance

Manuf./Distrib.: AC Ind.; Aldrich; Alemark; Allchem Ind.; Amber Syn.; Ashland; J.T. Baker; BASF; CPS; Ferro/Grant; Fluka; Houghton Chem.; Mallinckrodt; Osaka Org. Chem. Ind.; Sigma; Spectrum Chem. Mfg.; Toho Chem. Ind.; Union Carbide; Van Waters & Rogers

p-Dioxane. *See* 1,4-Dioxane

3,6-Dioxaoctane-1,8-diol. *See* Triethylene glycol

4,7-Dioxaundecan-2-ol. *See* 1-Butoxyethoxy-2-propanol

1,3-Dioxolan. *See* 1,3-Dioxolane

1,3-Dioxolane

CAS 646-06-0

UN No. 1166

Synonyms: Ethylene glycol formal; 1,3-Dioxacyclopentane; 1,3-Dioxolan; Formal glycol; Dihydro-1,3-dioxole; Glycoformal; Glycol formal

Classification: Sat. cyclic aliphatic ether

Empirical: $C_3H_6O_2$

Formula: $-CH_2OCH_2OCH_2-$ (5-membered ring)

Properties: Water-wh. liq.; ether-like odor; sol. in water, ethanol, ether, acetone, other org. solvs.; m.w. 74.09; m.p. -95 C; b.p. 74 C; flash pt. 2 C

Toxicology: LD50 (oral, rat) 3000 mg/kg, (skin, rabbit) 8480 mg/kg; mod. toxic by ing. and IP; mildly to mod. toxic by inh.; probable mild skin irritant; may be absorbed thru skin; vapor may be very irritating to eyes

Precaution: Flamm.; very dangerous fire hazard exposed to heat or flame; incompat. with acids (may form peroxides), oxidizers (may react vigorously and increase risk of fire/explosion); potentially explosive reaction with lithium perchlorate; Hazardous decomp. prods.: May form peroxides on prolonged exposure to air, in the absence of inhibitors; heated to decomp., emits acrid smoke and irritating fumes

NFPA: Health 2, Flammability 3, Reactivity 2

Uses: Low-boiling solvent and extractant for oils, waxes, dyes, cellulose derivs.;

crosslinking agent for phenolic novolac resins; in lithium batteries
Manuf./Distrib.: Aldrich; Amber Syn.; Fluka

1,3-Dioxolan-2-one. *See* Ethylene carbonate
1,3-Dioxolan-2-one, 4-methyl. *See* Propylene carbonate
Dioxolone-2. *See* Ethylene carbonate
Dioxyethylene ether. *See* 1,4-Dioxane
Dipentene. *See* d-Limonene; dl-Limonene; l-Limonene
Dipentylamine. *See* Di-n-amylamine
Di-n-pentylamine. *See* Di-n-amylamine
Dipropylene carbonate. *See* Propylene carbonate

Dipropylene glycol
CAS 110-98-5; 25265-71-8; EINECS 203-821-4; 246-770-3
Synonyms: DPG; Di-1,2-propylene glycol; Bis (hydroxypropyl) ether; 2,2´-Dihydroxydipropyl ether; 1,1´-Oxybis-2-propanol; 1,1´-Oxydi-2-propanol; 2,2´-Dihydroxyisopropyl ether; Methyl-2(methyl-2) oxybispropanol
Classification: Mixture of diols
Empirical: $C_6H_{14}O_3$
Formula: $CH_3CHOHCH_2OCH_2CHOHCH_3$
Properties: Colorless, sl. visc. liq., nearly odorless; hygroscopic; sol. in toluene, water; m.w. 134.18; dens. 1.023; b.p. 233 C; flash pt. (OC) 280 F; ref. index 1.4410
Toxicology: LD50 (oral, rat) 14,850 mg/kg; mildly toxic by ingestion; skin and eye irritant; mutagenic data; heated to decomp., emits acrid smoke and irritating fumes
Precaution: Combustible when exposed to heat and flame; can react vigorously with oxidizing materials
Uses: Solvent in hydraulic brake fluids, cutting oils, textile lubricants, inks; polyester and alkyd resins, reinforced plastics, plasticizers, solvents, fuel additives, in paints, cosmetics
Regulatory: FDA 21CFR §175.105, 176.170, 176.200, 178.3910
Manuf./Distrib.: Aldrich; Allchem Ind.; ARCO; Ashland; Berje; Brown; Chemcentral; Coyne; Fluka; Great Western; C.P. Hall; Houghton Chem.; Olin; PMC Spec.; Samson; Stanchem; Sunnyside; Texaco; Union Carbide; Van Waters & Rogers

Di-1,2-propylene glycol. *See* Dipropylene glycol

Dipropylene glycol butyl ether
CAS 29911-28-2
Synonyms: 2-Propanol, 1-(2-butoxy-1-methylethoxy)-; Butyl dipropasol solvent
Empirical: $C_{10}H_{22}O_3$
Properties: M.w. 190.32
Toxicology: LDLo (oral, rat) 2000 mg/kg; mod. toxic by ingestion; primary irritant; Hazardous decomp. prods.: Heated to decomp., emits acrid smoke and irritating fumes
Uses: Slow evaporating solvent with good solvency for coating resins, cleaners, cosmetics, agric., electronics, ink, textile and adhesives prods.
Manuf./Distrib.: Aldrich
Trade names: Arcosolv® DPNB; Dowanol® DPnB

Dipropylene glycol t-butyl ether
CAS 132739-31-2
Uses: Slow evaporating solvent with good solvency for coating resins
Manuf./Distrib.: Aldrich
Trade names: Arcosolv® DPTB

Dipropylene glycol dimethyl ether
CAS 111109-77-4; EINECS 404-640-5
Synonyms: Bis(methoxypropyl) ether
Empirical: $C_8H_{18}O_3$
Formula: $CH_3OCH_2CH(CH_3)OCH_2CH(CH_3)OCH_3$
Properties: Colorless liq., low odor; sol. 53 g/100 g water; m.w. 162.23; sp.gr. 0.898; f.p. < -71 C; b.p. 175 C; flash pt. 65 C; ref. index 1.407
Uses: Solvent used in coatings incl. PU coatings, agric. formulations, resin prod., cleaning prods.
Manuf./Distrib.: Dow; Fluka
Trade names: Proglyde DMM

Dipropylene glycol ethyl ether
CAS 15764-24-6
Synonyms: DPGEE; Dipropylene glycol monoethyl ether; 1-2-(Ethoxy-2-methylethoxy)-2-propanol
Classification: Aliphatic ether alcohol
Empirical: $C_8H_{18}O_3$
Formula: $CH_3CH(OH)CH_2OCH_2CH(CH_3)OCH_2CH_3$
Properties: Colorless liq.; sol. in water, variety of org. solvs.; m.w. 162.26; sp.gr. 0.927; b.p. 193-195 C; flash pt. (COC) 90.5 C
Toxicology: High concs. of vapor expected to cause nose/throat irritation; absorbed thru skin, but health effects unlikely; mild eye irritant; low oral toxicity
Precaution: Combustible; incompat. with oxidizing agents (increases fire/explosion hazard), strong acids (may cause decomp. of ethers)
Uses: Solvent for paints, lacquers, resins, inks, dyes, oils, greases, cleaners

Dipropylene glycol methyl ether. *See* PPG-2 methyl ether
Dipropylene glycol methyl ether acetate. *See* PPG-2 methyl ether acetate
Dipropylene glycol monoethyl ether. *See* Dipropylene glycol ethyl ether
Dipropylene glycol monomethyl ether. *See* PPG-2 methyl ether
Dipropylene glycol monomethyl ether acetate. *See* PPG-2 methyl ether acetate
Dipropylene glycol monopropyl ether. *See* Dipropylene glycol n-propyl ether
Dipropylene glycol propyl ether. *See* Dipropylene glycol n-propyl ether

Dipropylene glycol n-propyl ether
CAS 29911-27-1
Synonyms: Dipropylene glycol propyl ether; Dipropylene glycol monopropyl ether
Empirical: $C_9H_{20}O_3$
Formula: $C_3H_7O[CH_2(CH)CH_3O]_2H$
Properties: Colorless liq., low odor; mod. sol. in water; m.w. 176.2; sp.gr. 0.922; b.p. 212 C; surf. tens. 27.6 dynes/cm
Toxicology: May cause general depressed activity, diarrhea
Uses: Coalescent, solvent for water- or solv.-borne coatings
Manuf./Distrib.: Ashland; Dow
Trade names: Arcosolv® DPNP; Dowanol® DPnP

Dipropyl methane. *See* Heptane
Distillates (petroleum), hydrotreated middle. *See* Hydrocarbon solvent
Distilled spirits. *See* Alcohol
2,3-Dithiobutane. *See* Methyl disulfide
Dithiocarbonic anhydride. *See* Carbon disulfide
Divinylene sulfide. *See* Thiophene
Dizene. *See* o-Dichlorobenzene
DMA. *See* Dimethyl acetamide; Dimethylamine

DMAC. *See* Dimethyl acetamide
DMDS. *See* Methyl disulfide
DME. *See* Ethylene glycol dimethyl ether
DMEP. *See* Dimethoxyethyl phthalate
DMF. *See* Dimethyl formamide
DMFA. *See* Dimethyl formamide
DMI. *See* Dimethyl isosorbide
DMMP. *See* Dimethyl methylphosphonate
DMP. *See* Dimethyl phthalate
DMSO. *See* Dimethyl sulfoxide
DNBA. *See* Dibutylamine
DOA. *See* Dioctyl adipate
1-Docosanol. *See* Behenyl alcohol
13-Docosenamide. *See* Erucamide
cis 13-Docosenamide. *See* Erucamide
13-Docosenoic acid, 9-octadecenyl ester. *See* Oleyl erucate
Dodecafluoropentane. *See* Perfluoropentane
Dodecane. *See* n-Dodecane

n-Dodecane
CAS 112-40-3
Synonyms: Dodecane; Bihexyl; Dihexyl; Duodecane
Classification: Sat. aliphatic hydrocarbon
Empirical: $C_{12}H_{26}$
Formula: $CH_3(CH_2)_{10}CH_3$
Properties: Colorless liq.; gasoline-like odor; sol. in most org. solvs.; insol. in water; m.w. 170.34; sp.gr. 0.753; m.p. -9.6 C; b.p. 216 C; flash pt. 74 C
Toxicology: May cause mild skin irritation; conc. vapor may be irritating to eyes; inh. of conc. vapor may cause nose/throat irritation, headache, drowsiness, nausea, difficulty breathing
Precaution: Combustible; incompat. with strong oxidizing agents (increases fire and explosion hazard); Hazardous decomp. prods.: Heated to decomp., emits acrid smoke and irritating fumes
NFPA: Health 0, Flammability 2, Reactivity 0
Uses: Component of gasoline, jet fuel, kerosene, petrol. solvents such as wh. spirit; solvent; org. synthesis; copying fluids; jet fuel research; distillation chaser
Manuf./Distrib.: Aldrich; Fluka; Phillips; Sigma

Dodecanoic acid benzyl ester. *See* Benzyl laurate
Dodecanoic acid, 2-hydroxypropyl ester. *See* Propylene glycol laurate
Dodecanoic acid methyl ester. *See* Methyl laurate
Dodecanoic acid, 1-methylethyl ester. *See* Isopropyl laurate
Dodecanoic acid, monoester with 1,2-propanediol. *See* Propylene glycol laurate
Dodecanoic acid, phenylmethyl ester. *See* Benzyl laurate
1-Dodecanol, 2-octyl-. *See* Octyldodecanol
Dodecylbenzenesulfonic acid, comp. with 2-propanamine (1:1). *See* Isopropylamine dodecylbenzenesulfonate
N-Dodecyl-2-pyrrolidone. *See* Lauryl pyrrolidone

Dodoxynol-7
CAS 9014-92-0 (generic); 26401-47-8 (generic)
Synonyms: PEG-7 dodecyl phenyl ether; POE (7) dodecyl phenyl ether
Definition: Ethoxylated alkyl phenol
Formula: $C_{12}H_{25}C_6H_4(OCH_2CH_2)_nOH$, avg. n = 7
Uses: Emulsifier; surfactant for solvents; solvent and emulsion cleaning
Regulatory: FDA 21CFR §172.710, 175.105, 176.210, 178.3400

DOM. *See* Dioctyl maleate
DOP. *See* Dioctyl phthalate
DOS. *See* Dioctyl sebacate
DPG. *See* Dipropylene glycol
DPGEE. *See* Dipropylene glycol ethyl ether
DPGME. *See* PPG-2 methyl ether
DPM. *See* PPG-2 methyl ether
Duodecane. *See* n-Dodecane
Dutch liquid. *See* Ethylene dichloride
Dutch oil. *See* Ethylene dichloride

EA. *See* Ethylamine
Earthnut oil. *See* Peanut oil
EB. *See* Ethylbenzene
EDA. *See* Ethylenediamine
EDC. *See* Ethylene dichloride
EE. *See* Ethoxyethanol
EEA. *See* Ethoxyethanol acetate; Ethylacetoacetate
EEP. *See* Ethyl 3-ethoxypropionate
EGDS. *See* Glycol distearate
EGME. *See* Methoxyethanol
EGMEA. *See* Methoxyethanol acetate
EGMS. *See* Glycol stearate
2-EH. *See* 2-Ethylhexanol
Eicosanoic acid methyl ester. *See* Methyl eicosenate
Elaic acid. *See* Oleic acid
Enanthic alcohol. *See* Heptyl alcohol
Enanthyl alcohol. *See* Heptyl alcohol
Enzactin. *See* Triacetin
EOEOEA. *See* 2-(2-Ethoxyethoxy) ethyl acrylate
1,4-Epoxybutane. *See* Tetrahydrofuran
1,4-Epoxy-p-menthane. *See* 1,4-Cineole
1,8-Epoxy-p-menthane. *See* Eucalyptol
1,2-Epoxypropane. *See* Propylene oxide

Erucamide

CAS 112-84-5; EINECS 204-009-2

Synonyms: Erucic acid amide; Erucylamide; 13-Docosenamide; cis 13-Docosenamide

Classification: Aliphatic amide

Empirical: $C_{22}H_{43}ON$

Formula: $CH_3(CH_2)_7CH=CH(CH_2)_{11}CONH_2$

Properties: Solid; sol. in isopropanol; sl. sol. in alcohol, acetone; dens. 0.888; m.p. 75-80 C

Precaution: Combustible

Uses: Foam stabilizer; solvent for waxes and resins, emulsions; slip/antiblock agent for polyethylene; lubricant, mold release for rubber and plastics; lamination of polyethylene to cellophane and in polyethylene extrusion coatings; food contact applics.

Regulatory: FDA 21CFR §175.105, 177.1200, 177.1210, 178.3860
Manuf./Distrib.: Akzo Nobel; Aldrich; Chemax; Cookson Spec. Addit.; Croda Universal Ltd; Sigma; Syn. Prods.; Witco/Oleo-Surf.

Erucic acid amide. *See* Erucamide
Erucic acid, oleyl ester. *See* Oleyl erucate
Erucylamide. *See* Erucamide
Ethanamine. *See* Ethylamine
Ethane, 1-chloro-1,1-difluoro-. *See* 1,1,1-Chlorodifluoroethane
1,2-Ethanediamine. *See* Ethylenediamine
1,2-Ethanediamine, N-(2-aminoethyl)-N´-(2-((2-aminoethyl)amino)ethyl)-. *See* Tetraethylenepentamine
Ethane dichloride. *See* Ethylene dichloride
Ethane, 1,2-dichloro-1,1,2,2-tetrafluoro-. *See* Dichlorotetrafluoroethane
1,2-Ethanediol. *See* Glycol
Ethane-1,2-diol. *See* Glycol
1,2-Ethanediol carbonate. *See* Ethylene carbonate
1,2-Ethanediol diacetate. *See* Ethylene glycol diacetate
1,2-Ethanediol monoacetate. *See* Ethylene glycol acetate
2,2´-[1,2-Ethanediylbis(oxy)]bisethanol. *See* Triethylene glycol
Ethanenitrile. *See* Acetonitrile
Ethane, nitro-. *See* Nitroethane
Ethane trichloride. *See* 1,1,2-Trichloroethane
Ethane, 1,1,1-trichloro-. *See* 1,1,1-Trichloroethane
Ethane, 1,1,2-trichloro-1,2,2-trifluoro. *See* Trichlorotrifluoroethane
Ethanoic acid. *See* Acetic acid; Acetic acid, glacial
Ethanol. *See* Alcohol
Ethanol, 2-(2-aminoethoxy)-. *See* 2-(2-Aminoethoxy) ethanol
Ethanol, 2-butoxy-, acetate. *See* Butoxyethanol acetate
Ethanol, 2-(2-(2-butoxyethoxy)ethoxy)-. *See* Butoxytriglycol
Ethanol, 2-(2-ethoxyethoxy). *See* Ethoxydiglycol
Ethanol, 2-((2-ethylhexyl)oxy)-. *See* Ethylene glycol 2-ethylhexyl ether
Ethanol, 2-((2-hexyloxy)ethoxy)-. *See* Diethylene glycol n-hexyl ether
Ethanolic acid. *See* Acetic acid, glacial
Ethanol, 2-(2-methoxyethoxy)-. *See* Methoxydiglycol
Ethanol, 2(2-(2-methoxyethoxy)ethoxy)-. *See* Triglycol monomethyl ether
Ethanol, 2-propoxy. *See* Ethylene glycol propyl ether
Ethanol, 2,2´-thiodi-. *See* Thiodiglycol
1-Ethanol-2-thiol. *See* 2-Mercaptoethanol
Ethanol, undenatured. *See* Alcohol
Ether. *See* Ethyl ether
Ether, bis (2-butoxyethyl). *See* Diethylene glycol dibutyl ether
Ether chloratus. *See* Ethyl chloride
Ether cyanatus. *See* Propionitrile
Ether hydrochloric. *See* Ethyl chloride
Ether muriatic. *See* Ethyl chloride
2-Ethinylbutanol-2. *See* Methyl pentynol
Ethinylmethylethyl carbinol. *See* Methyl pentynol
Ethinyl trichloride. *See* Trichloroethylene
Ethohexadiol. *See* Ethyl hexanediol

Ethoxydiglycol
CAS 111-90-0; EINECS 203-919-7
Synonyms: DEGEE; Diethylene glycol monoethyl ether; Diethylene glycol ethyl ether; Diglycol monoethyl ether; 'Carbitol'; Ethanol, 2-(2-ethoxyethoxy); Eth-

ylene diglycol monoethyl ether; 2-(2-Ethoxyethoxy) ethanol
Classification: Aliphatic ether alcohol
Empirical: $C_6H_{14}O_3$
Formula: $CH_2OHCH_2OCH_2CH_2OC_2H_5$
Properties: Colorless liq., mild pleasant odor; sol. in water and org. solvs., e.g., acetone, benzene, chloroform, ethanol, ether; m.w. 134.20; sp.gr. 0.990 (20 C); m.p. -76 C; b.p. 195-202 C; flash pt. (OC) 96.1 C; ref. index 1.425 (25 C)
Toxicology: LD50 (oral, rat) 5500 mg/kg; moderately toxic by ingestion and other routes; skin and eye irritant; can be absorbed thru skin, but health effects unlikely
Precaution: Combustible liq. and vapor; peroxide former; Hazardous decomp. prods.: Heated to decomp., emits acrid smoke
NFPA: Health 1, Flammability 1, Reactivity 0
Uses: Solvent for dyes, nitrocellulose, resins, lacquers, quick-dry varnishes, enamels, wood stains, thinners, cosmetics; mutual solvent for min. oil-soap and min. oil-sulfonated oil mixts.; textiles, textile printing, soaps; organic synthesis; brake fluid diluent; mfg. of plasticizers; solvent, thinner in nail enamels; solvent, solubilizer, cosurfactant for pharmaceuticals
Regulatory: FDA 21CFR §175.105, 176.180
Manuf./Distrib.: Aldrich; Allchem Ind.; Ashland; Eastman; Fluka; Great Western; Occidental; Oxiteno; Sigma; Spectrum Chem. Mfg.; Union Carbide
Trade names: Carbitol®; Carbitol® Low Gravity; DE; DE; Dioxitol-Low Gravity; Eastman® DE; Ethyl Di-Icinol; Poly-Solv® DE (Low Gravity); Solvent APV Spec.; Transcutol
Trade names containing: Dioxitol-High Gravity; Eastman® DE-HG; Icinol EE22; Poly-Solv® DE (High Gravity)

Ethoxydiglycol acetate
CAS 112-15-2; EINECS 203-940-1
Synonyms: Diethylene glycol monoethyl ether acetate; 2-(2-Ethoxyethoxy) ethanol acetate; 2-(2-Ethoxyethoxy) ethyl acetate; Diethylene glycol ethyl ether acetate
Empirical: $C_8H_{16}O_4$
Formula: $CH_3COO(CH_2)_2O(CH_2)_2OCH_2CH_3$
Properties: Colorless liq.; sol. in water; m.w. 176.22; sp.gr. 1.011 (20 C); m.p. -25 C; b.p. 214 C; ref. index 1.42 (20 C)
Toxicology: Irritant
Uses: Solvent, coalescing aid for latex paints; reagent
Regulatory: FDA 21CFR §175.105
Manuf./Distrib.: Aldrich; ARCO; Eastman; Fluka; Spectrum Chem. Mfg.
Trade names: Carbitol® Acetate; DEA; DE Acetate; Eastman® DE Acetate

Ethoxyethanol
CAS 110-80-5; EINECS 203-804-1
UN No. 1171 (DOT)
Synonyms: EE; Ethyl glycol; Ethylene glycol ethyl ether; Ethylene glycol monoethyl ether; 2-Ethoxyethanol; 'Cellosolve'
Classification: Ether alcohol
Empirical: $C_4H_{10}O_2$
Formula: $CH_3CH_2OCH_2CH_2OH$
Properties: Colorless liq., pract. odorless; misc. with hydrocarbons, alcohol, ether, acetone, liq. esters, water; m.w. 90.14; dens. 0.9311 (20/20 C); m.p. -70 C; b.p. 135.6 C; flash pt. 48.9 C; ref. index 1.4060 (25 C)
Toxicology: TLV 5 ppm (skin); LD50 (oral, rat) 3 g/kg; toxic by skin absorption; moderately toxic by ingestion, IV, IP routes; mildly toxic by inhalation, subcut.

routes
Precaution: Flamm.; reacts with oxidizing materials; mod. explosion hazard in form of vapor when exposed to flames
NFPA: Health 2, Flammability 2, Reactivity 0
Uses: Solvent for nitrocellulose, alkyd resins, lacquer, lacquer thinners, dyeing and printing textiles, varnish removers, cleaning solutions
Regulatory: FDA 21CFR §73.1, 175.105, 177.2600; SARA reportable
Manuf./Distrib.: Aldrich; Allchem Ind.; ARCO; Ashland; Brown; Fluka; Great Western; Occidental; Oxiteno; Sigma; Spectrum Chem. Mfg.; Union Carbide
Trade names: Cellosolve®; EE; Ethyl Icinol; Oxitol; Poly-Solv® EE

2-Ethoxyethanol. *See* Ethoxyethanol

Ethoxyethanol acetate
CAS 111-15-9; EINECS 203-839-2
UN No. 1172
Synonyms: EEA; Ethylene glycol ethyl ether acetate; Ethylene glycol monoethyl ether acetate; 2-Ethoxyethanol acetate; Acetic acid 2-ethoxyethyl ester; 2-Ethoxyethyl acetate
Classification: Carboxylic acid ester
Definition: Ester of ethoxyethanol and acetic acid
Empirical: $C_6H_{12}O_3$
Formula: $CH_2COOCH_2CH_2OCH_2CH_3$
Properties: Colorless liq., pleasant odor; sol. in≈6 parts water; misc. with aromatic hydrocarbons; m.w. 132.16; sp.gr. 0.975 (20/20 C); m.p. -62 C; b.p. 156 C; flash pt. (CC) 47 C; ref. index 1.4030
Toxicology: ACGIH TLV/TWA 5 ppm (skin); LD50 (rat, oral) 5.1 g/kg; toxic by ingestion and skin absorp.; inh. may cause mild irritation of nose, throat, and eyes; high concs. may cause CNS depression
Precaution: Combustible; reacts with oxidizing materials; incompat. with acids (can hydrolyze to form acetic acid and alcohol), bases (can hydrolyze to form acetate salt and alcohol); Hazardous decomp. prods.: Heated to decomp., emits acrid smoke and irritating fumes
NFPA: Health 2, Flammability 2, Reactivity 0
Uses: Solvent for coatings (NC, cellulose acetate, ethyl cellulose, butyrate, acrylic, urethane), automobile lacquers; retards evaporation and imparts high gloss; varnish removers; blush retarder in lacquers; wood stains; textiles; leather finishing
Regulatory: FDA 21CFR §175.105
Manuf./Distrib.: Aldrich; Allchem Ind.; ARCO; Ashland; Fluka; Occidental; Spectrum Chem. Mfg.; Union Carbide
Trade names: Cellosolve® Acetate; EEA

2-Ethoxyethanol acetate. *See* Ethoxyethanol acetate
2-(2-Ethoxyethoxy) ethanol. *See* Ethoxydiglycol
2-(2-Ethoxyethoxy)ethanol acetate. *See* Ethoxydiglycol acetate
1-Ethoxy-2-(β-ethoxyethoxy) ethane. *See* Diethylene glycol diethyl ether
2-[2-(2-Ethoxyethoxy) ethoxy] ethanol. *See* Ethoxytriglycol
2-[2-(2-Ethoxyethoxy)ethoxy]ethene. *See* Diethylene glycol ethylvinyl ether
2-(2-Ethoxyethoxy) ethyl acetate. *See* Ethoxydiglycol acetate
Ethoxyethoxyethyl acrylate. *See* 2-(2-Ethoxyethoxy) ethyl acrylate

2-(2-Ethoxyethoxy) ethyl acrylate
Synonyms: EOEOEA; Ethoxyethoxyethyl acrylate
Uses: Monomer for radiation-curable ink and coating formulations
Manuf./Distrib.: Monomer-Polymer & Dajac Labs; Morton Int'l.
Trade names containing: SR-256

1-(2-Ethoxyethoxy)-2-vinyloxy ethane. *See* Diethylene glycol ethylvinyl ether
2-Ethoxyethyl acetate. *See* Ethoxyethanol acetate
Ethoxyethyl ether. *See* Diethylene glycol diethyl ether
2-Ethoxyethyl ether. *See* Diethylene glycol diethyl ether
2-Ethoxyethyl 2-(vinyloxy)ethyl ether. *See* Diethylene glycol ethylvinyl ether
1-[2-(2-Ethoxy-1-methylethoxy)-1-methylethoxy]-2-propanol. *See* Tripropylene glycol ethyl ether
1-2-(Ethoxy-2-methylethoxy)-2-propanol. *See* Dipropylene glycol ethyl ether
1-Ethoxy-2-propanol. *See* Propylene glycol ethyl ether
Ethoxypropionic acid ethyl ester. *See* Ethyl 3-ethoxypropionate
3-Ethoxypropionic acid ethyl ester. *See* Ethyl 3-ethoxypropionate

Ethoxytriglycol
CAS 112-50-5; EINECS 203-978-9
Synonyms: Triethylene glycol monoethyl ether; Triethylene glycol ethyl ether; PEG-3 ethyl ether; Triglycol monoethyl ether; 2-[2-(2-Ethoxyethoxy) ethoxy] ethanol
Empirical: $C_8H_{18}O_4$
Formula: $C_2H_5O(C_2H_4O)_3H$
Properties: Colorless liq.; sol. in water; m.w. 178.26; dens. 1.0208 (20/20 C); b.p. 255.4 C; f.p. -18.7 C; flash pt. (OC)135 C
Toxicology: LD50 (oral, rat) 10,610 mg/kg, (dermal, rabbit) 8 g/kg; mildly toxic by ingestion and skin contact; eye irritant; heated to decomp., emits acrid smoke and fumes
Precaution: Combustible; can react with oxidizing materials
Uses: Chemical intermediate; solvent
Manuf./Distrib.: Ashland; Fluka; Oxiteno
Trade names: Dowanol® TMH
Trade names containing: Icinol EE22; Icinol ME33

Ethyl acetate
CAS 141-78-6; EINECS 205-500-4
UN No. 1173 (DOT)
Synonyms: Acetic ether; Acetic ester; Acetoxyethane; Ethyl acetic ester; Acetic acid, ethyl ester; Ethyl ethanoate; Vinegar naphtha
Classification: Sat. aliphatic carboxylic acid ester
Definition: Ester of ethyl alcohol and acetic acid
Empirical: $C_4H_8O_2$
Formula: $CH_3COOC_2H_5$
Properties: Colorless liq., fragrant fruity odor, acetous burning taste; sol. in chloroform, alcohol, ether, acetone, benzene, fixed and volatile oils; sl. sol. in water; m.w. 88.12; dens. 0.902 (20/4 C); bulk dens. 0.8945 g/ml; f.p. -83.6 C; b.p. 77 C; flash pt. -4.4 C; ref. index 1.3723; surf. tens. 24 dynes/cm (20 C)
Toxicology: TLV 400 ppm in air; LD50 (oral, rat) 5620 mg/kg; poison by inhalation; mildly toxic by ingestion; irritant to eyes, skin, mucous membranes; mutagenic data; mildly narcotic; CNS depressant
Precaution: DOT: Flamm. liq.; very dangerous fire hazard exposed to heat or flame; can react vigorously with oxidizers; incompat. with strong oxidizers, strong acids, bases, potassium t-butoxide; Hazardous decomp. prods.: Ethanol, acetic acid; heated to decomp., emits acrid smoke and irritating fumes
NFPA: Health 1, Flammability 3, Reactivity 0
Uses: General solvent in coatings, plastics, lacquers, printing inks, adhesives, organic synthesis, smokeless powders, artificial leather, photographic films and plates, pharmaceuticals; synthetic fruit essences, flavoring agent; cleaning textiles; nail polish removers; color diluent, flavoring agent, solvent for food

use; reagent
Regulatory: FDA 21CFR § 73.1, 172.560, 173.228, 175.320, 177.1200, 182.60, GRAS; 27CFR §21.106; FEMA GRAS; Japan approved with restrictions; FDA approved for ophthalmics, orals, topicals; USP/NF, BP, Ph.Eur. compliance
Manuf./Distrib.: Aldrich; Allchem Ind.; Ashland; J.T. Baker; Baychem; Berje; BP Chems. Ltd; Brown; Chemcentral; Chisso Am.; Eastman; Fluka; Harcros USA; Hoechst Celanese; Houghton Chem.; Hüls AG; Lonza Ltd; Mallinckrodt; Monsanto; Penta Mfg.; Punda Mercantile; Samson; Sigma; Spectrum Chem. Mfg.; Sunnyside; Union Carbide; Van Waters & Rogers
Trade names containing: Filmex® C; Propsolv I-1; Propsolv I-2; Propsolv III-1; Propsolv III-2; Tecsol® 1; Tecsol® 3; Tecsol® C

Ethyl acetic ester. *See* Ethyl acetate

Ethylacetoacetate
CAS 141-97-9; EINECS 205-516-1
UN No. 1993 (DOT)
Synonyms: EEA; 3-Oxobutanoic acid ethyl ester; Diacetyl ether; Acetoacetic acid ethyl ester; Acetoacetic ester; Ethyl 3-oxobutanoate
Classification: Aliphatic organic compd.
Empirical: $C_6H_{10}O_3$
Formula: $CH_3COCH_2COOC_2H_5$
Properties: Colorless liq., fruity odor; sol. in ≈ 35 parts water; misc. with common org. solvs.; m.w. 130.14; dens. 1.0213 (25/4 C); m.p. -45 C; b.p. 180.8 C (760 mm); flash pt. (CC) 184 F; ref. index 1.4180-1.4195
Toxicology: LD50 (oral, rat) 3.98 g/kg; mod. toxic by ingestion; mod. irritating to skin, mucous membranes, eyes; heated to decomp., emits acrid smoke and irritating fumes
Precaution: DOT: Flamm. liq.; combustible liq. when exposed to heat or flame; can react with oxidizing materials
NFPA: Health 2, Flammability 2, Reactivity 0
Uses: Catalyst for esterification and olefin polymerization; resin crosslinking agent for automotive prods., coatings, elastomers, films/paints, graphic arts, plastics; syn. flavoring agent for foods and pharmaceuticals; solvent and extractant; intermediate for vitamins, feed additives, drugs, pesticides; prod. of azo dyes
Regulatory: FDA 21CFR §172.515; FEMA GRAS; Japan approved as flavoring
Manuf./Distrib.: Aceto; Aldrich; Berje; Eastman; Fluka; Lonza; Penta Mfg.; Sigma; Spectrum Chem. Mfg.; Wacker-Chemie GmbH

Ethyl acetone. *See* Methyl propyl ketone
Ethyl alcohol. *See* Alcohol
Ethyl alcohol, undenatured. *See* Alcohol

Ethylamine
CAS 75-04-7
UN No. 1036 (DOT), 2270 (DOT)
Synonyms: EA; MEA; Aminoethane; 1-Aminoethane; Ethanamine; Ethylamine anhydrous; Monoethylamine
Classification: Saturated aliphatic amine
Empirical: C_2H_7N
Formula: $CH_3CH_2NH_2$
Properties: Colorless liq. below 16.6 C; colorless gas above 17 C; strong ammonia-like odor; sol. in water, ethanol, acetone, ether, dimethyl sulfoxide; m.w. 45.09; sp.gr. 0.689 (15 C); vapor pressure 873 mm Hg (20 C); m.p. -81 C; b.p. 16.6 C; flash pt. (CC) < -18 C; pH 1.9 (0.1 M sol'n.)

Toxicology: LD50 (oral, rat) 400 mg/kg, (skin, rabbit) 390 mg/kg; ACGIH TLV/TWA 5 ppm (skin); STEL 15 ppm (skin); poison by ing., skin contact, IV routes; mod. toxic by inh.; skin irritant; severe eye irritant; lachrymator
Precaution: Flamm., corrosive, toxic gas or liq.; dangerous fire hazard; mod. explosion hazard; incompat. with mercury, acids, oxidizing agents, calcium or sodium hypochlorite, nitromethane, nitrosyl perchlorate; violent/exlosive reactions possible; Hazardous decomp. prods.: Emits toxic fumes of NO_x
NFPA: Health 3, Flammability 4, Reactivity 0
Uses: Intermediate; mfg. of dyes; pharmaceuticals; emulsifying agents; syn. resins; detergents; pesticides; as solvent; stabilizer for rubber latex; curing agent for epoxy resins
Manuf./Distrib.: Aldrich; Ashland; BASF AG; Elf Atochem N. Am.; Fluka; Sigma; Spectrum Chem. Mfg.; Union Carbide

Ethylamine anhydrous. *See* Ethylamine

Ethylbenzene
CAS 100-41-4; EINECS 202-849-4
UN No. 1175 (DOT)
Synonyms: EB; Phenylethane; Ethylbenzol
Classification: Aromatic hydrocarbon
Empirical: C_8H_{10}
Formula: $C_6H_5C_2H_5$
Properties: Colorless liq., aromatic odor; sol. in alcohol, benzene, CCl_4, ether; almost insol. in water; m.w. 106.17; sp.gr. 0.867 (20 C); f.p. -95 C; b.p. 136.187 C; flash pt. 15 C; ref. index 1.495 (20 C)
Toxicology: ACGIH TLV/TWA 100 ppm; STEL 125 ppm; LD50 (oral, rat) 3500 mg/kg, (dermal, rabbit) 17,800 mg/kg; moderately toxic by ingestion, intraperitoneal route; mildly toxic by inhalation and skin contact; eye irritant
Precaution: Highly flamm.; dangerous fire risk; incompat. with strong oxidizing agents (can react violently; increased risk of fire and explosion); Hazardous decomp. prods.: None
NFPA: Health 2, Flammability 3, Reactivity 0
Uses: Solvent and raw material for prod. of cellulose acetate, acetophenone, diethyl benzene, anthraquinones; chem. intermediate in mfg. of styrene, other plastics, syn. rubber, styrene latexes; minor component of gasoline and aviation fuels; reagent
Regulatory: SARA reportable
Manuf./Distrib.: Aldrich; Ashland; Fluka; Phibro Energy USA; Sigma; Spectrum Chem. Mfg.

Ethylbenzol. *See* Ethylbenzene
Ethylbis(2-hydroxyethyl)amine. *See* N-Ethyldiethanolamine
2-Ethylbutanol-1. *See* 2-Ethyl-1-butanol
2-Ethylbutanol. *See* 2-Ethyl-1-butanol

2-Ethyl-1-butanol
CAS 97-95-0
UN No. 2275 (DOT), 2282 (DOT)
Synonyms: 2-Ethylbutyl alcohol; 2-Ethyl butan-1-ol; 2-Ethylbutanol; 2-Ethylbutanol-1; s-Hexyl alcohol; sec-Hexyl alcohol; s-Hexanol; 3-Methylolpentane; 3-Pentylcarbinol; sec-Pentylcarbinol; Pseudohexyl alcohol
Classification: Primarly aliphatic alcohol
Empirical: $C_6H_{14}O$
Formula: $CH_3CH_2CH(CH_2CH_3)CH_2OH$
Properties: Colorless clear liq.; sweet musty odor; sol. in most org. solvs.; sl. sol. in

water; m.w. 102.17; sp.gr. 0.8328 (20 C); vapor pressure 0.9 mm Hg (20 C); m.p. -114 C; b.p. 150 C; flash pt. (CC) 58 C

Toxicology: LD50 (oral, rat) 1850 mg/kg, (skin, rabbit) 1260 mg/kg; mod. toxic by ing. and skin contact; skin and severe eye irritant; inh. may cause nose/throat irritation, CNS effects

Precaution: Flamm. exposed to heat or flame; incompat. with oxidizing agents (increase fire/explosion hazard); Hazardous decomp. prods.: Heated to decomp., emits acrid smoke and irritating fumes

NFPA: Health 1, Flammability 2, Reactivity 0

Uses: Solvent for oils, resins, waxes, dyes; synthesis of pharmaceuticals, surfactants, syn. lubricants, lubricant additives

Manuf./Distrib.: Aldrich; Fluka

2-Ethyl butan-1-ol. *See* 2-Ethyl-1-butanol
3-Ethylbutinol. *See* Methyl pentynol
2-Ethylbutyl alcohol. *See* 2-Ethyl-1-butanol
Ethyl butyl carbinol. *See* 3-Heptanol
3-Ethylbutynol. *See* Methyl pentynol
Ethyl carbinol. *See* n-Propyl alcohol

Ethyl chloride

CAS 75-00-3

UN No. 1037 (DOT)

Synonyms: Chlorethyl; Chloroethane; Chloryl; Ether chloratus; Ether hydrochloric; Ether muriatic; Monochlorethane; Monochloroethane; Muriatic ether; Hydrochloric ether

Classification: Halogenated aliphatic hydrocarbon

Empirical: C_2H_5Cl

Formula: CH_3CH_2Cl

Properties: Colorless gas @ R.T., liq. below 12 C or under pressure; ether-like pungent odor; misc. with methanol, ethanol, diethyl ether, ethyl acetate, methylene chloride, chloroform, CCl_4, benzene; sl. sol. in water; m.w. 64.52; sp.gr. 0.891 (liq.); vapor pressure 1000 mm Hg (20 C); m.p. -139 C; b.p. 12.3 C; flash pt. -50 C

Toxicology: LC50 (inh., rat, 2 h) 160 g/m^3; ACGIH TLV/TWA 100 ppm (skin); mildly toxic by inh.; may cause cooling of skin due to evaporation, allergic reactions; sl. eye irritant; skin, mucous membrane irritant

Precaution: Flamm.; compressed gas; very dangerous fire hazard exposed to heat, or flame; severe explosion hazard exposed to flame; incompat. with alkali metals, powd. metals, strong oxidizers; violent reactions, increased risk of fire or explosion possible; Hazardous decomp. prods.: Reacts with water or steam to produce toxic and corrosive fumes; heated to decomp., emits toxic fumes of phosgene and Cl^-

NFPA: Health 1, Flammability 4, Reactivity 0

Uses: Raw material in prod. of tetraethyl lead and ethyl cellulose; refrigerant; solvent; local anesthetic; intermediate in chem. synthesis, mfg. of dyes, drugs, perfumes; aerosol propellant

Manuf./Distrib.: Air Prods.; Aldrich; PPG Ind.; Sigma; Spectrum Chem. Mfg.

Ethyl citrate. *See* Triethyl citrate
Ethyl cyanide. *See* Propionitrile

Ethylcyclohexane

CAS 1678-91-7

Classification: Sat. alicyclic hydrocarbon

Empirical: C_8H_{16}

Properties: Colorless liq.; sol. in most org. solvs.; insol. in water; m.w. 112.21; sp.gr.

0.792 (20 C); m.p. -111 C; b.p. 132 C; flash pt. 35 C
Toxicology: May cause mild skin irritation; vapor may be irritating to eyes; probably a CNS depressant; sl. nose/throat irritant; high vapor concs. may cause headache, lightheadedness, nausea, drowsiness; very high concs. may cause unconsciousness, death
Precaution: Flamm.; incompat. with strong oxidizing agents (increased fire and explosion hazard); Hazardous decomp. prods.: None
NFPA: Health 1, Flammability 3, Reactivity 0
Uses: Solvent; org. synthesis
Manuf./Distrib.: Aldrich; Fluka

Ethylcyclopentane
CAS 1640-89-7
Classification: Sat. alicyclic hydrocarbon
Empirical: C_7H_{14}
Properties: Colorless liq.; gasoline-like odor; sol. in most org. solvs.; insol. in water; m.w. 98.19; sp.gr. 0.763 (20 C); m.p. -138 C; b.p. 103 C; flash pt. 15 C
Toxicology: May cause mild skin irritation; vapor may be irritating to eyes; probably an asphyxiant and CNS depressant; high vapor concs. may cause nose/throat irritation, lightheadedness, nausea, drowsiness
Precaution: Flamm.; very dangerous fire and explosion hazard exposed to heat or flame; incompat. with strong oxidizing agents (increased fire and explosion hazard); easily forms explosive mixts. in air; Hazardous decomp. prods.: Heated to decomp., emits acrid smoke and fumes
NFPA: Health 1, Flammability 3, Reactivity 0
Uses: Extraction solvent for essential oils; org. synthesis; component in gasoline
Manuf./Distrib.: Aldrich

Ethyldiethanolamine. *See* N-Ethyldiethanolamine

N-Ethyldiethanolamine
CAS 139-87-7
Synonyms: N,N-Bis(2-hydroxyethyl)ethylamine; Ethyldiethanolamine; Diethanolethylamine; 2,2´-(Ethylimino)bisethanol; 2,2´-(Ethylimino) diethanol; Ethylbis(2-hydroxyethyl)amine; N-Ethyl-2,2´-iminodiethanol
Classification: Amino alcohol
Empirical: $C_6H_{15}NO_2$
Formula: $CH_3CH_2N(CH_2CH_2OH)_2$
Properties: Water-wh. liq.; amine odor; sol. in water; m.w. 133.22; sp.gr. 1.015 (20 C); m.p. -50 C; sets to glasslike solid below -44 C; b.p. 246-252 C; flash pt. (CC) 123 C
Toxicology: LD50 (oral, rat) 4570 mg/kg; mildly toxic by ingestion; skin irritant; severe eye irritant; Hazardous decomp. prods.: Heated to decomp., emits toxic fumes of NO_x
NFPA: Health 2, Flammability 1, Reactivity 0
Uses: Solvent; detergents
Manuf./Distrib.: Aldrich; Elf Atochem N. Am.; Harcros USA

Ethyl diglyme. *See* Diethylene glycol diethyl ether
Ethyldimethyl methane. *See* Isopentane
Ethylene acetate. *See* Ethylene glycol diacetate
Ethylene alcohol. *See* Glycol

Ethylene carbonate
CAS 96-49-1; EINECS 202-510-0
Synonyms: Ethylene glycol carbonate; Ethylene glycol cyclic carbonate; Glycol carbonate; 1,3-Carbonyldioxyethane; 1,2-Ethanediol carbonate; Dioxolone-2;

1,3-Dioxolan-2-one; Ethylene carbonic acid; Cyclic ethylene carbonate; Carbonic acid, cyclic ethylene ester
Classification: Carbonic acid ester
Empirical: $C_3H_4O_3$
Formula: $(-CH_2O)_2CO$
Properties: Colorless solid (below 36.4 C) or liq., odorless; very sol. in water; sol. in ether, n-butanol, CCl_4; misc. with alcohol, ethyl acetate, benzene, chloroform; m.w. 88.07; sp.gr. 1.3218 (39/4 C); vapor pressure 0.01 mm Hg (20 C); m.p. 36.4 C; b.p. 248 C; flash pt. (OC) 143 C; ref. index 1.4158 (50 C)
Toxicology: LD50 (oral, rat) 10 g/kg; LDLo (IP, mouse) 500 mg/kg; moderately toxic by intraperitoneal route; mildly toxic by ingestion; skin and eye irritant
Precaution: Combustible exposed to heat or flame; incompat. with strong oxidizing agents (increases fire risk), acids, bases, and reducing agents (decomp. can occur); Hazardous decomp. prods.: May produce ethylene oxide and CO_2 if an acid, base, or salt is present in an aq. sol'n. of ethylene carbonate; heated to decomp., emits acrid smoke and irritating fumes
NFPA: Health 2, Flammability 1, Reactivity 1
Uses: Solvent for polymers and resins; plasticizer; intermediate for pharmaceuticals, hydroxyethylation reactions, rubber chemicals, textile finishing agents; extraction of aromatic hydrocarbons
Manuf./Distrib.: Aldrich; Fluka; Hüls AG; Monomer-Polymer & Dajac Labs
Trade names: Jeffsol™ EC; Texacar® EC

Ethylene carbonic acid. *See* Ethylene carbonate
Ethylene chloride. *See* Ethylene dichloride
Ethylene diacetate. *See* Ethylene glycol diacetate

Ethylenediamine
CAS 107-15-3 (anhyd.); 6780-13-8 (monohydrate); EINECS 203-468-6
UN No. 1604 (DOT)
Synonyms: EDA; 1,2-Diaminoethane; 1,2-Ethanediamine; Dimethylenediamine
Classification: Aliphatic organic compd.; alkaline compd.
Empirical: $C_2H_8N_2$
Formula: $NH_2CH_2CH_2NH_2$
Properties: Colorless to sl. yel. clear volatile liq., ammonia-like odor; hygroscopic; misc. with water, alcohol; m.w. 60.12; dens. 0.8994 (20/4 C); m.p. 8.5 C; b.p. 117.2 C; flash pt. (CC) 110 F; ref. index 1.4565
Toxicology: TLV:TWA 10 ppm; LD50 (oral, rat) 500 mg/kg; human irritant poison by inh.; mod. toxic by ingestion, skin contact; corrosive; severe skin and eye irritant; allergen, sensitizer; mutagenic data; heated to decomp., emits toxic fumes of NO_x and NH_3
Precaution: DOT: Corrosive material; flamm. exposed to heat, flame, oxidizers; can react violently with acetic acid, acetic anhydride, acrylic acid, epichlorohydrin, many others
NFPA: Health 3, Flammability 2, Reactivity 0
Uses: Fungicide, mfg. of chelating agents, dimethylolethylene-urea resins, chemical intermediate, solvent, emulsifier, textile lubricants, antifreeze inhibitor, agric. chems., corrosion inhibitors, surfactants, ion-exchange resins, rubber chems.; thermoplastics lubricants; stabilizer for dyes; solvent for shellac; flume wash water additive for food processing; antimicrobial for cane-sugar/beet-sugar mills; stabilizer, buffer for pharmaceuticals
Regulatory: FDA 21CFR §173.315, 173.320 (1 ppm max.), 178.3120, 556.270, 181.30; FDA approved for IV, injectables, orals, rectals, topicals; USP, BP, Ph.Eur., JP compliance
Manuf./Distrib.: Aldrich; Allchem Ind.; BASF; BASF AG; Coyne; Diamine &

Chems. Ltd; Fluka; Houghton Chem.; Nova Molecular Tech.; Sigma; Spectrum Chem. Mfg.; Texaco; Tosoh; Union Carbide

Ethylene dichloride
CAS 107-06-2; EINECS 203-458-1
UN No. 1184 (DOT)
Synonyms: 1,2-DCE; EDC; 1,2-Ethylene dichloride; 1,2-Bichloroethane; Dichloroethylene; Ethane dichloride; Dutch liquid; Dutch oil; Glycol dichloride; 1,2-Dichloroethane; α,β-Dichloroethane; sym-Dichloroethane; Dichloro-1,2-ethane; Ethylene chloride
Classification: Halogenated aliphatic hydrocarbon
Empirical: $C_2H_4Cl_2$
Formula: $Cl \cdot CH_2CH_2 \cdot Cl$
Properties: Colorless oily liq., chloroform-like odor, sweet taste; misc. with most common solvs.; sl. sol. in water; m.w. 98.96; dens. 1.2554 (20/4 C); vapor pressure 100 mm Hg (29.4 C); f.p. -35.5 C; b.p. 83.5 C; flash pt. 56 C; ref. index 1.445
Toxicology: LD50 (oral, rat) 670 mg/kg, (skin, rabbit) 3890 mg/kg; ACGIH TLV/TWA 10 ppm; human poison by ingestion; mod. toxic by inh., IP route; strong narcotic; skin/severe eye irritant; human systemic effects by ing., inh.
Precaution: Flamm.; dangerous fire hazard exposed to heat, flame, or oxidizers; mod. explosive as vapor exposed to flame; violent reaction with Al, NH_3; can react vigorously with oxidizers and emit vinyl chloride and HCl; Hazardous decomp. prods.: Heated to decomp., emits highly toxic fumes of Cl^-, phosgene
NFPA: Health 2, Flammability 3, Reactivity 0
Uses: Production of vinyl chloride, trichloroethylene, vinylidene chloride, trichloroethane; lead scavenger in gasoline; paint, varnish remover; metal degreasing; soaps; wetting/penetrating agents; organic synthesis; ore flotation; solvent; fumigant; pesticide; extraction solvent for foods, flume wash water additive, pesticide; intermediate for paints
Regulatory: FDA 21CFR §172.560, 172.710, 173.230, 173.315, 175.105, 573.440; SARA reportable
Manuf./Distrib.: Albemarle; Albright & Wilson Am.; Aldrich; Ashland; BASF; BP Chems. Ltd; Fluka; Georgia Gulf; Houghton Chem.; Norsk Hydro AS; Occidental; PPG Ind.; Sigma; Spectrum Chem. Mfg.; Vulcan

1,2-Ethylene dichloride. *See* Ethylene dichloride
Ethylene diethanoate. *See* Ethylene glycol diacetate
Ethylene diglycol. *See* Diethylene glycol
Ethylene diglycol monoethyl ether. *See* Ethoxydiglycol
Ethylene dihydrate. *See* Glycol
Ethylenedinitrilotetra-2-propanol. *See* Tetrahydroxypropyl ethylenediamine
1,1′,1′′,1′′′-(Ethylenedinitriolo)tetra-2-propanol. *See* Tetrahydroxypropyl ethylenediamine
2,2′-Ethylenedioxydiethanol. *See* Triethylene glycol
2,2′-Ethylenedioxyethanol. *See* Triethylene glycol
Ethylene glycol. *See* Glycol

Ethylene glycol acetate
CAS 542-59-6
Synonyms: Ethylene glycol monoacetate; Glycol monoacetate; Glycol monoacetin; Acetic acid, 2-hydroxyethyl ester; 1,2-Ethanediol monoacetate; 2-Hydroxyethyl acetate
Classification: Glycol ester
Empirical: $C_4H_8O_3$
Formula: $CH_3COOCH_2CH_2OH$

Properties: Colorless liq.; almost odorless; sol. in water, alcohol, ether, benzene, toluene; m.w. 104.22; sp.gr. 1.108; b.p. 181 C; flash pt. (OC) 102 C
Toxicology: LD50 (oral, rat) 8250 mg/kg, (IP, mouse) 1310 mg/kg; mod. toxic by ing., IP routes; eye irritant; inh. of vapor may irritate nose/throat; ing. expected to cause mild CNS depression
Precaution: Combustible exposed to heat or flame; incompat. with strong oxidizing agents (increases fire/explosion hazard), strong acids and strong bases (decomp. can occur); Hazardous decomp. prods.: Ethylene glycol, acetic acid; heated to decomp., emits acrid smoke and irritating fumes
NFPA: Health 0, Flammability 1, Reactivity 0
Uses: Solvent for camphor, dyes, nitrocellulose, cellulose acetate; in lacquers, paints, varnish removers; antifoaming agent for aq. sol'ns.
Manuf./Distrib.: Aldrich; Fluka

Ethylene glycol bis(2-hydroxyethyl ether). *See* Triethylene glycol
Ethylene glycol butyl ether. *See* Butoxyethanol
Ethylene glycol n-butyl ether. *See* Butoxyethanol
Ethylene glycol butyl ether acetate. *See* Butoxyethanol acetate
Ethylene glycol carbonate. *See* Ethylene carbonate
Ethylene glycol cyclic carbonate. *See* Ethylene carbonate

Ethylene glycol diacetate
CAS 111-55-7; EINECS 203-881-1
Synonyms: Glycol diacetate; 1,2-Ethanediol diacetate; Ethylene diacetate; Ethylene acetate; Ethylene diethanoate; 1,2-Diacetoxyethane
Classification: Glycol ester
Empirical: $C_6H_{10}O_4$
Formula: $CH_3COOCH_2CH_2OOCCH_3$
Properties: Colorless liq., faint fruity odor; sol. in alcohol, ether, benzene, acetone, CCl_4, acetic acid, CS_2; mod. sol. in water (10%); insol. in petrol.; m.w. 146.14; sp.gr. 1.1063 (20/20 C); f.p. -31 C; b.p. 190.5 C; flash pt. (OC) 96 C; ref. index 1.416
Toxicology: LD50 (oral, rat) 6850 mg/kg, (skin, rabbit) 8480 mg/kg; moderately toxic by intraperitoneal route; mildly toxic by ingestion and skin contact; eye irritant; inh. may cause nose/throat irritation
Precaution: Combustible; incompat. with strong oxidizing agents (increases fire/explosion hazard), strong acids, strong reducing agents (decomp. can occur); Hazardous decomp. prods.: Ethylene glycol, acetic acid; heated to decomp., emits acrid smoke and fumes
NFPA: Health 1, Flammability 1, Reactivity 0
Uses: Extraction solvent; foundry resins; perfume fixative; solvent for coatings, lacquers, cellulose esters, cellulose ethers, oils, resins; mfg. of printing inks; plasticizer for cellulosics, PVAc
Manuf./Distrib.: Aceto; Aldrich; Ashland; BP Chems. Inc; Chemoxy Int'l. plc; CPS; Eastman; Fluka; Houghton Chem.; Sigma

Ethylene glycol diethyl ether
CAS 629-14-1; EINECS 211-076-1
UN No. 1153 (DOT)
Synonyms: 1,2-Diethoxyethane; Ethyl glyme
Empirical: $C_6H_{14}O_2$
Formula: $C_2H_5OCH_2CH_2OC_2H_5$
Properties: Colorless liq., sl. odor; immiscible in water; m.w. 118.20; dens. 0.8417 (20/20 C); f.p. -74 C; b.p. 121.4 C; flash pt. 35 C; ref. index 1.393 (20 C)
Toxicology: LD50 (oral, rat) 4390 mg/kg; moderately toxic by ingestion; mildly toxic by inhalation; eye irritant

Precaution: Flamm.; mod. fire risk; can react with oxidizing materials
Uses: Organic synthesis (reaction medium); solvent and diluent for detergents; solvent for paints
Manuf./Distrib.: Aldrich; Brand-Nu Labs; Eastern Chem.; Ferro/Grant; Fluka; Sigma
Trade names: Ethyl Glyme

Ethylene glycol dihydroxydiethyl ether. *See* Triethylene glycol

Ethylene glycol dimethyl ether
CAS 110-71-4; EINECS 203-794-9
UN No. 2252 (DOT)
Synonyms: DME; 1,2-Dimethoxyethane; Dimethylglycol; Monoglyme; Glycol dimethyl ether
Empirical: $C_4H_{10}O_2$
Formula: $CH_3OCH_2CH_2OCH_3$
Properties: Water-wh. liq., sharp ethereal odor; misc. with water, alcohol; sol. in hydrocarbons; m.w. 90.12; dens. 0.86285 (20/4 C); m.p. -58 C; b.p. 82-83 (760 mm); flash pt. 4.5 C; ref. index 1.3813 (20 C)
Toxicology: Experimental reproductive effects
Precaution: DOT: Flamm. liq.; mod. fire hazard; may form explosive peroxides
NFPA: Health 2, Flammability 2, Reactivity 0
Uses: Solvent for paints; pharmaceutical solvent, vehicle
Regulatory: BP compliance
Manuf./Distrib.: Aldrich; Brand-Nu Labs; Ferro/Grant; Fluka; Sigma; Spectrum Chem. Mfg.
Trade names: Monoglyme

Ethylene glycol distearate. *See* Glycol distearate
Ethylene glycol ethyl ether. *See* Ethoxyethanol
Ethylene glycol ethyl ether acetate. *See* Ethoxyethanol acetate

Ethylene glycol 2-ethylhexyl ether
CAS 1559-35-9
Synonyms: Ethylene glycol mono-2-ethylhexyl ether; Ethanol, 2-((2-ethylhexyl)oxy)-
Empirical: $C_{10}H_{22}O_2$
Properties: M.w. 174.32
Toxicology: LD50 (oral, rat) 3080 mg/kg; primary irritant
Uses: Solvent for coatings and cleaning applics.
Trade names: EEH
Trade names containing: Eastman® EEH

Ethylene glycol formal. *See* 1,3-Dioxolane

Ethylene glycol hexyl ether
CAS 112-25-4; EINECS 203-951-1
Synonyms: Ethylene glycol monohexyl ether; Hexylglycol; Ethylene glycol n-hexyl ether; 2-(Hexyloxy)ethanol; n-Hexyl 'Cellosolve'
Classification: Glycol ether
Empirical: $C_8H_{18}O_2$
Formula: $C_6H_{13}OC_2H_4OH$
Properties: Water-wh. liq.; m.w. 146.26; sp.gr. 0.887 (20/20 C); f.p. -58 F; b.p. 208.1 C; flash pt. 90.5 C; ref. index 1.4290 (20 C)
Toxicology: LD50 (oral, rat) 830 mg/kg, (skin, rabbit) 810 mg/kg; primary irritant
Precaution: Combustible
Uses: High-boiling solvent for coatings

Manuf./Distrib.: Ashland; Eastman; Fluka; Union Carbide
Trade names: EH; Hexyl Cellosolve®

Ethylene glycol n-hexyl ether. *See* Ethylene glycol hexyl ether
Ethylene glycol methyl ether. *See* Methoxyethanol
Ethylene glycol methyl ether acetate. *See* Methoxyethanol acetate
Ethylene glycol monoacetate. *See* Ethylene glycol acetate
Ethylene glycol monobutyl ether. *See* Butoxyethanol
Ethylene glycol monobutyl ether acetate. *See* Butoxyethanol acetate
Ethylene glycol monoethyl ether. *See* Ethoxyethanol
Ethylene glycol monoethyl ether acetate. *See* Ethoxyethanol acetate
Ethylene glycol mono-2-ethylhexyl ether. *See* Ethylene glycol 2-ethylhexyl ether
Ethylene glycol monohexyl ether. *See* Ethylene glycol hexyl ether
Ethylene glycol monomethyl ether. *See* Methoxyethanol
Ethylene glycol monomethyl ether acetate. *See* Methoxyethanol acetate
Ethylene glycol monophenyl ether. *See* Phenoxyethanol
Ethylene glycol monopropyl ether. *See* Ethylene glycol propyl ether
Ethylene glycol monostearate. *See* Glycol stearate
Ethylene glycol phenyl ether. *See* Phenoxyethanol

Ethylene glycol propyl ether
CAS 2807-30-9; EINECS 220-548-6
Synonyms: Propyl 'Cellosolve'; 2-Propoxyethanol; Ethanol, 2-propoxy; Ethylene glycol monopropyl ether
Empirical: $C_5H_{12}O_2$
Formula: $CH_2OHCH_2OCH_2CH_2CH_3$
Properties: Liq.; sol. in water; m.w. 104.17; dens. 0.913 (20/20 C); b.p. 149.5 C (760 mm); f.p. -90 C; flash pt. (TCC) 49 C
Toxicology: LD50 (oral, rat) 3089 mg/kg; moderately toxic by ingestion and skin contact; mild toxicity by inhalation; skin/eye irritant
Precaution: Combustible; Hazardous decomp. prods.: Heated to decomp., emits acrid smoke, irritating fumes
Uses: Solvent for coatings, cosmetics, resins; coupling solvent for resin/water systems
Manuf./Distrib.: Aldrich; Ashland; Eastman
Trade names: Eastman® EP; EP; Propyl Cellosolve®

Ethylene glycol stearate. *See* Glycol stearate
Ethylene tetrachloride. *See* Perchloroethylene
Ethylene thioglycol. *See* 2-Mercaptoethanol
Ethylene trichloride. *See* Trichloroethylene

Ethyl 3-epoxypropionate
Uses: Let-down solvent for thinning syn. coating resins, esp. for PU coatings
Trade names: Eastman® EEP

N-Ethylethanamine. *See* Diethylamine
Ethyl ethanoate. *See* Ethyl acetate

Ethyl ether
CAS 60-29-7; EINECS 200-467-2
UN No. 1155
Synonyms: Diethyl ether; Diethyl oxide; Anesthetic ether; Ether; Ethyl oxide; Oxybis-1,1′-ethane; Sulfuric ether
Classification: Sat. aliphatic ether
Empirical: $C_4H_{10}O$
Formula: $CH_3CH_2OCH_2CH_3$
Properties: Colorless clear volatile liq., sweet pungent odor; sol. in water, chloro-

form, benzene, conc. HCl; misc. with alcohol, ether, most org. solvs.; m.w. 74.12; sp.gr. 0.7135 (20/4 C); m.p. -116.2 C; b.p. 34.6 C; flash pt. (CC) -45 C; pH neutral

Toxicology: ACGIH TLV/TWA 400 ppm; STEL 500 ppm; LD50 (oral, rat) 1215 mg/kg; mod. toxic to humans by ing., IP, IV routes; mildly toxic by inh.; human systemic effects by inh.; CNS depressant; anesthetic; mod. to severe skin and eye irritant

Precaution: DOT: Flamm. liq.; severe fire and explosion hazard when exposed to heat or flame; forms explosive peroxides; incompat. with sulfur compds., halogens, interhalogens, strong oxidizers; violent or explosive reactions possible; Hazardous decomp. prods.: Can form peroxides on prolonged exposure to air and light, in absence of inhibitors

NFPA: Health 2, Flammability 4, Reactivity 1

Uses: Organic synthesis; mfg. of monoethanolamine; smokeless powders; industrial solvent for paints; analytical chemistry; anesthetic; extractant; solvent for waxes, fats, and oils; fuel additive; pharmaceutical antiseptic, analgesic, expectorant

Regulatory: FDA approved for transmucosal pharmaceuticals

Manuf./Distrib.: Aldrich; J.T. Baker; Exxon; Fluka; Hüls AG; Mallinckrodt; Quantum; Sigma; Spectrum Chem. Mfg.; Van Waters & Rogers

Ethyl 3-ethoxypropionate

CAS 763-69-9

Synonyms: EEP; Ethyl-β-ethoxypropionate; Ethoxypropionic acid ethyl ester; 3-Ethoxypropionic acid ethyl ester

Empirical: $C_7H_{14}O_3$

Formula: $C_2H_5OC_3H_5O_2C_2H_5$

Properties: Liq.; sl. sol. in water; m.w. 146.19; dens. 0.9496 (20/20 C); b.p. 165-172 C; f.p. < -50 C; flash pt. (Seta) 58 C; ref. index 1.4074 (20 C)

Toxicology: LD50 (oral, rat) 5000 mg/kg, (skin, rabbit) 10 g/kg; mildly toxic by ingestion and skin contact; skin and eye irritant; heated to decomp., emits acrid smoke and irritating fumes

Precaution: Flamm. when exposed to heat or flame; can react with oxidizing materials

Uses: Solvent for polymerization, esp. high-solids acrylics; retarder solvent for coatings; let-down solvent for thinning synthetic coating resins, esp. moisture-sensitive resins for PU coatings; air-dry and baking finishes; intermediate for vitamin B_1

Manuf./Distrib.: Aldrich; Ashland; Eastman; Houghton Chem.; Union Carbide

Trade names: Ektapro® EEP Solvent; Ucar® Ester EEP

Ethyl-β-ethoxypropionate. *See* Ethyl 3-ethoxypropionate

Ethyl ethynyl methyl carbinol. *See* Methyl pentynol

Ethyl formate

CAS 109-94-4; EINECS 203-721-0

UN No. 1190 (DOT)

Synonyms: Ethyl formic ester; Formic acid ethyl ester; Ethyl methanoate; Formic ether

Classification: Aliphatic organic compd.; carboxylic acid ester

Definition: Ester of ethyl alcohol and formic acid

Empirical: $C_3H_6O_2$

Formula: $HCOOC_2H_5$

Properties: Colorless liq., sharp rum-like odor; very sol. in water; sol. in acetone, alcohol, benzene, ether, fixed oils, propylene glycol; sl. sol. in min. oil; m.w.

74.09; sp.gr. 0.9236 (20/20 C); vapor pressure 100 mm Hg (5.4 C); m.p. -79 C; b.p. 54 C; flash pt. (CC) -20 C; ref. index 1.359

Toxicology: LD50 (oral, rat) 1850 mg/kg, (skin, rabbit) 20 g/kg; ACGIH TLV/TWA 100 ppm; mod. toxic by ingestion, subcutaneous routes; mildly toxic by skin contact and inh.; skin and eye irritant

Precaution: DOT: Flamm. liq.; dangerous fire and explosion hazard exposed to heat, flame, oxidizers; incompat. with oxidizing materials, acids (decomp. can occur), bases (vigorous decomp. can occur), moisture; Hazardous decomp. prods.: Ethanol, formic acid; dec. slowly in water, releasing formic acid; heated to decomp., emits acrid smoke and irritating fumes

NFPA: Health 2, Flammability 3, Reactivity 0

Uses: Ethereal, fruity fragrance and flavoring, rum-like on dilution; intermediate for prod. of pharmaceuticals and crop protection agents; solvent for nitrocellulose; insecticide, fungicide, and larvicide for tobacco, cereals, dried fruit, etc.; acetone substitute; org. synthesis; syn. flavoring agent for foods and pharmaceuticals; reagent

Regulatory: FDA 21CFR §172.515, 184.1295, GRAS; 193.210, insecticide residue tolerance of 250 ppm in raisins and currants; FEMA GRAS

Manuf./Distrib.: AC Ind.; Aldrich; BASF; Fluka; Int'l. Chem. Group; Penta Mfg.; Rhône-Poulenc N. Am.; Sigma; Spectrum Chem. Mfg.

Ethyl formic ester. *See* Ethyl formate
Ethyl glycol. *See* Ethoxyethanol
Ethyl glyme. *See* Ethylene glycol diethyl ether

Ethyl hexanediol

CAS 94-96-2; EINECS 202-377-9

Synonyms: Ethohexadiol; 1,3-Octanediol; Octylene glycol; 1,3-Octylene glycol; 2-Ethyl-1,3-hexanediol; 2-Ethylhexane-1,3-diol; 2-Ethylhexanediol-1,3; Ethyl hexylene glycol; 2-Ethyl-3-propyl-1,3-propanediol; 3-Hydroxymethyl-n-yeptan-4-ol

Classification: Aliphatic alcohol

Empirical: $C_8H_{18}O_2$

Formula: $CH_3CH_2CH_2CHOHCHCH_2CH_3CH_2OH$

Properties: Colorless, sl. visc. liq., odorless; sol. in alcohol, ether, acetone, toluene, benzene, CCl_4; partly sol. in water; m.w. 146.26; dens. 0.9422 (20/20 C); bulk dens. 7.8 lb/gal (20 C); vapor pressure < 0.01 mm Hg (20 C); f.p. -40 C; b.p. 244 C; flash pt. 126 C; ref. index 1.4465-1.4515

Toxicology: LD50 (oral, rat) 1400 mg/kg, (skin rabbit) 2000 mg/kg; moderately toxic by ingestion and skin contact; skin and severe eye irritant

Precaution: Combustible exposed to heat or flame; can react with oxidizers; Hazardous decomp. prods.: Heated to decomp., emits acrid smoke and irritating fumes

Uses: Insecticide, insect repellent, raw material for oily cosmetics, vehicle and solvent in printing inks, medicine, chelating agent for boric acid; brake oil; pharmaceutical solvent for topical ointments

Regulatory: FDA approved for topicals

Manuf./Distrib.: Aldrich; Dixie; Fluka; Hüls Am.; Union Carbide

2-Ethylhexanediol-1,3. *See* Ethyl hexanediol
2-Ethyl-1,3-hexanediol. *See* Ethyl hexanediol
2-Ethylhexane-1,3-diol. *See* Ethyl hexanediol
2-Ethylhexanoic acid, cetyl/stearyl ester. *See* Cetearyl octanoate
2-Ethylhexanoic acid, 1-methyl-1,2-ethanediyl ester. *See* Propylene glycol dioctanoate

2-Ethylhexanol
CAS 104-76-7; EINECS 203-234-3
Synonyms: 2-EH; 2-Ethylhexyl alcohol; 2-Ethyl-1-hexanol; Octyl alcohol; Alcohol C_8
Empirical: $C_8H_{18}O$
Formula: $CH_3(CH_2)_3CHC_2H_5CH_2OH$
Properties: Colorless liq., mild oily sweet sl. rose fragrance; misc. with most org. solvs.; sl. sol. in water; m.w. 130.26; dens. 0.83 (20 C); b.p. 183.5 C; f.p. -76 C; flash pt. 81.1 C; ref. index 1.4300 (20 C)
Toxicology: LD50 (rat, oral) 12.46 ml/kg; moderately toxic by ingestion, skin contact, IP, subcut., parenteral routes; severe eye irritant; moderate skin irritant; heated to decomp., emits acrid smoke and fumes
Precaution: Dangerous fire hazard when exposed to heat or flame; can react vigorously with oxidizing materials
Uses: Plasticizer for PVC resins; defoaming agent, wetting agent, organic synthesis, solvent mix for nitrocellulose; solvent for paints, waterborne coatings; penetrant for plasticizing inks, etc.; syn. flavoring agent for foods and pharmaceuticals
Regulatory: FEMA GRAS
Manuf./Distrib.: Aldrich; Aristech; Ashland; BASF; BP Chems. Ltd; Coyne; Eastman; Fluka; Hoechst AG; Houghton Chem.; Penta Mfg.; Shell

2-Ethyl-1-hexanol. *See* 2-Ethylhexanol; Isooctyl alcohol
2-Ethyl-1-hexanol acetate. *See* 2-Ethylhexyl acetate
2-Ethyl-1-hexanol phosphate. *See* Trioctyl phosphate
2-Ethylhexanyl acetate. *See* 2-Ethylhexyl acetate

2-Ethylhexyl acetate
CAS 103-09-0
Synonyms: Acetic acid 2-ethylhexyl ester; β-Ethylhexyl acetate; 2-Ethylhexanyl acetate; 2-Ethylhexyl ethanoate; 2-Ethyl-1-hexanol acetate; 2-Ethyl-1-hexyl acetate; Octyl acetate
Classification: Carboxylic acid ester
Empirical: $C_{10}H_{20}O_2$
Formula: $CH_3COOCH_2CHC_2H_5C_4H_9$
Properties: Water-wh. stable liq.; sweet odor; sol. in ether, alcohol; very sl. sol. in water; m.w. 172.27; sp.gr. 0.873 (20/20 C); m.p. -93 C; b.p. 199-205 C; flash pt. (CC) 71 C; ref. index 1.4201 (20 C); pH neutral
Toxicology: LD50 (oral, rat) 3000 mg/kg; mod. toxic by ingestion; sl. to mod. skin and eye irritant; may be sl. absorbed thru skin; inh. may cause nose/throat irritation; unlikely to cause significant health effects
Precaution: Combustible; incompat. with strong acids or bases (decomp. can occur), oxidizing materials (increases fire/explosion risk); Hazardous decomp. prods.: Heated to decomp., emits acrid smoke and fumes
NFPA: Health 2, Flammability 2, Reactivity 0
Uses: Solvent for nitrocellulose, resins, lacquers, baking finishes, waxes, oils; in fragrances
Manuf./Distrib.: Aceto; Allan; Ashland; Eastman; Houghton Chem.; Hüls AG; Penta Mfg.; Union Carbide; Van Waters & Rogers
See also Octyl acetate

2-Ethyl-1-hexyl acetate. *See* 2-Ethylhexyl acetate
β-Ethylhexyl acetate. *See* 2-Ethylhexyl acetate
2-Ethylhexyl alcohol. *See* 2-Ethylhexanol
Ethyl hexylene glycol. *See* Ethyl hexanediol

2-Ethylhexyl ethanoate. *See* 2-Ethylhexyl acetate
2-Ethylhexylhydroxystearate. *See* Octyl hydroxystearate
2-Ethylhexyl isononanoate. *See* Octyl isononanoate
2-Ethylhexyl oxystearate. *See* Octyl hydroxystearate
2-Ethylhexyl sebacate. *See* Dioctyl sebacate
Ethyl 2-hydroxypropionate. *See* Ethyl lactate
Ethyl α-hydroxypropionate. *See* Ethyl lactate
Ethyl-S(-)-2-hydroxypropionate. *See* Ethyl lactate
Ethylidene diethyl ether. *See* Acetal
Ethylidenelactic acid. *See* Lactic acid
2,2´-(Ethylimino)bisethanol. *See* N-Ethyldiethanolamine
2,2´-(Ethylimino)diethanol. *See* N-Ethyldiethanolamine
N-Ethyl-2,2´-iminodiethanol. *See* N-Ethyldiethanolamine
Ethylisobutylmethane. *See* Isoheptane

Ethyl lactate
CAS 97-64-3; EINECS 202-598-0
UN No. 1192
Synonyms: Lactic acid ethyl ester; 2-Hydroxypropionic acid ethyl ester; Ethyl α-hydroxypropionate; Ethyl 2-hydroxypropionate; Ethyl-S(-)-2-hydroxypropionate
Definition: Ethyl ester of lactic acid
Empirical: $C_5H_{10}O_3$
Formula: $CH_3CHOHCOOC_2H_5$
Properties: Colorless liq., mild odor; misc. with water, alcohol, ketones, esters, hydrocarbons, oil; m.w. 118.13; dens. 1.020-1.036 (20/20 C); m.p. -26 C; b.p. 154 C; flash pt. 46.1 C; ref. index 1.410-1.420
Toxicology: LD50 (oral, mouse) 2500 mg/kg; moderately toxic by ingestion, intraperitoneal, subcutaneous, intravenous routes; heated to decomp., emits acrid smoke and irritating fumes
Precaution: Flamm. or combustible; can react with oxidizers; sl. explosion hazard in vapor form exposed to flame
Uses: Dipped latex prods., coatings, specialty finishes, inks, diluents, adhesives, intermediates, photoresist solvents, screen printing of electronic parts, solvent resins, safety glass; syn. flavoring and fragrance for foods and pharmaceuticals; non-ozone depleting solvent for electronics, coatings, cleaning agents, precision cleaning, inks, metal industry
Regulatory: FDA 21CFR §172.515; FEMA GRAS
Manuf./Distrib.: Aldrich; CPS; Farleyway Chem. Ltd; Int'l. Chem. Inc; Penta Mfg.; Purac Am.
Trade names: Purasolv® EL; Purasolv® ELECT; Purasolv® ELS

Ethyllic acid. *See* Acetic acid
Ethyl methanoate. *See* Ethyl formate
Ethylmethyl carbinol. *See* 2-Butanol
Ethyl methyl ketone. *See* Methyl ethyl ketone
Ethyl nitrile. *See* Acetonitrile
Ethyl 9-octadecenoate. *See* Ethyl oleate

Ethyl oleate
CAS 111-62-6; 85049-36-1; EINECS 203-889-5; 285-206-0
Synonyms: 9-Octadecenoic acid ethyl ester; Ethyl 9-octadecenoate
Definition: Ester of ethyl alcohol and oleic acid
Empirical: $C_{20}H_{38}O_2$
Formula: $CH_3(CH_2)_7CH{:}CH(CH_2)_7COOC_2H_5$
Properties: Ylsh. oily liq.; misc. with alcohol, ether, veg. oils; insol. in water; m.w.

310.52; dens. 0.870 (20/4 C); visc. 75.15 cp; m.p. -32 C; b.p. 216-218 (15 mm); acid no. 0.5 max.; iodine no. 75-85; sapon. no. 177-188; flash pt. 175 C; ref. index 1.451

Toxicology: No known toxicity

Precaution: Combustible

Uses: Syn. flavoring agent for foods; pharmaceutical vehicle, solvent

Regulatory: FDA 21CFR §172.515; FEMA GRAS; USP/NF, BP compliance

Manuf./Distrib.: Aldrich; Fluka; Penta Mfg.; Pfaltz & Bauer; Sigma

Ethyl oxide. *See* Ethyl ether
Ethyl 3-oxobutanoate. *See* Ethylacetoacetate
Ethyl phosphate. *See* Triethyl phosphate
Ethyl phthalate. *See* Diethyl phthalate
Ethyl propanoate. *See* Ethyl propionate

Ethyl propionate

CAS 105-37-3; EINECS 203-291-4

UN No. 1195

Synonyms: Propanoic acid ethyl ester; Ethyl propanoate; Propionic ether

Classification: Carboxylic acid ester

Definition: Ester of propionic acid and ethyl alcohol

Empirical: $C_5H_{10}O_2$

Formula: $CH_3CH_2COOC_2H_5$

Properties: Colorless liq., fruity odor; sol. in ≈ 60 parts water; sol. in most org. solvs.; misc. with alcohol, ether; m.w. 102.14; sp.gr. 0.890 (20/4 C); m.p. -73 C; b.p. 96-99 C; flash pt. (CC) 12 C; ref. index 1.384 (20 C)

Toxicology: LD50 (oral, rabbit) 3500 mg/kg; mod. toxic by ingestion and IP routes; skin and eye irritant; probable respiratory tract irritant; high concs. may cause dizziness, headache, extreme concs. may cause unconsciousness, death

Precaution: Highly flamm. liq.; dangerous fire risk; reacts with oxidizing materials; Hazardous decomp. prods.: May hydrolyze slowly to ethanol and propionic acid if water is present; heated to decomp., emits acrid smoke and fumes

NFPA: Flammability:; 3, Reactivity 0

Uses: Solvent for paints, various natural and synthetic resins, cellulose ethers and esters; syn. flavoring agent for foods, fruit syrups, pharmaceuticals; cutting agent for pyroxylin

Regulatory: FDA 21CFR §172.515; FEMA GRAS; Japan approved as flavoring

Manuf./Distrib.: Aldrich; Ashland; Berje; Fluka; Northwestern Flavors; Penta Mfg.; Sigma; Union Carbide

2-Ethyl-3-propyl-1,3-propanediol. *See* Ethyl hexanediol
1-Ethyl-2-pyrrolidinone. *See* N-Ethyl-2-pyrrolidone

N-Ethyl-2-pyrrolidone

CAS 2687-91-4

Synonyms: 1-Ethyl-2-pyrrolidinone

Properties: Colorless liq.; m.w. 113.16; b.p. 97 C (20 mm)

Uses: Reaction solvent for chemical reactions down to -70 C; antibiotic prod.; elec. capacitors; mfg. of polyamide fibers and film; ink; photoresist stripper; polymerization; synthesis

Trade names: NEP®

2-Ethynyl-2-butanol. *See* Methyl pentynol
Ethynyldimethyl carbinol. *See* Methyl butynol
EtOH. *See* Alcohol

Eucalyptol

CAS 470-82-6; EINECS 207-431-5

Synonyms: Cineol; Cineole; 1,8-Cineol; 1,8-Cineole; Cajeputol; Eucalyptole; 1,8-Epoxy-p-menthane; Limonene oxide; p-Menthane, 1,8-epoxy; 2-Oxabicyclo (2.2.2) octane, 1,3,3-trimethyl-; 1,8-Oxido-p-menthane; 1,3,3-Trimethyl-2-oxabicyclo[2.2.2]octane
Definition: Constituent of oil of eucalyptus
Empirical: $C_{10}H_{18}O$
Properties: Colorless liq., camphor-like odor, spicy cooling taste; misc. with alcohol, chloroform, ether, glac. acetic acid, fixed or volatile oils; pract. insol. in water; m.w. 154.25; dens. 0.924 (20/4 C); m.p. > 1.5 C; b.p. 175-179 C; flash pt. 49 C; ref. index 1.4550-1.4600 (20 C)
Toxicology: LD50 (oral, rat) 2480 mg/kg, (subcut., mouse) 1070 mg/kg; poison by subcut. and intramuscular routes; mod. toxic by ing.; experimental carcinogen, reproductive effects; mutagen
Precaution: Combustible liq.; Hazardous decomp. prods.: Heated to decomp., emits acrid smoke and fumes
Uses: Pharmaceuticals (cough syrups, expectorants), flavoring for bakery prods., beverages, chewing gum, confections, ice cream; perfumery
Regulatory: FDA 21CFR §172.515; 27CFR §21.65, 21.151; FEMA GRAS; Japan approved as flavoring; FDA approved for dentals
Manuf./Distrib.: Aldrich; Berje; Fluka; Penta Mfg.; Polarome Mfg.; Quest Int'l.; Ruger; SCM Glidco Organics; Sigma; Ungerer
Trade names containing: Acintene® AC-P; Acintene® DP 738

Eucalyptole. *See* Eucalyptol

F-11. *See* Trichlorofluoromethane
F 12. *See* Dichlorodifluoromethane
F 22. *See* Chlorodifluoromethane
F112. *See* 1,1,2,2-Tetrachloro-1,2-difluoroethane
F 114. *See* Dichlorotetrafluoroethane
Fatty acids, tallow, methyl esters. *See* Methyl tallowate
FC 11. *See* Trichlorofluoromethane
FC 12. *See* Dichlorodifluoromethane
FC-112. *See* 1,1,2,2-Tetrachloro-1,2-difluoroethane
FC 113. *See* Trichlorotrifluoroethane
FC 114. *See* Dichlorotetrafluoroethane
FC142b. *See* 1,1,1-Chlorodifluoroethane
Fluorane 114. *See* Dichlorotetrafluoroethane
Fluorocarbon-11. *See* Trichlorofluoromethane
Fluorocarbon-12. *See* Dichlorodifluoromethane
Fluorocarbon 22. *See* Chlorodifluoromethane
Fluorocarbon 112. *See* 1,1,2,2-Tetrachloro-1,2-difluoroethane
Fluorocarbon 113. *See* Trichlorotrifluoroethane
Fluorocarbon-114. *See* Dichlorotetrafluoroethane
Fluorocarbon FC142b. *See* 1,1,1-Chlorodifluoroethane
Fluorotrichloromethane. *See* Trichlorofluoromethane
Food freezant 12. *See* Dichlorodifluoromethane
Formal glycol. *See* 1,3-Dioxolane

Formamide
CAS 75-12-7; EINECS 200-842-0

Synonyms: Amide C_1; Methanamide; Methanoic acid, amide; Carbamaldehyde
Classification: Amide
Empirical: CH_3NO
Formula: $HCONH_2$
Properties: Colorless oily liq.; sol. in water, ethanol, methanol, phenol, acetic acid, dioxane, acetone, ethylene glycol; very sl. sol. in ether; m.w. 45.04; sp.gr. 1.134; vapor pressure 29.7 mm Hg (129.4 C); m.p. 2.6 C; b.p. 210 C (dec.); flash pt. (OC) 154 C; pH 7.3
Toxicology: LD50 (oral, rat) 5577 mg/kg, (subcut., rat) > 4 g/kg; ACGIH TLV/TWA 20 ppm (skin); poison by skin contact, subcut. routes; mod. toxic by ing., IP routes; primary irritant; skin and eye irritant
Precaution: Combustible exposed to heat or flame; can react vigorously with oxidizers; incompat. with Karl Fischer reagent (toluene/pyridine/sulfur trioxide); reaction may cause explosion of closed container; incompat. with I_2, pyridine, SO_3; Hazardous decomp. prods.: Heated to decomp., emits toxic fumes of NO_x
NFPA: Health 2, Flammability 1
Uses: Solvent; softener; intermediate in organic synthesis
Manuf./Distrib.: AC Ind.; Aldrich; Alfa Aesar Johnson Matthey; AMRESCO; BASF; BASF AG; Electron Microscopy Sciences; Fluka; Nat'l. Diagnostics; Penta Mfg.; Research Organics; Sigma; Spectrum Chem. Mfg.

Formic acid
CAS 64-18-6; EINECS 200-579-1
UN No. 1779 (DOT)
Synonyms: Hydrogen carboxylic acid; Methanoic acid; Aminic acid
Classification: Organic acid
Empirical: CH_2O_2
Formula: HCOOH
Properties: Colorless fuming liq., penetrating odor; sol. in water, alcohol, ether; m.w. 46.03; dens. 1.22 (20/4 C); m.p. 8.3 C; b.p. 100.8 C; flash pt. (OC) 69 C; ref. index 1.3714
Toxicology: LD50 (oral, rat) 1100 mg/kg; poison by IV route; mod. toxic by ingestion, intraperitoneal routes; skin and severe eye irritant; TLV 5 ppm in air; heated to decomp., emits acrid smoke and irritating fumes
Precaution: Flamm.; DOT: corrosive material; can react vigorously with oxidizers; explosive with furfuryl alcohol
NFPA: Health 3, Flammability 2, Reactivity 0
Uses: Reducing agent, dyeing and finishing of textiles, leather treatment, chemicals, mfg. of fumigants, insecticides, refrigerants, solvents for perfumes, lacquers; electroplating; silvering glass; ore flotation; paper mfg. aid; antibacterial preservative for foods; syn. flavoring adjunct for foods and pharmaceuticals
Regulatory: FDA 21CFR §172.515, 186.1316, GRAS as indirect additive, 573.480; FEMA GRAS; Europe listed; prohibited in UK
Manuf./Distrib.: Aldrich; Ashland; BASF; BASF AG; BP Chems. Ltd; Coyne; Fluka; Hoechst Celanese; Houghton Chem.; Mallinckrodt; Norsk Hydro AS; Sigma; Spectrum Chem. Mfg.

Formic acid ethyl ester. *See* Ethyl formate
Formic acid isobutyl ester. *See* Isobutyl formate
Formic acid methyl ester. *See* Methyl formate
Formic ether. *See* Ethyl formate
N-Formyldimethylamine. *See* Dimethyl formamide
Formyl trichloride. *See* Chloroform
Freon 22. *See* Chlorodifluoromethane

Freon 112. *See* 1,1,2,2-Tetrachloro-1,2-difluoroethane
Freon 113. *See* Trichlorotrifluoroethane
Freon 114. *See* Dichlorotetrafluoroethane
Freon 142. *See* 1,1,1-Chlorodifluoroethane
Freon 142b. *See* 1,1,1-Chlorodifluoroethane
Fural. *See* Furfural
2-Furaldehyde. *See* Furfural
Furale. *See* Furfural
2-Furanaldehyde. *See* Furfural
2-Furancarbinol. *See* Furfuryl alcohol
2-Furancarbonal. *See* Furfural
2-Furancarboxaldehyde. *See* Furfural
Furanidine. *See* Tetrahydrofuran
2-Furanmethanol. *See* Furfuryl alcohol
Furan, tetrahydro-. *See* Tetrahydrofuran

Furfural
CAS 98-01-1; EINECS 202-627-7
UN No. 1199 (DOT)
Synonyms: 2-Furaldehyde; Fural; Furale; 2-Furanaldehyde; 2-Furancarbonal; 2-Furfural; Furfuraldehyde; Furfurol; Furfurole; 2-Furancarboxaldehyde; Furole; α-Furole; 2-Furyl methanal; Pyromucic aldehyde; Artificial oil of ants; Artificial ant oil
Classification: Cyclic aldehyde
Empirical: $C_5H_4O_2$
Formula: C_4H_3OCHO
Properties: Colorless liq. (pure); reddish-brown (on exposure to air and light); almond-like odor; sol. in alcohol, ether, benzene, 8.3% in water; m.w. 96.08; dens. 1.1598 (20/4 C); f.p. -36.5 C; b.p. 161.7 C; flash pt. 60 C; ref. index 1.5260 (20 C)
Toxicology: LD50 (oral, rat) 65 mg/kg; poison by ingestion, intraperitoneal, subcutaneous, IV, IM routes; moderately toxic by inhalation; TLV 2 ppm in air; irritates mucous membranes, acts on CNS; heated to decomp., emits acrid smoke and irritating fumes
Precaution: Flamm. or combustible; can react with oxidizing materials; mod. explosion hazard exposed to heat or flame or by chem. reaction; incompat. with strong min. acids or alkalies
NFPA: Health 3, Flammability 2, Reactivity 0
Uses: Chemical intermediate for mfg. of derivs. (furan, THF), paints; solvent for petrol. lube, nitrocellulose; wetting agent; in mfg. of furfural-phenol plastics; vulcanization accelerator; insecticide, fungicide, germicide; reagent in analytical chemistry; syn. flavoring agent for foods and pharmaceuticals
Regulatory: FDA 21CFR §175.105; FEMA GRAS; SARA reportable; Japan approved as flavoring
Manuf./Distrib.: Aldrich; Allchem Ind.; Fluka; Great Lakes; Penta Mfg.; QO; Sigma; Spectrum Chem. Mfg.; Van Waters & Rogers
Trade names: QO® Furfural

2-Furfural. *See* Furfural
Furfural alcohol. *See* Furfuryl alcohol
Furfuralcohol. *See* Furfuryl alcohol
Furfuraldehyde. *See* Furfural
Furfurol. *See* Furfural

Furfuryl alcohol
CAS 98-00-0; EINECS 202-626-1

UN No. 2874 (DOT)
Synonyms: 2-Furancarbinol; 2-Furanmethanol; 2-Hydroxymethylfuran; α-Furylcarbinol; 2-Furylcarbinol; (2-Furyl) methanol; Furfural alcohol; Furfuralcohol; Furyl alcohol
Classification: Heterocyclic substituted alcohol
Empirical: $C_5H_6O_2$
Formula: $C_4H_3OCH_2OH$
Properties: Colorless mobile liq., brn.-dk. red (air/lt. exposed), low odor, cooked sugar taste; sol. in alcohol, ether, chloroform, benzene; misc. with water but unstable; insol. in petrol. hydrocarbons, most oils; m.w. 98.10; sp.gr. 1.1285 (20/4 C); m.p. -29 C; b.p. 170 C; flash pt. (OC) 75 C; ref. index 1.485
Toxicology: ACGIH TLV/TWA 10 ppm (skin); STEL 15 ppm (skin); LD50 (oral, rat) 275 mg/kg, (dermal, rabbit) 450 mg/kg; poison by ingestion, skin contact, subcut. route; mod. toxic by inh, IP routes; mild skin irritant; may cause severe eye irritation
Precaution: Combustible; can react with oxidizing materials; mod. explosion hazard when exposed to heat or flame; may react explosively with mineral and some organic acids; Hazardous decomp. prods.: Heated to decomp., emits acrid smoke and fumes
NFPA: Health 1, Flammability 2, Reactivity 1
Uses: Mfg. of phenolic and thermosetting resins; wetting agent; furan polymers; foundry sand binders; corrosion-resist. resins; intermediate for esterification and etherification, paints; solvent-plasticizer for phenolic resins; solvent for dyes and resins; visc. reducer, cure promoter, and carrier in amine-cured epoxy resins; polymer sealants/cements; syn. flavoring for foods and pharmaceuticals
Regulatory: FEMA GRAS
Manuf./Distrib.: AC Ind.; Aldrich; Allchem Ind.; Ashland; Browning; Fluka; Great Lakes; Penta Mfg.; QO; Schweizerhall; Spectrum Chem. Mfg.; Van Waters & Rogers
Trade names: QO® Furfuryl Alcohol (FA®)

Furole. *See* Furfural
α-Furole. *See* Furfural
Furyl alcohol. *See* Furfuryl alcohol
2-Furylcarbinol. *See* Furfuryl alcohol
α-Furylcarbinol. *See* Furfuryl alcohol
2-Furyl methanal. *See* Furfural
(2-Furyl) methanol. *See* Furfuryl alcohol

Gas. *See* Gasoline
Gasolene. *See* Gasoline

Gasoline
CAS 8006-61-9; 68425-31-0; 68514-15-8
UN No. 1203 (DOT), 1257 (DOT)
Synonyms: Automotive gasoline; Petrol; Motor fuel; Motor spirit; Motor spirits; Benzin; Natural gasoline; Gasolene; Gas
Classification: Mixed petroleum hydrocarbons
Definition: Common commercial unleaded automotive gasoline; may contain

additives and blending agents

Properties: Colorless liq.; char. odor; insol. in water; sol. in ether, chloroform, benzene, ethanol, other petrol. solvs.; m.w. 108 (avg.); sp.gr. 0.72-0.76; m.p. < -60 C; b.p. 50-200 C; flash pt. -43 C; pH 9 max.

Toxicology: LC50 (inh., rat, 5 min) 300 g/m^3; LCLo (inh., mammal, 5 min) 30,000 ppm; ACGIH TLV/TWA 300 ppm; STEL 500 ppm; mod. toxic by ing.; may cause burning in mouth, throat, and chest, stomach irritation, nausea, vomiting, cyanosis

Precaution: Extremely flamm.; very dangerous fire and explosion hazard exposed to heat or flame; incompat. with strong oxidizing agents (can cause fire or explosion); common air contaminant

NFPA: Health 1, Flammability 3, Reactivity 0

Uses: Fuel for motor vehicles; some use as diluent and solvent

Gingelly oil. *See* Sesame oil

Gingilli oil. *See* Sesame oil

Glacial acetic acid. *See* Acetic acid, glacial

Glicerol. *See* Glycerin

D-Glucitol, 1,4:3,6-dianhydro-2,5-di-o-methyl. *See* Dimethyl isosorbide

Glycereth-7 triacetate

CAS 57569-76-3

Synonyms: PEG-7 glyceryl ether triacetate; POE (7) glyceryl ether triacetate

Definition: Triester of acetic acid with a PEG ether of glycerin containing an avg. 7 moles ethylene oxide

Empirical: $C_{23}H_{42}O_{13}$

Properties: Clear pale yel. liq., mild typ. odor; sol. in water; m.w. 526; sp.gr. 1.15; b.p. > 200 C

Toxicology: May cause mild skin and eye irritation on prolonged contact

Precaution: Store away from strong oxidizing agents; avoid excessive heat

Uses: Humectant, emollient, solubilizer; solvent for sunscreen ingreds.

Trade names: Pelemol® G7A

Glycerides, lard mono-, acetates. *See* Acetylated lard glyceride

Glycerides, lard mono-, hydrogenated, acetates. *See* Acetylated hydrogenated lard glyceride

Glycerides, palm kernel oil mono-, di-, and tri-, acetates. *See* Acetylated palm kernel glycerides

Glycerin

CAS 56-81-5; EINECS 200-289-5

Synonyms: Glicerol; Glycerol; Glycyl alcohol; 1,2,3-Propanetriol; Propane-1,2,3-triol; Glycerine; Trihydroxypropane glycerol

Classification: Polyhydric alcohol

Empirical: $C_3H_8O_3$

Formula: $HOCH_2COHHCH_2OH$

Properties: Clear colorless syrupy liq., odorless, sweet taste; hygroscopic; sol. in water, alcohol; insol. in ether, benzene, chloroform; m.w. 92.09; dens. 1.26201 (25/25 C); m.p. 17.8 C; b.p. 290 C (dec.); flash pt. (OC) 176 C; ref. index 1.4730 (25 C)

Toxicology: LD50 (oral, rat) > 20 ml/kg, (IV, rat) 4.4 ml/kg; poison by subcutaneous route; mildly toxic by ingestion; human systemic and GI effects by ingestion; human mutagenic data; skin and eye irritant; nuisance dust; heated to dec., emits acrid smoke

Precaution: Combustible liq. exposed to heat, flame, strong oxidizers; highly explosive with hydrogen peroxide

NFPA: Health 1, Flammability 1, Reactivity 0

Uses: Alkyd resins, paints, dynamite, ester gums, cosmetics, perfumery, lubricants, softener, bacteriostat, penetrant, emollient; antifreeze; in prod. of antibiotics in OTC drugs; solvent, humectant, plasticizer, bodying agent for foods, food pkg. materials; plasticizer, solvent, solubilizer, preservative for pharmaceuticals

Use level: 0.2-65.7% (topicals), 1-50% (liq. orals), 50% (parenterals), 7-10% (dentifrices), 0.5-3.0% (ophthalmics)

Regulatory: FDA 21CFR §169.175, 169.176, 169.177, 169.178, 169.180, 169.181, 172.866, 175.300, 178.3500, 182.90, 182.1320, GRAS; FEMA GRAS; FDA approved for orals, parenterals, ophthalmics, dentifrices, intramuscular injectables, rectals; USP/NF compliance; Japan approved; Europe listed; UK approved; BP compliance

Manuf./Distrib.: Akzo Nobel; Alba Int'l.; Aldrich; Allchem Ind.; Alnor Oil; Asahi Denka Kogyo; Ashland; Baychem; Boehle; Chemcentral; Croda Universal Ltd; Dial; Farleyway Chem. Ltd; Fina Chems.; Fluka; C.P. Hall; Henkel/Emery; Houghton Chem.; Lonza; Penta Mfg.; Procter & Gamble; Samson; Sigma; Spectrum Chem. Mfg.; Unichema Int'l.; United Min. & Chem.; Van Waters & Rogers; Welch, Holme & Clark; Witco/Oleo-Surf.

Trade names: Emery® 912; Emery® 916; Emery® 917; Emery® 918; Kemstrene® 96.0% USP; Kemstrene® 99.0%; Kemstrene® 99.7% USP; Kemstrene® High Gravity

Glycerine. *See* Glycerin
Glycerine triacetate. *See* Triacetin
Glycerin monoacetate. *See* Glyceryl acetate
Glycerin monostearate. *See* Glyceryl stearate
Glycerol. *See* Glycerin
Glycerol-1-acetate. *See* Glyceryl acetate
Glycerol formal. *See* Glyceryl formal
Glycerol monoacetate. *See* Glyceryl acetate
Glycerol-1-monoacetate. *See* Glyceryl acetate
Glycerol-α-monoacetate. *See* Glyceryl acetate
Glycerol monostearate. *See* Glyceryl stearate
Glycerol stearate. *See* Glyceryl stearate
Glycerol triacetate. *See* Triacetin

Glyceryl acetate

CAS 106-61-6; 26446-35-5

Synonyms: 1-Acetate-1,2,3-propanetriol; Acetic acid, monoglyceride; Acetin; 2,3-Dihdyroxypropyl acetate; Glycerin monoacetate; Glycerol-1-acetate; Glycerol monoacetate; Glycerol-α-monoacetate; Glycerol-1-monoacetate; Glyceryl monoacetate; Monacetin; α-Monacetin; 1-Monacetin; Monacetyl glycerine; 1,2,3-Propanetriol monoacetate

Empirical: $C_5H_{10}O_4$

Formula: $C_3H_7O_2 \cdot OOCCH_3$

Properties: Colorless or pale yel. (commercial) liq.; char. odor; very hygroscopic; sol. in water, alcohol; sl. sol. in ether; insol. in benzene; m.w. 134.13; dens. 1.206 (20/4 C); b.p. 158 C (17 mm); ref. index 1.451

Toxicology: LD50 (subcut., rat) 5500 mg/kg, (subcut., mouse) 3500 mg/kg; mod. toxic by subcut. route; mutagen; Hazardous decomp. prods.: Heated to decomp., emits acrid smoke and irritating fumes

Uses: In mfg. of smokeless powd., dynamite; solvent for basic dyes; in tanning leather; plasticizer for cellulose acetate, cellulose nitrate

Manuf./Distrib.: C.P. Hall; Penta Mfg.; Sigma

Glyceryl caprylate
CAS 26402-26-6; EINECS 247-668-1
Synonyms: Glyceryl monocaprylate; Octanoic acid, monoester with 1,2,3-propanetriol; Monooctanoin
Definition: Monoester of glycerin and caprylic acid
Empirical: $C_{11}H_{22}O_4$
Formula: $CH_3(CH_2)_6COOCH_2COHHCH_2OH$
Properties: Crystals; m.p. 39.5-40.5 C
Uses: Solubilizer, emulsifier for pharmaceutical drugs, cosmetics; carrier/vehicle for drugs in capsules; antistat for syn. materials; solvent, solubilizer for active ingreds.
Regulatory: FDA 21CFR §176.210, 177.2800
Trade names: Imwitor® 908

Glyceryl caprylate/caprate
Definition: Mixture of monoglycerides of caprylic and capric acids
Uses: Cosolvent and coupler for org. compds.; w/o emulsifier; antistat and lubricant for syn. materials; emulsifier for cosmetics, foods, pharmaceuticals; solvent, dispersant, solubilizer, vehicle, penetrant for pharmaceuticals
Regulatory: FDA 21CFR §176.210, 177.2800
Trade names: Capmul® MCM; Capmul® MCM-90

Glyceryl dioleate
CAS 25637-84-7; EINECS 247-144-2
Synonyms: 9-Octadecenoic acid, diester with 1,2,3-propanetriol
Definition: Diester of glycerin and oleic acid
Empirical: $C_{39}H_{72}O_5$
Uses: Surfactant for cosmetics, pharmaceuticals, industrial applics.; emulsifier, thickener, stabilizer, emollient; lubricant for PVC processing; antifog for PVC film; solvent for dyes and pigments
Regulatory: FDA 21CFR §175.105, 176.210, 177.2800
Manuf./Distrib.: Sigma
Trade names: Emerest® 2419

Glyceryl formal
CAS 4740-78-7; EINECS 225-248-9
Synonyms: Glycerol formal
Empirical: $C_4H_8O_3$
Properties: Colorless liq., low odor; hygroscopic; m.w. 104.11; dens. 1.218 (20/4 C); b.p. 95-97 C (20 mm); ref. index 1.451 (20 C)
Toxicology: Low toxicity
Uses: Solvent for pharmaceutical and cosmetic compds.; vehicle for veterinary injectables (sulphadiazine and trimethoprin)
Manuf./Distrib.: Chemie Linz N. Am.; Fluka; Hüls AG; Occidental; Pentagon Chems. Ltd; Sigma

Glyceryl monoacetate. *See* Glyceryl acetate
Glyceryl monocaprylate. *See* Glyceryl caprylate
Glyceryl monooleate. *See* Glyceryl oleate
Glyceryl monostearate. *See* Glyceryl stearate

Glyceryl oleate
CAS 111-03-5; 25496-72-4; 37220-82-9; EINECS 203-827-7; 253-407-2
Synonyms: Glyceryl monooleate; Monoolein; 9-Octadecenoic acid, monoester with 1,2,3-propanetriol
Definition: Monoester of glycerin and oleic acid
Empirical: $C_{21}H_{40}O_4$

Glyceryl stearate

Formula: $CH_3(CH_2)_7CH{=}CH(CH_2)_7CHOHOOCH_2CCH_2OH$

Properties: Yel. oil or soft solid; insol. in water; somewhat sol. in alcohol, most org. solvs.; sp.gr. 0.940-0.960; m.p. 149-19 C; acid no. 3 max.; iodine no. 65-80; sapon. no. 166-174

Toxicology: Heated to decomp., emits acrid smoke and irritating fumes

Precaution: Combustible

Uses: Emulsifier, coemulsifier, stabilizer, wetting agent, lubricant, and antistat; used in cosmetic, pharmaceutical, industrial, food applics.; plasticizer; internal antistat for PE, PP, PVC; antifog, cling agent, lubricant for PVC film; flavor adjuvant, solvent, vehicle, defoamer, dispersant, emulsifier, plasticizer for food use (coffee whiteners, baking mixes, beverages, chewing gum, meat prods., pkg. materials, veg. oil); odorless paints; rust-preventive oils

Regulatory: FDA 21CFR §175.105, 175.300, 176.210, 177.2800, 181.22, 181.27, 182.4505, 184.1323, GRAS; FEMA GRAS; FDA approved for orals

Manuf./Distrib.: ABITEC; Am. Ingreds.; Aquatec Quimica SA; Calgene; Croda Surf. Ltd; Ferro/Keil; Fluka; Grindsted Prods.; Henkel/Emery; ICI Surf. Am.; Inolex; Lonza; Mona Ind.; Sigma; Spectrum Chem. Mfg.; Stepan; Unichema N. Am.; Witco/Oleo-Surf.

Glyceryl stearate

CAS 123-94-4; 11099-07-3; 31566-31-1; 85666-92-8; 85251-77-0; EINECS 250-705-4; 234-325-6; 204-664-4; 286-490-9

Synonyms: GMS; Monostearin; 1,2,3-Propanetriol octadecanoate; Octadecanoic acid monoester with 1,2,3-propanetriol; Glycerin monostearate; Glyceryl monostearate; Glycerol stearate; Glycerol monostearate; 2,3-Dihydroxypropyl octadecanoate; Stearic monoglyceride

Classification: Aliphatic carboxylic acid ester

Definition: Monoester of glycerin and stearic acid

Empirical: $C_{21}H_{42}O_4$

Formula: $CH_3(CH_2)_{16}COOCH_2COHHCH_2OH$

Properties: Wh. to cream wax-like flakes, sl. fatty odor and taste; sol. in hot org. solvs.; insol. in water, ethanol, glycerin, propylene glycol; disp. in min. oil; m.w. 358.57; sp.gr. 0.97; m.p. 56-59 C; b.p. 238-240 C; acid no. 6 max.; iodine no. 3 max.; sapon. no. 162-175; hyd. no. 300-330; flash pt. (OC) 230 C; pH 9.3 (3%)

Toxicology: ACGIH TLV/TWA 10 $mg/^3$ (stearates); LD50 (IP, mouse) 200 mg/kg; poison by IP route; heated to decomp., emits acrid smoke and irritating fumes

Precaution: Combustible dust; incompat. with oxidizing materials (increased risk of fire and explosion); Hazardous decomp. prods.: None

Uses: Nonionic sec. o/w emulsifier for creams and lotions; emulsifier for oils, waxes, solvents; thickener, visc. booster for emulsions; plasticizer for cellulose nitrate; antistat, antifog, lubricant, processing aid for plastics; protective coating for hygroscopic powds.; cosmetics, pharmaceuticals; opacifier; detackifier; food additive (thickener, emulsifier, antisticking, antistalant, solvent)

Regulatory: FDA 21CFR §139.110, 139.115, 139.117, 139.120, 139.121, 139.122, 139.125, 139.135, 139.138, 139.140, 139.150, 139.155, 139.160, 139.165, 139.180, 175.105, 175.210, 175.300, 176.200, 176.210, 177.2800, 184.1324, GRAS; FEMA GRAS; Europe listed; FDA approved for orals, ophthalmics, otics, rectals, topicals; USP/NF, BP, Ph.Eur. compliance

Manuf./Distrib.: ABITEC; Allan; Am. Ingreds.; Aquatec Quimica SA; Calgene; Croda Surf. Ltd; Eastman; Goldschmidt; Grindsted Prods.; Hart Prods.; Henkel/Emery; ICI Surf. Am.; Inolex; ISP Van Dyk; Koster Keunen; Lanaetex Prods.; Lipo; Lonza; Patco; Protameen; Rhône-Poulenc Surf. & Spec.; Spectrum Chem. Mfg.; Stepan; Witco/Oleo-Surf.

Glyceryl triacetate. *See* Triacetin
Glyceryl tri-(12-acetylricinolate). *See* Glyceryl triacetyl ricinoleate

Glyceryl triacetyl ricinoleate
CAS 101-34-8; EINECS 202-935-1
Synonyms: 9-Octadecenoic acid, 12-(acetyloxy)-, 1,2,3-propanetriol ester; Castor oil, acetylated; 1,2,3-Propanetriyl 12-(acetyloxy)-9-octadecenoate; Glyceryl tri-(12-acetylricinolate)
Definition: Triester of glycerin and acetyl ricinoleic acid
Formula: $C_3H_5(OOCC_{17}H_{32}OCOCH_3)_3$
Properties: Clear pale yel. oily liq., mild odor; sol. in most org. liqs.; insol. in water; dens. 0.967 (25/25 C); solidifies @ -40 C; iodine no. 76; sapon. no. 300
Precaution: Combustible
Uses: Plasticizer for nitrocellulose, ethylcellulose, and PVC; lubricants; protective coatings; mild emollient, pigment wetter, cosolvent; softener for waxes and resins
Manuf./Distrib.: CasChem
Trade names: Naturechem® GTR

Glyceryl tricaprylate/caprate
Uses: Bakery lubricant, release agent, glazing agent for confectionery; solvent carrier for flavors and fragrances
Trade names: Crodamol GTCC

Glyceryl triundecanoate. *See* Triundecanoin
Glycoformal. *See* 1,3-Dioxolane

Glycol
CAS 107-21-1; EINECS 203-473-3
Synonyms: Ethylene glycol; Ethylene dihydrate; Glycol alcohol; 1,2-Dihydroxyethane; 1,2-Ethanediol; Ethane-1,2-diol; Monoethylene glycol; Ethylene alcohol
Classification: Aliphatic dihydric alcohol
Empirical: $C_2H_6O_2$
Formula: $HOCH_2CH_2OH$
Properties: Clear liq., sweet taste (poisonous); very hygroscopic; sol. in water, lower aliphatic alcohols, glycerol, glycol ethers, acetone; poor sol. in toluene, benzene, chloroform; m.w. 62.07; sp.gr. 1.1135 (20/4 C); visc. 17.3 cps; m.p. -13 C; b.p. 197.6 C; flash pt. (OC) 115 C; ref. index 1.43063
Toxicology: ACGIH TLV/CL 39.4 ppm; LD50 (oral, rat) 4700 mg/kg; toxic by ingestion and inhalation; vapor can cause nose/throat irritation and burning; may cause skin and eye irritation; may be absorbed thru skin
Precaution: Combustible; incompat. with strong oxidizing agents, perchloric acid, strong bases, phosphorus (V) sulfide, strong acids, silvered copper wires carrying dc current; Hazardous decomp. prods.: Heated to decomp., emits acrid smoke and fumes
NFPA: Health 1, Flammability 1, Reactivity 0
Uses: Antifreeze in cooling and heating systems; in hydraulic brake fluids; industrial humectant; solvent in paints, plastics, printing inks; softening agent for cellophane; stabilizer; in explosives, alkyd resins, elastomers, waxes; asphalt; wood stains; raw material for prod. of syn. polyester fibers, latex paints, polyester and alkyd resins; plasticizer; heat transfer agent; low pressure laminates; textile processing; foam stabilizer; pharmaceutical solvent
Regulatory: FDA 21CFR §175.105, 176.300; FDA approved for topicals; BP compliance
Manuf./Distrib.: Aldrich; Ashland; J.T. Baker; BASF; Baychem; CasChem;

Chemcentral; Coyne; Eastman; Fluka; C.P. Hall; Hoechst Celanese; Houghton Chem.; Huntsman; Inspec Group plc; Mitsui Petrochem. Ind.; Mitsui Toatsu; Mobil; Occidental; Olin; Oxiteno; Shell; Sigma; Spectrum Chem. Mfg.; Sunnyside; Texaco; Union Carbide; Van Waters & Rogers

Trade names containing: Dioxitol-High Gravity; Eastman® DE-HG; Poly-Solv® DE (High Gravity)

Glycol alcohol. *See* Glycol
Glycol bis (hydroxyethyl) ether. *See* Triethylene glycol
Glycol butyl ether. *See* Butoxyethanol
Glycol carbonate. *See* Ethylene carbonate
Glycol diacetate. *See* Ethylene glycol diacetate
Glycol dichloride. *See* Ethylene dichloride
Glycol dimethyl ether. *See* Ethylene glycol dimethyl ether

Glycol distearate
CAS 627-83-8; EINECS 211-014-3
Synonyms: EGDS; Ethylene glycol distearate; Octadecanoic acid, 1,2-ethanediyl ester
Definition: Diester of ethylene glycol and stearic acid
Formula: $CH_3(CH_2)_{16}COOCH_2CH_2OCO(CH_2)_{16}CH_3$
Properties: Sp.gr. 0.97; m.p. 60 C; flash pt. (COC) 171 C
Uses: Pearlescent and opacifier; thickener, intermediate, lubricant, emulsifier, emollient; for emulsion shampoos and foam baths; plasticizer, lubricant, antistat for plastics and rubber; surfactant paints; pharmaceutical surfactant; specialized solvent
Regulatory: FDA 21CFR §73.1, 176.210
Manuf./Distrib.: Ashland; Cedar Concepts; Inolex; Lonza; PPG Ind.; Rhône-Poulenc Surf. & Spec.; Witco/Oleo-Surf.
Trade names: Kemester® EGDS

Glycol ether. *See* Diethylene glycol
Glycol ether EM acetate. *See* Methoxyethanol acetate
Glycol ethylene ether. *See* 1,4-Dioxane
Glycol ethyl ether. *See* Diethylene glycol
Glycol formal. *See* 1,3-Dioxolane

Glycolic acid-n-butyl ester
CAS 7397-62-8; EINECS 230-991-7
UN No. 8027
Empirical: $C_6H_{12}O_3$
Formula: $CH_3CH_2CH_2CH_2OOCCH_2OH$
Properties: Colorless liq.; very faint ester-like odor; misc. with common org. solvs.; only partly misc. with water; dens. 1.015-1.023 g/cc (20 C); vapor pressure 1.33 mbar (20 C); f.p. ≈ -26 C; b.p. 82-97 C; flash pt. ≈ 75 C; ref. index 1.423-1.426 (20 C)
Toxicology: LD50 (oral, rat) 4595 mg/kg; irritant to mucous membranes; risk of serious eye damage
Environmental: Readily biodeg.
Uses: Paint additive, flow agent for nitrocellulose lacquers, brush-applied paints, alkyd resin and oil paints; high-boiling solvent in stoving finishes; mfg. of printing inks
Manuf./Distrib.: Fluka
Trade names: Polysolvan® O

Glycol monoacetate. *See* Ethylene glycol acetate
Glycol monoacetin. *See* Ethylene glycol acetate

Glycol monobutyl ether acetate. *See* Butoxyethanol acetate
Glycol monomethyl ether. *See* Methoxyethanol
Glycol monomethyl ether acetate. *See* Methoxyethanol acetate
Glycol monostearate. *See* Glycol stearate

Glycol stearate
CAS 111-60-4; 97281-23-7; EINECS 203-886-9; 306-522-8
Synonyms: EGMS; Ethylene glycol monostearate; Ethylene glycol stearate; PEG-1 stearate; Glycol monostearate; POE (1) stearic acid; 2-Hydroxyethyl octadecanoate
Definition: Ester of ethylene glycol and stearic acid
Empirical: $C_{20}H_{40}O_3$
Formula: $CH_3(CH_2)_{16}COOCH_2CH_2OH$
Properties: Yel. waxy solid; sol. in alcohol, hot ether, acetone; insol. in water; m.w. 328.60; dens. 0.96 (25 C); m.p. 57-60 C
Toxicology: LD50 (IP, rat) 200 mg/kg; primary irritant; poison by intraperitoneal route; skin irritant; heated to decomp., emits acrid smoke and fumes
Precaution: Combustible
Uses: Opacifier and pearling agent for cream shampoos, other cosmetics; plasticizer for cellulose nitrate; lubricant for plasticized PVC; surfactant for industrial and cosmetic applics., pulp/paper, pharmaceuticals, metalworking, lubricants, textiles; agric., paints, adhesives; specialized solvent
Regulatory: FDA 21CFR §176.210; FDA approved for topicals
Manuf./Distrib.: Ashland; C.P. Hall; Inolex; ISP Van Dyk; PPG Ind.; Rhône-Poulenc Surf. & Spec.; Ruger; Stepan; Witco/Oleo-Surf.
Trade names: Kemester® EGMS

Glycyl alcohol. *See* Glycerin
Glyme-3. *See* PEG-3 dimethyl ether
GMS. *See* Glyceryl stearate
Gomenoleo oil. *See* Olive oil
Groundnut oil. *See* Peanut oil
Gum turpentine. *See* Turpentine oil

Halocarbon 11. *See* Trichlorofluoromethane
Halocarbon 112. *See* 1,1,2,2-Tetrachloro-1,2-difluoroethane
Halocarbon 113. *See* Trichlorotrifluoroethane
Halocarbon 114. *See* Dichlorotetrafluoroethane
Halon. *See* Dichlorodifluoromethane
Heavy mineral oil. *See* Mineral oil
Hendecane. *See* n-Undecane
1-Heptadecanol, 16-methyl-. *See* Isostearyl alcohol
2,2,4,4,6,6,8-Heptamethylnonane. *See* Isohexadecane
2,2,4,4,6,8.8-Heptamethylnonane. *See* Isohexadecane

Heptane
CAS 142-82-5; 64742-89-8; EINECS 205-563-8
UN No. 1206 (DOT)
Synonyms: n-Heptane; Alkane C_7; Dipropyl methane; Heptyl hydride
Classification: Sat. aliphatic hydrocarbon
Empirical: C_7H_{16}

n-Heptane

Formula: $CH_3(CH_2)_5CH_3$

Properties: Volatile colorless liq.; gasoline-like odor; sol. in alcohol, ether, chloroform; insol. in water; m.w. 100.21; sp.gr. 0.68368 (20 C); m.p. -91 C; b.p. 98.428 C; flash pt. (CC) -3.89 C; ref. index 1.38764 (20 C)

Toxicology: ACGIH TLV/TWA 400 ppm; STEL 500 ppm; LD50 (IV, mouse) 222 mg/kg; lethal conc. for mice 15,900 ppm in air; irritating to respiratory tract; narcotic in high concs.; poison by IV route

Precaution: Flamm. when exposed to heat or flame; dangerous fire risk; can react vigorously with oxidizing materials; mod. explosive exposed to heat or flame; violent reaction with phosphorus + chlorine; Hazardous decomp. prods.: Heated to decomp., emits acrid smoke and fumes

NFPA: Health 1, Flammability 3, Reactivity 0

Uses: Standard for octane rating determinations; anesthetic; solvent for rubber compounding, cements/sealants, inks, extraction of oils and fats; organic synthesis; replacement for n-hexane; reagent; major ingred. in gasoline, aviation fuel, petrol. solvents such as petrol. naphtha and rubber solvent

Regulatory: FDA 21CFR §175.105, 177.1580, 27CFR §21.111

Manuf./Distrib.: Aldrich; Ashland; Burdick & Jackson; Coyne; Exxon; Fluka; Great Western; Houghton Chem.; Humphrey; Phibro Energy USA; Phillips; Sigma; Spectrum Chem. Mfg.; Texaco; Triple Crown Am.

Trade names: Exxsol® Heptane; Shell Tolu-Sol® A HT Solv.; Texsolve C

Trade names containing: Shell Sol B HT; Shell Tolu-Sol® W HT Solv.; Tecsol® 1; Tecsol® 3; Texsolve E; Texsolve H

n-Heptane. *See* Heptane

Heptanol-2. *See* 2-Heptanol

1-Heptanol. *See* Heptyl alcohol

2-Heptanol

CAS 543-49-7

UN No. 1987

Synonyms: Amyl methyl carbinol; sec-Heptyl alcohol; s-Heptyl alcohol; 2-Hydroxyheptane; Methyl-n-amyl carbinol; Methyl amyl carbinol; Heptanol-2; Heptan-2-ol

Classification: Sec. aliphatic alcohol

Empirical: $C_7H_{16}O$

Formula: $CH_3CH_2CH_2CH_2CH_2CH(OH)CH_3$

Properties: Colorless liq.; mild odor; sol. in ethanol, ether, benzene; very sl. sol. in water; m.w. 116.20; sp.gr. 0.8193 (20 C); vapor pressure 1 mm Hg (14.6 C); b.p. 160-162 C; flash pt. (CC) 64 C

Toxicology: LD50 (oral, rat) 2580 mg/kg, (skin, rabbit) 1780 mg/kg; mod. toxic by ing. and skin contact; severe eye and sl. skin irritant; may be absorbed thru skin; inh. may cause nose/throat irritation

Precaution: Flamm. exposed to heat and flame; incompat. with oxidizing agents (increases risk of fire and explosion)

NFPA: Health 0, Flammability 2, Reactivity 0

Uses: Solvent for syn. resins; frothing agent in ore flotation

Manuf./Distrib.: Aldrich; Fluka; Penta Mfg.

Heptan-2-ol. *See* 2-Heptanol

3-Heptanol

CAS 589-82-2; EINECS 209-661-1

Synonyms: 3-Heptyl alcohol; 3-Hydroxyheptane; Butyl ethyl carbinol; Ethyl butyl carbinol; Heptan-3-ol

Classification: Sec. aliphatic alcohol

Empirical: $C_7H_{16}O$
Formula: $CH_3(CH_2)_3CH(OH)CH_2CH_3$
Properties: Colorless oily liq., herbaceous odor, pungent sl. bitter taste; sol. in alcohol, ether; sl. sol. in water; m.w. 116.20; sp.gr. 0.820 (20/4 C); vapor pressure 0.5 mm Hg (20 C); m.p. -70 C; b.p. 155-157 C; flash pt. (CC) 60 C; ref. index 1.422 (20 C)
Toxicology: LD50 (oral, rat) 1870 mg/kg, (skin, rabbit) 4360 mg/kg; mod. toxic by ing. and skin contact; skin and severe eye irritant; can be absorbed thru skin; inh. may cause nose/throat irritation; high concs. of vapor may cause CNS depression
Precaution: Flamm. exposed to heat or flame; incompat. with oxidizing agents (increase fire/explosion risk); Hazardous decomp. prods.: Heated to decomp., emits acrid smoke and fumes
NFPA: Health 0, Flammability 2, Reactivity 0
Uses: Flotation frother; solvent in org. coatings; intermediate in chem. synthesis; syn. flavoring agent for foods and pharmaceuticals
Regulatory: FDA 21CFR §172.515; FEMA GRAS
Manuf./Distrib.: Aldrich; Fluka; Penta Mfg.

Heptan-3-ol. *See* 3-Heptanol
n-Heptanol. *See* Heptyl alcohol
2-Heptanone. *See* Methyl n-amyl ketone
3,6,9,12,15,18,21-Heptaoxatricosane-1,2,3-diol. *See* PEG-8

Heptyl alcohol
CAS 111-70-6; EINECS 203-897-9
Synonyms: 1-Heptanol; n-Heptanol; 1-Hydroxyheptane; Hydroxyheptane; pri-Heptyl alcohol; Alcohol C-7; Enanthyl alcohol; Enanthic alcohol
Classification: Primary aliphatic alcohol
Empirical: $C_7H_{16}O$
Formula: $CH_3(CH_2)_6OH$
Properties: Colorless liq., citrus odor; sl. sol. in water @ 175 C; misc. with alcohol, fixed oils, ether; m.w. 116.23; sp.gr. 0.824 (20/4 C); m.p. -34.6 C; b.p. 175.8 C; flash pt. 70 C; ref. index 1.423-.1427
Toxicology: LD50 (oral, rat) 500 mg/kg, (skin, rabbit) 2 g/kg; mod. toxic by ingestion, skin contact; mildly toxic by inh.; probably an eye irritant; inh. can probably cause nose/throat irritation; high concs. may cause headache, nausea, drowsiness
Precaution: Flamm. exposed to heat, flames; reactive with oxidizing materials; Hazardous decomp. prods.: Heated to decomp., emits acrid smoke and fumes
Uses: Organic intermediate; solvent; cosmetic formulations; mfg. of plastics; food additive; syn. flavoring agent for foods and pharmaceuticals
Regulatory: FDA 21CFR §172.515; FEMA GRAS
Manuf./Distrib.: Aldrich; Elf Atochem N. Am.; Fluka; Penta Mfg.; Sigma; Spectrum Chem. Mfg.; Suchema AG

3-Heptyl alcohol. *See* 3-Heptanol
s-Heptyl alcohol. *See* 2-Heptanol
sec-Heptyl alcohol. *See* 2-Heptanol
Heptyl carbinol. *See* Caprylic alcohol
Heptyl hydride. *See* Heptane
Hexadecafluoroheptane. *See* Perfluoroheptane
Hexadecanoic acid isopropyl ester. *See* Isopropyl palmitate
Hexadecanoic acid, methyl ester. *See* Methyl palmitate
Hexadecanoic acid, 1-methylethyl ester. *See* Isopropyl palmitate
Hexadecanoic acid, 2-methylpropyl ester. *See* Isobutyl palmitate

1-Hexadecanol. *See* Cetyl alcohol
1-Hexadecanol, acetate. *See* Cetyl acetate
1-Hexadecanol lactate. *See* Cetyl lactate
Hexadecyl acetate. *See* Cetyl acetate
Hexadecyl alcohol. *See* Cetyl alcohol
n-Hexadecyl 2-ethylhexanoate. *See* Cetyl octanoate
n-Hexadecyl lactate. *See* Cetyl lactate
Hexaethylene glycol. *See* PEG-6
Hexahydrobenzene. *See* Cyclohexane
Hexahydrocresol. *See* Methylcyclohexanol
Hexahydrocumene. *See* Isopropylcyclohexane
Hexahydromethylphenol. *See* Methylcyclohexanol
Hexahydrophenol. *See* Cyclohexanol
Hexahydropyridine. *See* Piperidine
Hexahydrotoluene. *See* Methyl cyclohexane
Hexahydroxylene. *See* 1,3-Dimethylcyclohexane
Hexahydro-m-xylene. *See* 1,3-Dimethylcyclohexane
Hexahydro-o-xylene. *See* 1,2-Dimethylcyclohexane
Hexahydro-p-xylene. *See* 1,4-Dimethylcyclohexane
Hexalin. *See* Cyclohexanol
Hexalin acetate. *See* Cyclohexyl acetate
Hexaline acetate. *See* Cyclohexyl acetate
Hexamethylene. *See* Cyclohexane
Hexanaphthene. *See* Cyclohexane

Hexane
CAS 110-54-3; 64742-49-0; EINECS 203-777-6
UN No. 1208
Synonyms: n-Hexane; Normal hexane; Alkane C-6; Hexyl hydride
Classification: Aliphatic hydrocarbon
Empirical: C_6H_{14}
Formula: $CH_3(CH_2)_4CH_3$
Properties: Colorless volatile liq., faint petrol. odor; sol. in alcohol, acetone, ether; insol. in water; m.w. 86.20; sp.gr. 0.65937 (20/4 C); m.p. -95 C; b.p. 68.742 C; flash pt. (CC) -21.7 C; ref. index 1.37486 (20 C)
Toxicology: ACGIH TLV/TWA 50 ppm; LD50 (oral, rat) 28.710 mg/kg; sl. toxic by ingestion, inh.; human systemic effects by inh. (nausea, headache, dizziness, drowsiness); irritating to respiratory tract; skin and eye irritant; narcotic in high concs.
Precaution: Flamm.; very dangerous fire/explosion hazard exposed to heat or flame; can react vigorously with oxidizers; Hazardous decomp. prods.: Heated to decomp., emits acrid smoke and fumes
NFPA: Health 1, Flammability 3, Reactivity 0
Uses: Solvent and thinner for veg. oil and pharmaceutical extraction, compounding rubber cements, shoe, cosmetics industries; alcohol denaturant; paint diluent; polymerization reaction medium; filling for thermometers; cleaning agent for textiles, furniture, leather; in adhesives, cements, paints; laboratory reagent; extraction solvent for hops extract, spice oleoresins
Regulatory: FDA 21CFR §173.270; Japan approved with restrictions
Manuf./Distrib.: Aldrich; Ashland; BP Chems. Ltd; Burdick & Jackson; Coyne; Exxon; Fluka; Great Western; Houghton Chem.; Humphrey; Mitsui Petrochem. Ind.; Phibro Energy USA; Phillips; Shell; Sigma; Spectrum Chem. Mfg.; Texaco
Trade names: Exxsol® Hexane; Texsolve B
Trade names containing: Shell Sol B HT; Texsolve E; Texsolve H

n-Hexane. *See* Hexane
Hexanedinitrile. *See* Adiponitrile
Hexanedioic acid, bis (2-ethylhexyl) ester. *See* Dioctyl adipate
Hexanedioic acid, bis (2-hexyl decyl) ester. *See* Diisocetyl adipate
Hexanedioic acid, bis (1-methylethyl) ester. *See* Diisopropyl adipate
Hexanedioic acid, bis (2-methylpentadecyl) ester. *See* Diisocetyl adipate
Hexanedioic acid, dibutyl ester. *See* Dibutyl adipate
Hexanedioic acid diisobutyl ester. *See* Diisobutyl adipate
Hexanedioic acid dinitrile. *See* Adiponitrile
Hexanoic acid, 2-ethyl-, hexadecyl ester. *See* Cetyl octanoate
Hexanoic acid, 3,5,5-trimethyl-, isodecyl ester. *See* Isodecyl isononanoate
Hexanoic acid, 3,5,5-trimethyl-, isooctadecyl ester. *See* Isostearyl isononanoate
1-Hexanol. *See* Hexyl alcohol
n-Hexanol. *See* Hexyl alcohol
s-Hexanol. *See* 2-Ethyl-1-butanol
1-Hexanol, 2-ethyl-, phosphate. *See* Trioctyl phosphate
Hexanone. *See* Methyl isobutyl ketone
Hexazane. *See* Piperidine
Hexone. *See* Methyl isobutyl ketone
n-Hexoxyethoxyethanol. *See* Diethylene glycol n-hexyl ether

Hexyl acetate
CAS 142-92-7; EINECS 205-572-7
Synonyms: Hexyl ethanoate; Acetic acid hexyl ester; n-Hexyl acetate; 1-Hexyl acetate; Hexyl alcohol, acetate
Empirical: $C_8H_{16}O_2$
Formula: $CH_3COO(CH_2)_5CH_3$
Properties: Colorless oily liq., pleasant fruity odor, bittersweet taste; sol. in alcohol, ether; insol. in water; m.w. 144.22; sp.gr. 0.873 (20/4 C); m.p. -81 C; b.p. 167-169 C; flash pt. 41 C; ref. index 1.409 (20 C)
Toxicology: LD50 (oral, rat) 42 g/kg; mildly toxic by ingestion; heated to decomp., emits acrid smoke and fumes
Precaution: Flamm.
Uses: Solvent for cellulosic and acrylic coatings; syn. flavoring agent for foods and pharmaceuticals
Regulatory: FDA 21CFR §172.515; FEMA GRAS
Manuf./Distrib.: Aldrich; Berje; Fluka; Penta Mfg.; Sigma

1-Hexyl acetate. *See* Hexyl acetate
n-Hexyl acetate. *See* Hexyl acetate
s-Hexyl acetate. *See* Methyl amyl acetate

Hexyl alcohol
CAS 111-27-3; 68526-79-4; EINECS 203-852-3
UN No. 2282 (DOT)
Synonyms: n-Hexyl alcohol; 1-Hexanol; n-Hexanol; Alcohol C-6; 1-Hydroxyhexane; Caproyl alcohol; Pentylcarbinol; Amylcarbinol
Classification: Primary aliphatic alcohol
Empirical: $C_6H_{14}O$
Formula: $CH_3(CH_2)_4CH_2OH$
Properties: Colorless liq., fruity odor, aromatic flavor; sol. in alcohol, ether, acetone, chloroform, benzene; sl. sol. in water; m.w. 102.20; sp.gr. 0.8186; f.p. -51.6 C; b.p. 157.2 C; flash pt. (TOC) 65 C; ref. index 1.1469 (25 C)
Toxicology: LD50 (rat, oral) 4.59 g/kg, (skin, rabbit) 3100 mg/kg; poison by intravenous route; moderately toxic by ingestion, skin contact; skin and severe eye irritant; can be absorbed thru skin

Precaution: Flamm. or combustible liq.; reactive with oxidizing materials, strong acids
NFPA: Health 1, Flammability 2, Reactivity 0
Uses: Solvent for waterborne coatings; pharmaceuticals (antiseptics, perfume esters); plasticizer; syn. flavoring agent for foods and pharmaceuticals; emulsion polymerization; intermediate for textile and leather finishing agents, mfg. of food additives
Regulatory: FDA 21CFR §172.515, 172.864, 178.3480; FEMA GRAS
Manuf./Distrib.: Albemarle; Aldrich; Ashland; Condea Vista; Fluka; Penta Mfg.; Sigma; Spectrum Chem. Mfg.

n-Hexyl alcohol. *See* Hexyl alcohol
s-Hexyl alcohol. *See* 2-Ethyl-1-butanol
sec-Hexyl alcohol. *See* 2-Ethyl-1-butanol
Hexyl alcohol, acetate. *See* Hexyl acetate
Hexyl 'Carbitol'. *See* Diethylene glycol n-hexyl ether
n-Hexyl 'Cellosolve'. *See* Ethylene glycol hexyl ether
Hexyldecanol. *See* 2-Hexyl-1-decanol

2-Hexyl-1-decanol
CAS 2425-77-6; 36311-34-9; EINECS 219-370-1
Synonyms: Hexyldecanol
Classification: Aliphatic alcohol
Empirical: $C_{16}H_{34}O$
Formula: $CH_3(CH_2)_7CHCH_2OHCH_3(CH_2)_4CH_2$
Uses: Extender and solvent for dyes and fragrance oils; intermediate for prod. of compds. for applics. in lubricants, emulsifiers, metal processing, and textiles; solubilizer for pharmaceuticals
Trade names: Exxal® 16; Primarol 1006

Hexyldiglycol. *See* Diethylene glycol n-hexyl ether

Hexylene glycol
CAS 107-41-5; EINECS 203-489-0
Synonyms: 2,4-Dihydroxy-2-methylpentane; 4-Methyl-2,4-pentanediol; 2-Methyl-2,4-pentanediol; 2-Methylpetnane-2,4-diol; α,α,α´-Trimethyltrimethyleneglycol
Classification: Aliphatic dihydric alcohol
Empirical: $C_6H_{14}O_2$
Formula: $(CH_3)_2COHCH_2CHOHCH_3$
Properties: Colorless liq., nearly odorless; sol. in water, aliphatic and aromatic hydrocarbons, alcohols, ethers, fatty acids; m.w. 118.18; sp.gr. 0.9216 (20/4 C); m.p. -50 C; sets to glass; b.p. 198.3 C; flash pt. (TCC) 98 C; ref. index 1.4276 (20 C)
Toxicology: ACGIH TLV/CL 25 ppm; LD50 (rat, oral) 4.70 g/kg; skin and eye irritant; inh. may cause nose/respiratory irritation; very high concs. may cause headache, dizziness, nausea, incoordination; very low oral toxicity
Precaution: Combustible; can react vigorously with oxidizing materials; incompat. with strong acids (decomp. occurs); Hazardous decomp. prods.: Heated to decomp., emits acrid smoke and fumes
NFPA: Health 0, Flammability 1, Reactivity 0
Uses: Hydraulic brake fluids; printing inks; coupling agent and penetrant for textiles, cosmetics; ice inhibitor in carburetors; fuel and lubricant emulsifier; solvent for printing inks, textile dyes, in petrol. refining; cement additive; component of soaps, cosmetics; intermediate in chem. prod.; defoaming agent; pharmaceutic aid, humectant, solvent

Regulatory: FDA 21CFR §175.105, 176.180, 176.200, 176.210, 177.1210, 177.2800; FDA approved for topicals; USP/NF compliance
Manuf./Distrib.: Aldrich; Allchem Ind.; Ashland; BP Chems. Ltd; Coyne; Elf Atochem SA; Fluka; Great Western; Houghton Chem.; Mitsui Petrochem. Ind.; Penta Mfg.; Shell; Sigma; Spectrum Chem. Mfg.; Union Carbide

Hexyl ethanoate. *See* Hexyl acetate
Hexylglycol. *See* Ethylene glycol hexyl ether
Hexyl hydride. *See* Hexane
Hexyl methyl carbinol. *See* 2-Octanol
Hexyl methyl ketone. *See* Methyl hexyl ketone
2-(Hexyloxy)ethanol. *See* Ethylene glycol hexyl ether
2-(2-Hexyloxyethoxy)ethanol. *See* Diethylene glycol n-hexyl ether
High flash naphtha. *See* Stoddard solvent
Huile de ricini. *See* Castor oil

Hybrid safflower oil
Synonyms: Safflower oil, hybrid
Definition: Oil derived from safflower seeds; contains predominantly oleic acid triglyceride
Properties: Bland liq.; insol. in water; sol. in oil
Uses: Solubilizer, emollient, solvent for cosmetics and pharmaceuticals and in cooking oils

Hydrobromic acid
CAS 10035-10-6; EINECS 233-113-0
UN No. 1048 (DOT), 1788 (DOT)
Synonyms: Hydrogen bromide anhydrous; Anhydrous hydrobromic acid; Hydrogen bromide
Empirical: BrH
Formula: HBr
Properties: Colorless gas or pale yel. liq.; misc. with water, alcohol; m.w. 80.92; dens. 3.50 g/l (0 C); m.p. -87 C; b.p.-66.5 C; nonflamm.
Toxicology: LD50 (IP, rat) 76 mg/kg; ACGIH TLV:CL 3 ppm; mildly toxic by inh.; corrosive irritant to eyes, skin, mucous membranes
Precaution: Corrosive gas; protect from light; reacts violently with F_2, Fe_wO_3, NH_3, O_3; Hazardous decomp. prods.: Heated to decomp. or reacting with water or steam, emits toxic and corrosive fumes of Br^- and HBr
NFPA: Health 3, Flammability 0, Reactivity 0
Uses: Analytical chemistry, solvent for ore minerals, manufacture of inorganic and some alkyl bromides, alkylation catalyst
Manuf./Distrib.: Advance Research Chems.; Albemarle; Aldrich; Allchem Ind.; AmeriBrom; Assoc. Octel Co Ltd; EM Ind.; Fluka; Great Lakes; Kalama; Sigma; Spectrum Chem. Mfg.

Hydrochloric ether. *See* Ethyl chloride
Hydrocyanic ether. *See* Propionitrile
Hydrofuran. *See* Tetrahydrofuran

Hydrogenated castor oil
CAS 8001-78-3; EINECS 232-292-2
Synonyms: Opalwax; Castorwax; Castor oil, hydrogenated
Definition: End prod. of controlled hydrogenation of castor oil, consisting mainly of the triglyceride of hydroxystearic acid
Properties: Wh. hard wax; very insol. in water and in the more common org. solvs.; m.w. ≈ 932; m.p. 86-88 C; iodine no. 5 max.; sapon. no. 176-182; hyd. no. 154-162

Toxicology: Ingestion of large amts. may cause pelvic congestion
Uses: In water-repellent coatings, candles, polishes, ointments, cosmetics; impregnant for paper, wood, cloth; as lubricant, mold release in mfg. of formed plastics and rubber goods; wax for food applics., pharmaceutical ointments; solvent for pharmaceutical injectables; tablet coatings
Regulatory: FDA 21CFR §175.105, 175.300, 176.170, 176.210, 177.1200, 177.1210, 177.2420, 177.2800, 178.3280; USP/NF compliance
Manuf./Distrib.: Akzo Nobel; Amber Syn.; Arista Ind.; Ashland; Hoechst AG; Southern Clay Prods.; Welch, Holme & Clark

Hydrogen bromide. *See* Hydrobromic acid
Hydrogen bromide anhydrous. *See* Hydrobromic acid
Hydrogen carboxylic acid. *See* Formic acid
1-Hydroxybutane. *See* Butyl alcohol
2-Hydroxybutane. *See* 2-Butanol
4-Hydroxybutanoic acid lactone. *See* Butyrolactone
γ-Hydroxybutyric acid cyclic ester. *See* Butyrolactone
4-Hydroxybutyric acid γ-lactone. *See* Butyrolactone
γ-Hydroxybutyrolactone. *See* Butyrolactone
Hydroxycyclohexane. *See* Cyclohexanol
1-Hydroxydecane. *See* n-Decyl alcohol
Hydroxydimethylbenzene. *See* Xylenol
4-Hydroxydiphenyldimethylmethane. *See* Cumyl phenol
1-Hydroxyethane 1-carboxylic acid. *See* Lactic acid
2-Hydroxy-1-ethanethiol. *See* 2-Mercaptoethanol
β-(β-Hydroxyethoxy) ethylamine. *See* 2-(2-Aminoethoxy) ethanol
N-(2-Hydroxyethyl)acetamide. *See* Acetamide MEA
2-Hydroxyethyl acetate. *See* Ethylene glycol acetate
Hydroxyethylmercaptan. *See* 2-Mercaptoethanol
2-Hydroxyethylmercaptan. *See* 2-Mercaptoethanol
2-Hydroxyethyl octadecanoate. *See* Glycol stearate
N-(2-Hydroxyethyl) peanut acid amide. *See* Peanutamide MEA

N-Hydroxyethylpyrrolidone
CAS 3445-11-2
Uses: Solvent for coating and cleaning applics.
Trade names: Gaftex® HI-20H; HEP®

β-Hydroxyethyl sulfide. *See* Thiodiglycol
Hydroxyheptane. *See* Heptyl alcohol
1-Hydroxyheptane. *See* Heptyl alcohol
2-Hydroxyheptane. *See* 2-Heptanol
3-Hydroxyheptane. *See* 3-Heptanol
1-Hydroxyhexane. *See* Hexyl alcohol
α-Hydroxy-ω-hydroxy poly(oxy-1,2-ethanediyl). *See* Polyethylene glycol
4-Hydroxy-1-keto-4-methylpentane. *See* Diacetone alcohol
2-Hydroxy-2-methyl-3-butyne. *See* Methyl butynol
2-Hydroxymethylfuran. *See* Furfuryl alcohol
3-Hydroxymethyl-n-heptan-4-ol. *See* Ethyl hexanediol
4-Hydroxy-4-methylpentanone-2. *See* Diacetone alcohol
4-Hydroxy-4-methyl-2-pentanone. *See* Diacetone alcohol
4-Hydroxy-4-methyl pentan-2-one. *See* Diacetone alcohol
1-Hydroxymethylpropane. *See* Isobutyl alcohol
12-Hydroxyoctadecanoic acid, 2-ethylhexyl ester. *See* Octyl hydroxystearate
12-Hydroxy-9-octadecenoic acid, monoester with 1,2-propanediol. *See* Propylene glycol ricinoleate

1-Hydroxyoctane. *See* Caprylic alcohol
2-Hydroxyoctane. *See* 2-Octanol
2-Hydroxypentane. *See* s-Amyl alcohol
1-Hydroxypropane. *See* n-Propyl alcohol
2-Hydroxy-1,2,3-propanetricarboxylic acid, tributyl ester. *See* Tributyl citrate
2-Hydroxy-1,2,3-propanetricarboxylic acid, triethyl ester. *See* Triethyl citrate
2-Hydroxypropanoic acid. *See* Lactic acid
2-Hydroxypropanoic acid butyl ester. *See* Butyl lactate
2-Hydroxypropanoic acid hexadecyl ester. *See* Cetyl lactate
2-Hydroxypropanoic acid, isostearyl ester. *See* Isostearyl lactate
2-Hydroxypropanol. *See* Propylene glycol
2-Hydroxypropionic acid. *See* Lactic acid
α-Hydroxypropionic acid. *See* Lactic acid
2-Hydroxypropionic acid ethyl ester. *See* Ethyl lactate
Hydroxypropyl alginate. *See* Propylene glycol alginate
N-(2-Hydroxypropyl)peanut acid amide. *See* Peanutamide MIPA
α-Hydroxytoluene. *See* Benzyl alcohol
Hydroxyxylene. *See* Xylenol
Hypnone. *See* Acetophenone

2,2´-Iminobisethylamine. *See* Diethylenetriamine
2,2´-Iminodiethylamine. *See* Diethylenetriamine
Inactive limonene. *See* dl-Limonene
Indigenous peanut oil. *See* Peanut oil
IPA. *See* Isopropyl alcohol
IPM. *See* Isopropyl myristate
IPP. *See* Isopropyl palmitate
Isoacetophorone. *See* Isophorone

Isoamyl acetate
CAS 123-92-2; EINECS 204-662-3
UN No. 1104
Synonyms: Banana oil; 3-Methyl-1-butyl acetate; 3-Methyl-1-butanol acetate; β-Methylbutyl acetate; 3-Methylbutyl acetate; Isopentyl alcohol acetate; Acetic acid isopentyl ester; Isoamyl ethanoate; 3-Methylbutyl ethanoate; Isopentyl acetate; Pear oil
Classification: Carboxylic acid ester
Definition: Ester of isoamyl alcohol and acetic acid
Empirical: $C_7H_{14}O_2$
Formula: $CH_3COOCH_2CH_2CH(CH_3)_2$
Properties: Colorless liq., banana-like odor; sl. sol. in water; misc. with alcohol, ether, ethyl acetate, fixed oils; insol. in glycerin, propylene glycol; m.w. 130.21; sp.gr. 0.876 (20 C); m.p. -78.5 C; b.p. 142 C; flash pt. (CC) 25 C; ref. index 1.400
Toxicology: LD50 (oral, rat) 16,600 mg/kg; ACGIH TLV/TWA 100 ppm; mildly toxic by ingestion, inh., subcutaneous routes; eye and mild skin irritant; inh. can cause nose/throat irritation, CNS depression
Precaution: Highly flamm.; dangerous fire hazard exposed to heat or flame; can react vigorously with reducing materials; also incompat. with oxidizing agents (increases fire/explosion risk), strong acids, strong bases (decomp. can occur);

Hazardous decomp. prods.: Heated to decomp., emits acrid smoke and fumes
NFPA: Health 1, Flammability 3, Reactivity 0
Uses: Fresh fruity fragrance and flavoring, pear- or banana-like on dilution, for foods, pharmaceuticals, perfumery; artificial leather, photographic film; solvent for nitrocellulose, lacquers, resins, waxes; masking undesirable odors
Regulatory: FDA 21CFR §172.515; FEMA GRAS; Japan approved as flavoring
Manuf./Distrib.: Aceto; Aldrich; Berje; Fluka; Northwestern Flavors; Oxiteno; Penta Mfg.; Rhône-Poulenc N. Am.; Sigma; Spectrum Chem. Mfg.

Isoamyl alcohol
CAS 123-51-3; EINECS 204-633-5
UN No. 1105
Synonyms: Isoamylol; 3-Methylbutanol; 3-Methyl-1-butanol; 2-Methyl-4-butanol; Primary isoamyl alcohol; Isopentyl alcohol; Isopentanol; Isobutyl carbinol
Classification: Primary aliphatic alcohol
Empirical: $C_5H_{12}O$
Formula: $(CH_3)_2CHCH_2CH_2OH$
Properties: Colorless liq.; disagreeable sweet odor; pungent taste; sl. sol. in water; sol. in alcohol, ether, acetone, many org. solvs.; m.w. 88.15; sp.gr. 0.813 (15/4 C); f.p. -117.2 C; b.p. 132 C; flash pt. (CC) 42.7 C; ref. index 1.407 (20 C)
Toxicology: ACGIH TLV/TWA 100 ppm; STEL 125 ppm; highly toxic; ing. has caused human deaths from respiratory failure; may cause heart, lung, kidney damage; CNS depressant
Precaution: DOT: Flamm. liq.; explosive limits in air 1.2-9%; incompat. with strong oxidizing agents, reducing agents, strong acids, alkalies, hydrogen trisulfide; potential explosive hazard; Hazardous decomp. prods.: None known
NFPA: Health 1, Flammability 2, Reactivity 0
Uses: Syn. flavor and fragrance for foods and pharmaceuticals; fine chemicals; extraction solvent; herbicides, plasticizers, mining chemicals, solv. for gums, resins, and lacquers, oil additives, intermediates, photographic chemicals; organic synthesis, microscopy
Regulatory: FDA 21CFR §172.515; FEMA GRAS
Manuf./Distrib.: Aldrich; Allchem Ind.; AMRESCO; BASF; Coyne; CPS; Fluka; Hoechst Celanese; Oxiteno; Penta Mfg.; Sigma; Spectrum Chem. Mfg.
Trade names: Isoamyl Alcohol 95%; Isoamyl Alcohol 99%
See also 3-Pentanol

Isoamyl ethanoate. *See* Isoamyl acetate
Isoamylhydride. *See* Isopentane
Isoamyl methyl ketone. *See* Methyl isoamyl ketone
Isoamylol. *See* Isoamyl alcohol
Isobutanol. *See* Isobutyl alcohol
Isobutenyl methyl ketone. *See* Mesityl oxide
Isobutoxy-2-ethoxy-2-ethanol. *See* Diethylene glycol isobutyl ether
2-(2-Isobutoxyethoxy) ethanol. *See* Diethylene glycol isobutyl ether

Isobutyl acetate
CAS 110-19-0; EINECS 203-745-1
UN No. 1123 (DOT), 1213 (DOT)
Synonyms: 2-Methylpropyl acetate; 2-Methyl-1-propyl acetate; β-Methylpropyl ethanoate; Acetic acid isobutyl ester; Acetic acid 2-methylpropyl ester
Classification: Carboxylic acid ester
Definition: Ester of isobutyl alcohol and acetic acid
Empirical: $C_6H_{12}O_2$
Formula: $CH_3COOCH_2CH(CH_3)_2$

Properties: Colorless liq., fruit-like odor; very sol. in alcohol, fixed oils, propylene glycol; sol. in ether, hydrocarbons; sl. sol. in water; m.w. 116.18; sp.gr. 0.8685 (15 C); m.p. -98.9 C; b.p. 118 C; flash pt. (CC) 18 C; ref. index 1.389

Toxicology: ACGIH TLV/TWA 150 ppm; LD50 (oral, rat) 13,400 mg/kg; mildly toxic by ingestion and inh.; vapors can probably irritate nose and throat; high concs. may cause CNS depression

Precaution: Highly flamm.; very dangerous fire and mod. explosion hazard on exposure to heat, flame, oxidizers; incompat. with strong oxidizers, strong bases, strong acids; increased risk of fire/explosion; decomp. (hydrolysis) can occur; Hazardous decomp. prods.: Heated, emits acrid smoke and fumes

NFPA: Health 1, Flammability 3, Reactivity 0

Uses: Solvent for nitrocellulose; in thinners, varnishes, sealants, topcoat lacquers; perfumery; syn. flavoring agent for foods and pharmaceuticals

Regulatory: FDA 21CFR §172.515; FEMA GRAS; BP compliance

Manuf./Distrib.: Aldrich; Allchem Ind.; Ashland; BASF; Coyne; Eastman; Fluka; Hoechst Celanese; Houghton Chem.; Janssen Chimica; Penta Mfg.; Union Carbide

Isobutylacetone. *See* Methyl isoamyl ketone

Isobutyl adipate. *See* Diisobutyl adipate

Isobutyl alcohol

CAS 78-83-1; EINECS 201-148-0

UN No. 1213

Synonyms: Isobutanol; 1-Hydroxymethylpropane; Isopropylcarbinol; 2-Methyl-1-propanol; 2-Methylpropanol

Classification: Primary aliphatic alcohol

Empirical: $C_4H_{10}O$

Formula: $(CH_3)_2CHCH_2OH$

Properties: Colorless liq., sweet odor; partly sol. in water; sol. in alcohol, ether; m.w. 74.12; sp.gr. 0.806 (15 C); m.p. -108 C; b.p. 106-109 C; f.p. -108 C; flash pt. (TCC) 29 C; ref. index 1.396

Toxicology: ACGIH TLV/TWA 50 ppm; LD50 (oral, rat) 2460 mg/kg; poison by IV, intraperitoneal route; mod. toxic by ingestion, skin contact; severe skin/eye irritant; inh. of high concs. or ing. may cause CNS depression effects

Precaution: Flamm.; dangerous fire hazard with heat, flame; mod. explosive as vapor with heat, flame, oxidizers; also incompat. with chromium trioxide; explosion potential with various chems.; Hazardous decomp. prods.: Heated to decomp., emits acrid smoke and fumes

NFPA: Health 1, Flammability 3, Reactivity 0

Uses: Flavors and fragrances; organic synthesis; latent solvent; intermediate; solvent for paints, lacquers, paint removers, cleaners, hydraulic fluids; fluorometric determinations; liq. chromatography; mfg. of isobutyl esters for use in solvents, plasticizers, flavorings, and perfumes; syn. flavoring agent for foods and pharmaceuticals

Regulatory: FDA 21CFR §172.515; FEMA GRAS

Manuf./Distrib.: Aldrich; Allchem Ind.; Ashland; BASF; CPS; Eastman; Fluka; Hoechst Celanese; Houghton Chem.; Penta Mfg.; Shell; Sigma; Spectrum Chem. Mfg.; Union Carbide

Trade names containing: Filmex® B; Solvenon® I

Isobutyl carbinol. *See* Isoamyl alcohol

Isobutylcyclohexane

CAS 1678-98-4

Classification: Sat. alicyclic hydrocarbon

Isobutylcyclopentane

Empirical: $C_{10}H_{20}$
Formula: $C_6H_{11}CH_2CH(CH_3)CH_3$
Properties: Colorless liq.; sol. in most org. solvs.; insol. in water; m.w. 140.27; sp.gr. 0.795 (20 C); m.p. -95 C; b.p. 169 C
Toxicology: May cause mild skin irritation; conc. vapor may cause eye irritation, nose/throat irritation, headache, lightheadedness, nausea, drowsiness; probably no significant CNS effects
Precaution: Combustible; incompat. with strong oxidizing agents (increased risk of fire and explosion); Hazardous decomp. prods.: None
NFPA: Health 0, Reactivity 0
Uses: Solvent; org. synthesis

Isobutylcyclopentane
CAS 3788-32-7
Classification: Sat. alicyclic hydrocarbon
Empirical: C_9H_{18}
Formula: $C_5H_9CH_2CH(CH_3)CH_3$
Properties: Colorless liq.; sol. in most org. solvs.; insol. in water; m.w. 126.23
Toxicology: May cause mild skin irritation; conc. vapor may be irritating to eyes; probably a CNS depressant; high vapor concs. may cause nose/throat irritation, lightheadedness, nausea; very high concs. may cause unconsciousness and death
Precaution: Combustible; incompat. with strong oxidizing agents (increased fire and explosion hazard); Hazardous decomp. prods.: None
Uses: Solvent; org. synthesis

Isobutyl formate
CAS 542-55-2
UN No. 2393 (DOT)
Synonyms: Tetryl formate; Formic acid isobutyl ester
Empirical: $C_5H_{10}O_2$
Formula: $HCOOCH_2CH(CH_3)_2$
Properties: Liq., fruity ether-like odor, rum-like taste; sol. in 100 parts water; misc. with alcohol, ether; m.w. 102.13; dens. 0.885 (20/4 C); m.p. -95 C; b.p. 98-99 C; flash pt. 50 F; ref. index 1.3858 (20 C)
Toxicology: Mod. toxic by ingestion
Precaution: Flamm. liq.; very dangerous fire hazard
Uses: Syn. flavoring agent for foods and pharmaceuticals
Regulatory: FDA 21CFR §172.515; FEMA GRAS
Manuf./Distrib.: Aldrich
Trade names containing: Solvenon® I

Isobutyl heptyl ketone
CAS 123-18-2
Synonyms: 2,6,8-Trimethyl-4-nonanone
Empirical: $C_{12}H_{24}O$
Formula: $(CH_3)_2CHCH_2COCH_2CH(CH_3)CH_2CH(CH_3)_2$
Properties: Water-wh. liq., pleasant odor; insol. in water; m.w. 184.32; sp.gr. 0.822 (20/20 C); b.p. 224 C; flash pt. (CC) 189 F
Uses: Solvent for coatings (NC emulsions, vinyls); dispersant; intermediate; high solvating power for vinyl resins, cellulose esters and ethers
Manuf./Distrib.: Union Carbide

Isobutyl isobutyrate
CAS 97-85-8; EINECS 202-612-5
UN 2528 (DOT)

Synonyms: 2-Methylpropanoic acid 2-methylpropyl ester
Empirical: $C_8H_{16}O_2$
Formula: $(CH_3)_2CHCOOCH_2CH(CH_3)_2$
Properties: Liq., fruity odor; misc. with alcohol; insol. in water; m.w. 144.22; dens. 0.854 (20/4 C); m.p. -81 C; b.p. 149-151 C; flash pt. 40 C; ref. index 1.399 (20 C)
Toxicology: LD50 (oral, rat) 12,800 mg/kg; mildly toxic by ingestion and inhalation; irritant
Precaution: Flamm.
Uses: Solvent for paints; syn. flavoring agent for foods and pharmaceuticals
Regulatory: FDA 21CFR §172.515; FEMA GRAS
Manuf./Distrib.: Aldrich; Ashland; Eastman; Fluka; Houghton Chem.; Penta Mfg.

Isobutyl ketone. *See* Diisobutyl ketone
Isobutyl methyl carbinol. *See* Methyl amyl alcohol
Isobutyl methyl ketone. *See* Methyl isobutyl ketone
Isobutyl methyl methanol. *See* Methyl amyl alcohol

Isobutyl palmitate
CAS 110-34-9
Synonyms: 2-Methylpropyl hexadecanoate; Palmitic acid, isobutyl ester; Hexadecanoic acid, 2-methylpropyl ester
Definition: Ester of isobutyl alcohol and palmitic acid
Formula: $CH_3(CH_2)_{14}COOCH_2CH(CH_3)_2$
Uses: Emollient, lubricant, solvent; for cosmetics, textiles, metalworking

Isobutyltrimethylethane. *See* Isooctane

Isocetyl alcohol
CAS 36311-34-9; EINECS 252-964-9
Synonyms: Isohexadecanol; Isohexadecyl alcohol; Isopalmityl alcohol
Definition: Mixture of branched chain C16 aliphatic alcohols
Empirical: $C_{16}H_{34}O$
Formula: $C_8H_{17}CHC_6H_{13}CH_2OH$
Properties: Liq.; sol. in min. oil, 95% ethanol, IPM, oleyl alcohol, castor oil, cyclomethicone; insol. in water, glycerin, propylene glycol; m.w. 242; sp.gr. 0.830-0.840; flash pt. (COC) 156 C
Toxicology: LD50 (rat, oral) > 50 g/kg; mildly irritating to skin and eyes
Uses: Emollient for emulsions, pigment dispersant and binder, coupling agent for min. oil and castor oil; carrier and extender for flavor and fragrance oils in foods and pharmaceuticals; solubilizer; extender and solvent for pigments, lipsticks dyes
Trade names: Ceraphyl® ICA; Standamul® G-16

Isocetyl salicylate
CAS 138208-68-1
Classification: Ester
Empirical: $C_{23}H_{38}O_3$
Properties: Water-white liq., odorless; insol. in water; sp.gr. 1.02; flash pt. (PMCC) > 200 F
Toxicology: May cause mild skin and eye irritation on prolonged contact
Precaution: Store away from strong oxidizing agents; avoid excessive heat
Uses: Emollient; solvent-plasticizer for benzophenone in sunscreen formulations, waxes, resins, polymers
Trade names: Dermol ICSA

Isocineole. *See* 1,4-Cineole
Isocumene. *See* Propylbenzene

Isodecyl isononanoate
CAS 41395-89-5
Synonyms: 3,5,5-Trimethylhexanoic acid, isodecyl ester; Hexanoic acid, 3,5,5-trimethyl-, isodecyl ester
Definition: Ester of branched chain decyl alcohols and a branched chain nonanoic acid
Empirical: $C_{19}H_{38}O_2$
Formula: $(CH_3)_3CCH_2CCH_3HCH_2COOC_{10}H_{21}$
Properties: Sol. in alcohol, animal and vegetable oils, min. oil; insol. in water, glycerin, propylene glycol
Uses: Emollient with solvency chars. for skin and hair care prods.; pigment wetter
Trade names: Wickenol® 152

Isodecyl oleate
CAS 59231-34-4; EINECS 261-673-6
Synonyms: 9-Octadecenoic acid, isodecyl ester
Definition: Ester of branched chain decyl alcohols and oleic acid
Empirical: $C_{28}H_{54}O_2$
Formula: $CH_3(CH_2)_7CH=CH(CH_2)_7COOC_{10}H_{21}$
Properties: Wh. to straw liq., char. mild odor; sol. in peanut oil, 95% ethanol, IPM, oleyl alcohol; insol. in water, glycerin, propylene glycol; sp.gr. 0.858-0.864; ref. index 1.4540-1.4560
Toxicology: LD50 (rat, oral) > 40 ml/kg; nonirritating to eyes; mildly irritating to skin
Uses: Emollient, cosolvent for cosmetics, pharmaceutical topicals; emulsifier, lubricant, antistat, defoamer for metalworking, textiles, plastics, paper
Trade names: Ceraphyl® 140-A

Isododecane
CAS 141-70-8; 31807-55-3; 13475-82-6; EINECS 205-495-9
Synonyms: 2,2,6,6-Tetramethyl-4-methyleneheptane
Definition: Branched chain aliphatic hydrocarbon with 12 carbons
Empirical: $C_{12}H_{26}$
Formula: $H_3C(C(CH_3)_2CH_2)_2CH(CH_3)_2$
Properties: m.w. 170
Uses: Rheological additive; cosolubilizer for nonhydrocarbon materials; solvent for debris on skin
Manuf./Distrib.: Fluka
Trade names: Permethyl® 99A

Isoeicosyl alcohol. *See* Octyldodecanol
Isoforone. *See* Isophorone

Isoheptane
CAS 591-76-4
UN No. 1206
Synonyms: 2-Methylhexane; Ethylisobutylmethane
Classification: Sat. aliphatic hydrocarbon
Empirical: C_7H_{16}
Formula: $CH_3CHCH_3CH_2CH_2CH_2CH_3$
Properties: Colorless liq.; gasoline-like odor; sol. in most org. solvs.; insol. in water; m.w. 100.2; sp.gr. 0.683 (20 C); vapor pressure 40 mm Hg (14.9 C); m.p. -118 C; b.p. 90 C; flash pt. (CC) < -18 C
Toxicology: ACGIH TLV/TWA 400 ppm; STEL 500 ppm; may cause mild skin irritation; conc. vapor may cause sl. eye irritation; conc. vapor may cause sl. nose/throat irritation, lightheadedness, dizziness, nausea, difficulty breath-

ing; narcotic @ 0.8-1.1%

Precaution: Flamm.; very dangerous fire hazard exposed to heat, flame; explosive as vapor exposed to heat, flame; incompat. with strong oxidizing agents (increases fire and explosion hazard); Hazardous decomp. prods.: Heated to decomp., emits acrid smoke and irritating fumes

NFPA: Health 0, Flammability 3, Reactivity 0

Uses: Ingred. in gasoline, aviation fuels; solvent; org. synthesis; raw material for prod. of plasticizers

Manuf./Distrib.: Aldrich; Fluka

Isohexadecane

CAS 4390-04-9; EINECS 224-506-8

Synonyms: 2,2,4,4,6,6,8-Heptamethylnonane; 2,2,4,4,6,8.8-Heptamethylnonane

Classification: Liquid isoparaffin

Empirical: $C_{16}H_{34}$

Formula: $H_3C(C(CH_3)_2CH_2)_4CH(CH_3)_2$

Properties: m.w. 226

Uses: Emollient, moisture barrier, solvent

Manuf./Distrib.: Aldrich; Sigma

Isohexadecanol. *See* Isocetyl alcohol

Isohexadecyl alcohol. *See* Isocetyl alcohol

Isohexane

CAS 107-83-5; EINECS 203-523-4

UN No. 1208 (DOT), 2462 (DOT)

Synonyms: Dimethylpropylmethane; Methyl pentane; 2-Methylpentane

Classification: Sat. aliphatic hydrocarbon

Empirical: C_6H_{14}

Formula: $CH_3CH(CH_3)CH_2CH_2CH_3$

Properties: Colorless liq., faint petrol. odor; misc. with most org. solvs.; insol. in water; m.w. 86.18; sp.gr. 0.65 (20 C); m.p. -153 C; b.p. 60 C; flash pt. (CC) -32 C

Toxicology: ACGIH TLV/TWA 500 ppm; STEL 1000 ppm; may cause skin irritation; human eye irritant; inh. of high concs. can cause respiratory tract irritation, sl. nausea, headache, lightheadedness

Precaution: Highly flamm.; very dangerous fire hazard exposed to heat, flame, or oxidizers; severe explosion hazard exposed to heat, flame; incompat. with strong oxidizing agents (increases risk of fire and explosion); Hazardous decomp. prods.: Heated to decomp., emits acrid smoke and irritating fumes

NFPA: Health 1, Flammability 3, Reactivity 0

Uses: Solvent for extraction of veg. oils; chem. reaction medium; f.p. depressant

Manuf./Distrib.: Aldrich; Fluka

Isohexyl alcohol. *See* 2-Methyl-1-pentanol

Isonitropropane. *See* 2-Nitropropane

Isononanoic acid, cetyl/stearyl ether. *See* Cetearyl isononanoate

Isononanoic acid, 2-ethylhexyl ester. *See* Octyl isononanoate

Isononanoic acid isononyl ester. *See* Isononyl isononanoate

Isononanoic acid, isotridecyl ester. *See* Isotridecyl isononanoate

Isononyl isononanoate

CAS 42131-25-9

Synonyms: 3,5,5-Trimethylhexanoic acid, 3,5,5-trimethylhexyl ester; Isononanoic acid isononyl ester

Definition: Ester of a branched chain nonyl alcohol with a branched chain nonanoic acid

Empirical: $C_{18}H_{36}O_2$

Isooctadecanoic acid, diester with triglycerol

Properties: Sol. in alcohol, animal and vegetable oils, min. oil; insol. in water, glycerin, propylene glycol
Toxicology: No known toxicity
Uses: Emollient, moisturizer, pigment wetter with solvency chars. for hair care prods.
Trade names: Wickenol® 151

Isooctadecanoic acid, diester with triglycerol. *See* Polyglyceryl-3 diisostearate
Isooctadecanoic acid, isooctadecyl ester. *See* Isostearyl isostearate
Isooctadecanol. *See* Isostearyl alcohol
Isooctadecyl 2-hydroxypropanoate. *See* Isostearyl lactate

Isooctane
CAS 540-84-1
UN No. 1262 (DOT)
Synonyms: 2,4,-TMP; 2,2,4-Trimethylpentane; Isobutyltrimethylethane
Classification: Sat. aliphatic hydrocarbon
Empirical: C_8H_{18}
Formula: $CH_3C(CH_3)_2CH_2CH(CH_3)CH_3$
Properties: Colorless clear liq.; gasoline-like odor; completely misc. with most org. solvs.; insol. in water; m.w. 114.26; sp.gr. 0.692 (20 C); vapor pressure 40.6 mm Hg (21 C); m.p. -107 C; b.p. 99 C; flash pt. 4.5 C
Toxicology: ACGIH TLV/TWA 300 ppm; STEL 375 ppm; mild skin irritant; conc. vapor irritating to eyes; inh. of conc. vapor may cause nose/throat irritation, headache, drowsiness, nausea, difficulty breathing; very high concs. may cause unconsciousness, death
Precaution: Flamm.; very dangerous fire hazard exposed to heat, flame, oxidizers; explosive as vapor exposed to heat, flame; can react vigorously with reducing agents; incompat. with strong oxidizing agents; Hazardous decomp. prods.: Heated to decomp., emits acrid smoke and irritating fumes
NFPA: Health 0, Flammability 3, Reactivity 0
Uses: Std. reference fuel for octane rating of fuels; reagent in spectrophotometric analysis; org. synthesis; as solvent and thinner
Manuf./Distrib.: Aldrich; Fluka; Phillips; Sigma; Spectrum Chem. Mfg.; Triple Crown Am.

Isooctanol. *See* Isooctyl alcohol

Isooctyl alcohol
CAS 68526-83-0; 26952-21-6; EINECS 248-133-5
Synonyms: Isooctanol; 2-Ethyl-1-hexanol; Alcohol C_8
Empirical: $C_8H_{18}O$
Formula: $CH_3(CH_2)_3CH(C_2H_5)CH_2OH$
Properties: Clear liq.; m.w. 130.26; dens. 0.832 (20/20 C); b.p. 182-185 C; flash pt. (TOC) 180 F; ref. index 1.431
Toxicology: LD50 (oral, rat) 1480 mg/kg, (skin, rabbit) 2520 mg/kg; moderately toxic by skin absorption and ingestion; severe eye irritant; TLV 50 ppm
Precaution: Combustible
Uses: In plasticizers; intermediate for nonionic detergents and surfactants, hydraulic fluids; resin, solvent, emulsifier, antifoaming agent; in coatings
Manuf./Distrib.: Allchem Ind.; Chemcentral

Isooctyl stearate
CAS 91031-48-0; EINECS 292-951-5
Definition: Ester of isooctyl alcohol and stearic acid
Uses: Emollient, solvent for cosmetics; plasticizer and lubricant for plastics, paper industry

Isopalmityl alcohol. *See* Isocetyl alcohol

Isopentane
CAS 78-78-4; EINECS 201-142-8
UN No. 1265 (DOT)
Synonyms: 2-Methylbutane; Butane, 2-methyl-; Ethyldimethyl methane; Isoamylhydride; 1,1,2-Trimethylethane
Classification: Sat. aliphatic hydrocarbon
Empirical: C_5H_{12}
Formula: $CH_3CH_2CH(CH_3)_2$
Properties: Colorless liq.; gasoline-like odor; sol. in hydrocarbons, alcohol, oils, ether; insol. in water; m.w. 72.17; sp.gr. 0.625 (20 C); vapor pressure 595 mm Hg (21 C); f.p. -159.89 C; b.p. 27.85 C; flash pt. -51 C
Toxicology: LCLo (inh., mouse, 2 h) 419 g/m^3; ACGIH TLV/TWA 600 ppm; STEL 750 ppm; mildly toxic and narcotic by inhalation; mild skin irritation on evaporative cooling; may cause temporary eye irritation
Precaution: Highly flamm.; very dangerous fire and explosion hazard exposed to heat, flame, or oxidizers; incompat. with strong oxidizing agents (increases fire and explosion hazard); Hazardous decomp. prods.: Heated to decomp., emits acrid smoke and irritating fumes
NFPA: Health 1, Flammability 4, Reactivity 0
Uses: Solvent; blowing agent for polystyrene and other polymers; synthesis of org. chems. (e.g., chlorinated hydrocarbons; component of motor vehicle and aviation fuel
Manuf./Distrib.: Aldrich; Exxon; Exxon Europe; Fluka; Phillips; Sigma; Spectrum Chem. Mfg.; Triple Crown Am.
Trade names: Exxsol® Isopentane

Isopentanol. *See* Isoamyl alcohol
Isopentyl acetate. *See* Isoamyl acetate
Isopentyl alcohol. *See* Isoamyl alcohol
Isopentyl alcohol acetate. *See* Isoamyl acetate
Isopentyl methyl ketone. *See* Methyl isoamyl ketone

Isophorone
CAS 78-59-1; EINECS 201-126-0
Synonyms: 3,3,5-Trimethyl-2-cyclohexen-1-one; 3,3,5-Trimethyl-2-cyclohexene-1-one; Isoacetophorone; Isoforone; 1,1,3-Trimethyl-3-cyclohexene-5-one
Classification: Cyclic aliphatic ketone
Empirical: $C_9H_{14}O$
Formula: $OCHC:C(CH_3)CH_2C(CH_3)_2CH_2$
Properties: Colorless to wh. liq.; peppermint or camphor-like odor; sol. in alcohol, ether, acetone; misc. with org. solvs.; insol. in water; m.w. 138.20; sp.gr. 0.922 (20/20 C); m.p. -8 C; b.p. 210-218 C; flash pt. (CC) 84 C; ref. index 1.4781 (20 C)
Toxicology: ACGIH TLV/CL 5 ppm; LD50 (oral, rat) 2330 mg/kg; mildly toxic by inhalation; irritating to eyes, skin, respiratory system
Precaution: Combustible; incompat. with strong oxidizing agents such as peroxides, nitrates, and perchlorates
NFPA: Health 2, Flammability 2, Reactivity 0
Uses: Solvent for lacquers, plastics, oils, fats, gums, polyvinyl and nitrocellulose resins; mfg. of pesticides, coatings; flavoring agent for foods and pharmaceuticals
Regulatory: FEMA GRAS; SARA reportable
Manuf./Distrib.: Aceto; Aldrich; Allchem Ind.; Ashland; BP Chems. Inc;

Chemcentral; Coyne; Elf Atochem N. Am.; Exxon; Fabrichem; Fluka; Great Western; Houghton Chem.; Hüls AG; Sigma; Stanchem; Sunnyside; Union Carbide; Van Waters & Rogers

Isopropanol. *See* Isopropyl alcohol
4-Isopropenyl-1-methylcyclohexene. *See* d-Limonene
4-Isopropenyl-1-methyl-1-cyclohexene. *See* dl-Limonene
(R)-4-Isopropenyl-1-methyl-1-cyclohexene. *See* d-Limonene
2-Isopropoxypropane. *See* Isopropyl ether

1-Isopropoxy-2-propanol
Uses: Solvent for resins and dyes, surface coatings
Trade names containing: Solvenon® IPP

2-Isopropoxy-1-propanol
Uses: Solvent for resins and dyes, surface coatings
Trade names containing: Solvenon® IPP

Isopropyl acetate
CAS 108-21-4; EINECS 203-561-1
UN No. 1220 (DOT)
Synonyms: Acetic acid 1-methylethyl ester; Acetic acid isopropyl ester; 2-Propyl ethanoate; 2-Propyl acetate; 2-Acetoxypropane; 1-Methylethyl acetate
Classification: Sat. aliphatic carboxylic acid ester
Empirical: $C_5H_{10}O_2$
Formula: $CH_3COOCH(CH_3)2$
Properties: Colorless aromatic liq., fruity odor; sol. in acetone, most org. solvs.; mod. sol. in water; misc. with alcohol, ether, fixed oils; m.w. 102.15; dens. 0.874 (20/20 C); m.p. 073 C; f.p. -69.3 C; b.p. 88.4 C; flash pt. 40 F; ref. index 1.377; surf. tens. 26 dynes/cm (20 C)
Toxicology: TLV 250 ppm; STEL 310 ppm; LD50 (oral, rat) 3000 mg/kg; mod. toxic by ingestion; mildly toxic by inh.; high concs. of vapor irritating to nose and throat; human systemic effects on inh.; narcotic in high conc.
Precaution: Highly flamm.; dangerous fire hazard with heat, flame, oxidizers; mod. explosive with heat or flame; incompat. with strong oxidizers, strong acids, bases, potassium t-butoxide; Hazardous decomp. prods.: Isopropanol, acetic acid
NFPA: Health 1, Flammability 3, Reactivity 0
Uses: Solvent for paint mfg., syn. resins (nitrocellulose, ethylcellulose, CAB, some vinyl copolymers, PS, methacrylate resins), printing inks, gums, lacquers, plastics, oils and fats; organic synthesis; odorant in perfumes; syn. flavoring agent for foods and pharmaceuticals
Regulatory: FDA 21CFR §172.515, 175.105, 177.1200; FEMA GRAS
Manuf./Distrib.: Aldrich; Allchem Ind.; Ashland; J.T. Baker; BP Chems. Ltd; Chemcentral; Coyne; Eastman; Fluka; Harcros USA; Houghton Chem.; Samson; Stanchem; Sunnyside; Union Carbide; Van Waters & Rogers

Isopropylacetone. *See* Methyl isobutyl ketone
Isopropyl adipate. *See* Diisopropyl adipate

Isopropyl alcohol
CAS 67-63-0; EINECS 200-661-7
UN No. 1219
Synonyms: IPA; Isopropanol; Petrohol; 2-Propanol; s-Propyl alcohol; sec-Propyl alcohol; Dimethyl carbinol
Classification: Aliphatic alcohol
Empirical: C_3H_8O
Formula: $(CH_3)_2CHOH$

Properties: Colorless volatile liq., pleasant odor, sl. bitter taste; sol. in water, alcohol, ether, chloroform, most org. solvs.; m.w. 60.11; sp.gr. 0.7863 (20/20 C); f.p. -86 C; b.p. 82.4 C (760 mm); flash pt. (TOC) 11.7 C; ref. index 1.3756 (20 C)

Toxicology: ACGIH TLV/TWA 400 ppm; STEL 500 ppm; LD50 (oral, rat) 5045 mg/kg; poison by ingestion, subcutaneous routes; human systemic effects by ingestion/inhalation (headache, nausea, vomiting, narcosis); 100 ml can be fatal

Precaution: Flamm.; very dangerous fire hazard with heat, flame, oxidizers; reacts with air to form dangerous peroxides; incompat. with strong oxidizers, phosgene, iron salts, hydrogen-palladium, potassium t-butoxide, nitroform; Hazardous decomp. prods.: Heated to decomp., emits acrid smoke and fumes

NFPA: Health 1, Flammability 3, Reactivity 0

Uses: Solvent for essential oils, alkaloids, gums, resins, cellulose derivs., coatings; deicing agent for liq. fuels; lacquers; dehydrating agent; denaturing ethyl alcohol; in cosmetics; mfg. of acetone, glycerol, isopropyl acetate; preservative in extraction processes; cosolvent for pesticides; syn. flavoring agent, color diluent for foods and pharmaceuticals; antiseptic in topical anti-infectives

Regulatory: FDA 21CFR §73.1 (no residue), 73.1001, 172.515, 172.560, 172.712, 173.240 (limitation 50 ppm in spice oleoresins, 6 ppm in lemon oil, 2% in hops extract), 173.340; 175.105, 176.200, 176.210, 177.1200, 177.2800, 178.1010, 178.3910; 27CFR §21.112; use in bread is permitted in Ireland and Japan; FEMA GRAS; FDA approved for orals, topicals; USP/NF, BP compliance

Manuf./Distrib.: Aldrich; Allchem Ind.; ARCO; Ashland; J.T. Baker; Baychem; BP Chems. Ltd; R.E. Carroll; Chemcentral; Coyne; Eastman; Exxon; Fluka; General Chem.; C.P. Hall; Harcros USA; Houghton Chem.; Hüls AG; Mallinckrodt; Mitsui Toatsu; Olin; Samson; Shell; Sigma; Spectrum Chem. Mfg.; Union Carbide; Van Waters & Rogers; Veckridge

Trade names containing: Filmex® A-1; Filmex® D-1; MS-195/CO2 Safezone® Cleaning Solvent & Flux Remover; Tecsol® A; Tecsol® B; Tecsol® D; Tecsol® H

Isopropylamine

CAS 75-31-0

UN No. 1221 (DOT)

Synonyms: MIPA; 2-Aminopropane; 1-Methylethylamine; Monoisopropylamine; 2-Propanamine; 2-Propylamine; sec-Propylamine

Classification: Sat. aliphatic amine

Empirical: C_3H_9N

Formula: $(CH_3)_2CHNH_2$

Properties: Colorless liq.; ammonia-like odor; sol. in water, ethanol, ether, benzene, chloroform, many org. solvs.; very sol. in acetone; m.w. 59.11; sp.gr. 0.6870 (20/20 C); vapor pressure 472 mm Hg (20 C); m.p. -101 C; b.p. 32. 4 C; flash pt. (CC) -37 C; pH 11.8 (0.1M aq. sol'n.)

Toxicology: LD50 (oral, rat) 820 mg/kg, (skin, rabbit) 550 mg/kg; ACGIH TLV/TWA 5 ppm; STEL 10 ppm; corrosive, very toxic; skin and eye irritant; can be absorbed thru skin in toxic amts.; possible permanent eye injury; inh. can cause nose/throat/lung irritation

Precaution: Flamm.; incompat. with oxidizers, acids, halogenated hydrocarbons, nitroparaffins, nitromethane, calcium or sodium hypochlorite, nitrosyl perchlorate, chloromethyloxirane, mercury, perchloryl fluoride; violent/explosive reactions possible; Hazardous decomp. prods.: Emits toxic fumes of NO_x

NPFA; Health 3, Flammability 4, Reactivity 0

Uses: Intermediate; solvent; corrosion inhibitor; dehairing agent; mfg. of pharmaceuticals, dyes, insecticides, bactericides

Isopropylamine dodecylbenzenesulfonate

Manuf./Distrib.: Aldrich; Ashland; Elf Atochem N. Am.; Fluka; Sigma; Union Carbide

Isopropylamine dodecylbenzenesulfonate
CAS 26264-05-1; 68584-24-7; EINECS 247-556-2
Synonyms: Dodecylbenzenesulfonic acid, comp. with 2-propanamine (1:1)
Classification: Aromatic compd.
Definition: Salt of isopropylamine and doecylbenzene sulfonic acid
Empirical: $C_{21}H_{39}O_3NS$
Formula: $C_{18}H_{30}O_3S \cdot C_3H_9N$
Uses: Emulsifier, solubilizer, solvent, detergent, and wetting agent for oil-based systems; dispersant in oil and water-based systems; used in dry-cleaning surfactants; hydrotrope for liq. detergents; latex emulsifier; emulsion polymerization
Regulatory: FDA 21CFR §176.210
Trade names: Rhodacal® 330

Isopropylbenzene. *See* Cumene
Isopropyl benzol. *See* Cumene
Isopropylbiphenyl. *See* 1,1´-Biphenyl (1-methylethyl)
Isopropylcarbinol. *See* Isobutyl alcohol
3-Isopropylcumene. *See* 1,3-Diisopropyl benzene
4-Isopropylcumene. *See* 1,4-Diisopropyl benzene

Isopropylcyclohexane
CAS 696-29-7
Synonyms: Hexahydrocumene; (1-Methylethyl)cyclohexane; Normanthane
Classification: Sat. alicyclic hydrocarbon
Empirical: C_9H_{18}
Properties: Colorless liq.; sol. in most org. solvs.; insol. in water; m.w. 126.24; sp.gr. 0.802 (20 C); m.p. -95 C; b.p. 155 C; flash pt. 35 C
Toxicology: May cause mild skin irritation; vapor may be irritating to eyes; probably a CNS depressant; sl. nose/throat irritant; high vapor concs. may cause headache, lightheadedness, nausea, drowsiness; very high concs. may cause unconsciousness, death
Precaution: Flamm.; incompat. with strong oxidizing agents (increased fire and explosion hazard); Hazardous decomp. prods.: None
NFPA: Health 1, Reactivity 0
Uses: Solvent; org. synthesis
Manuf./Distrib.: Aldrich; Fluka

Isopropylcyclopentane
CAS 3875-51-2
Synonyms: 2-Cyclopentylpropane; Cyclopentyl-2 propane
Classification: Sat. alicyclic hydrocarbon
Empirical: C_8H_{16}
Formula: $C_5H_9CH(CH_3)CH_3$
Properties: Colorless liq.; sol. in most org. solvs.; insol. in water; sp.gr. 0.7765 (20 C); m.p. -111 C; b.p. 126 C
Toxicology: May cause mild skin irritation; conc. vapor may be irritating to eyes; probably a CNS depressant; high vapor concs. may cause nose/throat irritation, lightheadedness, nausea, dizziness; very high concs. may cause unconsciousness and death
Precaution: Flamm.; incompat. with strong oxidizing agents (increased fire and explosion hazard); Hazardous decomp. prods.: None
Uses: Solvent; org. synthesis

Isopropyl dimethyl carbinol. *See* 2-Methyl-1-pentanol
Isopropyl diphenyl. *See* 1,1′-Biphenyl (1-methylethyl)

Isopropyl ether
CAS 108-20-3; EINECS 203-560-6
UN No. 1159 (DOT)
Synonyms: Diisopropyl oxide; 2-Isopropoxypropane
Empirical: $C_6H_{14}O$
Formula: $(CH_3)_2CHOCH(CH_3)_2$
Properties: Colorless liq., ethereal odor; misc. with water, ethanol, ether; m.w. 102.20; dens. 0.719; m.p. -60 C; b.p. 68.5 C; flash pt. (CC) -18 F; ref. index 1.368 (20 C)
Toxicology: LD50 (oral, rat) 8470 mg/kg, (dermal, rabbit) 20 g/kg; TLV 250 ppm in air; mod. toxic by IP and other routes; mildly toxic by ingestion, inh., skin contact; skin irritant; heated to decomp., emits acrid smoke and fumes
Precaution: DOT: Flamm. liq.; severe explosion hazard when exposed to heat, flame, sparks, or oxidizers; may form explosive peroxides; shock may cause explosion; violent reaction with chlorosulfonic acid, HNO_3
NFPA: Health 1, Flammability 3, Reactivity 1
Uses: Solvent for paints, animal, vegetable, and min. oils, waxes, and resins; extraction solvent; rubber cements; reagent
Manuf./Distrib.: Allchem Ind.; Ashland; J.T. Baker; Exxon; Fluka; Shell; Spectrum Chem. Mfg.

Isopropyl hexadecanoate. *See* Isopropyl palmitate
Isopropyl n-hexadecanoate. *See* Isopropyl palmitate
Isopropyl-S-(-)-2-hydroxy propionate. *See* Isopropyl lactate
Isopropylideneacetone. *See* Mesityl oxide

Isopropyl lactate
CAS 617-51-6
Synonyms: Isopropyl-S-(-)-2-hydroxy propionate; Propanoic acid, 2-hydroxy-, 1-methylethyl ester; Lactic acid, isopropyl ester; 1-Methylethyl 2-hydroxypropanoate
Definition: Isopropyl ester of lactic acid
Empirical: $C_6H_{12}O_3$
Formula: $CH_3CH(OH)COOCH(CH_3)_2$
Properties: M.w. 132.18
Toxicology: LDLo (IM, guinea pig) 2500 mg/kg
Uses: Solvent for paints and coatings, electronics, printed circuit boards, metal industry
Manuf./Distrib.: Purac Am.
Trade names: Purasolv® IPL

Isopropyl laurate
CAS 10233-13-3; EINECS 233-560-1
Synonyms: Dodecanoic acid, 1-methylethyl ester; 1-Methylethyl dodecanoate
Definition: Ester of isopropyl alcohol and lauric acid
Formula: $CH_3(CH_2)_{10}COOCH(CH_3)_2$
Uses: Emollient, lubricant, plasticizer, and cosolvent in cosmetic and pharmaceutical industries
Regulatory: FDA 21CFR §176.210, 177.2800
Trade names: Crodamol IPL

Isopropyl methyl ketone. *See* Methyl isopropyl ketone

Isopropyl myristate
CAS 110-27-0; EINECS 203-751-4

Isopropyl myristate/palmitate.

Synonyms: IPM; 1-Methylethyl tetradecanoate; Tetradecanoic acid, 1-methylethyl ester; Tetradecanoic acid isopropyl ester; Myristic acid isopropyl ester
Classification: Synthetic fatty alcohol
Definition: Ester of isopropyl alcohol and myristic acid
Empirical: $C_{17}H_{34}O_2$
Formula: $CH_3(CH_2)_{12}COOCH(CH_3)_2$
Properties: Colorless oily low-visc. liq., odorless; sol. in most org. solvs., acetone, chloroform, ethyl acetate, ethanol, min. oil, veg. oil; dissolves waxes; insol. in water; m.w. 270.44; dens. 0.850-0.860; f.p. 3 C; b.p. 192.6 C (20 mm); dec. 208 C; acid no. 1 max.; iodine no. 1 max.; sapon. no. 202-212; ref. index 1.432-1.436 (20 C)
Toxicology: Suspected tumorigen; human skin irritant; causes blackheads
Precaution: Light-sensitive; Hazardous decomp. prods.: Heated to decomp., emits toxic smoke and fumes
Uses: Emollient, lubricant, vehicle, solvent in cosmetic creams, topical medicinals; plasticizer for cellulosics
Regulatory: FDA 21CFR §176.210, 177.2800; FEMA GRAS; FDA approved for topicals; USP/NF, BP, Ph.Eur. compliance
Manuf./Distrib.: Aldrich; Amerchol; Aquatec Quimica SA; Burdick & Jackson; Fluka; Goldschmidt; Henkel/Emery; Inolex; Lanaetex Prods.; Penta Mfg.; Pentagon Chems. Ltd; Sigma; Spectrum Chem. Mfg.; Stepan; Unichema Int'l.
Trade names: Crodamol IPM; Estol IPM 1509 (BIO-IPM); Kessco® IPM; Lexol® IPM; Lexol® IPM-NF; Promyr; Schercemol IPM; Tegosoft® M
Trade names containing: Crodamol CAP; Lexol® 3975; Wickenol® 136

Isopropyl myristate/palmitate. *See* Isopropyl myristopalmitate

Isopropyl myristopalmitate
Synonyms: Isopropyl palmitate/myristate; Isopropyl myristate/palmitate
Definition: Ester of isopropyl alcohol and mixture of myristic and palmitic acids
Properties: Sp.gr. 0.852; flash pt. (COC) 182 C; ref. index 1.4340
Uses: Cosmetics emollient, solvent; plasticizer for PVC; lubricant for PS
Manuf./Distrib.: Inolex; Union Camp

Isopropyl palmitate
CAS 142-91-6; EINECS 205-571-1
Synonyms: IPP; Isopropyl n-hexadecanoate; Isopropyl hexadecanoate; Hexadecanoic acid, 1-methylethyl ester; Hexadecanoic acid isopropyl ester; 1-Methylethyl hexandecanoate
Definition: Ester of isopropyl alcohol and palmitic acid
Empirical: $C_{19}H_{38}O_2$
Formula: $CH_3(CH_2)_{14}COOCH(CH_3)_2$
Properties: Colorless mobile liq., very sl. odor; sol. in 4 parts 90% alcohol, min. oil, fixed oils, acetone, castor oil, chloroform, cottonseed oil, ethyl acetate; insol. in water, glycerin, propylene glycol; m.w. 298.57; dens. 0.850-0.855; m.p. 14 C; acid no. 1 max.; iodine no. 1 max.; sapon. no. 183-193; flash pt. > 230 F; ref. index 1.4350-1.4390 (20 C)
Toxicology: LD50 (IP, mouse) 100 mg/kg; poison by intraperitoneal route; human skin irritant
Precaution: Combustible; Light-sensitive; Hazardous decomp. prods.: Heated to decomp., emits acrid smoke and fumes
Uses: Emollient, emulsifier, solvent, cosolvent, stabilizer in lotions, creams, similar cosmetics, deodorants; vehicle in topical pharmaceuticals; plasticizer for cellulosics
Regulatory: FDA 21CFR §176.210, 177.2800; FDA approved for topicals; USP/NF,

BP, Ph.Eur. compliance
Manuf./Distrib.: Aldrich; Amerchol; Aquatec Quimica SA; Fluka; Goldschmidt; Henkel/Emery; Inolex; Lanaetex Prods.; Penta Mfg.; Spectrum Chem. Mfg.; Stepan; Unichema Int'l.
Trade names: Crodamol IPP; Kessco® IPP; Propal; Tegosoft® P
Trade names containing: Lexol® 3975; Wickenol® 136

Isopropyl palmitate/myristate. *See* Isopropyl myristopalmitate

Isopropyl stearate
CAS 112-10-7; EINECS 203-934-9
Synonyms: 1-Methylethyl octadecanoate; Octadecanoic acid, 1-methylethyl ester
Definition: Ester of isopropyl alcohol and stearic acid
Empirical: $C_{21}H_{42}O_2$
Formula: $CH_3(CH_2)_{16}COOCH(CH_3)_2$
Properties: M.w. 326.63
Toxicology: LDLo (oral, rat) 8 g/kg; primary irritant
Uses: Emollient, cosolvent, and lubricant for cosmetics and topical pharmaceuticals
Regulatory: FDA 21CFR §176.210, 177.2800; FDA approved for topicals
Manuf./Distrib.: Henkel/Emery
Trade names: Wickenol® 127
Trade names containing: Lexol® 3975; Wickenol® 136

Isopropyltrimethylmethane. *See* 2,2,3-Trimethylbutane
Isosorbide dimethyl ether. *See* Dimethyl isosorbide

Isostearyl alcohol
CAS 27458-93-1; 70693-04-8; EINECS 248-470-8
Synonyms: 1-Heptadecanol, 16-methyl-; Isooctadecanol; 16-Methyl-1-heptadecanol
Definition: Mixture of branched chain 18 carbon aliphatic alcohols
Properties: m.w. 267-285; flash pt. (COC) 170 C
Uses: Coemulsifier, lubricant, foam control agent, cosolvent, plasticizer, stabilizer, emollient, intermediate; for metal lubricants, inks, textiles, emulsions, paper, cosmetics, mineral processing, oil field chemicals, fabric softeners
Manuf./Distrib.: Ashland; M. Michel; Witco/Oleo-Surf.
Trade names: Adol® 66; Witcohol™ 66

Isostearyl isononanoate
Synonyms: Hexanoic acid, 3,5,5-trimethyl-, isooctadecyl ester
Definition: Ester of isostearyl alcohol and isononanoic acid
Empirical: $C_{27}H_{54}O_2$
Formula: $CH_3(CH_2)_5CHCH_3COOC_{18}H_{37}$
Uses: Emollient, superfatting agent, solvent for lipophilic components
Trade names: Tegosoft® 189

Isostearyl isostearate
CAS 41669-30-1; EINECS 255-485-3
Synonyms: Isooctadecanoic acid, isooctadecyl ester
Definition: Ester of isostearyl alcohol and isostearic acid
Empirical: $C_{36}H_{62}O_2$
Properties: Liq.; insol. in water, sol. in hydrocarbons
Toxicology: No known toxicity
Uses: Nonionic emollient; used in cosmetics imparting slip and lubricity, luster, and sheen; cosolvent and solubilizer in perfumes; emollient, lubricant for pharmaceuticals

Isostearyl lactate

Manuf./Distrib.: Gattefosse
Trade names: Schercemol 1818

Isostearyl lactate
CAS 42131-28-2; EINECS 255-674-0
Synonyms: 2-Hydroxypropanoic acid, isostearyl ester; Isooctadecyl 2-hydroxypropanoate
Definition: Ester of isostearyl alcohol and lactic acid
Empirical: $C_{21}H_{42}O_3$
Formula: $CH_3CHOHCOOC_{18}H_{37}$
Uses: Emollient, substantivity agent, softener, slip agent, lubricant, cosolvent, solubilizer in cosmetics and pharmaceuticals

Isotridecyl isononanoate
CAS 42131-27-1; 59231-37-7
Synonyms: 3,5,5-Trimethylhexanoic acid, isotridecyl ester; Isononanoic acid, isotridecyl ester
Definition: Ester of isotridecyl alcohol and isononanoic acid
Empirical: $C_{22}H_{44}O_2$
Properties: Sol. in alcohol, animal and vegetable oils, min. oil; insol. in water, glycerin, propylene glycol
Uses: Emollient with solvency chars. for skin and hair care prods.; pigment wetter
Trade names: Wickenol® 153

Isovalerone. *See* Diisobutyl ketone

Katchung oil. *See* Peanut oil
Kautschin. *See* d-Limonene

Kerosene
CAS 8008-20-6; EINECS 232-366-4
UN No. 1223 (DOT)
Synonyms: Kerosine; Coal oil; Straight-run kerosene
Classification: Petroleum hydrocarbon
Definition: Mixt. of petroleum hydrocarbons, chiefly of the methane series having 10-16 carbon atoms per molecule
Properties: Water-wh. to pale yel. oily liq.; mild char. petrol. odor; sol. in other petrol. solvs.; insol. in water; sp.gr. 0.8; vapor pressure 1.5 mm Hg (20 C); f.p. -45.6 C; b.p. 151-301 C; flash pt. 122 F; KB value 36
Toxicology: LD50 (oral, rabbit) 28 g/kg, (IP, rabbit) 6600 mg/kg, (IV, rabbit) 180 mg/kg; LDLo (oral, human male) 500 mg/kg; poison by IV and intratracheal routes; primary irritant; mod. to severe skin irritant; human systemic effects by ing., IV routes
Environmental: HAPS contg.
Precaution: Combustible exposed to heat or flame; mod. explosive as vapor exposed to heat or flame; incompat. with strong oxidizing agents; risk of fire and explosion; Hazardous decomp. prods.: Heated to decomp., emits acrid smoke and fumes
NFPA: Health 0, Flammability 2, Reactivity 0
Uses: Fuel for lamps, flares, stoves; diesel fuel for automobiles; vehicle for paints, thinners, enamels, polishes, varnishes; carrier for pesticides; cleaning

degreasing solvent; mold release agent for ceramics, pottery; solvent in asphalt coating
Manuf./Distrib.: Aldrich; Houghton Chem.; Sigma; Spectrum Chem. Mfg.

Kerosine. *See* Kerosene
Ketohexamethylene. *See* Cyclohexanone
Ketone C-7. *See* Methyl n-amyl ketone
Ketone methyl phenyl. *See* Acetophenone
Ketone propane. *See* Acetone
β-Ketopropane. *See* Acetone

Lactamide DGA
Synonyms: Lactamide N-(1-ethoxy-2-hydroxyethyl); Propanamide, 2-hydroxy-N-(1-ethoxy-2-hydroxyethyl)
Definition: Ethoxylated monoethanolamine (DGA) of lactic acid
Uses: Humectant, emollient, cosolvent for skin and hair prods.
Trade names: Foamid LM2E

Lactamide N-(1-ethoxy-2-hydroxyethyl). *See* Lactamide DGA

Lactic acid
CAS 50-21-5; 598-82-3 (DL); 79-33-4 (L); 10326-41-7 (D); EINECS 200-018-0; 209-954-4 (DL); 201-296-2 (L); 233-713-2 (D)
UN No. 1760
Synonyms: 2-Hydroxypropanoic acid; 1-Hydroxyethane 1-carboxylic acid; 2-Hydroxypropionic acid; α-Hydroxypropionic acid; Milk acid; Acetonic acid; Ethylidenelactic acid
Classification: Organic acid
Definition: Prod. of the metabolism of glucose and glycogen
Empirical: $C_3H_6O_3$
Formula: $CH_3CHOHCOOH$
Properties: Colorless to yellowish cryst. or syrupy liq., nearly odorless; hygroscopic; misc. with water, alcohol, glycerol, furfural; insol. in chloroform; m.w. 90.09; dens. 1.249; m.p. 18 C; b.p. 122 C (15 mm); flash pt. > 230 F; ref. index 1.4251
Toxicology: LD50 (oral, rat) 3730 mg/kg; mod. toxic by ingestion, rectal routes; mutagenic data; severe skin and eye irritant; heated to decomp., emits acrid smoke and irritating fumes
Precaution: DOT: Corrosive material; mixts. with nitric acid + hydrofluoric acid may react vigorously
Uses: Cultured dairy prods., as acidulant, chemicals (salts, plasticizers, adhesives, pharmaceuticals), mordant in wool dyeing, food additive, mfg. of lactates; sequestrant in metal plating, cataphoresis paints; leather tanning delimer; acidulant, antimicrobial, preservative, flavoring for foods, pharmaceuticals; solvent, vehicle for foods
Use level: Up to 10%
Regulatory: FDA 21CFR §131.144, 133, 150.141, 150.161, 172.814, 184.1061, GRAS; USDA 9CFR §318.7, 381.147; BATF 27CFR §240.1051, GRAS; not for use in infant foods; FEMA GRAS; Japan approved; Europe listed; UK approved; FDA approved for injectables, parenterals, orals, topicals, vaginals; USP/NF, BP, Ph.Eur., JP compliance

Manuf./Distrib.: Aldrich; Am. Biorganics; Balchem; Chemical; Coyne; Fluka; Lohmann; Penta Mfg.; Pfanstiehl Labs; Purac Biochem; Research Organics; Rhône-Poulenc N. Am.; San Yuan; Sigma; Spectrum Chem. Mfg.; Wilke Int'l.

Lactic acid butyl ester. *See* Butyl lactate
Lactic acid ethyl ester. *See* Ethyl lactate
Lactic acid, isopropyl ester. *See* Isopropyl lactate
Lactic acid methyl ester. *See* Methyl lactate

Lactol spirits
Classification: Petroleum naphtha
Properties: b.p. 200-265 F
Uses: Low-boiling solvent

LAM. *See* 2-Pyrrolidone

Laneth-5
CAS 61791-20-6 (generic)
Synonyms: PEG-5 lanolin ether; POE (5) lanolin ether; PEG-5 lanolin alcohol
Definition: PEG ether of lanolin alcohol with avg. ethoxylation value of 5
Uses: Nonionic emulsifier, emollient, stabilizer, solubilizer, cosolvent, wetting agent, dispersant, conditioner for cosmetics, pharmaceuticals, creams and lotions; bromo dye solvent
Trade names: Nikkol BWA-5; Polychol 5
Trade names containing: Solulan® 5

Laneth-10
CAS 61791-20-6 (generic)
Synonyms: PEG-10 lanolin ether; POE (10) lanolin ether; PEG 500 lanolin ether
Definition: PEG ether of lanolin alcohol with avg. ethoxylation value of 10
Uses: Nonionic emulsifier, emollient, stabilizer, solubilizer, cosolvent, wetting agent, dispersant, conditioner for cosmetics, creams and lotions, pharmaceuticals
Trade names: Nikkol BWA-10

Laneth-15
CAS 61791-20-6 (generic); 84650-19-1
Synonyms: PEG-15 lanolin ether; POE (15) lanolin ether
Definition: PEG ether of lanolin alcohol with avg. ethoxylation value of 15
Uses: Nonionic emulsifier, emollient, stabilizer, solubilizer, wetting agent, dispersant, conditioner for cosmetics, topical pharmaceuticals, creams and lotions; cosolvent for nonmiscible oils
Trade names: Polychol 15

Laneth-16
CAS 61791-20-6 (generic)
Synonyms: PEG-16 lanolin ether; POE (16) lanolin ether
Definition: PEG ether of lanolin alcohol with avg. ethoxylation value of 16
Uses: Emulsifier for cosmetics and pharmaceuticals
Trade names containing: Solulan® 16

Lanolin
CAS 8006-54-0 (anhyd.); 8020-84-6 (hyd.); EINECS 232-348-6
Synonyms: Anhydrous lanolin; Adeps lanae; Wool wax; Wool fat
Definition: Deriv. of unctuous fatty sebaceous secretion of sheep, *Ovis aries*, consistg. of complex mixt. of esters of high m.w. aliphatic, steroid, or triterpenoid alcohol and fatty acids
Properties: Yel.-wh. semisolid; sol. in chloroform, ether; insol. in water; m.p. 38-42 C; iodine no. 18-36; flash pt. > 230 F

Toxicology: Can cause allergic reactions, contact dermatitis
Precaution: Hazardous decomp. prods.:; Heated to decomp., emits acrid smoke and irritating fumes
NFPA: Health 0, Flammability 1, Reactivity 0
Uses: Ointments, soaps, face creams, facial tissues, hair set and sun-tan preps.; plasticizer for rubber; lubricant for textiles, metalworking compds.; EP and slip agent for metalworking/rust preventative coatings; in printing inks; waterproofing agent for leather; chewing gum base; pharmaceutical emollient, ointment base, filler, vehicle; cosolvent, emollient, plasticizer, emulsifier, wetting agent for cosmetics
Regulatory: FDA 21CFR §172.615, 175.300, 176.170, 176.210, 177.1200, 177.2600, 178.3910; Japan approved; FDA approved for ophthalmics, topicals; USP/NF, BP, Ph.Eur. compliance
Manuf./Distrib.: Aldrich; Amerchol; Charkit; Croda Chem. Ltd; Fluka; Henkel/Emery; Integra; Lanaetex Prods.; Penta Mfg.; R.I.T.A.; Ruger; Sigma; Spectrum Chem. Mfg.; Stevenson Cooper; Westbrook Lanolin
Trade names: Corona Lanolin; Coronet Lanolin

Lanolin alcohol
CAS 8027-33-6; EINECS 232-430-1
Synonyms: Alcohols, lanolin; Wool wax alcohol
Definition: Mixture of organic alcohols obtained from hydrolysis of lanolin
Properties: Amber waxy solid, char. odor; sol. in ether, chloroform; sl. sol. in alcohol; insol. in water; m.p. 56 C; acid no. 2 max.; sapon. no. 12 max.
Toxicology: Less likely to cause allergic reaction than lanolin
Uses: W/o emulsifier, stabilizer, softener, emollient, gelling agent, thickener, plasticizer, moisturizer for absorption bases, cosmetics, cleansing preps.
Regulatory: FDA approved for ophthalmics, topicals; USP/NF compliance
Manuf./Distrib.: Ashland; ChemMark Development; Croda Chem. Ltd; Henkel/Emery; Heterene
Trade names containing: Argobase 125T

Lanolin, alcohols, acetates. *See* Acetylated lanolin alcohol

Lanolin linoleate
Synonyms: 9,12-Octadienoic acid, lanolin alcohol ester
Definition: Ester of lanolin alcohol and linoleic acid
Trade names containing: Polylan®

Lanolin oil
CAS 8038-43-5; 70321-63-0; 8006-54-0; EINECS 274-559-6
Synonyms: Dewaxed lanolin; Oils, lanolin
Definition: Liq. fraction of lanolin obtained by physical means from whole lanolin
Toxicology: LD50 (rat, oral) > 20 g/kg; may cause allergic reactions, contact skin rashes
Uses: W/o emulsifier; dispersant for pigments; conditioning agent, cosolvent; emollient, lubricant, moisturizer, penetrant, superfatting agent for lipsticks, baby oils, brilliantines, cleansing lotions; pharmaceutical ointment base
Manuf./Distrib.: Aldrich; ChemMark Development; Croda Inc; Fluka; Henkel/Emery; Lanaetex Prods.; R.I.T.A.; Sigma
Trade names: Lipolan R

Lauric acid, benzyl ester. *See* Benzyl laurate

Lauryl pyrrolidone
CAS 2687-96-9
Synonyms: N-Dodecyl-2-pyrrolidone; 2-Pyrrolidinone, 1-dodecyl-
Classification: Substituted heterocyclic compd.

Light mineral oil

Empirical: $C_{16}H_{31}NO$
Properties: M.w. 253.48
Toxicology: LDLo (oral, rat) 5 g/kg, (IP, mouse) 1 g/kg; primary irritant
Uses: Conditioner, foam stabilizer, wetting agent; specialty solvent for commercial cleaning, textile processing, lithographic printing prods., water-borne coatings, inks; VOC replacement
Manuf./Distrib.: Aldrich
Trade names: Gaftex® HI-12; Surfadone® LP-300

Light mineral oil. *See* Mineral oil
Ligroin. *See* Mineral spirits
(±)-Limonene. *See* dl-Limonene
(-)-Limonene. *See* l-Limonene

d-Limonene
CAS 5989-27-5; EINECS 227-813-5
UN No. 1993 (DOT)
Synonyms: (D)-Limonene; Dextro-limonene; d-p-Mentha-1,8-diene; p-Mentha-1,8-diene; Cinene; Dipentene; Cajeputene; Kautschin; R(+)-Limonene; D-(+)-Limonene; (+)-Carvene; (R)-1-Methyl-4-(1-methylethenyl)-cyclohexene; (R)-4-Isopropenyl-1-methyl-1-cyclohexene; 4-Isopropenyl-1-methylcyclohexene
Classification: Unsat. alicylic hydrocarbon; terpene
Definition: Naturally occurring C10 cycloolefinic hydrocarbon; major component in citrus oils
Empirical: $C_{10}H_{16}$
Formula: $CH_3C_6H_8C(CH_3)=CH_2$
Properties: Colorless liq., citrus odor; insol. in water, propylene glycol; sol. in alcohol, ether; m.w. 136.26; sp.gr. 0.8411 (20 C); m.p. -74 C; b.p. 176-176.4 C; flash pt. (CC) 48 C; ref. index 1.471; KB value 63 (tech.); VOC 837 g/l
Toxicology: LD50 (oral, rat) 4400 mg/kg, (IV, rat) 110 mg/kg; poison by IV route; mod. toxic by intraperitoneal route; mildly toxic by ingestion; skin irritant; inh. of vapor may cause nose/throat irritation
Environmental: Fully biodeg.
Precaution: Combustible; incompat. with strong oxidizers, iodine pentafluoride, tetrafluoroethylene; explosive reactions possible; Hazardous decomp. prods.: Heated to decomp., emits acrid smoke and irritating fumes
NFPA: Health 0, Flammability 2, Reactivity 0
Uses: Syn. flavoring for foods and pharmaceuticals; fragrance and perfume ingred.; chem. intermediate; wetting agent, dispersant in terpene resin mfg.; industrial solvent for degreasing, in household and I&I cleaning prods.; minor component of turpentine; component of essential oils; inhibitor of tetrafluoroethylene polymerization; gallstone solubilizer
Regulatory: FDA 21CFR §182.60, GRAS; FEMA GRAS
Manuf./Distrib.: Aldrich; Allchem Ind.; Charkit; Coyne; Florida Chem.; Fluka; Int'l. Flavors & Fragrances; Langley Smith Ltd; Penta Mfg.; SCM Glidco Organics; Sigma; Spectrum Chem. Mfg.

(D)-Limonene. *See* d-Limonene
D-(+)-Limonene. *See* d-Limonene

dl-Limonene
CAS 138-86-3; EINECS 205-341-0
UN No. 2052 (DOT)
Synonyms: d,l-Limonene; Dipentene; Cinene; Cajeputene; p-Mentha-1,8-diene; dl-p-Mentha-1,8-diene; Racemic limonene; Inactive limonene; 1-Methyl-4-

isopropenyl-1-cyclohexene; 1-Methyl-4(1-methylethenyl)cyclohexene; (±)-Limonene; 4-Isopropenyl-1-methyl-1-cyclohexene
Classification: Unsat. alicyclic hydrocarbon
Empirical: $C_{10}H_{16}$
Properties: Colorless liq., pleasant lemon-like odor; sol. in alcohol, ether; pract. insol. in water; m.w. 136.23; sp.gr. 0.847 (15.5/15.5 C); m.p. -96.9 C; b.p. 175-176 C (763 mm); flash pt. (CC) 45 C; ref. index 1.4744
Toxicology: LD50 (oral, rat) 5000 mg/kg; skin irritant and sensitizer; inh. of vapor may cause nose and throat irritation; probable eye irritant; not very toxic by ing.
Precaution: DOT: Flamm. liq.; can react vigorously with oxidizers; also incompat. with iodine pentafluoride and tetrafluoroethylene; Hazardous decomp. prods.: Heated to decomp., emits acrid smoke and irritating fumes
NFPA: Health 0, Flammability 2, Reactivity 0
Uses: Solvent for oleoresinous prods., rosin, ester gum, alkyd resins, waxes, rubber.; rubber compding. and reclaiming; wetting agent, dispersant for oils, resins and combinations, pigments, driers, paints, enamels, lacquers; printing inks, perfumes, floor waxes, furniture polishes; flavor and fragrance ingred. for foods, pharmaceuticals; minor component of turpentine
Regulatory: FDA 21CFR §175.105, 177.2600, 182.60, GRAS; FDA approved for topicals
Manuf./Distrib.: Aldrich; Arizona; Fluka; Hercules; Penta Mfg.; SCM Glidco Organics; Spectrum Chem. Mfg.; Veitsiluoto Oy
Trade names containing: Acintene® AC-P; Acintene® B Beta Pinene; Acintene® DP 738

d,l-Limonene. *See* dl-Limonene

l-Limonene
CAS 138-86-3; 5989-54-8; EINECS 205-341-0
UN No. 2052 (DOT)
Synonyms: l-p-Mentha-1,8-diene; (S)-(-)-p-Mentha-1,8-diene; Dipentene; Cyclohexene, 1-methyl-4-(1-methylethenyl)-, (S)-; Carvene; (-)-Limonene; 1-Methyl-4-(1-methylethenyl)-(S)-cyclohexene; Cajeputene; p-Mentha-1,8-diene, (S)-(-)-
Empirical: $C_{10}H_{16}$
Properties: Liq.; misc. with alcohol; insol. in water; m.w. 136.23; dens. 0.8407 (20.5/4 C); b.p. 175.5-176.5 C (763 mm); flash pt. 168 F; ref. index 1.474 (21 C)
Toxicology: Skin irritant, sensitizer; Hazardous decomp. prods.: Heated to decomp., emits acrid smoke and irritating fumes
NFPA: Health 0, Flammability 2, Reactivity 0
Uses: Syn. flavoring agent for foods; solvent
Regulatory: FDA 21CFR §182.60, GRAS
Manuf./Distrib.: Polarome Mfg.; SCM Glidco Organics; Spectrum Chem. Mfg.
Trade names: Dipentene No. 122

R(+)-Limonene. *See* d-Limonene
Limonene oxide. *See* Eucalyptol
Linoleic acid, oleyl ester. *See* Oleyl linoleate
Liquid paraffin. *See* Mineral oil
Liquid petrolatum. *See* Mineral oil
Liquified petroleum gas. *See* Propane
Lye. *See* Potassium hydroxide; Sodium hydroxide

MAA. *See* Methyl acetoacetate
MAAC. *See* Methyl amyl acetate
Macrogol 300. *See* PEG-6
Macrogol 400. *See* PEG-8
Macrogol 600. *See* PEG-12
Macrogol 1000. *See* PEG-20
Macrogol 1540. *See* PEG-32
Macrogol 6000. *See* PEG-150
Maize oil. *See* Corn oil
MAK. *See* Methyl n-amyl ketone
MAOH. *See* Methyl amyl alcohol
MB. *See* Methyl bromide
MBX. *See* Methyl bromide
MDS. *See* Methyl disulfide
2-ME. *See* 2-Mercaptoethanol
MEA. *See* Ethylamine
MEBR. *See* Methyl bromide
MEK. *See* Methyl ethyl ketone
dl-p-Mentha-1,8-diene. *See* dl-Limonene
d-p-Mentha-1,8-diene. *See* d-Limonene
l-p-Mentha-1,8-diene. *See* l-Limonene
p-Mentha-1,3-diene. *See* α-Terpinene
p-Mentha-1,8-diene. *See* d-Limonene; dl-Limonene
(S)-(-)-p-Mentha-1,8-diene. *See* l-Limonene
p-Mentha-1,8-diene, (S)-(-)-. *See* l-Limonene
p-Menthane, 1,4-epoxy-. *See* 1,4-Cineole
p-Menthane, 1,8-epoxy. *See* Eucalyptol
p-Menth-1,4(8)-diene. *See* Terpinolene
p-Menthen-1-en-8-ol. *See* α-Terpineol
p-Menth-1-en-8-ol. *See* α-Terpineol
Meparfynol. *See* Methyl pentynol
1-Mercaptobutane. *See* n-Butyl mercaptan
Mercaptoethanol. *See* 2-Mercaptoethanol

2-Mercaptoethanol
CAS 60-24-2; EINECS 200-464-6
UN No. 2966 (DOT)
Synonyms: 2-ME; Mercaptoethanol; β-Mercaptoethanol; Thioethylene glycol; Ethylene thioglycol; 1-Ethanol-2-thiol; 2-Hydroxy-1-ethanethiol; Thioglycol; Monothioethyleneglycol; 2-Thioethanol; Thiomonoglycol; Hydroxyethylmercaptan; 2-Hydroxyethylmercaptan
Classification: Aliphatic organic compd.; alkyl mercapto alcohol
Empirical: C_2H_6OS
Formula: $HSCH_2CH_2OH$
Properties: Water-wh. mobile liq.; disagreeable odor; sol. in water, most org. solvs. incl. alcohol, benzene, ether; m.w. 78.13; sp.gr. 1.114; m.p. < -100 C; b.p. 157 C; flash pt. 74 C; ref. index 1.5006
Toxicology: LD50 (oral, rat) 244 mg/kg, (skin, rabbit) 150 mg/kg; poison by ingestion, skin contact, IP routes; mod. toxic by IV route; severely irritating to eyes, skin, nose; massive exposures can be fatal; ing. may irritate GI tract
Precaution: Combustible; flamm. exposed to heat, flame, oxidizers; incompat. with oxidizers, reducing agents, strong acids, nitric acid, strong bases, alkali metals, moisture; vigorous or explosive reactions possible; Hazardous decomp. prods.: Heated to decomp., emits highly toxic fumes of SO_x

NFPA: Health 2, Flammability 2
Uses: Solvent for dyestuffs; intermediate for producing dyestuffs, pharmaceuticals, rubber chemicals, flotation agents, insecticides, plasticizers; reducing agent; biochemical reagent; PVC stabilizers; agricultural chemicals; textile auxiliary; chain transfer agent in mfg. of acrylic resins for high-solids automotive topcoats; mfg. of polythioethers for high-solids elastomers, coatings, and adhesives
Manuf./Distrib.: Aldrich; BASF; BASF AG; Elf Atochem N. Am.; Fluka; Morton Int'l.; Research Organics; Rhône-Poulenc Surf. & Spec.; Sigma; Spectrum Chem. Mfg.; Toray Thiokol

β-Mercaptoethanol. *See* 2-Mercaptoethanol
Mercaptopriopionic acid, dibutyltin salt. *See* Dibutyltin β-mercaptopropionate
Mesitylene. *See* 1,3,5-Trimethylbenzene

Mesityl oxide
CAS 141-79-7; EINECS 205-502-5
UN No. 1229 (DOT)
Synonyms: 4-Methyl-3-penten-2-one; 4-Methyl-3-pentene-2-one; 2-Methyl-2-penten-4-one; Isobutenyl methyl ketone; Isopropylideneacetone; Methyl isobutenyl ketone
Empirical: $C_6H_{10}O$
Formula: $(CH_3)_2C:CHCOCH_3$
Properties: Colorless oily liq., strong odor; somewhat water-sol. @ 20 C; misc. with alcohol, ether, most org. liqs.; m.w. 98.15; dens. 0.854 (20/4 C); m.p. -59 C; b.p. 128-130 C; flash pt. 31 C; ref. index 1.446 (20 C)
Toxicology: LD50 (oral, rat) 1120 mg/kg, (skin, rabbit) 5150 mg/kg; ACGIH TLV:TWA 15 ppm; STEL 25 ppm; poison by IP route; mod. toxic by ing.; mildly toxic by inh., skin contact; human systemic effects; highly irritating to tissue; narcotic in high concs.
Precaution: DOT: Flamm. liq.; dangerous fire hazard exposed to heat, sparks, flame; reactive with oxidizers; violent reactions possible; Hazardous decomp. prods.: Heated to decomp., emits acrid smoke, irritating fumes
Uses: Solvent for paints; insect repellent
Manuf./Distrib.: Fluka; Penta Mfg.; Van Waters & Rogers

Metacetone. *See* Diethyl ketone
Methacetone. *See* Diethyl ketone
Methanamide. *See* Formamide
Methanecarbonitrile. *See* Acetonitrile
Methanecarboxylic acid. *See* Acetic acid
Methane, chloro-. *See* Methyl chloride
Methane dichloride. *See* Methylene chloride
Methane-, dichloro-. *See* Methylene chloride
Methane trichloride. *See* Chloroform
Methane, trichloro-. *See* Chloroform
Methane, trichlorofluoro-. *See* Trichlorofluoromethane
Methanoic acid. *See* Formic acid
Methanoic acid, amide. *See* Formamide
Methanol. *See* Methyl alcohol
Methenyl chloride. *See* Chloroform
Methenyl tribromide. *See* Bromoform
1-Methoxy-2-acetoxypropane. *See* Propylene glycol methyl ether acetate
1-Methoxybutane-1,3-dione. *See* Methyl acetoacetate

3-Methoxybutanol
CAS 2517-43-3
UN No. 1987
Formula: $CH_3CH(OCH_3)CH_2CH_2OH$
Properties: Colorless liq.; mild odor; sol. in water; misc. with commonly used org. solvs.; dens. 0.920-0.922 g/cc; vapor pressure 0.9 mm Hg (20 C); sets to glass @ -85 C; b.p. 161 C; flash pt. 74 C; ref. index 1.415-1.416 (20 C)
Toxicology: LD50 (oral, rat) 4000 mg/kg; toxic by ingestion
Environmental: Good biological eliminability
Precaution: Combustible
Uses: High-boiling solvent for lacquers; good dissolving power for many nat. resins, nitrocellulose, PVB, aldehyde, ketone, indene, phenol-formaldehyde, alkyd, and maleic resins, commonly used plasticizers; coupling agent for brake fluids; intermediate for plasticizers, herbicides; film-forming additive in PVA emulsions; solvent for pharmaceuticals; provides antiskinning in conventional and aq. paints
Manuf./Distrib.: Hoechst AG

Methoxybutyl acetate
CAS 4435-53-4; EINECS 224-644-9
UN No. 2708 (DOT)
Synonyms: Acetic acid (3-methoxy-n-butyl) ester; Acetic acid-3-methoxybutyl ester; Methyl-1,3-butylene glycol acetate; 'Butoxyl'; 3-Methoxybutyl acetate
Empirical: $C_7H_{14}O_3$
Formula: $CH_3COOCH_2CH_2CH(OCH_3)CH_3$
Properties: Colorless liq.; acrid odor; bitter taste; sol. in water, org. solvs.; m.w. 146.21; dens. 0.952-0.958 (20/20 C); b.p. 171 C; flash pt. 62.5 C
Toxicology: LD50 (oral, rat) 4210 mg/kg; mildly toxic by ingestion; skin and eye irritant
Precaution: Combustible exposed to heat or flame; can react with oxidizing materials; Hazardous decomp. prods.: Heated to decomp., emits acrid smoke and irritating fumes
Uses: Solvent, thinner for coatings, nitrocellulose, acryl, urethane coatings, epoxy resin, aminoalkyd resin
Manuf./Distrib.: Daicel Chem. Ind.
Trade names: Butoxyl®

3-Methoxybutyl acetate. *See* Methoxybutyl acetate

Methoxydiglycol
CAS 111-77-3; EINECS 203-906-6
Synonyms: Diethylene glycol methyl ether; Diethylene glycol monomethyl ether; Ethanol, 2-(2-methoxyethoxy)-; 2-(2-Methoxyethoxy) ethanol; (2-β-Methyl 'Carbitol'), methoxyethoxy ethanol
Classification: Aliphatic ether alcohol
Empirical: $C_5H_{12}O_3$
Formula: $CH_3OCH_2CH_2OCH_2CH_2OH$
Properties: Water-wh. liq.; sol. in water; m.w. 120.17; dens. 1.0211 (20/4 C); b.p. 194 C; flash pt. 93.3 C; ref. index 1.4264 (27 C)
Toxicology: LD50 (oral, rat) 5500 mg/kg; moderately toxic by skin contact, intraperitoneal route; mildly toxic by ingestion
Precaution: Combustible
NFPA: Health 1, Flammability 1, Reactivity 0
Uses: Solvent; brake fluid component; intermediate; solvent for coating formulations incl. nongrain-raising wood stains, printing inks, dye pastes for textiles; coalescing aid for PVAc latex paints; solvent in stamp pad and stencil inks

Regulatory: FDA 21CFR §175.105
Manuf./Distrib.: Aldrich; Ashland; Brown; Eastman; Fluka; Great Western; Harcros USA; Hoechst AG; Occidental; Oxiteno; Spectrum Chem. Mfg.; Union Carbide
Trade names: DM; DM; Dowanol® DM; Eastman® DM; Ektasolve® DM; Hisolve DM; Methyl Carbitol®; Methyl Di-Icinol; Poly-Solv® DM

Methoxy dipropylene glycol. *See* PPG-2 methyl ether

Methoxyethanol
CAS 109-86-4; EINECS 203-713-7
UN No. 1188 (DOT)
Synonyms: EGME; Ethylene glycol methyl ether; Ethylene glycol monomethyl ether; Glycol monomethyl ether; 2-Methoxyethanol; β-Methoxyethanol; Methoxyhydroxyethane; Methyl 'Cellosolve'
Classification: Aliphatic ether alcohol
Empirical: $C_3H_8O_2$
Formula: $CH_3OCH_2CH_2OH$
Properties: Colorless liq., mild agreeable odor.; sol. in water, alcohols, ketones, glycols, most hydrocarbons; misc. with ether, benzene, glycerol, acetone, dimethylformamide; m.w. 76.10; sp.gr. 0.964 (20/4 C); f.p. -86.5 C; b.p. 123-124 C (760 mm); flash pt. (OC) 43 C; ref. index 1.4028 (20 C); pH neutral
Toxicology: ACGIH TLV/TWA 5 ppm (skin); LD50 (oral, rat) 2460 mg/kg; mod. toxic to humans by ingestion; human systemic effects; skin and eye irritant; inh. can cause headache, confusion, weakness, nausea, difficulty breathing, increased heart rate
Precaution: Flamm. on exposure to heat and flame; mod. explosion hazard; can react with oxidizers to form explosive peroxides; incompat. with acid chlorides, acid anhydrides (forms ester compds.); Hazardous decomp. prods.: Acetaldehyde, methanol; heated to dec., emits acrid smoke, irritating fumes
NFPA: Health 2, Flammability 2, Reactivity 0
Uses: Solvent for nitrocellulose, cellulose acetate, natural resins, some synthetic resins, paints, inks, dyes, leather dyeing, nail polishes, varnishes, enamels, wood stains; plastics and plasticizers; in modified Karl Fischer reagent; jet fuel deicing additive; sealing moisture-proof cellophane; fixative in perfumes; chem. intermediate; in semiconductor and microfilm industries; binder, color diluent, extender in foods
Regulatory: FDA 21CFR §73.1 (no residue), 175.105; USDA 9CFR §381.147; SARA reportable
Manuf./Distrib.: Aldrich; ARCO; Ashland; Brown; Fluka; Great Western; Harcros USA; Hoechst AG; Occidental; Oxiteno; Sigma; Spectrum Chem. Mfg.; Union Carbide
Trade names: EM; EM; Hisolve MC; Methyl Cellosolve®; Methyl Icinol; Poly-Solv® EM

2-Methoxyethanol. *See* Methoxyethanol
β-Methoxyethanol. *See* Methoxyethanol

Methoxyethanol acetate
CAS 110-49-6; EINECS 203-772-9
UN No. 1189
Synonyms: EGMEA; Ethylene glycol methyl ether acetate; Ethylene glycol monomethyl ether acetate; Glycol monomethyl ether acetate; Methoxyethyl acetate; 2-Methoxyethyl acetate; 2-Methoxyethanol acetate; Glycol ether EM acetate; 1-Acetoxy-2-methoxyethane; Methyl glycol monoacetate; Methyl glycol acetate; Acetic acid 2-methoxyethyl ester; β-Methoxyethyl acetate

2-Methoxyethanol acetate

Classification: Aliphatic ether alcohol ester
Empirical: $C_5H_{10}O_3$
Formula: $CH_3COOC(CH_2)_2OCH_3$
Properties: Colorless liq.; pleasant sweet odor; sol. in water, ethanol, ether; m.w. 118.14; sp.gr. 1.007 (20/20 C); f.p. -70 C; b.p. 143 C; flash pt. (CC) 49 C; ref. index 1.402 (20 C)
Toxicology: ACGIH TLV/TWA 5 ppm (skin); LD50 (oral, rat) 3390 mg/kg, (dermal, rabbit) 5250 mg/kg; very toxic; mod. toxic by ing., IP, subcut. routes; mildly toxic by inh., skin contact; human systemic effects; eye irritant; inh. irritant
Precaution: DOT: Flamm. liq.; explosive limits 1.7-8.2%; mod. explosion hazard; can react with oxidizing materials ; may form peroxides which can be explosive; decomp. (hydrolysis) can occur with strong acids and bases; Hazardous decomp. prods.: 2-Methoxyethanol, acetic acid; heated to decomp., emits acrid smoke and irritating fumes
NFPA: Health 1, Flammability 2
Uses: Solvent for coatings, waxes, oils, various gums, resins, cellulose acetate, nitrocellulose; in photographic films, lacquers, textile printing
Manuf./Distrib.: Aldrich; Fluka; Sigma; Spectrum Chem. Mfg.
Trade names: EMA; Methyl Cellosolve® Acetate

2-Methoxyethanol acetate. *See* Methoxyethanol acetate
Methoxy ether of propylene glycol. *See* Methoxyisopropanol
2-(2-Methoxyethoxy) ethanol. *See* Methoxydiglycol
2-(2-(2-Methoxyethoxy)ethoxy)ethanol. *See* Triglycol monomethyl ether
Methoxyethyl acetate. *See* Methoxyethanol acetate
2-Methoxyethyl acetate. *See* Methoxyethanol acetate
β-Methoxyethyl acetate. *See* Methoxyethanol acetate
2-Methoxyethyl ether. *See* Diethylene glycol dimethyl ether
2-Methoxyethyl phthalate. *See* Dimethoxyethyl phthalate
Methoxyhydroxyethane. *See* Methoxyethanol
1-Methoxy-2-hydroxypropane. *See* Methoxyisopropanol

Methoxyisopropanol
CAS 107-98-2; EINECS 203-539-1
UN No. 3092 (DOT)
Synonyms: PGME; Monopropylene glycol methyl ether; Polypropylene glycol monomethyl ether; Methoxy ether of propylene glycol; Propylene glycol methyl ether; Propylene glycol monomethyl ether; α-Propylene glycol monomethyl ether; 1-Methoxy-2-hydroxypropane; 1-Methoxy-2-propanol; 2-Methoxy-1-methylethanol; 2-Propanol, 1-methoxy
Classification: Aliphatic ether alcohol
Empirical: $C_4H_{10}O_2$
Formula: $CH_3OCH_2CH(OH)CH_3$
Properties: Colorless clear liq.; ether-like sweet odor; sol. in water, acetone, benzene, CCl_4, ethyl ether, petrol. ether, methanol; m.w. 90.14; sp.gr. 0.919; m.p. -96.7 C; b.p. 120 C; flash pt. (CC) 32 C
Toxicology: ACGIH TLV/TWA 100 ppm; STEL 150 ppm; LD50 (oral, rat) 5660 mg/kg, (dermal, rabbit) 13 g/kg; moderately toxic by intravenous route; mildly toxic by ingestion, inhalation, skin contact; vapor irritates eyes and causes tearing; skin irritant
Precaution: Flamm.; dangerous fire hazard when exposed to heat or flame; can react with oxidizing materials; Hazardous decomp. prods.: Heated to decomp., emits acrid smoke and irritating fumes
NFPA: Health 0, Flammability 3, Reactivity 0

Uses: Antifreeze and coolant for diesel engines; solvent for paints, lacquers, resins, dyes, oils/greases, cellulosics, acrylics, cleaners, dyes, inks, electronics, textile dyes, agric. prods., cosmetics, floor polish, adhesives, fuel additives; solvent sealing of cellophane; oil field, mining, epoxy laminates; chem. intermediate applics.; heat transfer fluid

Regulatory: FDA 21CFR §175.105, 181.22, 181.30

Manuf./Distrib.: Aldrich; Allchem Ind.; Ashland; Fluka; Harcros USA; Norman, Fox; Spectrum Chem. Mfg.

Trade names: Arcosolv® PM; Dowanol® PM; Eastman® PM; Icinol PM; Methyl Propasol®; Poly-Solv® MPM; Solvenon® PM; Solvent PM

1-(2-Methoxyisopropoxy)-2-propanol. *See* PPG-2 methyl ether
2-Methoxy-1-methylethanol. *See* Methoxyisopropanol
1-(2-(2-Methoxy-1-methylethoxy)-1methylethoxy)-2-propanol. *See* PPG-3 methyl ether
2-(2-Methoxymethylethoxy)methylethoxy)propanol. *See* PPG-3 methyl ether
1-(2-Methoxy-1-methylethoxy)-2-propanol. *See* PPG-2 methyl ether
(2-Methoxymethylethoxy) propanol. *See* PPG-2 methyl ether

4-Methoxy-4-methylpentanone-2

CAS 107-70-0; EINECS 203-512-4

UN No. 2293 (DOT)

Synonyms: 4-Methoxy-4-methylpentan-2-one; 4-Methoxy-4-methyl-2-pentanone; 2-Pentanone, 4-methoxy-4-methyl-; 4-Methyl-4-methoxy-2-pentanone; Methoxy-4-methyl-4-pentanone-2; Pentoxane

Classification: Aliphatic ketone

Empirical: $C_7H_{14}O_2$

Formula: $CH_3C(CH_3)(OCH_3)CH_2COCH_3$

Properties: Colorless to water-wh. liq.; mod. sol. in water; sol. in ethanol, acetone; m.w. 130.21; sp.gr. 0.899; b.p. 147-163 C; flash pt. (CC) 60.5 C

Toxicology: LD50 (oral, mouse) 2050 mg/kg; irritant to skin and eyes; mod. toxic by ing. and skin contact; mildly toxic by inhalation

Precaution: Flamm. exposed to heat or flame; moderate fire risk; can react vigorously with oxidizers; Hazardous decomp. prods.: Heated to decomp., emits acrid smoke and irritating fumes

Uses: Solvent for resin coatings

Trade names: PentoXone

Methoxy-4-methyl-4-pentanone-2. *See* 4-Methoxy-4-methylpentanone-2
4-Methoxy-4-methylpentan-2-one. *See* 4-Methoxy-4-methylpentanone-2
4-Methoxy-4-methyl-2-pentanone. *See* 4-Methoxy-4-methylpentanone-2
2-Methoxy-2-methylpropane. *See* Methyl t-butyl ether
Methoxy polyethylene glycol. *See* Polyethylene glycol monomethyl ether
1-Methoxy-2-propanol. *See* Methoxyisopropanol
1-Methoxy-2-propanol acetate. *See* Propylene glycol methyl ether acetate
2-(2-(2-Methoxypropoxy)ethoxy)ethanol. *See* Diethylene glycol isobutyl ether
2-(2-(2-Methoxypropoxy)propoxy)-1-propanol. *See* PPG-3 methyl ether
1-Methoxy-2-propyl acetate. *See* Propylene glycol methyl ether acetate
2-Methoxypropyl acetate. *See* Propylene glycol methyl ether acetate
Methoxytriglycol. *See* Triglycol monomethyl ether

Methyl acetate

CAS 79-20-9; EINECS 201-185-2

UN No. 1231 (DOT)

Synonyms: Acetic acid methyl ester; Methyl acetic ester; Methyl ethanoate

Classification: Sat. aliphatic carboxylic acid ester

Methyl acetic ester

Definition: Ester of methyl alcohol and acetic acid
Empirical: $C_3H_6O_2$
Formula: CH_3COOCH_3
Properties: Colorless volatile liq., fragrant odor, sl. bitter flavor; very sol. in water; sol. in ethanol, diethyl ether, acetone, benzene, chloroform; m.w. 74.09; dens. 0.92438; vapor pressure 100 mm Hg (9.4 C); f.p. -98.05 C; b.p. 54.05 C; flash pt. (CC) -10 C; ref. index 1.3614 (20 C); surf. tens. 24 dynes/cm (20 C)
Toxicology: LD50 (oral, rabbit) 3705 mg/kg; ACGIH TLV/TWA 200 ppm in air; STEL 250 ppm; mod. toxic by several routes; human systemic irritant by inh.; mod. skin and severe eye irritant; irritant to respiratory tract; narcotic in high concs.
Precaution: Flamm.; dangerous fire and explosion risk; explosive limits 3-16% in air; incompat. with oxidizing agents, strong acids, bases, potassium t-butoxide; Hazardous decomp. prods.: Methanol, acetic acid; heated to decomp., emits acrid smoke and fumes
NFPA: Health 1, Flammability 3, Reactivity 0
Uses: Solvent for paint removers, lacquers, resins, oils, cellulose nitrate, nitrocellulose, acetylcellulose; intermediate; syn. flavoring agent in foods, pharmaceuticals; odorant; mfg. of artificial leather; catalyst for biodeg. of org. materials; mfg. of acetic anhydride
Regulatory: FDA 21CFR §172.515, 175.105; FEMA GRAS
Manuf./Distrib.: Akzo Nobel; Aldrich; ARCO; Ashland; J.T. Baker; Eastern Chem.; Eastman; Fluka; Hampshire; Hoechst Celanese; Monsanto; Penta Mfg.; Spectrum Chem. Mfg.; Van Waters & Rogers; Wacker-Chemie GmbH

Methyl acetic ester. *See* Methyl acetate

Methyl acetoacetate
CAS 105-45-3; EINECS 203-299-8
Synonyms: MAA; Acetoacetic acid methyl ester; 1-Methoxybutane-1,3-dione; Methyl acetylacetate; Methyl acetylacetonate; Methyl 3-oxobutanoate; Methyl 3-oxobutyrate; 3-Oxobutanoic acid methyl ester
Classification: Carboxylic acid ester
Empirical: $C_5H_8O_3$
Formula: $CH_3COCH_2COOCH_3$
Properties: Colorless liq.; very sol. in water; sol. in alcohol, ether; m.w. 116.1; sp.gr. 1.076 (20/4 C); m.p. -80 C; b.p. 167-170 C; flash pt. 62 C; pH neutral
Toxicology: LD50 (oral, rat) 3228 mg/kg; mod. toxic by ing.; severe eye and mild skin irritant; high concs. of vapor may cause nose/throat irritation, drowsiness, headache, lightheadedness
Precaution: Flamm. exposed to heat or flame; incompat. with oxidizing agents (increases fire/explosion hazard), bases (decomp. can occur); Hazardous decomp. prods.: Methanol, acetic acid; heated to decomp., emits acrid smoke and irritating fumes
NFPA: Health 2, Flammability 2, Reactivity 0
Uses: Copromoter for cobalt-contg. polyester-accelerator systems; solvent for cellulose ethers; extractant; ingred. of solvent mixts. for cellulose esters; synthesis of nifedipine (a cardiac drug); org. synthesis; intermediate for vitamins, feed additives, drugs, pesticides; component for flavors and fragrances; prod. of azo dyes
Manuf./Distrib.: Aceto; Aldrich; Eastman; Fluka; Lonza; Penta Mfg.; Sigma; Wacker-Chemie GmbH

Methyl acetone. *See* Methyl ethyl ketone
Methyl 12-acetoxy-9-octadecenoate. *See* Methyl acetyl ricinoleate
Methyl 12-acetoxyoleate. *See* Methyl acetyl ricinoleate

Methyl acetylacetate. *See* Methyl acetoacetate
Methyl acetylacetonate. *See* Methyl acetoacetate

Methyl acetyl ricinoleate
CAS 140-03-4; EINECS 205-392-9
Synonyms: Methyl 12-acetoxyoleate; Ricinoleic acid, methyl ester, acetate; Methyl 12-acetoxy-9-octadecenoate
Empirical: $C_{21}H_{38}O_4$
Properties: Cryst.; sol. in most organic solvents; insol. in water; pale-yellow liquid; mild odor; m.w. 1306.59; sp.gr. 0.938; m.p. -15 C; flash pt. (COC) 196 C; ref. index 1.4545
Toxicology: LD50 (oral, mouse) 34,900 mg/kg; probably mod. toxic by ing.; primary irritant to eyes; probable skin and mucous membrane irritant
Precaution: Combustible exposed to heat or flame; can react with oxidizing materials; Hazardous decomp. prods.: Heated to decomp., emits toxic fumes of NO_x
Uses: Plasticizer, lubricant, protective coatings, synthetic rubbers, vinyl compds. and cellulosics; chewing gum base; solubilizer for benzophenone-3; cosolvent props.
Regulatory: FDA 21CFR §175.105; Japan approved
Manuf./Distrib.: CasChem
Trade names: Naturechem® MAR

Methyl adipate. *See* Dimethyl adipate

Methyl alcohol
CAS 67-56-1; EINECS 200-659-6
UN No. 1230 (DOT)
Synonyms: Methanol; Wood alcohol; Wood naphtha; Wood spirit; Carbinol; Columbian spirits; Methyl hydroxide; Methylol
Empirical: CH_4O
Formula: CH_3OH
Properties: Clear colorless liq., alcoholic odor (pure), pungent odor (crude); highly polar; misc. with water, alcohol, ether, benzene, ketones; m.w. 32.05; dens. 0.7924; m.p. -97.8 C; b.p. 64.5 C (760 mm); flash pt.(OC) 54 F; ref. index 1.3292 (20 C)
Toxicology: LD50 (oral, rat) 5628 mg/kg; toxic (causes blindness); poisonous by ingestion, inhalation, or percutaneous absorption; experimental teratogen, reproductive effects; eye and skin irritant; narcotic; usual fatal dose 100-250 ml; TLV 200 ppm in air
Precaution: DOT: Flamm. liq.; dangerous fire risk; explosive limits 6.0-36.5% vol. in air; reacts vigorously with oxidizers; Hazardous decomp. prods.: Heated to decomp., emits acrid smoke and irritating fumes
NFPA: Health 1, Flammability 3, Reactivity 0
Uses: Industrial solvent; mfg. of formaldehyde, acetic acid, dimethyl terephthalate; chemical synthesis; antifreeze; solvent for nitrocellulose, polyvinyl butyral, shellac, rosin, manila resin, dyes, paints, waterborne coatings; denaturant; extraction solvent for hops extract, spice oleoresins (foods); pharmaceutical solvent, excipient
Regulatory: FDA 21CFR §172.560, 173.250, 173.385, 175.105, 176.180, 176.200, 176.210, 177.2420, 177.2460, 27CFR §21.115; FDA approved for orals; USP/NF, BP compliance
Manuf./Distrib.: Air Prods.; Albright & Wilson Am.; Aldrich; Allchem Ind.; AlliedSignal; Ashland; J.T. Baker; Baychem; Brown; Chemcentral; Coyne; CPS; DuPont; Eastman; Fluka; General Chem.; C.P. Hall; Harcros USA; Hoechst Celanese; Houghton Chem.; Mitsui Toatsu; Nissan Chem. Ind.; Norsk Hydro

AS; Quantum; Samson; Sigma; Spectrum Chem. Mfg.; Sunnyside; Van Waters & Rogers; Veckridge

Trade names containing: Filmex® A-1; Filmex® A-2; Filmex® B; Filmex® C; Filmex® D-1; Filmex® D-2; Tecsol® A-2; Tecsol® B; Tecsol® D-2

Methyl amyl acetate

CAS 108-84-9

UN No. 1233 (DOT)

Synonyms: MAAC; Methylisoamyl acetate; Methylisobutyl carbinol acetate; Methylisobutylcarbinyl acetate; s-Hexyl acetate; 1,3-Dimethylbutyl acetate; Acetic acid-1,3-dimethylbutyl ester; 4-Methyl-2-pentanol acetate; 4-Methyl-2-pentyl acetate; 2-Pentanol, 4-methyl-, acetate

Empirical: $C_8H_{16}O_2$

Formula: $CH_3COOCH(CH_3)C_4H_9$

Properties: Colorless liq., mild odor; insol. in water; sol. in alcohol; m.w. 144.21; sp.gr. 0.858 (20/20 C); vapor pressure 3.8 mm Hg (20 C); f.p. -64 C; b.p. 146-150 C; flash pt. (TCC) 96 F; ref. index 1.4008 (20 C)

Toxicology: LD50 (oral, rat) 6160 mg/kg, (skin, rabbit) 20 g/kg; LCLo (inh., rat, 4 h) 2000 ppm; TLV 50 ppm in air; toxic by inhalation; mildly toxic by ing., skin contact; skin and human eye irritant; human systemic effects by inh.

Precaution: Flamm. exposed to heat or flames; moderate fire risk; can react with oxidizing materials

Uses: Solvent for coatings, nitrocellulose, and other lacquers

Manuf./Distrib.: CTC Organics; Eastman; Penta Mfg.

Methyl amyl alcohol

CAS 108-11-2; EINECS 203-551-7

UN No. 2053 (DOT)

Synonyms: MAOH; MIBC; MIC; Methylisobutyl carbinol; Isobutyl methyl carbinol; Isobutyl methyl methanol; 4-Methylpentanol-2; 4-Methyl-2-pentanol; 2-Methyl-4-pentanol; 2-Pentanol, 4-methyl-

Empirical: $C_6H_{14}O$

Formula: $(CH_3)_2CHCH_2CH(CH_3)OH$

Properties: Colorless clear liq.; misc. with most common org. solvs.; sl. sol. in water; m.w. 102.18; sp.gr. 0.8079 (20/20 C); vapor pressure 2.8 mm Hg (20 C); f.p. < -9 C (sets to a glass); b.p. 131.8 C; flash pt. (OC) 40.5 C; ref. index 1.4089

Toxicology: LD50 (oral, rat) 2590 mg/kg, (IP, mouse) 812 mg/kg; ACGIH TLV/TWA 25 ppm; STEL 40 ppm (skin); mod. toxic by ing., skin contact, IP routes; mildly toxic by inh.; irritating to respiratory system; skin and severe eye irritant

Precaution: Flamm. exposed to heat or flame; moderate fire risk; mod. explosion hazard exposed to heat or flame; explosive limits in air 1-5.5%; can react with oxidizers; Hazardous decomp. prods.: Heated to decomp., emits acrid smoke and fumes

Uses: Solvent for dyestuffs, oils, gums, resins, waxes, nitrocellulose, ethylcellulose; latent solvent for coatings; organic synthesis; froth flotation; brake fluids

Manuf./Distrib.: Aldrich; Allchem Ind.; Ashland; Fluka; Houghton Chem.; O'Brien Ind.; Shell; Union Carbide; Zinkan Enterprises

Methylamyl alcohol. *See* 2-Methyl-1-pentanol
Methyl amyl carbinol. *See* 2-Heptanol
Methyl-n-amyl carbinol. *See* 2-Heptanol
Methyl amyl ketone. *See* Methyl n-amyl ketone

Methyl n-amyl ketone

CAS 110-43-0; EINECS 203-767-1

UN No. 1110 (DOT)
Synonyms: MAK; 2-Heptanone; Ketone C-7; Methyl amyl ketone; Amyl methyl ketone; n-Amyl methyl ketone; Butylacetone; Methyl pentyl ketone; n-Pentyl methyl ketone
Classification: Sat. aliphatic ketone
Empirical: $C_7H_{14}O$
Formula: $CH_3CH_2CH_2CH_2CH_2COCH_3$
Properties: Water-wh. liq., penetrating fruity odor; sol. in ethanol, diethyl ether; almost insol. in water; misc. with org. solvs.; m.w. 114.18; sp.gr. 0.8166 (20/20 C); m.p. -26.9 C; b.p. 150.6 C; flash pt. 49 C; ref. index 1.4110 (20 C)
Toxicology: LD50 (oral, rat) 1670 mg/kg, (skin, rabbit) 12,600 mg/kg; ACGIH TLV/TWA 50 ppm; mod. toxic by ingestion; mildly toxic by inhalation, skin contact; skin irritant; narcotic in high concs.
Precaution: Flamm. exposed to heat or flames; moderate fire risk; reactive with oxidizing materials; incompat. with strong acids, alkalies, hydrogen peroxides; violent reactions possible; Hazardous decomp. prods.: Explosive peroxides; heated to decomp., emits acrid smoke and fumes
NFPA: Health 1, Flammability 2, Reactivity 0
Uses: Industrial solvent for nitrocellulose lacquers, syn. resin finishes, high solids coatings; syn. flavoring and perfumery; fragrance ingred. in creams, lotions, soaps, detergents; food and pharmaceutical flavoring; inert reaction medium
Regulatory: FDA 21CFR §172.515; FEMA GRAS
Manuf./Distrib.: Aldrich; Allchem Ind.; Ashland; Eastman; Fluka; Houghton Chem.; Shell; Sigma; Union Carbide; Van Waters & Rogers

Methyl arachidate. *See* Methyl eicosenate
1-Methylazacyclopentan-2-one. *See* N-Methyl-2-pyrrolidone

Methyl behenate
CAS 929-77-1; EINECS 213-207-8
Synonyms: Methyl docosanoate
Definition: Methyl ester of behenic acid
Empirical: $C_{23}H_{46}O_2$
Formula: $CH_3(CH_2)_{20}COOCH_3$
Properties: Wax-like solid; insol. in water; sol. in alcohol and ether; m.w. 354.62; m.p. 54 C; b.p. 215.5 C (3.75 mm)
Precaution: Combustible
Uses: Special synthesis; intermediate for pure behenic acid, prod. of superamides; biochemical and medical research; reference standard in gas chromatography; in metalworking lubricants, specialized solvents; opacifier, visc. control agent; as solvent
Manuf./Distrib.: Aldrich; Fluka; Sigma
Trade names: Kemester® 9022

Methylbenzene. *See* Toluene
Methyl benzenecarboxylate. *See* Methyl benzoate
Methylbenzenesulfonic acid, sodium salt. *See* Sodium toluenesulfonate
4-Methylbenzenesulfonic acid, sodium salt. *See* Sodium toluenesulfonate

Methyl benzoate
CAS 93-58-3; EINECS 202-259-7
UN No. 2938 (DOT)
Synonyms: Benzoic acid methyl ester; Niobe oil; Oil of niobe; Methyl benzenecarboxylate
Classification: Ester
Empirical: $C_8H_8O_2$

Methylbenzol

Formula: $C_6H_5COOCH_3$

Properties: Colorless liq., fragrant odor; sol. in alcohol, fixed oils, propylene glycol, water @ 30 C; misc. with alcohol, ether; insol. in glycerin; m.w. 136.15; dens. 1.082-1.088; vapor pressure 1 mm Hg (39 C); m.p. -12.5 C; b.p. 199.6 C; flash pt. 181 F; ref. index 1.515

Toxicology: LD50 (oral, rat) 1350 mg/kg; mod. toxic by ingestion; mildly toxic by skin contact; skin and eye irritant; heated to decomp., emits acrid smoke and irritating fumes

Precaution: Flamm. exposed to heat or flame; reactive with oxidizing materials; Hazardous decomp. prods.: Heated to decomp., emits acrid smoke and irritating fumes

NFPA: Health 0, Flammability 2, Reactivity 0

Uses: Perfumery, solvent for cellulose esters and ethers, resins, rubber; syn. flavoring for foods, pharmaceuticals

Regulatory: FDA 21CFR §172.515; FEMA GRAS

Manuf./Distrib.: Aldrich; Berje; Fluka; Hüls AG; Morflex; Pentagon Chems. Ltd; Penta Mfg.; Schweizerhall; Sigma; Spectrum Chem. Mfg.; Sybron

Methylbenzol. *See* Toluene

Methyl bromide

CAS 74-83-9

UN No. 1062 (DOT)

Synonyms: MB; MBX; MEBR; Bromomethane; Monobromomethane

Empirical: CH_3Br

Properties: Colorless odorless gas or liquified gas; chloroform-like odor; burning taste; readily sol. in lower alcohols, ethers, esters, ketones, halogenated hydrocarbons, aromatic hydrocarbons, carbon disulfide; m.w. 94.95; sol. 1.75 g/100 g water; sp.gr. 1.732 (0 C); vapor pressure 1824 mm Hg; b.p. 3.6 C; flash pt. none

Toxicology: LD50 (oral, rat) 214 mg/kg; LCLo (human, inh., 2 h) 60,000 ppm; ACGIH TLV/TWA 5 ppm (skin); human poison by inh.; can cause respiratory distress, cardiac arrest, CNS effects, neurotoxic effects; corrosive to skin

Precaution: 10-15% mixts. with air may be ignited with difficulty; mod. explosive exposed to sparks or flame; incompat. with metals, dimethyl sulfoxide, ethylene oxide; Hazardous decomp. prods.: Heated to decomp., emits toxic fumes of Br^-

Uses: Soil and space fumigant; disinfestation of potatoes, tomatoes, other crops; organic synthesis; extraction solvent for vegetable oils

Regulatory: Restricted use; SARA reportable

Manuf./Distrib.: Akzo Nobel; Albemarle; Aldrich; Allchem Ind.; AmeriBrom; Great Lakes; Sigma

2-Methylbutane. *See* Isopentane
2-Methyl butanol-2. *See* t-Amyl alcohol
1-Methyl-1-butanol. *See* s-Amyl alcohol
2-Methyl butanol. *See* 2-Methyl-1-butanol

2-Methyl-1-butanol

CAS 137-32-6

UN No. 1105

Synonyms: s-Butyl carbinol; 2-Methyl butanol; Active primary amyl alcohol; dl-sec-Butylcarbinol

Classification: Primary aliphatic alcohol

Empirical: $C_5H_{12}O$

Formula: $CH_3CH_2CH(CH_3)CH_2OH$

Properties: Colorless liq.; sour sharp odor; m.w. 88.15; very sol. in acetone; sol. in alcohols, ether; sl. sol. in water; m.w. 88.15; sp.gr. 0.816 (20 C); f.p. < -70 C; b.p. 128 C; flash pt. (CC) 43 C

Toxicology: LD50 (oral, rat) 4920 mg/kg; LDLo (skin, rabbit) 3540 mg/kg; toxic; eye, skin, mucous membrane irritant; can be absorbed thru skin; inh. may cause nose/throat irritation, headache, dizziness, dyspnea, nausea, vomiting

Precaution: Flamm. exposed to heat, flame, oxidizers; explosive as vapor exposed to heat or flame; incompat. with oxidizing agents (increases fire/explosion hazard), hydrogen trisulfide (explosive reaction possible); Hazardous decomp. prods.: Heated to decomp., emits acrid smoke and irritating fumes

NFPA: Health 2, Flammability 2, Reactivity 0

Uses: Solvent; org. synthesis; lubricants; plasticizers; additive for oils and paints; frothing agent

Manuf./Distrib.: Aldrich

2-Methyl-2-butanol. *See* t-Amyl alcohol
2-Methyl-4-butanol. *See* Isoamyl alcohol
3-Methylbutanol. *See* Isoamyl alcohol
3-Methyl-1-butanol. *See* Isoamyl alcohol
3-Methylbutan-3-ol. *See* t-Amyl alcohol
3-Methyl-1-butanol acetate. *See* Isoamyl acetate
3-Methylbutan-2-one. *See* Methyl isopropyl ketone
3-Methyl-2-butanone. *See* Methyl isopropyl ketone
1-Methylbutyl acetate. *See* s-Amyl acetate
3-Methylbutyl acetate. *See* Isoamyl acetate
3-Methyl-1-butyl acetate. *See* Isoamyl acetate
β-Methylbutyl acetate. *See* Isoamyl acetate
1-Methyl-4-t-butylbenzene. *See* p-t-Butyl toluene
p-Methyl-t-butylbenzene. *See* p-t-Butyl toluene
Methyl-1,3-butylene glycol acetate. *See* Methoxybutyl acetate
3-Methylbutyl ethanoate. *See* Isoamyl acetate

Methyl t-butyl ether

CAS 1634-04-4; EINECS 216-653-1

UN No. 2398 (DOT)

Synonyms: MTBE; Methyl tertiary butyl ether; Methyl 1,1-dimethylethyl ether; Propane, 2-methoxy-2-methyl; t-Butyl methyl ether; 2-Methoxy-2-methylpropane

Classification: Aliphatic ether

Empirical: $C_5H_{12}O$

Formula: $CH_3OC(CH_3)_3$

Properties: Clear liq., terpene-like odor; sl. sol. in water; misc. with all gasoline-type hydrocarbons; m.w. 88.15; dens. 0.7335; b.p. 91.1 C; f.p. -75 C; flash pt. –25.6 C

Toxicology: Slight skin and eye irritant; heated to decomp., emits acrid smoke and irritating fumes

Precaution: Flamm. exposed to heat or flame

Uses: Octane booster in gasoline; intermediate; solvent for paints; extraction solvent; solvent, reaction medium in pharmaceuticals, for polymerizations, Grignard reactions

Manuf./Distrib.: Aldrich; Allchem Ind.; ARCO; Ashland; Fluka; Hüls Am.; Sigma; Spectrum Chem. Mfg.; Texas Petrochem.

Trade names: High Purity MTBE

Methyl butynol

CAS 115-19-5; EINECS 204-070-5

UN No. 1993

2-Methyl-3-butyn-2-ol

Synonyms: 2-Methyl-3-butyn-2-ol; 2-Hydroxy-2-methyl-3-butyne; Dimethyl acetylene carbinol; Dimethyl ethynyl carbinol; Ethynyldimethyl carbinol; α,α-Dimethylpropargyl alcohol; 1,1-Dimethylpropargyl alcohol
Classification: Aliphatic organic compd.
Definition: Tertiary acetylenic alcohol
Empirical: C_5H_8O
Formula: $(CH_3)_2COHCCH$
Properties: Colorless liq., fragrant odor; misc. with water, acetone, benzene, CCl_4; sol. in most org. solvs.; m.w. 84.13; dens. 0.8672 (20/20 C); m.p. 2.6 C; b.p. 104-105 C; flash pt. (TOC) 25 C; ref. index 0.861 (20C)
Toxicology: LD50 (oral, rat) 1950 mg/kg, (IP, mouse) 3600 mg/kg, (subcut., mouse) 2340 mg/kg; moderately toxic by ingestion, intraperitoneal, subcutaneous routes
Precaution: Flamm.; very dangerous fire hazard exposed to heat or flame; reactive with oxidizing agents; Hazardous decomp. prods.: Heated to decomp., emits acrid smoke and irritating fumes
Uses: Corrosion inhibitor; reactive intermediate in mfg. of pharmaceuticals, plastics, rubbers, fragrances, agric.; as solvent, acid inhibitor, viscosity reducer, and stabilizer in vinyl plastisols, platinum catalyst blocker for silicones
Manuf./Distrib.: Air Prods.; Aldrich; BASF; Fluka

2-Methyl-3-butyn-2-ol. *See* Methyl butynol

Methyl canolate
Uses: Solvent for pesticides and herbicides
Trade names: Emery® 2231; Emery® 2232

Methyl caprylate/caprate
CAS 67762-39-4
Definition: Mixture of esters of methyl alcohol and caprylic and capric acids
Uses: In lubricants for automotive, textile, metal rolling operations; food coatings; solvent for pesticides and herbicides; detergent intermediate
Regulatory: FDA 21CFR §172.225, 176.200, 176.210, 177.2800
Trade names: Emery® 2209

(2-β-Methyl 'Carbitol'), methoxyethoxy ethanol. *See* Methoxydiglycol
Methyl 'Cellosolve'. *See* Methoxyethanol

Methyl chloride
CAS 74-87-3
UN No. 1063 (DOT)
Synonyms: Chloromethane; Monochloromethane; Methane, chloro-
Empirical: CH_3Cl
Properties: Colorless gas; ethereal odor; sweet taste; sl. sol. in water; sol. in alcohol; misc. with chloroform, ether, glac. acetic acid; m.w. 50.49; sp.gr. 0.918 (20/4 C); m.p. -97 C; b.p. -23.7 C; flash pt. < 32 F
Toxicology: LD50 (oral, rat) 1800 mg/kg; LCLo (inh., human, 2 h) 20,000 ppm; ACGIH TLV/TWA 50 ppm; STEL 100 ppm (skin); mildly toxic by inh.; human systemic effects; can cause convulsions, nausea, vomiting; sl. irritant; some narcotic effect
Precaution: Flamm. gas; very dangerous fire hazard exposed to heat, flame, or powerful oxidizers; mod. explosion hazard; explodes on contact with interhalogens, magnesium, potassium, sodium, zinc, or their alloys; Hazardous decomp. prods.: Heated to decomp., emits highly toxic fumes of Cl^-
Uses: Catalyst carrier in low-temp. polymerization, tetramethyl lead, silicones, refrigerant, fluid for thermometric/thermostatic equipment, methylating agent in organic synthesis, extractant and low-temp. solvent, herbicide, topical

anesthetic; permitted as food additive
Regulatory: SARA reportable
Manuf./Distrib.: Air Prods.; Aldrich; Alexander; ARC; ICI Am.; Mitsui Toatsu; Occidental; Sigma; Vulcan

Methylchloroform. *See* 1,1,1-Trichloroethane

Methyl cocoate
CAS 61788-59-8; EINECS 262-988-1
Definition: Ester of methyl alcohol and coconut fatty acids
Formula: $RCO\text{-}OCH_3$, RCO^- represents the fatty acids derived from coconut oil
Uses: Emollient, plasticizer, lubricant for cosmetics, pharmaceuticals, plastics, lubricating oils, textiles, leather, cutting oils; chemical and detergent intermediate; food coatings; solvent for pesticides and herbicides
Regulatory: FDA 21CFR §172.225, 175.105, 176.200, 176.210, 177.2260, 177.2800, 178.3910
Trade names: Emery® 2253; Emery® 2254

Methyl cyanide. *See* Acetonitrile

Methyl cyclohexane
CAS 108-87-2; EINECS 203-624-3
UN No. 2296 (DOT)
Synonyms: Hexahydrotoluene; Cyclohexylmethane; Cyclohexane, methyl-; Toluene hexahydride
Classification: Sat. alicyclic hydrocarbon
Empirical: C_7H_{14}
Formula: $C_6H_{11}CH_3$
Properties: Colorless liq.; faint benzene-like odor; sol. in most org. solvs.; insol. in water; m.w. 98.19; sp.gr. 0.77 (20 C); vapor pressure 40 mm Hg (20 C); m.p. –126 C; b.p. 101 C; flash pt.–4 C
Toxicology: LD50 (oral, mouse) 2250 mg/kg; LDLo (oral, rabbit) 4 g/kg; ACGIH TLV/TWA 400 ppm; mod. toxic by ing.; may cause mild skin irritation; vapor may irritate eyes; probably a CNS depressant; sl. nose/throat irritant
Precaution: Flamm.; incompat. with strong oxidizing agents (increases fire and explosion hazard); Hazardous decomp. prods.: Heated to decomp., emits acrid smoke and fumes
NFPA: Health 2, Flammability 3, Reactivity 0
Uses: Solvent for cellulose ethers, organic synthesis; reagent
Manuf./Distrib.: Aldrich; Fluka; Janssen Chimica; Penta Mfg.; Phillips; Spectrum Chem. Mfg.; Triple Crown Am.

Methylcyclohexanol
CAS 25639-42-3; EINECS 247-152-6
UN No. 2617 (DOT)
Synonyms: Hexahydrocresol; Hexahydromethylphenol; Methylcyclohexyl alcohol
Classification: Alicyclic alcohol
Empirical: $C_7H_{14}O$
Formula: $CH_3C_6H_{10}OH$ (C_6H_{10} = cyclohexane ring)
Properties: Colorless visc. oily liq.; aromatic menthol-like odor; sol. in common org. solvs.; mod. sol. in water; m.w. 114.2; sp.gr. 0.913; m.p. -50 C; b.p. 155-180 C; flash pt. (CC) 68 C
Toxicology: LD50 (oral, rat) 1660 mg/kg, (subcut., rat) 2900 mg/kg; LDLo (skin, rabbit) 6800 mg/kg; ACGIH TLV/TWA 50 ppm; toxic; liq. may cause skin irritation; vapor can irritate eyes; liq. splashes may be very irritating
Precaution: Flamm. exposed to heat, flame, or oxidizers; incompat. with strong

oxidizing agents (increases fire/explosion hazard); Hazardous decomp. prods.: On heating, emits acrid fumes
NFPA: Health 0, Flammability 2, Reactivity 0
Uses: Solvent for lacquers, oils, gums, waxes, resins; blending agent for soaps and detergents; lubricant antioxidant; degreasing agent
Manuf./Distrib.: Fluka; Penta Mfg.

Methylcyclohexyl alcohol. *See* Methylcyclohexanol

Methylcyclopentane
CAS 96-37-7
UN No. 2298 (DOT)
Classification: Sat. alicyclic hydrocarbon
Empirical: C_6H_{12}
Properties: Colorless liq. or solid; sweetish gasoline-like odor; sol. in ether, most org. solvs.; insol. in water; m.w. 84.18; sp.gr. 0.749 (20 C); vapor pressure 100 mm Hg (17.9 C); m.p. -142 C; b.p. 72 C; flash pt. -23.9 C
Toxicology: LCLo (inh., mouse) 9500 mg/m^3; mildly toxic by inh.; may cause mild skin irritation; vapor can temporarily irritate eyes; probably a CNS depressant; high vapor concs. may cause nose/throat irritation, lightheadedness, nausea, drowsiness
Precaution: Flamm.; very dangerous fire hazard exposed to heat, flame, oxidizers; incompat. with strong oxidizing agents (increased fire and explosion hazard); Hazardous decomp. prods.: Heated to decomp., emits acrid smoke and irritating fumes
NFPA: Health 2, Flammability 3, Reactivity 0
Uses: Extraction solvent for essential oils; org. synthesis; azeotropic distillation agent
Manuf./Distrib.: Aldrich; Fluka

3-Methyl-N,N-diethylbenzamide. *See* Diethyl toluamide
Methyl diglyme. *See* Diethylene glycol dimethyl ether
Methyl 1,1-dimethylethyl ether. *See* Methyl t-butyl ether
8-Methyl-3,6-dioxa-1-undecanol. *See* Diethylene glycol methylpentyl ether
4-Methyl-1,3-dioxolan-2-one. *See* Propylene carbonate

Methyl disulfide
CAS 624-92-0; EINECS 210-871-0
UN No. 2381 (DOT)
Synonyms: MDS; DMDS; Dimethyl disulfide; Methyldithiomethane; 2,3-Dithiobutane
Classification: Alkyl disulfide
Empirical: $C_2H_6S_2$
Formula: CH_3SSCH_3
Properties: Pale yel. liq.; disagreeable onion odor; very sol. in alcohol, ether; sl. sol. in water; m.w. 94.20; sp.gr. 1.062 (20/4 C); vapor pressure 28.6 mm Hg; m.p. -84.7 C; b.p. 108-110 C; flash pt. (CC) 16 C; ref. index 1.527 (20 C)
Toxicology: LD50 (mammal) 138 mg/kg; skin and eye irritant; very toxic by inh. in animal tests; breaks down to methyl mercaptan in body; inh. of vapors probably causes headache, dizziness, nausea, vomiting; massive exposure can be fatal
Precaution: Highly flamm.; very dangerous fire hazard exposed to heat, flame, oxidizers; incompat. with oxidizing agents (can react violently), reducing materials (can produce methyl mercaptan), copper and alloys (corrosive); Hazardous decomp. prods.: Hydrogen sulfide, methyl mercaptan
Uses: Intermediate in mfg. of chemicals; prep. of catalysts for petrol. refining;

solvent for sulfur; syn. flavoring in foods and pharmaceuticals
Regulatory: FDA 21CFR §172.515; FEMA GRAS
Manuf./Distrib.: Aldrich; Elf Atochem N. Am.; Fluka; Sigma

Methyldithiomethane. *See* Methyl disulfide
Methyl docosanoate. *See* Methyl behenate
Methyl dodecanoate. *See* Methyl laurate

Methyl eicosenate
CAS 1120-28-1; EINECS 214-304-8
Synonyms: Methyl arachidate; Eicosanoic acid methyl ester
Classification: Ester
Empirical: $C_{21}H_{42}O_2$
Formula: $CH_3(CH_2)_{18}COOCH_3$
Properties: m.w. 326.57
Uses: Intermediate in prod. of alkanolamides, in metalworking lubricants, as specialized solvents; foam depressant and nutrient in fermentation
Manuf./Distrib.: Aldrich; Fluka; Sigma
Trade names: Kemester® 2050

Methylene bromide
CAS 74-95-3
UN No. 2664
Synonyms: Dibromomethane; Methylene dibromide
Classification: Halogenated aliphatic hydrocarbon
Empirical: CH_2Br_2
Formula: $BrCH_2Br$
Properties: Colorless clear liq.; sol. in alcohol, ether, chloroform, acetone; mod. sol. in water; m.w. 173.85; sp.gr. 2.49; m.p. -52 C; b.p. 97 C; nonflamm.
Toxicology: Toxic; sl. skin and eye irritant; may be absorbed thru skin; inh. may cause sl. nose/throat irritation, CNS depression (lightheadedness, nausea, fatigue, incoordination); prolonged/repeated high-level exposure may cause liver/kidney damage
Precaution: Incompat. with strong oxidizers, strong bases, metals (may react vigorously or increase risk of fire and explosion)
Uses: Solvent; intermediate in org. synthesis, prod. of herbicides
Manuf./Distrib.: Albemarle; Aldrich; AmeriBrom; Assoc. Octel Co Ltd; Fluka; Great Lakes; Sigma

Methylene chloride
CAS 75-09-2; EINECS 200-838-9
UN No. 1593
Synonyms: DCM; Dichloromethane; Methylene dichloride; Methane dichloride; Methane-, dichloro-
Classification: Halogenated aliphatic hydrocarbon
Empirical: CH_2Cl_2
Properties: Colorless clear volatile liq., penetrating ether-like odor; sol. in alcohol, ether, phenols, aldehydes, ketones; misc. with fixed and volatile oils; mod. sol. in water; m.w. 84.93; sp.gr. 1.335 (15/4 C); f.p. -97 C; b.p. 40.1 C; ref. index 1.4244 (20 C); KB value 136; nonflamm.
Toxicology: ACGIH TLV/TWA 50 ppm; LD50 (oral, rat) 2136 mg/kg; very toxic; poison by IV route; mod. toxic by ingestion, subcutaneous, IP routes; human systemic effects; eye and severe skin irritant; narcotic in high concs.
Environmental: HAPS contg.
Precaution: Explosive as vapor exposed to heat or flame; incompat. with methanol, Al powd., amines, quat. ion exchange resins (azide form), DMSO, perchlo-

ric acid, alkali metals, nitric acid, etc.; Hazardous decomp. prods.: Contact with hot surfaces cause decomp., yielding toxic fumes; heated to decomp., emits highly toxic fumes of phosgene and Cl^-

NFPA: Health 2, Flammability 1, Reactivity 0

Uses: Paint removers, solvent degreasing, plastics processing, blowing agent in foams, solvent for cellulose acetate, propellant for paint aerosols, refrigerant, pharmaceutic aid; extraction solvent for industrial, food use (decaffeinated coffee, hops extract, spice oleoresins); color diluent for foods; pharmaceutical solvent, anesthetic

Regulatory: FDA 21CFR §73.1 (no residue), 173.255; FDA approved for orals; USP/NF compliance; SARA reportable

Manuf./Distrib.: Aldrich; Allchem Ind.; Ashland; R.E. Carroll; Chemcentral; Coyne; Elf Atochem N. Am.; Farleyway Chem. Ltd; Fluka; C.P. Hall; Harcros USA; Houghton Chem.; Hüls Am.; ICI Spec. Chems.; Mallinckrodt; Mitsui Toatsu; Occidental; Primachem; Samson; Sigma; Spectrum Chem. Mfg.; Stanchem; TR-AMC; Van Waters & Rogers; Vulcan

Trade names: Dynasolve 200; Dynasolve 210; Striptron Stripper

Methylene dibromide. *See* Methylene bromide
Methylene dichloride. *See* Methylene chloride
Methyl ester stearic acid. *See* Methyl stearate
Methyl ethanoate. *See* Methyl acetate
1-Methylethyl acetate. *See* Isopropyl acetate
Methylethylacetylenyl carbinol. *See* Methyl pentynol
1-Methylethylamine. *See* Isopropylamine
(1-Methylethyl) benzene. *See* Cumene
(1-Methylethyl) 1,1′-biphenyl. *See* 1,1′-Biphenyl (1-methylethyl)
Methyl ethyl carbinol. *See* 2-Butanol
(1-Methylethyl)cyclohexane. *See* Isopropylcyclohexane
1-Methylethyl dodecanoate. *See* Isopropyl laurate
1-Methyl ethylene carbonate. *See* Propylene carbonate
Methyl ethylene glycol. *See* Propylene glycol
Methyl ethylene oxide. *See* Propylene oxide
Methyl ethyl glycol. *See* Propylene glycol
1-Methylethyl hexandecanoate. *See* Isopropyl palmitate
1-Methylethyl 2-hydroxypropanoate. *See* Isopropyl lactate

Methyl ethyl ketone

CAS 78-93-3; EINECS 201-159-0

UN No. 1193

Synonyms: MEK; Ethyl methyl ketone; 2-Butanone; 3-Butanone; Butanone; 2-Oxobutane; Methyl acetone; Methyl-2-propanone

Classification: Sat. aliphatic ketone

Empirical: C_4H_8O

Formula: $CH_3COCH_2CH_3$

Properties: Colorless liq., acetone-like odor; sol. in 4 parts water, benzene, alcohol, ether; misc. with oils; m.w. 72.10; sp.gr. 0.8255 (0/4 C); m.p. -86 C; b.p. 79.6 C; flash pt. (TOC) 24 F; visc. 0.40 cp (25 C); ref. index 1.3814 (15 C)

Toxicology: ACGIH TLV/TWA 200 ppm; STEL 300 ppm; LD50 (oral, rat) 2737 mg/kg; mod. toxic by ingestion, skin contact, IP routes; toxic by inhalation; mod. to severe eye irritant; experimental teratogen, reproductive effects

Precaution: DOT: Flamm. liq; dangerous fire risk; explosive limits in air 2-10%; incompat. with oxidizers, strong acids, mixts. of haloforms and strong bases, mixts. of hydrogen peroxide and nitric acid; Hazardous decomp. prods.: Explosive peroxides; heated to decomp., emits acrid smoke and fumes

NFPA: Health 1, Flammability 3, Reactivity 0

Uses: Solvent in nitrocellulose coatings, acrylic coatings, vinyl films, paint removers, lacquers, varnishes, cements, adhesives, magnetic tapes, printing inks, polymerization, organic synthesis; metal degreasing; syn. leathers; chem. intermediate; mfg. of smokeless powder; cleaning fluids; priming, catalyst carrier; extraction solvent in food processing; syn. flavoring agent

Regulatory: FDA 21CFR §172.515, 175.105, 175.320, 177.1200

Manuf./Distrib.: Aldrich; Allchem Ind.; AlliedSignal; Ashland; J.T. Baker; Baychem; BP Chems. Ltd; R.E. Carroll; Chemcentral; Elf Atochem N. Am.; Exxon; Fluka; General Chem.; C.P. Hall; Harcros USA; Hoechst Celanese; Houghton Chem.; Mallinckrodt; Olin; Oxiteno; Penta Mfg.; Primachem; Samson; Shell; Spectrum Chem. Mfg.; Sunnyside; Texaco; Union Carbide; Van Waters & Rogers

Methylethylmethane. *See* Butane

1-Methylethyl octadecanoate. *See* Isopropyl stearate

1-Methylethyl tetradecanoate. *See* Isopropyl myristate

Methyl formate

CAS 107-31-3

UN No. 1243 (DOT)

Synonyms: Formic acid methyl ester; Methyl methanoate

Classification: Carboxylic acid ester

Empirical: $C_2H_4O_2$

Formula: $HCOOCH_3$

Properties: Colorless liq.; ether-like odor; very sol. in water; sol. in alcohol, ether; m.w. 60.06; sp.gr. 0.975 (20 C); vapor pressure 400 mm Hg (16/0 C); m.p. -100 C; b.p. 32 C; solidifies @ 100 C; flash pt. (CC) -19 C; pH neutral

Toxicology: LD50 (oral, rabbit) 1622 mg/kg; ACGIH TLV/TWA 100 ppm; STEL 150 ppm; probably skin and mod. to severe eye irritant; can be absorbed thru skin; inh. can cause nose/throat/lung irritation, CNS depression

Precaution: Flamm.; very dangerous fire hazard; incompat. with oxidizing materials (increases fire/explosion hazard), acids (decomp. can occur), bases (vigorous decomp. can occur), moisture or water vapor (slowly hydrolyzes to metahnol, formic acid); Hazardous decomp. prods.: Methanol, formic acid; breaks down slowly in water releasing formic acid; heated to decomp., emits acrid smoke and irritating fumes

NFPA: Health 2, Flammability 4, Reactivity 0

Uses: Solvent for cellulose acetate; fumigant and larvicide for tobacco, dried fruits, cereals, etc.; intermediate for dimethylformamide, other org. synthesis; in prep. of antileukemic agents

Manuf./Distrib.: Aldrich; BASF; BASF AG; Fluka; Hoechst Celanese; Sigma; Spectrum Chem. Mfg.

Methyl gluceth-10

CAS 68239-42-9

Synonyms: PEG-10 methyl glucose ether; POE (10) methyl glucose ether

Definition: PEG ether of methyl glucose

Formula: $CH_3.C_6H_{10}O_5.(OCH_2CH_2)_nOH$, avg. n = 10

Uses: Humectant for personal care prods.; freezing pt. depressant; emollient; moisturizer; foam modifier in detergent and shampoo systems; solvent and solubilizer for topical pharmaceuticals

Trade names: Glucam® E-10

Methyl gluceth-20

CAS 68239-43-0

Synonyms: PEG-20 methyl glucose ether; POE (20) methyl glucose ether

Methyl glycol

Definition: PEG ether of methyl glucose
Formula: $CH_3\text{-}C_6H_{10}O_5\text{-}(OCH_2CH_2)_nOH$, avg. n = 20
Uses: Humectant for personal care prods.; freezing pt. depressant; emollient; moisturizer; foam modifier in detergent and shampoo systems; solvent and solubilizer for topical pharmaceuticals
Trade names: Glucam® E-20

Methyl glycol. *See* Propylene glycol
Methyl glycol acetate. *See* Methoxyethanol acetate
Methyl glycol monoacetate. *See* Methoxyethanol acetate
16-Methyl-1-heptadecanol. *See* Isostearyl alcohol
Methyl hexadecanoate. *See* Methyl palmitate
2-Methylhexane. *See* Isoheptane
5-Methyl-2-hexanone. *See* Methyl isoamyl ketone
Methyl hexyl carbinol. *See* 2-Octanol

Methyl hexyl ketone
CAS 111-13-7; EINECS 203-837-1
Synonyms: 2-Octanone; Hexyl methyl ketone
Empirical: $C_8H_{16}O$
Formula: $CH_3(CH_2)_5COCH_3$
Properties: Colorless liq., apple odor, camphor taste; misc. with alcohol, ether, esters, hydrocarbons; insol. in water; m.w. 128.22; dens. 0.818 (20/4 C); m.p. –16 C; b.p. 170-172 C; flash pt. 56 C; ref. index 1.416 (20 C)
Toxicology: LD50 (mouse) 1600 mg/kg; mod. toxic; skin irritant; heated to decomp., emits acrid smoke and irritating fumes
Precaution: Flamm. exposed to heat, flame, oxidizers
Uses: Solvent for paints; syn. flavoring agent for foods and pharmaceuticals
Regulatory: FDA 21CFR §172.515; FEMA GRAS
Manuf./Distrib.: Aldrich; Fluka; Penta Mfg.; Union Camp; UOP

Methyl hydroxide. *See* Methyl alcohol
Methyl-S-(-)-2-hydroxy propionate. *See* Methyl lactate
Methylisoamyl acetate. *See* Methyl amyl acetate

Methyl isoamyl ketone
CAS 110-12-3; EINECS 203-737-8
UN No. 2302 (DOT)
Synonyms: MIAK; 5-Methyl-2-hexanone; Isoamyl methyl ketone; Methyl isopentyl ketone; Isobutylacetone; Isopentyl methyl ketone
Empirical: $C_7H_{14}O$
Formula: $CH_3COC_2H_4CH(CH_3)_2$
Properties: Colorless liq., pleasant odor; sl. sol. in water; misc. with most org. solvs.; m.w. 114.19; sp.gr. 0.8164; f.p. -101 F; b.p. 286 F; flash pt. (TCC) 96 F; ref. index 1.4078
Toxicology: LD50 (oral, rat) 4760 mg/kg, (skin, rabbit) 10 g/kg; TLV 50 ppm; mod. toxic by ingestion; mildly toxic by inh., skin contact; heated to decomp., emits acrid smoke and irritating fumes
Precaution: Flamm.; combustible exposed to heat, flame, oxidizers
Uses: Solvent for paints
Manuf./Distrib.: Aldrich; Allchem Ind.; Ashland; Burdick & Jackson; Eastman; Fluka; Sigma

Methyl isobutenyl ketone. *See* Mesityl oxide
Methylisobutyl carbinol. *See* Methyl amyl alcohol; 2-Methyl-1-pentanol
Methylisobutyl carbinol acetate. *See* Methyl amyl acetate
Methylisobutylcarbinyl acetate. *See* Methyl amyl acetate

Methyl isobutyl ketone
CAS 108-10-1; EINECS 203-550-1
UN No. 1245
Synonyms: MIBK; 4-Methyl-2-pentanone; 2-Methyl-4-pentanone; Hexanone; Hexone; 2-Methylpropyl methyl ketone; Isopropylacetone; Isobutyl methyl ketone; 4-Methyl-2-oxopentane
Classification: Sat. aliphatic ketone
Empirical: $C_6H_{12}O$
Formula: $CH_3COCH_2CH(CH_3)_2$
Properties: Colorless volatile liq., faint ketonic/camphoraceous odor; sl. sol. in water; misc. with alcohol, ether, benzene, most org. solvs.; m.w. 100.18; sp.gr. 0.8042 (20/20 C); m.p. -85 C; b.p. 115.8 C; flash pt. (CC) 13 C; ref. index 1.396
Toxicology: ACGIH TLV/TWA 50 ppm; STEL 75 ppm; LD50 (oral, rat) 2080 mg/kg; poison by IP route; mod. toxic by ingestion; mildly toxic by inh.; very irritating to skin, eyes, mucous membranes; narcotic in high conc.
Precaution: Flamm.; dangerous fire risk; explosive limits 1.4-7.5% in air; incompat. with oxidizing/reducing agents, strong bases, potassium t-butoxide; violent reactions possible; Hazardous decomp. prods.: Explosive peroxides (e.g., methyl isobutyl peroxide)
NFPA: Health 2, Flammability 3, Reactivity 1
Uses: Solvent for paints, varnishes, nitrocellulose, lacquers, resin-based protective coatings, adhesives, mfg. of methyl amyl alcohol, extraction processes (metal, uranium from fission prods.), pharmaceuticals; organic synthesis; raw material in prod. of antioxidants; denaturant for alcohol; prod. of surfactants for inks, paints, pesticides; syn. flavoring adjuvant for foods, pharmaceuticals; pharmaceutical solvent
Regulatory: FDA 21CFR §172.515; FEMA GRAS; USP/NF compliance; SARA reportable
Manuf./Distrib.: Aldrich; Allchem Ind.; Ashland; J.T. Baker; Baychem; Chemcentral; Eastman; Elf Atochem SA; Exxon; Fluka; C.P. Hall; Harcros USA; Houghton Chem.; Hüls Am.; Primachem; Samson; Shell; Sigma; Spectrum Chem. Mfg.; Union Carbide; Van Waters & Rogers
Trade names containing: Filmex® A-1; Filmex® A-2; Filmex® B; Filmex® C; Filmex® D-1; Filmex® D-2; Propsolv III-1; Propsolv III-2; Tecsol® 3; Tecsol® A; Tecsol® A-2; Tecsol® B; Tecsol® C; Tecsol® D; Tecsol® D-2

Methyl isopentyl ketone. *See* Methyl isoamyl ketone
1-Methyl-4-isopropenyl-1-cyclohexene. *See* dl-Limonene
1-Methyl-4-isopropylcyclohexadiene-1,3. *See* α-Terpinene
1-Methyl-4-isopropyl-1,3-cyclohexadiene. *See* α-Terpinene
1-Methyl-4-isopropyl-1-cyclohexen-8-ol. *See* α-Terpineol

Methyl isopropyl ketone
CAS 563-80-4
UN No. 2397 (DOT)
Synonyms: MIPK; 2-Butanone, 3-methyl-; Isopropyl methyl ketone; 3-Methylbutan-2-one; 2-Acetyl propane; 3-Methyl-2-butanone
Classification: Ketone
Empirical: $C_5H_{10}O$
Formula: $CH_3COCH(CH_3)_2$
Properties: Colorless liq.; sol. in alcohol, ether; very sl. sol. in water; m.w. 86.15; sp.gr. 0.805 (20/20 C); dens. 6.64 lb/gal; f.p. -92 C; b.p. 94.3 C; flash pt. (Seta) 0.5 C; ref. index 1.38788 (16 C)
Toxicology: LD50 (oral, rat) 148 mg/kg, (skin, rabbit) 6350 mg/kg, (IP, rat) 800 mg/kg; ACGIH TLV/TWA 200 ppm; primary irritant; poison by ing.; mildly

toxic by inh. and skin contact; skin and eye irritant; mutagen
Precaution: Flamm. exposed to heat or flame; can react vigorously with oxidizing materials; Hazardous decomp. prods.: Heated to decomp., emits acrid smoke and irritating fumes
Uses: Solvent for nitrocellulose lacquers
Manuf./Distrib.: Allchem Ind.; Chisso Am.; Eastern Chem.; Eastman

Methyl ketone. *See* Acetone

Methyl lactate
CAS 547-64-8; EINECS 208-930-0
Synonyms: Methyl-S-(-)-2-hydroxy propionate; Lactic acid methyl ester; Methyl DL-lactate
Definition: Methyl ester of lactic acid
Empirical: $C_4H_8O_3$
Formula: $CH_3CH(OH)COOCH_3$
Properties: M.w. 104.11; sp.gr. 1.085-1.098 (20 C); b.p. 140-150 C; flash pt. 51 C; ref. index 1.410-1.415
Toxicology: LDLo (IP, rat) > 2 g/kg; mild irritant
Precaution: Flamm.
Uses: Solvent for paints and coatings, electronics, printed circuit boards, metal industry
Manuf./Distrib.: Fluka; Purac Am.; Rhône-Poulenc N. Am.
Trade names: Purasolv® ML

Methyl DL-lactate. *See* Methyl lactate

Methyl lardate
Uses: Solvent, carrier for agric. spray prods.; defoaming component in metalworking, paper deinking, pharmaceutical fermentation; wetting agent, lubricant ingred. for metalworking and lubricating oils
Manuf./Distrib.: Anar; Chemol; Norman, Fox
Trade names: Emery® Methyl Lardate

Methyl laurate
CAS 111-82-0; 67762-40-7; EINECS 203-911-3
Synonyms: Methyl dodecanoate; Dodecanoic acid methyl ester
Definition: Ester of methyl alcohol and lauric acid
Empirical: $C_{13}H_{26}O_2$
Formula: $CH_3(CH_2)_{10}COOCH_3$
Properties: Water-white liq., fatty floral odor; insol. in water; m.w. 214.35; dens. 0.8702 (20/4 C); m.p. 4.8 C; b.p. 262 C (766 mm); flash pt. > 230 F; ref. index 1.4320
Precaution: Combustible; noncorrosive
Uses: Intermediate for detergents, emulsifiers, wetting agents, stabilizers, lubricants, plasticizers, textiles, plastics, cosmetics; syn. flavoring agent for foods, pharmaceuticals; solvent for pesticides and herbicides
Regulatory: FDA 21CFR §172.225, 172.515, 176.200, 176.210, 177.2260, 177.2800; FEMA GRAS; BP compliance
Manuf./Distrib.: Aldrich; Fluka; Henkel/Emery; Penta Mfg.; Procter & Gamble; Sigma; Stepan
Trade names: Emery® 2270; Emery® 2290; Emery® 2296
Trade names containing: Degreez

N-Methylmethanamine. *See* Dimethylamine
Methyl methanoate. *See* Methyl formate
4-Methyl-4-methoxy-2-pentanone. *See* 4-Methoxy-4-methylpentanone-2
1-Methyl-4(1-methylethenyl)cyclohexene. *See* dl-Limonene

1-Methyl-4-(1-methylethenyl)-(S)-cyclohexene. *See* l-Limonene
(R)-1-Methyl-4-(1-methylethenyl)-cyclohexene. *See* d-Limonene
Methyl-2(methyl-2) oxybispropanol. *See* Dipropylene glycol
Methylmorpholine. *See* p-Methyl morpholine
4-Methyl morpholine. *See* p-Methyl morpholine
N-Methyl morpholine. *See* p-Methyl morpholine

p-Methyl morpholine
CAS 109-02-4; EINECS 203-640-0
UN No. 2535 (DOT)
Synonyms: Methylmorpholine; 4-Methyl morpholine; Morpholine, 4-methyl-; N-Methyl morpholine
Empirical: $C_5H_{11}NO$
Formula: $CH_2CH_2OCH_2CH_2NCH_3$
Properties: Water-white liq., ammonia odor; forms constant-boiling mixture with 25% water and boiling at 97 C; misc. with benzene, water; m.w. 101.17; dens. 0.921; b.p. 115.4 C; f.p. -66 C; flash pt. (TOC) 75 F; ref. index 1.4349
Toxicology: LD50 (oral, rat) 1970 mg/kg, (skin, rabbit) 1242 mg/kg; mod. toxic by ing. and skin contact; mildly toxic by inh.; corrosive irritant to skin, eyes, mucous membranes
Precaution: Flamm. exposed to heat or flames; dangerous fire risk; corrosive; can react vigorously with oxidizing materials; Hazardous decomp. prods.: Heated to decomp., emits toxic fumes of NO_x
Uses: Catalyst in polyurethane foams; extraction solvent; stabilizer for chlorinated hydrocarbons; self-polishing waxes; corrosion inhibitor; pharmaceuticals
Manuf./Distrib.: Aldrich; BASF; BASF AG; Fluka; Sigma

Methyl myristate
CAS 124-10-7; EINECS 204-680-1
Synonyms: Methyl tetradecanoate; Tetradecanoic acid, methyl ester
Definition: Ester of methyl alcohol and myristic acid
Empirical: $C_{15}H_{30}O_2$
Formula: $CH_3(CH_2)_{12}COOCH_3$
Properties: Colorless liq., honey and orris-like odor; insol. in water; m.w. 242.40; dens. 0.866 (20/4 C); m.p. 17.8 C; b.p. 186.8 C (30 mm); flash pt. > 112 C; ref. index 1.438 (20 C)
Precaution: Combustible
Uses: Intermediate for myristic acid detergents; emulsifiers, wetting agents, stabilizers, resins, lubricants, plasticizers, textiles, animal feeds; standard for gas chromatography; syn. flavoring agent for foods, pharmaceuticals; solvent for pesticides and herbicides
Regulatory: FDA 21CFR §172.225, 172.515, 176.200, 176.210, 177.2260, 177.2800; FEMA GRAS; BP compliance
Manuf./Distrib.: Aldrich; Fluka; Henkel/Emery; Penta Mfg.; Sigma; Stepan
Trade names: Emery® 2214
Trade names containing: Degreez

Methyl octadecanoate. *See* Methyl stearate
Methyl 9-octadecenoate. *See* Methyl oleate
Methyl-cis-9-octadecenoate. *See* Methyl oleate
Methylol. *See* Methyl alcohol

Methyl oleate
CAS 112-62-9; 67762-38-3; EINECS 203-992-5; 267-015-4
Synonyms: Methyl 9-octadecenoate; Methyl-cis-9-octadecenoate; 9-Octadecenoic

acid, methyl ester; Oleic acid, methyl ester, cis-
Definition: Ester of methyl alcohol and oleic acid
Empirical: $C_{19}H_{36}O_2$
Formula: $CH_3(CH_2)_7CH=CH(CH_2)_7COOCH_3$
Properties: Clear to amber liq., faint fatty odor; sol. in alcohols, most org. solvs.; insol. in water; m.w. 296.55; dens. 0.8739 (20 C); f.p. -19.9 C; b.p. 218.5 C (20 mm); ref. index 1.4510 (26 C)
Toxicology: ACGIH TLV/TWA 10 mg/m^3 (total dust); low oral toxicity; mildly irritating to skin; experimental tumorigen which causes leukemia and skin tumors by skin contact
Precaution: Combustible; Hazardous decomp. prods.: Heated to decomp., emits acrid smoke and irritating fumes
Uses: Intermediate for detergents, emulsifiers, wetting agents, stabilizers, paints, textile treatment, plasticizers for PS, cellulosics, duplicating inks, rubbers, waxes, etc.; chromatographic reference standard; defoamer for pharmaceutical fermentation; solvent-carrier for pesticides and herbicides
Regulatory: FDA 21CFR §172.225, 175.105, 176.200, 176.210, 177.2260, 177.2800; BP compliance
Manuf./Distrib.: Aldrich; Calgene; Ferro/Bedford; Fluka; Henkel/Emery; Norman, Fox; Sigma; Stepan; Unichema Int'l.; Union Camp; Witco/Oleo-Surf.
Trade names: Emery® 2219; Emery® 2301; Emery® Methyl Oleate; Kemester® 115; Kemester® 205
Trade names containing: Degreez

Methyl oleate/linoleate
Definition: Ester of methyl alcohol and a mixture of oleic and linoleic acids
Uses: Emulsifier, emollient for cosmetics; lubricant for leather; carrier for agric. spray prods.; as solvent
Trade names: Kemester® 213

3-Methylolpentane. *See* 2-Ethyl-1-butanol
Methylolpropane. *See* Butyl alcohol
Methyloxacyclopropane. *See* Propylene oxide
Methyloxirane. *See* Propylene oxide
S(-)-Methyloxirane. *See* Propylene oxide
Methyl 3-oxobutanoate. *See* Methyl acetoacetate
Methyl 3-oxobutyrate. *See* Methyl acetoacetate
4-Methyl-2-oxopentane. *See* Methyl isobutyl ketone

Methyl palmitate
CAS 112-39-0; EINECS 203-966-3
Synonyms: Methyl hexadecanoate; Hexadecanoic acid, methyl ester
Definition: Ester of methyl alcohol and palmitic acid
Empirical: $C_{17}H_{34}O_2$
Formula: $CH_3(CH_2)_{14}COOCH_3$
Uses: Chemical and detergent intermediate, chemical synthesis; lubricant in min., cutting, lamination, textile oils, and rust inhibitors; textile and leather applic.; solvent for pesticides and herbicides
Regulatory: FDA 21CFR §172.225, 175.105, 176.200, 176.210, 177.2260, 177.2800, 178.3910
Manuf./Distrib.: Aldrich; Fluka; Henkel/Emery; Penta Mfg.; Sigma; Stepan
Trade names: Emery® 2216
Trade names containing: Degreez

Methylparafynol. *See* Methyl pentynol
Methyl pentane. *See* Isohexane

2-Methylpentane. *See* Isohexane
2-Methyl-2,4-pentanediol. *See* Hexylene glycol
2-Methylpentane-2,4-diol. *See* Hexylene glycol
4-Methyl-2,4-pentanediol. *See* Hexylene glycol
2-Methylpentanol-1. *See* 2-Methyl-1-pentanol
4-Methylpentanol-2. *See* Methyl amyl alcohol

2-Methyl-1-pentanol
CAS 105-30-6
Synonyms: Amyl methyl alcohol; Methylamyl alcohol; Isopropyl dimethyl carbinol; Isohexyl alcohol; Methyl isobutyl carbinol; 2-Methyl pentan-1-ol; 2-Methylpentanol-1; 2-Methyl-2-propylethanol; 1,3-Dimethyl butanol
Classification: Primary aliphatic alcohol
Empirical: $C_6H_{14}O$
Formula: $CH_3CH_2CH_2CH(CH_3)CH_2OH$
Properties: Colorless liq.; sol. in ethanol, ether, acetone; very sl. sol. in water; m.w. 102.18; sp.gr. 0.8263 (20 C); b.p. 148 C; flash pt. 54 C
Toxicology: LD50 (oral, rat) 1410 mg/kg, (skin, rabbit) 3560 mg/kg; toxic; may cause mild skin and severe eye irritation; skin absorp. can occur; inh. of mist may cause nose/throat irritation, headaches, nausea, dizziness, drowsiness, incoordination
Precaution: Mod. flamm.; incompat. with oxidizing agents (increases fire/explosion risk); Hazardous decomp. prods.: Heated to decomp., emits smoke and acrid fumes
NFPA: Health 0, Flammability 2, Reactivity 0
Uses: Solvent for metallurgy, metal processing, surf. coatings; chem. intermediate for plastics, other org. compds.
Manuf./Distrib.: Aldrich; BASF AG; Fluka; Penta Mfg.; Sigma; Union Carbide

2-Methyl pentan-1-ol. *See* 2-Methyl-1-pentanol
2-Methyl-4-pentanol. *See* Methyl amyl alcohol
4-Methyl-2-pentanol. *See* Methyl amyl alcohol
4-Methyl-2-pentanol acetate. *See* Methyl amyl acetate
2-Methyl-2-pentanol-4-one. *See* Diacetone alcohol
2-Methyl-4-pentanone. *See* Methyl isobutyl ketone
4-Methyl-2-pentanone. *See* Methyl isobutyl ketone
4-Methyl-3-pentene-2-one. *See* Mesityl oxide
2-Methyl-2-penten-4-one. *See* Mesityl oxide
4-Methyl-3-penten-2-one. *See* Mesityl oxide
3-Methylpentin-3-ol. *See* Methyl pentynol
4-Methyl-2-pentyl acetate. *See* Methyl amyl acetate
2-Methylpentyl 'Carbitol'. *See* Diethylene glycol methylpentyl ether
Methyl pentyl ketone. *See* Methyl n-amyl ketone
2-[2-((2-Methylpentyl)oxy)ethoxy]ethanol. *See* Diethylene glycol methylpentyl ether

Methyl pentynol
CAS 77-75-8; EINECS 201-055-5
UN No. 1993
Synonyms: 3-Methylpentin-3-ol; 3-Methylpent-1-yn-3-ol; 3-Methyl-1-pentyn-3-ol; Ethyl ethynyl methyl carbinol; Ethinylmethylethyl carbinol; Methylparafynol; 2-Ethynyl-2-butanol; 2-Ethinylbutanol-2; 3-Ethylbutinol; 3-Ethylbutynol; Meparfynol; m-Pentynol; Methylethylacetylenyl carbinol
Classification: Aliphatic organic compd.
Empirical: $C_6H_{10}O$
Formula: $CH_3CH_2CH_3COHCCH$

3-Methylpent-1-yn-3-ol

Properties: Colorless liq.; acrid odor; burning taste; sol. in water, ether; misc. with acetone, benzene, CCl_4, ethyl acetate; m.w. 98.16; dens. 0.8721 (20/20 C); f.p. –30.6 C; b.p. 121.4 C; flash pt. (TOC) 38.3 C; ref. index 1.4310

Toxicology: LD50 (oral, rat) 300 mg/kg, (IP, mouse) 625 mg/kg; toxic by ingestion; moderately toxic by intraperitoneal and subcutaneous routes; a soporific; may cause dermatitis, psychoses, CNS effects

Precaution: Flamm. exposed to heat or flame; can react with oxidizing materials; Hazardous decomp. prods.: Heated to decomp., emits acrid smoke and fumes

Uses: Corrosion inhibitor; reactive intermediate in the mfg. of pharmaceuticals, perfumes, plastics, rubbers, specialty chemicals; as solvent, acid inhibitor, visc. reducer and stabilizer, electroplating brightener, halogenated solvent stabilizer

Manuf./Distrib.: Air Prods.; Aldrich; Fluka

3-Methylpent-1-yn-3-ol. *See* Methyl pentynol
3-Methyl-1-pentyn-3-ol. *See* Methyl pentynol
Methyl phenyl ketone. *See* Acetophenone
Methyl phenyl polysiloxane. *See* Phenyl trimethicone
Methylphosphonic acid, dimethyl ester. *See* Dimethyl methylphosphonate
Methyl phthalate. *See* Dimethyl phthalate

2-Methyl-1,3-propanediol

CAS 2163-42-0

Empirical: $C_4H_{10}O_2$

Formula: $HOCH_2CH(CH_3)CH_2OH$

Properties: M.w. 90.12; dens. 1.015; m.p. -91 C; b.p. 123-125 C (20 mm); flash pt. > 110 C; ref. index 1.4450 (20 C)

Toxicology: Irritant

Uses: Intermediate used in prod. of solvents, urethanes, unsat. polyesters, gel coats, sat. polyester and alkyd coatings, polymeric plasticizers; nonvolatile solvent, compatibilizer for personal care, adjuvants, fabric softener carriers, nonionic surfactants, pigment dispersions

Manuf./Distrib.: Aldrich; ARCO; Fluka

Trade names: MPDiol® Glycol

2-Methylpropanoic acid 2-methylpropyl ester. *See* Isobutyl isobutyrate
1-Methyl propanol. *See* 2-Butanol
2-Methylpropanol. *See* Isobutyl alcohol
2-Methyl-1-propanol. *See* Isobutyl alcohol
2-Methyl-2-propanol. *See* t-Butyl alcohol
Methyl-2-propanone. *See* Methyl ethyl ketone
1-Methylpropyl acetate. *See* s-Butyl acetate
2-Methylpropyl acetate. *See* Isobutyl acetate
2-Methyl-1-propyl acetate. *See* Isobutyl acetate
1-Methylpropylamine. *See* s-Butylamine
Methyl propyl carbinol. *See* s-Amyl alcohol
β-Methylpropyl ethanoate. *See* Isobutyl acetate
2-Methyl-2-propylethanol. *See* 2-Methyl-1-pentanol
2-Methylpropyl hexadecanoate. *See* Isobutyl palmitate

Methyl propyl ketone

CAS 107-87-9; EINECS 203-528-1

UN No. 1249 (DOT)

Synonyms: MPK; 2-Pentanone; Ethyl acetone; Methyl n-propyl ketone

Empirical: $C_5H_{10}O$

Formula: $CH_3COCH_2CH_2CH_3$

Properties: Water-wh. liq., fruity ethereal odor; sl. sol. in water; misc. with alcohol, ether; m.w. 86.14; dens. 0.801-0.806; m.p. -78 C; b.p. 216 F; flash pt. 45 F
Toxicology: LD50 (oral, rat) 3730 mg/kg, (IP, rat) 800 mg/kg; ACGIH TLV:TWA 200 ppm; STEL 250 ppm; mod. toxic by ingestion, IP; mildly toxic by skin contact, inh.; human systemic effects; skin irritant; mutagenic data
Precaution: Highly flamm.; very dangerous fire hazard exposed to heat or flame; reacts vigorously with oxidizers; explosion hazard in vapor form exposed to heat or flame; mixts. with bromine trifluoride may explode; Hazardous decomp. prods.: Heated to decomp., emits acrid smoke and irritating fumes
NFPA: Health 2, Flammability 3, Reactivity 0
Uses: Solvent for paints; substitute for diethyl ketone; syn. flavoring agent for foods, pharmaceuticals; in org. synthesis
Regulatory: FDA 21CFR §172.515; FEMA GRAS
Manuf./Distrib.: Aldrich; Ashland; Chisso Am.; Eastman; Fluka; Houghton Chem.; Janssen Chimica; Penta Mfg.; Spectrum Chem. Mfg.

Methyl n-propyl ketone. *See* Methyl propyl ketone
2-Methylpropyl methyl ketone. *See* Methyl isobutyl ketone
2-Methylpyridine. *See* α-Picoline
3-Methylpyridine. *See* β-Picoline
4-Methylpyridine. *See* γ-Picoline
α-Methylpyridine. *See* α-Picoline
1-Methylpyrrolidinone. *See* N-Methyl-2-pyrrolidone
1-Methyl-2-pyrrolidinone. *See* N-Methyl-2-pyrrolidone
N-Methylpyrrolidinone. *See* N-Methyl-2-pyrrolidone
N-Methyl-2-pyrrolidinone. *See* N-Methyl-2-pyrrolidone
Methylpyrrolidone. *See* N-Methyl-2-pyrrolidone
1-Methylpyrrolidone. *See* N-Methyl-2-pyrrolidone
1-Methyl-2-pyrrolidone. *See* N-Methyl-2-pyrrolidone
N-Methylpyrrolidone. *See* N-Methyl-2-pyrrolidone

N-Methyl-2-pyrrolidone
CAS 872-50-4; EINECS 212-828-1
DOT No. NA 1993
Synonyms: NMP; Methylpyrrolidone; N-Methylpyrrolidone; 1-Methylpyrrolidone; 1-Methyl-2-pyrrolidone; 1-Methylpyrrolidinone; N-Methylpyrrolidinone; N-Methyl-2-pyrrolidinone; 1-Methyl-2-pyrrolidinone; 1-Methylazacyclopentan-2-one
Classification: Sat. nitrogen heterocyclic compd.
Empirical: C_5H_9NO
Properties: Colorless liq., mild amine odor; sol. in water, alcohols, ketones, ethers, ethyl acetate, aromatic hydrocarbons; mod. sol. in aliphatic hydrocarbons; misc. with castor oil; m.w. 99.13; sp.gr. 1.032 (20/4 C); vapor pressure 0.3 mm Hg (20 C); f.p. -24 C; b.p. 202 C; flash pt. 95 C; ref. index 1.470 (20 C); pH 7.7-8.0 (10% aq.)
Toxicology: LD50 (oral, rat) 7000 mg/kg; severely irritating to eyes, skin; mod. toxic by IP and IV routes; mildly toxic by ing., skin contact; inh. of high concs. may cause fatigue, difficulty breathing; ing. may cause nausea, vomiting
Environmental: readily biodeg.; EC50 (rainbow trout, 96 h) ≥ 500 mg/l, (algae, 72 h) ≥ 500 mg/l
Precaution: Combustible exposed to heat, open flame, powerful oxidizers; incompat. with oxidizing agents (may spontaneously ignite), strong acids and strong bases (can cause hydrolysis); Hazardous decomp. prods.: May form hydroperoxides on exposure to air and sunlight; heated to decomp., emits toxic fumes of NO_x

HMIS: Health 2, Flammability 1, Reactivity 0
Uses: Solvent for resins, paints, acetylene, industrial cleaning, agric., mold cleaning; petrochem. and petrol. processing; spinning agent for PVC; microelectronics industry plastic solvent; solvent for CO_2 removal in ammonia synthesis; polymer solvent for PVAc, PVDF, PS, vinyl copolymers, nylon and aromatic polyamides and polyimides, polyesters, acrylics, PC, cellulose derivs., syn. elastomers, waxes; veterinary pharmaceuticals; intermediate; pigment dispersant
Regulatory: SARA reportable
Manuf./Distrib.: Aldrich; Allchem Ind.; ARCO; Ashland; BASF; Coyne; Dynaloy; Fluka; ISP; Janssen Chimica; Olin; Sigma; Spectrum Chem. Mfg.
Trade names: GAFsolver™ TMP; M-Pyrol®; NMP; NMP-EL; Pharmasolve™
Trade names containing: Ecco Strip LS-W; FoamFlush® Urethane Remover; Partsprep™ Degreaser; PrepRite™ Coatings Remover; Printsolve™ Ink Remover; Ship Shape® Resin Cleaner

Methyl soyate
Definition: Ester of methyl alcohol and soya acid
Uses: Intermediate in prod. of superamides, in metalworking lubricants, specialized solvents; opacifier, visc. control agent
Trade names: Emery® 2224; Kemester® 226

Methyl stearate
CAS 112-61-8; 85586-21-6; EINECS 203-990-4; 287-824-6
Synonyms: Methyl octadecanoate; Stearic acid methyl ester; Octadecanoic acid, methyl ester; Methyl ester stearic acid
Definition: Ester of methyl alcohol and stearic acid
Empirical: $C_{19}H_{38}O_2$
Formula: $CH_3(CH_2)_{16}COOCH_3$
Properties: Wh. crystals; insol. in water; sol. in ether, alcohol; m.w. 298.57; m.p. 37.8 C; b.p. 234.5 C (30 mm); flash pt. 307 F
Toxicology: Experimental tumorigen; heated to decomp., emits acrid smoke and irritating fumes
Precaution: Combustible exposed to heat or flame; can react with oxidizing materials
Uses: Intermediate for stearic acid detergents, emulsifiers, wetting agents, stabilizers, resins, lubricants, plasticizers, paints; foam depressant and nutrient in fermentation; topical pharmaceuticals; specialized solvent; solvent for pesticides and herbicides
Regulatory: FDA 21CFR §172.225, 176.200, 176.210, 177.2260, 177.2800, 178.3910; FDA approved for topicals; BP compliance
Manuf./Distrib.: Aldrich; Ashland; Ferro/Bedford; Fluka; Henkel/Emery; Penta Mfg.; Sea-Land; Sigma; Union Camp; Witco/Oleo-Surf.
Trade names: Emery® 2218; Kemester® 4516; Kemester® 9018
Trade names containing: Degreez

Methyl succinate. *See* Dimethyl succinate
Methyl sulfinylmethane. *See* Dimethyl sulfoxide
Methyl sulfoxide. *See* Dimethyl sulfoxide

Methyl sunflowerate
Uses: Solvent for pesticides and herbicides
Trade names: Emery® 2230

Methyl tallowate
CAS 61788-61-2; EINECS 262-989-7
Synonyms: Fatty acids, tallow, methyl esters

Definition: Ester of methyl alcohol and tallow acid
Uses: Wetting/oiliness agent for metalworking and lubricating oils; lubricant, plasticizer for cosmetics, leather, rubber prods.; solvent for pesticides and herbicides
Trade names: Emery® 2203; Kemester® 143

Methyl tertiary butyl ether. *See* Methyl t-butyl ether
Methyl tetradecanoate. *See* Methyl myristate
Methyl toluene. *See* Xylene
3-Methyltoluene. *See* m-Xylene
4-Methyltoluene. *See* p-Xylene
m-Methyltoluene. *See* m-Xylene
o-Methyltoluene. *See* o-Xylene
p-Methyltoluene. *See* p-Xylene
Methyltrichloromethane. *See* 1,1,1-Trichloroethane
Methyltriglycol. *See* Triglycol monomethyl ether
Methyltrimethylene glycol. *See* Butylene glycol
MIAK. *See* Methyl isoamyl ketone
MIBC. *See* Methyl amyl alcohol
MIBK. *See* Methyl isobutyl ketone
MIC. *See* Methyl amyl alcohol
Milk acid. *See* Lactic acid
Mineral naphtha. *See* Benzene

Mineral oil
CAS 8012-95-1; 8020-83-5 (wh.); 8042-47-5; EINECS 232-384-2; 232-455-8
Synonyms: Heavy mineral oil; Light mineral oil; White mineral oil; Paraffin oil; Liquid paraffin; Petrolatum liquid; Liquid petrolatum
Definition: Liq. mixture of hydrocarbons obtained from petroleum
Properties: Colorless transparent oily liq., odorless, tasteless; insol. in water, alcohol; sol. in benzene, chloroform, ether, petrol. ether, volatile oils; dens. 0.83-0.86 (light), 0.875-0.905 (heavy); flash pt. (OC) 444 F; surf. tens. < 35 dynes/cm
Toxicology: Eye irritant; human carcinogen and teratogen by inhalation; heated to decomp., emits acrid smoke and fumes
Precaution: Combustible
NFPA: Health 0, Flammability 1, Reactivity 0
Uses: Cathartic; laxative; protectant; lubricant; binder; carrier; mold release for foods, coating for fruits and vegetables, food pkg. materials; plasticizer, lubricant for plastics; pharmaceutical vehicle, solvent, lubricant, laxative; in cosmetics, pharmaceuticals, plastics, agric., paper, textiles, etc.
Regulatory: FDA 21CFR §172.878, 173.340 (limitation 0.008% in wash water for sliced potatoes, 150 ppm in yeast), 175.105, 175.210, 175.230, 175.300, 176.170, 176.200, 176.210, 177.1200, 177.2260, 177.2600, 177.2800, 178.3570, 178.3620, 178.3740, 178.3910, 179.45; 573.680; ADI not specified (FAO/WHO); FDA approved for ophthalmics, orals, topicals; USP/NF, BP, Ph.Eur. compliance
Manuf./Distrib.: Air-Scent Int'l.; Aldrich; Amoco/Lubricants; Chemisphere; Coyne; Exxon; Fluka; Magie Bros. Oil; Mobil; Penreco; Penta Mfg.; Ruger; San Yuan; Sea-Land; Sigma; Spectrum Chem. Mfg.; Surco Prods.; Total Petrol.; Witco/Golden Bear; Witco/Petrol. Spec.
Trade names: Drakeol® 7; PD-23
Trade names containing: Argobase 125T; Gilugel MIN

Mineral spirits
CAS 8032-32-4; 64475-85-0; EINECS 232-453-7
UN No. 1271 (DOT)

Synonyms: White spirits; Ligroin; Petroleum spirits
Definition: Mixture of hydrocarbons from petroleum with distillation range of 300-415 F; avail. in type I (reg.), II (high flash), III (odorless), IV (low dry pt.)
Properties: Clear colorless, volatile, nonfluorescent liq.; dens. @ 15.6/15.6 C: 0.654-0.820 (I), 0.768-0.820 (II), 0.775 max. (III), 0.754-0.800 (IV); i.b.p. 149 C (I), 177 C (II), 149 C (III), 149 C (IV); flash pt. 38 C min. except 60 C (II); KB value 38 (non-rule 66 type)
Toxicology: LD50 (oral, rat) > 34,600 mg/kg; TLV:TWA 300 ppm; STEL 400 ppm; mod. toxic (human); mildly toxic by inh., IP routes; defatting on skin contact; vapor can cause headache, dizziness, unconsciousness with high concs.; heated to dec., emits acrid smoke
Environmental: HAPS contg.
Precaution: Combustible
NFPA: Health 0, Flammability 2, Reactivity 0
Uses: Solvent; paint thinner; in coatings, dry-cleaning, adhesives, agric., automotive and chemical specialties, cleaning and polishing, pharmaceuticals, inks, textile coatings, asphalt, wood treating; charcoal lighter fluid, concrete curing
Regulatory: FDA 21CFR §178.3800
Manuf./Distrib.: Aldrich; Ashland; Houghton Chem.; Spectrum Chem. Mfg.; Texaco
Trade names: Shell Mineral Spirits 135; Shell Mineral Spirits 145-EC; Shell Mineral Spirits 146 HT; Shell Mineral Spirits 200 HT; Shell Sol 71; Shell Sol 142 HT; Shell Sol 340 HT; Texsolve S; Texsolve S-66; Texsolve S-LO
See also Stoddard solvent

MIPA. *See* Isopropylamine
MIPK. *See* Methyl isopropyl ketone

Mixed terpenes
Definition: Mixt. of hydrocarbons distilled from certain plants and trees such as pine and citrus; consists of terpenes, sesquiterpenes, diterpenes, polyterpenes
Uses: Hydrocarbon solvent; constituent of industrial cleaners, deodorants
Trade names: Solvenol® No. 2

Mixed xylenes. *See* Xylene
Monacetin. *See* Glyceryl acetate
1-Monacetin. *See* Glyceryl acetate
α-Monacetin. *See* Glyceryl acetate
Monacetyl glycerine. *See* Glyceryl acetate
Monoamylamine. *See* Pentylamine
Monobromomethane. *See* Methyl bromide
Monobutyl glycol ether. *See* Butoxyethanol
Monochlorethane. *See* Ethyl chloride
Monochlorodifluoromethane. *See* Chlorodifluoromethane
Monochloroethane. *See* Ethyl chloride
Monochloromethane. *See* Methyl chloride

Monochlorotoluene
CAS 25168-05-2
UN No. 2238 (DOT)
Synonyms: Toluene, ar-chloro-; Chloromethylbenzene
Empirical: C_7H_7Cl
Properties: Liq.; m.w. 126.59
Toxicology: LC50 (inh., unspecified mammal) 48 g/m^3
Precaution: Flamm.
Uses: Solvent, extender, diluent for dye carrier, fuel oil additive, sludge solvent,

component of paint thinners and strippers, metal parts cleaners, adhesives
Trade names: Halso® 99
Trade names containing: Oxsol® 73; Oxsol® 253; Oxsol® 325; Oxsol® 550

Mono- and diglycerides of fatty acids
CAS 67701-32-0; 67701-33-1; 68990-53-4
Definition: Mixt. of glycerol mono- and di-esters, with minor amts. of tri-esters, of fatty acids from edible oils
Properties: Yel. liqs. to ivory plastics to hard solids, bland odor and taste; sol. in alcohol, ethyl acetate, chloroform, other chlorinated hydrocarbons; insol. in water; acid no. 4 max.
Toxicology: Heated to decomp., emits acrid smoke and irritating fumes
Uses: Antistat for PE, flexible PVC, plastisols; food and pharmaceutical emulsifier; flavoring agent; lubricant, release agent, softener, solvent, stabilizer, surfactant, texturizer, vehicle in foods
Regulatory: FDA 21CFR §172.863, 182.4505, 184.1505, GRAS; USDA 9CFR §318.7, 381.147; Europe listed; UK approved; USP/NF compliance
Manuf./Distrib.: Int'l. Chem. Inc

Monoethanolamine peanut acid amide. *See* Peanutamide MEA
Monoethylamine. *See* Ethylamine
Monoethylene glycol. *See* Glycol
Monofluorotrichloromethane. *See* Trichlorofluoromethane
Monoglyme. *See* Ethylene glycol dimethyl ether
Monoisopropylamine. *See* Isopropylamine
Monoisopropyl biphenyl. *See* 1,1´-Biphenyl (1-methylethyl)
Monooctanoin. *See* Glyceryl caprylate
Monoolein. *See* Glyceryl oleate
Monopropylene glycol. *See* Propylene glycol
Monopropylene glycol methyl ether. *See* Methoxyisopropanol
Monostearin. *See* Glyceryl stearate
Monothioethyleneglycol. *See* 2-Mercaptoethanol

Morpholine
CAS 110-91-8; EINECS 203-815-1
UN No. 1760 (DOT), 2054 (DOT)
Synonyms: Tetrahydro-p-isoxazine; Tetrahydro-1,4-isoxazine; Tetrahydro-p-oxazine; Tetrahydro-1,4-oxazine; 1-Oxa-4-azacyclohexane; Tetrahydro-2H-1,4-oxazine; Tetrahydro-4H-1,4-oxazine; Diethylene imidoxide; Diethylene oximide; Diethyleneimide oxide; Diethylenimide oxide
Classification: Heterocyclic sec. amine
Empirical: C_4H_9NO
Formula: C_4H_8ONH
Properties: Colorless clear oily liq., amine-like odor; sol. in water, acetone, benzene, ether, ethyl acetate, chloroform, ethylene glycol, castor oil, alcohols; m.w. 87.14; sp.gr. 1.002 (20/20 C); vapor pressure 10 mm Hg (23 C); f.p. -4.9 C; b.p. 128.9 C; flash pt. (OC) 37.7 C; autoignition temp. 590 F; ref. index 1.4540 (20 C)
Toxicology: LD50 (oral, rat) 1050 mg/kg, (skin, rabbit) 500 mg/kg; ACGIH TLV/TWA 20 ppm; STEL 30 ppm (skin); mod. toxic by ing., inh., skin contact, IP; corrosive irritant to skin, eyes, mucous membranes; very high concs. may be fatal
Precaution: Flamm.; dangerous fire hazard exposed to flame, heat or oxidizers; reactive with oxidizers; explosive with nitromethane; incompat. with strong acids, acid anhydrides, nitrites, cellulose nitrate, isocyanates, org. peroxides, epoxides; Hazardous decomp. prods.: Emits highly toxic fumes of NO_x

NFPA: Health 3, Flammability 3, Reactivity 0
Uses: Org. intermediate for vulcanization accelerators, textile lubricants, softeners, sizing agents, antioxidants, surfactants, plasticizers, insecticides, dyes, bactericides, analgesics, anesthetics; solvent for resins, waxes, casein, dyes; neutralizing agent; emulsifier for cosmetics, rubless waxes, polishes; defoaming agent; corrosion inhibitor in steam boiler systems; optical brightener for detergents; lubricant, coolant for machine tools; protective food coatings
Regulatory: FDA 21CFR §172.235, 173.310 (10 ppm max. in steam), 175.105, 176.210, 178.3300; BP compliance
Manuf./Distrib.: Air Prods.; Aldrich; Allchem Ind.; BASF; BASF AG; Coyne; Fluka; Houghton Chem.; Huntsman; Nippon Nyukazai; PMC Spec.; Sigma; Spectrum Chem. Mfg.; Texaco

Morpholine, 4-methyl-. *See* p-Methyl morpholine
Motor benzol. *See* Benzene
Motor fuel. *See* Gasoline
Motor spirit(s). *See* Gasoline
MPA. *See* Propylene glycol methyl ether acetate
MPK. *See* Methyl propyl ketone
MTBE. *See* Methyl t-butyl ether
Muriatic ether. *See* Ethyl chloride
Myristic acid isopropyl ester. *See* Isopropyl myristate

Myristyl propionate
CAS 6221-95-0; EINECS 226-300-9
Synonyms: 1-Tetradecanol, propanoate
Definition: Ester of myristyl alcohol and propionic acid
Empirical: $C_{17}H_{34}O_2$
Formula: $CH_3CH_2COOCH_2(CH_2)_{12}CH_3$
Uses: Emollient, penetrant, and spreading agent in cosmetics, antiperspirants; perfume solvent in personal care prods.; humectant
Trade names: Lonzest® 143-S

Naphtha
CAS 8030-30-6; 68920-06-9; 64742-95-6 (lt. aromatic)
UN No. 1255 (DOT), 1270 (DOT), 2553 (DOT)
Synonyms: Coal tar naphtha; Benzin; Benzine; Petroleum naphtha; VM&P naphtha; Petroleum benzin; Petroleum ether; Petroleum spirit
Definition: Petroleum distillate
Properties: Dark straw-colored to colorless liq.; sol. in benzene, toluene, xylene; dens. 0.862-0.892; b.p. 149-216 C; flash pt. (CC) 107 F
Toxicology: LD50 (oral, rat) > 5 g/kg, (skin, rabbit) 3 g/kg; mildly toxic by inhalation; human poison and systemic effects by IV route; irritating to eyes, skin, respiratory tract; vapor inh. may cause nausea, dizziness, unconsciousness; tumorigen
Precaution: Flamm. when exposed to heat or flame; sl. explosion hazard; can react with oxidizing materials; keep containers tightly closed
NFPA: Health 1, Flammability 3, Reactivity 0
Uses: Solvent, thinner in paint, dry-cleaning fluid, rubber compounding, sealants, chem. absorption; blending with natural gas; color diluent, protective coating solvent in foods

Regulatory: FDA 21CFR §73.1, 172.250
Manuf./Distrib.: Ashland; J.T. Baker; Baychem; R.E. Carroll; Chemcentral; Crowley Chem.; Exxon; Harcros USA; Houghton Chem.; Kerr-McGee; Mobil; Monsanto; Norsk Hydro AS; Phillips; Samson; Shell; Spectrum Chem. Mfg.; Sunnyside; Texaco; Van Waters & Rogers
Trade names: Aromatic 150; Hi-Sol® 10; Hi-Sol® 15; Hi-Sol® 70; Shell VM&P Naphtha HT; Shell Sol B; Shell Super VM&P Naphtha EC; Shell Tolu-Sol® 5; Texsolve V

Naphthalane. *See* Decahydronaphthalene
Naphthalene-1,2,3,4-tetrahydride. *See* Tetrahydronaphthalene
Naphtha safety solvent. *See* Stoddard solvent

Naphthenic acid
CAS 1338-24-5; EINECS 215-662-8
UN No. 9137 (DOT)
Synonyms: Carboxylic acids, naphthenic; Naphthenic acids
Classification: Alicyclic and aliphatic carboxylic acids
Properties: Yel. visc. liq., mild odor (purified), dk. red visc. liq., strong offensive odor (crude); sol. in most org. solvs.; insol. in water; m.w. 235-280; sp.gr. 0.94-0.96; b.p. 132-243 C; flash pt. (OC) 149 C
Toxicology: LD50 (oral, rat) 3000 mg/kg, (IP, rat) 640 mg/kg; moderately toxic by ingestion, IP routes; probable sl. to mod. skin irritant and eye irritant; inh. may cause CNS depression (headache, dizziness, nausea)
Precaution: Incompat. with strong oxidizers (increases fire and explosion risk), strong bases (violent reactions possible); Hazardous decomp. prods.: Heated to decomp., emits acrid smoke and irritating fumes
Uses: Paint dryers, fungicides, metal catalysts, corrosion inhibitors, lubricants, fracturing fluids, cellulose preservatives; solvent; detergents; rubber reclaiming agent; intermediate for synthesis of metal napthenates and succinic acid
Manuf./Distrib.: Crowley Chem.; Esprit; Fabrichem; Fluka; OM Group

Naphthenic acids. *See* Naphthenic acid
Naphthol. *See* Naphthol spirits

Naphthol spirits
CAS 1321-67-1 (generic)
Synonyms: Naphthol
Empirical: $C_{10}H_8O$
Properties: Monoclinic; sol. in cold and hot water, chloroform; very sol. in alcohol, benzene; m.w. 144.18; dens. 1.217 (4 C); m.p. 122-123 C; b.p. 285-286 C
Toxicology: LDLo (subcut., rabbit) 1000 mg/kg, (subcut., dog) 333 mg/kg; poison by subcutaneous route; Hazardous decomp. prods.: Heated to decomp., emits acrid smoke and irritating fumes
Uses: Med.-boiling, combustible liq. solvent

Natural gasoline. *See* Gasoline
NE. *See* Nitroethane
Neoamyl alcohol. *See* 2,2-Dimethylpropanol

Neoheptane
CAS 590-35-2
UN No. 1206
Synonyms: 2,2-Dimethylpentane; Dimethyl-2,2-pentane
Classification: Sat. aliphatic hydrocarbon
Empirical: C_7H_{16}
Formula: $CH_3CH_2CH_2C(CH_3)_3$
Properties: Colorless liq.; gasoline-like odor; sol. in most org. solvs.; insol. in water;

m.w. 100.2; sp.gr. 0.674 (20 C); m.p. -124 C; b.p. 79 C; flash pt. (CC) 15 C
Toxicology: ACGIH TLV/TWA 400 ppm; STEL 500 ppm; may cause mild skin irritation; conc. vapor may cause sl. eye irritation; conc. vapor may cause sl. nose/throat irritation, lightheadedness, dizziness, nausea, difficulty breathing
Precaution: Flamm.; incompat. with strong oxidizing agents (increases fire and explosion hazard); Hazardous decomp. prods.: None
Uses: Ingred. in gasoline, aviation fuels; solvent; org. synthesis
Manuf./Distrib.: Aldrich; Fluka

Neopentanoic acid, tridecyl ester. *See* Tridecyl neopentanoate
Neopentyl alcohol. *See* 2,2-Dimethylpropanol

Neopentyl glycol dicaprate
CAS 27841-06-1
Synonyms: Decanoic acid, 2,2-dimethyl-1,3-propanediol diester
Definition: Diester of neopentyl glycol and decanoic acid
Empirical: $C_{25}H_{48}O_4$
Formula: $CH_3(CH_2)_8COOCH_2C(CH_3)_2CH_2OCO(CH_2)_8CH_3$
Uses: Solvency properties for use in make-up removers
Trade names: Schercemol NGDC

Neopentyl glycol dioctanoate
CAS 28510-23-8
Uses: Low freeze pt. emollient; solvent for makeup removers
Trade names: Schercemol NGDO

Niobe oil. *See* Methyl benzoate
Nitration benzene. *See* Benzene
Nitrocarbol. *See* Nitromethane

Nitroethane
CAS 79-24-3; EINECS 201-188-9
UN No. 2842 (DOT)
Synonyms: NE; Ethane, nitro-
Classification: Aliphatic organic compd.; nitroparaffin
Empirical: $C_2H_5NO_2$
Formula: $CH_3CH_2NO_2$
Properties: Oily colorless liq., agreeable odor; sol. in water, alcohols, esters, ketones, ethers, aromatic hydrocarbons, chloroform, acid, alkali; m.w. 75.07; sp.gr. 1.048 (20/4 C); f.p. -50 C; m.p. -90 C; b.p. 112-116 C; decomp. pt. 335-382 C; flash pt. (TCC) 30 C; ref. index 1.391 (20 C)
Toxicology: LD50 (oral, rat) 1100 mg/kg; ACGIH TLV/TWA 100 ppm; poison by intraperitoneal route; moderately toxic by ingestion; irritating to eyes, mucous membranes; may cause headache, nausea, coughing, sneezing, CNS depression
Precaution: DOT: Flamm. liq.; explodes when heated; incompat. with strong oxidizers, strong acids, reducing agents, alkali metals, inorg. bases, amines, heavy metal oxides, hydrocarbons, activated carbon; Hazardous decomp. prods.: Heated to decomp., emits toxic fumes of NO_x
NFPA: Health 1, Flammability 3, Reactivity 3
Uses: Intermediate for synthesis; stabilizer for chlorinated solvents; fuel additive; specialty solvent for coating, adhesive, and ink formulations; solvent for cellulose esters, vinyl, alkyd, other resins and waxes; propellant; raw material for pharmaceutical synthesis
Regulatory: BP compliance

Manuf./Distrib.: Aldrich; ANGUS; Fluka; Spectrum Chem. Mfg.
Trade names: NE™ 101

Nitroisopropane. *See* 2-Nitropropane

Nitromethane
CAS 75-52-5; EINECS 200-876-6
UN No. 1261 (DOT)
Synonyms: NM; Nitrocarbol
Classification: Aliphatic nitro compd.; nitroparaffin
Empirical: CH_3NO_2
Properties: Colorless oily liq., mild fruity to disagreeable odor; sol. in water, alcohol, ether, aromatic hydrocarbons, esters, ketones; m.w. 61.04; sp.gr. 1.139 (20/20 C); m.p. -29 C; b.p. 100-103 C; flash pt. (TCC) 35-35.6 C; ref. index 1.382 (20 C)
Toxicology: ACGIH TLV/TWA 20 ppm; LD50 (oral, rat) 940 mg/kg, (IP, mouse) 110 mg/kg; poison by ingestion, inhalation, intraperitoneal routes; vapor inh. may cause mild respiratory tract irritation
Precaution: Flamm.; dangerously reactive; dangerous fire/explosion risk; may detonate under high temps. and pressures; incompat. with acids, alkali metals, inorg. bases, amines, strong oxidizers, hydrocarbons, metal oxides, aluminum chloride, etc.; Hazardous decomp. prods.: Emits toxic fumes of NO_x
NFPA: Health 1, Flammability 3, Reactivity 4
Uses: Stabilizer for chlorinated solvents; chemical intermediate; solvent for cellulosic compds., polymers, waxes and in chem. synthesis; rocket fuel, gasoline additive; in coatings industry; raw material for pharmaceutical synthesis
Regulatory: BP compliance
Manuf./Distrib.: Aldrich; Allchem Ind.; ANGUS; Fluka; Sigma; Spectrum Chem. Mfg.
Trade names: NM-55®

Nitropropane
CAS 108-03-2; EINECS 203-544-9
UN No. 2608
Synonyms: 1-NP; 1-Nitropropane
Classification: Aliphatic nitro compd.; nitroparaffin
Empirical: $C_3H_7NO_2$
Formula: $CH_3CH_2CH_2NO_2$
Properties: Colorless oily liq.; mild fruity odor; very sl. sol. in water; sol. in aromatic hydrocarbons, alcohols, esters, ketones, ethers, lower carboxylic acids; m.w. 89.09; sp.gr. 1.003 (20/20 C); m.p. -108 C; b.p. 129-133 C; flash pt. (TCC) 35.6 C; ref. index 1.4018 (20 C)
Toxicology: LD50 (oral, rat) 455 mg/kg; ACGIH TLV/TWA 25 ppm; poison by IP route and ing.; irritating to mucous membranes; inh. may cause eye/nose/throat irritation, headaches, nausea, methemoglobinemia; mild skin irritant; human eye irritant
Precaution: Flamm. liq.; reacts violently; incompat. with strong oxidizers, metal oxides, strong acids, amines, inorg. bases, hydrocarbons, activated carbon; Hazardous decomp. prods.: Heated to decomp., emits toxic fumes of NO_x
NFPA: Health 1, Flammability 3, Reactivity 2
Uses: Solvent for inks and coatings, cellulose acetate, vinyl resins, lacquers, synthetic rubbers, fats, oils, dyes; chemical intermediate; rocket propellant; gasoline additive; pharmaceutical intermediate
Manuf./Distrib.: Aldrich; ANGUS; Fluka
Trade names: NiPar S-10™

1-Nitropropane. *See* Nitropropane

2-Nitropropane
CAS 79-46-9; EINECS 201-209-1
UN No. 2608
Synonyms: 2-NP; s-Nitropropane; sec-Nitropropane; β-Nitropropane; Propane, 2-nitro-; Dimethylnitromethane; Nitroisopropane; Isonitropropane
Classification: Aliphatic nitro compd.; nitroparaffin
Empirical: $C_3H_7NO_2$
Formula: $CH_3CH(NO_2)CH_3$
Properties: Colorless clear oily liq.; mild fruity odor; sl. sol. in water; sol. in most aromatic hydrocarbons, alcohols, esters, ketones, chloroform, ethers, lower carboxylic acids; m.w. 89.09; sp.gr. 0.992 (20/20 C); m.p. -93 C; b.p. 119-122 C; flash pt. (TOC) 24 C; ref. index 1.394 (20 C)
Toxicology: LD50 (oral, rat) 725 mg/kg, (IP, rat) 75 mg/kg; ACGIH TLV/TWA 10 ppm; very toxic; poison by ingestion, inhalation, and intraperitoneal routes (nausea, vomiting, diarrhea, anorexia, severe headaches, cyanosis)
Precaution: Flamm. when exposed to heat, open flame, oxidizers; may explode on heating; incompat. with strong oxidizers, acids, amines, inorg. bases, activated carbon, hydrocarbons; Hazardous decomp. prods.: Heated to decomp., emits toxic fumes of NO_x
NFPA: Health 1, Flammability 3, Reactivity 2
Uses: Solvent for coatings, varnishes, lacquers, inks, polymers, cellulose acetate, vinyl resins, synthetic rubbers, fats, oils, dyes; chemical intermediate in prod. of pharmaceuticals, dyes, insecticides, textile chems.; rocket propellant; gasoline additive; solvent for fractionating edible fats and oils
Regulatory: SARA reportable
Manuf./Distrib.: Aldrich; ANGUS; Ashland; Fluka

β-Nitropropane. *See* 2-Nitropropane
s-Nitropropane. *See* 2-Nitropropane
sec-Nitropropane. *See* 2-Nitropropane
NM. *See* Nitromethane
NMP. *See* N-Methyl-2-pyrrolidone

n-Nonane
CAS 111-84-2
UN No. 1920 (DOT)
Classification: Sat. aliphatic hydrocarbon
Empirical: C_9H_{20}
Formula: $CH_3(CH_2)_7CH_3$
Properties: Colorless liq.; gasoline-like odor; misc. with most org. solvs.; sol. in abs. alcohol, ether; insol. in water; m.w. 128.26; sp.gr. 0.714; vapor pressure 10 mm Hg (38 C); m.p. -53.5 C; b.p. 151 C; flash pt. 31 C
Toxicology: LD50 (IV, mouse) 218 mg/kg; ACGIH TLV/TWA 200 ppm; toxic; skin irritant; conc. vapor probably irritating to eyes; inh. of conc. vapor may cause nose/throat irritation, headache, drowsiness, nausea, tremors, difficulty breathing
Precaution: Flamm.; very dangerous fire hazard exposed to heat or flame; explosive as vapor exposed to heat or flame; incompat. with strong oxidizing agents (increases fire and explosion hazard); Hazardous decomp. prods.: Heated to decomp., emits acrid smoke and irritating fumes
NFPA: Health 0, Flammability 3, Reactivity 0
Uses: Solvent; org. synthesis; component of gasoline and petrol. solvents such as VM&P naphtha and Stod. solvent; mfg. of paraffin prods.; paper processing; rubber industry; synthesis of biodeg. detergents; jet fuel research; distillation

chaser
Manuf./Distrib.: Aldrich; Fluka; Sigma

Nonanoic acid, 1-methyl-1,2-ethanediyl ester. *See* Propylene glycol dipelargonate
Noncarbinol. *See* n-Decyl alcohol

Nonoxynol-4
CAS 7311-27-5; 9016-45-9 (generic); 26027-38-3 (generic); 37205-87-1 (generic); 27176-97-2; EINECS 230-770-5
Synonyms: PEG-4 nonyl phenyl ether; POE (4) nonyl phenyl ether; PEG 200 nonyl phenyl ether
Classification: Ethoxylated alkyl phenol
Empirical: $C_{23}H_{40}O_5$
Formula: $C_9H_{19}C_6H_4(OCH_2CH_2)_nOH$, avg. n = 4
Properties: Yel. to almost colorless liq.; sol. in oil
Toxicology: Moderately toxic by ingestion, skin contact; severe eye and mild skin irritant in humans; heated to dec., emits acrid smoke and fumes
Uses: Nonionic surfactant; as detergent, emulsifier, wetting agent, dispersant, stabilizer, defoamer; intermediate in synthesis of anionic surfactants; pharmaceutic aids; plasticizer, antistat for plastics; in cosmetics, fat liquoring, cutting oils, agric., petrol. oils; solvent cleaner
Regulatory: FDA 21CFR §175.105, 176.180, 176.210, 178.3400; FDA approved for ophthalmics, topicals, vaginals
Manuf./Distrib.: Aldrich; Fluka; Sigma
Trade names: T-Det® N-4

Nonoxynol-5
CAS 9016-45-9 (generic); 26027-38-3 (generic); 37205-87-1 (generic); 26264-02-8; 20636-48-0; EINECS 247-555-7
Synonyms: PEG-5 nonyl phenyl ether; POE (5) nonyl phenyl ether; 14-(Nonylphenoxy)-3,6,9,12-tetraoxatetradecan-1-ol
Classification: Ethoxylated alkyl phenol
Empirical: $C_{25}H_{44}O_6$
Formula: $C_9H_{19}C_6H_4(OCH_2CH_2)_nOH$, avg. n = 5
Properties: Yel. to almost colorless liq.; sol. in oil
Toxicology: Moderately toxic by ingestion, skin contact; severe eye and mild skin irritant in humans; heated to dec., emits acrid smoke and fumes
Uses: Nonionic surfactant; as detergent, emulsifier, wetting agent, dispersant, stabilizer, defoamer; intermediate in synthesis of anionic surfactants; emulsion polymerization; paint; pharmaceutical surfactant; solvent cleaner
Regulatory: FDA 21CFR §175.105, 176.180, 176.210, 178.3400
Manuf./Distrib.: Aldrich; Fluka; Sigma
Trade names: Renex® 648

Nonoxynol-6
CAS 9016-45-9 (generic); 26027-38-3 (generic); 37205-87-1 (generic); 27177-01-1; 27177-05-5
Synonyms: PEG-6 nonyl phenyl ether; POE (6) nonyl phenyl ether; PEG 300 nonyl phenyl ether
Classification: Ethoxylated alkyl phenol
Empirical: $C_{27}H_{48}O_7$
Formula: $C_9H_{19}C_6H_4(OCH_2CH_2)_nOH$, avg. n = 6
Properties: Yel. to almost colorless liq.
Toxicology: Moderately toxic by ingestion, skin contact; severe eye and mild skin irritant in humans; heated to dec., emits acrid smoke and fumes
Uses: Nonionic surfactant; as detergent, emulsifier, wetting agent, dispersant,

stabilizer, defoamer; intermediate in synthesis of anionic surfactants; plasticizer, antistat for plastics; prep. of emulsified paints; food and pharmaceutical applics.; solvent cleaner
Regulatory: FDA 21CFR §175.105, 176.180, 176.210, 178.3400
Manuf./Distrib.: Aldrich; Fluka; Sigma
Trade names: T-Det® N-6

Nonoxynol-9 phosphate
CAS 51609-41-7 (generic); 51811-79-1; 66197-78-2; 68412-53-3; EINECS 266-231-6
Synonyms: PEG-9 nonyl phenyl ether phosphate; 26-(Nonylphenoxy)-3,6,9,12,15,18,21,24-octaoxahexacosan-1-ol, dihydrogenphosphate; POE (9) nonyl phenyl ether phosphate; PEG 450 nonyl phenyl ether phosphate
Definition: Complex mixture of esters of phosphoric acid and nonoxynol-9
Uses: Emulsifier, lubricant, antistat, detergent, corrosion inhibitor for agric., industrial use; solvent, antisoil redeposition for dry-cleaning; emulsion polymerization; household and industrial detergents; fabric finishes; paints, textile wet processing, metals
Regulatory: FDA 21CFR §175.105, 178.3400
Trade names: Schercophos NP-9

Nonylcarbinol. *See* n-Decyl alcohol
26-(Nonylphenoxy)-3,6,9,12,15,18,21,24-octaoxahexacosan-1-ol, dihydrogen phosphate. *See* Nonoxynol-9 phosphate
14-(Nonylphenoxy)-3,6,9,12-tetraoxatetradecan-1-ol. *See* Nonoxynol-5
Nopinen. *See* β-Pinene
Nopinene. *See* β-Pinene
Normal hexane. *See* Hexane
Normal pentane. *See* n-Pentane
normal Propyl alcohol. *See* n-Propyl alcohol
Normanthane. *See* Isopropylcyclohexane
1-NP. *See* Nitropropane
2-NP. *See* 2-Nitropropane

Oceol. *See* Oleyl alcohol
9,12-Octadecadienoic acid, 9-octadecenyl ester. *See* Oleyl linoleate
Octadecanoic acid butyl ester. *See* Butyl stearate
Octadecanoic acid, 1,2-ethanediyl ester. *See* Glycol distearate
Octadecanoic acid, methyl ester. *See* Methyl stearate
Octadecanoic acid, 1-methylethyl ester. *See* Isopropyl stearate
Octadecanoic acid monoester with 1,2,3-propanetriol. *See* Glyceryl stearate
Octadecanol. *See* Stearyl alcohol
1-Octadecanol. *See* Stearyl alcohol
n-Octadecanol. *See* Stearyl alcohol
9-Octadecenoic acid. *See* Oleic acid
cis-9-Octadecenoic acid. *See* Oleic acid
9-Octadecenoic acid, 12-(acetyloxy)-, 1,2,3-propanetriol ester. *See* Glyceryl triacetyl ricinoleate
9-Octadecenoic acid decyl ester. *See* Decyl oleate
9-Octadecenoic acid, diester with 1,2,3-propanetriol. *See* Glyceryl dioleate

9-Octadecenoic acid ethyl ester. *See* Ethyl oleate
9-Octadecenoic acid, isodecyl ester. *See* Isodecyl oleate
9-Octadecenoic acid, methyl ester. *See* Methyl oleate
9-Octadecenoic acid, monoester with 1,2,3-propanetriol. *See* Glyceryl oleate
9-Octadecenoic acid, 9-octadecenyl ester. *See* Oleyl oleate
9-Octadecen-1-ol. *See* Oleyl alcohol
cis-9-Octadecen-1-ol. *See* Oleyl alcohol
(Z)-9-Octadecen-1-ol. *See* Oleyl alcohol
9-Octadecenyl 13-docosenoate. *See* Oleyl erucate
9-Octadecenyl 9,12-octadecadienoate. *See* Oleyl linoleate
Octadecyl alcohol. *See* Stearyl alcohol
n-Octadecyl alcohol. *See* Stearyl alcohol
9,12-Octadienoic acid, lanolin alcohol ester. *See* Lanolin linoleate

Octane
CAS 111-65-9; EINECS 203-892-1
UN No. 1262 (DOT)
Synonyms: n-Octane
Classification: Sat. aliphatic hydrocarbon
Empirical: C_8H_{18}
Formula: $CH_3(CH_2)_6CH_3$
Properties: Colorless liq.; gasoline-like odor; completely misc. with most org. solvs.; insol. in water; sl. sol. in alcohol, ether; m.w. 114.23; sp.gr. 0.70 (20 C); vapor pressure 10 mm Hg (19 C); m.p. -56.7 C; b.p. 125-126 C; flash pt. 13 C
Toxicology: LD50 (IV, mouse) 428 mg/kg; ACGIH TLV/TWA 300 ppm; STEL 375 ppm; may cause skin irritation; conc. vapor is irritating to eyes; mild narcotic at high vapor concs.; may cause nose/throat irritation, headache, drowsiness, nausea, difficulty breathing
Precaution: Highly flamm.; severe explosion hazard exposed to heat, flame, oxidizers; incompat. with strong oxidizing agents (increases fire and explosion hazard); Hazardous decomp. prods.: Heated to decomp., emits acrid smoke and irritating fumes
NFPA: Health 0, Flammability 3, Reactivity 0
Uses: Solvent; component of high-octane motor and aviation fuel; ingred. of VM&P naphtha and other petrol. solvents; mfg. of benzene, toluene, xylene aromatics; azeotropic distillations; syn. flavoring for foods
Manuf./Distrib.: Aldrich; Fluka; Phibro Energy USA; Phillips; Sigma; Texaco; Triple Crown Am.
Trade names containing: Texsolve E

n-Octane. *See* Octane
1,3-Octanediol. *See* Ethyl hexanediol
Octanoic acid, 1-methyl-1,2-ethanediyl ester. *See* Propylene glycol dicaprylate
Octanoic acid, monoester with 1,2,3-propanetriol. *See* Glyceryl caprylate
Octanoic acid, 1,3-propanediyl ester. *See* Propylene glycol dioctanoate
Octanoic/decanoic acid triglyceride. *See* Caprylic/capric triglyceride
Octanol. *See* Caprylic alcohol
1-Octanol. *See* Caprylic alcohol

2-Octanol
CAS 123-96-6
UN No. 1987 (DOT)
Synonyms: 2-n-Octanol; sec-n-Octyl alcohol; Capryl alcohol; Hexyl methyl carbinol; Methyl hexyl carbinol; 2-Hydroxyoctane; Secondary caprylic alcohol
Classification: Sec. aliphatic alcohol
Empirical: $C_8H_{18}O$

2-n-Octanol

Formula: $CH_3(CH_2)_5CH(OH)CH_3$
Properties: Colorless clear oily liq.; pungent aromatic odor; sol. in ethanol, ether, acetone, aromatic and aliphatic hydrocarbons; pract. insol. in water; m.w. 130.23; sp.gr. 0.819 (20/4 C); m.p. -39 C; b.p. 178.5 C; flash pt. 76 C; ref. index 1.426 (20 C)
Toxicology: Toxic; skin and eye irritant; inh. may cause nose/throat irritation, CNS depression symptoms; ing. may cause 'alcohol' intoxication symptoms
Precaution: Combustible; incompat. with oxidizing agents (increases risk of fire or explosion)
NFPA: Health 1, Flammability 2, Reactivity 0
Uses: Intermediate in mfg. of plasticizers, disinfectant soaps; antifoaming agent; solvent for fats and waxes; in mfg. of wetting agents, hydraulic oils, petrol. additives; in lacquers, enamels; masking of industrial odors; syn. fragrance and flavoring agent for foods and pharmaceuticals
Regulatory: FDA 21CFR §172.515; FEMA GRAS
Manuf./Distrib.: Aldrich; Fluka; Henkel/Emery; Penta Mfg.; Sigma; Spectrum Chem. Mfg.

2-n-Octanol. *See* 2-Octanol
n-Octanol. *See* Caprylic alcohol
1-Octanol acetate. *See* Octyl acetate
2-Octanone. *See* Methyl hexyl ketone
n-Octanyl acetate. *See* Octyl acetate

Octyl acetate
CAS 112-14-1
Synonyms: 1-Octyl acetate; n-Octyl acetate; 2-Ethylhexyl acetate; 1-Octanol acetate; Octyl alcohol acetate; Acetate C-8; n-Octanyl acetate; Acetic acid octyl ester; Caprylyl acetate
Classification: Carboxylic acid ester
Empirical: $C_{10}H_{20}O_2$
Formula: $CH_3(CO_2(CH_2)_7CH_3$
Properties: Colorless liq.; floral-fruity odor; sol. in alcohol, ether; sl. sol. in water; m.w. 172.27; sp.gr. 0.868; m.p. -38.5 C; b.p. 199 C; flash pt. (CC) 86 C; ref. index 1.4180 (20 C)
Toxicology: LD50 (oral, rat) 3000 mg/kg; low oral toxicity; eye and sl. skin irritant; inh. of mist may cause nose/throat irritation; high concs. may cause CNS depression, drowsiness, headache, lightheadedness
Precaution: Incompat. with strong oxidizing agents (increases fire/explosion hazard), strong bases, strong acids (decomp. can occur); Hazardous decomp. prods.: n-Octyl alcohol, acetic acid; heated to decomp., emits acrid smoke and irritating fumes
Uses: Solvent; flavor and fragrance additive for foods and pharmaceuticals
Regulatory: FDA 21CFR §172.515; FEMA GRAS
Manuf./Distrib.: Aldrich; Hüls AG; Penta Mfg.; Sigma
See also 2-Ethylhexyl acetate

1-Octyl acetate. *See* Octyl acetate
n-Octyl acetate. *See* Octyl acetate
Octyl alcohol. *See* 2-Ethylhexanol
n-Octyl alcohol. *See* Caprylic alcohol
sec-n-Octyl alcohol. *See* 2-Octanol
Octyl alcohol acetate. *See* Octyl acetate
n-Octyldecyl alcohol. *See* Stearyl alcohol

Octyldodecanol
CAS 5333-42-6; EINECS 226-242-9
Synonyms: 2-Octyl dodecanol; 1-Dodecanol, 2-octyl-; Isoeicosyl alcohol
Classification: Aliphatic alcohol
Empirical: $C_{20}H_{42}O$
Formula: $CH_3(CH_2)_9CHCH_2OHCH_3(CH_2)_6CH_2$
Properties: Water-wh.clear liq.; sol. in alcohol, ether; insol. in water; m.w. 298; acid no. 0.5 max.; iodine no. 8 max.; sapon. no. 5 max.; hyd. no. 175-190; flash pt. (COC) 180 C
Uses: Carrier for oil-sol. prods.; lubricant, emollient for cosmetics and pharmaceuticals; pigment dispersant, solubilizer, coupling agent; intermediate for lubricants, emulsifiers, metal processing and textiles; pharmaceutical vehicle
Regulatory: FDA approved for topicals, vaginal creams; USP/NF compliance
Trade names: Standamul® G
Trade names containing: Argobase 125T

2-Octyl dodecanol. *See* Octyldodecanol
Octylene glycol. *See* Ethyl hexanediol
1,3-Octylene glycol. *See* Ethyl hexanediol

Octyl hydroxystearate
CAS 29383-26-4; 29710-25-6
Synonyms: 2-Ethylhexyl oxystearate; 12-Hydroxyoctadecanoic acid, 2-ethylhexyl ester; 2-Ethylhexylhydroxystearate
Definition: Ester of 2-ethylhexyl alcohol and 12-hydroxystearic acid
Empirical: $C_{26}H_{52}O_3$
Formula: $CH_3(CH_2)_5COHH(CH_2)_{10}COOCH_2CCH_2CH_3H(CH_2)_3CH_3$
Uses: Emollient, solubilizer, cosolvent for cosmetics, pharmaceuticals; refatting additive for soaps, cleansers
Trade names: Wickenol® 171

Octyl isononanoate
CAS 71566-49-9; EINECS 275-637-2
Synonyms: 2-Ethylhexyl isononanoate; Isononanoic acid, 2-ethylhexyl ester
Definition: Ester of 2-ethylhexyl alcohol and a branched chain nonanoic acid
Empirical: $C_{17}H_{34}O_2$
Formula: $CH_3(CH_3)_2CCH_2CH_3CHCH_2COOCH_2(CH_2CH_3)CH(CH_2)_3CH_3$
Properties: Clear water-white liq., typ. mild odor; sp.gr. 0.85; f.p. -34 C; b.p. 200 C; flash pt. (COC) 127 C
Toxicology: Nontoxic
Precaution: Store away from strong oxidizing agents
Uses: Emollient, lubricant, solvent; cosmetics, pharmaceutical topicals; resin plasticizer

Octyl phosphate. *See* Trioctyl phosphate
1-Octyl-2-pyrrolidinone. *See* N-Octyl pyrrolidone
N-Octyl pyrrolidinone. *See* N-Octyl pyrrolidone

N-Octyl pyrrolidone
CAS 7425-87-8
Synonyms: N-Octyl pyrrolidinone; 1-Octyl-2-pyrrolidinone; 2-Pyrrolidinone, 1-octyl-
Empirical: $C_{12}H_{23}NO$
Properties: M.w. 197.36
Toxicology: LD50 (oral, rat) 2050 mg/kg; primary irritant
Trade names: Gaftex® HI-8

N-Octyl-2-pyrrolidone. *See* Caprylyl pyrrolidone

Octyl sebacate. *See* Dioctyl sebacate
ODB. *See* o-Dichlorobenzene
ODCB. *See* o-Dichlorobenzene
Oil of niobe. *See* Methyl benzoate
Oil of Palma Christi. *See* Castor oil
Oil of pine. *See* Pine oil
Oils, lanolin. *See* Lanolin oil
Oils, rice bran. *See* Rice bran oil
Olea europaea oil. *See* Olive oil

Oleic acid
CAS 112-80-1; EINECS 204-007-1
Synonyms: cis-9-Octadecenoic acid; Red oil; Elaic acid; Oleinic acid; 9-Octadecenoic acid
Classification: Aliphatic carboxylic acid; unsaturated fatty acid
Empirical: $C_{18}H_{34}O_2$
Formula: $CH_3(CH_2)_7CH{:}CH(CH_2)_7COOH$
Properties: Colorless liq., odorless; darkens when exposed to oxygen; insol. in water; sol. in alcohol, ether, benzene, chloroform, fixed/volatile oils; m.w. 282.47; dens. 0.895 (25/25 C); m.p. 6 C; b.p. 286 C (100 mm); acid no. 196-204; flash pt. 100 C; ref. index 1.463 (18 C)
Toxicology: LD50 (oral, rat) 74 g/kg; poison by intravenous route; mildly toxic by ingestion; experimental tumorigen; irritant to skin, mucous membranes; heated to decomp., emits acrid smoke and irritating fumes
Precaution: Combustible when exposed to heat or flame; incompat. with Al and perchloric acid; light-sensitive
Uses: Soap base, mfg. of oleates, ointments, cosmetics, polishing compds., lubricants, food-grade additives, Turkey red oil, driers; intermediates for paints; waterproofing textiles; oiling wool; pharmaceutic aid, solvent; activator, plasticizer, softener, in rubbers; flume wash water additive for food processing
Regulatory: FDA 21CFR §172.210, 172.860, 172.862, 173.315 (0.1 ppm max. in wash water), 173.340, 175.105, 175.320, 176.170, 176.200, 176.210, 177.1010, 177.1200, 177.2260, 177.2600, 177.2800, 178.3570, 178.3910, 182.70, 182.90; FEMA GRAS; FDA approved for inhalants, orals, topicals; USP/NF, BP compliance
Manuf./Distrib.: Akzo Nobel; Aldrich; Allchem Ind.; Am. Biorganics; Anar; Arizona; Ashland; J.T. Baker; Baychem; Brown; Browning; R.E. Carroll; Climax Perf. Materials; Ferro/Keil; Fluka; C.P. Hall; Henkel/Emery; Hercules; Int'l. Chem. Inc; Penta Mfg.; Rhône-Poulenc N. Am.; Schweizerhall; Sigma; Spectrum Chem. Mfg.; Unichema Int'l.; Union Derivan SA; Van Waters & Rogers; Welch, Holme & Clark; Witco/Oleo-Surf.

Oleic acid, methyl ester, cis-. *See* Methyl oleate
Oleinic acid. *See* Oleic acid
Oleol. *See* Oleyl alcohol

Oleth-2
CAS 9004-98-2 (generic); 5274-65-7
Synonyms: PEG-2 oleyl ether; POE (2) oleyl ether; PEG 100 oleyl ether
Definition: PEG ether of oleyl alcohol
Formula: $CH_3(CH_2)_7CH{=}CH(CH_2)_7CH_2(OCH_2CH_2)_nOH$, avg. n = 2
Uses: Solubilizer, emulsifier, dispersant, stabilizer, spreading agent, emollient, lipophilic cosolvent for cosmetics, detergents, fragrance prods., textile, leather, paper, pharmaceutical applics.
Manuf./Distrib.: Aldrich; Fluka; Sigma
Trade names: Ameroxol® OE-2; Eumulgin® PWM2

Oleth-5

CAS 9004-98-2 (generic); 5353-27-5

Synonyms: PEG-5 oleyl ether; POE (5) oleyl ether; 3,6,9,12,15-Pentaoxatriacont-24-en-1-ol

Definition: PEG ether of oleyl alcohol

Empirical: $C_{28}H_{56}O_6$

Formula: $CH_3(CH_2)_7CH{=}CH(CH_2)_7CH_2(OCH_2CH_2)_nOH$, avg. n = 5

Uses: Emulsifier, lubricant, solubilizer for pesticides, cosmetics, pharmaceuticals, floor polishes; dispersant for paints

Manuf./Distrib.: Aldrich; Fluka; Sigma

Trade names containing: Solulan® 5

Oleth-16

CAS 9004-98-2 (generic); 25190-05-0 (generic)

Synonyms: PEG-16 oleyl ether; POE (16) oleyl ether

Definition: PEG ether of oleyl alcohol

Formula: $CH_3(CH_2)_7CH{=}CH(CH_2)_7CH_2(OCH_2CH_2)_nOH$, avg. n = 16

Uses: Emulsifier for cosmetics and pharmaceuticals

Regulatory: FDA 21CFR §176.200, 177.2800

Manuf./Distrib.: Aldrich; Fluka; Sigma

Trade names containing: Solulan® 16

Oleum abietis. *See* Pine oil

Oleum Gossypii seminis. *See* Cottonseed oil

Oleum olivae. *See* Olive oil

Oleum ricini. *See* Castor oil

Oleyl alcohol

CAS 143-28-2; EINECS 205-597-3

Synonyms: 9-Octadecen-1-ol; (Z)-9-Octadecen-1-ol; cis-9-Octadecen-1-ol; Oceol; Oleol

Classification: Unsaturated fatty alcohol

Empirical: $C_{18}H_{36}O$

Formula: $CH_3(CH_2)_7CH{=}CH(CH_2)_8OH$

Properties: Colorless to pale yel. oily visc. liq., faint char. odor, bland taste; insol. in water; sol. in alcohol, ether; m.w. 268.49; dens. 0.84; m.p. 13-19 C; b.p. 207 C (13 mm); acid no. 1 max.; iodine no. 85-95; hyd. no. 205-215; cloud pt. ≤ 10 C; flash pt. > 110 C; ref. index 1.4582 (27.5 C)

Toxicology: Human skin irritant; heated to decomp., emits acrid smoke and irritating fumes

Uses: Plasticizer, emulsion stabilizer, antifoam, coupler, cosolvent, pigment dispersant, detergent in cutting oils, inks, textile finishing, petrochem., pulp/paper, paints, plastics, food applics., pharmaceuticals, cosmetics; mfg. of sulfuric esters used as surfactants

Regulatory: FDA 21CFR §176.170, 176.210, 177.1010, 177.1210, 177.2800, 178.3910; FDA approved for topicals; USP/NF compliance

Manuf./Distrib.: Aldrich; Croda Inc; Croda Chem. Ltd; Fluka; R.W. Greeff; Henkel/Emery; Lanaetex Prods.; M. Michel; Rhône-Poulenc Surf. & Spec.; Ronsheim & Moore; Sigma; Witco/Oleo-Surf.

Trade names: Adol® 80; Adol® 85 NF; Fancol OA-95; Novol NF; Witcohol™ 350

Trade names containing: Witcohol™ 42; Witcohol™ 320; Witcohol™ 330; Witcohol™ 340

Oleyl erucate

CAS 17673-56-2; EINECS 241-654-9

Synonyms: 9-Octadecenyl 13-docosenoate; Erucic acid, oleyl ester; 13-Docosenoic

acid, 9-octadecenyl ester
Definition: Ester of oleyl alcohol and erucic acid
Empirical: $C_{40}H_{76}O_2$
Uses: Emollient for cosmetics, pharmaceuticals; jojoba oil substitute; pharmaceutical solubilizer, solvent, carrier/vehicle
Trade names: Dynacerin® 660

Oleyl linoleate
CAS 17673-59-3
Synonyms: Linoleic acid, oleyl ester; 9,12-Octadecadienoic acid, 9-octadecenyl ester; 9-Octadecenyl 9,12-octadecadienoate
Definition: Ester of oleyl alcohol and linoleic acid
Empirical: $C_{36}H_{66}O_2$
Trade names containing: Polylan®

Oleyl oleate
CAS 3687-45-4; EINECS 222-980-4
Synonyms: 9-Octadecenoic acid, 9-octadecenyl ester
Definition: Ester of oleyl alcohol and oleic acid
Empirical: $C_{36}H_{68}O_2$
Uses: Emollient for cosmetic and pharmaceutical formulations contributing luster, softness, and high degree of lubricity in skin and hair preparations; cosolvent and solubilizer in perfumes; lubricant for metalworking, wire drawing; sperm oil substitute
Regulatory: FDA approved for topicals
Manuf./Distrib.: Sigma
Trade names: Schercemol OLO

Olive oil
CAS 8001-25-0; EINECS 232-277-0
Synonyms: Olea europaea oil; Gomenoleo oil; Azeite; Oleum olivae
Definition: Fixed oil obtained from the ripe fruit of *Olea europaea*, contg. glycerides of oleic acid, palmitic acid, linoleic acid, stearic acid, and arachidic acid
Properties: Yel. to lt. grnsh. liq., pleasant odor, sl. char. taste; sl. sol. in alcohol; misc. with ether, chloroform, carbon disulfide; dens. 0.909-0.915 (25/25 C); m.p. -6 C; solid. pt. 17-26 C; iodine no. 79-88; sapon. no. 187-196; flash pt. (CC) 437 F; ref. index 1.466-1.468; becomes rancid on exposure to air
Toxicology: LD50 (IV, mouse) 1320 mg/kg; mod. toxic by IP route; human skin irritant; devoid of side effects
Precaution: Combustible exposed to heat or flame; can react with oxidizers; some spontaneous heating; photosensitive; Hazardous decomp. prods.: Heated to decomp., emits acrid smoke and irritating fumes
Uses: Lubricant, emollient in cosmetics, pharmaceuticals; pharmaceutical vehicle, solvent for ointments, liniments, soaps; salad dressing and other foods
Regulatory: FDA 21CFR §175.105, 176.200, 176.210, GRAS; Japan approved (olive); USP/NF, BP, Ph.Eur., JP compliance
Manuf./Distrib.: ABITEC; Aldrich; Arista Ind.; Charkit; Croda Inc; Croda Chem. Ltd; Fluka; Freedom Textile; Int'l. Chem. Inc; Penta Mfg.; Sigma; Spectrum Chem. Mfg.; Tri-K Ind.; Welch, Holme & Clark
Trade names: EmCon Olive; Super Refined® Olive Oil

Opalwax. *See* Hydrogenated castor oil
Optal. *See* n-Propyl alcohol
Orthodichlorobenzene. *See* o-Dichlorobenzene
Orthodichlorobenzol. *See* o-Dichlorobenzene
1-Oxa-4-azacyclohexane. *See* Morpholine

7-Oxabicyclo (2.2.1) heptane, 1-methyl-4-(1-methylethyl)-. *See* 1,4-Cineole
2-Oxabicyclo (2.2.2) octane, 1,3,3-trimethyl-. *See* Eucalyptol
Oxacyclopentane. *See* Tetrahydrofuran
3-Oxa-1-heptanol. *See* Butoxyethanol
3-Oxapentane-1,5-diol. *See* Diethylene glycol
3-Oxa-1,5-pentanediol. *See* Diethylene glycol
1,8-Oxido-p-menthane. *See* Eucalyptol
Oxirane, methyl, polymer and oxibane, butyl ether. *See* PPG-2-buteth-3
2-Oxobutane. *See* Methyl ethyl ketone
3-Oxobutanoic acid ethyl ester. *See* Ethylacetoacetate
3-Oxobutanoic acid methyl ester. *See* Methyl acetoacetate

Oxo-heptyl acetate
CAS 90438-79-2
Synonyms: C7 alkyl acetate
Properties: Sp.gr. 0.87; flash pt. (TCC) 150 F
Uses: Solvent for high solids, maintenance, marine, and industrial waterborne coatings, acrylic resin polymerization media, metalworking fluids, urethane foam and industrial cleaners and degreasers, paint strippers, pigment dispersions; pesticide emulsifiable concs., textile scouring aids, NC or screen printing inks, penetrating and lubricating oils, vinyl plastisols; good solvency, low surf. tens. and interfacial tens., good wetting, lubricity; visc. modifier; latex coalescer
Manuf./Distrib.: Chemcentral; Houghton Chem.
Trade names: Exxate® 700

Oxo-hexyl acetate
CAS 88230-35-7
Synonyms: C6 alkyl acetate
Properties: Sp.gr. 0.87; flash pt. (TCC) 134 F
Uses: Solvent for high solids, maintenance, marine, and industrial waterborne coatings, acrylic resin polymerization media, metalworking fluids, urethane foam and industrial cleaners and degreasers, paint strippers, pigment dispersions; pesticide emulsifiable concs., textile scouring aids, NC or screen printing inks, penetrating and lubricating oils, vinyl plastisols; good solvency, low surf. tens. and interfacial tens., good wetting, lubricity; visc. modifier; latex coalescer
Manuf./Distrib.: Houghton Chem.
Trade names: Exxate® 600

Oxolane. *See* Tetrahydrofuran
2-Oxopyrrolidine. *See* 2-Pyrrolidone
1,1´-Oxybis (butane). *See* Dibutyl ether
Oxybis-1,1´-ethane. *See* Ethyl ether
1,1´-(Oxybis(2,1-ethanediyloxy))bisbutane. *See* Diethylene glycol dibutyl ether
2,2´-[Oxybis(2,1-ethanediyloxy)]bisethanol. *See* PEG-4
2,2´-Oxybisethanol. *See* Diethylene glycol
1,1´-Oxybis(2-methoxyethane). *See* Diethylene glycol dimethyl ether
1,1´-Oxybis-2-propanol. *See* Dipropylene glycol
2,2´-Oxydiethanol. *See* Diethylene glycol
1,1´-Oxydi-2-propanol. *See* Dipropylene glycol

Palmitic acid, isobutyl ester. *See* Isobutyl palmitate
Palmityl acetate. *See* Cetyl acetate
Palmityl alcohol. *See* Cetyl alcohol
Parachlorobenzotrifluoride. *See* p-Chlorobenzotrifluoride

Paraffin, chlorinated
CAS 63449-39-8
Synonyms: Chlorocosane; Chlorcosane; Chlorinated paraffin; Paraffin waxes and hydrocarbon waxes, chlorinated
Definition: Paraffin hydrocarbons treated with chlorine; contains about 50% Cl
Properties: Pale to amber, neutral, light viscous oils or soft waxes; insol. in water; sl. sol. in alcohol; sol. in benzene, chloroform; dens. 0.900-1.50
Toxicology: Primary irritant
Precaution: Nonflamm.
Uses: In mfg. of textiles, draperies, etc.; solvent for dichloramine-T; high-pressure lubricants; flame retardants in plastics; as plasticizer for PVC in PE sealants; plasticizer in paints
Manuf./Distrib.: AC Ind.; Allchem Ind.; R.E. Carroll; Chemcentral; Dover; Ferro/Keil; C.P. Hall; Harwick; Hoechst AG; Morton Int'l.; Occidental; Punda Mercantile; Sea-Land; StanChem; Tri-Iso; Witco/Polymer Addit.

Paraffin oil. *See* Mineral oil
Paraffin waxes and hydrocarbon waxes, chlorinated. *See* Paraffin, chlorinated
PCBTF. *See* p-Chlorobenzotrifluoride
PDMS. *See* Dimethicone

Peanutamide MEA
Synonyms: N-(2-Hydroxyethyl) peanut acid amide; Monoethanolamine peanut acid amide; Peanut fatty acid monoethanolamide
Definition: Mixture of ethanolamides of fatty acids derived from peanut oil
Formula: $RCO\text{-}NHCH_2CH_2OH$, RCO- represents fatty acids derived from peanut oil
Uses: Emulsifier, detergent intermediate; solvent; specialty cosmetic prod.

Peanutamide MIPA
Synonyms: N-(2-Hydroxypropyl)peanut acid amide
Definition: Mixture of isopropanolamides of fatty acids derived from peanut oil
Formula: $RCO\text{-}NHCH_2CCH_3HOH$, RCO- represents fatty acids derived from peanut oil
Uses: Emulsifier, detergent intermediate; solvent; specialty cosmetic prod.

Peanut fatty acid monoethanolamide. *See* Peanutamide MEA

Peanut oil
CAS 8002-03-7; EINECS 232-296-4
Synonyms: Arachis oil; Groundnut oil; Indigenous peanut oil; Katchung oil; Earthnut oil; Pecan shell powder
Classification: Fixed oil
Definition: Refined fixed oil obtained from seed kernels of one or more cultivated varieties of *Arachis hypogaea*
Properties: Pale yel. liq., nutty odor, bland taste; sol. in benzene, ether, chloroform, CCl_4, oils; misc. with carbon disulfide; sl. sol. in alcohol; insol. in alkalies; dens. 0.916-0.922; solid. pt. -5 C; iodine no. 84-100; sapon. no. 185-195; flash pt. 540 F; ref. index 1.466-1.470
Toxicology: Experimental tumorigen; human skin irritant, mild allergen; mutagenic data
Precaution: Combustible exposed to heat or flame; can react with oxidizing materials; sl. spontaneous heating; light-sensitive; Hazardous decomp. prods.:

Heated to decomp., emits acrid smoke and irritating fumes
NFPA: Health 0, Flammability 1, Reactivity 0
Uses: Edible oil; substitute for olive oil; vehicle for medicine; in mfg. of margarine, soaps, paints; heat transfer medium in laboratory; dietary supplement; food coatings; pharmaceutical solvent, vehicle
Regulatory: FDA 21CFR §175.105, 176.200, 176.210, 177.2800, 182.70; GRAS; FDA approved for injecables, orals, vaginals; USP/NF, BP compliance
Manuf./Distrib.: ABITEC; Alnor Oil; Arista Ind.; Charkit; Croda Chem. Ltd; Penta Mfg.; Spectrum Chem. Mfg.; Sigma; Tri-K Ind.; Welch, Holme & Clark
Trade names: Super Refined® Peanut Oil

Pear oil. *See* Amyl acetate; Isoamyl acetate
Pecan shell powder. *See* Peanut oil
PEG. *See* Polyethylene glycol

PEG-4
CAS 25322-68-3 (generic); 112-60-7; EINECS 203-989-9
Synonyms: PEG 200; POE (4); 2,2´-[Oxybis(2,1-ethanediyloxy)]bisethanol
Definition: Polymer of ethylene oxide
Empirical: $C_8H_{18}O_5$
Formula: $H(OCH_2CH_2)_nOH$, avg. n = 4
Properties: Visc. liq., sl. char. odor; hygroscopic; m.w. 190-210; dens. 1.127 (25/25 C); visc. 4.3 cSt (210 F); supercools on freezing
Toxicology: LD50 (oral, rat) 28,900 mg/kg; mildly toxic by ingestion; heated to decomp., emits acrid smoke and irritating fumes
Precaution: Solvent action on some plastics
NFPA: Health 0, Flammability 1, Reactivity 0
Uses: Lubricant for rubber molds, textile fibers, metalworking; in food and food pkg.; in cosmetics and hair preps.; pharmaceutic aid; in gas chromatography; in paints, paper coatings, polishes, ceramics; food coating, binder, plasticizer, lubricant, defoamer; pharmaceutical solvent, vehicle, bulking agent
Regulatory: FDA 21CFR §73.1, 172.210, 172.770, 172.820, 173.310, 173.340, 175.105, 175.300, 178.3750; FDA approved for topicals
Manuf./Distrib.: Aldrich; Ashland; Fluka; C.P. Hall; Harwick; Henkel; Sigma; Spectrum Chem. Mfg.; Union Carbide
Trade names: Dow E200; Pluracol® E200; Teric PEG 200

PEG-6
CAS 25322-68-3 (generic); 2615-15-8; EINECS 220-045-1
Synonyms: Polyethylene glycol 300; PEG 300; Hexaethylene glycol; Macrogol 300
Definition: Polymer of ethylene oxide
Empirical: $C_{12}H_{26}O_7$
Formula: $H(OCH_2CH_2)_nOH$, avg. n = 6
Properties: Sp.gr. 1.124-1.127; visc. 5.4-6.4 cSt (99 C); pour pt. -15 to -8 C; flash pt. (COC) 196 C; ref. index 1.463-1.4641 (20 C); pH 4.5-7.5 (5%)
Toxicology: LD50 (oral, rat) 27,500 mg/kg; mildly toxic by ingestion; heated to decomp., emits acrid smoke and irritating fumes
NFPA: Health 0, Flammability 1, Reactivity 0
Uses: Lubricant for rubber molds, textile fibers, metalworking; in food and food pkg.; in cosmetics and hair preps.; pharmaceutic aid; in gas chromatography; in paints, paper coatings, polishes, ceramics; plasticizer for cellulose nitrate; food coating, binder, plasticizer, lubricant, flavor adjuvant, bodying agent, defoamer; pharmaceutical coating, suppository base, lubricant, solvent
Regulatory: FDA 21CFR §172.210, 172.770, 172.820, 173.310, 173.340, 175.105, 175.300, 178.3750, 178.3910; FDA approved for parenterals, ophthalmics, topicals; USP/NF, BP compliance

Manuf./Distrib.: Aldrich; Ashland; Fluka; Harwick; Henkel; Sigma; Spectrum Chem. Mfg.; Union Carbide
Trade names: Carbowax® PEG 300; Carbowax® Sentry® PEG 300; Dow E300; Dow E300 NF; Emery® 6687; Lutrol® E 300; Pluracol® E300; Renex® PEG 300; Rhodasurf® PEG 300; Teric PEG 300
Trade names containing: Carbowax® Sentry® PEG 540 Blend; Pluracol® E1500

PEG-8
CAS 25322-68-3 (generic); 5117-19-1; EINECS 225-856-4
Synonyms: Macrogol 400; Polyethylene glycol 400; PEG 400; POE (8); 3,6,9,12,15,18,21-Heptaoxatricosane-1,2,3-diol
Definition: Polymer of ethylene oxide
Empirical: $C_{16}H_{34}O_9$
Formula: $H(OCH_2CH_2)_nOH$, avg. n = 8
Properties: Visc. liq., sl. char. odor, minimal taste; sl. hygroscopic; m.w. 380-420; dens. 1.128 (25/25 C); visc. 7.3 cSt (210 F); m.p. 4-8 C; pH 4.5-7.5 (5%)
Toxicology: LD50 (rat, oral) 30 ml/kg; low toxicity by ingestion, IV, IP routes; heated to decomp., emits acrid smoke and irritating fumes
NFPA: Health 0, Flammability 1, Reactivity 0
Uses: Lubricant for rubber molds, textile fibers, metalworking; in food and food pkg.; in cosmetics and hair preps.; pharmaceutic aid; in gas chromatography; in paints, paper coatings, polishes, ceramics; plasticizer for cellulose nitrate; food coating, binder, plasticizer, lubricant, flavor adjuvant, bodying agent, defoamer; pharmaceutical coating, suppository base, bulking agent, solvent
Regulatory: FDA 21CFR §172.210, 172.770, 172.820, 173.310, 173.340, 175.105, 175.300, 178.3750, 178.3910, 181.22, 181.30; USP/NF, BP, Ph.Eur. compliance
Manuf./Distrib.: Aldrich; Ashland; Fluka; C.P. Hall; Harwick; Henkel; Sigma; Spectrum Chem. Mfg.; Union Carbide
Trade names: Carbowax® Sentry® PEG 400; Dow E400; Dow E400 NF; Emery® 6709; Lutrol® E 400; Pluracol® E400; Pluracol® E400 NF; Renex® PEG 400; Teric PEG 400

PEG-9
CAS 25322-68-3 (generic); 3386-18-3; EINECS 222-206-1
Synonyms: PEG 450; POE (9)
Definition: Polymer of ethylene oxide
Empirical: $C_{18}H_{38}O_{10}$
Formula: $H(OCH_2CH_2)_nOH$, avg. n = 9
Uses: Lubricant for rubber molds, textile fibers, metalworking; in food and food pkg.; in cosmetics and hair preps.; pharmaceutic aid; in gas chromatography; in paints, paper coatings, polishes, ceramics; food coating, binder, plasticizer, lubricant, flavor adjuvant, bodying agent, defoamer; pharmaceutical bulking agent, solvent
Regulatory: FDA 21CFR §172.210, 172.770, 172.820, 173.310, 173.340, 175.105, 175.300, 178.3750, 178.3910
Manuf./Distrib.: Aldrich; Fluka; Sigma
Trade names: Rhodasurf® PEG 400

PEG-12
CAS 25322-68-3 (generic); 6790-09-6; EINECS 229-859-1
Synonyms: Polyethylene glycol 600; PEG 600; POE (12); Macrogol 600
Definition: Polymer of ethylene oxide
Empirical: $C_{24}H_{50}O_{13}$
Formula: $H(OCH_2CH_2)_nOH$, avg. n = 12
Properties: Visc. liq., char. odor; sl. hygroscopic; m.w. 570-630; dens. 1.128 (25/25 C); m.p. 20-25 C; visc. 10.5 cSt (210 F); pH 4.5-7.5 (5%)

Toxicology: LD50 (oral, rat) 38,100 mg/kg; low toxicity by ingestion; eye irritant; heated to decomp., emits acrid smoke and irritating fumes
NFPA: Health 0, Flammability 1, Reactivity 0
Uses: Lubricant for rubber molds, textile fibers, metalworking; in food and food pkg.; in cosmetics and hair preps.; pharmaceutic aid; in gas chromatography; in paints, paper coatings, polishes, ceramics; plasticizer for cellulose nitrate; food coating, binder, plasticizer, lubricant, flavor adjuvant, bodying agent, defoamer; pharmaceutical coating, suppository base, bulking agent, solvent
Regulatory: FDA 21CFR §172.210, 172.770, 172.820, 173.310, 173.340, 175.105, 175.300, 178.3750, 178.3910; USP/NF compliance
Manuf./Distrib.: Aldrich; Ashland; Fluka; C.P. Hall; Harwick; Sigma; Spectrum Chem. Mfg.; Union Carbide
Trade names: Carbowax® Sentry® PEG 600; Dow E600; Dow E600 NF; Emery® 6686; Pluracol® E600; Pluracol® E600 NF; Renex® PEG 600

PEG-14
CAS 25322-68-3 (generic)
Synonyms: POE (14); PEG 700
Definition: Polymer of ethylene oxide
Empirical: $C_{28}H_{58}O_{15}$
Formula: $H(OCH_2CH_2)_nOH$, avg. n = 14
Properties: Visc. 11.5-13.0 cSt
Uses: Lubricant for rubber molds, textile fibers, metalworking; in food and food pkg.; in cosmetics and hair preps.; pharmaceutic aid; in gas chromatography; in paints, paper coatings, polishes, ceramics; food coating, binder, plasticizer, lubricant, flavor adjuvant, bodying agent, defoamer; pharmaceutical bulking agent, solvent
Regulatory: FDA 21CFR §172.210, 172.770, 172.820, 173.310, 173.340, 175.105, 175.300, 178.3750, 178.3910
Manuf./Distrib.: Aldrich; Fluka; Sigma
Trade names: Rhodasurf® PEG 600

PEG-16
CAS 25322-68-3 (generic)
Synonyms: PEG 800; POE (16)
Definition: Polymer of ethylene oxide
Empirical: $C_{32}H_{66}O_{17}$
Formula: $H(OCH_2CH_2)_nOH$, avg. n = 16
Properties: Visc. 12.5-14.5 cSt
Uses: Lubricant for rubber molds, textile fibers, metalworking; in food and food pkg.; in cosmetics and hair preps.; pharmaceutic aid; in gas chromatography; in paints, paper coatings, polishes, ceramics; food coating, binder, plasticizer, lubricant, flavor adjuvant, bodying agent, defoamer; pharmaceutical bulking agent, solvent
Regulatory: FDA 21CFR §172.210, 172.770, 172.820, 173.310, 173.340, 175.105, 175.300, 178.3750, 178.3910
Manuf./Distrib.: Aldrich; Fluka; Sigma

PEG-20
CAS 25322-68-3 (generic); EINECS 203-989-9
Synonyms: Polyethylene glycol 1000; PEG 1000; Macrogol 1000; POE (20)
Definition: Polymer of ethylene oxide
Empirical: $C_{40}H_{82}O_{21}$
Formula: $H(OCH_2CH_2)_nOH$, avg. n = 20
Properties: Solid; sp.gr. 1.085; visc. 16-19 cSt (99 C); pour pt. 37-40 C; flash pt. (COC) 265 C; pH 4.5-7.5 (5%)

Toxicology: LD50 (oral, rat) 42 g/kg; mod. toxic by IP, IV routes; mildly toxic by ingestion; experimental tumorigen; heated to decomp., emits acrid smoke and irritating fumes
NFPA: Health 0, Flammability 1, Reactivity 0
Uses: Lubricant for rubber molds, textile fibers, metalworking; in food and food pkg.; in cosmetics and hair preps.; pharmaceutic aid; in gas chromatography; in paints, paper coatings, polishes, ceramics; plasticizer for cellulose nitrate; food coating, binder, plasticizer, lubricant, flavor adjuvant, bodying agent, defoamer; pharmaceutical coating, ointment and suppository base, bulking agent, solvent, plasticizer, lubricant
Regulatory: FDA 21CFR §172.210, 172.770, 172.820, 173.310, 173.340, 175.105, 175.300, 178.3750, 178.3910; USP/NF, BP, Ph.Eur. compliance
Manuf./Distrib.: Aldrich; Ashland; Fluka; Harwick; Sigma; Spectrum Chem. Mfg.; Union Carbide
Trade names: Carbowax® Sentry® PEG 900; Carbowax® Sentry® PEG 1000; Dow E1000; Dow E1000 NF; Renex® PEG 1000

PEG-32
CAS 25322-68-3 (generic); EINECS 203-989-9
Synonyms: PEG 1540; Macrogol 1540; POE (32)
Definition: Polymer of ethylene oxide
Empirical: $C_{64}H_{130}O_{33}$
Formula: $H(OCH_2CH_2)_nOH$, avg. n = 32
Properties: Wh. powd.
Toxicology: LD50 (oral, rat) 44,200 mg/kg; mildly toxic by ingestion; human skin irritant; heated to decomp., emits acrid smoke and irritating fumes
Uses: Lubricant for rubber molds, textile fibers, metalworking; in food and food pkg.; in cosmetics and hair preps.; pharmaceutic aid; in gas chromatography; in paints, paper coatings, polishes, ceramics; food coating, binder, plasticizer, lubricant, flavor adjuvant, bodying agent, defoamer; pharmaceutical coating, ointment base, bulking agent; solvent, solubilizer, emulsifier
Regulatory: FDA 21CFR §172.210, 172.770, 172.820, 173.310, 173.340, 175.105, 175.300, 178.3750, 178.3910; FDA approved for dentals, orals, rectals, topicals; BP compliance
Manuf./Distrib.: Aldrich; Fluka; Sigma
Trade names: Carbowax® Sentry® PEG 1450; Dow E1450; Dow E1450 NF; Lutrol® E 1500; Pluracol® E1450 NF
Trade names containing: Carbowax® Sentry® PEG 540 Blend; Pluracol® E1500

PEG-75
CAS 25322-68-3 (generic); EINECS 203-989-9
Synonyms: Polyethylene glycol 3350; PEG 3350; POE (75)
Definition: Polymer of ethylene oxide
Empirical: $C_{150}H_{302}O_{76}$
Formula: $H(OCH_2CH_2)_nOH$, avg. n = 75
Properties: Wh. powd. or creamy-white flakes; m.w. 3000-3700; dens. 1.212 (25/25 C); visc. 76-110 cSt (210 F); m.p. 54-58 C; pH 4.5-7.5 (5%)
Toxicology: LD50 (oral, rat) 50 g/kg; mildly toxic by ingestion; skin irritant; heated to decomp., emits acrid smoke and irritating fumes
NFPA: Health 0, Flammability 1, Reactivity 0
Uses: Lubricant for rubber molds, textile fibers, metalworking; in food and food pkg.; in cosmetics and hair preps.; pharmaceutic aid; in gas chromatography; in paints, paper coatings, polishes, ceramics; food coating, binder, plasticizer, lubricant, flavor adjuvant, bodying agent, defoamer; pharmaceutical coating, ointment and suppository base, bulking agent, solvent, plasticizer

Regulatory: FDA 21CFR §172.210, 172.770, 172.820, 173.310, 173.340, 175.105, 175.300, 178.3750, 178.3910; FDA approved for injectables, orals, rectals, topicals, vaginals; USP/NF compliance
Manuf./Distrib.: Aldrich; Fluka; Sigma; Spectrum Chem. Mfg.
Trade names: Carbowax® Sentry® PEG 3350; Dow E3350; Dow E3350 NF; Lutrol® E 4000; Pluracol® E4000

PEG-100
CAS 25322-68-3 (generic); EINECS 203-989-9
Synonyms: PEG (100); POE (100)
Definition: Polymer of ethylene oxide
Empirical: $C_{200}H_{402}O_{101}$
Formula: $H(OCH_2CH_2)_nOH$, avg. n = 100
Uses: Lubricant for rubber molds, textile fibers, metalworking; in food and food pkg.; in cosmetics and hair preps.; pharmaceutic aid; in gas chromatography; in paints, paper coatings, polishes, ceramics; food coating, binder, plasticizer, lubricant, flavor adjuvant, bodying agent, defoamer; pharmaceutical coating, ointment base, bulking agent, tablet binder; solvent, vehicle
Regulatory: FDA 21CFR §172.210, 172.770, 172.820, 173.310, 173.340, 175.105, 175.300, 178.3750, 178.3910
Manuf./Distrib.: Aldrich; Fluka; Sigma
Trade names: Carbowax® Sentry® PEG 4600; Dow E4500; Dow E4500 NF

PEG (100). *See* PEG-100

PEG-150
CAS 25322-68-3 (generic); EINECS 203-989-9
Synonyms: PEG 6000; Macrogol 6000; POE (150)
Definition: Polymer of ethylene oxide
Formula: $H(OCH_2CH_2)_nOH$, avg. n = 150
Properties: Powd. or creamy-white flakes; water-sol.; m.w. 7000-9000; dens. 1.21 (25/25 C); visc. 470-900 cSt (210 F); m.p. 56-63 C; flash pt. > 887 F
Toxicology: LD50 (oral, rat) > 50 g/kg; mildly toxic by ingestion; mutagenic data; skin irritant; heated to decomp., emits acrid smoke and irritating fumes
Precaution: Combustible exposed to heat or flame; NPFA; Health 0, Flammability 1, Reactivity 0
Uses: Lubricant for rubber molds, textile fibers, metalworking; in food and food pkg.; in cosmetics and hair preps.; pharmaceutic aid; in gas chromatography; in paints, paper coatings, polishes, ceramics; food coating, binder, plasticizer, lubricant, flavor adjuvant, bodying agent, defoamer; pharmaceutical coating, ointment and suppository base, bulking agent, solvent, plasticizer
Regulatory: FDA 21CFR §172.210, 172.770, 172.820, 173.310, 173.340, 175.300, 177.2420, 178.3750, 178.3910
Manuf./Distrib.: Aldrich; Fluka; Sigma; Spectrum Chem. Mfg.
Trade names: Carbowax® Sentry® PEG 8000; Dow E8000; Dow E8000 NF; Lutrol® E 6000; Pluracol® E6000

PEG 200. *See* PEG-4
PEG 300. *See* PEG-6
PEG 400. *See* PEG-8
PEG 450. *See* PEG-9
PEG 600. *See* PEG-12
PEG 700. *See* PEG-14
PEG 800. *See* PEG-16
PEG 1000. *See* PEG-20
PEG 1540. *See* PEG-32

PEG 3350. *See* PEG-75
PEG 6000. *See* PEG-150

PEG-6 abietate
Synonyms: POE (6) abietate; PEG 300 abietate
Uses: Additive for cutting oils; solvent, emulsifier for solvents, creams, cleansing milks and pesticides; dispersant
Trade names: Secoster® MA 300

PEG 300 abietate. *See* PEG-6 abietate

PEG-13 betanaphthol ether
CAS 35545-57-4
Uses: Emollient, emulsifier, visc. control agent, lubricant, pigment dispersant, perfume solubilizer, used in cosmetics, household, textile industry, metal treating and plating; intermediate
Trade names: Hetoxide BN-13

PEG-2 butyl ether. *See* Butoxydiglycol
PEG-3 butyl ether. *See* Butoxytriglycol

PEG-4 butyl ether
Synonyms: Tetraethylene glycol butyl ether; Tetraethylene glycol monobutyl ether
Trade names containing: Icinol BE33

PEG 400 caprylate/caprate glycerides. *See* PEG-8 caprylic/capric glycerides

PEG-8 caprylic/capric glycerides
CAS 57307-99-0; 85536-08-9
Synonyms: PEG 400 caprylate/caprate glycerides
Definition: PEG deriv. of a mixt. of mono, di, and triglycerides of caprylic and capric acids with an avg. of 8 moles of ethylene oxide
Uses: Hydrophilic oil; nonionic surfactant for drug delivery systems; bioavailability enhancer, emulsifier, solv., solubilizer for pharmaceuticals
Trade names: Labrafac® CM10; Labrasol

PEG-12 castor oil
CAS 61791-12-6 (generic)
Synonyms: POE (12) castor oil
Definition: PEG deriv. of castor oil with avg. 12 moles of ethylene oxide
Uses: Detergent, emulsifier, coemulsifier in sol. oils, solvent cleaners, coatings, industrial/institutional cleaners, metalworking; mold release in plastics; textile fiber lubricant
Manuf./Distrib.: Fluka; Sigma

PEG-33 castor oil
CAS 61791-12-6 (generic)
Synonyms: POE (33) castor oil
Definition: PEG deriv. of castor oil with avg. 33 moles of ethylene oxide
Uses: Solvent, emulsifier for pharmaceuticals
Regulatory: FDA 21CFR §175.105, 175.300, 1716.210, 177.2800
Manuf./Distrib.: Fluka; Sigma
Trade names: Ricinion

PEG-35 castor oil
CAS 61791-12-6 (generic)
Synonyms: POE (35) castor oil; Polyoxyl 35 castor oil
Definition: PEG deriv. of castor oil with avg. 35 moles of ethylene oxide
Properties: Yel. oily liq., faint char. odor, sl. bitter taste; very sol. in water; sol. in

alcohol, ethyl acetate; insol. in min. oils; sp.gr. 1.105-1.06; visc. 650-850 cps; acid no. 2 max.; iodine no. 25-35; sapon. no. 60-76; hyd. no. 65-80
Toxicology: May cause a hepatitis-like reaction in large doses
Uses: Solubilizer, emulsifier used for essential oils, pharmaceuticals, cosmetics, veterinary medicine; pharmaceutical solvent
Regulatory: FDA 21CFR §175.105, 175.300, 177.2800; USP/NF compliance
Manuf./Distrib.: Fluka; Sigma
Trade names: Cremophor® EL; Incrocas 35

PEG-5 cetyl ether. *See* Ceteth-5
PEG-16 cetyl ether. *See* Ceteth-16
PEG-24 cetyl ether. *See* Ceteth-24
PEG-2 cetyl/oleyl ether. *See* Cetoleth-2
PEG-3 cetyl/oleyl ether. *See* Cetoleth-3
PEG-6 cetyl/oleyl ether. *See* Cetoleth-6
PEG-24 cholesteryl ether. *See* Choleth-24
PEG-6 decyl ether. *See* Deceth-6
PEG 300 decyl ether. *See* Deceth-6

PEG-4 dilaurate
CAS 9005-02-1 (generic)
Synonyms: PEG 200 dilaurate; POE (4) dilaurate; Polyoxyl 4 dilaurate
Definition: PEG diester of lauric acid
Empirical: $C_{32}H_{62}O_7$
Formula: $CH_3(CH_2)_{10}CO(OCH_2CH_2)_nOCO(CH_2)_{10}CH_3$, avg. n = 4
Uses: Surfactant used as emulsifier, dispersant, solubilizer, visc. control agent, defoamer, lubricant, wetting agent, cosolvent for cosmetics, pharmaceuticals, plastics, textiles, metalworking, agric., food
Regulatory: FDA 21CFR §175.105, 175.300, 176.170, 176.180, 176.200, 176.210; FDA approved for topicals
Trade names: Cithrol 2DL

PEG-8 dilaurate
CAS 9005-02-1 (generic)
Synonyms: POE (8) dilaurate; PEG 400 dilaurate
Definition: PEG diester of lauric acid
Empirical: $C_{40}H_{78}O_{11}$
Formula: $CH_3(CH_2)_{10}CO(OCH_2CH_2)_nOCO(CH_2)_{10}CH_3$, avg. n = 8
Properties: Sp.gr. 1.030; m.p. 15 C; flash pt. (COC) 249 C; ref. index 1.459
Uses: Emulsifier, solubilizer, cosolvent, dispersing agent, wetting agent, lubricant, softener, release agent, coupling agent used in personal care prods., pharmaceutical, industrial applics., agric. chemical sprays, industrial and textile lubricants; plasticizer
Regulatory: FDA 21CFR §175.105, 175.300, 176.210, 177.1210, 177.2260, 177.2800, 178.3520
Manuf./Distrib.: C.P. Hall; Inolex; Velsicol
Trade names: Cithrol 4DL

PEG-12 dilaurate
CAS 9005-02-1 (generic)
Synonyms: POE (12) dilaurate; PEG 600 dilaurate
Definition: PEG diester of lauric acid
Empirical: $C_{48}H_{94}O_{15}$
Formula: $CH_3(CH_2)_{10}CO(OCH_2CH_2)_nOCO(CH_2)_{10}CH_3$, avg. n = 12
Uses: Dispersant and emulsifier for metalworking lubricants, fiber lubricants and softeners, solubilizers, defoamers, antistats, cosmetics, pharmaceuticals, plas-

tics, agric., and chemical intermediates; cosolvent
Regulatory: FDA 21CFR §175.105, 175.300, 176.210, 177.2260, 177.2800
Trade names: Cithrol 6DL

PEG-20 dilaurate
CAS 9005-02-1 (generic)
Synonyms: POE (20) dilaurate; PEG 1000 dilaurate
Definition: PEG diester of lauric acid
Empirical: $C_{64}H_{126}O_{23}$
Formula: $CH_3(CH_2)_{10}CO(OCH_2CH_2)_nOCO(CH_2)_{10}CH_3$, avg. n = 20
Uses: Dispersant, emulsifier, wetting agent, cosolvent, solubilizer, thickener, emollient, opacifier, spreading agent; used in cosmetic, pharmaceutical, food, agric., plastic, other industries
Regulatory: FDA 21CFR §175.300, 176.210, 177.2260, 177.2800
Trade names: Cithrol 10DL

PEG 200 dilaurate. *See* PEG-4 dilaurate
PEG 400 dilaurate. *See* PEG-8 dilaurate
PEG 600 dilaurate. *See* PEG-12 dilaurate
PEG 1000 dilaurate. *See* PEG-20 dilaurate

PEG-3 dimethyl ether
CAS 112-49-2; 24991-55-7
Synonyms: Triethylene glycol dimethyl ether; 2,5,8,11-Tetraoxadodecane; Triglyme; Glyme-3
Empirical: $C_{18}H_{18}O_4$
Properties: Liq.; misc. with water, hydrocarbon solvs.; m.w. 178.26; dens. 0.990 (20/4 C); m.p. -45 C; b.p. 215 C; flash pt. 111 C; ref. index 1.4233
Toxicology: TDLo (oral, mouse) 7500 mg/kg, (oral, rabbit) 3500 mg/kg; reproductive effects
Precaution: Flamm. exposed to heat or flame; Hazardous decomp. prods.: Heated to decomp., emits acrid smoke and irritating fumes
Uses: Complexing agent, solvent; in electrochem., polymer/boron chem., gas absorp., extraction, stabilization, fuels, lubricants, textiles, pharmaceuticals, pesticides
Manuf./Distrib.: Aldrich; Brand-Nu Labs; Fluka; Hoechst AG; Sigma
Trade names: Triglyme

PEG-4 dimethyl ether
CAS 24991-55-7 (generic)
Uses: Complexing agent; pharmaceutical solvent, drug carrier, stabilizer
Manuf./Distrib.: Aldrich; Fluka
Trade names: Tetraglyme

PEG-4 dioleate
CAS 9005-07-6 (generic); 52688-97-0 (generic); 134141-38-1
Synonyms: POE (4) dioleate; PEG 200 dioleate
Definition: PEG diester of oleic acid
Toxicology: LD50 (oral, rat) 1900 mg/kg
Uses: Surfactant; coemulsifier, thickener, solubilizer, cosolvent, emollient, opacifier, spreading agent, wetting agent, dispersant for cosmetics, pharmaceuticals, food, agric., plastics, etc.
Regulatory: FDA 21CFR §173.340, 175.105, 175.300, 176.210
Trade names: Cithrol 2DO

PEG-8 dioleate
CAS 9005-07-6 (generic); 52688-97-0 (generic)
Synonyms: POE (8) dioleate; PEG 400 dioleate

Definition: PEG diester of oleic acid
Toxicology: LD50 (oral, rat) 1900 mg/kg
Uses: Emulsifier, solubilizer, cosolvent, lubricant, dispersant, defoamer, anticorrosive, wetting agent for cosmetic, pharmaceutical, textile, metalworking, agric., plastics, foods; defoamer for beet sugar and yeast processing
Regulatory: FDA 21CFR §173.340, 175.105, 175.300, 176.170, 176.200, 176.210, 177.1210, 177.2260, 177.2800
Trade names: Cithrol 4DO; Emerest® 2648

PEG-12 dioleate
CAS 9005-07-6 (generic); 52688-97-0 (generic); 85736-49-8; EINECS 288-459-5
Synonyms: POE (12) dioleate; PEG 600 dioleate
Definition: PEG diester of oleic acid
Toxicology: LD50 (oral, rat) 1900 mg/kg
Uses: Emulsifier, dispersant; additive for cutting oils; component of paper defoamers, softeners, lubricants; cosmetics, pharmaceuticals, foods, agric., plastics, solv. and solventless coatings; lubricant for textile spin finishes; defoamer for beet sugar and yeast processing; cosolvent
Regulatory: FDA 21CFR §173.340, 175.105, 175.300, 176.200, 176.210, 177.2260, 177.2800
Trade names: Cithrol 6DO; Secoster® DO 600

PEG-20 dioleate
CAS 9005-07-6 (generic); 52688-97-0 (generic)
Synonyms: POE (20) dioleate; PEG 1000 dioleate
Definition: PEG diester of oleic acid
Toxicology: LD50 (oral, rat) 1900 mg/kg
Uses: Surfactant, solubilizer, cosolvent, thickener, emollient, opacifier, spreading agent, wetting agent, dispersant for cosmetic, pharmaceutical, perfume, food, agric., plastic, other industries
Regulatory: FDA 21CFR §175.300, 176.210, 177.2260, 177.2800
Trade names: Cithrol 10DO

PEG 200 dioleate. *See* PEG-4 dioleate
PEG 400 dioleate. *See* PEG-8 dioleate
PEG 600 dioleate. *See* PEG-12 dioleate
PEG 1000 dioleate. *See* PEG-20 dioleate
PEG-7 dodecyl phenyl ether. *See* Dodoxynol-7
PEG-3 ethyl ether. *See* Ethoxytriglycol
PEG-4 ethyl ether. *See* Tetraethylene glycol ethyl ether
PEG-7 glyceryl ether triacetate. *See* Glycereth-7 triacetate

PEG-15 hydroxystearate
Synonyms: POE (15) hydroxystearate
Definition: PEG ester of hydroxystearic acid
Formula: $CH_3(CH_2)_5CHOH(CH_2)_{10}CO(OCH_2CH_2)_nOH$, avg. n = 15
Uses: Solvent for injection sol'ns.
Trade names: G-4972; Solutol® HS 15

PEG-6 isostearate
CAS 56002-14-3 (generic)
Synonyms: PEG 300 monoisostearate; POE (6) monoisostearate
Definition: PEG ester of isostearic acid
Formula: $C_{17}H_{35}CO(OCH_2CH_2)_nOH$, avg. n = 6
Uses: Pharmaceutical solvent, emulsifier
Trade names: Olepal ISO

PEG-2 lactamide
Synonyms: Diethylene glycol lactamide
Classification: Amide
Formula: $CH_3CHOHCONH(CH_2CH_2O)_nH$, avg. n = 2
Uses: Coupling agent, lubricant, emollient, humectant for personal care prods.; solvent in cosmetic creams and lotions
Trade names: Naetex-L

PEG-3 lanolate
CAS 68459-50-7 (generic)
Synonyms: POE (3) lanolate
Definition: PEG ester of lanolin acid
Formula: $RCO(OCH_2CH_2)_nOH$, RCO^- represents fatty acids from lanolin, avg. n = 3
Uses: Emulsifier, solvent
Trade names: Agnosol 3

PEG-5 lanolate
CAS 68459-50-7 (generic)
Synonyms: PEG-5 lanolin acids; POE (5) lanolate
Definition: PEG ester of lanolin acid
Formula: $RCO\text{-}(OCH_2CH_2)_nOH$, RCO- rep. fatty acids from lanolin, avg. n = 5
Uses: O/w emulsifier, solvent for bromo acid dyes in lipsticks, skin care and makeup prods., solubilizer, wetting agent, dispersant
Trade names: Agnosol 5; Lanpol 5; Ritalafa® 5

PEG-10 lanolate
CAS 68459-50-7 (generic)
Synonyms: POE (10) lanolate; PEG 500 lanolate; PEG-10 lanolin acid
Definition: PEG ester of lanolin acid
Formula: $RCO\text{-}(OCH_2CH_2)_nOH$, RCO- rep. fatty acids from lanolin, avg. n = 10
Uses: O/w emulsifier, solvent for bromo acid dyes in lipsticks, skin care and makeup prods., solubilizer, wetting agent, dispersant
Trade names: Ritalafa® 10

PEG-20 lanolate
CAS 68459-50-7 (generic)
Synonyms: POE (20) lanolate; PEG 1000 lanolate
Definition: PEG ester of lanolin acid
Formula: $RCO\text{-}(OCH_2CH_2)_nOH$, RCO- rep. fatty acids from lanolin, avg. n = 20
Uses: O/w emulsifier, solvent for bromo acid dyes in lipsticks, skin care and makeup prods., solubilizer, wetting agent, dispersant

PEG 500 lanolate. *See* PEG-10 lanolate
PEG 1000 lanolate. *See* PEG-20 lanolate

PEG-75 lanolin
CAS 8039-09-6; 61790-81-6 (generic)
Synonyms: POE (75) lanolin; PEG 4000 lanolin; Polyoxyl 75 lanolin
Definition: PEG deriv. of lanolin with avg. 75 moles of ethylene oxide
Uses: Emulsifier for cosmetic and pharmaceutical emulsions; solvent, emollient, superfatting agent, conditioner for skin and hair care prods., household detergents; solubilizer, wetting agent, dispersing aid
Regulatory: FDA approved for topicals
Trade names: Solulan® 75; Solulan® L-575

PEG 4000 lanolin. *See* PEG-75 lanolin
PEG-10 lanolin acid. *See* PEG-10 lanolate

PEG-5 lanolin acids. *See* PEG-5 lanolate
PEG-5 lanolin alcohol. *See* Laneth-5
PEG-5 lanolin ether. *See* Laneth-5
PEG-10 lanolin ether. *See* Laneth-10
PEG-15 lanolin ether. *See* Laneth-15
PEG-16 lanolin ether. *See* Laneth-16
PEG 500 lanolin ether. *See* Laneth-10
PEG-3 methyl ether. *See* Triglycol monomethyl ether
PEG-4 methyl ether. *See* Tetraethylene glycol methyl ether
PEG-5 methyl ether. *See* Pentaethylene glycol methyl ether

PEG-6 methyl ether
CAS 9004-74-4 (generic)
Synonyms: POE (6) methyl ether; PEG 300 methyl ether; PEG-6 monomethyl ether
Definition: PEG ether of methyl alcohol
Formula: $CH_3(OCH_2CH_2)_nOH$, avg. n = 6
Uses: Intermediate, lubricant for adhesives, inks, mining, soaps, detergents; provides solvency, low odor, low irritation, low toxicity to personal care prods.
Manuf./Distrib.: Aldrich; Fluka; Sigma
Trade names: Dow MPEG350; Upiwax 350

PEG-10 methyl ether
CAS 9004-74-4 (generic)
Synonyms: POE (10) methyl ether; PEG 500 methyl ether
Definition: PEG ether of methyl alcohol
Formula: $CH_3(OCH_2CH_2)_nOH$, avg. n = 10
Uses: Intermediate, lubricant for adhesives, inks, mining, soaps and detergents; provides solvency, low odor, low irritation, low toxicity to personal care prods.
Manuf./Distrib.: Aldrich; Fluka; Sigma
Trade names: Dow MPEG550

PEG-16 methyl ether
CAS 9004-74-4 (generic)
Synonyms: POE (16) methyl ether
Definition: PEG ether of methyl alcohol
Formula: $CH_3(OCH_2CH_2)_nOH$, avg. n = 16
Uses: Chemical intermediate, lubricant for adhesives; provides solvency, low odor, low irritation, low toxicity to personal care prods.
Manuf./Distrib.: Aldrich; Fluka; Sigma
Trade names: Dow MPEG750

PEG 300 methyl ether. *See* PEG-6 methyl ether
PEG 500 methyl ether. *See* PEG-10 methyl ether
PEG-10 methyl glucose ether. *See* Methyl gluceth-10
PEG-20 methyl glucose ether. *See* Methyl gluceth-20
PEG 300 monoisostearate. *See* PEG-6 isostearate
PEG-6 monomethyl ether. *See* PEG-6 methyl ether
PEG 200 monooleate. *See* PEG-4 oleate
PEG 300 monooleate. *See* PEG-6 oleate
PEG 400 monooleate. *See* PEG-8 oleate
PEG 1000 monooleate. *See* PEG-20 oleate
PEG-4 nonyl phenyl ether. *See* Nonoxynol-4
PEG-5 nonyl phenyl ether. *See* Nonoxynol-5
PEG-6 nonyl phenyl ether. *See* Nonoxynol-6
PEG 200 nonyl phenyl ether. *See* Nonoxynol-4
PEG 300 nonyl phenyl ether. *See* Nonoxynol-6

PEG-9 nonyl phenyl ether phosphate. *See* Nonoxynol-9 phosphate
PEG 450 nonyl phenyl ether phosphate. *See* Nonoxynol-9 phosphate

PEG-4 oleate
CAS 9004-96-0 (generic); 10108-25-5; EINECS 233-293-0
Synonyms: POE (4) monooleate; PEG 200 monooleate
Definition: PEG ester of oleic acid
Empirical: $C_{26}H_{50}O_6$
Formula: $CH_3(CH_2)_7CHCH(CH_2)_7CO(OCH_2CH_2)_nOH$, avg. n = 4
Uses: Wetting agent, penetrant, spreading agent, defoamer, detergent, emulsifier, solubilizer, thickening agent, dispersant, textile aux., softener, lubricant for textiles, cosmetics, metalworking, food, agric., plastics; solvent in metal cleaning, degreasing
Regulatory: FDA 21CFR §175.105, 175.300, 176.210
Trade names: Emerest® 2624

PEG-5 oleate
CAS 9004-96-0 (generic); 23336-36-9
Synonyms: POE (5) monooleate
Definition: PEG ester of oleic acid
Formula: $CH_3(CH_2)_7CHCH(CH_2)_7CO(OCH_2CH_2)_nOH$, avg. n = 5
Uses: Surfactant in cutting oils, degreasing solvents, metal cleaners; emulsifier; dyeing assistant; cosmetics, textile processing
Regulatory: FDA 21CFR §175.105, 175.300
Trade names: Chemax E-200 MO

PEG-6 oleate
CAS 9004-96-0 (generic); 60344-26-5
Synonyms: POE (6) monooleate; PEG 300 monooleate
Definition: PEG ester of oleic acid
Empirical: $C_{30}H_{58}O_8$
Formula: $CH_3(CH_2)_7CHCH(CH_2)_7CO(OCH_2CH_2)_nOH$, avg. n = 6
Uses: Emulsifier, lubricant, chemical intermediate, antifoam, dispersant; for cosmetics, food, agriculture, plastics, textiles; solvent for lipids
Regulatory: FDA 21CFR §175.105, 175.300, 176.210
Trade names: Olepal 1

PEG-8 oleate
CAS 9004-96-0 (generic)
Synonyms: POE (8) monooleate; PEG 400 monooleate
Definition: PEG ester of oleic acid
Formula: $CH_3(CH_2)_7CHCH(CH_2)_7CO(OCH_2CH_2)_nOH$, avg. n = 8
Properties: Dk. red oil; sol. in alcohol; disp. in water; misc. with cottonseed oil
Uses: Emulsifier, dispersant, lubricant, chemical intermediate, solubilizer, solvent, visc. control agent; for cosmetics, pharmaceuticals, food, agric., plastics, coatings
Regulatory: FDA 21CFR §175.105, 175.300, 176.170, 176.200, 177.1200, 177.1210, 177.2260, 177.2800
Trade names: Secoster® MO 400

PEG-20 oleate
CAS 9004-96-0 (generic)
Synonyms: POE (20) monooleate; PEG 1000 monooleate
Definition: PEG ester of oleic acid
Formula: $CH_3(CH_2)_7CHCH(CH_2)_7CO(OCH_2CH_2)_nOH$, avg. n = 20
Toxicology: LD50 (IV, mouse) 500 mg/kg; mod. toxic by IV route; skin and eye irritant; heated to decomp., emits acrid smoke and irritating fumes

Uses: Raw material for finishing agents in the syn. fiber industry; nonionic surfactant, emulsifier, thickener, solubilizer, solvent, emollient, opacifier, wetting agent, dispersant for cosmetics, pharmaceuticals, food, agric., plastics, paints, etc.
Regulatory: FDA 21CFR §175.300, 176.200, 177.2260, 177.2800
Trade names: Chemax E-1000 MO

PEG-2 oleyl ether. *See* Oleth-2
PEG-5 oleyl ether. *See* Oleth-5
PEG-16 oleyl ether. *See* Oleth-16
PEG 100 oleyl ether. *See* Oleth-2

PEG-8 palmitostearate
Uses: Pharmaceutical solvent, emulsifier
Trade names: Stearate 400 WL 817

PEG-150 palmitostearate
Uses: Pharmaceutical solvent, emulsifier
Trade names: Stearate 6000 WL 1644

PEG/PPG-17/6 copolymer
CAS 9003-11-6 (generic)
Definition: Random copolymer produced by interaction of 17 moles ethylene oxide with 6 moles of propylene oxide
Uses: Lubricant for antiperspirants; emollient, solvent, bodying agent; lubricant for textile spin finishes; intermediate for prep. of resins, plasticizers, modifiers, and surfactants; demulsifier
Manuf./Distrib.: Aldrich; Fluka; Sigma
Trade names: Ucon® 75-H-450

PEG/PPG-24/24 glycerin. *See* PPG-24-glycereth-24

PEG-20 sorbitan isostearate
CAS 66794-58-9 (generic)
Synonyms: PEG 1000 sorbitan monoisostearate; POE (20) sorbitan monoisostearate; Polysorbate 120
Definition: Ethoxylated sorbitan monoester of isostearic acid with avg. 20 moles ethylene oxide
Uses: Emulsifier, solubilizer for cosmetics; emulsifier, solvent, wetting agent for pharmaceuticals
Trade names: Crillet 6

PEG 1000 sorbitan monoisostearate. *See* PEG-20 sorbitan isostearate
PEG-20 sorbitan oleate. *See* Polysorbate 80
PEG-20 sorbitan stearate. *See* Polysorbate 60
PEG-1 stearate. *See* Glycol stearate
PEG-5 stearyl ether. *See* Steareth-5
PEG-16 stearyl ether. *See* Steareth-16

PEG-3 triethanolamine
CAS 36936-60-4
Uses: Solvent neutralizer for acid dyes
Manuf./Distrib.: Aldrich
Trade names: Emery® 5794

1,4,7,10,13-Pentaazatridecane. *See* Tetraethylenepentamine
Pent-acetate. *See* Amyl acetate

3-(n-Penta-8′-decenyl) phenol
CAS 8007-24-7

3-(n-Pentadecyl) phenol

Formula: $C_6H_4 \cdot OH \cdot C_{15}H_{29}$
Uses: Raw material for surfactants, antioxidants, anticorrosives; lubricant additive; resin modifier; cosolvent for insecticides
Trade names: Cardolite® NC-511

3-(n-Pentadecyl) phenol
CAS 501-24-6; 3158-56-3
Empirical: $C_{21}H_{36}O$
Formula: $CH_3(CH_2)_{14}C_6H_4OH$
Properties: m.w. 304.52; m.p. 50-53 C; b.p. 190-195 C (1 mm); flash pt. > 110 C
Uses: Raw material for surfactants, antioxidants, anticorrosives; lubricant additive; cosolvent for insecticides, germicides; resin modifier; coupling agent
Manuf./Distrib.: Aldrich; Sigma
Trade names: Cardolite® NC-507; Cardolite® NC-510

Pentaethylene glycol methyl ether
Synonyms: Pentaethylene glycol monomethyl ether; PEG-5 methyl ether
Trade names containing: Icinol ME45

Pentaethylene glycol monomethyl ether. *See* Pentaethylene glycol methyl ether
Pentamethylene. *See* Cyclopentane
Pentamethyleneimine. *See* Piperidine
1-Pentanamine. *See* Pentylamine

n-Pentane
CAS 109-66-0; EINECS 203-692-4
UN No. 1265 (DOT)
Synonyms: Amyl hydride; Alkane C_5; Normal pentane
Classification: Aliphatic hydrocarbon
Empirical: C_5H_{12}
Formula: $CH_3(CH_2)_3CH_3$
Properties: Colorless liq.; gasoline-like odor; sol. in alcohol and most org. solvs.; insol. in water; m.w. 72.17; sp.gr. 0.62624; vapor pressure 400 mm Hg (18.5 C); f.p. -129.7 C; b.p. 36 C; flash pt. (CC) -49 C
Toxicology: LD50 (IV, mouse) 446 mg/kg; ACGIH TLV/TWA 600 ppm; STEL 750 ppm; narcotic in high concs. (dizziness, headache); extreme exposure can cause nausea, confusion, persistent gasoline taste, unconsciousness; lethal @ 130,000 ppm
Precaution: Highly flamm.; severe explosion hazard exposed to heat or flame; explosive limits 1.4-8%; shock can shatter metal containers; incompat. with strong oxidizing agents (increases fire and explosion hazard); Hazardous decomp. prods.: Heated to decomp., emits acrid smoke and irritating fumes
NFPA: Health 1, Flammability 4, Reactivity 0
Uses: Solvent; solvent extraction; reagent; artificial ice mfg.; low-temp. thermometers; solvent extract processes; blowing agent in plastics (expandable polystyrene); pesticides
Manuf./Distrib.: Aldrich; Ashland; Burdick & Jackson; Fluka; Phillips; Sigma; Spectrum Chem. Mfg.; Triple Crown Am.

1,3-Pentanediamine
CAS 589-37-7
UN No. 2734
Synonyms: DAMP; 1,3-Diaminopentane
Empirical: $C_5H_{14}N_2$
Formula: $H_2NCH_2CH_2CH(C_2H_5)NH_2$
Properties: Liq., sl. odor; misc. with water; m.w. 102; dens. 0.855 g/ml; visc. 1.89 cp; vapor pressure 1 mm Hg (20 C); b.p. 164 C; flash pt. (CC) 59 C; surf. tens.

32.2 dyne/cm
Precaution: Corrosive, flamm.; incompat. with strong oxidants; Hazardous decomp. prods.: Thermal decomp. may emit NO_x
Uses: Epoxy curative; solvent for gas treatment; surfactant (asphalt emulsifiers, textiles); scale/corrosion inhibitor; additive for fuel, oil, plastic, petrol. and chem. processing; extender, catalyst for PU; in coatings, adhesives, sealants, elastomers
Manuf./Distrib.: Aldrich
Trade names: Dytek® EP

1,3-Pentanediol, 2,2,4-trimethyl-, monoisobutyrate. *See* 2,2,4-Trimethyl-1,3-pentanediol monoisobutyrate
Pentanedione-2,4. *See* Acetylacetone
2,4-Pentanedione. *See* Acetylacetone
Pentane, dodecafluoro-. *See* Perfluoropentane
Pentanol-2. *See* s-Amyl alcohol
Pentanol-3. *See* 3-Pentanol
1-Pentanol. *See* n-Amyl alcohol
Pentan-1-ol. *See* n-Amyl alcohol
2-Pentanol. *See* s-Amyl alcohol
Pentan-2-ol. *See* s-Amyl alcohol

3-Pentanol
CAS 584-02-1; EINECS 209-526-7
UN No. 1105
Synonyms: Diethyl carbinol; Pentanol-3; Isoamyl alcohol; Pentan-3-ol
Classification: Sec. aliphatic alcohol
Empirical: $C_5H_{12}O$
Formula: $(C_2H_5)_2CHOH$
Properties: Colorless visc. liq.; char. odor; sl. sol. in water; sol. in alcohol, ether; m.w. 88.15; sp.gr. 0.819 (20/4 C); m.p. < -75 C; b.p. 114-116 C; flash pt. (CC) 41 C; ref. index 1.410
Toxicology: LD50 (oral, rat) 1.87 g/kg, (skin, rabbit) 2520 mg/kg; ACGIH TLV/TWA 100 ppm; STEL 125 ppm; irritant to eyes, nose, and throat; severe eye irritant, sl. skin irritant; inh. may cause dizziness, coughing, nausea, vomiting
Precaution: Flamm.; dangerous fire and explosion hazard exposed to heat, flame, oxidizers; incompat. with oxidizing agents (increases fire/explosion hazard), hydrogen trisulfide (explosive reaction possible); Hazardous decomp. prods.: Heated to decomp., emits acrid smoke and irritating fumes
NFPA: Health 1, Flammability 2, Reactivity 0
Uses: As flotation agent, solvent, in organic synthesis, mfg. of pharmaceuticals
Manuf./Distrib.: Aceto; Aldrich; Fluka; Hüls AG; Penta Mfg.

Pentan-3-ol. *See* 3-Pentanol
n-Pentanol. *See* n-Amyl alcohol
t-Pentanol. *See* t-Amyl alcohol
1-Pentanol acetate. *See* Amyl acetate
2-Pentanol acetate. *See* s-Amyl acetate
2-Pentanol, 4-methyl-. *See* Methyl amyl alcohol
2-Pentanol, 4-methyl-, acetate. *See* Methyl amyl acetate
Pentanone-3. *See* Diethyl ketone
2-Pentanone. *See* Methyl propyl ketone
3-Pentanone. *See* Diethyl ketone
2-Pentanone, 4-hydroxy-4-methyl-. *See* Diacetone alcohol
2-Pentanone, 4-methoxy-4-methyl-. *See* 4-Methoxy-4-methylpentanone-2
3,6,9,12,15-Pentaoxahentriacontan-1-ol. *See* Ceteth-5

3,6,9,12,15-Pentaoxatriacont-24-en-1-ol. *See* Oleth-5
Pentoxane. *See* 4-Methoxy-4-methylpentanone-2
Pentyl acetate. *See* Amyl acetate
1-Pentyl acetate. *See* Amyl acetate
2-Pentyl acetate. *See* s-Amyl acetate
n-Pentyl acetate. *See* Amyl acetate
Pentyl alcohol. *See* n-Amyl alcohol
n-Pentyl alcohol. *See* n-Amyl alcohol
sec-Pentyl alcohol. *See* s-Amyl alcohol
t-Pentyl alcohol. *See* t-Amyl alcohol

Pentylamine
CAS 110-58-7; EINECS 203-780-2
UN No. 1106
Synonyms: Amylamine; n-Amylamine; Monoamylamine; Amine C_5; n-Pentylamine; 1-Pentylamine; 1-Pentanamine; 1-Aminopentane
Classification: Sat. aliphatic amine
Empirical: $C_5H_{13}N$
Formula: $CH_3(CH_2)_4NH_2$
Properties: Colorless liq.; ammoniacal odor; very sol. in water, ethanol, acetone, ether, benzene; sol. in dimethylsulfoxide; sp.gr. 0.753 (20/4 C); m.p. -55 C; b.p. 102-104 C; flash pt. (CC) -1 C; ref. index 1.412 (20 C); mod. strong base
Toxicology: corrosive, toxic; can cause burns of skin; can be absorbed thru skin in toxic amts.; eye irritant; possible permanent eye injury, 'halo vision'; inh. of vapor can cause nose/throat irritation, coughing, chest pain, difficulty breathing
Precaution: Highly flamm.; incompat. with oxidizing agents, acids, acid chlorides, acid anhydrides, calcium or sodium hypochlorite, nitromethane, nitrosyl perchlorate; violent/explosive reactions possible
NFPA: Health 2, Flammability 3, Reactivity 0
Uses: Prod. of dyes, rubber chems., topical pharmaceuticals, insecticides, syn. detergents, other chems.; flotation agent; solvent; corrosion inhibitor; gasoline additive
Regulatory: FDA approved for topicals
Manuf./Distrib.: Aldrich; BASF; Elf Atochem N. Am.; Fluka; Sigma

1-Pentylamine. *See* Pentylamine
n-Pentylamine. *See* Pentylamine
Pentylcarbinol. *See* Hexyl alcohol
3-Pentylcarbinol. *See* 2-Ethyl-1-butanol
sec-Pentylcarbinol. *See* 2-Ethyl-1-butanol
n-Pentyl methyl ketone. *See* Methyl n-amyl ketone
N-Pentyl-1-pentanamine. *See* Di-n-amylamine
Pentyl propanoate. *See* n-Pentyl propionate

n-Pentyl propionate
CAS 624-54-4
Synonyms: Propanoic acid, pentyl ester; Pentyl propanoate; Amyl propanoate
Empirical: $C_8H_{16}O_2$
Formula: $CH_3(CH_2)_4OCOCH_2CH_3$
Properties: Insol. in water; m.w. 144.12; sp.gr. 0.874 (20/20 C); vapor pressure 1.5 mm Hg (20 C); b.p. 164.9 C; flash pt. (CC) 135 F
Toxicology: LD50 (oral, rat) > 14 g/kg, (skin, rabbit) > 14 g/kg
Uses: Solvent for high solids coatings, cellulosic and urethane paints
Manuf./Distrib.: Aldrich; Penta Mfg.; Union Carbide
Trade names: Ucar® n-Pentyl Propionate

m-Pentynol. *See* Methyl pentynol

Perchloroethylene

CAS 127-18-4; EINECS 204-825-9

UN No. 1897 (DOT)

Synonyms: Perc; Perchloroethylene; Tetrachlorethylene; Tetrachloroethylene; Tetrachloroethene; Ethylene tetrachloride

Classification: Chlorinated hydrocarbon

Empirical: C_2Cl_4

Formula: $Cl_2C=CCl_2$

Properties: Colorless liq., strong chloroform-like odor; misc. with alcohol, ether, chloroform, benzene, oil; insol. in water; m.w. 165.82; sp.gr. 1.6311 (15/4 C); m.p. -23 C; b.p. 121 C; ref. index 1.5055 (20 C); flash pt. none; KB value 90; nonflamm.

Toxicology: ACGIH TLV:TWA 25 ppm; STEL 100 ppm; LD50 (oral, mice) 8.85 g/kg; poison by IV route; mod. toxic by inh.; narcotic in high concs.; vapor inh. affects CNS; severe acute/chronic exposure causes liver damage

Environmental: HAPS contg.

Precaution: Incompat. with bases, metals (aluminum, barium, beryllium, lithium, magnesium, potassium), liq. oxygen, dinitrogen tetraoxide; Hazardous decomp. prods.: Phosgene

NFPA: Health 2, Flammability 0, Reactivity 0

Uses: Dry-cleaning solvent; degreasing solvent; heat exchange fluid; vermifuge; drying agent; degreasing metals; scouring, sizing, and desizing in textile applics.

Regulatory: SARA reportable

Manuf./Distrib.: Aldrich; Allchem Ind.; Asahi-Penn; Ashland; Coyne; Cross World Sales; Elf Atochem SA; Fluka; General Chem.; Houghton Chem.; ICI Am.; Occidental; PPG Ind.; Primachem; Sigma; Spectrum Chem. Mfg.; Universal Preserv-A-Chem; Vulcan

Trade names: PPG Perchlor

Trade names containing: Oxsol® 73; Oxsol® 253; Oxsol® 325

Perchloroethylene. *See* Perchloroethylene

Perchloromethane. *See* Carbon tetrachloride

Perfluoroacetic acid. *See* Trifluoroacetic acid

Perfluoroheptane

CAS 335-57-9; EINECS 206-392-1

Synonyms: Hexadecafluoroheptane; Perfluoro-n-heptane

Empirical: C_7F_{16}

Formula: $CF_3(CF_2)_5CF_3$

Properties: M.w. 388.07; dens. 1.745 (20/4 C); b.p. 82-84 C

Toxicology: LDLo (IP, mouse) 100 mg/kg; poison by IP route; heated to decomp., emits very toxic fumes of F^-

Uses: Dispersant for lubricants, mold release agents, protective coatings; reaction media for polymerizations, purification, separation processes; solvent for removal of halogenated lubricants, oils, greases; carrier for halogenated material; zero ozone depletion potential

Manuf./Distrib.: Aldrich; Fluka

Trade names: PF-5070

Perfluoro-n-heptane. *See* Perfluoroheptane

Perfluorohexane

CAS 355-42-0; EINECS 206-585-0

Synonyms: Tetradecafluorohexane; n-Tetradecafluorohexane; Perfluoro-n-hexane

Perfluoro-n-hexane

Empirical: C_6F_{14}
Formula: $CF_3(CF_2)_4CF_3$
Properties: M.w. 338.04; dens. 1.684 (20/4 C); b.p. 54-58 C; ref. index 1.2520
Toxicology: LC (inh., rat, 6 h) > 20,000 ppm
Uses: Dispersant for lubricants, mold release agents, protective coatings; reaction media for polymerizations, purification, separation processes; solvent for removal of halogenated lubricants, oils, greases; carrier for halogenated material; for quick-cooling or freezing foods; zero ozone depletion potential
Regulatory: FDA 21CFR §173.342
Manuf./Distrib.: Aldrich; Fluka
Trade names: PF-5060

Perfluoro-n-hexane. *See* Perfluorohexane

Perfluoro-N-methylmorpholine
Uses: Dispersant for lubricants, mold release agents, protective coatings; reaction media for polymerizations, purification, separation processes; solvent for removal of halogenated lubricants, oils, greases; carrier for halogenated material; zero ozone depletion potential
Trade names: PF-5052

Perfluoropentane
CAS 678-26-2
Synonyms: Dodecafluoropentane; Perfluoro-n-pentane; Pentane, dodecafluoro-
Empirical: C_5F_{12}
Properties: M.w. 288.04
Toxicology: LC50 (inh., rat) 120 g/m^3
Uses: Dispersant for lubricants, mold release agents, protective coatings; reaction media for polymerizations, purification, separation processes; solvent for removal of halogenated lubricants, oils, greases; carrier for halogenated material; zero ozone depletion potential
Trade names: PF-5050

Perfluoro-n-pentane. *See* Perfluoropentane
Perhydronaphthalene. *See* Decahydronaphthalene
Petrohol. *See* Isopropyl alcohol
Petrol. *See* Gasoline

Petrolatum
CAS 8009-03-8 (NF); 8027-32-5 (USP); EINECS 232-373-2
Synonyms: Petroleum jelly; White soft paraffin; Petrolatum amber; Vaseline; Petrolatum white; White petrolatum
Classification: Petroleum hydrocarbons
Definition: Semisolid mixture of hydrocarbons obtained from petroleum
Properties: Ylsh. to lt. amber or white semisolid, unctuous mass; pract. odorless and tasteless; sol. in benzene, chloroform, ether, petrol. ether, oils; pract. insol. in water; dens. 0.820-0.865 (60/25 C); m.p. 38-54 C; ref. index 1.460-1.474 (60 C)
Toxicology: Can cause allergic skin reactions in hypersensitive persons; generally nontoxic; heated to decomp., emits acrid smoke and irritating fumes
Uses: Laxative; dispersant; as ointment base, emollient, solvent in pharmaceuticals and cosmetics; leather grease; shoe polish; rust preventives; plastics, rubber, textiles lubricant; defoamer in beet sugar and yeast processing; protective food coatings; release agent/lubricant in bakery prods., confectionery
Regulatory: FDA 21CFR §172.880, 172.884, 173.340, 175.105, 175.125, 175.176, 175.300, 176.170, 176.200, 176.210, 177.2600, 177.2800, 1787.3570, 178.3700, 178.3910, 573.720; FDA approved for ophthalmics, orals, topicals, otics; USP/NF, BP compliance

Manuf./Distrib.: Barco Chem. Prods.; Exxon; Harcros USA; Magie Bros. Oil; Mobil; Penreco; Sea-Land; Spectrum Chem. Mfg.; Stevenson Cooper; Witco/Oleo-Surf.; Witco/Petrol. Spec.
Trade names: Penreco Snow

Petrolatum amber. *See* Petrolatum
Petrolatum liquid. *See* Mineral oil
Petrolatum white. *See* Petrolatum
Petroleum benzin. *See* Naphtha

Petroleum distillates
CAS 8002-05-9; 64742-14-9; 64742-47-8; EINECS 232-298-5
UN No. 1268 (DOT)
Classification: Petroleum hydrocarbons
Definition: Mixture of volatile hydrocarbons obtained from petroleum
Toxicology: Human systemic effects by parenteral route (cough, dyspnea, nausea, vomiting); skin contact may cause defatting; vapor inh. can cause CNS depression, headache, unconsciousness
Precaution: Combustible exposed to heat or flame; Hazardous decomp. prods.: Heated to decomp., emits acrid smoke and irritating fumes
Uses: Hydrocarbon processing solvent, foam control agent, in waterless hand cleaners, agric. sprays, polishes, fruit and veg. processing, cleaning oils; vehicle for pesticides
Trade names: Penreco 2251 Oil; Penreco 2263 Oil

Petroleum ether. *See* Naphtha
Petroleum jelly. *See* Petrolatum
Petroleum naphtha. *See* Naphtha
Petroleum spirit. *See* Naphtha
Petroleum spirits. *See* Mineral spirits
PGEE. *See* Propylene glycol ethyl ether
PGME. *See* Methoxyisopropanol
PGMEA. *See* Propylene glycol methyl ether acetate
Phene. *See* Benzene
Phenethanol. *See* Phenethyl alcohol
β-Phenethanol. *See* Phenethyl alcohol

Phenethyl alcohol
CAS 60-12-8; EINECS 200-456-2
Synonyms: Benzeneethanol; Benzyl carbinol; Phenethanol; 2-Phenylethanol; β-Phenylethanol; β-Phenethanol; Phenylethyl alcohol; β-Phenylethyl alcohol; 2-Phenylethyl alcohol
Classification: Aromatic alcohol
Empirical: $C_8H_{10}O$
Formula: $C_6H_5CH_2CH_2OH$
Properties: Colorless liq., floral rose odor, burning taste; misc. with alcohol, ether; sol. in fixed oils, glycerin, propylene glycol; sl. sol. in water; m.w. 122.18; dens. 1.0245 (15 C); m.p. -27 C; b.p. 220 C; flash pt. 102 C; ref. index 1.532 (20 C)
Toxicology: LD50 (oral, rat) 1790 mg/kg; poison by ingestion, IP routes; mod. toxic by skin contact; skin and eye irritant; experimental teratogenic effects; severe CNS injury
Precaution: Combustible when exposed to heat or flame; reactive with oxidizing materials; Hazardous decomp. prods.: Heated to decomp., emits acrid smoke and irritating fumes
NFPA: Health 1, Flammability 1, Reactivity 0
Uses: Antiseptic for topical use; solvent (pharmaceuticals); local anesthetic;

fragrance and flavoring for foods and pharmaceuticals; antimicrobial preservative for pharmaceuticals
Regulatory: FDA 21CFR §172.515, CIR approved, EPA reg.; FEMA GRAS; JSCI listed; FDA approved for ophthalmics, otics, topicals, vaginals; USP/NF compliance
Manuf./Distrib.: Aldrich; ARCO; BASF; Fluka; Penta Mfg.; Polarome Mfg.; Sigma; Spectrum Chem. Mfg.

Phenolcarbinol. *See* Benzyl alcohol

Phenoxydiglycol
Uses: Solvent for resins, in metal cleaners, paint strippers, cleaning compds., as ink vehicle
Trade names containing: Igepal® OD-410

Phenoxyethanol
CAS 122-99-6; EINECS 204-589-7
Synonyms: 2-Phenoxyethanol; Phenylglycol; Ethylene glycol monophenyl ether; Ethylene glycol phenyl ether; Phenoxytol
Classification: Aromatic ether alcohol
Definition: Phenol polyglycol ether
Empirical: $C_8H_{10}O_2$
Formula: $C_6H_5OCH_2CH_2OH$
Properties: Clear liq., faint aromatic odor, burning taste; sl. sol. in water; sol. in alcohol, ether, NaOH sol'ns.; m.w. 138.18; dens. 1.1094 (20/20 C); m.p. 14 C; b.p. 242 C; flash pt. 121 C; ref. index 1.534 (20 C)
Toxicology: LD50 (rat, oral) 1.26 g/kg; moderately toxic by ingestion and skin contact; skin and severe eye irritant
NFPA: Health 0, Flammability 1, Reactivity 0
Uses: Solvent for resins, dyes, inks, in organic synthesis; perfume fixative, bactericidal agent, plasticizer, germicide, as insect repellent; coalescent in adhesives, water-based architectural and industrial coatings; plasticizer for cellulosics; antimicrobial preservative for pharmaceuticals
Regulatory: FDA 21CFR §175.105; CIR approved, EPA reg.; JSCI listed 1.0% max.; Europe listed 1% max.; BP, Ph.Eur. compliance
Manuf./Distrib.: Aldrich; Amber Syn.; Ashland; Fluka; Jan Dekker BV; Hüls AG; Oxiteno; Penta Mfg.; Sigma; Spectrum Chem. Mfg.; Tri-K Ind.
Trade names: Dowanol® EPh; Emery® 6705; Rewopal® MPG 10; Rewopal® MPG 40
Trade names containing: Igepal® OD-410

2-Phenoxyethanol. *See* Phenoxyethanol

Phenoxyglycol
Trade names containing: Igepal® OD-410

1-Phenoxypropanol-2
Trade names containing: Solvenon® PP

2-Phenoxypropanol-1
Uses: Solvent
Trade names containing: Solvenon® PP

1-Phenoxy-2-propanol. *See* Propylene glycol phenyl ether
2-Phenoxypropanol. *See* Propylene glycol phenyl ether
2-Phenoxypropyl alcohol. *See* Propylene glycol phenyl ether
Phenoxytol. *See* Phenoxyethanol
Phenylcarbinol. *See* Benzyl alcohol
Phenyl cyanide. *See* Benzonitrile

Phenylethane. *See* Ethylbenzene
2-Phenylethanol. *See* Phenethyl alcohol
β-Phenylethanol. *See* Phenethyl alcohol
1-Phenylethanone. *See* Acetophenone
Phenylethyl alcohol. *See* Phenethyl alcohol
2-Phenylethyl alcohol. *See* Phenethyl alcohol
β-Phenylethyl alcohol. *See* Phenethyl alcohol
Phenylglycol. *See* Phenoxyethanol
Phenyl hydride. *See* Benzene
4-(2-Phenylisopropyl)phenol. *See* Cumyl phenol
Phenylmethanal. *See* Benzaldehyde
Phenylmethane. *See* Toluene
Phenylmethanol. *See* Benzyl alcohol
Phenylmethyl acetate. *See* Benzyl acetate
Phenylmethyl alcohol. *See* Benzyl alcohol
Phenylmethyl benzoate. *See* Benzyl benzoate
Phenylmethyl dodecanoate. *See* Benzyl laurate
Phenyl methyl ketone. *See* Acetophenone
1-Phenylpropane. *See* Propylbenzene
2-Phenylpropane. *See* Cumene

Phenyl trimethicone
CAS 2116-84-9; EINECS 218-320-6
Synonyms: Methyl phenyl polysiloxane; Polyphenylmethyl siloxane; Phenyl tristrimethyl siloxysilane
Classification: Siloxane polymer
Empirical: $C_{15}H_{32}O_3Si_4$
Properties: M.w. 372.8; dens. 0.97; m.p. < -60 C; b.p. 264-6 C; ref. index 1.459; flash pt. 127 C
Uses: Oil-phase ingred. in alcoholic milky-white lotions; hydrophobing agent; co-binder for primers, esp. for facade coatings; lubricant with solvency props.
Regulatory: FDA 21CFR §175.105, 175.300
Manuf./Distrib.: Fluka
Trade names: T4250

Phenyl tristrimethyl siloxysilane. *See* Phenyl trimethicone
Phosphonic acid, methyl-, dimethyl ester. *See* Dimethyl methylphosphonate
Phosphoric acid tributyl ester. *See* Tributyl phosphate
o-Phosphoric acid, triethyl ester. *See* Triethyl phosphate
Phosphoric acid, tris(2-ethylhexyl) ester. *See* Trioctyl phosphate
Phthalic acid bis(2-methoxyethyl) ester. *See* Dimethoxyethyl phthalate
Phthalic acid, butyl octyl ester. *See* Butyl octyl phthalate
Phthalic acid, diethyl ester. *See* Diethyl phthalate
Phthalic acid diisobutyl ester. *See* Diisobutyl phthalate
Phthalic acid, di(methoxyethyl) ester. *See* Dimethoxyethyl phthalate
Phthalic acid dimethyl ester. *See* Dimethyl phthalate
Phthalol. *See* Diethyl phthalate
2-Picoline. *See* α-Picoline
3-Picoline. *See* β-Picoline
4-Picoline. *See* γ-Picoline

α-Picoline
CAS 109-06-8; EINECS 203-643-7
UN No. 2313 (DOT)
Synonyms: α-Methylpyridine; 2-Methylpyridine; 2-Picoline; o-Picoline
Empirical: C_6H_7N

β-Picoline

Formula: $C_5H_4N(CH_3)$
Properties: Colorless liq., strong unpleasant odor; very sol. in water; misc. in alcohol, ether; m.w. 93.13; dens. 0.943; vapor pressure 10 mm Hg (24.4 C); m.p. -70 C; b.p. 128-129 C; flash pt. 26 C; ref. index 1.500
Toxicology: LD50 (oral, rat) 790 mg/kg, (IP, rat) 200 mg/kg, (skin, rabbit) 410 mg/kg; poison by IP route; mod. toxic by ing. and skin contact; primary irritant; skin and severe eye irritant; mutagen
Precaution: Flamm. exposed to heat and flame; mixts. with hydrogen peroxide + iron (II) sulfate + sulfuric acid may ignite and explode; Hazardous decomp. prods.: Heated to decomp., emits toxic fumes of NO_x
Uses: Organic intermediate for pharmaceuticals, dyes, rubber chemicals, solvent, source for vinyl pyridine, laboratory reagent
Manuf./Distrib.: Aldrich; Fluka; Lonza; Nepera; Raschig; Reilly Ind.; Schweizerhall

β-Picoline
CAS 108-99-6; EINECS 203-636-9
UN No. 2313 (DOT)
Synonyms: 3-Picoline; m-Picoline; 3-Methylpyridine; Pyridine, 3-methyl-
Empirical: C_6H_7N
Properties: Colorless liq.; sweetish, not unpleasant odor; sol. in water, alcohol, ether; m.w. 93.13; dens. 0.957; f.p. -18 C; b.p. 144 C; flash pt. 36 C; ref. index 1.5050
Toxicology: LD50 (inh., rat, 2 h) 8700 ppm, (IP, rat) 150 mg/kg; poison by IV and IP routes; mod. toxic by ingestion; irritant
Precaution: Flamm. exposed to heat or flame; moderate fire risk; can react vigorously with oxidizers; Hazardous decomp. prods.: Heated to decomp., emits toxic fumes of NO_x
Uses: Solvent in synthesis of pharmaceuticals, resins, dyestuffs, rubber accelerators, insecticides; prep. of nicotinic acid, nicotinic acid amide, waterproofing agents; laboratory reagent
Manuf./Distrib.: Aldrich; Fabrichem; Fluka; Nepera; Raschig; Reilly Ind.; Schweizerhall

γ-Picoline
CAS 108-89-4; EINECS 203-626-4
UN No. 2313 (DOT)
Synonyms: 4-Methylpyridine; 4-Picoline; p-Picoline; Pyridine, 4-methyl-
Empirical: C_6H_7N
Formula: $NCHCHC(CH_3)CHCH$
Properties: Colorless, moderately volatile liq.; disagreeable odor; sol. in water, alcohol, ether; m.w. 93.13; dens. 0.957; m.p. 2.4 C; b.p. 145 C; flash pt. 56 C; ref. index 1.5050
Toxicology: LD50 (oral, rat) 1290 mg/kg, (IP, rat) 163 mg/kg; poison by skin contact, IP routes; mod. toxic by ing.; mildly toxic by inh.; severe primary irritant to eyes and skin
Precaution: Flamm. exposed to heat, flames, oxidizers; Hazardous decomp. prods.: Heated to decomp., emits toxic fumes of NO_x
Uses: Solvent in synthesis of pharmaceuticals, resins, dyestuffs, rubber accelerators, pesticides, waterproofing agents; laboratory reagent; mfg. of isoniazid; catalyst; curing agent
Manuf./Distrib.: Aldrich; Fluka; Lonza; Raschig; Reilly Ind.; Schweizerhall

m-Picoline. *See* β-Picoline
o-Picoline. *See* α-Picoline
p-Picoline. *See* γ-Picoline

Pimelic ketone. *See* Cyclohexanone
Pimelin ketone. *See* Cyclohexanone
2-Pinene. *See* α-Pinene
2(10)-Pinene. *See* β-Pinene

α-Pinene
CAS 80-56-8; 7785-26-4 (L, -); 7785-70-8 (+); EINECS 232-077-3 (-); 232-087-8 (+)
UN No. 2368 (DOT)
Synonyms: 2-Pinene; 2,6,6-Trimethylbicyclo(3.1.1)-2-hept-2-ene; 2,6,6-Trimethylbicyclo(3.1.1)-2-heptene
Classification: Terpene hydrocarbon
Empirical: $C_{10}H_{16}$
Properties: Colorless liq., turpentine odor; insol. in water; sol. in alcohol, chloroform, ether, glac. acetic acid; m.w. 136.26; dens. 0.8592 (20/4 C); m.p. -55 C; b.p. 155 C; flash pt. 91 F; ref. index 1.464-1.468
Toxicology: LD50 (oral, rat) 3700 mg/kg; deadly poison by inh.; mod. toxic by ingestion; eye, mucous membrane, and severe skin irritant; toxic effects similar to turpentine
Precaution: DOT: Flamm. liq.; dangerous fire hazard exposed to heat, flame, oxidizers; explodes on contact with nitrosyl perchlorate
NFPA: Health 1, Flammability 3, Reactivity 0
Uses: Solvent for protective coatings; synthesis of camphene, pine oil; odorant; insecticides; additive for lubricants; raw material in prod. of camphor; syn. flavoring agent for foods and pharmaceuticals
Regulatory: FDA 21CFR §172.515; FEMA GRAS
Manuf./Distrib.: Aldrich; Arizona; Fluka; Hercules; Penta Mfg.; SCM Glidco Organics; Sigma; Spectrum Chem. Mfg.; Veitsiluoto Oy
Trade names containing: Acintene® AC-P; Acintene® B Beta Pinene; Acintene® DP 738; Acintene® P-AC Alpha-Pinene

β-Pinene
CAS 127-91-3; 18172-67-3 (-); EINECS 242-060-2 (-)
Synonyms: 6,6-Dimethyl-2-methylenebicyclo[3.1.1]heptane; Nopinen; Nopinene; 2(10)-Pinene; Pseudopinen; Pseudopinene
Classification: Terpene hydrocarbon
Empirical: $C_{10}H_{16}$
Properties: Colorless liq., terpene odor; insol. in water; sol. in alcohol, chloroform, ether; m.w. 136.24; dens. 0.859; m.p. -61 C; b.p. 165-167 C; soften. pt. 112-118 C; flash pt. 32 C
Toxicology: LD50 (oral, rat) 4700 mg/kg; mildly toxic by ingestion; irritating to eyes, skin, respiratory system
Precaution: Flamm.; fire risk; Hazardous decomp. prods.: Heated to decomp., emits acrid smoke and irritating fumes
Uses: Raw material for prod. of org. chems.; polyterpene resins; intermediate for perfumes and flavorings; syn. flavoring agent for foods and pharmaceuticals
Regulatory: FDA 21CFR §172.280, 172.515; FEMA GRAS
Manuf./Distrib.: Aldrich; Arizona; Fluka; Penta Mfg.; SCM Glidco Organics
Trade names containing: Acintene® B Beta Pinene

Pine oil
CAS 8002-09-3
UN No. 1272 (DOT)
Synonyms: Yarmor; Yarmor pine oil; Oil of pine; Oleum abietis
Definition: Volatile oil obtained from distillation of the species *Pinus*
Properties: Colorless to pale yel. liq., penetrating pine odor; insol. in water; sol. in org. solvs.; dens. 0.86; b.p. 200-220 C; flash pt. (CC) 172 F

Toxicology: LD50 (oral, rat) 3200 mg/kg, (skin, rabbit) 5 g/kg; moderately toxic by ingestion; mildly toxic by skin contact; irritating to eyes, skin, mucous membranes; weak allergen; large doses may cause CNS depression
Precaution: Flamm. exposed to heat or flame; can react with oxidizers; mod. spontaneous heating
NFPA: Health 0, Flammability 2, Reactivity 0
Uses: Odorant, disinfectant, penetrant, wetting agent, solvent, frothing agent, preservative, laboratory reagent, fragrances; natural flavoring agent for foods
Regulatory: FDA 21CFR §172.510, 175.105, 176.180, 176.200, 176.210, 177.2800; 27CFR §21.65, 21.151
Manuf./Distrib.: Allchem Ind.; Arizona; Hercules; Langley Smith Ltd; Lebermuth; Penta Mfg.; SCM Glidco Organics; Sea-Land; Spectrum Chem. Mfg.; T&R Chem.
Trade names: Herco® Pine Oil

Piperidine
CAS 110-89-4; EINECS 203-813-0
UN No. 2401 (DOT)
Synonyms: Hexahydropyridine; Hexazane; Pentamethyleneimine; Azacyclohexane; Cyclopentimine
Empirical: $C_5H_{11}N$
Formula: $NH(CH_2)_4CH_2$
Properties: Colorless clear liq.; amine-like odor; sol. in alcohol, benzene, chloroform; misc. with water; m.w. 85.15; dens. 0.862 (20/4 C); vapor pressure 40 mm Hg (29.2 C); m.p. -11 to -9 C; b.p. 104-106 C; flash pt. 16 C; ref. index 1.453 (20 C)
Toxicology: LD50 (oral, rat) 0.52 ml/kg, (skin, rabbit) 320 mg/kg, (inh., mammal) 6500 mg/m^3; toxic by ing., inh., skin contact, IP routes; causes burns; experimental carcinogen, reproductive effects; mutagenic data
Precaution: Highly flamm.; very dangerous fire hazard exposed to heat, flame, or oxidizers; can react vigorously with oxidizers; explodes on contact with 1-perchloryl piperidine, dicyanofurazan, N-nitrosoacetanilide; Hazardous decomp. prods.: Heated to decomp., emits highly toxic fumes of NO_x
NFPA: Health 3, Flammability 3, Reactivity 0
Uses: Solvent and intermediate, curing agent for rubber and epoxy resins, catalyst for condensation reactions, ingredient in oils and fuels, complexing agent; intermediate for rubber accelerators; in agric. and pharmaceuticals; syn. flavoring agent for foods, pharmaceuticals
Regulatory: FDA 21CFR §172.515; FEMA GRAS
Manuf./Distrib.: Aldrich; Am. Biorganics; BASF AG; Fluka; Janssen Chimica; Nepera; Penta Mfg.; Raschig; Reilly Ind.; Schweizerhall; Sigma; Spectrum Chem. Mfg.

PM acetate. *See* Propylene glycol methyl ether acetate
POE (4). *See* PEG-4
POE (8). *See* PEG-8
POE (9). *See* PEG-9
POE (12). *See* PEG-12
POE (14). *See* PEG-14
POE (16). *See* PEG-16
POE (20). *See* PEG-20
POE (32). *See* PEG-32
POE (75). *See* PEG-75
POE (100). *See* PEG-100
POE (150). *See* PEG-150

POE (6) abietate. *See* PEG-6 abietate
POE (12) castor oil. *See* PEG-12 castor oil
POE (33) castor oil. *See* PEG-33 castor oil
POE (35) castor oil. *See* PEG-35 castor oil
POE (5) cetyl ether. *See* Ceteth-5
POE (16) cetyl ether. *See* Ceteth-16
POE (24) cetyl ether. *See* Ceteth-24
POE (2) cetyl/oleyl ether. *See* Cetoleth-2
POE (3) cetyl/oleyl ether. *See* Cetoleth-3
POE (6) cetyl/oleyl ether. *See* Cetoleth-6
POE (24) cholesteryl ether. *See* Choleth-24
POE (6) decyl ether. *See* Deceth-6
POE (4) dilaurate. *See* PEG-4 dilaurate
POE (8) dilaurate. *See* PEG-8 dilaurate
POE (12) dilaurate. *See* PEG-12 dilaurate
POE (20) dilaurate. *See* PEG-20 dilaurate
POE (4) dioleate. *See* PEG-4 dioleate
POE (8) dioleate. *See* PEG-8 dioleate
POE (12) dioleate. *See* PEG-12 dioleate
POE (20) dioleate. *See* PEG-20 dioleate
POE (7) dodecyl phenyl ether. *See* Dodoxynol-7
POE (7) glyceryl ether triacetate. *See* Glycereth-7 triacetate
POE (15) hydroxystearate. *See* PEG-15 hydroxystearate
POE (3) lanolate. *See* PEG-3 lanolate
POE (5) lanolate. *See* PEG-5 lanolate
POE (10) lanolate. *See* PEG-10 lanolate
POE (20) lanolate. *See* PEG-20 lanolate
POE (75) lanolin. *See* PEG-75 lanolin
POE (5) lanolin ether. *See* Laneth-5
POE (10) lanolin ether. *See* Laneth-10
POE (15) lanolin ether. *See* Laneth-15
POE (16) lanolin ether. *See* Laneth-16
POE (6) methyl ether. *See* PEG-6 methyl ether
POE (10) methyl ether. *See* PEG-10 methyl ether
POE (16) methyl ether. *See* PEG-16 methyl ether
POE (10) methyl glucose ether. *See* Methyl gluceth-10
POE (20) methyl glucose ether. *See* Methyl gluceth-20
POE (6) monoisostearate. *See* PEG-6 isostearate
POE (4) monooleate. *See* PEG-4 oleate
POE (5) monooleate. *See* PEG-5 oleate
POE (6) monooleate. *See* PEG-6 oleate
POE (8) monooleate. *See* PEG-8 oleate
POE (20) monooleate. *See* PEG-20 oleate
POE (4) nonyl phenyl ether. *See* Nonoxynol-4
POE (5) nonyl phenyl ether. *See* Nonoxynol-5
POE (6) nonyl phenyl ether. *See* Nonoxynol-6
POE (9) nonyl phenyl ether phosphate. *See* Nonoxynol-9 phosphate
POE (2) oleyl ether. *See* Oleth-2
POE (5) oleyl ether. *See* Oleth-5
POE (16) oleyl ether. *See* Oleth-16
POE (24) POP (24) glyceryl ether. *See* PPG-24-glycereth-24
POE (3) POP (2) monobutyl ether. *See* PPG-2-buteth-3
POE (5) POP (3) monobutyl ether. *See* PPG-3-buteth-5
POE (7) POP (5) monobutyl ether. *See* PPG-5-buteth-7

POE (10) POP (7) monobutyl ether. *See* PPG-7-buteth-10
POE (12) POP (9) monobutyl ether. *See* PPG-9-buteth-12
POE (16) POP (12) monobutyl ether. *See* PPG-12-buteth-16
POE (30) POP (20) monobutyl ether. *See* PPG-20-buteth-30
POE (35) POP (28) monobutyl ether. *See* PPG-28-buteth-35
POE (45) POP (33) monobutyl ether. *See* PPG-33-buteth-45
POE (3) POP (3) myristyl ether. *See* PPG-3-myreth-3
POE (20) sorbitan monoisostearate. *See* PEG-20 sorbitan isostearate
POE (20) sorbitan monooleate. *See* Polysorbate 80
POE (20) sorbitan monostearate. *See* Polysorbate 60
POE (1) stearic acid. *See* Glycol stearate
POE (5) stearyl ether. *See* Steareth-5
POE (16) stearyl ether. *See* Steareth-16

Poloxamer 181
CAS 9003-11-6 (generic); 53637-25-5
Classification: Polyoxyethylene, polyoxypropylene block polymer
Formula: $HO(CH_2CH_2O)_x(CCH_3HCH_2O)_y(CH_2CH_2O)_zH$, avg. x=3, y=30, z=3
Toxicology: LD50 (oral, rat) 2300 mg/kg; moderately toxic by ingestion, intraperitoneal route
Uses: Nonionic surfactant; as food additives, defoamers, antistats, demulsifiers, detergents, wetting agents, gellants, emulsifiers, dispersants, dye levelers; paints, paper, agric., foods, pharmaceuticals, etc.; provides solvency, low odor, low irritation, low toxicity to personal care prods.
Regulatory: FDA 21CFR §172.808, 173.340, 175.105, 176.180, 176.200, 176.210, 177.1200
Manuf./Distrib.: Aldrich; Fluka; Sigma
Trade names: Dow EP530

Polyacetal. *See* Acetal
Polydimethylsiloxane. *See* Dimethicone
Polydimethyl siloxy cyclics. *See* Cyclomethicone
Polyether glycol. *See* Polyethylene glycol

Polyethylene glycol
CAS 25322-68-3; EINECS 203-473-3
Synonyms: PEG; α-Hydroxy-ω-hydroxy poly(oxy-1,2-ethanediyl); Polyglycol; Poly(ethylene oxide); Polyether glycol
Classification: Aliphatic organic compd.
Definition: Condensation polymer of ethylene glycol
Formula: $H(OC_2H_4)_nOH$, $n \geq 4$
Properties: Clear liq. or wh. solid; sol. in org. solvs., aromatic hydrocarbons; dens. 1.110-1.140 (20 C); m.p. 4-10 C; flash pt. 471 F; ref. index 1.4590
Toxicology: LD50 (oral, rat) 33,750 mg/kg; sl. toxic by ingestion; skin and eye irritant; contact dermatitis sensitization; may cause hives
Precaution: Combustible exposed to heat or flames; Hazardous decomp. prods.: Heated to decomp., emits acrid smoke and irritating fumes
Uses: Chemical intermediates, plasticizers, softeners, humectants, ointments, polishes, paper coating, mold lubricants, bases for cosmetics and pharmaceuticals, solvents, binders, metal and rubber processing, food additives, laboratory reagent; antistat; paints; textile fibers; food pkg.; boiler water additive for food processing; pharmaceutical ointment/suppository base, coating agent, plasticizer, solvent
Regulatory: FDA 21CFR §172.210, 172.820, 173.340; FDA approved for orals, topicals; USP/NF compliance
Manuf./Distrib.: Aldrich; BASF; BP Chems. Ltd; Calgene; Dow; DuPont; Fluka;

Harcros USA; Henkel; Hüls; Inolex; Olin; Rhône-Poulenc Surf. & Spec.; Sigma; Texaco; Union Carbide;
See also PEG…

Polyethylene glycol 300. *See* PEG-6
Polyethylene glycol 400. *See* PEG-8
Polyethylene glycol 600. *See* PEG-12
Polyethylene glycol 1000. *See* PEG-20
Polyethylene glycol 3350. *See* PEG-75

Polyethylene glycol monomethyl ether
CAS 9004-74-4 (generic)
Synonyms: Methoxy polyethylene glycol
Definition: Addition polymer of ethylene oxide and methanol
Formula: $CH_3(OCH_2CH_2)_nOH$, n = no. of oxyethylene groups
Properties: Liq. grades: Colorless or pract. colorless clear to sl. hazy sl. hygroscopic visc. liqs., sl. char. odor; sol. in acetone, alcohol, chloroform, ethyl acetate, toluene; misc. with water; sp.gr. 1.09-1.10; Solid grades: wh. waxy plastic, flakes, beads, or powds., pract. odorless and tasteless; sol. in water, acetone, alcohol, chloroform,ethyl acetate, toluene
Uses: Pharmaceutical ointment base, solvent, plasticizer
Manuf./Distrib.: Aldrich; Fluka; Sigma

Poly(ethylene oxide). *See* Polyethylene glycol

Polyglyceryl-3 diisostearate
CAS 66082-42-6; 85404-84-8; EINECS 291-548-1
Synonyms: Triglyceryl diisostearate; Isooctadecanoic acid, diester with triglycerol
Definition: Diester of isostearic acid and a glycerin polymer with avg. 3 glycerin units
Uses: Dye and pigment wetter; emollient; thickener; solvent; for creams, lotions, lip prods.; food emulsifier; pharmaceutical emollient, thickener, solvent
Trade names: Emerest® 2452

Polyglycol. *See* Polyethylene glycol
Poly(oxy(dimethylsilylene)). *See* Dimethicone
Polyoxyl 35 castor oil. *See* PEG-35 castor oil
Polyoxyl 4 dilaurate. *See* PEG-4 dilaurate
Polyoxyl 75 lanolin. *See* PEG-75 lanolin
Polyoxypropylene (9). *See* PPG-9
Polyoxypropylene (17). *See* PPG-17
Polyoxypropylene (20). *See* PPG-20
Polyoxypropylene (26). *See* PPG-26
Polyoxypropylene (30). *See* PPG-30
Polyoxypropylene (3) methyl ether. *See* PPG-3 methyl ether
Polyphenylmethyl siloxane. *See* Phenyl trimethicone

Polypropylene glycol
CAS 25322-69-4; EINECS 200-338-0
Synonyms: PPG
Classification: Aliphatic organic compd.
Empirical: $(C_3H_8O_2)_n$
Formula: $HO(C_3H_6O)_nH$
Properties: Colorless clear visc. liq., sl. bitter taste; sol. in water, aliphatic ketones, alcohol; insol. in ether, aliphatic hydrocarbons; m.w. 400-2000; dens. 1.001-1.007; m.p. does not cryst.; flash pt. > 390 F
Toxicology: LD50 (oral, rat) 4190 mg/kg; mildly toxic by ingestion; skin and eye

irritant; linked to sensitive reactions; heated to decomp., emits acrid smoke and irritating fumes

Precaution: Combustible exposed to heat or flame; reactive with oxidizers

Uses: Hydraulic fluids, rubber lubricants, antifoam agents, intermediates for urethane foams, adhesives, coatings, elastomers, plasticizers, paint formulations, lab reagent; defoamer for beet sugar and yeast processing, foods; pharmaceutical solvent for fats, oils, waxes, resins

Regulatory: FDA 21CFR §173.340; FDA approved for ophthalmics, orals, topicals

Manuf./Distrib.: Aldrich; ARCO; Ashland; BASF; Bayer; BP Chems. Ltd; Calgene; Coyne; Dow; Eastman; Fluka; Harcros USA; Hüls AG; Monomer-Polymer & Dajac Labs; Olin; PPG Ind.; Rhône-Poulenc Surf. & Spec.; Texaco; Witco/Oleo-Surf.

See also PPG...

Polypropylene glycol (9). *See* PPG-9
Polypropylene glycol (12). *See* PPG-17
Polypropylene glycol (20). *See* PPG-20
Polypropylene glycol (26). *See* PPG-26
Polypropylene glycol (30). *See* PPG-30
Polypropylene glycol monomethyl ether. *See* Methoxyisopropanol

Polysorbate 60

CAS 9005-67-8 (generic)

Synonyms: POE (20) sorbitan monostearate; PEG-20 sorbitan stearate; Sorbimacrogol stearate 300

Definition: Mixture of stearate esters of sorbitol and sorbitol anhydrides, with ≈ 20 moles ethylene oxide

Empirical: $C_{64}H_{126}O_{26}$

Properties: Lemon to orange oily liq., faint char. odor, bitter taste; sol. in water, aniline, ethyl acetate, toluene; insol. in min. and veg. oils; m.w. 1311.90; acid no. 2 max.; sapon. no. 45-55; hyd. no. 81-96

Toxicology: LD50 (IV, rat) 1220 mg/kg; moderately toxic by intravenous route; experimental tumorigen, reproductive effects; Hazardous decomp. prods.: Heated to decomp., emits acrid smoke and irritating fumes

Uses: Emulsifier, solvent for polymerization, cosmetics, pharmaceuticals, industrial chemicals, agric., textiles, plastics additive, paints, pulp/paper; pharmaceutical and veterinary drug; solubilizer; foaming agent in beverage mixes; food emulsifier

Regulatory: FDA 21CFR §73.1001, 172.515, 172.836, 172.878, 172.886, 173.340, 175.105, 175.300, 178.3400; USDA CFR9 §318.7, 381.147 (limitation 1% max., 1% total combined with polysorbate 80, 0.0175% in scald water); FEMA GRAS; Europe listed; UK approved; FDA approved for orals, rectals, topicals, vaginals; USP/NF, BP, Ph.Eur. compliance

Manuf./Distrib.: Aldrich; Fluka; Sigma; Spectrum Chem. Mfg.

Polysorbate 80

CAS 9005-65-6 (generic); 37200-49-0; 61790-86-1; EINECS 215-665-4; 200-849-9

Synonyms: POE (20) sorbitan monooleate; PEG-20 sorbitan oleate; Sorbimacrogol oleate 300

Definition: Mixture of oleate esters of sorbitol and sorbitol anhydrides, with ≈ 20 moles ethylene oxide

Properties: Amber visc. liq.; nonionic; very sol. in water; sol. in alcohol, cottonseed oil, corn oil, ethyl acetate, methanol, toluene; insol. in min. oil; dens. 1.06-1.10; visc. 270-430 cSt; acid no. 2 max.; sapon. no. 45-55; hyd. no. 65-80; pH 5-7 (5% aq.)

Toxicology: LD50 (oral, mouse) 25 g/kg; mod. toxic by intravenous route; mildly toxic by ingestion; eye irritant; experimental tumorigen, reproductive effects; mutagenic data; Hazardous decomp. prods.: Heated to decomp., emits acrid smoke and irritating fumes

Uses: Pharmaceutic aid (surfactant); as emulsifier and dispersant in medicinal prods.; as defoamer and emulsifier in foods; adjuvant in herbicides; polymerization, paints; color diluent, flavoring

Regulatory: FDA 21CFR §73.1, 73.1001, 172.515, 172.840, 173.340, 175.105, 175.300, 178.3400; USDA 9CFR §318.7, 381.147 (limitation 1% alone, 1% total combined with polysorbate 60); FEMA GRAS; Europe listed; UK approved; FDA approved for buccals, intramuscular injectables, intravenous, parenterals, ophthalmics, orals, otics, rectals, topicals, vaginals; USP/NF, BP, Ph.Eur. compliance

Manuf./Distrib.: Aldrich; Fluka; Sigma; Spectrum Chem. Mfg.

Trade names: Capmul® POE-O

Trade names containing: Ritawax AEO

Polysorbate 120. *See* PEG-20 sorbitan isostearate
POP (5) butyl ether. *See* PPG-5 butyl ether
POP (9) butyl ether. *See* PPG-9 butyl ether
POP (14) butyl ether. *See* PPG-14 butyl ether
POP (15) butyl ether. *See* PPG-15 butyl ether
POP (18) butyl ether. *See* PPG-18 butyl ether
POP (22) butyl ether. *See* PPG-22 butyl ether
POP (24) butyl ether. *See* PPG-24 butyl ether
POP (33) butyl ether. *See* PPG-33 butyl ether
POP (40) butyl ether. *See* PPG-40 butyl ether
POP (53) butyl ether. *See* PPG-53 butyl ether
POP (10) cetyl ether. *See* PPG-10 cetyl ether
POP (30) cetyl ether. *See* PPG-30 cetyl ether
POP (50) cetyl ether. *See* PPG-50 cetyl ether
POP (55) glyceryl ether. *See* PPG-55 glyceryl ether
POP (40) methyl diethyl ammonium chloride. *See* PPG-40 diethylmonium chloride
POP (2) methyl ether. *See* PPG-2 methyl ether
POP (10) methyl glucose ether. *See* PPG-10 methyl glucose ether
POP (20) methyl glucose ether. *See* PPG-20 methyl glucose ether
POP (3) myristyl ether. *See* PPG-3 myristyl ether
POP (2) myristyl ether propionate. *See* PPG-2 myristyl ether propionate
POP (23) oleyl ether. *See* PPG-23 oleyl ether
POP (37) oleyl ether. *See* PPG-37 oleyl ether
POP (24) POE (24) glyceryl ether. *See* PPG-24-glycereth-24
POP (2) POE (3) monobuty ether. *See* PPG-2-buteth-3
POP (3) POE (5) monobutyl ether. *See* PPG-3-buteth-5
POP (5) POE (6) monobutyl ether. *See* PPG-5-buteth-7
POP (7) POE (10) monobutyl ether. *See* PPG-7-buteth-10
POP (9) POE (12) monobutyl ether. *See* PPG-9-buteth-12
POP (12) POE (16) monobutyl ether. *See* PPG-12-buteth-16
POP (20) POE (30) monobutyl ether. *See* PPG-20-buteth-30
POP (28) POE (35) monobutyl ether. *See* PPG-28-buteth-35
POP (33) POE (45) monobutyl ether. *See* PPG-33-buteth-45
POP (3) POE (3) myristyl ether. *See* PPG-3-myreth-3
POP (11) stearyl ether. *See* PPG-11 stearyl ether
POP (15) stearyl ether. *See* PPG-15 stearyl ether
Potash lye. *See* Potassium hydroxide
Potassa. *See* Potassium hydroxide

Potassium hydrate. *See* Potassium hydroxide

Potassium hydroxide
CAS 1310-58-3; EINECS 215-181-3
UN No. 1813 (DOT), 1814 (DOT)
Synonyms: Caustic potash; Potassium hydrate; Lye; Potash lye; Potassa
Classification: Inorganic base
Empirical: HKO
Formula: KOH
Properties: Wh. flakes, lumps or pellets; sol. in water, alcohol, glycerol; sl. sol. in ether; m.w. 56.11; dens. 2.044; m.p. 405 C; b.p. 1320 C
Toxicology: LD50 (oral, rat) 365 mg/kg; ACGIH TLV:CL 2 mg/m^3; toxic by ingestion, inh.; strong caustic; eye and severe skin irritant; ing. may cause violent throat pain and stricture of esophagus
Precaution: DOT: Corrosive material; releases heat in contact with water and acids; corrodes many metals when wet; violent exothermic reactions with water; potentially explosive reactions with chlorine dioxide, nitrobenzene, nitromethane; Hazardous decomp. prods.: Heated to decomp., emits toxic fumes of K_2O; above 84 C reacts with reducing sugars to form CO
NFPA: Health 3, Flammability 0, Reactivity 1
Uses: Soap mfg., bleaching, mfg. of chemicals, electrolyte, absorbent, liq. fertilizers, food additive, herbicides, electroplating, mercerizing, paint removers, reagent, water treatment; direct food additive, pH control, processing aid, stabilizer; pharmaceutical alkalizing agent, cuticle solvent, caustic for wart removal; reagent
Regulatory: FDA 21CFR §163.110, 175.210, 184.1631, GRAS; USDA 9CFR §381.147; Europe listed; UK approved; FDA approved for intravenous, parenterals, orals, topicals; USP/NF, BP, JP compliance
Manuf./Distrib.: Aldrich; Alfa Aesar Johnson Matthey; AMRESCO; ASHTA; EM Ind.; Fluka; General Chem.; Hüls Am.; ICI Spec. Chems.; Int'l. Chem. Inc; K3; Occidental; Olin; San Yuan; Sigma; Spectrum Chem. Mfg.; Universal Preserv-A-Chem

Potato alcohol. *See* Alcohol denatured
PPG. *See* Polypropylene glycol

PPG-9
CAS 25322-69-4 (generic)
Synonyms: Polyoxypropylene (9); Polypropylene glycol (9); PPG 400
Definition: Polymer of propylene oxide
Formula: $H(OCH_2CHCH_3)_nOH$, avg. n = 9
Uses: Chemical intermediate, antifoam agent in fermentation and in paint formulations, antiblooming agent for pentachlorophenol-treated wood; binder and lubricant for ceramics; plasticizer of resin-treated papers; mold release applics.; provides solvency, low odor, low irritation, low toxicity to personal care prods.
Regulatory: FDA 21CFR §173.310, 175.105, 176.200, 176.210
Manuf./Distrib.: Aldrich; Fluka
Trade names: Dow P425

PPG-17
CAS 25322-69-4 (generic)
Synonyms: Polyoxypropylene (17); Polypropylene glycol (12)
Definition: Polymer of propylene oxide
Formula: $H(OCH_2CHCH_3)_nOH$, avg. n = 17
Uses: Chemical intermediate, antifoam agent in fermentation and in paint

formulations, antiblooming agent for pentachlorophenol-treated wood; binder and lubricant for ceramics; plasticizer of resin-treated papers; defoamer for food-grade coatings and paper prods.; provides solvency, low odor, low irritation, low toxicity to personal care prods.

Regulatory: FDA 21CFR §173.310, 175.105, 176.170, 176.200, 176.210
Manuf./Distrib.: Aldrich; Fluka
Trade names: Dow P1000TB

PPG-20

CAS 25322-69-4 (generic)
Synonyms: Polyoxypropylene (20); Polypropylene glycol (20); PPG 1200
Definition: Polymer of propylene oxide
Formula: $H(OCH_2CHCH_3)_nOH$, avg. n = 20
Uses: Defoamer, mold release applics., chemical intermediates for fatty acid esters, components for urethane resins; textile lubricants, metalworking compds., cosmetics, paints, urethane foams, hydraulic fluids, plasticizers, release agents; provides solvency, low odor, low irritation, low toxicity to personal care prods.
Regulatory: FDA 21CFR §173.310, 173.340, 175.105, 176.170, 176.200, 176.210, 178.3740
Manuf./Distrib.: Aldrich; Fluka
Trade names: Dow P1200

PPG-26

CAS 25322-69-4 (generic)
Synonyms: Polyoxypropylene (26); Polypropylene glycol (26); PPG 2000
Definition: Polymer of propylene oxide
Formula: $H(OCH_2CHCH_3)_nOH$, avg. n = 26
Uses: Defoamer, mold release applics., chemical intermediates for fatty acid esters, components for urethane resins; boiler water additive for prep. of steam that will contact food; defoamer for beet sugar and yeast processing; defoamer for food-grade coatings and paper prods.; provides solvency, low odor, low irritation, low toxicity to personal care prods.
Regulatory: FDA 21CFR §173.310, 173.340, 175.105, 176.170, 176.200, 176.210, 178.3740
Manuf./Distrib.: Aldrich; Fluka
Trade names: Dow P2000

PPG-30

CAS 25322-69-4 (generic)
Synonyms: Polyoxypropylene (30); Polypropylene glycol (30); PPG 4000
Definition: Polymer of propylene oxide
Formula: $H(OCH_2CHCH_3)_nOH$, avg. n = 30
Uses: Defoamer, mold release applics., chemical intermediates for fatty acid esters, components for urethane resins; antifoam agent in fermentation and in paint formulations, antiblooming agent for pentachlorophenol-treated wood; binder and lubricant for ceramics; plasticizer of resin-treated papers; provides solvency, low odor, low irritation, low toxicity to personal care prods.
Regulatory: FDA 21CFR §173.310, 173.340, 175.105, 176.170, 176.200, 178.3740
Manuf./Distrib.: Aldrich; Fluka
Trade names: Dow P4000

PPG-400

Uses: Intermediate yielding esters; useful as lubricants, defoaming agents in rubber and pharmaceuticals, solvents and humectant modifiers for inks, plasticizers, and functional fluids
Trade names: Jeffox™ PPG-400

PPG 400. *See* PPG-9
PPG 1200. *See* PPG-20

PPG-2000
Uses: Intermediate yielding esters; useful as lubricants, defoaming agents in rubber and pharmaceuticals, solvents and humectant modifiers for inks, plasticizers, and functional fluids
Trade names: Jeffox™ PPG-2000

PPG 2000. *See* PPG-26
PPG 4000. *See* PPG-30

PPG-10 butanediol
Definition: POP ether of butanediol
Formula: $HO(CH_2CHCH_3O)_xCH_2CH_2CH_2CH_2(OCHCH_3CH_2)_yOH$, avg. (x+y)=10
Uses: Emollient for antiperspirants; solvent for sunscreen actives; crystal growth inhibitor
Trade names: Macol® 57

PPG-2-buteth-3
CAS 9038-95-3 (generic); 9065-63-8 (generic)
Synonyms: POE (3) POP (2) monobutyl ether; POP (2) POE (3) monobuty ether; Oxirane, methyl, polymer and oxibane, butyl ether
Definition: Polyoxypropylene, polyoxyethylene ether of butyl alcohol
Empirical: $(C_7H_{14}O_2 \cdot C_6H_{12}O_2)_x$
Formula: $C_4H_9(OCHCH_3CH_2)_x(OCH_2CH_2)_yOH$, avg. x = 2, avg. y = 3
Uses: Emollient for cosmetics; lubricant for compressor, rubber, greases, textiles; intermediate for prep. of resins, plasticizers, surfactants; solvent for inks and dyes
Regulatory: FDA 21CFR §173.310, 173.340, 175.105, 176.210, 178.3570
Manuf./Distrib.: Aldrich; Fluka; Sigma
Trade names: Ucon® 50-HB-55

PPG-3-buteth-5
CAS 9038-95-3 (generic); 9065-63-8 (generic)
Synonyms: POE (5) POP (3) monobutyl ether; POP (3) POE (5) monobutyl ether
Definition: Polyoxypropylene, polyoxyethylene ether of butyl alcohol
Empirical: $(C_7H_{14}O_2 \cdot C_6H_{12}O_2)_x$
Formula: $C_4H_9(OCHCH_3CH_2)_x(OCH_2CH_2)_yOH$, avg. x = 3, avg. y = 5
Uses: Emollient for cosmetics; lubricant for compressor, rubber, greases, textiles; intermediate for prep. of resins, plasticizers, surfactants; solvent for inks and dyes
Regulatory: FDA 21CFR §173.310, 175.105, 176.210, 178.3570
Manuf./Distrib.: Aldrich; Fluka; Sigma
Trade names: Ucon® 50-HB-100

PPG-5-buteth-7
CAS 9038-95-3 (generic); 9065-63-8 (generic); 74623-31-7
Synonyms: POE (7) POP (5) monobutyl ether; POP (5) POE (6) monobutyl ether
Definition: Polyoxypropylene, polyoxyethylene ether of butyl alcohol
Empirical: $(C_7H_{14}O_2 \cdot C_6H_{12}O_2)_x$
Formula: $C_4H_9(OCH_3CHCH_2)_x(OCH_2CH_2)_yOH$, avg. x = 5, avg. y = 7
Toxicology: LD50 (oral, rat) 5370 mg/kg
Uses: Emollient, component in demulsifying and wetting formulations; brake and metalworking fluids; rubber and fiber lubricant; textile applics.; defoamer for hot and cold applics., food and chemical processing; cosmetic formulations; boiler water additive for food processing; solvent for inks and dyes

Regulatory: FDA 21CFR §173.310, 175.105, 176.210, 178.3570
Manuf./Distrib.: Aldrich; Fluka; Sigma
Trade names: Ucon® 50-HB-170

PPG-7-buteth-10
CAS 9038-95-3 (generic); 9065-63-8 (generic)
Synonyms: POE (10) POP (7) monobutyl ether; POP (7) POE (10) monobutyl ether
Definition: POP, POE ether of butyl alcohol
Formula: $C_4H_9(OCHCH_3CH_2)_x(OCH_2CH_2)_yOH$, avg. x = 7, avg. y = 10
Uses: Detergent for toilet cleaners, laundry detergents, emulsion polymerization, defoamers, metalworking fluids, hydraulic fluids; emollient for cosmetics; lubricant for compressor, rubber, greases, textiles; intermediate for prep. of resins, plasticizers, surfactants; solvent for inks and dyes
Regulatory: FDA 21CFR §173.310, 175.105, 176.210, 178.3570
Manuf./Distrib.: Aldrich; Fluka; Sigma
Trade names: Ucon® 50-HB-260

PPG-9-buteth-12
CAS 9038-95-3 (generic); 9065-63-8 (generic)
Synonyms: POE (12) POP (9) monobutyl ether; POP (9) POE (12) monobutyl ether
Definition: Polyoxypropylene, polyoxyethylene ether of butyl alcohol
Empirical: $(C_7H_{14}O_2 \cdot C_6H_{12}O_2)_x$
Formula: $C_4H_9(OCH_3CHCH_2)_x(OCH_2CH_2)_yOH$, avg. x = 9, avg. y = 12
Uses: Emollient for cosmetics; lubricant for compressor, rubber, greases, textiles; intermediate for prep. of resins, plasticizers, surfactants; solvent for inks and dyes
Regulatory: FDA 21CFR §173.310, 175.105, 176.210, 178.3570
Manuf./Distrib.: Aldrich; Fluka; Sigma
Trade names: Ucon® 50-HB-400

PPG-12-buteth-16
CAS 9038-95-3 (generic); 9065-63-8 (generic); 74623-31-7
Synonyms: POE (16) POP (12) monobutyl ether; POP (12) POE (16) monobutyl ether
Definition: Polyoxypropylene, polyoxyethylene ether of butyl alcohol
Empirical: $(C_7H_{14}O_2 \cdot C_6H_{12}O_2)_x$
Formula: $C_4H_9(OCH_3CHCH_2)_x(OCH_2CH_2)_yOH$, avg. x = 12, avg. y = 16
Toxicology: LD50 (oral, rat) 5370 mg/kg
Uses: Defoaming agent, lubricant, emollient; in demulsifying and wetting formulations; brake fluids; metalworking; rubber lubricant; textiles; fiber lubricant; chemical processing; cosmetics; boiler water additive for food processing; solvent for inks and dyes
Regulatory: FDA 21CFR §173.310, 175.105, 176.210, 178.3570
Manuf./Distrib.: Aldrich; Fluka; Sigma
Trade names: Ucon® 50-HB-660

PPG-20-buteth-30
CAS 9038-95-3 (generic); 9065-63-8 (generic)
Synonyms: POE (30) POP (20) monobutyl ether; POP (20) POE (30) monobutyl ether
Definition: Polyoxypropylene, polyoxyethylene ether of butyl alcohol
Empirical: $(C_7H_{14}O_2 \cdot C_6H_{12}O_2)_x$
Formula: $C_4H_9(OCH_3CHCH_2)_x(OCH_2CH_2)_yOH$, avg. x = 20, avg. y = 30
Uses: Emollient; defoamer for food processing; lubricant for compressor, rubber, greases, textiles; intermediate for prep. of resins, plasticizers, surfactants; solvent for inks and dyes

Regulatory: FDA 21CFR §173.310, 175.105, 176.210, 178.3570
Manuf./Distrib.: Aldrich; Fluka; Sigma
Trade names: Ucon® 50-HB-2000

PPG-28-buteth-35
CAS 9038-95-3 (generic); 9065-63-8 (generic)
Synonyms: POE (35) POP (28) monobutyl ether; POP (28) POE (35) monobutyl ether
Definition: Polyoxypropylene, polyoxyethylene ether of butyl alcohol
Empirical: $(C_7H_{14}O_2 \cdot C_6H_{12}O_2)_x$
Formula: $C_4H_9(OCH_3CHCH_2)_x(OCH_2CH_2)_yOH$, avg. x = 28, avg. y = 35
Uses: Emollient; detergent; defoamer; metalworking and hydraulic fluids; mold release agent; emulsion polymerization; defoamer for food processing; lubricant for compressor, rubber, greases, textiles; intermediate for prep. of resins, plasticizers, surfactants; solvent for inks and dyes
Regulatory: FDA 21CFR §173.310, 175.105, 176.210, 178.3570
Manuf./Distrib.: Aldrich; Fluka; Sigma
Trade names: Ucon® 50-HB-3520

PPG-33-buteth-45
CAS 9038-95-3 (generic); 9065-63-8 (generic)
Synonyms: POE (45) POP (33) monobutyl ether; POP (33) POE (45) monobutyl ether
Definition: Polyoxypropylene, polyoxyethylene ether of butyl alcohol
Empirical: $(C_7H_{14}O_2 \cdot C_6H_{12}O_2)_x$
Formula: $C_4H_9(OCH_3CHCH_2)_x(OCH_2CH_2)_yOH$, avg. x = 33, avg. y = 45
Toxicology: LD50 (oral, rat) 5370 mg/kg
Uses: Defoaming agent, lubricant, emollient; in demulsifying and wetting formulations; brake fluids; metalworking; rubber lubricant; textiles; fiber lubricant; food processing; chemical processing; cosmetics; boiler water additive for food processing; solvent for inks and dyes
Regulatory: FDA 21CFR §173.310, 173.340, 175.105, 176.210, 178.3570
Manuf./Distrib.: Aldrich; Fluka; Sigma
Trade names: Ucon® 50-HB-5100

PPG-3 butyl ether
Uses: Solvent for coatings
Trade names: Arcosolv® TPNB

PPG-5 butyl ether
CAS 9003-13-8 (generic)
Synonyms: POP (5) butyl ether
Definition: Polypropylene glycol ether of butyl alcohol
Formula: $C_4H_9(OCHCH_3CH_2)_nOH$, avg. n = 5
Uses: Lubricant for two-stroke cycle engines, rubber, greases, textiles; reactive intermediate for resins, plasticizers, surfactants; antifoam in boiler water, latex processing; solvent for inks and dyes
Manuf./Distrib.: Aldrich
Trade names: Ucon® LB-65

PPG-9 butyl ether
CAS 9003-13-8 (generic)
Synonyms: POP (9) butyl ether
Definition: Polypropylene glycol ether of butyl alcohol
Formula: $C_4H_9(OCHCH_3CH_2)_nOH$, avg. n = 9
Uses: Lubricant for two-stroke cycle engines, rubber, greases, textiles; reactive intermediate for resins, plasticizers, surfactants; antifoam in boiler water, latex

processing; solvent for inks and dyes
Manuf./Distrib.: Aldrich
Trade names: Ucon® LB-135

PPG-14 butyl ether
CAS 9003-13-8 (generic)
Synonyms: POP (14) butyl ether
Definition: Polypropylene glycol ether of butyl alcohol
Formula: $C_4H_9(OCHCH_3CH_2)_nOH$, avg. n = 14
Uses: Emollient, solvent, bodying agent for personal care prods.; lubricant for antiperspirants
Manuf./Distrib.: Aldrich
Trade names: Dow L910; Ucon® Fluid AP

PPG-15 butyl ether
CAS 9003-13-8 (generic)
Synonyms: POP (15) butyl ether
Definition: Polypropylene glycol ether of butyl alcohol
Formula: $C_4H_9(OCHCH_3CH_2)_nOH$, avg. n = 15
Uses: Lubricant for two-stroke cycle engines, rubber, greases, textiles; reactive intermediate for resins, plasticizers, surfactants; antifoam in boiler water, latex processing; solvent for inks and dyes
Manuf./Distrib.: Aldrich
Trade names: Ucon® LB-285

PPG-18 butyl ether
CAS 9003-13-8 (generic)
Synonyms: POP (18) butyl ether; PPG (18) butyl ether
Definition: Polypropylene glycol ether of butyl alcohol
Formula: $C_4H_9(OCH_3CHCH_2)_nOH$, avg. n = 18
Properties: Liq.; insol. in water; sol. in alcohol
Uses: Emollient; lubricant for two-stroke cycle engines, rubber, greases, textiles; reactive intermediate for resins, plasticizers, surfactants; antifoam in boiler water, latex processing; solvent for inks and dyes
Regulatory: FDA 21CFR §175.105, 176.210
Manuf./Distrib.: Aldrich
Trade names: Dow L1150; Ucon® LB-385

PPG (18) butyl ether. *See* PPG-18 butyl ether

PPG-22 butyl ether
CAS 9003-13-8 (generic)
Synonyms: POP (22) butyl ether
Definition: Polypropylene glycol ether of butyl alcohol
Formula: $C_4H_9(OCH_3CHCH_2)_nOH$, avg. n = 22
Uses: Lubricant for two-stroke cycle engines, rubber, greases, textiles; reactive intermediate for resins, plasticizers, surfactants; antifoam in boiler water, latex processing; solvent for inks and dyes
Regulatory: FDA 21CFR §175.105, 176.200,176.210
Manuf./Distrib.: Aldrich
Trade names: Ucon® LB-525

PPG-24 butyl ether
CAS 9003-13-8 (generic)
Synonyms: POP (24) butyl ether; PPG (24) butyl ether
Definition: Polypropylene glycol ether of butyl alcohol
Formula: $C_4H_9(OCH_3CHCH_2)_nOH$, avg. n = 24
Properties: Liq.; insol. in water; sol. in alcohol

PPG (24) butyl ether

Uses: Emollient; lubricant for two-stroke cycle engines, rubber, greases, textiles; reactive intermediate for resins, plasticizers, surfactants; antifoam in boiler water, latex processing; solvent for inks and dyes
Regulatory: FDA 21CFR §175.105, 176.210, 178.3570
Manuf./Distrib.: Aldrich
Trade names: Ucon® LB-625

PPG (24) butyl ether. *See* PPG-24 butyl ether

PPG-33 butyl ether
CAS 9003-13-8 (generic)
Synonyms: POP (33) butyl ether; PPG (33) butyl ether
Definition: Polypropylene glycol ether of butyl alcohol
Formula: $C_4H_9(OCH_3CHCH_2)_nOH$, avg. n = 33
Properties: Liq.; insol. in water; sol. in alcohol
Uses: Emollient; lubricant for two-stroke cycle engines, rubber, greases, textiles; reactive intermediate for resins, plasticizers, surfactants; antifoam in boiler water, latex processing; solvent for inks and dyes
Regulatory: FDA 21CFR §166.110, 172.615, 175.105, 175.25, 175.300, 175.380, 175.390, 175.170, 176.210, 177.1010, 177.1210, 177.1350, 184.1660, 178.3570
Manuf./Distrib.: Aldrich
Trade names: Ucon® LB-1145

PPG (33) butyl ether. *See* PPG-33 butyl ether

PPG-40 butyl ether
CAS 9003-13-8 (generic)
Synonyms: POP (40) butyl ether; PPG (40) butyl ether
Definition: Polypropylene glycol ether of butyl alcohol
Formula: $C_4H_9(OCH_3CHCH_2)_nOH$, avg. n = 40
Properties: Liq.; insol. in water; sol. in alcohol
Uses: Emollient; lubricant for two-stroke cycle engines, rubber, greases, textiles; reactive intermediate for resins, plasticizers, surfactants; antifoam in boiler water, latex processing; solvent for inks and dyes
Regulatory: FDA 21CFR §175.105, 176.200, 176.210, 178.3570
Manuf./Distrib.: Aldrich
Trade names: Ucon® LB-1715

PPG (40) butyl ether. *See* PPG-40 butyl ether

PPG-53 butyl ether
CAS 9003-13-8 (generic)
Synonyms: POP (53) butyl ether; PPG (53) butyl ether
Definition: Polypropylene glycol ether of butyl alcohol
Formula: $C_4H_9(OCH_3CHCH_2)_nOH$, avg. n = 53
Properties: Liq.; insol. in water; sol. in alcohol
Uses: Emollient; lubricant for two-stroke cycle engines, rubber, greases, textiles; reactive intermediate for resins, plasticizers, surfactants; antifoam in boiler water, latex processing; solvent for inks and dyes
Regulatory: FDA 21CFR §178.3570
Manuf./Distrib.: Aldrich
Trade names: Ucon® LB-3000

PPG (53) butyl ether. *See* PPG-53 butyl ether

PPG-10 cetyl ether
CAS 9035-85-2 (generic)
Synonyms: POP (10) cetyl ether; PPG (10) cetyl ether
Definition: Polypropylene glycol ether of cetyl alcohol

Formula: $CH_3(CH_2)_{14}CH_2(OCH_3CHCH_2)_nOH$, avg. n = 10
Uses: Emollient, coupler, cosolvent, plasticizer, superfatting, wetting and spreading agent, penetrant; lubricant in cosmetics and personal care prods.
Trade names: Procetyl 10

PPG (10) cetyl ether. *See* PPG-10 cetyl ether

PPG-30 cetyl ether
CAS 9035-85-2 (generic)
Synonyms: POP (30) cetyl ether; PPG (30) cetyl ether
Definition: Polypropylene glycol ether of cetyl alcohol
Formula: $CH_3(CH_2)_{14}CH_2(OCH_3CHCH_2)_nOH$, avg. n = 30
Properties: Liq.; sol. in oils, alcohol; insol. in water
Toxicology: No known toxicity
Uses: Emollient, coupler, cosolvent, plasticizer, superfatting agent; in cosmetics, pharmaceuticals
Trade names: Procetyl 30

PPG (30) cetyl ether. *See* PPG-30 cetyl ether

PPG-50 cetyl ether
CAS 9035-85-2 (generic)
Synonyms: POP (50) cetyl ether; PPG (50) cetyl ether
Definition: Polypropylene glycol ether of cetyl alcohol
Formula: $CH_3(CH_2)_{14}CH_2(OCH_3CHCH_2)_nOH$, avg. n = 50
Properties: Liq.; sol. in lanolin oil, alcohol
Uses: Emollient, coupler, cosolvent, plasticizer, superfatting agent for cosmetics
Trade names: Procetyl 50

PPG (50) cetyl ether. *See* PPG-50 cetyl ether

PPG-40 diethylmonium chloride
CAS 9076-43-1
Synonyms: POP (40) methyl diethyl ammonium chloride; Quaternium-21
Classification: Quaternary ammonium salt
Formula: $[CH_3N(CH_2CH_3)_2(CH_2CH_3CHO)_nH]^+Cl^-$, avg. n = 40
Uses: Pigment dispersant, particle suspension aid, emulsifier, solvent, conditioner, antistat, lubricant, corrosion inhibitor for toiletries, cosmetics, germicides, syn. fibers and plastics, textiles, industrial processes; ore flotation additive
Trade names: Emcol® CC-42

PPG-24-glycereth-24
CAS 51258-15-2 (generic); 9082-00-2
Synonyms: PEG/PPG-24/24 glycerin; POE (24) POP (24) glyceryl ether; POP (24) POE (24) glyceryl ether
Definition: Polyoxypropylene, polyoxyethylene ether of glycerin
Formula: $C_3H_7O_2(OCH_3CHCH_2)_x(OCH_2CH_2)_yOH$, avg. x = 24, avg. y = 24
Uses: Emollient; provides solvency, low odor, low irritation, low toxicity to personal care prods.
Manuf./Distrib.: Aldrich
Trade names: Dow 15-200

PPG-66-glycereth-12
CAS 51258-15-2 (generic)
Definition: POP, POE ether of glycerin
Uses: Provides solvency, low odor, low irritation, low toxicity to personal care prods.

PPG-10 glyceryl ether

Manuf./Distrib.: Aldrich
Trade names: Dow 112-2

PPG-10 glyceryl ether
CAS 25791-96-2
Definition: PPG ether of glycerin
Uses: Provides solvency, low odor, low irritation, low toxicity to personal care prods.
Manuf./Distrib.: Aldrich
Trade names: Dow PT700

PPG-55 glyceryl ether
CAS 25791-96-2 (generic)
Synonyms: POP (55) glyceryl ether
Definition: Polypropylene glycol ether of glycerin
Formula: $C_3H_7O_2(OCH_3CHCH_2)_nOH$, avg. n = 55
Uses: Provides solvency, low odor, low irritation, low toxicity to personal care prods.
Manuf./Distrib.: Aldrich
Trade names: Dow PT3000

PPG-2 methyl ether
CAS 13429-07-7; 34590-94-8; EINECS 236-547-9; 252-104-2
Synonyms: DPM; DPGME; Dipropylene glycol methyl ether; Dipropylene glycol monomethyl ether; 1,4-Dimethyl-3,6-dioxa-1-heptanol; Methoxy dipropylene glycol; (2-Methoxymethylethoxy) propanol; 1-(2-Methoxyisopropoxy)-2-propanol; POP (2) methyl ether; 1-(2-Methoxy-1-methylethoxy)-2-propanol
Classification: Aliphatic ether alcohol
Definition: PPG ether of methyl alcohol
Empirical: $C_7H_{16}O_3$
Formula: $CH_3(OCH_3CHCH_2)_nOH$, avg. n = 2
Properties: Colorless clear liq.; mild ethereal sweet odor; sol. in water, acetone, ethanol, benzene, CCl_4, ether, methanol, monochlorobenzene, petrol, ether, VM&P naphtha; m.w. 148.23; sp.gr. 0.951; f.p. -83 C; b.p. 190 C; flash pt. (CC) 86 C; ref. index 1.4205
Toxicology: ACGIH TLV/TWA 100 ppm (skin); STEL 150 ppm (skin); LD50 (oral, rat) 5660 mg/kg, (skin, rabbit) 9500 mg/kg; mildly toxic by ing. and skin contact; skin and eye irritant; causes tears; mild allergen
Precaution: Combustible; flamm. when exposed to heat or flame; can react with oxidizing materials; Hazardous decomp. prods.: Heated to decomp., emits acrid smoke, irritating fumes
NFPA: Health 0, Flammability 2, Reactivity 0
Uses: Solvent for paints, lacquers, resins, dyes, oils/greases, cellulose, cleaners, inks, cosmetics, agric., epoxy laminates, adhesives, floor polish, fuel additives, oilfield, mining, and electronic chems.; chem. intermediate applics.; heat transfer fluid
Regulatory: FDA 21CFR §175.105, 181.22, 181.30
Manuf./Distrib.: Aldrich; Ashland; Fluka; Hoechst AG; Oxiteno; Van Waters & Rogers
Trade names: Arcosolv® DPM; Dowanol® DPM; Icinol DPM; Methyl Dipropasol; Poly-Solv® DPM
Trade names containing: Printsolve™ Ink Remover

PPG-3 methyl ether
CAS 25498-49-1; 10213-77-1; 20324-33-8; 37286-64-9 (generic)
Synonyms: TPGME; Tripropylene glycol monomethyl ether; Tripropylene glycol

methyl ether; 2-(2-Methoxymethylethoxy)methylethoxy)propanol; Polyoxypropylene (3) methyl ether; 1-(2-(2-Methoxy-1-methylethoxy)-1methylethoxy)-2-propanol; PPG (3) methyl ether; 2-(2-(2-Methoxypropoxy)propoxy)-1-propanol

Classification: Aliphatic ether alcohol
Definition: PPG ether of methyl alcohol
Empirical: $C_{10}H_{22}O_4$
Formula: $CH_3(OCH_3CHCH_2)_nOH$, avg. n = 3
Properties: Colorless clear liq.; sl. sweetish ether odor; sol. in water, acetone, ethanol, benzene, CCl_4, ether, methanol, monochlorobenzene, VM&P naphtha; m.w. 206.32; sp.gr. 0.967 (25/25 C); f.p. -60 C; b.p. 243 C; flash pt. (CC) 121 C
Toxicology: Low oral toxicity; mild eye irritant; vapor inh. nonirritating; high concs. of mists can irritate nose/throat, cause CNS depression, nausea, headache, weakness, fatigue
Precaution: Incompat. with oxidizing agents; increases risk of fire and explosion
NFPA: Health 0, Flammability 1, Reactivity 0
Uses: High-boiling solvent for paints, lacquers, resins, dyes, oils, greases, cellulose, cleaners, inks, functional fluids, agric., cosmetics, epoxy laminates, adhesives, floor polish, fuel additives, oilfield, mining, and electronic chems.; chemical intermediate; component of hydraulic brake fluids
Regulatory: FDA 21CFR §181.22, 181.30
Manuf./Distrib.: Aldrich; Oxiteno
Trade names: Arcosolv® TPM; Dowanol® TPM; Poly-Solv® TPM

PPG (3) methyl ether. *See* PPG-3 methyl ether

PPG-2 methyl ether acetate
CAS 88917-22-0
Synonyms: Dipropylene glycol methyl ether acetate; Dipropylene glycol monomethyl ether acetate
Formula: $CH_3(OC_3H_6)_2OAc$
Properties: M.w. 190.2
Uses: Solvent for paints, inks, cleaners, epoxy laminates, agric., adhesives, floor polish, oilfield, mining, and electronic chems.
Manuf./Distrib.: Aldrich; Ashland
Trade names: Arcosolv® DPM Acetate; Dowanol® DPMA

PPG-10 methyl glucose ether
CAS 61849-72-7
Synonyms: POP (10) methyl glucose ether; PPG (10) methyl glucose ether
Definition: PPG ether of methyl glucose
Formula: $CH_3(CH_6H_{10}O_5)(OCH_3CHCH_2)_nOH$, avg. n = 10
Uses: Humectant for personal care prods.; freezing pt. depressant; emollient, moisturizer, foam modifier in detergent and shampoo systems; solvent and solubilizer for topical pharmaceuticals
Trade names: Glucam® P-10

PPG (10) methyl glucose ether. *See* PPG-10 methyl glucose ether

PPG-20 methyl glucose ether
CAS 61849-72-7
Synonyms: POP (20) methyl glucose ether; PPG (20) methyl glucose ether
Definition: PPG ether of methyl glucose
Formula: $CH_3(CH_6H_{10}O_5)(OCH_3CHCH_2)_nOH$, avg. n = 20
Uses: Humectant for personal care prods.; freezing pt. depressant; emollient, moisturizer, foam modifier in detergent and shampoo systems; solvent and

PPG (20) methyl glucose ether

solubilizer for topical pharmaceuticals
Trade names: Glucam® P-20; Prox-onic MG-020 P

PPG (20) methyl glucose ether. *See* PPG-20 methyl glucose ether

PPG-3-myreth-3
CAS 37311-04-9 (generic)
Synonyms: POE (3) POP (3) myristyl ether; POP (3) POE (3) myristyl ether
Definition: Polyoxypropylene, polyoxyethylene ether of myristyl alcohol
Formula: $CH_3(CH_2)_{12}CH_2(OCH_3CHCH_2)_x(OCH_2CH_2)_yOH$, avg. x=3, avg. y=3
Uses: Lubricant, emulsifier, wetting agent, penetrant, dye dispersant, scouring aid, antistat, solvent coupler; syn. oils for personal care prods.; metal processing; textile surfactant
Trade names: Witconol™ APEM

PPG-3 myristyl ether
CAS 63793-60-2 (generic)
Synonyms: Tripropylene glycol myristyl ether; POP (3) myristyl ether; PPG (3) myristyl ether
Definition: PPG ether of myristyl alcohol
Formula: $CH_3(CH_2)_{12}CH_2(OCH_3CHCH_2)_nOH$, avg. n = 3
Properties: Liq.; insol. in water; sol. in alcohol, min. oil
Uses: Emollient for cosmetics and toiletries; lubricant, emulsifier, wetting agent, penetrant, dye dispersant, scouring aid for metal processing, textile use; solvent coupler
Trade names: Witconol™ APM

PPG (3) myristyl ether. *See* PPG-3 myristyl ether
PPG-1 myristyl ether acetate. *See* Propylene glycol myristyl ether acetate

PPG-2 myristyl ether propionate
Synonyms: POP (2) myristyl ether propionate; PPG (2) myristyl ether propionate
Definition: Ester of propionic acid and the PPG ether of myristyl alcohol
Formula: $CH_3(CH_2)_{12}CH_2(OCH_3CHCH_2)_2OCOCH_2CH_3$
Uses: Emollient, lubricant with dry, lt. greaseless feel; coupling agent; solvent for sunscreen actives; for bath oils, creams, moisturizers, emulsions
Trade names: Crodamol PMP

PPG (2) myristyl ether propionate. *See* PPG-2 myristyl ether propionate

PPG-23 oleyl ether
CAS 52581-71-2 (generic)
Synonyms: POP (23) oleyl ether
Definition: Polypropylene glycol ether of oleyl alcohol
Formula: $CH_3(CH_2)_7CH=CH(CH_2)_7CH_2(OCH_3CHCH_2)_nOH$, avg. n = 23
Uses: Emollient, cosolvent

PPG-37 oleyl ether
CAS 52581-71-2 (generic)
Synonyms: POP (37) oleyl ether
Definition: PPG ether of oleyl alcohol
Formula: $CH_3(CH_2)_7CH=CH(CH_2)_7CH_2(OCH_3CHCH_2)_nOH$, avg. n = 37
Uses: Emollient, cosolvent

PPG-11 stearyl ether
CAS 25231-21-4 (generic)
Synonyms: POP (11) stearyl ether; PPG (11) stearyl ether
Definition: PPG ether of stearyl alcohol
Formula: $CH_3(CH_2)_{16}CH_2(OCH_3CHCH_2)_nOH$, avg. n = 11
Properties: Liq.; insol. in water; sol. in alcohol, min. oil

Uses: Emollient, lubricant, emulsifier, wetting agent, penetrant, dye dispersant, scouring aid, antistat; in cosmetics, toiletries, metal processing, as textile surfactant; solvent coupler
Trade names: Witconol™ APS

PPG (11) stearyl ether. *See* PPG-11 stearyl ether

PPG-15 stearyl ether
CAS 25231-21-4 (generic)
Synonyms: POP (15) stearyl ether; PPG (15) stearyl ether
Definition: PPG ether of stearyl alcohol
Formula: $CH_3(CH_2)_{16}CH_2(OCH_3CHCH_2)_nOH$, avg. n = 15
Uses: Surfactant used as emulsifier, dispersant, solubilizer, visc. control agent, emollient, solvent, lubricant for cosmetics, pharmaceuticals, and industrial applics.; coupler for fragrances
Regulatory: FDA approved for topicals
Trade names: Arlamol® E

PPG (15) stearyl ether. *See* PPG-15 stearyl ether
pri-Heptyl alcohol. *See* Heptyl alcohol
Primary amyl acetate. *See* Amyl acetate
Primary isoamyl alcohol. *See* Isoamyl alcohol
pri-Octyl alcohol. *See* Caprylic alcohol
Propanamide, 2-hydroxy-N-(1-ethoxy-2-hydroxyethyl). *See* Lactamide DGA
2-Propanamine. *See* Isopropylamine

Propane
CAS 74-98-6; EINECS 200-827-9
UN No. 1075 (DOT), 1978 (DOT)
Synonyms: Dimethylmethane; Propyl hydride; Liquified petroleum gas
Classification: Sat. aliphatic hydrocarbon
Empirical: C_3H_8
Formula: $CH_3CH_2CH_3$
Properties: Colorless gas, nat. gas odor; easily liquefied under pressure at R.T.; noncorrosive; sol. in ether, alcohol, chloroform; sl. sol. in water, acetone; m.w. 44.09; sp.gr. 0.513 (0 C, as liq.), 1.56 (0 C, as vapor); f.p. -189.9 C; b.p. -42.5 C; flash pt. -104 C
Toxicology: OSHA PEL/TWA 1000 ppm; asphyxiant; narcotic in high concs.; liq. propane may cause frostbite to skin and eyes
Precaution: Flamm.; dangerous fire hazard exposed to heat, flame; explosive limits in air 2.4-9.5%; vigorously reactive with oxidizers; explosive as vapor exposed to heat or flame; explosive with ClO_2; violent exothermic reaction with barium peroxide, heat
NFPA: Health 1, Flammability 4, Reactivity 0
Uses: Aerosol propellant; paint aerosols; organic synthesis, household and industrial fuel, mfg. of ethylene, extractant, solvent, refrigerant, gas enricher; direct food additive, propellant, aerating agent; pharmaceutical propellant
Regulatory: FDA 21CFR §173.350, 184.1655, GRAS; FDA approved for topicals; USP/NF compliance
Manuf./Distrib.: Air Prods.; Aldrich; Exxon; Fina Chems.; Marathon Oil; Phillips; Stanchem

Propane, 1,2-dichloro. *See* Propylene dichloride
1,2-Propanediol. *See* Propylene glycol
Propane-1,2-diol. *See* Propylene glycol
Propane-1,2-diol alginate. *See* Propylene glycol alginate
1,2-Propanediol carbonate. *See* Propylene carbonate

1,2-Propanediol monomethyl ether acetate. *See* Propylene glycol methyl ether acetate
1,2-Propanediyl carbonate. *See* Propylene carbonate
Propane, 1,2-epoxy-. *See* Propylene oxide
Propane, 2-methoxy-2-methyl. *See* Methyl t-butyl ether
Propanenitrile. *See* Propionitrile
Propane, 2-nitro-. *See* 2-Nitropropane
1,2,3-Propanetriol. *See* Glycerin
Propane-1,2,3-triol. *See* Glycerin
1,2,3-Propanetriol monoacetate. *See* Glyceryl acetate
1,2,3-Propanetriol octadecanoate. *See* Glyceryl stearate
1,2,3-Propanetriol triacetate. *See* Triacetin
1,2,3-Propanetriyl 12-(acetyloxy)-9-octadecenoate. *See* Glyceryl triacetyl ricinoleate
Propanoic acid butyl ester. *See* n-Butyl propionate
Propanoic acid ethyl ester. *See* Ethyl propionate
Propanoic acid, 2-hydroxy-, 1-methylethyl ester. *See* Isopropyl lactate
Propanoic acid, pentyl ester. *See* n-Pentyl propionate
Propanoic acid, propyl ester. *See* Propyl propionate
Propanol-1. *See* n-Propyl alcohol
1-Propanol. *See* n-Propyl alcohol
Propan-1-ol. *See* n-Propyl alcohol
2-Propanol. *See* Isopropyl alcohol
n-Propanol. *See* n-Propyl alcohol
2-Propanol, 1-(2-butoxy-1-methylethoxy)-. *See* Dipropylene glycol butyl ether
Propanol, (1,1-dimethylethoxy)-. *See* Propylene glycol t-butyl ether
2-Propanol, 1-ethoxy-. *See* Propylene glycol ethyl ether
2-Propanol, 1-methoxy. *See* Methoxyisopropanol
2-Propanol, 2-methyl-. *See* t-Butyl alcohol
1-Propanol, 2-phenoxy-. *See* Propylene glycol phenyl ether
2-Propanol, 1-propoxy-. *See* Propylene glycol n-propyl ether
Propanone. *See* Acetone
2-Propanone. *See* Acetone
Propellant 11. *See* Trichlorofluoromethane
Propellant 12. *See* Dichlorodifluoromethane
Propellant 22. *See* Chlorodifluoromethane
Propellant 114. *See* Dichlorotetrafluoroethane
Propene oxide. *See* Propylene oxide
Propione. *See* Diethyl ketone
Propionic acid, 2-methyl-, monoester with 2,2,4-trimethyl-1,3-pentanediol. *See* 2,2,4-Trimethyl-1,3-pentanediol monoisobutyrate
Propionic acid, propyl ester. *See* Propyl propionate
Propionic ether. *See* Ethyl propionate
Propionic nitrile. *See* Propionitrile

Propionitrile
CAS 107-12-0
UN No. 2404 (DOT)
Synonyms: Cyanoethane; Ether cyanatus; Ethyl cyanide; Hydrocyanic ether; Propanenitrile; Propionic nitrile; Propylnitrile
Classification: Nitrile/cyano compd.
Empirical: C_3H_5N
Formula: CH_3CH_2CN
Properties: Colorless liq.; sweet ether-like odor; sol. in water, dimethylformamide,

ether, alcohol; m.w. 55.08; sp.gr. 0.783 (20 C); m.p. -93 C; b.p. 97 C; flash pt. (CC) 2 C

Toxicology: LD50 (oral, rat) 39 mg/kg, (skin, rabbit) 210 mg/kg; poison by ing., skin contact, IV, IP routes; mod. toxic by inh.; mild skin irritant; can be absorbed thru skin; mod. eye irritant; vapor can cause nose/throat irritation

Precaution: Flamm.; dangerous fire hazard exposed to heat, flame, oxidizers; incompat. with oxidizing agents (can react explosively), strong acids or bases (can react violently); explosive with N-bromosuccinimide and heat; Hazardous decomp. prods.: Hydrogen cyanide; heated to decomp., emits toxic fumes of NO_x and CN^-

NFPA: Health 4, Flammability 3, Reactivity 1

Uses: Solvent in petrol. refining; dielectric fluid; chem. intermediate; raw material for drug mfg.

Manuf./Distrib.: Aldrich; BASF AG; Fluka; Lonza

2-Propoxyethanol. *See* Ethylene glycol propyl ether
2-(2-Propoxyethoxy) ethanol. *See* Diethylene glycol propyl ether
1-Propoxy-2-propanol. *See* Propylene glycol n-propyl ether
n-Propoxypropanol. *See* Propylene glycol n-propyl ether

Propyl acetate

CAS 109-60-4; EINECS 203-686-1

UN No. 1276 (DOT)

Synonyms: Acetic acid n-propyl ester; Acetic acid propyl ester; n-Propyl acetate; 1-Propyl acetate; 1-Acetoxypropane; n-Propyl ethanoate

Classification: Sat. aliphatic carboxylic acid ester

Definition: Ester of propyl alcohol and acetic acid

Empirical: $C_5H_{10}O_2$

Formula: $CH_3COOCH_2CH_2CH_3$

Properties: Colorless liq., pear-like odor; sol. in alcohols, ketones, esters, hydrocarbons, diethyl ether, castor oil, linseed oil; mod. sol. in water; m.w. 102.14; dens. 0.887 (20/20 C); m.p. -92 C; b.p. 99-102 C; flash pt. (CC) 14 C; ref. index 1.384 (20 C); surf. tens. 24.3 dynes/cm (20 C)

Toxicology: LD50 (oral, rat) 9370 mg/kg, (IP, mouse) 1420 mg/kg; ACGIH TLV/TWA 200 ppm; STEL 250 ppm; mod. toxic by IP and subcut. routes; may be irritating to skin, mucous membranes; narcotic in high concs.

Precaution: Highly flamm.; dangerous fire hazard exposed to heat, flame, oxidizers; explosive as vapor exposed to heat, flame; incompat. with oxidizing agents, strong acids, potassium t-butoxide, reactive nitrogen compds.; Hazardous decomp. prods.: n-Propanol, acetic acid; heated to decomp., emits acrid smoke and irritating fumes

NFPA: Health 1, Flammability 3, Reactivity 0

Uses: Syn. flavoring agent for foods, pharmaceuticals; fragrance ingred. (perfumery); solvent for nitrocellulose and other cellulose derivs., natural and syn. resins, lacquers, plastics, printing inks, chlorinated rubber, heat-reactive phenolics; organic synthesis, lab reagent

Regulatory: FDA 21CFR §172.515, 177.1200; FEMA GRAS

Manuf./Distrib.: Aldrich; BASF; Berje; BP Chems. Ltd; Coyne; Eastman; Fluka; Hoechst Celanese; Houghton Chem.; Penta Mfg.; Union Carbide

1-Propyl acetate. *See* Propyl acetate
2-Propyl acetate. *See* Isopropyl acetate
n-Propyl acetate. *See* Propyl acetate
Propyl alcohol. *See* n-Propyl alcohol
1-Propyl alcohol. *See* n-Propyl alcohol

n-Propyl alcohol
CAS 71-23-8; EINECS 200-746-9
UN No. 1274 (DOT)
Synonyms: 1-Propanol; Propanol-1; Propan-1-ol; n-Propanol; normal Propyl alcohol; Propylic alcohol; Albacol; Optal; Propyl alcohol; 1-Propyl alcohol; Ethyl carbinol; 1-Hydroxypropane; Ethyl carbinol; Alcohol C-3
Classification: Aliphatic alcohol
Empirical: C_3H_8O
Formula: $CH_3CH_2CH_2OH$
Properties: Colorless clear liq., alcoholic and sl. stupefying odor; very sol. in benzene; sol. in water, ethyl alcohol, ether, acetone, propylene glycol; dissolves fat; m.w. 60.10; sp.gr. 0.804 (20/4 C); vapor pressure 10 mm Hg (14.7 C); m.p. -127 C; b.p. 97-98 C; flash pt. (TCC) 23 C; ref. index 1.385 (20 C)
Toxicology: LD50 (oral, rat) 1.87 g/kg; ACGIH TLV/TWA 200 ppm (skin); STEL 250 ppm (skin); toxic by skin absorption; drying effect on skin may lead to cracking, fissuring, and infections; mildly irritating to eyes, mucous membranes
Precaution: Highly flamm.; dangerous fire hazard exposed to heat, flame, oxidizers; explosive limits in air 2-13%; incompat. with oxidizing agents (increases risk of fire and explosion); ignites on contact with potassium t-butoxide; Hazardous decomp. prods.: Emits acrid smoke and irritating fumes
NFPA: Health 1, Flammability 3, Reactivity 0
Uses: Solvent for natural and syn. resins, cellulose esters and ethers, paints, waxes, veg. oils; organic synthesis; chemical intermediate; in polishes; brake fluids; degreasing solvent; antiseptic; syn. flavoring for foods, pharmaceuticals
Regulatory: FDA 21CFR §172.515, 175.105, 177.1200, 573.880; FEMA GRAS; FDA approved for topicals; BP compliance
Manuf./Distrib.: Aldrich; Allchem Ind.; ARCO; Ashland; J.T. Baker; Baychem; Burdick & Jackson; Chemcentral; Coyne; Eastman; Fluka; Hoechst Celanese; Houghton Chem.; Mallinckrodt; Penta Mfg.; Samson; Sigma; Spectrum Chem. Mfg.; Sunnyside; Union Carbide; Van Waters & Rogers

s-Propyl alcohol. *See* Isopropyl alcohol
sec-Propyl alcohol. *See* Isopropyl alcohol
2-Propylamine. *See* Isopropylamine
sec-Propylamine. *See* Isopropylamine

Propylbenzene
CAS 103-65-1; EINECS 203-132-9
UN No. 2364 (DOT)
Synonyms: Isocumene; 1-Phenylpropane; 1-Propylbenzene; n-Propylbenzene
Classification: Aromatic hydrocarbon
Empirical: C_9H_{12}
Formula: $C_6H_5CH_2CH_2CH_3$
Properties: Colorless clear liq.; sol. in acetone, benzene, petrol. ether; pract. insol. in water; m.w. 120.19; sp.gr. 0.852 (20 C); vapor pressure 10 mm Hg (43.4 C); m.p. -99 C; b.p. 159 C; flash pt. (CC) 30 C
Toxicology: LD50 (oral, rat) 6040 mg/kg; LCLo (inh., mouse) 20 g/m^3; mildly toxic by ing. and inh.
Precaution: Flamm.; very dangerous fire hazard exposed to heat, flame, oxidizers; reactive with oxidizers; mod. explosive hazard as vapor exposed to heat or flame; Hazardous decomp. prods.: Heated to decomp., emits acrid smoke and fumes
NFPA: Health 2, Flammability 3, Reactivity 0

Uses: Solvent for cellulose acetate; in textile printing and dyeing; mfg. of methylstyrene; constituent of asphalt and naphtha
Manuf./Distrib.: Aldrich; Fluka; Phillips; Sigma

1-Propylbenzene. *See* Propylbenzene
n-Propylbenzene. *See* Propylbenzene
Propyl carbinol. *See* Butyl alcohol
Propyl 'Cellosolve'. *See* Ethylene glycol propyl ether

Propylcyclohexane
CAS 1678-92-8
Classification: Sat. alicyclic hydrocarbon
Empirical: C_9H_{18}
Formula: $C_3H_7C_6H_{11}$
Properties: Colorless liq.; sol. in most org. solvs.; insol. in water; m.w. 126.24; sp.gr. 0.794 (20 C); m.p. -95 C; b.p. 156-157 C; flash pt. 35 C
Toxicology: May cause mild skin irritation; conc. vapor can irritate eyes; probably a CNS depressant; high vapor concs. may cause sl. nose/throat irritation, headache, lightheadedness, nausea, drowsiness; very high concs. may cause unconsciousness, death
Precaution: Flamm.; incompat. with strong oxidizing agents (increased fire and explosion hazard); Hazardous decomp. prods.: None
NFPA: Health 0, Reactivity 0
Uses: Solvent; org. synthesis
Manuf./Distrib.: Aldrich

Propylcyclopentane
CAS 2040-96-2
Classification: Sat. alicyclic hydrocarbon
Empirical: C_8H_{16}
Formula: $C_3H_7C_5H_9$
Properties: Colorless liq.; sol. in most org. solvs.; insol. in water; m.w. 112.22; sp.gr. 0.776 (20 C); m.p. -117 C; b.p. 131 C
Toxicology: May cause mild skin irritation; conc. vapor may be irritating to eyes; probably a CNS depressant; high vapor concs. may cause nose/throat irritation, lightheadedness, nausea, dizziness; very high concs. may cause unconsciousness, death
Precaution: Flamm.; incompat. with strong oxidizing agents (increased risk of fire and explosion); Hazardous decomp. prods.: None
NFPA: Health 0, Reactivity 0
Uses: Solvent; org. synthesis
Manuf./Distrib.: Fluka; Sigma

Propylene carbonate
CAS 108-32-7; EINECS 203-572-1
Synonyms: 1,3-Dioxolan-2-one, 4-methyl; 1,3-Carbonyl dioxypropane; Cyclic propylene carbonate; Dipropylene carbonate; Cyclic methylethylene carbonate; 4-Methyl-1,3-dioxolan-2-one; 1-Methyl ethylene carbonate; 1,2-Propanediol carbonate; 1,2-Propanediyl carbonate; 1,2-Propylene carbonate; Propylene glycol cyclic carbonate; Carbonic acid, 1,2-propylene glycol ester; Carbonic acid, cyclic propylene ester; Cyclic 1,2-propylene carbonate
Classification: Organic compd.; carbonic acid ester
Empirical: $C_4H_6O_3$
Properties: Colorless clear liq.; odorless; very sol. in water; misc. with acetone, benzene, chloroform, ether, ethyl acetate; m.w. 102.10; sp.gr. 1.2069 (20/20 C); vapor pressure 0.03 mm Hg (20 C); m.p. -48.8 C; b.p. 242.1 C; flash pt. (OC) 135

C; ref. index 1.422 (20 C); pH 8.8; can remain liq. (supercools) below its f.p.
Toxicology: LD50 (oral, rat) 29 g/kg, (skin, rabbit) 20 mL/kg; mildly toxic by ingestion; human skin irritant; eye irritant
Precaution: Combustible exposed to heat or flame; incompat. with strong oxidizing agents (increases fire risk), acids, bases, reducing agents (decomp. can occur); Hazardous decomp. prods.: May produce propylene oxide if an acid, base, or salt is present in an aq. sol'n. of propylene carbonate; heated to decomp., emits acrid smoke and irritating fumes
NFPA: Health 1, Flammability 1, Reactivity 0
Uses: Solvent for inorg. salts, plasticizers, syn. fibers, polymers, pigments, dyes, paints; extraction solvent; intermediate for organic syntheses; reactive diluent for urethane foams and coatings, foundry sand binders, textiles, natural gas purification; cosmetics lubricant; flavoring agent for pharmaceuticals
Regulatory: FDA 21CFR §175.105; FDA approved for topicals
Manuf./Distrib.: Aldrich; Allchem Ind.; ARCO; Ashland; BASF AG; Fluka; Great Western; Hüls AG; Sigma; Spectrum Chem. Mfg.
Trade names: Arconate® 1000; Arconate® HP; Arconate® PC; Jeffsol™ PC; Solvenon® PC; Texacar® PC

1,2-Propylene carbonate. *See* Propylene carbonate
Propylene chloride. *See* Propylene dichloride

Propylene dichloride
CAS 78-87-5; EINECS 201-152-2
UN No. 1279 (DOT)
Synonyms: 1,2-Dichloropropane; Propane, 1,2-dichloro; Propylene chloride; α,β-Dichloropropane; α,β-Propylene dichloride
Empirical: $C_3H_6Cl_2$
Formula: $CH_3CHClCH_2Cl$
Properties: Colorless liq.; m.w. 112.99; dens. 1.1593 (20/20 C); vapor pressure 40 mm Hg (19.4 C); b.p. 96.8 C; flash pt. 60 F
Toxicology: LD50 (oral, rat) 1947 mg/kg, (skin, rabbit) 8750 mg/kg, (IP, rat) 1100 mg/kg; ACGIH TLV/TWA 75 ppm; STEL 110 ppm; mod. toxic by inh. and ing.; mildly toxic by skin contact; primary irritant; eye irritant; can cause dermatitis
Precaution: Very dangerous fire hazard exposed to heat or flame; reacts with aluminum to form aluminum chloride (reaction can be explosive); can react vigorously with oxidizers; Hazardous decomp. prods.: Heated to decomp., emits toxic fumes of Cl^-
Uses: Intermediate for perchloroethylene, CCl_4; lead scavenger for antiknock fluids; solvents for fats, oils, waxes, gums, resins, cellulose esters and ethers; scouring compounds; metal degreasers; soil fumigant for nematodes
Manuf./Distrib.: Aldrich; Fluka; Sigma

α,β-Propylene dichloride. *See* Propylene dichloride
Propylene epoxide. *See* Propylene oxide

Propylene glycol
CAS 57-55-6; 4254-15-3 (+); 4254-14-2 (-); 4254-16-4 (±); EINECS 200-338-0
Synonyms: 1,2-Propanediol; Propane-1,2-diol; α-Propylene glycol; 1,2-Propylene glycol; Monopropylene glycol; 1,2-Dihydroxypropane; 2-Hydroxypropanol; Methyl glycol; Methyl ethylene glycol; Methyl ethyl glycol; Trimethyl glycol
Classification: Aliphatic dihydric alcohol
Empirical: $C_3H_8O_2$
Formula: $CH_3CHOHCH_2OH$
Properties: Colorless clear visc. liq., odorless, sl. acrid taste; sol. in water, essential

oils, acetone, chloroform, ether, ethanol; m.w. 76.11; sp.gr. 1.0362; m.p. -60 C; b.p. 188.2 C; flash pt. (OC) 99 C

Toxicology: LD50 (oral, rat) 25 ml/kg, (skin, rabbit) 20,800 mg/kg; sl. toxic by ingestion, IP,IV, subcut. routes; eye/human skin irritant; human systemic effects by ing.

Precaution: Combustible exposed to heat or flame; explosive as vapor exposed to heat or flames; explosive limits 2.6-12.6%; reactive with oxidizers; can increase risk of fire and explosion; Hazardous decomp. prods.: Heated to decomp., emits acrid smoke and irritating fumes

NFPA: Health 0, Flammability 1, Reactivity 0

Uses: Intermediate in synthesis of org. compds., polyester and alkyd resins, polypropylene glycol; solvent, emulsifier in production paints, foods, drugs; substitute for ethylene glycol and glycerol; hygroscopic agent; energy source in animal feed; raw material for resinous plasticizers; plasticizer for adhesives, cork, paper prods.; mold growth and fermentation inhibitor; bactericide (air sterilization); component of antifreeze, hydraulic and brake fluids; pharmaceutical vehicle, solvent

Use level: 10-25% (oral sol'ns.), 10-80% (parenterals), 5-80% (topicals)

Regulatory: FDA 21CFR §169.175, 169.176, 169.177, 169.178, 169.180, 169.181, 175.300, 177.2600, 178.3300, 184.1666, 582.4666, GRAS; USDA 9CFR §318.7, 381.147; BATF 27CFR §240.1051; EPA reg., approved for some drugs; Japan approved with limitations; Europe listed; FEMA GRAS; FDA approved for orals, parenterals, topicals; USP/NF, BP, Ph.Eur., JP compliance

Manuf./Distrib.: Aldrich; Allchem Ind.; ARCO; Asahi Denka Kogyo; Ashland; J.T. Baker; Baychem; BP Chems. Ltd; Chemcentral; Coyne; Eastman; Fluka; Focus; C.P. Hall; Harcros USA; Houghton Chem.; Hüls AG; Huntsman; Olin; Primachem; Samson; Seeler Ind.; Sigma; Spectrum Chem. Mfg.; Sunnyside; Texaco; Union Carbide; Van Waters & Rogers; Veckridge; Westco

Trade names: Propylene Glycol USP/FCC Ultra Grade

1,2-Propylene glycol. *See* Propylene glycol
α-Propylene glycol. *See* Propylene glycol

Propylene glycol alginate

CAS 9005-37-2

Synonyms: Hydroxypropyl alginate; Alginic acid, ester with 1,2-propanediol; Propane-1,2-diol alginate

Definition: Mixture of propylene glycol esters of alginic acid

Formula: $(C_9H_{14}O_7)_8$

Properties: Wh. to ylsh. fibrous or gran. powd., pract. odorless and tasteless; sol. in water, dil. organic acids, hydroalcoholic mixts.; m.w. 1873.6

Toxicology: LD50 (oral, rat) 7200 mg/kg; mildly toxic by ingestion; Hazardous decomp. prods.: Heated to decomp., emits acrid smoke and irritating fumes

Uses: Food additive (human); gellant, film-former, emulsifier, suspending aid; reactive with milk; for foods (dairy prods., extruded foods, beverages, salad dressings, dessert gels, sauces, frozen desserts, bakery prods.); cosmetics/pharmaceuticals (lotions, vitamin suspension), industrial (paper coatings, adhesives, textile printing, dyeing, explosives); pharmaceutical solvent

Regulatory: FDA 21CFR §133.133, 133.134, 133.162, 133.178. 133.179, 172.210, 172.820, 172.858, 173.340, 176.170, GRAS; FEMA GRAS; Japan approved (1% max.); Europe listed; UK approved; FDA approved for orals; USP/NF compliance

Manuf./Distrib.: Ashland; Kelco Int'l.; Meer; Spectrum Chem. Mfg.

Propylene glycol butoxyethyl ether. *See* 1-Butoxyethoxy-2-propanol
Propylene glycol butyl ether. *See* Butoxypropanol

Propylene glycol 1-t-butyl ether. *See* Propylene glycol t-butyl ether
Propylene glycol n-butyl ether. *See* Butoxypropanol

Propylene glycol t-butyl ether
CAS 57018-52-7
Synonyms: PTB; 1-t-Butoxy-2-propanol; Propylene glycol 1-t-butyl ether; Propanol, (1,1-dimethylethoxy)-
Empirical: $C_7H_{16}O_2$
Formula: $C_4H_9OCH_2CHOHCH_3$
Properties: Partly sol. in water; m.w. 132.2; dens. 0.874 (20/4 C); f.p. -69 F; b.p. 151 C; flash pt. (TCC) 113 F; ref. index 1.4116
Toxicology: Skin and eye irritant; tumorigen
Precaution: Flamm.
Uses: Solvent for water-based paints, cleaners, inks, cutting fluids; in polyester and alkyd resin prod.
Manuf./Distrib.: Aldrich; Eastman; Fluka
Trade names: Arcosolv® PTB; Butyl Propasol

Propylene glycol caprylate
CAS 31565-12-5
Synonyms: Caprylic acid, monoester with 1,2-propanediol
Definition: Ester of caprylic acid and propylene glycol
Empirical: $C_{11}H_{22}O_3$
Formula: $CH_3(CH_2)_6COOCH_2CHCH_3OH$
Uses: Emollient, solubilizer, solvent for cosmetics, pharmaceuticals
Trade names: Imwitor® 408; Nikkol Sefsol 218

Propylene glycol cyclic carbonate. *See* Propylene carbonate

Propylene glycol dicaprylate
CAS 7384-98-7; EINECS 230-962-9
Synonyms: Octanoic acid, 1-methyl-1,2-ethanediyl ester
Definition: Diester of propylene glycol and caprylic acid
Empirical: $C_{19}H_{36}O_4$
Formula: $CH_3(CH_2)_6COOCH_2CHCH_3OCO(CH_2)_6CH_3$
Uses: Emollient, solvent, solubilizer for cosmetics, pharmaceuticals
Regulatory: FDA 21CFR §172.856, 173.340, 175.300, 176.170, 176.210, 177.2800
Trade names: Nikkol Sefsol 228

Propylene glycol dicaprylate/dicaprate
CAS 9062-04-8; 58748-27-9; 68583-51-7; 68988-72-7
Synonyms: Decanoic acid, 1-methyl-1,2-ethanediyl ester mixed with 1-methyl-1,2-ethanediyl dioctanoate
Definition: Mixture of the propylene glycol diesters of caprylic and capric acids
Uses: Carrier, coupler, solvent for flavors, fragrance oil, sol. colorants, vitamins, medicinals, cosmetics; emollient for creams, lotions, makeup; food additive, emulsifier, stabilizer, defoamer
Regulatory: FDA 21CFR §172.856, 173.340, 175.300, 176.170, 176.210, 177.2800
Trade names: Captex® 200; Lexol® PG-865; Liponate PC; Miglyol® 840; Myritol® PC; Neobee® 20; Neobee® 1054; Neobee® M-20

Propylene glycol dinonanoate. *See* Propylene glycol dipelargonate

Propylene glycol dioctanoate
CAS 7384-98-7; 56519-71-1
Synonyms: 2-Ethylhexanoic acid, 1-methyl-1,2-ethanediyl ester; Octanoic acid, 1,3-propanediyl ester
Definition: Diester of propylene glycol and 2-ethylhexanoic acid

Empirical: $C_{19}H_{36}O_4$
Formula: $CH_3(CH_2)_3CHCH_3CH_2COOCH_2CHCH_3OCOCHCH_2CH_3(CH_2)_3CH_3$
Properties: Sol. in alcohol, min. oil; sp.gr. 0.921; flash pt. (COC) 159 C; ref. index 1.4350
Uses: Emollient for creams, lotions, topicals, makeup, bath preps., after-shave lotions; vehicle for fragrance; plasticizer; carrier for flavors; vehicle for vitamins, nutritional prods.; pharmaceutical emollient, lubricant, carrier/vehicle, humectant; solvent, carrier, vehicle in flavors, fragrances, vitamins, antibiotics, medicinals
Manuf./Distrib.: Inolex
Trade names: Lexol® PG-800

Propylene glycol dipelargonate
CAS 41395-83-9; EINECS 255-350-9
Synonyms: Propylene glycol dinonanoate; Nonanoic acid, 1-methyl-1,2-ethanediyl ester
Definition: Diester of propylene glycol and pelargonic acid
Empirical: $C_{21}H_{40}O_4$
Formula: $CH_3(CH_2)_7COOCH_2CHCH_3OCO(CH_2)_7CH_3$
Properties: Sp.gr. 0.918; flash pt. (COC) 197 C; ref. index 1.4404
Uses: Emollient, lubricant, cosolvent for bath oils, preshave lotions, aerosol systems, lipsticks, glosses, makeup bases, pharmaceuticals; carrier for fragrances; plasticizer
Manuf./Distrib.: Henkel; Inolex
Trade names: Schercemol PGDP

Propylene glycol ethyl ether
CAS 1569-02-4; 52125-53-8
Synonyms: PGEE; 1-Ethoxy-2-propanol; Propylene glycol monoethyl ether; 2-Propanol, 1-ethoxy-
Classification: Aliphatic ether alcohol
Empirical: $C_5H_{12}O_2$
Formula: $CH_3CH(OH)CH_2OCH_2CH_3$
Properties: Colorless clear liq.; sol. in water, ethanol, ether, other org. solvs.; m.w. 104.17; sp.gr. 0.896; b.p. 131 C; flash pt. (COC) 43 C
Toxicology: LD50 (oral, rat) 4400 mg/kg, (skin, rabbit) 8100 mg/kg; mildly toxic by ingestion and skin contact; vapor irritating to eyes, nose, and throat; absorbed thru skin, but health effects unlikely; mod. to severe eye irritant; probable lachrymator
Precaution: Combustible; incompat. with oxidizing agents (increases fire/explosion hazard), strong acids (may cause decomp. of ethers); Hazardous decomp. prods.: Heated to decomp., emits acrid smoke and irritating fumes
Uses: Fast evaporating solvent for coatings, lacquers, resins, dyes, oils, greases, cleaners, cosmetics, agric., electronics, inks, textile and adhesive prods.
Trade names: Arcosolv® PE

Propylene glycol ethyl ether acetate
CAS 98516-30-4
Synonyms: Propylene glycol monoethyl ether acetate
Uses: Highly misc. solvent with broad solvency for coatings, cleaners, agric., cosmetics, electronics, ink, textile and adhesive prods.
Manuf./Distrib.: Ashland; Harcros USA
Trade names: Arcosolv® PE Acetate

Propylene glycol laurate
CAS 142-55-2; 27194-74-7; EINECS 205-542-3

Propylene glycol methyl ether

Synonyms: Dodecanoic acid, 2-hydroxypropyl ester; Dodecanoic acid, monoester with 1,2-propanediol; Propylene glycol monolaurate
Definition: Ester of propylene glycol and lauric acid
Empirical: $C_{15}H_{30}O_3$
Formula: $CH_3(CH_2)_{10}COOCH_2CHCH_3OH$
Properties: Sp.gr. 0.911; m.p. 0-12 C; flash pt. (COC) 188 C
Toxicology: Nontoxic but can cause allergic reactions in hypersensitive persons
Uses: Surfactant, emulsifier, coemulsifier, stabilizer, wetting agent, lubricant, plasticizer, emollient, solvent, and antistat; used in cosmetic, pharmaceutical, industrial, food applics., pulp/paper, metalworking, lubricants, textiles, agric., paints; adhesives; food additive, emulsifier, stabilizer, defoamer; pharmaceutical emulsifier, stabilizer, lubricant, solvent
Regulatory: FDA 21CFR §172.856, 173.340, 175.105, 175.300, 176.170, 176.210, 177.2800
Manuf./Distrib.: Calgene; Inolex; Stepan; Velsicol
Trade names: Imwitor® 412; Schercemol PGML

Propylene glycol methyl ether. *See* Methoxyisopropanol

Propylene glycol methyl ether acetate
CAS 108-65-6; 84540-57-8; EINECS 203-603-9
UN No. 1993 (DOT)
Synonyms: MPA; PGMEA; PM acetate; 1-Methoxy-2-acetoxypropane; 2-Methoxypropyl acetate; Acetic acid, 2-methoxy-1-methyl ethyl ester; Propylene glycol monomethyl ether acetate; 1-Methoxy-2-propanol acetate; 1-Methoxy-2-propyl acetate; 1,2-Propanediol monomethyl ether acetate
Classification: Aliphatic ether alcohol
Empirical: $C_6H_{12}O_3$
Formula: $CH_3COOCHCH_3CH_2OCH_3$
Properties: Colorless liq.; mild fruity odor; sol. 20% in water; probably sol. in many org. solvs.; m.w. 132.2; sp.gr. 0.97 (20/20 C); m.p. < -67 C; b.p. 146 C; flash pt. (Seta) 45 C; ref. index 1.402 (20 C)
Toxicology: LD50 (oral, rat) 8532 mg/kg, (skin, rabbit) 5 g/kg, (IP, mouse) 750 mg/kg; inh. of vapor may cause nose/throat irritation; low oral toxicity; very high exposures unlikely but may cause CNS depression; low toxicity by skin absorption
Precaution: Flamm.; incompat. with oxidizing agents (increases fire/explosion hazard)
NFPA: Health 0, Flammability 2, Reactivity 0
Uses: Solvent for electronics, inks, acrylics, nitrocellulose, paints, urethane coatings, lacquers, water-based paints, oils, gums
Manuf./Distrib.: Aldrich; Allchem Ind.; Ashland; Fluka; Harcros USA; Oxiteno; Spectrum Chem. Mfg.; Van Waters & Rogers
Trade names: Arcosolv® PM Acetate; Dowanol® PMA; Eastman® PM Acetate; Ektasolve® PM Acetate; Methyl Propasol® Acetate

Propylene glycol mono-n-butyl ether. *See* Butoxypropanol
Propylene glycol monoethyl ether. *See* Propylene glycol ethyl ether
Propylene glycol monoethyl ether acetate. *See* Propylene glycol ethyl ether acetate
Propylene glycol monolaurate. *See* Propylene glycol laurate
Propylene glycol monomethyl ether. *See* Methoxyisopropanol
α-Propylene glycol monomethyl ether. *See* Methoxyisopropanol
Propylene glycol monomethyl ether acetate. *See* Propylene glycol methyl ether acetate
Propylene glycol monoricinoleate. *See* Propylene glycol ricinolcatc

Propylene glycol myristyl ether acetate
Synonyms: PPG-1 myristyl ether acetate
Definition: Ester of propylene glycol myristyl ether and acetic acid
Uses: Emollient, solvent, and plasticizer for anhyd. oil systems, emulsions
Trade names: Hetester PMA

Propylene glycol phenyl ether
CAS 770-35-4; 4169-04-4
Synonyms: 1-Phenoxy-2-propanol; 2-Phenoxypropanol; 1-Propanol, 2-phenoxy-; 2-Phenoxypropyl alcohol; Propylene phenoxetol
Classification: Ether alcohol
Empirical: $C_9H_{12}O_2$
Formula: $C_6H_5OC_3H_6OH$
Properties: Colorless clear liq.; sl. 'rotten orange' odor; sol. in a variety of org. solvs.; mod. sol. in water; m.w. 152.20; sp.gr. 1.063; f.p. 11.4 C; b.p. 242.7 C; flash pt. (TCC) 116 C
Toxicology: LD50 (oral, rat) 2830 mg/kg; inh. may cause nose/throat irritation; low toxicity by inh. unless material is heated or a mist formed; repeated skin contact may cause mild irritation
Precaution: Combustible; incompat. with oxidizing agents (increases fire risk), strong acids (may cause decomp. of ethers)
Uses: Solvent for coatings, paints, lacquers, resins, inks, dyes, oils, greases, cleaners; bactericide agent; fixative for soaps and perfumes; intermediate for plasticizers
Manuf./Distrib.: Aldrich; Fluka
Trade names: Dowanol® PPh

Propylene glycol propyl ether. *See* Propylene glycol n-propyl ether

Propylene glycol n-propyl ether
CAS 1569-01-3
Synonyms: Propylene glycol propyl ether; 1-Propoxy-2-propanol; n-Propoxypropanol; 2-Propanol, 1-propoxy-
Classification: Aliphatic ether alcohol
Empirical: $C_6H_{14}O_2$
Formula: $C_3H_7OCH_2CH(CH_3)OH$
Properties: Colorless clear liq.; mild char. odor; very sol. in water; sol. in a variety of org. solvs.; m.w. 118.18; sp.gr. 0.886 (20/20 C); f.p. -112 F; b.p. 149.8 C; flash pt. (TCC) 48 C; ref. index 1.4121
Toxicology: LD50 (oral, rat) 3250 mg/kg, (skin, rabbit) 3560 mg/kg; toxic; low toxicity absorbed thru skin; low toxicity by ing. and short-term inh.; mild to mod. skin irritant; mod. to severe eye irritant; may cause corneal injury
Precaution: Flamm.; incompat. with oxidizing agents (increases fire/explosion hazard), strong acids (may cause decomp. of ethers); Hazardous decomp. prods.: Heated to decomp., emits acrid smoke and irritating fumes
Uses: Solvent for coatings, cleaners, cosmetics, agric., electronics, ink, textiles, adhesives
Manuf./Distrib.: Aldrich
Trade names: Arcosolv® PNP; Dowanol® PnP

Propylene glycol ricinoleate
CAS 26402-31-3; EINECS 247-669-7
Synonyms: 12-Hydroxy-9-octadecenoic acid, monoester with 1,2-propanediol; Propylene glycol monoricinoleate
Definition: Ester of propylene glycol and ricinoleic acid
Empirical: $C_{21}H_{40}O_4$

Propylene oxide

Properties: Sp.gr. 0.960; m.p. < -16 C; flash pt. (COC) 221 C; ref. index 1.469
Uses: Emulsifier, coemulsifier, stabilizer, wetting agent, lubricant, emollient, and antistat; used in cosmetic, pharmaceutical, industrial, textile, food applics.; dye solvent; plasticizer for cellulosics
Manuf./Distrib.: CasChem
Trade names: Flexricin® 9

Propylene oxide
CAS 75-56-9; EINECS 200-879-2
UN No. 1280 (DOT)
Synonyms: 1,2-Propylene oxide; 1,2-Epoxypropane; Propane, 1,2-epoxy-; Methyloxirane; S(-)-Methyloxirane; Methyloxacyclopropane; Methyl ethylene oxide; Propene oxide; Propylene epoxide
Empirical: C_3H_6O
Properties: Colorless liq., ethereal odor; sl. sol. in water; sol. in alcohol and ether; m.w. 58.08; dens. 0.8304 (20/20 C); vapor pressure 400 mm Hg (17.8 C); f.p. –104.4 C; b.p. 33.9 C; flash pt. –35 F; ref. index 1.366 (20C)
Toxicology: LD50 (oral, rat) 380 mg/kg, (skin, rabbit) 1245 mg/kg; ACGIH TLV/TWA 20 ppm; primary irritant; poison by IP route; mod. toxic by ing., inh., skin contact; severe skin and eye irritant
Precaution: Highly flamm.; explosive limits in air 2-22%; very dangerous fire and explosion hazard exposed to heat or flame; explosive reaction with epoxy resin and sodium hydroxide; forms explosive mixts. with oxygen; reacts vigorously with oxidizers; Hazardous decomp. prods.: Emits acrid smoke and fumes
NFPA: Health 3, Flammability 4, Reactivity 2
Uses: Intermediate; polyols for urethane foams, propylene glycols, surfactants, detergents, isopropanolamines, fumigant, synthetic lubricants, synthetic elastomers, solvent
Regulatory: SARA reportable
Manuf./Distrib.: Aldrich; ARC; ARCO; Ashland; Electron Microscopy Sciences; Fluka; Hüls AG; Spectrum Chem. Mfg.; Texaco

1,2-Propylene oxide. *See* Propylene oxide
Propylene phenoxetol. *See* Propylene glycol phenyl ether
2-Propyl ethanoate. *See* Isopropyl acetate
n-Propyl ethanoate. *See* Propyl acetate
Propyl hydride. *See* Propane
Propylic alcohol. *See* n-Propyl alcohol
Propylmethanol. *See* Butyl alcohol
Propylnitrile. *See* Propionitrile
Propyl propanoate. *See* Propyl propionate

Propyl propionate
CAS 106-36-5; EINECS 203-389-7
Synonyms: Propanoic acid, propyl ester; n-Propyl propionate; Propyl propanoate; Propionic acid, propyl ester
Empirical: $C_6H_{12}O_2$
Formula: $CH_3CH_2COOCH_2CH_2CH_3$
Properties: Clear liq., complex fruity odor; sol. in 200 parts water; misc. with alcohol, ether; m.w. 116.16; dens. 0.881 (20/4 C); vapor pressure 10 mm Hg (19.4 C); m.p. -76 C; b.p. 120-122 C; flash pt. 22 C; ref. index 1.3935 (20 C)
Toxicology: LD50 (oral, rat) 10,331 mg/kg; LC50 (inh., unspecified mammal) 16,900 mg/m^3; mod. toxic by ing.; very mildly toxic by inh.; skin irritant
Precaution: Flamm. exposed to heat or flame; can react with oxidizers; Hazardous decomp. prods.: Heated to decomp., emits acrid smoke and irritating fumes
Uses: Solvent for paints; syn. flavoring agent for foods and pharmaceuticals

Regulatory: FDA 21CFR §172.515; FEMA GRAS
Manuf./Distrib.: Aldrich; Ashland; Fluka; Penta Mfg.; Union Carbide

n-Propyl propionate. *See* Propyl propionate
Pseudohexyl alcohol. *See* 2-Ethyl-1-butanol
Pseudopinen. *See* β-Pinene
Pseudopinene. *See* β-Pinene
PTB. *See* Propylene glycol t-butyl ether
PTBT. *See* p-t-Butyl toluene
Purified gum spirits. *See* Turpentine oil

Pyridine
CAS 110-86-1; EINECS 203-809-9
UN No. 1282 (DOT)
Synonyms: Azabenzene; Azine
Classification: Heterocyclic aromatic compd.
Empirical: C_5H_5N
Formula: NCHCHCHCHCH
Properties: Colorless liq., char. disagreeable odor, sharp burning taste; sol. in water, alcohol, ether, chloroform, petrol. ether, oils, other org. liqs.; m.w. 79.10; sp.gr. 0.98272 (20/4 C); vapor pressure 10 mm Hg (13.2 C); m.p. -41.6 C; b.p. 115-116 C; flash pt. (CC) 20 C; ref. index 1.510 (20 C); pH 8.5 (0.2 molar sol'n. in water)
Toxicology: LD50 (oral, rat) 1.58 g/kg, (skin, rabbit) 1121 mg/kg; ACGIH TLV/TWA 5 ppm; poison by IP route; mod. toxic by ing., IV, subcut.; may cause human CNS depression, irritation of skin and respiratory tract; may cause severe corrosive damage to eyes
Precaution: Highly flamm.; dangerous fire hazard exposed to heat, flame, oxidizers; volatile with steam; incompat. with oxidizing materials, strong acids or bases, formamide, iodine, maleic anhydride, silver perchlorate, perchromates, etc.; Hazardous decomp. prods.: Emits highly toxic fumes of NO_x
NFPA: Health 2, Flammability 3, Reactivity 0
Uses: Chem. intermediate, acid scavenger; in agrochemicals, sedative, surfactants, photographic materials, coatings, curing agents, rubber chemicals, vulcanization accelerator, plastics, antidandruff shampoos; denaturant for alcohols; solvent for many org. and inorg. materials; synthesis of vitamins, drugs, sedative; syn. flavoring agent for foods, pharmaceuticals; processing aid in textile dyeing
Regulatory: FDA 21CFR §172.515; FEMA GRAS; BP compliance
Manuf./Distrib.: AC Ind.; Aldrich; Allchem Ind.; Berje; Daicel Chem. Ind.; Fabrichem; Fluka; Koei Chem.; Nepera; Penta Mfg.; Raschig; Reilly Ind.; Schweizerhall; Sigma; Spectrum Chem. Mfg.

Pyridine, 3-methyl-. *See* β-Picoline
Pyridine, 4-methyl-. *See* γ-Picoline
Pyroacetic acid. *See* Acetone
Pyroacetic ether. *See* Acetone
Pyrobenzol. *See* Benzene
Pyrobenzole. *See* Benzene
Pyroligneus acid. *See* Acetic acid
Pyromucic aldehyde. *See* Furfural
2-Pyrrolidinone. *See* 2-Pyrrolidone
2-Pyrrolidinone, 1-cyclohexyl-. *See* N-Cyclohexyl pyrrolidone
2-Pyrrolidinone, 1-dodecyl-. *See* Lauryl pyrrolidone
2-Pyrrolidinone, 1-octyl-. *See* N-Octyl pyrrolidone
Pyrrolidone-2. *See* 2-Pyrrolidone

2-Pyrrolidone
CAS 616-45-5; EINECS 204-648-7
Synonyms: LAM; Butyrolactam; 2-Pyrrolidinone; Pyrrolidone-2; 2-Oxopyrrolidine; γ-Aminobutyric acid lactam
Empirical: C_4H_7NO
Formula: $CH_2CH_2CH_2C(O)NH$
Properties: Lt. yel. liq.; sol. in water, ethanol, ethyl ether, chloroform, benzene, ethyl acetate, carbon disulfide; m.w. 85.12; dens. 1.1; b.p. 245 C; flash pt. 265 F
Toxicology: LD50 (oral, rat) 328 mg/kg, (IP, rat) 160 mg/kg; mildly toxic by ingestion, subcutaneous route; mutagen; Hazardous decomp. prods.: Heated to decomp., emits toxic fumes of NO_x
Uses: Plasticizer and coalescing agent for acrylic latices; solvent for veterinary medicine; pharmaceutical solvent, solubilizer, intermediate
Manuf./Distrib.: Aldrich; Allchem Ind.; BASF; Fluka; ISP; UCB SA; Unitex
Trade names: 2-Pyrol®; Soluphor® P

Quaternium-21. *See* PPG-40 diethylmonium chloride

R 11. *See* Trichlorofluoromethane
R 12. *See* Dichlorodifluoromethane
R 22. *See* Chlorodifluoromethane
R 113. *See* Trichlorotrifluoroethane
R 114. *See* Dichlorotetrafluoroethane
Racemic limonene. *See* dl-Limonene
Red oil. *See* Oleic acid
Refrigerant 12. *See* Dichlorodifluoromethane
Refrigerant 22. *See* Chlorodifluoromethane
Refrigerant 112. *See* 1,1,2,2-Tetrachloro-1,2-difluoroethane
Refrigerant 113. *See* Trichlorotrifluoroethane
Refrigerant 114. *See* Dichlorotetrafluoroethane

Rice bran oil
CAS 68553-81-1; 84696-37-7; EINECS 271-397-8
Synonyms: Oils, rice bran; Rice oil
Definition: Oil expressed from rice bran
Properties: Golden yel. oil; misc. with hexane and other fat solvs.; dens. 0.916-0.921; cloud pt. < -7 C; acid no. < 0.1; iodine no. 92-115; sapon. no. 181-189; ref. index 1.470-1.473
Uses: Soaps, hydrogenated shortenings, edible oil; also suitable for cosmetics, pharmaceuticals; pharmaceutical ointment solvent
Regulatory: FDA 21CFR §175.105, 176.200, 176.210, 177.2260, 177.2800
Manuf./Distrib.: Arista Ind.; Penta Mfg.; Pokonobe Ind.; Tri-K Ind.; Welch, Holme & Clark
Trade names: EmCon Rice Bran

Rice oil. *See* Rice bran oil
Ricini oleum. *See* Castor oil
Ricinoleic acid, methyl ester, acetate. *See* Methyl acetyl ricinoleate
Ricinus oil. *See* Castor oil

Saccharose palmitate. *See* Sucrose palmitate
Safety solvent naphtha. *See* Stoddard solvent

Safflower oil
CAS 8001-23-8; EINECS 232-276-5
Synonyms: Carthanus tinctorious oil
Definition: Oily liq. obtained from seeds of *Carthanus tinctorius* consisting principally of triglycerides of linoleic acid
Properties: Lt. yel. oil; sol. in oil and fat solvs.; insol. in water; misc. with ether, chloroform; dens. 0.9211-0.9215 (25/25 C); iodine no. 135-150; sapon. no. 188-194; ref. index 1.472-1.475
Toxicology: Human skin irritant; ingestion in large volumes produces vomiting
Precaution: Light-sensitive; becomes rancid on exposure to air; Hazardous decomp. prods.: Heated to decomp., emits acrid smoke and irritating fumes
Uses: Alkyd resins; drying oil for paints, varnishes, medicine; salad oil blends; food coatings, food emulsifier, texturizer, flavoring agent; pharmaceutical emollient, vehicle, solvent, laxative
Regulatory: FDA 21CFR §175.105, 175.300, 176.200, 176.210, GRAS; Japan approved (safflower); USP/NF compliance
Manuf./Distrib.: ABITEC; Aldrich; Alnor Oil; Arista Ind.; Calgene; Charkit; Croda Chem. Ltd; Lipo; Penta Mfg.; Sigma; Spectrum Chem. Mfg.; Traco Labs; Tri-K Ind.; Welch, Holme & Clark
Trade names: Neobee® 18; Super Refined® Safflower Oil

Safflower oil, hybrid. *See* Hybrid safflower oil
SBA. *See* 2-Butanol

SD alcohol 1
Definition: Ethyl alcohol denatured with methyl alcohol and one of the following: denatonium benzoate, MIBK, mixed isomers of nitropropane, or methyl n-butyl ketone
Regulatory: FDA 27CFR §20.11, 21.32
Trade names containing: Tecsol® 1; Tecsol® 3

SD alcohol 3A
Definition: Ethyl alcohol denatured with methyl alcohol
Uses: Solvent used in coating systems for pharmaceutical tablets
Regulatory: FDA 27CFR §20.11, 21.35
Manuf./Distrib.: Eastman
Trade names containing: Tecsol® A; Tecsol® A-2; Tecsol® B; Tecsol® C; Tecsol® D; Tecsol® D-2; Tecsol® H

Secondary butyl alcohol. *See* 2-Butanol
Secondary caprylic alcohol. *See* 2-Octanol

Sesame oil
CAS 8008-74-0; EINECS 232-370-6
Synonyms: Gingilli oil; Gingelly oil; Teel oil; Sesame seed oil; Benne oil

Sesame seed oil

Definition: Refined fixed oil obtained from the seeds of *Sesamum indicum*
Properties: Bland ylsh. oily liq., pract. odorless, bland taste; sl. sol. in alcohol; misc. with ether, chloroform, hexane, carbon disulfide; dens. 0.916-0.921; iodine no. 103-116; sapon. no. 188-195; flash pt. 491 F; ref. index 1.4575-1.4598 (60 C)
Toxicology: LD50 (IV, rabbit) 678 μg/kg; poison by intravenous route; human skin irritant; may cause allergic reactions, primarily contact dermatitis; experimental carcinogen and tumorigen
Precaution: Combustible exposed to heat or flame; Hazardous decomp. prods.: Heated to decomp., emits acrid smoke and irritating fumes
NFPA: Health 0, Flammability 1, Reactivity 0
Uses: Cosmetics, lotions, injectables; natural flavorant for foods; pharmaceutical solvent, vehicle, keratin softener
Regulatory: FDA 21CFR §175.105, 175.300, 176.200, 176.210; FDA approved for parenterals, orals, topicals; USP/NF, BP, Ph.Eur., JP compliance
Manuf./Distrib.: Arista Ind.; Charkit; Croda Inc; Fluka; Penta Mfg.; Pokonobe Ind.; Quest Int'l.; Sigma; Spectrum Chem. Mfg.; Traco Labs; Tri-K Ind.; Vitamins, Inc; Welch, Holme & Clark
Trade names: Lipovol SES; Super Refined® Sesame Oil

Sesame seed oil. *See* Sesame oil

Shea butter
CAS 68424-60-2; 977026-99-5
Definition: Natural fat obtained from fruit of the Karite tree, *Butyrospermum parkii*
Properties: Iodine no. 53-65; sapon. no.178-190
Uses: Emollient, consistency giving agent for o/w and w/o creams and emulsions; native fatting agent for creams, lotions, anhyd. creams, pharmaceutical dermatologicals, suppositories; skin conditioner, occlusive agent, solvent for suntan preps., body lotions, soaps, shampoos
Manuf./Distrib.: ABITEC
Trade names: Fancol Karite Butter

Silicone rubber. *See* Dimethicone
Soda lye. *See* Sodium hydroxide
Sodium aluminosilicate. *See* Sodium silicoaluminate
Sodium aluminum silicate. *See* Sodium silicoaluminate
Sodium hydrate. *See* Sodium hydroxide

Sodium hydroxide
CAS 1310-73-2; EINECS 215-185-5
UN No. 1823 (DOT); 1824
Synonyms: Caustic soda; Sodium hydrate; White caustic; Soda lye; Lye
Classification: Alkali hydroxide
Empirical: HNaO
Formula: NaOH
Properties: Wh. deliq. solid beads or pellets; absorbs water and CO_2 from air; sol. in water, alcohol, glycerin; m.w. 40; dens. 2.12 (20/4 C); m.p. 318 C; b.p. 1390 C
Toxicology: TLV:Cl 2 mg/m^3 of air; LD50 (IP, mouse) 40 mg/kg; poison by IP route; mod. toxic by ingestion; mutagenic data; corrosive irritant to eye, skin, mucous membrane; mists and dusts cause small burns; heated to decomp., emits toxic fumes of Na_2O
Precaution: DOT: Corrosive material; strong base; incompat. with acids and water; may ignite or react violently with many org. compds.; dangerous material to handle
NFPA: Health 3, Flammability 0, Reactivity 1

Uses: Chemical mfg., rayon, cellophane, neutralizing agent in petrol. refining; textile processing; pulp, paper, soaps; vegetable oil refining; reclaiming rubber; food additive, pH control, solvent for colorants, processing aid; etching and electroplating; boiler water treatment, coagulation, corrosion/scale control, pH control, softening; pharmaceutical alkalizing agent, buffering agent

Regulatory: FDA 21CFR §163.110, 172.560, 172.814, 172.892, 173.310, 184.1763, GRAS; USDA 9CFR §318.7, 381.147; BATF 27CFR §21.101, 240.1051a; Japan restricted; Europe listed; UK approved; FDA approved for injectables, dentals, parenterals, inhalants, ophthalmics, orals, rectals, topicals, vaginals; USP/NF, BP, Ph.Eur. compliance; SARA reportable

Manuf./Distrib.: Advance Research Chems.; Akzo Nobel; Aldrich; Am. Biorganics; Asahi Chem. Ind.; Asahi Denka Kogyo; Elf Atochem N. Am.; EM Ind.; Fluka; General Chem.; Georgia Gulf; Georgia-Pacific; Hüls AG; ICI Spec. Chems.; LaRoche Ind.; Nissan Chem. Ind.; Norsk Hydro AS; Occidental; Olin; PPG Ind.; Primachem; Rasa Ind.; Sigma; Spectrum Chem. Mfg.; Universal Preserv-A-Chem; Wright

Sodium iodide

CAS 7681-82-5; EINECS 231-679-3

Synonyms: Sodium monoiodide; Sodium iodine

Empirical: INa

Formula: NaI

Properties: Wh. or colorless cryst. or powd.; very sol. in water; sol. in alcohol and glycerin; m.w. 149.89; dens. 3.667; vapor pressure 1 mm Hg (767 C); m.p. 661 C; b.p. 1300 C

Toxicology: LD50 (oral, rat) 4340 mg/kg, (oral, mouse) 1 g/kg, (IP, mouse) 430 mg/kg; primary irritant; mod. toxic by ing., IV, IP routes; human teratogenic and reproductive effects by ing.; skin and eye irritant

Precaution: Reacts violently with BrF_3, $HClO_4$, oxidants; Hazardous decomp. prods.: Heated to decomp., emits toxic fumes of I^- and Na_2O

Uses: Photography, pharmaceuticals, solvent for iodine, organic chemicals, reagent, feed additive, cloud seeding, scintillation, expectorant; antiseptic in topical anti-infective prods.

Regulatory: FDA approved for parenterals, topicals, vaginals; USP/NF, BP, Ph.Eur. compliance

Manuf./Distrib.: Advance Research Chems.; Ajay; Aldrich; Allan; CM Chem. Prods.; EM Ind.; Fluka; Mallinckrodt; Research Organics; Sigma; Spectrum Chem. Mfg.; West Agro

Sodium iodine. *See* Sodium iodide

Sodium p-methylbenzenesulfonate. *See* Sodium toluenesulfonate

Sodium monoiodide. *See* Sodium iodide

Sodium paratoluene sulfonate. *See* Sodium toluenesulfonate

Sodium rhodanate. *See* Sodium thiocyanate

Sodium rhodanide. *See* Sodium thiocyanate

Sodium silicoaluminate

CAS 1318-02-1; 1344-00-9; EINECS 215-684-8

Synonyms: Sodium aluminosilicate; Sodium aluminum silicate; Aluminum sodium silicate

Definition: Series of hydrated sodium aluminum silicates

Empirical: $Na_2O : Al_2O_3 : SiO_2$ with mole ratio ≈ 1:1:13.2

Properties: Wh. fine amorphous powd. or beads, odorless and tasteless; insol. in water, alcohol, org. solvs.; partly sol. in strong acids and alkali hydroxides @ 80-100 C; pH 6.5-10.5 (20% slurry)

Toxicology: Irritant to skin, eyes, mucous membranes; heated to decomp., emits

toxic fumes of Na_2O
Precaution: Noncombustible
Uses: Anticaking agent in detergents, desiccants, food preps.; reinforcing filler for rubbers; extender for paints; oral pharmaceuticals; solvent; detergent builder; gas separation; wh. pigment in paper, paint, plastics
Regulatory: FDA 21CFR §133.146, 160.105, 160.185, 182.2727 (2% max.), 582.2727, GRAS; Europe listed; UK approved; FDA approved for orals
Manuf./Distrib.: Albemarle; Crosfield; Fluka; J.M. Huber/Chems.; PQ; Punda Mercantile; Sigma; Spectrum Chem. Mfg.
See also Zeolite

Sodium sulfocyanate. *See* Sodium thiocyanate
Sodium sulfocyanide. *See* Sodium thiocyanate

Sodium thiocyanate
CAS 540-72-7; EINECS 208-754-4
Synonyms: Sodium rhodanate; Sodium rhodanide; Sodium sulfocyanate; Sodium sulfocyanide; Sodium thiocyanide; Thiocyanate sodium
Classification: Inorganic compd.
Empirical: CNNaS
Formula: NaSCN
Properties: Colorless cryst.; sol. in water, alcohol, acetone; m.w. 81.07; m.p. 287 C
Toxicology: LD50 (oral, rat) 764 mg/kg, (IP, rat), 540 mg/kg, (IV, mouse) 484 mg/kg; poison by ing., IV, subcut. routes; mod. toxic by IP route; irritant; ing. of large doses may cause vomiting, convulsions; Hazardous decomp. prods.: Heated to decomp., emits very toxic fumes of NO_x, SO_x, Na_2O
Uses: Intermediate for agric. chems., slimicides for paper, pharmaceuticals; brightener in electroplating; analytical reagent; photographic emulsion stabilizer; textile dye and printing assistant; corrosion inhibitor for water treatment; solvent for textile baths
Manuf./Distrib.: Akzo Nobel bv; Aldrich; Alemark; Allchem Ind.; Amber Syn.; Am. Int'l.; Barker Ind.; Degussa; EM Ind.; Fluka; Sigma; Witco/Polymer Addit.; Sanko Kagaku; Spectrum Chem. Mfg.; Toyo Kasei Kogyo; Witco/Polymer Addit.

Sodium thiocyanide. *See* Sodium thiocyanate

Sodium toluenesulfonate
CAS 657-84-1; 12068-03-0; EINECS 235-088-1
Synonyms: 4-Methylbenzenesulfonic acid, sodium salt; Methylbenzenesulfonic acid, sodium salt; Sodium tosylate; Sodium p-toluenesulfonate; 4-Toluenesulfonic acid sodium salt; p-Toluenesulfonic acid, sodium salt; Sodium p-methylbenzenesulfonate; Sodium paratoluene sulfonate; Benzenesulfonic acid, 4-methyl-, sodium salt
Classification: Substituted aromatic compd.
Empirical: $C_7H_8O_3S \cdot Na$
Formula: $CH_3CH_6H_4SO_3Na$
Properties: Crystals; sol. in water; m.w. 194.19
Toxicology: LD50 (IV, mouse) 1700 mg/kg; mod. toxic by IV route; Hazardous decomp. prods.: Heated to decomp., emits toxic fumes of SO_x and Na_2O
Uses: Hydrotropic solvent
Regulatory: FDA 21CFR §178.1010
Manuf./Distrib.: Aldrich; Ashland; Fabrichem; Fluka; Howard Hall; Hüls AG; Monomer-Polymer & Dajac Labs; Ruetgers-Nease; Witco/Oleo-Surf.

Sodium p-toluenesulfonate. *See* Sodium toluenesulfonate
Sodium tosylate. *See* Sodium toluenesulfonate

Sodium xylenesulfonate
CAS 1300-72-7; EINECS 215-090-9
Synonyms: Dimethylbenzene sulfonic acid, sodium salt
Definition: Sodium salt of ring sulfonated mixed xylene isomers
Empirical: $C_8H_{10}O_3S \cdot Na$
Formula: $(CH_3)_2C_6H_3SO_3Na$
Properties: m.w. 208.21
Toxicology: No known toxicity; irritant
Uses: Hydrotropic solvent, used in detergents; pharmaceutical solubilizer
Regulatory: FDA 21CFR §175.105, 176.180; FDA approved for topicals
Manuf./Distrib.: Aldrich; Ashland; Browning; Burlington Chem.; Fluka; Howard Hall; Huntsman; Mitsubishi Gas; Pilot; Ruetgers-Nease; Stepan; Witco/Oleo-Surf.
Trade names: Pilot SXS-40

Solvent naphtha. *See* Stoddard solvent
Sorbimacrogol oleate 300. *See* Polysorbate 80
Sorbimacrogol stearate 300. *See* Polysorbate 60
Sorbitan, esters, monododecanoate. *See* Sorbitan laurate
Sorbitan, esters, monohexadecanoate. *See* Sorbitan palmitate

Sorbitan laurate
CAS 1338-39-2; 5959-89-7; EINECS 215-663-3; 227-729-9
Synonyms: Sorbitan monolaurate; Sorbitan, esters, monododecanoate; Anhydrosorbitol monolaurate; Sorbitan monododecanoate
Definition: Monoester of lauric acid and hexitol anhydrides derived from sorbitol
Empirical: $C_{18}H_{34}O_6$
Properties: Yel. to amber oily liq., bland char. odor; sol. in methanol, alcohol, min. oil; sl. sol. in cottonseed oil, ethyl acetate; insol. in water; m.w. 346.47; dens. 1.032; acid no. 8 max.; sapon. no. 158-170; hyd. no. 330-358; flash pt. > 230 F; ref. index 1.4740
Toxicology: Experimental neoplastigen, reproductive effects; Hazardous decomp. prods.: Heated to decomp., emits acrid smoke and irritating fumes
Uses: Emulsifier for foods, cosmetics, household prods., industrial applics., coatings, household (polishes, cleaners), pharmaceutical uses; lubricant, antifog, antistat for PVC and other plastics; antifog for food pkg. films; polymerization emulsifier for thermoplastics; emollient for skin care prods.; lubricant, emulsifier in textile spin finishes; cosolvent
Regulatory: FDA 21CFR §175.320, 178.3400; Europe listed; UK approved; FDA approved for ophthalmics, orals; USP/NF, BP compliance
Manuf./Distrib.: Aldrich; Fluka; Sigma; Spectrum Chem. Mfg.
Trade names: Crill 1

Sorbitan monododecanoate. *See* Sorbitan laurate
Sorbitan monolaurate. *See* Sorbitan laurate
Sorbitan monopalmitate. *See* Sorbitan palmitate

Sorbitan palmitate
CAS 26266-57-9; EINECS 247-568-8
Synonyms: Sorbitan monopalmitate; 1,4-Anhydro-D-glucitol, 6-hexadecanoate; Sorbitan, esters, monohexadecanoate
Definition: Partial ester of palmitic acid with sorbitol mono- and dianhydrides
Empirical: $C_{22}H_{42}O_6$
Properties: Tan gran. waxy solid, faint tallow-like odor; sol. in ethyl acetate, warm abs. alcohol; sol. hazy in warm peanut or min. oil; insol. in water; m.w. 1109.59; dens. 0.989; acid no. 8 max.; sapon. no. 140-150; hyd. no. 275-305; flash pt. > 230

F; ref. index 1.4780
Toxicology: No known toxicity; irritant
Uses: Nonionic surfactant, emulsifier for foods, cosmetics, pharmaceuticals, household prods., industrial applics., polymerization, pulp/paper, metalworking, lubricants, textiles, agric., paints, adhesives; oil additive, corrosion inhibitor; cosolvent
Regulatory: FDA 21CFR §175.320, 178.3400; Europe listed; UK approved; FDA approved for intramuscular injectables; USP/NF compliance
Manuf./Distrib.: Aldrich; Fluka; Sigma; Spectrum Chem. Mfg.
Trade names: Crill 2

Sorbitan tri-9-octadecenoate. *See* Sorbitan trioleate

Sorbitan trioleate
CAS 26266-58-0; 85186-88-5; EINECS 247-569-3; 286-074-7
Synonyms: STO; Anhydrosorbitol trioleate; Sorbitan tri-9-octadecenoate
Definition: Triester of oleic acid and hexitol anhydrides derived from sorbitol
Empirical: $C_{60}H_{108}O_8$
Properties: Yel. to amber oily liq.; sol. in alcohol, veg. oil, min. oil; insol. in water, ethylene glycol, propylene glycol; m.w. 957.52; dens. 0.956; acid no. 17 max.; iodine no. 77-85; sapon. no. 170-190; hyd. no. 50-75; flash pt. > 230 F; ref. index 1.4760
Toxicology: Irritant
Uses: Emulsifier for foods, cosmetics, household prods., industrial applics., polymerization; antifog, lubricant, cling additive for plastic film; emollient, solvent
Regulatory: FDA 21CFR §175.320, 178.3400; FDA approved for inhalants, orals, topicals
Manuf./Distrib.: Aldrich; Fluka; Sigma; Spectrum Chem. Mfg.
Trade names: Nikkol SO-30; Nikkol SO-30R

Soya bean oil. *See* Soybean oil
Soya oil. *See* Soybean oil

Soybean oil
CAS 8001-22-7; EINECS 232-274-4
Synonyms: Soya oil; Soya bean oil; Chinese bean oil
Classification: Fixed oil
Definition: Oil obtained from seeds of soya plant, *Glycine soja*, by extraction or expression; consists of triglycerides of oleic, linoleic, linolenic, and saturated acids
Properties: Pale yel. to brnsh. yel. oil, sl. char. odor and taste; sol. in alcohol, ether, chloroform, carbon disulfide; insol. in water; dens. 0.924-0.929; visc. 50.09 cps; m.p. 22-31 C; iodine no. 120-141; sapon. no. 189-195; flash pt. 540 F; ref. index 1.471
Toxicology: May cause allergic reactions incl. hair damage and acne-like pimples; heated to decomp., emits acrid smoke and irritating fumes
Precaution: Combustible; light-sensitive
Uses: Soap mfg., plasticizer (epoxidized), high-protein foods, drying oil for inks, paints, varnishes, cattle feeds; dietary supplement; food coatings, emulsifier, texturizer, lubricant; pharmaceutical vehicle, solvent
Regulatory: FDA 21CFR §175.105, 175.300, 176.200, 176.210, 177.2800, GRAS; FDA approved for orals, topicals; USP/NF, BP compliance
Manuf./Distrib.: ABITEC; ADM; Akzo Nobel; Alba Int'l.; Aldrich; Alnor Oil; Arista Ind.; Ashland; Central Soya; Charkit; Croda Chem. Ltd; Degen; Lomas Int'l.; Reichhold; Sigma; Spectrum Chem. Mfg.; Werner G. Smith; Tri-K Ind.;

Welch, Holme & Clark
Trade names: Super Refined® Soybean Oil USP

Spirit of Hartshorn. *See* Ammonium hydroxide
Spirits of turpentine. *See* Turpentine oil
Spirit of turpentine. *See* Turpentine oil

Steareth-5
CAS 9005-00-9 (generic); 71093-13-5
Synonyms: PEG-5 stearyl ether; POE (5) stearyl ether
Definition: PEG ether of stearyl alcohol
Empirical: $C_{28}H_{58}O_6$
Formula: $CH_3(CH_2)_{16}CH_2(OCH_2CH_2)_nOH$, avg. n = 5
Uses: Detergent, wetting agent, emulsifier, dispersant, solubilizer, stabilizer, coupling agent for cosmetics, textiles, metalworking lubricants, household prods., industrial uses
Manuf./Distrib.: Aldrich; Fluka; Sigma
Trade names containing: Solulan® 5

Steareth-16
CAS 9005-00-9 (generic)
Synonyms: PEG-16 stearyl ether; POE (16) stearyl ether
Definition: PEG ether of stearyl alcohol
Empirical: $C_{50}H_{102}O_{17}$
Formula: $CH_3(CH_2)_{16}CH_2(OCH_2CH_2)_nOH$, avg. n = 16
Uses: Pharmaceutical surfactant, emulsifier
Regulatory: FDA 21CFR §177.2800
Manuf./Distrib.: Aldrich; Fluka; Sigma
Trade names containing: Solulan® 16

Stearic acid methyl ester. *See* Methyl stearate
Stearic monoglyceride. *See* Glyceryl stearate
Stearol. *See* Stearyl alcohol

Stearyl alcohol
CAS 112-92-5; EINECS 204-017-6
Synonyms: Octadecanol; n-Octadecanol; Octadecyl alcohol; n-Octadecyl alcohol; Stearol; 1-Octadecanol; n-Octyldecyl alcohol; C18 linear alcohol; Decyl octyl alcohol
Classification: Fatty alcohol
Empirical: $C_{18}H_{38}O$
Formula: $CH_3(CH_2)_{16}CH_2OH$
Properties: Wh. unctuous flakes or gran., faint odor, bland taste; sol. in alcohol, acetone, ether; insol. in water; m.w. 270.56; dens. 0.8124 (59/4 C); m.p. 55-60 C; b.p. 210.5 C (15 mm); acid no. 2 max.; iodine no. 2 max.; hyd. no. 195-220
Toxicology: LD50 (oral, rat) 20 g/kg; mildly toxic by ingestion; may cause skin sensitivity; nonallergenic; experimental neoplastigen
Precaution: Flamm. when exposed to heat or flame; can react with oxidizers; Hazardous decomp. prods.: Heated to decomp., emits acrid smoke and irritating fumes
NFPA: Health 0, Reactivity 0
Uses: Perfumery, cosmetics, intermediate, surface active agents, lubricants, resins, antifoam agent; food additive; intermediate in mfg. of food additives; pharmaceutical surfactant, emulsion stabilizer, emulsifier, astringent; cosolvent
Regulatory: FDA 21CFR §172.755, 172.864, 175.105, 175.300, 176.200, 176.210, 177.1010, 177.1200, 177.2800, 178.3480, 178.3910; FDA approved for orals,

topicals, vaginals; USP/NF, BP, Ph.Eur. compliance
Manuf./Distrib.: Aarhus Oliefabrik A/S; AC Ind.; Albemarle; Aldrich; Allchem Ind.; Amerchol; Brown; Chemron; Condea Vista; Croda Inc; Croda Chem. Ltd; Fluka; R.W. Greeff; Henkel/Emery; Kraft Chem.; Lipo; Lonza; M. Michel; Procter & Gamble; Sea-Land; Sigma; Spectrum Chem. Mfg.
Trade names: Adol® 620 NF
Trade names containing: Witcohol™ 42; Witcohol™ 320; Witcohol™ 330; Witcohol™ 340

STO. *See* Sorbitan trioleate

Stoddard solvent
CAS 8052-41-3; EINECS 232-489-3
UN No. 1256 (DOT)
Synonyms: White spirits; Mineral spirits; Naphtha safety solvent; Safety solvent naphtha; Solvent naphtha; High flash naphtha
Classification: Mixed petroleum hydrocarbons
Definition: Mixed isomer contg. 85% nonane and 15% trimethylbenzene
Properties: Colorless clear liq.; petrol. odor; misc. with abs. alcohol, benzene, ether, chloroform, CCl_4, CS_2, most org. solvs.; insol. in water; m.w. ≈ 150; sp.gr. 0.78; f.p. -70 C; b.p. 157-196 C; flash pt. (CC) 40 C; ref. index 1.43 (20 C); KB value 32-37
Toxicology: LD50 (oral, rat) > 5 g/kg, (skin, rabbit) > 3 g/kg; ACGIH TLV/TWA 100 ppm; mildly toxic by inhalation (irritation, CNS effects, headache, dizziness, confusion, fatigue); human eye irritant; probably a mild to mod. skin irritation
Precaution: Combustible; flamm. exposed to heat or flame; explosive in vapor form when exposed to heat or flame; reactive with oxidizing materials; Hazardous decomp. prods.: Heated to decomp., may explode; heated to decomp., emits acrid fumes
NFPA: Health 0, Flammability 2, Reactivity 0
Uses: Solvent and thinner for paints, varnishes, other coatings; dry-cleaning solvent; spot and stain removal; solvent in textile industry; mfg. of pesticides, cleaners, aerosol prods.; metal and leather cleaning/degreasing; extraction of fats and oils; in rubber cements and polishes; tackifying agent for rubber
Manuf./Distrib.: Ashland; R.E. Carroll; Spectrum Chem. Mfg.
Trade names: 3135 Naphtha; Varsol® 1; Varsol® 18

Straight-run kerosene. *See* Kerosene
Strong ammonia solution. *See* Ammonium hydroxide

Sucrose palmitate
CAS 26446-38-8; EINECS 247-706-7
Synonyms: Saccharose palmitate
Definition: Monoester of palmitic acid and sucrose
Uses: O/w and w/o emulsifier, softener, conditioner, and aerating agent in foods; protective fruit coatings; pharmaceutical emulsifier, solubilizer, stabilizer, tablet lubricant; solvent, emulsifier for cosmetics
Regulatory: FDA 21CFR §172.859
Trade names: Sucro Ester 15

Sulfinylbis (methane). *See* Dimethyl sulfoxide

Sulfolane
CAS 126-33-0; EINECS 204-783-1
Synonyms: Tetrahydrothiophene 1,1-dioxide; Dapsone; Tetramethylene sulfone; Thiophan sulfone
Empirical: $C_4H_8O_2S$

Formula: $CH_2CH_2CH_2CH_2SO_2$
Properties: Wh. or creamy wh. cryst. powd., odorless, sl. bitter taste; very sl. sol. in water; sol. in alcohol, acetone, dilute min. acids; m.w. 120.16; dens. 1.27 (20 C); m.p. 25-28 C; b.p. 285 C; flash pt. 330 F; ref. index 1.485
Toxicology: LD50 (oral, rat) 1941 mg/kg; toxic by ingestion; irritant
Precaution: Combustible
Uses: Curing agent for epoxy resins; medicine (antibacterial); solvent for extraction of benzene, toluene, other aromatic hydrocarbons from oil refinery streams; used in Sulfinol process to remove acid gases; used for separation of low boiling alcohols, min. oils, tars; plasticizer; polymerization solvent; dielec. in elec. equipment; solv. in surf. coatings; component in hydraulic fluids
Manuf./Distrib.: Aldrich; Fluka; Phillips; Shell; Sigma; Syn. Chems. Ltd
Trade names: Sulfolane W

Sulfur dioxide
CAS 7446-09-5; EINECS 231-195-2
UN No. 1079 (DOT)
Synonyms: Sulfurous anhydride; Sulfurous acid anhydride; Sulfur oxide; Sulfurous oxide; Sulphur dioxide; Bisulfate; Bisulfite
Classification: Inorganic sulfur compd.
Empirical: O_2S
Formula: O=S=O
Properties: Colorless gas, strong suffocating odor; condenses @ -10 C to colorless liq.; sol. 8.5% in water, 25% in alcohol, 32% in methanol; sol. in ether, chloroform; m.w. 64.06; sp.gr. 1.5 (liq.); vapor pressure 2538 mm Hg (21.1 C); m.p. –72 C; b.p. –10 C (liq.); nonflamm.
Toxicology: LCLo (inh., human, 5 min) 3000 ppm; ACGIH TLV/TWA 2 ppm; STEL 5 ppm; poison gas; mildly toxic to humans by inh.; corrosive irritant to eyes, respiratory tract; gas causes skin irritation, liq. may cause burns due to freezing
Precaution: Incompat. with bases, chlorates, fluorine, interhalogens, powd. metals, metal oxides, metal acetylides, sodium hydroxide, diethyl zinc; violent reactions, ignition, explosions possible; common air contaminant; Hazardous decomp. prods.: Forms an acid sol'n. (sulfurous acid) on contact with moisture; heated to decomp., emits toxic fumes of SO_x
NFPA: Health 3, Flammability 0, Reactivity 0
Uses: Chem. intermediate; bleaching of paper pulp; ore and metal refining; soybean protein; extraction solvent; catalyst; bleaching oils and starch; sulfonation of oils; reducing agent, antioxidant, fungicide; oxygen scavenger; disinfectant in breweries; dechlorination and silica activation in water treatment (potable, sewage, effluent); bleaching agent, antimicrobial, fumigant, preservative for foods; antioxidant, preservative for pharmaceuticals; air pollutant, major contributor to acid rain
Regulatory: FDA 21CFR §182.3862, GRAS; BATF 27CFR §240.1051; FEMA GRAS; Japan approved (0.03-5 g/kg); Europe listed; UK approved; USP/NF, BP compliance; SARA reportable
Manuf./Distrib.: Air Prods.; Aldrich; Alexander; Boliden Intertrade; Coulton; Coyne; Hoechst Celanese; Outokumpu Oy; Peridot; Rhône-Poulenc Basic; Scott Spec. Gases; Thatcher

Sulfuric ether. *See* Ethyl ether
Sulfurous acid anhydride. *See* Sulfur dioxide
Sulfurous anhydride. *See* Sulfur dioxide
Sulfurous oxide. *See* Sulfur dioxide
Sulfur oxide. *See* Sulfur dioxide

Sulphur dioxide. *See* Sulfur dioxide

Sweet almond oil
CAS 8007-69-0
Synonyms: Almond oil, sweet; Almond oil
Definition: Fixed oil obtained from the ripe seed kernel of *Prunus amygdalus*, contg. chiefly glyceryl oleate
Properties: Colorless or pale yel. oily liq., almost odorless, bland taste; sl. sol. in alcohol; misc. with benzene, chloroform, ether, petrol. ether; insol. in water; dens. 0.910-0.915; iodine no. 93-100; sapon. no. 191-200; ref. index 1.4593-1.4646 (40 C)
Precaution: Keep cool, well closed; protect from light
Uses: Emollient; food ingred.; pharmaceutical vehicle, solvent, flavoring agent
Regulatory: USP/NF, BP compliance
Manuf./Distrib.: Arista Ind.; Sigma; Tri-K Ind.
Trade names: Super Refined® Almond Oil

Synthetic oil of bitter almond. *See* Benzaldehyde

Tallow alcohol
EINECS 308-980-4
Synonyms: **Alcohols, tallow**
Definition: **Mixt. of fatty alcohols derived from tallow**
Uses: **Emollient, consistency giving agent, cosolvent for skin creams and lotions**
Regulatory: **FDA 21CFR §176.170, 176.210**
Manuf./Distrib.: **Witco/Oleo-Surf.**
Trade names: Adol® 42

Tangantangan oil. *See* Castor oil
TBA. *See* t-Butyl alcohol
TBC. *See* Tributyl citrate
TBP. *See* Tributyl phosphate
TBT. *See* p-t-Butyl toluene
TCA. *See* Trichloroacetic acid
TCE. *See* 1,1,2,2-Tetrachloroethane; 1,1,1-Trichloroethane; Trichloroethylene
1,1,1-TCE. *See* 1,1,1-Trichloroethane
1,1,2-TCE. *See* 1,1,2-Trichloroethane
TCTFE. *See* Trichlorotrifluoroethane
TEA. *See* Triethylamine
TEBAC. *See* Benzyltriethyl ammonium chloride
TEC. *See* Triethyl citrate
Teel oil. *See* Sesame oil
TEG. *See* Triethylene glycol
TEN. *See* Triethylamine
TEP. *See* Triethyl phosphate
TEPA. *See* Tetraethylenepentamine
1,4(8)-Terpadiene. *See* Terpinolene

α-Terpinene
CAS 99-86-5
Synonyms: p-Mentha-1,3-diene; 1-Methyl-4-isopropylcyclohexadiene-1,3; 1-

Methyl-4-isopropyl-1,3-cyclohexadiene; Tirpilene; 1,3-Cyclohexadiene, 1-methyl-4-isopropyl-
Empirical: $C_{10}H_{16}$
Properties: Liq.; misc. in alcohol, ether; insol. in water; m.w. 136.24; dens. 0.834 (20/4 C); b.p. 181.5 C; flash pt. 115 F
Toxicology: LD50 (oral, rat) 1680 mg/kg; mod. toxic by ing.; Hazardous decomp. prods.: Heated to decomp., emits acrid smoke and irritating fumes
Uses: Lemon flavoring agent in pharmaceuticals
Regulatory: FEMA GRAS
Manuf./Distrib.: Aldrich; Fluka; SCM Glidco Organics; Sigma
Trade names containing: Acintene® DP 738

α-Terpineol
CAS 98-55-5; 8000-41-7 (mixt.)
Synonyms: α,α,4-Trimethyl-3-cyclohexene-1-methanol; 1-Methyl-4-isopropyl-1-cyclohexen-8-ol; p-Menth-1-en-8-ol; p-Menthen-1-en-8-ol
Empirical: $C_{10}H_{18}O$
Formula: $C_{10}H_{17}OH$
Properties: Colorless visc. liq. or solid, lilac-like odor; very sol. in alcohol, ether; sol. in propylene glycol; very sl. sol. in water; insol. in min. oil; m.w. 154.24; dens. 0.930; m.p. 40-41 C; f.p. 193 F; b.p. 214-224 C
Toxicology: LD50 (intramuscular, mouse) 2000 mg/kg; mod. toxic by intramuscular route; can be a sensitizer
Precaution: Combustible; Hazardous decomp. prods.: Heated to decomp., emits acrid smoke and irritating fumes
NFPA: Health 0, Flammability 2, Reactivity 0
Uses: Solvent for hydrocarbon materials, mutual solvent for resins, cellulose esters and ethers, perfumes, soaps, disinfectant, antioxidant; syn. flavoring agent for foods and pharmaceuticals; antiseptic, denaturant; pharmaceutical solvent
Regulatory: FDA 21CFR §172.515, 175.105; FEMA GRAS; FDA approved for topicals; BP compliance
Manuf./Distrib.: Aldrich; Arakawa Chem. Ind.; Berje; Hercules; Penta Mfg.; Quest Int'l.; SCM Glidco Organics; Sigma; Soda Aromatic; Spectrum Chem. Mfg.; Toyotama Perfumery; T&R Chem.

Terpinolene
CAS 68956-56-9
Synonyms: p-Menth-1,4(8)-diene; 1,4(8)-Terpadiene
Uses: Syn. flavoring agent in foods and pharmaceuticals
Regulatory: FDA 21CFR §172.515; FEMA GRAS
Manuf./Distrib.: Penta Mfg.; SCM Glidco Organics
Trade names containing: Acintene® AC-P; Acintene® DP 738

Tetrachlorethane. *See* 1,1,2,2-Tetrachloroethane
Tetrachlorethylene. *See* Perchloroethylene
Tetrachlorodifluoroethane. *See* 1,1,2,2-Tetrachloro-1,2-difluoroethane

1,1,2,2-Tetrachloro-1,2-difluoroethane
CAS 76-12-0
Synonyms: F112; FC-112; CFC 112; 1,2-Difluoro-1,1,2,2-tetrachloroethane; Tetrachlorodifluoroethane; Fluorocarbon 112; Freon 112; Halocarbon 112; Refrigerant 112
Classification: Halogenated aliphatic hydrocarbon
Empirical: $C_2Cl_4F_2$
Formula: Cl_2FCCCl_2F

1,1,2,2-Tetrachloroethane

Properties: Colorless liq. or wh. solid; sl. ether-like odor; sol. in alcohol, ether, chloroform; pract. insol. in water; m.w. 203.83; sp.gr. 1.6447; m.p. 25 C; b.p. 93 C; noncombustible

Toxicology: LD50 (oral, mouse) 800 mg/kg; LC50 (inh., mouse, 2 h) 123 g/m^3; ACGIH TLV/TWA 500 ppm; mod. toxic by ing.; low toxicity by inh.; may cause drowsiness and difficulty breathing

Precaution: Incompat. with reactive metals (sodium, potassium, magnesium, zinc, powd. aluminum); can react violently or explosively; Hazardous decomp. prods.: Heated to decomp., emits toxic fumes of F^- and Cl^-

Uses: Degreasing solvent in veterinary medicine to destroy parasitic worms

Manuf./Distrib.: Harris Spec. Chems.

1,1,2,2-Tetrachloroethane

CAS 79-34-5; EINECS 211-135-1

UN No. 1702 (DOT)

Synonyms: TCE; 1,1-Dichloro-2,2-dichloroethane; Tetrachlorethane; sym-Tetrachloroethane

Classification: Halogenated aliphatic hydrocarbon

Empirical: $C_2H_2Cl_4$

Formula: $CHCl_2CHCl_2$

Properties: Colorless liq.; sickly sweet chloroform-like odor; sol. in alcohol, ether; misc. with most org. solvs.; sl. sol. in water; m.w. 167.84; sp.gr. 1.596; m.p. -36 C; b.p. 146 C; flash pt. none; ref. index 1.4918

Toxicology: ACGIH TLV/TWA 1 ppm (skin); highly toxic; mild skin irritant; absorption thru skin may cause acute poisoning; may cause severe injury or death by inh., skin absorption, and ingestion; inh. causes nausea, vomiting, weight loss, anemia, liver injury

Precaution: Incompat. with strong bases (may react explosively), reactive metals (may react explosively with sodium, magnesium, aluminum); Hazardous decomp. prods.: Hydrogen chloride, phosgene

Uses: Intermediate in prod. of trichloroethylene; solvent; insecticide; cleansing and degreasing metals; paint removers; varnishes; lacquers; photographic film; resins; waxes; extraction of oils and fats; alcohol denaturant; org. synthesis; weed killer; fumigant

Manuf./Distrib.: AC Ind.; Aldrich; Fluka; Sigma; Spectrum Chem. Mfg.

sym-Tetrachloroethane. *See* 1,1,2,2-Tetrachloroethane
Tetrachloroethene. *See* Perchloroethylene
Tetrachloroethylene. *See* Perchloroethylene
Tetrachloromethane. *See* Carbon tetrachloride
Tetradecafluorohexane. *See* Perfluorohexane
n-Tetradecafluorohexane. *See* Perfluorohexane
Tetradecanoic acid isopropyl ester. *See* Isopropyl myristate
Tetradecanoic acid, methyl ester. *See* Methyl myristate
Tetradecanoic acid, 1-methylethyl ester. *See* Isopropyl myristate
1-Tetradecanol, propanoate. *See* Myristyl propionate
Tetraethylene glycol butyl ether. *See* PEG-4 butyl ether

Tetraethylene glycol ethyl ether

CAS 5650-20-4

Synonyms: Tetraethylene glycol monoethyl ether; PEG-4 ethyl ether

Trade names containing: Icinol EE22

Tetraethylene glycol methyl ether

CAS 23783-42-8

Synonyms: Tetraethylene glycol monomethyl ether; PEG-4 methyl ether

Empirical: $C_9H_{20}O_5$
Trade names containing: Icinol ME33; Icinol ME45

Tetraethylene glycol monobutyl ether. *See* PEG-4 butyl ether
Tetraethylene glycol monoethyl ether. *See* Tetraethylene glycol ethyl ether
Tetraethylene glycol monomethyl ether. *See* Tetraethylene glycol methyl ether

Tetraethylenepentamine
CAS 112-57-2; EINECS 203-986-2
UN No. 2320 (DOT)
Synonyms: TEPA; 1,4,7,10,13-Pentaazatridecane; 1,2-Ethanediamine, N-(2-aminoethyl)-N´-(2-((2-aminoethyl)amino)ethyl)-
Empirical: $C_8H_{23}N_5$
Formula: $NH_2(CH_2CH_2NH)_3CH_2CH_2NH_2$
Properties: Viscous liq., hygroscopic; sol. in most org. solvs. and water; m.w. 189.31; dens. 0.9980 (20/20 C); vapor pressure < 0.01 mm Hg (20 C); m.p. -30 C; b.p. 333 C; flash pt. (OC) 325 F; ref. index 1.5055
Toxicology: LD50 (oral, rat) 3990 mg/kg, (dermal, rabbit) 660 mg/kg, (IV, mouse) 320 mg./kg; poison by ing. and IV route; mod. toxic by skin contact; strong primary irritant to skin, eyes, mucous membranes; causes burns; mutagen
Precaution: Combustible exposed to heat or flame; corrosive; can react with oxidizing materials; Hazardous decomp. prods.: Heated to decomp., emits toxic fumes of NO_x
Uses: Solvent for sulfur, acid gases, various resins and dyes; saponifying agent for acidic materials; mfg. of syn. rubber; dispersant in motor oils; intermediate for oil additives; ion-exchange resins; surfactants; R.T. curing agent often used in two-pkg. protective coating systems
Manuf./Distrib.: Aldrich; Allchem Ind.; Ashland; Coyne; Fluka; Great Western; Houghton Chem.; Tosoh; Union Carbide

Tetrafluorodichloroethane. *See* Dichlorotetrafluoroethane
Tetrahydro-1,4-dioxin. *See* 1,4-Dioxane
Tetrahydro-p-dioxin. *See* 1,4-Dioxane

Tetrahydrofuran
CAS 109-99-9; EINECS 203-726-8
UN No. 2056 (DOT)
Synonyms: THF; Butylene oxide; 1,4-Epoxybutane; Oxacyclopentane; Oxolane; Diethylene oxide; Tetramethylene oxide; Cyclotetramethylene oxide; Furanidine; Hydrofuran; Furan, tetrahydro-
Classification: Sat. cyclic aliphatic ether
Empirical: C_4H_8O
Formula: $CH_2CH_2CH_2CH_2O$
Properties: Water-wh. liq., ethereal odor; very sol. in water; sol. in alcohols, ketones, esters, ethers, hydrocarbons.; m.w. 72.11; sp.gr. 0.888 (20/4 C); f.p. –65 C; b.p. 66 C; flash pt. (OC) –15 C; ref. index 1.4070
Toxicology: LD50 (IP, rat) 2900 mg/kg; LDLo (oral, rat) 3000 mg/kg; ACGIH TLV/TWA 200 ppm; STEL 250 ppm; toxic by ingestion and inhalation; irritant to eyes and mucous membranes; essentially nonirritating to skin; CNS depressant
Precaution: Extremely flamm.; flamm. limits in air 2-11.8%; incompat. with strong oxidizing agents (increases fire/explosion risk), bromine (reacts vigorously with gas evolution), caustic alkalies (violent reactions possible); Hazardous decomp. prods.: Peroxides may form and accumulate on exposure to air and light, in the absence of inhibitors; heated to decomp., emits acrid smoke and irritating fumes

NFPA: Health 2, Flammability 3, Reactivity 1
Uses: Solvent for Grignard reactions, reductions, polymerizations, resins, paints, adhesives, magnetic tapes, printing inks; chemical intermediate and monomer; pharmaceutical intermediate; extraction solvent (pharmaceuticals); steroid hormone prod. for use in birth control pills; preservative for histological samples
Regulatory: BP compliance
Manuf./Distrib.: Aldrich; Allchem Ind.; ARCO Europe; Ashland; BASF; J.T. Baker; Burdick & Jackson; Chemcentral; Coyne; DuPont; Great Lakes; Fluka; Harcros USA; Houghton Chem.; Hüls UK; ISP; Janssen Chimica; QO; Richman; Sigma; Spectrum Chem. Mfg.; Stanchem; Van Waters & Rogers
Trade names: Dynasolve 150; QO® Tetrahydrofuran (THF); THF; THF

Tetrahydro-2-furancarbinol. *See* Tetrahydrofurfuryl alcohol
Tetrahydro-2-furanmethanol. *See* Tetrahydrofurfuryl alcohol
Tetrahydro-2-furanone. *See* Butyrolactone

Tetrahydrofurfuryl alcohol
CAS 97-99-4; EINECS 202-625-6
Synonyms: THFA; Tetrahydro-2-furanmethanol; Tetrahydro-2-furancarbinol; Tetrahydro-2-furylmethanol
Classification: Cyclic alcohol
Empirical: $C_5H_{10}O_2$
Formula: $C_4H_7OCH_2OH$
Properties: Liq., mild odor; hygroscopic; misc. with water, alcohol, ether, acetone, chloroform, benzene; m.w. 102.14; dens. 1.053 (20/4 C); visc. 6.24 cp (20 C); m.p. < -80 C; b.p. 173-177 C; flash pt. 75 C; ref. index 1.453 (20 C); surf. tens. 37 dyne/cm
Toxicology: LD50 (oral, rat) 2500 mg/kg, (IV, rabbit) 725 mg/kg; mod. toxic by ing., IV, IP routes; severe eye irritant; moderately irritating to skin, mucous membranes
Precaution: Combustible exposed to heat, flame; explosive limit 1.5-9.7% by vol.; explosive as vapor exposed to heat or flame; incompat. with oxidizing materials; violent or explosive reaction with 3-nitro-N-bromophthalimide; Hazardous decomp. prods.: Heated to decomp., emits acrid smoke and irritating fumes
Uses: Solvent for vinyl resins, paints, dyes for leather, chlorinated rubber, cellulose esters, coupling agent, solvent-softener for nylon; syn. flavoring agent for foods and pharmaceuticals
Regulatory: FDA 21CFR §172.515, 175.105, 176.210; FEMA GRAS
Manuf./Distrib.: Aldrich; Allchem Ind.; Ashland; Fluka; Great Lakes; Great Western; Penta Mfg.; QO; Schweizerhall; Sigma; Van Waters & Rogers
Trade names: QO® Tetrahydrofurfuryl Alcohol (THFA®)

Tetrahydro-2-furylmethanol. *See* Tetrahydrofurfuryl alcohol
Tetrahydro-1,4-isoxazine. *See* Morpholine
Tetrahydro-p-isoxazine. *See* Morpholine

Tetrahydronaphthalene
CAS 119-64-2; EINECS 204-340-2
Synonyms: 1,2,3,4-Tetrahydronaphthalene; Tetralin; Naphthalene-1,2,3,4-tetrahydride
Empirical: $C_{10}H_{12}$
Properties: Colorless liq., pungent odor; misc. with most solvs.; insol. in water; m.w. 132.21; dens. 0.981 (13 C); m.p. -25 C; b.p. 206 C; flash pt. 160 F; ref. index 1.5410
Toxicology: LD50 (oral, rat) 2860 mg/kg, (dermal, rabbit) 17 g/kg; primary irritant

to eyes, skin, respiratory system; narcotic in high concs.
Uses: Chemical intermediate, solvent for greases, fats, oils, waxes, rubber, asphalt, coatings; substitute for turpentine
Manuf./Distrib.: Aldrich; DuPont; Fluka; Hüls AG; Sigma; Spectrum Chem. Mfg.
Trade names: Tetralin

1,2,3,4-Tetrahydronaphthalene. *See* Tetrahydronaphthalene
Tetrahydro-1,4-oxazine. *See* Morpholine
Tetrahydro-2H-1,4-oxazine. *See* Morpholine
Tetrahydro-4H-1,4-oxazine. *See* Morpholine
Tetrahydro-p-oxazine. *See* Morpholine

Tetrahydrothiophene
CAS 110-01-0; EINECS 203-728-9
UN No. 2412 (DOT)
Synonyms: Thiophane; Thiophene, tetrahydro-; Tetramethylene sulfide; Thiolane; Thiacyclopentane
Empirical: C_4H_8S
Formula: $S(CH_2)_3CH_2$
Properties: Water-white liq.; m.w. 88.18; dens. 1.00 (13.6/15.6 C); b.p. 115-124.4 C
Toxicology: LC50 (inh., mouse, 2 h) 27 g/m^3
Precaution: Flamm.; potentially explosive reaction with hydrogen peroxide; Hazardous decomp. prods.: Heated to decomp., emits toxic fumes of SO_x
Uses: Solvent, intermediate, fuel gas odorant
Manuf./Distrib.: Aldrich; Elf Atochem N. Am.; Fluka; Natural Gas Odorizing

Tetrahydrothiophene 1,1-dioxide. *See* Sulfolane

Tetrahydroxypropyl ethylenediamine
CAS 102-60-3; EINECS 203-041-4
Synonyms: N,N,N´,N´-Tetrakis(2-hydroxypropyl) ethylenediamine; Ethylenedinitrilotetra-2-propanol; 1,1´,1´´,1´´´-(Ethylenedinitriolo)tetra-2-propanol
Classification: Substituted amine
Empirical: $C_{14}H_{32}N_2O_4$
Formula: $(HOC_3H_6)_2NCH_2CH_2N(C_3H_6OH)$
Properties: Water-wh. visc. liq.; sol. in ethanol, toluene, ethylene glycol; misc. with water; m.w. 292.42; dens. 1.013; b.p. 175-181 C (0.8 mm); flash pt. > 230 F; ref. index 1.4812
Toxicology: LD50 (mammal) 3900 mg/kg; mod. toxic by unspecified route; may be irritating to skin and mucous membranes; may cause skin sensitization
Precaution: Combustible; Hazardous decomp. prods.: Heated to decomp., emits toxic fumes of NO_x
Uses: Chelating agent, intermediate, emulsifier, antistat, rewetting agent, grease additive, textile lubricant; for pharmaceuticals, herbicides, fungicides, insecticides, adhesives, latexes, resins, plasticizers, inks, cosmetics, resins; pharmaceutical solvent, preservative
Manuf./Distrib.: Aldrich; Fluka; Sigma

N,N,N´,N´-Tetrakis(2-hydroxypropyl) ethylenediamine. *See* Tetrahydroxypropyl ethylenediamine
Tetralin. *See* Tetrahydronaphthalene
Tetramethylene cyanide. *See* Adiponitrile
Tetramethylene oxide. *See* Tetrahydrofuran
Tetramethylene sulfide. *See* Tetrahydrothiophene
Tetramethylene sulfone. *See* Sulfolane
2,2,6,6-Tetramethyl-4-methyleneheptane. *See* Isododecane
2,5,8,11-Tetraoxadodecane. *See* PEG-3 dimethyl ether

Tetryl formate. *See* Isobutyl formate
Theobroma oil. *See* Cocoa butter
THF. *See* Tetrahydrofuran
THFA. *See* Tetrahydrofurfuryl alcohol
Thiacyclopentadiene. *See* Thiophene
Thiacyclopentane. *See* Tetrahydrothiophene
Thiaphene. *See* Thiophene
Thiobutylalcohol. *See* n-Butyl mercaptan
Thiocyanate sodium. *See* Sodium thiocyanate
2,2´-Thiodiethanol. *See* Thiodiglycol
Thiodiethylene glycol. *See* Thiodiglycol

Thiodiglycol
CAS 111-48-8; EINECS 203-874-3
Synonyms: β-Thiodigylcol; Thiodiethylene glycol; β-Hydroxyethyl sulfide; 2,2´-Thiodiethanol; Ethanol, 2,2´-thiodi-; Bis(2-hydroxyethyl) sulfide; Bis(β-hydroxyethyl)sulfide; Dihydroxyethyl sulfide; β,β´-Dihydroxydiethyl sulfide; β,β´-Dihydroxyethyl sulfide
Classification: Aliphatic alcohol
Empirical: $C_4H_{10}O_2S$
Formula: $HOCH_2CH_2 \cdot S \cdot CH_2CH_2OH$
Properties: Syrupy colorless liq., char. odor; sol. in acetone, alcohol, chloroform, water; sl. sol. in benzene, CCl_4, ether; m.w. 122.18; dens. 1.1852 (20/20 C); m.p. -16 C; b.p. 283 C; flash pt. (OC) 320 F
Toxicology: LD50 (oral, rat) 6610 mg/kg, (skin, rabbit) 20 ml/kg, (subcut., rat) 4 g/kg; mod. toxic by ing., IV, subcut. routes; skin and eye irritant
Precaution: Combustible exposed to heat or flame; reacts violently with acetone and sodium peroxide; can react with oxidizers; Hazardous decomp. prods.: Heated to decomp., emits highly toxic fumes of SO_x
Uses: Intermediate for elastomers and antioxidants, solvent for dyes in textile printing
Manuf./Distrib.: Aldrich; Fluka; Morton Int'l.; Sigma

β-Thiodigylcol. *See* Thiodiglycol
2-Thioethanol. *See* 2-Mercaptoethanol
Thioethylene glycol. *See* 2-Mercaptoethanol
Thiofuram. *See* Thiophene
Thiofuran. *See* Thiophene
Thiofurfuran. *See* Thiophene
Thioglycol. *See* 2-Mercaptoethanol
Thiolane. *See* Tetrahydrothiophene
Thiole. *See* Thiophene
Thiomonoglycol. *See* 2-Mercaptoethanol
Thiophane. *See* Tetrahydrothiophene
Thiophan sulfone. *See* Sulfolane
Thiophen. *See* Thiophene

Thiophene
CAS 110-02-1; EINECS 203-729-4
UN No. 2414 (DOT)
Synonyms: Thiaphene; Thiofuran; Thiofuram; Thiofurfuran; Thiole; Thiophen; Thiotetrole; Thiacyclopentadiene; Divinylene sulfide
Empirical: C_4H_4S
Properties: Colorless clear liq.; sl. aromatic odor similar to benzene; misc. with most org. solvs.; sol. in alcohol, ether; insol. in water; m.w. 84.14; dens. 1.051; vapor pressure 40 mm Hg (12.5 C); m.p. -38 C; b.p. 84 C; flash pt. 21.2 F; ref.

index 1.5270; may be heated to 850 C without decomp.

Toxicology: LD50 (oral, rat) 1400 mg/kg, (subcut., rabbit) 830 mg/kg, (IP, mouse) 100 mg/kg; poison by IP route; mildly toxic by inh.

Precaution: DOT: Flamm. liq.; dangerous fire hazard exposed to heat or flame; explosive reaction with N-nitrosoacetanilide; violent or explosive reaction with nitric acid; incompat. with oxidizers; Hazardous decomp. prods.: Heated to decomp., emits highly toxic fumes of SO_x

NFPA: Health 2, Flammability 3, Reactivity 0

Uses: Organic synthesis (condenses with phenol and formaldehyde, copolymerizes with maleic anhydride), solvent, dye, pharmaceutical mfg.

Manuf./Distrib.: Aldrich; Biddle Sawyer; Elf Atochem N. Am.; Fluka; Penta Mfg.; Sigma; Spectrum Chem. Mfg.

Thiophene, tetrahydro-. *See* Tetrahydrothiophene

Thiotetrole. *See* Thiophene

Tirpilene. *See* α-Terpinene

TMB. *See* 1,3,5-Trimethylbenzene

TMBAC. *See* Benzyl trimethyl ammonium chloride

2,4,-TMP. *See* Isooctane

TNBA. *See* Tributylamine

Toluene

CAS 108-88-3; EINECS 203-625-9

UN No. 1294 (DOT)

Synonyms: Methylbenzene; Methylbenzol; Phenylmethane; Toluol

Classification: Aromatic hydrocarbon

Empirical: C_7H_8

Formula: $C_6H_5CH_3$

Properties: Colorless clear liq., benzene odor; sol. in alcohol, benzene, ether, acetone, chloroform, petrol. ether, ethyl acetate; very sl. sol. in water; m.w. 92.13; sp.gr. 0.866 (20/4 C); vapor pressure 36.7 mm Hg (30 C); m.p. -94.5 C; b.p. 110.7 C; flash pt. (CC) 4.4 C; ref. index 1.4967 (20 C); KB value 105

Toxicology: LD50 (oral, rat) 7.53 g/kg, (skin, rabbit) 12,124 mg/kg; ACGIH TLV/TWA 100 ppm; STEL 150 ppm; toxic by ingestion, inhalation, IP, and skin absorption; irritant to eyes, skin, respiratory tract

Precaution: Highly flamm.; incompat. with strong oxidizers, nitric acid, sulfuric acid, nitrogen tetroxide, silver perchlorate, sulfur dichloride, etc.; fire and explosion risk; potential violent reactions; common air contaminant; Hazardous decomp. prods.: Emits acrid smoke and irritating fumes

NFPA: Health 2, Flammability 3, Reactivity 0

Uses: Aviation gasoline additive; solvent for paint, gums, resins; diluent and thinner in nitrocellulose lacquers; adhesive solvent in plastic toys; mfg. of benzoic acid, benzaldehyde; in extraction of various principles from plants; starting material for TDI for PU resins, TNT for explosives, toluene sulfonates for detergents and dyestuffs

Regulatory: FDA 21CFR §175.105, 175.320, 176.180, 177.1010, 177.1200, 177.1440, 177.1580, 177.1650, 177.2460, 178.3010, 27CFR §21.131; SARA reportable

Manuf./Distrib.: Aldrich; ARCO; Ashland; J.T. Baker; Baychem; BP Chems. Ltd; Burdick & Jackson; R.E. Carroll; Chemcentral; Chevron; Coyne; Exxon; Fina Chems.; Fluka; Harcros USA; Houghton Chem.; Marathon Oil; Mitsubishi Petrochem.; Mitsui Petrochem. Ind.; Mobil; Olin; Phillips; Shell; Sigma; Spectrum Chem. Mfg.; Texaco; Triple Crown Am.; Van Waters & Rogers

Trade names: Shell Toluene

Trade names containing: Texsolve E

Toluene, ar-chloro-. *See* Monochlorotoluene

Toluene, p-tert-butyl-. *See* p-t-Butyl toluene
Toluene, p-chloro-α,α,α,-trifluoro. *See* p-Chlorobenzotrifluoride
Toluene hexahydride. *See* Methyl cyclohexane
4-Toluenesulfonic acid sodium salt. *See* Sodium toluenesulfonate
p-Toluenesulfonic acid, sodium salt. *See* Sodium toluenesulfonate
α-Toluenol. *See* Benzyl alcohol
m-Toluic acid diethylamide. *See* Diethyl toluamide
Toluol. *See* Toluene
TPGEE. *See* Tripropylene glycol ethyl ether
TPGME. *See* PPG-3 methyl ether

Triacetin
CAS 102-76-1; EINECS 203-051-9
Synonyms: Glycerine triacetate; Glycerol triacetate; Glyceryl triacetate; Acetin; Enzactin; 1,2,3-Propanetriol triacetate; Triacetyl glycerol; Triacetyl glycerin
Definition: Triester of glycerin and acetic acid
Empirical: $C_9H_{14}O_6$
Formula: $C_3H_5(OCOCH_3)_3$
Properties: Colorless oily liq., sl. fatty odor, bitter taste; sol. in water, alcohol, ether, other org. solvs.; sl. sol. in CS_2; m.w. 218.20; dens. 1.160 (20 C); m.p. -78 C; b.p. 258-260 C; flash pt. 300 F; ref. index 1.4307 (20 C)
Toxicology: LD50 (oral, rat) 3000 mg/kg, (IV, mouse) 1600 ± 81 mg/kg; poison by ingestion; mod. toxic by IP, subcutaneous, IV routes; eye irritant
Precaution: Combustible exposed to heat, flame, or powerful oxidizers; Hazardous decomp. prods.: Heated to decomp., emits acrid smoke and irritating fumes
NFPA: Health 1, Flammability 1, Reactivity 0
Uses: Plasticizer for vinyl compds.; fixative in perfumery; mfg. of cosmetics; specialty solvent; mfg. of celluloid, photographic films; topical antifungal; adhesives, coatings, paper; syn. flavoring agent in foods, pharmaceuticals; food additive, nutrient, solvent, vehicle, humectant, formulation aid; OTC drug; topical antifungal
Regulatory: FDA 21CFR §175.300, 175.320, 181.22, 181.27, 184.1901, GRAS; FEMA GRAS; FDA approved for orals; USP/NF, BP compliance
Manuf./Distrib.: Aldrich; Allan; Eastman; Fluka; K3; Penta Mfg.; Sigma; Spectrum Chem. Mfg.; Unichema Int'l.; Yuki Gosei Kogyo
Trade names: Eastman® Triacetin

Triacetyl glycerin. *See* Triacetin
Triacetyl glycerol. *See* Triacetin
Tribromomethane. *See* Bromoform
Tributyl acetyl citrate. *See* Acetyl tributyl citrate

Tributylamine
CAS 102-82-9
UN No. 2542 (DOT)
Synonyms: TNBA; N,N-Dibutyl-1-butanamine; Tri-n-butylamine; Tris-n-butylamine
Classification: Sat. aliphatic amine
Empirical: $C_{12}H_{27}N$
Formula: $N(CH_2CH_2CH_2CH_3)_3$
Properties: Colorless to pale yel. liq.; ammonia-like odor; sol. in many org. solvs. incl. ethanol, acetone, ether, benzene; sl. sol. in water; m.w. 185.36; sp.gr. 0.78 (20 C); m.p. -70 C; b.p. 216-27 C; flash pt. (CC) 63 C
Toxicology: LD50 (oral, rat) 540 mg/kg, (skin, rabbit) 250 mg/kg; very toxic; corrosive; prolonged skin contact may cause severe irritation, absorption thru

skin in toxic amts.; inh. can cause nose/throat/lung irritation, chest pain, difficulty breathing
Precaution: Flamm. exposed to heat, flame, oxidizers; incompat. with oxidizing agents, strong acids, mercury; violent reactions and fire/explosion hazards possible; Hazardous decomp. prods.: Heated to decomp., emits toxic fumes of NO_x
NFPA: Health 3, Flammability 2, Reactivity 0
Uses: Inhibitor in hydraulic fluids; solvent; mfg. of pesticides, other org. chems.
Manuf./Distrib.: Aldrich; Ashland; BASF; BASF AG; Elf Atochem N. Am.; Fluka

Tri-n-butylamine. *See* Tributylamine

Tributyl citrate
CAS 77-94-1; EINECS 201-071-2
Synonyms: TBC; 2-Hydroxy-1,2,3-propanetricarboxylic acid, tributyl ester; Butyl citrate; Citric acid, tributyl ester
Definition: Triester of butyl alcohol and citric acid
Empirical: $C_{18}H_{32}O_7$
Formula: $C_3H_5O(COOC_4H_9)_3$
Properties: Colorless or pale yel., odorless; insol. in water; m.w. 360.45; dens. 1.042 (25/25 C); m.p. -20 C; b.p. 233.5 C (22.5 mm); flash pt. (COC) 315 F; ref. index 1.4453
Toxicology: LD50 (IP, mouse) 2900 mg/kg
Precaution: Combustible
Uses: Plasticizer, antifoam agent, solvent for nitrocellulose; plasticizer for aq. pharmaceutical coatings
Regulatory: FDA 21CFR §175.105
Manuf./Distrib.: Morflex; Unitex
Trade names: Citroflex® 4

Tributyl phosphate
CAS 126-73-8; EINECS 204-800-2
Synonyms: TBP; Tri-n-butyl phosphate; Butyl phosphate; Phosphoric acid tributyl ester
Classification: Organophosphate
Empirical: $C_{12}H_{27}O_4P$
Formula: $(C_4H_9)_3PO_4$
Properties: Stable colorless liq., odorless; misc. with alcohol, ether, most solvs. and diluents; sol. in water; m.w. 266.32; dens. 0.978 (20/20 C); m.p. < -80 C; b.p. 292 C (dec.); flash pt. (COC) 146 C; ref. index 1.4215 (25 C)
Toxicology: LD50 (oral, rat) 1400 mg/kg; ACGIH TLV/TWA 0.2 ppm; toxic by ingestion, inhalation, IP, IV routes; irritant to skin, mucous membranes, eyes
Precaution: Combustible exposed to heat or flame; Hazardous decomp. prods.: Heated to decomp., emits toxic fumes of PO_x
NFPA: Health 2, Flammability 1, Reactivity 0
Uses: Heat-exchange medium; solvent extraction of metal ions; plasticizer for cellulose esters, lacquers, plastics, vinyl resins; antifoam agent; dielectric; flame retardant
Manuf./Distrib.: AC Ind.; Akzo Nobel; Albright & Wilson Am.; Aldrich; Ashland; Bayer; Chemron; EM Ind.; Fluka; FMC; C.P. Hall; Harwick; Houghton Chem.; Sigma; Spectrum Chem. Mfg.; Velsicol
Trade names: Albrite® TBP; Amgard® $TBPO_4$; Kronitex® TBP; Reomol® TBP; TBP

Tri-n-butyl phosphate. *See* Tributyl phosphate
Tricarballylic acid-β-acetoxytributyl ester. *See* Acetyl triethyl citrate
Trichloracetic acid. *See* Trichloroacetic acid

Trichloran. *See* Trichloroethylene

Trichloroacetic acid

CAS 76-03-9

UN No. 1839 (DOT), 2564 (DOT)

Synonyms: TCA; Aceto-caustin; Trichloracetic acid; Trichloroethanoic acid

Classification: Halogenated carboxylic acid

Empirical: $C_2HCl_3O_2$

Formula: CCl_3COOH

Properties: Colorless cryst.; sharp, pungent odor; sol. in water; very sol. in alcohol, ether; m.w. 163.4; sp.gr. 1.62; vapor pressure 1 mm Hg (51 C); m.p. 58 C; b.p. 197 C; flash pt. none

Toxicology: LD50 (subcut., mouse) 270 mg/kg; ACGIH TLV/TWA 1 ppm; poison by subcut. route; corrosive irritant to skin, eyes, mucous membranes; inh. may cause mild to severe irritation or tissue damage, sore throat, nasal irritation, coughing

Precaution: Incompat. with dimethylsulfoxide, copper wool, reactive metals, bases, strong oxidizing agents; violent/explosive reactions possible; Hazardous decomp. prods.: May produce flamm. and explosive hydrogen gas with reactive metals; heated to decomp., emits toxic fumes of Cl^- and Na_2O

Uses: Prod. of its sodium salt; used as an herbicide; pickling or etching agent in metal surf. finishing; solvent in plastics industry; protein precipitating agent in laboratories; additive in min. lubricating oils; intermediate in chem. synthesis of esters; medical agent in treating skin disorders

Manuf./Distrib.: Aldrich; Alemark; Fluka; Hoechst Celanese; Rhône-Poulenc N. Am.; Sigma; Spectrum Chem. Mfg.

Trichlorobenzene. *See* 1,2,4-Trichlorobenzene

1,2,4-Trichlorobenzene

CAS 120-82-1; EINECS 204-428-0

UN No. 2321 (DOT)

Synonyms: Trichlorobenzene; unsym-Trichlorobenzene; Benzene, 1,2,4-trichloro-

Empirical: $C_6H_3Cl_3$

Properties: Colorless stable liq.; misc. with most org. solvs., ether, benzene, petrol. ether, carbon disulfide; insol. in water; m.w. 181.45; dens. 1.4634; vapor pressure 1 mm Hg (38.4 C); m.p. 17 C; b.p. 213 C; flash pt. 98.9 C; ref. index 1.5710

Toxicology: LD50 (oral, rat) 756 mg/kg, (oral, mouse) 300 mg/kg; ACGIH TLV:CL 5 ppm; toxic by ingestion and inhalation; mod. toxic by IP route; primary irritant; skin irritant

Precaution: Combustible exposed to heat or flame; can react vigorously with oxidizing materials; Hazardous decomp. prods.: Heated to decomp., emits toxic fumes of Cl^-

NFPA: Health 2, Flammability 1, Reactivity 0

Uses: Solvent in chemical manufacture, dyes, intermediates, dielectric fluid, synthetic transformer oils, lubricants, heat-transfer media, insecticides; herbicide for aquatic weed control in irrigation canals, lakes, ponds

Manuf./Distrib.: Aldrich; Allchem Ind.; Ashland; Bayer/Fibers, Orgs., Rubbers; Fabrichem; Fluka; Sigma; Spectrum Chem. Mfg.; Stan Chem Int'l. Ltd

1,3,5-Trichlorobenzene

CAS 108-70-3; EINECS 203-608-6

UN No. 2321

Synonyms: sym-Trichlorobenzene; Trichloro-1,3,5-benzene

Classification: Aromatic halogenated hydrocarbon
Empirical: $C_6H_3Cl_3$
Formula: $ClC_6H_3Cl_2$
Properties: Wh. to colorless needle-like cryst.; aromatic odor; freely sol. in diethyl ether, benzene, petrol. ether, glac. acetic acid; sparingly sol. in ethanol; pract. insol. in water; m.w. 181.45; m.p. 63-65 C; b.p. 208.5 C; flash pt. 107 C
Uses: Component in trichlorobenzene mixt. as dye solvent, degreasing solvent, dielectric fluid, lubricant additive; termite control; org. synthesis
Manuf./Distrib.: Aldrich; Fluka; Ishihara USA

Trichloro-1,3,5-benzene. *See* 1,3,5-Trichlorobenzene
sym-Trichlorobenzene. *See* 1,3,5-Trichlorobenzene
unsym-Trichlorobenzene. *See* 1,2,4-Trichlorobenzene
Trichloroethane. *See* 1,1,1-Trichloroethane

1,1,1-Trichloroethane
CAS 71-55-6; EINECS 200-756-3; 201-166-9
UN No. 2831 (DOT)
Synonyms: TCE; 1,1,1-TCE; Trichloroethane; α-Trichloroethane; Chloroethene; Chlorothene; Chloroform, methyl-; Ethane, 1,1,1-trichloro-; Trichloromethylmethane; Methylchloroform; Methyltrichloromethane
Classification: Halogenated aliphatic hydrocarbon
Empirical: $C_2H_3Cl_3$
Formula: CH_3CCl_3
Properties: Colorless liq., char. chloroform-like odor; pract. insol. in water; sol. in alcohol, ether, acetone, benzene, chloroform, methanol, CCl_4; m.w. 133.42; sp.gr. 1.3376 (20/4 C); vapor pressure 100 mm Hg (20 C); m.p. -32.5 C; b.p. 74.1 C; flash pt. none; ref. index. 1.43838 (20 C); KB value 124
Toxicology: LD50 (oral, rat) 10,300 mg/kg, (IP, rat) 3593 mg/kg; ACGIH TLV/TWA 350 ppm; STEL 450 ppm; poison by IV; mod. toxic by ing., inh., skin contact; primary irritant to eyes, tissue, skin, respiratory tract; vapor inh. produces CNS effects
Environmental: Ozone-depleting
Precaution: Nonflamm., but vapor may burn in presence of oxidizers or ignition source; incompat. with aluminum/magnesium/potassium and alloys, sodium, NaOH, KOH, nitrogen dioxide, oxygen (gas or liq.); Hazardous decomp. prods.: Hydrogen chloride, hydrochloric acid, phosgene; heated to decomp., emits toxic fumes of Cl^-
NFPA: Health 2, Flammability 1, Reactivity 0
Uses: Solvent for cleaning precision instruments, metal degreasing, dry-cleaning, resins, oils, waxes, adhesives, pesticides, textile processing, cutting fluids, printing inks, shoe polishes, spot cleaners; extraction solvent; diluent for paints; chem. intermediate for prod. of vinylidene chloride; catalyst for PU foams; carrier for aerosols; as aerosol propellant; reagent
Regulatory: FDA 21CFR §175.105, 177.1650; SARA reportable
Manuf./Distrib.: Aldrich; Allchem Ind.; Asahi Chem. Ind.; Asahi-Penn; Ashland; Chemoxy Int'l. plc; Elf Atochem N. Am.; Fabrichem; Fluka; Houghton Chem.; ICI Am.; Olin; PPG Ind.; Schumacher; Sigma; Spectrum Chem. Mfg.; Stanchem; Universal Preserv-A-Chem; Vulcan
Trade names: Ethana®
Trade names containing: MS-195/CO2 Safezone® Cleaning Solvent & Flux Remover

1,1,2-Trichloroethane
CAS 79-00-5; EINECS 201-166-9
Synonyms: 1,1,2-TCE; Ethane trichloride; β-Trichloroethane; Vinyl trichloride;

1,2,2-Trichloroethane
Classification: Halogenated aliphatic hydrocarbon
Empirical: $C_2H_3Cl_3$
Formula: Cl_2CHCH_2Cl
Properties: Colorless clear volatile liq.; char. sweet but sl. irritating odor; sol. in ethanol, chloroform, diethyl ether, esters, ketones, other chlorinated solvs.; poorly sol. in water; m.w. 133.41; sp.gr. 1.44 (20 C); m.p. -36.5 C; b.p. 113.8 C; surf. tens. 32.5 dynes/cm (20 C)
Toxicology: LD50 (oral, rat) 580 mg/kg, (skin, rabbit) 3730 mg/kg; ACGIH TLV/TWA 10 ppm (skin); very toxic; poison by ing., IV, subcut. routes; mod. toxic by inh., skin contact, IP routes; eye and severe skin irritant; narcotic
Precaution: Nonflamm., but vapor can be ignited by high-energy source; incompat. with strong oxidizers, strong alkalies, chemically active metals, potassium and alloys, sodium amide; fire and explosion risk; Hazardous decomp. prods.: Hydrogen chloride, hydrochloric acid; heated to decomp., emits toxic fumes of Cl^-
NFPA: Health 2, Flammability 1, Reactivity 0
Uses: Intermediate in prod. of vinylidene chloride; industrial solvent for chlorinated rubbers, pharmaceutical preps., mfg. of electronic components, in chem. synthesis
Manuf./Distrib.: Aldrich; Fluka; Sigma

1,2,2-Trichloroethane. *See* 1,1,2-Trichloroethane
α-Trichloroethane. *See* 1,1,1-Trichloroethane
β-Trichloroethane. *See* 1,1,2-Trichloroethane
Trichloroethanoic acid. *See* Trichloroacetic acid
Trichloroethene. *See* Trichloroethylene

Trichloroethylene
CAS 79-01-6; EINECS 201-167-4
UN No. 1710 (DOT)
Synonyms: TCE; 1,1,2-Trichloroethylene; Trichloran; Trichloroethene; 1,1-Dichloro-2-chloroethylene; Ethinyl trichloride; Ethylene trichloride; 1-Chloro-2,2-dichloroethylene; Acetylene trichloride
Classification: Aliphatic organic compd.
Empirical: C_2HCl_3
Formula: $CHCl:CCl_2$
Properties: Colorless, photoreactive liq., chloroform odor; misc. with common org. solvs.; sl. sol. in water; m.w. 131.40; dens. 1.456-1.462; vapor pressure 100 mm Hg (32 C); f.p. -86.8 C; m.p. -73 C; b.p. 86.7 C; flash pt. 89.6 F; ref. index 1.45560 (25 C); KB value 130
Toxicology: LD50 (oral, rat) 4.92 ml/kg, (IV, mouse) 34 mg/kg; ACGIH TLV/TWA 50 ppm; STEL 200 ppm; mildly toxic to humans by ingestion, inhalation; human systemic effects; eye, severe skin irritant; inh. of high concs. causes narcosis, anesthesia
Precaution: Flamm.; very dangerous fire hazard exposed to heat or flame; high concs. of vapor in high-temp. air can be made to burn mildly under strong flame; light-sensitive; reacts with alkali, epoxides; violent reactions possible; can form HCl with water; Hazardous decomp. prods.: Emits toxic fumes of Cl^-
NFPA: Health 2, Flammability 1, Reactivity 0
Uses: Metal degreasing; extraction solvent for oils, fats, waxes, mfg. of decaffeinated coffee, spice oleoresins; solvent for cellulose esters and ethers; dry-cleaning; solvent dyeing; mfg. of organic chemicals, pharmaceuticals, insecticides; analgesic, anesthetic
Regulatory: FDA 21CFR §173.290; SARA reportable; BP compliance

Manuf./Distrib.: Aldrich; Allchem Ind.; Ashland; Asahi-Penn; Coyne; Elf Atochem N. Am.; Fluka; General Chem.; Harcros USA; Houghton Chem.; ICI Am.; Kanto; Olin; PPG Ind.; Sigma; Spectrum Chem. Mfg.; Universal Preserv-A-Chem; Vulcan
Trade names: PPG Trichlor

1,1,2-Trichloroethylene. *See* Trichloroethylene

Trichlorofluoromethane
CAS 75-69-4; EINECS 200-892-3
Synonyms: F-11; FC 11; R 11; Fluorocarbon-11; Halocarbon 11; Propellant 11; Fluorotrichloromethane; Methane, trichlorofluoro-; Trichloromonofluoromethane; Monofluorotrichloromethane
Classification: Halogenated aliphatic hydrocarbon
Empirical: CCl_3F
Formula: Cl_3CF
Properties: Colorless clear volatile liq. or gas; nearly odorless @ < 20% conc., faint sweet odor at high concs.; sol. in alcohol, ether; insol. in water; m.w. 137.4; sp.gr. 1.494 (liq., 17.2 C); f.p. -111 C; b.p. 23.7 C; noncombustible
Toxicology: LD50 (IP, mouse) 1743 mg/kg; TDLo (inh., human, 30 min) 50,000 ppm; ACGIH TLV/CL 1000 ppm in air; poison by inhalation; CNS effects @ > 1% conc.; severe exposure can cause nausea, vomiting, irregular heartbeat, cardiac arrest, death
Precaution: Incompat. with reactive metals, e.g., powd. aluminum, sodium, potassium (can react violently), lithium (can detonate); Hazardous decomp. prods.: None known
Uses: Refrigerant; solvent; fire extinguishing agent; chemical intermediate; blowing agent; aerosol propellant; degreasing agent; dry-cleaning agent; pharmaceutical propellants
Regulatory: FDA approved for inhalants; USP/NF, BP compliance
Manuf./Distrib.: Aldrich; Allchem Ind.; Elf Atochem N. Am.; Elf Atochem SA; Fluka; Sigma

Trichloroform. *See* Chloroform
Trichloromethane. *See* Chloroform
Trichloromethylmethane. *See* 1,1,1-Trichloroethane
Trichloromonofluoromethane. *See* Trichlorofluoromethane

Trichlorotrifluoroethane
CAS 76-13-1; EINECS 200-936-1
Synonyms: FC 113; CFC 113; R 113; TCTFE; TTE; 1,1,2-Trichloro-1,2,2-trifluoroethane; Ethane, 1,1,2-trichloro-1,2,2-trifluoro; Chlorofluorocarbon 113; Freon 113; Halocarbon 113; Fluorocarbon 113; Refrigerant 113
Classification: Halogenated aliphatic hydrocarbon
Empirical: $C_2Cl_3F_3$
Formula: $ClCF_2CCl_2F$
Properties: Colorless volatile gas; nearly odorless to sweet ether-like odor at high concs.; sol. in alcohol, ether, benzene; very sl. sol. in water; m.w. 187.4; sp.gr. 1.42 (25 C); f.p. -35 C; m.p. 13.2 C; b.p. 47.6 C; noncombustible
Toxicology: LD50 (oral, rat) 43 g/kg; ACGIH TLV/TWA 1000 ppm; STEL 1250 ppm; primary irritant; may cause eye irritation, drying/cracking of skin; mildly toxic by ing. and inh.; high concs. may cause CNS symptoms, dizziness, headache
Precaution: Combustible exposed to heat or flame; incompat. with chemically active metals (calcium, powd. aluminum, zinc, magnesium, beryllium, titanium, samarium, lithium, barium); may react explosively

1,1,2-Trichloro-1,2,2-trifluoroethane

Uses: Refrigerant; dry-cleaning solvent; fire extinguishing agent; mfg. of chlorotrifluoroethylene; foaming/blowing agent; polymer intermediate for dechlorination of chems.; solvent drying; drying electronic parts and precision equipment; degreasing agent; quick-cooling or freezing of foods
Regulatory: FDA 21CFR §173.342
Manuf./Distrib.: Air Prods.; Aldrich; Allchem Ind.; AlliedSignal; Burdick & Jackson; Elf Atochem N. Am.; Fluka; Harris Spec. Chems.; LaRoche Ind.; PCR; Sigma; Spectrum Chem. Mfg.

1,1,2-Trichloro-1,2,2-trifluoroethane. *See* Trichlorotrifluoroethane
Tridecanol. *See* Tridecyl alcohol
1-Tridecanol. *See* Tridecyl alcohol
n-Tridecanol. *See* Tridecyl alcohol

Tridecyl alcohol
CAS 112-70-9; 68526-86-3; EINECS 203-998-8
Synonyms: Tridecanol; 1-Tridecanol; n-Tridecanol; C13 linear primary alcohol; n-Tridecyl alcohol
Classification: Synthetic alcohol; commercial mixt. of isomers
Empirical: $C_{13}H_{28}O$
Formula: $C_{12}H_{25}CH_2OH$
Properties: Low melting white solid, pleasant odor; m.w. 200.41; dens. 0.845 (20/20 C); m.p. 31 C; b.p. 274 C; flash pt. (TOC) 180 F
Toxicology: LD50 (oral, rat) 17,200 mg/kg, (skin, rabbit) 5600 mg/kg; mildly toxic by ingestion and skin contact; skin irritant
Precaution: Combustible exposed to heat or flame; Hazardous decomp. prods.: Heated to decomp., emits acrid smoke and irritating fumes
Uses: Esters for synthetic lubricants, detergents, antifoam agents, other tridecyl compds.; perfumery; solvent for paints
Regulatory: FDA 21CFR §175.105, 175.300, 176.210
Manuf./Distrib.: Aldrich; Allchem Ind.; Ashland; Browning; Chemcentral; Exxon; Fluka; Hoechst Celanese; Penta Mfg.; Sigma

n-Tridecyl alcohol. *See* Tridecyl alcohol

Tridecyl neopentanoate
CAS 106436-39-9
Synonyms: Neopentanoic acid, tridecyl ester
Definition: Ester of tridecyl alcohol and neopentanoic acid
Empirical: $C_{18}H_{36}O_2$
Formula: $(CH_3)_3CCOOC_{13}H_{27}$
Properties: Liq., char. mild odor; sol. in min. oil, ethanol (> 95%), IPM, cyclomethicone; insol. in water; m.w. 284; sp.gr. 0.850-0.860; ref. index 1.4345-1.4365
Toxicology: LD50 (oral, rat) > 5 g/kg; minimal eye irritation; mild skin irritation
Uses: Emollient imparting lubricity, skin feel, gloss, spreading to creams and lotions; cosolvent for oils and waxes in cosmetics sticks; pharmaceutical emollient, lubricant, gloss and spreading aid
Trade names: Trivent NP-13

Triethyl acetylcitrate. *See* Acetyl triethyl citrate

Triethylamine
CAS 121-44-8; EINECS 204-469-4
UN No. 1296 (DOT)
Synonyms: TEA; TEN; N,N-Diethylethanamine; (Diethylamino) ethane
Classification: Sat. aliphatic amine
Empirical: $C_6H_{15}N$
Formula: $(C_2H_5)_3N$

Properties: Colorless liq., strong ammoniacal odor; sol. in water, alcohol, acetone, ether, benzene, other org. solvs.; very sol. in chloroform; m.w. 101.22; sp.gr. 0.7293 (20/20 C); f.p. -115.3 C; b.p. 89.7 C; flash pt. (CC) -9 C; ref. index 1.4000
Toxicology: ACGIH TLV/TWA 10 ppm; STEL 15 ppm; LD50 (oral, rat) 460 mg/kg; LDLo (inh., rat, 4 h) 1000 ppm; very toxic, corrosive; mod. toxic by ingestion and inhalation; strong irritant to tissue; lachrymator
Precaution: Flamm.; dangerous fire risk; explosive limits in air 1.2-8%; incompat. with oxidizers, strong acids, maleic anhydride, nitrosating agents, nitrogen dioxide, nitroparaffins; violent reactions possible; Hazardous decomp. prods.: Heated to decomp., emits toxic fumes of NO_x; can react with nitrosating agents to form carcinogenic nitrosamines; dec. maleic anhydride @ temps. > 150 C generating heat and CO_2
NFPA: Health 2, Flammability 3, Reactivity 0
Uses: Propellant; corrosion inhibitor; catalytic solvent in chemical synthesis; accelerator activators for rubber; wetting, penetrating and waterproofing agents of quat. ammonium types; curing and hardening of polymers; solvent for paints; catalyst for PU foams, epoxy resins; stabilizer for chlorinated solvents; flux for copper soldering; mfg. of pesticides, pharmaceuticals, polymers, paints, rubber, dyes
Manuf./Distrib.: Air Prods.; Aldrich; Allchem Ind.; Ashland; BASF; BASF AG; Coyne; Elf Atochem N. Am.; Fluka; Focus; Houghton Chem.; Sigma; Spectrum Chem. Mfg.; Union Carbide

N,N,N-Triethylbenzenemethanaminium chloride. *See* Benzyltriethyl ammonium chloride

Triethyl citrate
CAS 77-93-0; EINECS 201-070-7
Synonyms: TEC; 2-Hydroxy-1,2,3-propanetricarboxylic acid, triethyl ester; Ethyl citrate
Definition: Triester of ethyl alcohol and citric acid
Empirical: $C_{12}H_{20}O_7$
Formula: $C_3H_5O(COOC_2H_5)_3$
Properties: Colorless mobile oily liq., odorless, bitter taste; sol. 65 g/100 cc water; sol. 0.8 g/100 cc oil; misc. with alcohol, ether; m.w. 276.32; dens. 1.136 (25 C); vapor pressure 1 mm Hg (107 C); b.p. 294 C; flash pt. (COC) 303 F; ref. index 1.4420
Toxicology: LD50 (oral, rat) 5900 mg/kg, (IP, mouse) 1750 mg/kg; mod. toxic by IP route; mildly toxic by ingestion, inh.; no known skin toxicity; heated to decomp., emits acrid smoke and irritating fumes
Precaution: Combustible exposed to heat or flame; Hazardous decomp. prods.: Heated to decomp., emits acrid smoke and irritating fumes
Uses: Solvent and plasticizer for nitrocellulose and natural resins; softener; paint removers; agglutinant; perfume base; food additive, sequestrant; flavoring agent, plasticizer in pharmaceuticals
Regulatory: FDA 21CFR §175.300, 175.320, 181.22, 181.27, 182.1911, GRAS; FEMA GRAS; FDA approved for orals; USP/NF compliance
Manuf./Distrib.: Aldrich; Fluka; Great Western; Int'l. Chem. Inc; Morflex; Penta Mfg.; Unitex
Trade names: Citroflex® 2

Triethylene glycol
CAS 112-27-6; EINECS 203-953-2
Synonyms: TEG; Di-β-hydroxyethoxyethane; 3,6-Dioxaoctane-1,8-diol; 2,2´-[1,2-Ethanediylbis(oxy)]bisethanol; 2,2´-Ethylenedioxydiethanol; 2,2´-Ethylenedioxyethanol; Ethylene glycol bis(2-hydroxyethyl ether); Ethylene glycol

dihydroxydiethyl ether; Glycol bis (hydroxyethyl) ether; Triglycol
Classification: Aliphatic alcohol
Empirical: $C_6H_{14}O_4$
Formula: $HOCH_2CH_2OCH_2CH_2OCH_2CH_2OH$
Properties: Colorless clear liq., odorless; sol. in water; misc. with alcohol, benzene; immisc. with benzene, toluene, gasoline; very sl. sol. in ether; m.w. 150.17; sp.gr. 1.12 (20 C); vapor pressure 1 mm Hg (114 C); f.p. -7.2 C; b.p. 278 C; flash pt. (PMCC) 335 F; ref. index 1.4550
Toxicology: LD50 (oral, rat) 17 g/kg, (IV, rat) 11,700 μg/kg; poison by IV route; mildly toxic by ing.; skin irritant
Precaution: Combustible exposed to heat or flame; reactive with oxidizing materials; explosive as vapor exposed to heat, flame; Hazardous decomp. prods.: Heated to decomp., emits acrid smoke and irritating fumes
NFPA: Health 1, Flammability 1, Reactivity 0
Uses: Dehydrating agent for natural gas; cork humectant; solvent and lubricant for textile dyeing/printing; plasticizer; humectant for tobacco; chem. intermediate
Regulatory: FDA 21CFR §175.105, 175.300, 177.1200, 178.3740, 178.3910, 179.45
Manuf./Distrib.: Aldrich; Allchem Ind.; Ashland; BASF; Coyne; Eastman; Fluka; Hoechst Celanese; Houghton Chem.; Marcus Int'l.; Occidental; Oxiteno; Shell; Spectrum Chem. Mfg.; Texaco; Union Carbide

Triethylene glycol butyl ether. *See* Butoxytriglycol
Triethylene glycol n-butyl ether. *See* Butoxytriglycol
Triethylene glycol dimethyl ether. *See* PEG-3 dimethyl ether
Triethylene glycol ethyl ether. *See* Ethoxytriglycol
Triethylene glycol monobutyl ether. *See* Butoxytriglycol
Triethylene glycol monoethyl ether. *See* Ethoxytriglycol
Triethylene glycol monomethyl ether. *See* Triglycol monomethyl ether
Triethylhexyl phosphate. *See* Trioctyl phosphate
Tri (2-ethylhexyl) phosphate. *See* Trioctyl phosphate

Triethyl phosphate
CAS 78-40-0; EINECS 201-114-5
Synonyms: TEP; o-Phosphoric acid, triethyl ester; Ethyl phosphate
Empirical: $C_6H_{15}O_4P$
Formula: $(C_2H_5O)_3PO$
Properties: Clear liq., mild odor; misc. with water, ethanol, ethyl acetate, acetone, chloroform, benzene, xylene, castor oil, isobutanol; m.w. 182.16; dens. 1.067-1.072 (20/20 C); vapor pressure 1 mm Hg (39.6 C); m.p. -56 C; b.p. 215 C; flash pt. (PMCC) 99 C; ref. index 1.4019
Toxicology: LD50 (oral, rat) 1311 mg/kg, (dermal, guinea pig) > 20 mL/kg; LDLo (oral, rat) 1600 mg/kg; moderately toxic by ing., IP, IV routes; causes eye irritation
Precaution: Combustible exposed to heat or flames; can react vigorously with oxidizing agents; Hazardous decomp. prods.: Heated to decomp., emits toxic fumes of phosphorous oxides
Uses: Intermediate for agric. insecticides; in floor polishes, unsaturated polyesters, lubricants; as plasticizer, solvent, catalyst, flame retardant
Manuf./Distrib.: Aldrich; Allchem Ind.; Ashland; Bayer; Eastman; Fluka; Focus
Trade names: Eastman® TEP

Trifluoracetic acid. *See* Trifluoroacetic acid

Trifluoroacetic acid
CAS 76-05-1; EINECS 200-929-3
UN No. 2699 (DOT)

Synonyms: Trifluoracetic acid; Acetic acid, trifluoro-; Perfluoroacetic acid; Trifluoroethanoic acid
Empirical: $C_2HF_3O_2$
Formula: CF_3COOH
Properties: Colorless liq.; strong pungent odor; m.w. 114.03; dens. 1.535 (0 C); m.p.-15.25 C; b.p. 71.1 C (734 mm)
Toxicology: LD50 (oral, rat) 200 mg/kg, (IV, mouse) 1200 mg/kg; poison by ing. and IP routes; mod. toxic by IV route; mildly toxic by inh.; corrosive irritant to skin, eyes, mucous membranes; mutagen
Precaution: Corrosive; Hazardous decomp. prods.: Heated to decomp., emits toxic fumes of F
Uses: Strong nonoxidizing acid, laboratory reagent, solvent, catalyst
Manuf./Distrib.: Advance Research Chems.; Aldrich; Aldrich Ltd; Allchem Ind.; AlliedSignal; Atomergic Chemetals; Fluka; Halocarbon Prods.; Janssen Chimica; Monomer-Polymer & Dajac Labs; PCR; Rhône-Poulenc N. Am.; Sigma; Solvay GmbH; Spectrum Chem. Mfg.

α,α,α-Trifluoro-4-chlorotoluene. *See* p-Chlorobenzotrifluoride
Trifluoroethanoic acid. *See* Trifluoroacetic acid
p-(Trifluoromethyl)chlorobenzene. *See* p-Chlorobenzotrifluoride
p-Trifluoromethylphenyl chloride. *See* p-Chlorobenzotrifluoride
Triglyceryl diisostearate. *See* Polyglyceryl-3 diisostearate
Triglycol. *See* Triethylene glycol
Triglycol monobutyl ether. *See* Butoxytriglycol
Triglycol monoethyl ether. *See* Ethoxytriglycol

Triglycol monomethyl ether
CAS 112-35-6
Synonyms: PEG-3 methyl ether; Ethanol, 2(2-(2-methoxyethoxy)ethoxy)-; 2-(2-(2-Methoxyethoxy)ethoxy)ethanol; Triethylene glycol monomethyl ether; Methyltriglycol; Methoxytriglycol
Empirical: $C_7H_{16}O_4$
Formula: $CH_3O(CH_2CH_2O)_2CH_2CH_2OH$
Properties: Misc. with water; m.w. 164.20; dens. 1.048 (20/4 C); b.p. 249 C; flash pt. 118 C; ref. index 1.439 (20 C)
Toxicology: LD50 (oral, rat) 11,300 mg/kg, (dermal, rabbit) 7100 mg/kg; mildly toxic by ingestion and skin contact; skin irritant
Precaution: Combustible exposed to heat or flame; Hazardous decomp. prods.: Heated to decomp., emits acrid smoke and irritating fumes
Uses: Solvent for brake fluids, hard-surf. cleaners, leather dyeing, paints, coatings, printing inks, textile vat dyeing and printing, adhesives, antifreeze, floor waxes/polishes, insect repellents; solubilizer for dyes; plasticizer
Manuf./Distrib.: Aldrich; Ashland; Fluka; Oxiteno; Sigma
Trade names: Hisolve TM; Poly-Solv® TM

Triglyme. *See* PEG-3 dimethyl ether

Trihexyl citrate
CAS 16544-70-0
Uses: Solvent for household, I&I cleaners
Trade names: Citroflex® 6

Trihydroxypropane glycerol. *See* Glycerin

Triisostearin PEG-6 esters
Definition: Mixt. formed from the transesterification of triisostearin and PEG-6
Uses: Solvent, amphiphilic agent for improving drug delivery; excipient
Trade names: Labrafil® Isostearique

1,3,5-Trimethylbenzene
CAS 108-67-8; EINECS 203-604-4
UN No. 2325
Synonyms: TMB; Trimethylbenzol; Trimethyl-1,3,5-benzene; Mesitylene; sym-Trimethylbenzene
Classification: Aromatic hydrocarbon
Empirical: C_9H_{12}
Formula: $C_6H_3(CH_3)_3$
Properties: Colorless clear liq.; char. aromatic odor; insol. in water; sol. in alcohol, ether, benzene, acetone, CCl_4, petrol. ether; m.w. 120.21; sp.gr. 0.863; f.p. -52.7 C; b.p. 164.6 C; flash pt. 50 C
Toxicology: ACGIH TLV/TWA 25 ppm; LC50 (inh., rat, 4 h) 24 mg/m^3; toxic by inhalation (can cause nose, throat, and lung irritation, headache, dizziness, nausea, CNS depression); probable sl. to mod. skin irritant and mild eye irritant
Precaution: Combustible; mod. fire hazard; incompat. with strong oxidizing agents (increased risk of fire and explosion); reacts violently with nitric acid
NFPA: Health 0, Flammability 2, Reactivity 0
Uses: Intermediate in prod. of dyestuffs and uv oxidation stabilizers for plastics; solvent and paint thinner
Manuf./Distrib.: Aldrich; B.I. Chems.; Fluka; Sigma

Trimethyl-1,3,5-benzene. *See* 1,3,5-Trimethylbenzene
sym-Trimethylbenzene. *See* 1,3,5-Trimethylbenzene
Trimethylbenzol. *See* 1,3,5-Trimethylbenzene
Trimethylbenzylammonium chloride. *See* Benzyl trimethyl ammonium chloride
1,7,7-Trimethylbicyclo[2.2.1]heptan-2-ol acetate. *See* Bornyl acetate
2,6,6-Trimethylbicyclo(3.1.1)-2-heptene. *See* α-Pinene
2,6,6-Trimethylbicyclo(3.1.1)-2-hept-2-ene. *See* α-Pinene

2,2,3-Trimethylbutane
CAS 464-06-2
UN No. 1206
Synonyms: Isopropyltrimethylmethane; Triptane
Classification: Sat. aliphatic hydrocarbon
Empirical: C_7H_{16}
Formula: $CH_3C(CH_3)_2CH(CH_3)CH_3$
Properties: Colorless liq.; gasoline-like odor; sol. in most org. solvs.; insol. in water; m.w. 100.20; sp.gr. 0.69 (20 C); m.p. -25 C; b.p. 81 C; flash pt. -6 C
Toxicology: ACGIH TLV/TWA 400 ppm; STEL 500 ppm; may cause mild skin irritation; conc. vapor may cause sl. eye irritation; probably a CNS depressant; conc. vapor may cause sl. nose/throat irritation, lightheadedness, dizziness, nausea, difficulty breathing
Precaution: Flamm.; incompat. with strong oxidizers (increases fire and explosion risk); Hazardous decomp. prods.: None
NFPA: Health 0, Flammability 3, Reactivity 0
Uses: Ingred. in gasoline and aviation fuel; org. synthesis; solvent
Manuf./Distrib.: Aldrich; Fluka

Trimethyl carbinol. *See* t-Butyl alcohol

Trimethyl citrate
CAS 1587-20-8
Empirical: $C_9H_{14}O_7$
Properties: Solid; sol. 8.2 g/100 ml in water; insol. in heptane; m.p. 76 C
Uses: Solvent for household, I&I cleaners
Trade names: Citroflex® 1

α,α,4-Trimethyl-3-cyclohexene-1-methanol. *See* α-Terpineol
1,1,3-Trimethyl-3-cyclohexene-5-one. *See* Isophorone
3,3,5-Trimethyl-2-cyclohexene-1-one. *See* Isophorone
3,3,5-Trimethyl-2-cyclohexen-1-one. *See* Isophorone

1,1,2-Trimethylcyclopentane
CAS 4259-00-1
Synonyms: Camphocean
Classification: Sat. alicyclic hydrocarbon
Empirical: C_8H_{16}
Formula: $(CH_3)_3C_5H_7$
Properties: Colorless liq.; sol. in most org. solvs.; insol. in water; m.w. 112.21; sp.gr. 0.7661 (20 C); b.p. 113-114 C
Toxicology: May cause mild skin irritation; conc. vapor may be irritating to eyes; probably a CNS depressant; high vapor concs. may cause nose/throat irritation, headache, lightheadedness, nausea, dizziness; very high concs. may cause unconsciousness, death
Precaution: Flamm.; incompat. with strong oxidizing agents (increased fire and explosion hazard); Hazardous decomp. prods.: None
Uses: Solvent; org. synthesis

1,1,3-Trimethylcyclopentane
CAS 4516-69-2
Classification: Sat. alicyclic hydrocarbon
Empirical: C_8H_{16}
Formula: $(CH_3)_3C_5H_7$
Properties: Colorless liq.; sol. in most org. solvs.; insol. in water; m.w. 112.21; sp.gr. 0.7703 (20 C); b.p. 115-116 C
Toxicology: May cause mild skin irritation; conc. vapor may be irritating to eyes; probably a CNS depressant; high vapor concs. may cause nose/throat irritation, headache, lightheadedness, nausea, dizziness; very high concs. may cause unconsciousness, death
Precaution: Flamm.; incompat. with strong oxidizing agents (increased fire and explosion hazard); Hazardous decomp. prods.: None
Uses: Solvent; org. synthesis

1,1,2-Trimethylethane. *See* Isopentane
Trimethyl glycol. *See* Propylene glycol
3,5,5-Trimethylhexanoic acid, isodecyl ester. *See* Isodecyl isononanoate
3,5,5-Trimethylhexanoic acid, isotridecyl ester. *See* Isotridecyl isononanoate
3,5,5-Trimethylhexanoic acid, 3,5,5-trimethylhexyl ester. *See* Isononyl isononanoate
2,6,8-Trimethyl-4-nonanone. *See* Isobutyl heptyl ketone
1,3,3-Trimethyl-2-oxabicyclo[2.2.2]octane. *See* Eucalyptol
2,2,4-Trimethylpentane. *See* Isooctane

2,2,4-Trimethyl-1,3-pentanediol monoisobutyrate
CAS 25265-77-4; EINECS 246-771-9
Synonyms: Propionic acid, 2-methyl-, monoester with 2,2,4-trimethyl-1,3-pentanediol; 2,2,4-Trimethyl-1,3-pentanediol mono (2-methylpropanoate); 1,3-Pentanediol, 2,2,4-trimethyl-, monoisobutyrate
Classification: Ester alcohol
Empirical: $C_{12}H_{24}O_3$
Formula: $(CH_3)_2CHCH(OH)C(CH_3)_2CH_2OOCCH(CH_3)_2$
Properties: Liq.; insol. in water; sol. in benzene, alcohol, acetone, CCl_4; m.w. 216.3; dens. 0.945-0.955 (20/20 C); b.p. 180-182 C (125 mm); flash pt. (COC) 245 F; ref.

index 1.4423
Toxicology: LDLo (oral, rat) 3200 mg/kg; mod. toxic by ingestion
Precaution: Combustible; Hazardous decomp. prods.: Heated to decomp., emits acrid smoke and irritating fumes
Uses: Solvent, intermediate; mfg. of plasticizers, surfactants, pesticides, resins, inks, latexes, drilling muds, lubricants, syn. detergents, herbicides; slow-evaporating solvent used as coalescing agent in latex finishes and water-base inks, PVAc latices, PVAc-acrylic latices, and acrylic, EVA, and B/S latices; solvent in electrodeposition coatings, lacquer coatings
Manuf./Distrib.: Aldrich; Ashland; Chisso Am.; Eastman
Trade names: Texanol® Ester-Alcohol; Ucar® Filmer IBT

2,2,4-Trimethyl-1,3-pentanediol mono (2-methylpropanoate). *See* 2,2,4-Trimethyl-1,3-pentanediol monoisobutyrate
α,α,α´-Trimethyltrimethyleneglycol. *See* Hexylene glycol

Trioctyl phosphate
CAS 78-42-2; EINECS 201-116-6
Synonyms: Tris(2-ethylhexyl) phosphate; Triethylhexyl phosphate; Tri (2-ethylhexyl) phosphate; Octyl phosphate; 2-Ethyl-1-hexanol phosphate; 1-Hexanol, 2-ethyl-, phosphate; Phosphoric acid, tris(2-ethylhexyl) ester
Empirical: $C_{24}H_{51}O_4P$
Formula: $[CH_3(CH_2)_3CH(C_2H_5)CH_2O]_3PO$
Properties: Liq.; sol. in alcohol, acetone, ether; m.w. 434.72; dens. 0.924 (26 C); b.p. 220-230 (8 mm); flash pt.(OC) 405 F; ref. index 1.4440
Toxicology: LD50 (oral, rat) 37 g/kg; LDLo (dermal, rabbit) 20 g/kg; mildly toxic by ingestion, skin contact; skin and eye irritant; suspected carcinogen; tumorigen
Precaution: Combustible exposed to heat or flame; can react with oxidizing materials; Hazardous decomp. prods.: Heated to decomp., emits toxic fumes of PO_x
Uses: Solvent, antifoaming agent for paints; plasticizer for vinyl resins, rubbers, nitrocellulose; flame retardant
Manuf./Distrib.: Akzo Nobel; Albright & Wilson Am.; Aldrich; Ashland; Bayer; Ferro/Keil; Fluka; FMC; Rhône-Poulenc; Sigma
Trade names: Kronitex® TOF; Reomol® TOP

Trioxanonane. *See* Diethylene glycol dimethyl ether
5,8,11-Trioxapentadecane. *See* Diethylene glycol dibutyl ether
3,6,9-Trioxaundecane. *See* Diethylene glycol diethyl ether

Tripropylene glycol n-butyl ether
CAS 55934-93-5
Formula: $C_4H_9OC_3H_6OC_3H_6OCH_2CHOHCH_3$
Uses: Solvent for heavy-duty cleaners, floor wax strippers; plasticizer for resin systems; coupling agent, grease penetrant in oven cleaners; solvent for inks
Manuf./Distrib.: Aldrich
Trade names: Dowanol® TPnB

Tripropylene glycol ethyl ether
CAS 20178-34-1
Synonyms: TPGEE; Tripropylene glycol monoethyl ether; 1-[2-(2-Ethoxy-1-methylethoxy)-1-methylethoxy]-2-propanol
Classification: Aliphatic ether alcohol
Empirical: $C_{11}H_{24}O_4$
Formula: $CH_3CH(OH)CH_2OCH(CH_3)CH_2OCH(CH_3)CH_2O$
Properties: Colorless clear liq.; sl. pleasant ether-like odor; sol. in water, variety of

org. solvs.; m.w. 220.3; sp.gr. 0.948; b.p. 252 C; flash pt. (COC) 132 C
Toxicology: Low oral toxicity; unlikely to irritate skin; may cause mild eye irritation; readily absorbed thru skin, but health effects unlikely; extreme/prolonged contact may cause CNS depression, possible unconsciousness
Precaution: Incompat. with oxidizing agents (increases fire/explosion hazard)
Uses: Solvent for paints, lacquers, resins, inks, dyes, oils, greases, cleaners

Tripropylene glycol methyl ether. *See* PPG-3 methyl ether
Tripropylene glycol monoethyl ether. *See* Tripropylene glycol ethyl ether
Tripropylene glycol monomethyl ether. *See* PPG-3 methyl ether
Tripropylene glycol myristyl ether. *See* PPG-3 myristyl ether
Triptane. *See* 2,2,3-Trimethylbutane
Tris-n-butylamine. *See* Tributylamine
Tris(2-ethylhexyl) phosphate. *See* Trioctyl phosphate
Triticum aestivum germ oil. *See* Wheat germ oil

Triundecanoin
CAS 13552-80-2; EINECS 236-935-8
Synonyms: Glyceryl triundecanoate; Undecanoic acid, 1,2,3-propanetriyl ester
Empirical: $C_{36}H_{68}O_6$
Formula: $HCO(CH_2OCO(CH_2)_9CH_3)_2CO(CH_2)_9CH_3$
Uses: Emollient, solvent, fixative, and extender in nutritional applics.; carrier for flavors and fragrances; pharmaceutical emollient, lubricant
Manuf./Distrib.: Fluka; Sigma
Trade names: Captex® 8227

TTE. *See* Trichlorotrifluoroethane
Turpentine. *See* Turpentine oil

Turpentine oil
CAS 8006-64-2
UN No. 1299
Synonyms: Turpentine; Purified gum spirits; Turpentine, steam distilled; Gum turpentine; Turpentine, purified; Spirit of turpentine; Spirits of turpentine; Wood turpentine; Turpentine, rectified; Turpentine oil, rectifier
Classification: Unsat. alicyclic hydrocarbon
Definition: Volatile essential oil obtained by distillation and rectification from turpentine, an oleoresin obtained from *Pinus* spp., contg. pinene and diterpene
Empirical: $\approx C_{10}H_{16}$
Properties: Colorless liq., char. aromatic or pine penetrating odor; insol. in water; sol. in alcohol, benzene, chloroform, ether, CS_2, petrol. ether, oils; m.w. ≈ 136; sp.gr. 0.860-0.875 (15 C); m.p. -50 to -60 C; b.p. 154-170 C; flash pt. (CC) 32-46 C; ref. index 1.463-1.483 (20 C)
Toxicology: ACGIH TLV/TWA 100 ppm; LD50 (oral, rat) 5760 mg/kg; toxic by ingestion; irritating to skin and mucous membranes; can cause allergic reactions, serious kidney irritation; CNS depressant; death due to respiratory failure
Precaution: DOT: Flamm. liq.; mod. fire risk and explosion hazard; incompat. with oxidizers, halogens, stannic chloride, hexachloromelamine, diatomaceous earth; Hazardous decomp. prods.: Heated to decomp., emits acrid smoke and irritating fumes
NFPA: Health 1, Flammability 3, Reactivity 0
Uses: Solvent in shoe polishes, printing inks, cleaning compds., waxes, paper prods.; solvent, thinner for paints, lacquers; solvent and reclaiming agent for rubber; synthesis of camphor and menthol; medicines (liniments); perfumery; pesticide mfg.; pharmaceutical solvent, rubefacient, diuretic

Turpentine oil, rectifier

Regulatory: FDA approved for inhalants; BP compliance
Manuf./Distrib.: Berje; Penta Mfg.; Quest Int'l.; Spectrum Chem. Mfg.

Turpentine oil, rectifier. *See* Turpentine oil
Turpentine, purified. *See* Turpentine oil
Turpentine, rectified. *See* Turpentine oil
Turpentine, steam distilled. *See* Turpentine oil

Undecane. *See* n-Undecane

n-Undecane
CAS 1120-21-4
UN No. 2330 (DOT)
Synonyms: Hendecane; Undecane
Classification: Sat. aliphatic hydrocarbon
Empirical: $C_{11}H_{24}$
Properties: Colorless liq.; gasoline-like odor; sol. in most org. solvs.; insol. in water; m.w. 156.31; sp.gr. 0.744 (20 C); m.p. -26 C; b.p. 196 C; flash pt. (CC) 60 C
Toxicology: LD50 (IV, mouse) 517 mg/kg; toxic; conc. vapor may be irritating to eyes; conc. vapor can probably cause nose/throat irritation, headache, drowsiness, nausea, difficulty breathing, however no significant narcotic effects likely
Precaution: Flamm. exposed to heat, flame, oxidizers; incompat. with strong oxidizing agents (increases fire and explosion hazard); Hazardous decomp. prods.: Heated to decomp., emits acrid smoke and irritating fumes
NFPA: Health 0, Flammability 2, Reactivity 0
Uses: Component of gasoline, jet fuel, kerosene, photocopying fluids, petrol. solvents such as wh. spirit; solvent; org. synthesis; petrol. research; distillation chaser
Manuf./Distrib.: Aldrich; Fluka; Sigma; Triple Crown Am.

Undecanoic acid, 1,2,3-propanetriyl ester. *See* Triundecanoin

Valerone. *See* Diisobutyl ketone
Vaseline. *See* Petrolatum
Vinegar acid. *See* Acetic acid; Acetic acid, glacial
Vinegar naphtha. *See* Ethyl acetate
Vinyl ether 'Carbitol'. *See* Diethylene glycol ethylvinyl ether
Vinyl trichloride. *See* 1,1,2-Trichloroethane
VM&P naphtha. *See* Naphtha

Walnut oil
CAS 8024-09-7; 84604-00-2
Definition: Oil derived from the nut meats of walnuts, *Juglans* spp.
Properties: Iodine no. 145-155; sapon. no. 190; ref. index 1.4691 (40 C)

Uses: Emollient for makeup, skin, and hair care prods.; occlusive solvent for suntan gels, topical pharmaceuticals, ointments; natural flavoring for foods; pharmaceutical lubricant
Regulatory: FDA 21CFR §172.510, 175.300
Manuf./Distrib.: Arista Ind.; Charkit; Penta Mfg.; Tri-K Ind.; Welch, Holme & Clark
Trade names: EmCon Walnut

Wheat germ oil
CAS 8006-95-9
Synonyms: Triticum aestivum germ oil
Definition: Oil obtained by expression or extraction of wheat germ
Properties: Lt. yel. oil; fat-sol.; sp.gr. 0.93-0.94; iodine no. 120-140; sapon. no. 179-194; ref. index 1.469-1.479
Uses: Emollient for skin and hair care prods.; gloss agent in makeups; moisturizer; food ingred., nutritive additive, dietary supplement; pharmaceutical nutritional source, topical moisturizer, lubricant; occlusive solvent
Manuf./Distrib.: Arista Ind.; Charkit; Croda Inc; Croda Chem. Ltd; Marine Bioprods.; Penta Mfg.; Pokonobe Ind.; Provital; Spectrum Chem. Mfg.; Tri-K Ind.; Vitamins, Inc
Trade names: EmCon W

White caustic. *See* Sodium hydroxide
White mineral oil. *See* Mineral oil
White petrolatum. *See* Petrolatum
White soft paraffin. *See* Petrolatum
White spirits. *See* Mineral spirits; Stoddard solvent
Wood alcohol. *See* Methyl alcohol
Wood naphtha. *See* Methyl alcohol
Wood spirit. *See* Methyl alcohol
Wood turpentine. *See* Turpentine oil
Wool fat. *See* Lanolin
Wool wax. *See* Lanolin
Wool wax alcohol. *See* Lanolin alcohol

Xylene
CAS 1330-20-7; EINECS 215-535-7
UN No. 1307 (DOT)
Synonyms: Methyl toluene; Mixed xylenes; Dimethylbenzene; Xylol
Classification: Aromatic hydrocarbon
Definition: Commercial mixture of 3 isomers: o-, m-, and p-xylene (1,3-dimethylbenzene, 1,2-dimethylbenzene, 1,4-dimethylbenzene)
Empirical: C_8H_{10}
Formula: $C_6H_4(CH_3)_2$
Properties: Colorless clear liq., char. sweet odor; pract. insol. in water; sol. in abs. alcohol, ethanol, diethyl ether, many org. liqs.; m.w. 106.16; sp.gr. 0.86; vapor pressure 6.72 mm Hg (21 C); b.p. 137-144 C; flash pt. 29 C; ref. index 1.4970; KB value 98
Toxicology: ACGIH TLV/TWA 100 ppm; STEL 150 ppm; LD50 (oral, rat) 4300 mg/kg; toxic by ing., inh.; irritant to eyes, skin, respiratory tract; may be narcotic in high concs. causing CNS depression, dizziness, nausea,

unconsciousness

Precaution: Highly flamm.; moderate fire risk; incompat. with strong oxidizing agents, nitric acid, dichlorohydrantoin; increased fire and explosion risk; Hazardous decomp. prods.: Heated to decomp., emits acrid smoke and irritating fumes

NFPA: Health 2, Flammability 3, Reactivity 0

Uses: Solvent for alkyd resins, lacquers, enamels, rubber cements, printing inks, leather, gums, resins, rubber, castor/linseed oils; raw material for prod. of benzoic acid, phthalic anhydride, dyes, paints, other organics; aviation gasoline; protective coating; degreaser, cleaning agent; carrier in epoxy resin prod.; chem. synthesis; mfg. of quartz crystal oscillators, perfumes, insect repellents

Regulatory: FDA 21CFR §175.105, 176.180, 177.1010, 177.1650; SARA reportable

Manuf./Distrib.: Aldrich; ARCO; Ashland; J.T. Baker; Baychem; BP Chems. Ltd; R.E. Carroll; Chemcentral; Coyne; Crowley Chem.; Exxon; Fina Chems.; Fluka; General Chem.; Harcros USA; Houghton Chem.; Mallinckrodt; Marathon Oil; Mitsubishi Petrochem.; Mitsui Petrochem. Ind.; Mobil; Olin; Phillips; Samson; Shell; Sigma; Spectrum Chem. Mfg.; Texaco; Van Waters & Rogers

Trade names: Shell Xylene

1,2-Xylene. *See* o-Xylene
1,3-Xylene. *See* m-Xylene
1,4-Xylene. *See* p-Xylene
3-Xylene. *See* m-Xylene
4-Xylene. *See* p-Xylene

m-Xylene

CAS 108-38-3

UN No. 1307 (DOT)

Synonyms: 1,3-Dimethylbenzene; m-Dimethylbenzene; 3-Methyltoluene; m-Methyltoluene; 1,3-Xylene; 3-Xylene; m-Xylol

Classification: Aromatic hydrocarbon

Empirical: C_8H_{10}

Formula: $CH_3C_6H_4CH_3$

Properties: Colorless clear liq.; char. aromatic odor; sol. in ethanol, diethyl ether, acetone, benzene, other org. solvs.; pract. insol. in water; m.w. 106.16; sp.gr. 0.8642 (20 C); vapor pressure 10 mm Hg (28.3 C); m.p. -47 C; b.p. 139 C; flash pt. (CC) 25 C

Toxicology: LD50 (IP, mouse) 1739 mg/kg, (skin, rabbit) 14,100 mg/kg; ACGIH TLV/TWA 100 ppm; STEL 150 ppm; toxic; severe skin irritant; sl. toxic by ing.; human systemic effects; ing. of large amts. may cause CNS depression

Precaution: Flamm.; very dangerous fire hazard exposed to heat, flame; incompat. with oxidizers, nitric acid, sulfuric acid, sulfur dichloride (increases fire and explosion hazard); common air contaminant; Hazardous decomp. prods.: Heated to decomp., emits acrid smoke and irritating fumes

NFPA: Health 2, Flammability 3, Reactivity 0

Uses: Chem. intermediate in mfg. of isophthalic acid for polyester resins, mfg. of alkyd resins; chem. intermediate for dyes, insecticides, org. synthesis; as solvent; in aviation fuel

Manuf./Distrib.: Aldrich; Fluka; Sigma; Spectrum Chem. Mfg.

o-Xylene

CAS 95-47-6; EINECS 202-422-2

UN No. 1307 (DOT)

Synonyms: o-Dimethylbenzene; 1,2-Dimethylbenzene; 1,2-Xylene; o-Xylol; o-Methyltoluene; Benzene, 1,2-dimethyl

Empirical: C_8H_{10}

Formula: $C_6H_4(CH_3)_2$
Properties: Colorless liq.; misc. with abs. alcohol, ether; insol. in water; m.w. 106.17; dens. 0.8802 (20 C); m.p. -25 C; b.p. 141.5-144.5 C; flash pt. 17 C
Toxicology: LDLo (oral, rat) 8800 mg/kg; LD50 (IP, mouse) 1364 mg/kg; OSHA PEL:TWA 100 ppm; mod. toxic by IP route; mildly toxic by ing., inh.; experimental teratogen; common air contaminant; heated to decomp., emits acrid smoke and irritating fumes
Precaution: DOT: Flamm. liq.; dangerous fire hazard exposed to heat or flame; explosive as vapor exposed to heat or flame
NFPA: Health 2, Flammability 3, Reactivity 0
Uses: Solvent for paints
Manuf./Distrib.: Mobil; Shell; Spectrum Chem. Mfg.

p-Xylene
CAS 106-42-3; EINECS 203-396-5
UN No. 1307 (DOT)
Synonyms: 1,4-Dimethylbenzene; p-Dimethylbenzene; 4-Methyltoluene; p-Methyltoluene; 1,4-Xylene; 4-Xylene; p-Xylol
Classification: Aromatic hydrocarbon
Empirical: C_8H_{10}
Formula: $CH_3C_6H_4CH_3$
Properties: Colorless clear liq.; char. aromatic odor; colorless cryst. solid below 13 C; sol. in ethanol, diethyl ether, acetone, benzene, other org. solvs.; pract. insol. in water; m.w. 106.16; sp.gr. 0.861 (20 C); vapor pressure 10 mm Hg (27.3 C); m.p. 13.3 C; b.p. 138.3 C; flash pt. (CC) 25 C
Toxicology: LD50 (IP, rat) 3810 mg/kg; ACGIH TLV/TWA 100 ppm; STEL 150 ppm; toxic; sl. toxic by ing.; ing. of large amts. causes CNS effects; inh. causes nose/throat irritation, CNS depression, headache, dizziness, nausea, vomiting
Precaution: Flamm.; very dangerous fire hazard exposed to heat or flame; explosive as vapor exposed to heat or flame; incompat. with oxidizers, nitric acid, sulfuric acid, sulfur dichloride (increases fire and explosion hazard); Hazardous decomp. prods.: Heated to decomp., emits acrid smoke and irritating fumes
NFPA: Health 2, Flammability 3, Reactivity 0
Uses: Chem. intermediate for synthesis of terephthalic acid for polyester resins and fibers, in mfg. of plastics, vitamins, pharmaceuticals, insecticides; solvent for adhesives and coatings
Manuf./Distrib.: Aldrich; Fluka; Sigma; Spectrum Chem. Mfg.

Xylenol
CAS 1300-71-6
UN No. 2261 (DOT)
Synonyms: Dimethylphenol; Hydroxydimethylbenzene; Hydroxyxylene; Dimethylhydroxybenzene
Empirical: $C_8H_{10}O$
Formula: $(CH_3)_2C_6H_3OH$
Properties: M.w. 122.17; Commercial mixt.: wh. cryst. solid; sol. in most org. solvs.; sl. sol. in water; dens. 1.02-1.03 (15 C); m.p. 20-76 C; b.p. 203-225 C
Toxicology: Toxic by ingestion, skin absorption; heated to decomp., emits acrid smoke and irritating fumes
Precaution: Combustible
Uses: Disinfectants, solvents, pharmaceuticals, insecticides, fungicides, herbicides, plasticizers, rubber chemicals, paints, additives to lubricants and gasoline, mfg. of polyphenylene oxide, wetting agents, dyestuffs, synthetic resin

Xylidine

Manuf./Distrib.: J.T. Baker; Coalite Chems. Div.; Crowley Chem.; Crowley Tar Prods.; Merichem

Xylidine. *See* n,n-Dimethylaniline
Xylol. *See* Xylene
m-Xylol. *See* m-Xylene
o-Xylol. *See* o-Xylene
p-Xylol. *See* p-Xylene

Yarmor. *See* Pine oil
Yarmor pine oil. *See* Pine oil

Zea mays oil. *See* Corn oil

Part III
Applications Cross-Reference

Application Cross-Reference

Trade name products and chemicals from the first two parts of this reference are grouped by broad application areas derived from manufacturers' specifications and research. The following products may not be limited to the application areas represented here.

Adhesives

Trade names: Agrisynth® BLO; Agsol Ex BLO; Arcosolv® DPM; Arcosolv® DPM Acetate; Arcosolv® DPNB; Arcosolv® PE; Arcosolv® PE Acetate; Arcosolv® PM; Arcosolv® PNB; Arcosolv® PNP; Arcosolv® TPM

BLO®

Carbowax® PEG 300; Citroflex® 1; Citroflex® 2; Citroflex® 4; Citroflex® 6; Citroflex® A-1; Citroflex® A-2; Citroflex® A-4

Dynasolve 200; Dynasolve M-35

Eastman® DBP; Eccoflex 090; Ecco Strip LS-W; Emery® 912; Emery® 916; Emery® 917; Emery® 918

FoamFlush® Urethane Remover

GAFsolver™ TMP

Halso® 99; Hisolve MC

Isopar® V

Jeffsol™ PC

Mod Acid

NE™ 101; Neosolv™ 2; Neosolv™ 4; Neosolv™ 6; Neosolv™ 8; Neosolv™ 24; Neosolv™ 150; NMP; Nusolv™ ABP-62; Nusolv™ ABP-74; Nusolv™ ABP-87; Nusolv™ ABP-103; Nusolv™ ABP-164

#15 Oil; #30 Oil; #40 Oil; Oxsol® 73; Oxsol® 100; Oxsol® 253; Oxsol® 325; Oxsol® 550; Oxsol® 1000

Pale 4; Pale 16; Pale 170; Pale 1000; Pilot SXS-40; Poly-Solv® DM; Poly-Solv® DPM; Poly-Solv® EM; Poly-Solv® MPM; Poly-Solv® TM; Poly-Solv® TPM

QO® Tetrahydrofurfuryl Alcohol (THFA®)

Shell VM&P Naphtha HT; Shell Mineral Spirits 135; Shell Mineral Spirits 146 HT; Shell Sol B HT

Tebol™; Tebol™ 99; THF; THF; Tolu-Sol® A HT Solv.; Tolu-Sol® W HT Solv.; Tri-Ethane

Uniplex 150

Vorite 105; Vorite 110; Vorite 115; Vorite 120; Vorite 125

Chemicals: Acetone; Acetyl tributyl citrate; Acetyl triethyl citrate; Acetyl trimethyl citrate

Adhesives *(cont'd.)*

Bis(methylethyl)-1,1′-biphenyl; 2-Butanol; t-Butyl alcohol
Castor oil, polymerized; p-Chlorobenzotrifluoride
3,4-Dichlorobenzotrifluoride; 1,4-Dioxane; Dipropylene glycol butyl ether
Ethyl acetate
Methyl ethyl ketone; Methyl isobutyl ketone; Mineral spirits; Monochlorotoluene
Naphtha; Nitroethane
PPG-2 methyl ether; PPG-3 methyl ether; PPG-2 methyl ether acetate; Propylene glycol ethyl ether; Propylene glycol ethyl ether acetate; Propylene glycol n-propyl ether
Tetrahydrofuran; Toluene; 1,1,1-Trichloroethane; Triglycol monomethyl ether
p-Xylene

Agriculture

Trade names: Agrisol PX401; Agrisol PX413; Agrisynth® BLO; Agsol Ex BLO; Arcosolv® DPM; Arcosolv® DPM Acetate; Arcosolv® DPNB; Arcosolv® PE; Arcosolv® PE Acetate; Arcosolv® PNB; Arcosolv® PNP; Arcosolv® TPM; Aromatic 100; Aromatic 150; Aromatic 150; Aromatic 200
BLO®; Butyl Diglyme; Butyl Di-Icinol; BVA ND-205 Solv.
Carbowax® PEG 300; Cardolite® NC-507; Cardolite® NC-510; Cardolite® NC-511; Crill 1; Crill 2
Diglyme; DMSO; Dowanol® EB; Dowanol® PM
Eastman® DBP; Emery® 912; Emery® 916; Emery® 917; Emery® 2203; Emery® 2209; Emery® 2214; Emery® 2216; Emery® 2218; Emery® 2219; Emery® 2224; Emery® 2230; Emery® 2231; Emery® 2232; Emery® 2253; Emery® 2254; Emery® 2270; Emery® 2290; Emery® 2296; Emery® 2301; Emery® Methyl Oleate; Ethyl Diglyme; Ethyl Di-Icinol; Ethyl Icinol; Exxate® 600; Exxate® 700; Exxate® 800; Exxate® 900; Exxate® 1000; Exxate® 1300; Exxsol® D 60; Exxsol® D 110; Exxsol® D-130
GBL
Icinol DPM; Icinol EE22; Icinol ME33; Icinol ME45; Icinol PM; Isopar® M; Isopar® V
Kodaflex® DBP
Methyl Di-Icinol; Methyl Icinol; Monoglyme; M-Pyrol®
Neosolv™ 2; Neosolv™ 4; Neosolv™ 6; Neosolv™ 8; Neosolv™ 24; Neosolv™ 150; NMP; Norpar® 12; Norpar® 13
PD-23; PD-25; PD-28; Penreco 2259 Oil; Penreco 2260 Oil; Penreco 2263 Oil; 2-Pyrol®
QO® Tetrahydrofurfuryl Alcohol (THFA®)
Renex® PEG 300; Renex® PEG 400; Renex® PEG 600; Renex® PEG 1000; Rhodacal® 330
Schercemol PGML; Secoster® DO 600; Secoster® MA 300; Shell VM&P Naphtha HT; Shell Cyclo Sol 53; Shell Cyclo Sol 63; Shell Mineral Spirits 146 HT; Shell Mineral Spirits 200 HT; Shell Sol 71; Shell Sol 142 HT; Shell Sol B HT

Agriculture *(cont'd.)*

Tebol™; Tebol™ 99; T-Mulz® 426; Tolu-Sol® A HT Solv.; Tolu-Sol® W HT Solv.; Triglyme
Vista LPA; Vista LPA-110; Vista LPA-142; Vista LPA-170; Vista LPA-210

Chemicals: t-Butyl alcohol; n-Butyl mercaptan
C9 alkyl acetate; C10 alkyl acetate; C13 alkyl acetate; Cyclohexane
Dipropylene glycol butyl ether; Dipropylene glycol dimethyl ether
Ethylacetoacetate; Ethyl formate
Isopropyl alcohol
Methoxyisopropanol; Methyl canolate; Methyl caprylate/caprate; Methyl cocoate; Methyl lardate; Methyl laurate; Methyl myristate; Methyl oleate; Methyl oleate/linoleate; Methyl palmitate; N-Methyl-2-pyrrolidone; Methyl stearate; Methyl sunflowerate; Methyl tallowate; Mineral oil; Mineral spirits
Oxo-heptyl acetate; Oxo-hexyl acetate
PEG-4 dilaurate; PEG-8 dilaurate; PEG-20 dilaurate; PEG-3 dimethyl ether; PEG-4 dioleate; PEG-8 dioleate; PEG-20 dioleate; PEG-8 oleate; PEG-20 oleate; 3-(n-Penta-8′-decenyl) phenol; 3-(n-Pentadecyl) phenol; Petroleum distillates; β-Picoline; γ-Picoline; Polysorbate 60; PPG-3 methyl ether; PPG-2 methyl ether acetate; Propylene glycol ethyl ether; Propylene glycol ethyl ether acetate; Propylene glycol n-propyl ether
1,2,4-Trichlorobenzene; 1,1,1-Trichloroethane; Triglycol monomethyl ether

Antifreeze

Trade names: Adol® 42; Adol® 80
Dowanol® DM
Eastman® CDA-19; Eastman® DM
Jeffsol™ PC
Poly-Solv® DM; Poly-Solv® DPM; Poly-Solv® EM; Poly-Solv® MPM; Poly-Solv® TM; Poly-Solv® TPM; Propylene Glycol USP/FCC Ultra Grade
Reomol® TBP

Chemicals: Alcohol denatured
Glycol
Methoxyisopropanol
Triglycol monomethyl ether

Brake Fluids

Trade names: Butyl Dioxitol®
Eastman® DB; Eastman® DE; Eastman® DP; Ektasolve® DM
Hisolve DM; Hisolve PM; Hisolve TM
Jeffsol™ EC; Jeffsol™ PC
Neosolv™ 2; Neosolv™ 4; Neosolv™ 6; Neosolv™ 8; Neosolv™ 24; Neosolv™ 150; Neosolvent NSG
Polyglyme; Poly-Solv® DM; Poly-Solv® DPM; Poly-Solv® MPM;

Brake Fluids *(cont'd.)*

Poly-Solv® TM; Poly-Solv® TPM
QO® Furfural

Chemicals: Alcohol denatured
2-Butanol; Butoxydiglycol
Diethylene glycol propyl ether; Dipropylene glycol
Ethoxydiglycol
Glycol
Hexylene glycol
Methoxydiglycol; Methyl amyl alcohol
Propylene glycol
Triglycol monomethyl ether

Ceramics

Trade names: Arconate® 1000; Arconate® HP; Arconate® PC
Emery® 912; Emery® 917; Emery® 918
Jeffsol™ PC
Santosol™ DMA; Santosol™ DME; Santosol™ DME-2; Santosol™ DME-3; Santosol™ DMG; Santosol™ DMS
Uniplex 155

Chemical Processing

Trade names: Arcosolv® PM; Arcosolv® PTB; Arcosolv® TPM
Diglycolamine® Agent (DGA®); Diglyme; Dowanol® DPnB; Dytek® EP
Eastman® DB Acetate; Ethyl Diglyme; Ethyl Glyme
Fancol OA-95; Filmex® A-2; Filmex® B; Filmex® C; Filmex® D-1; Filmex® D-2
GBL
High Purity MTBE; Hisolve MDM
Monoglyme
NEP®
QO® Furfural; QO® Tetrahydrofuran (THF)
Solvenon® PC
Tetraglyme; THF; Triglyme
Vista LPA; Vista LPA-110; Vista LPA-142; Vista LPA-170; Vista LPA-210

Chemicals: Acetonitrile; Alcohol; n-Amyl alcohol
Butylcyclohexane; s-Butylcyclohexane; t-Butylcyclohexane; Butylcyclopentane; s-Butylcyclopentane; t-Butylcyclopentane
1-Chloro-1-nitropropane; Cumene; Cyclohexane; N-Cyclohexyl pyrrolidone; Cyclopentane
Dibromodifluoromethane; Dibutyl ether; Diethylene glycol dibutyl ether; Diethylene glycol dimethyl ether; 1,4-Dioxane
Ethylacetoacetate; Ethylcyclopentane; Ethylene glycol diacetate; N-Ethyl-2-pyrrolidone
Heptane; Hydrobromic acid
Isoamyl alcohol; Isobutylcyclohexane; Isobutylcyclopentane; Isopropyl ether
Methyl acetoacetate; Methyl t-butyl ether; Methyl cyclohexane;

Chemical Processing *(cont'd.)*

Methylcyclopentane; Methyl ethyl ketone; N-Methyl-2-pyrrolidone
Nitromethane
PEG-3 dimethyl ether; 3-Pentanol
Sodium iodide
Tetrahydrofuran; 1,2,4-Trichlorobenzene; 1,1,2-Trichloroethane

Construction and Building Materials

Trade names: Arconate® 1000; Arconate® HP; Arconate® PC
Crodamol GTCC
Decalin; Dytek® EP
Eastman® DE
Nusolv™ ABP-62; Nusolv™ ABP-74; Nusolv™ ABP-87; Nusolv™ ABP-103; Nusolv™ ABP-164
Pluracol® E200; Pluracol® E300; Pluracol® E400; Pluracol® E400 NF; Pluracol® E600; Pluracol® E1450 NF
Shell Cyclo Sol 53; Shell Cyclo Sol 63; Shell Mineral Spirits 135; Shell Mineral Spirits 146 HT; Shell Mineral Spirits 200 HT
Tetralin

Chemicals: Bis(methylethyl)-1,1′-biphenyl
Decahydronaphthalene; Diacetone alcohol
Heptane
Kerosene
Methyl ethyl ketone; Mineral spirits
Tetrahydronaphthalene
Xylene

Cosmetic and Personal Care Products

Trade names: Acetamide MEA; Acetol® 1706; Acetulan®; Adol® 52 NF; Adol® 80; Adol® 85 NF; Antarox® P-400; Antarox® P-700; Arcosolv® DPM; Arcosolv® PE; Arcosolv® PE Acetate; Arcosolv® PM; Arcosolv® PNB; Arcosolv® PNP; Arcosolv® TPM; Argobase 125T; Arlamol® E; Arlasolve® DMI; Axol® E 61
Bernel® Ester DOM
Capmul® MCM; Capmul® MCM-90; Captex® 200; Captex® 300; Captex® 350; Captex® 810A; Captex® 810D; Captex® 8227; Carbopol® 940; Carbopol® 980; Carbowax® PEG 300; Carbowax® Sentry® PEG 300; Carbowax® Sentry® PEG 400; Carbowax® Sentry® PEG 540 Blend; Carbowax® Sentry® PEG 600; Carbowax® Sentry® PEG 900; Carbowax® Sentry® PEG 1000; Carbowax® Sentry® PEG 1450; Carbowax® Sentry® PEG 3350; Carbowax® Sentry® PEG 4600; Carbowax® Sentry® PEG 8000; Ceraphyl® 28; Ceraphyl® 140; Ceraphyl® 140-A; Ceraphyl® ICA; Cithrol 2DL; Cithrol 2DO; Cithrol 4DL; Cithrol 4DO; Cithrol 6DL; Cithrol 6DO; Cithrol 10DL; Cithrol 10DO; Citroflex® 2; Comperlan® COD; Corona Lanolin; Cosmetol® X; Cremophor® EL; Crill 1; Crill 2; Crodamol AB; Crodamol DA; Crodamol GTCC; Crodamol IPL; Crodamol IPM; Crodamol IPP; Crodamol PMP; Crystal® O; Crystal® Crown

Cosmetic and Personal Care Products *(cont'd.)*

Dermol EB; Dermol ICSA; Dow 15-200; Dow 112-2; Dow B100-1000; Dow B100-2000; Dow B100-4800; Dow E200; Dow EP530; Dow L910; Dow L1150; Dow MPEG350; Dow MPEG550; Dow MPEG750; Dow P425; Dow P1000TB; Dow P1200; Dow P2000; Dow P4000; Dow PT250; Dow PT700; Dow PT3000; Dowanol® DPM; Dowanol® EPh; Drakeol® 7; Dynacerin® 660

Ektasolve® PM Acetate; Emcol® CC-42; Emcol® CC-55; Emcol® CC-57; EmCon CO; EmCon Olive; EmCon Rice Bran; EmCon W; EmCon Walnut; Emerest® 2452; Emery® 912; Emery® 916; Emery® 917; Emery® 918; Estalan 430; Estasan GT 8-40 3578; Estasan GT 8-60 3575; Estasan GT 8-60 3580; Estasan GT 8-65 3577; Estasan GT 8-65 3581; Estasan GT 8-70 3579; Estol IPM 1509 (BIO-IPM)

Fancol ALA; Fancol CB; Fancol CB Extra; Fancol Karite Butter; Fancol OA-95; Fancorsil A; Flexricin® 9; Foamid AME-70; Foamid AME-75; Foamid AME-100; Foamid LM2E

GBL; Gilugel MIG; Gilugel MIN; Gilugel SIL5; Glucam® E-10; Glucam® E-20; Glucam® P-10; Glucam® P-20

Hallcomid® M-8-10; Hetoxide BN-13

Imwitor® 742; Incromectant AMEA-70; Isopar® G; Isopar® H; Isopar® K; Isopar® L; Isopar® M

Jeffsol™ PC

Kemstrene® 96.0% USP; Kemstrene® 99.0%; Kemstrene® 99.7% USP; Kemstrene® High Gravity; Kessco® IPM; Kessco® IPP; Kodaflex® DEP; Kodaflex® DMP

Labrafac® Hydro; Labrafac® LIPO; Labrafil® Isostearique; Labrasol; Lanpol 5; Lanpol 520; Lexol® 3975; Lexol® GT-865; Lexol® IPM; Lexol® IPM-NF; Lexol® PG-800; Lexol® PG-865; Lipolan R; Lipovol SES; Lonzest® 143-S; Lowenol 915

Macol® 57; Mazon® EE-1; Miglyol® 810; Miglyol® 812; Miglyol® 840; Miglyol® 8810; Monolan® PPG440, PPG1100, PPG2200; MPDiol® Glycol; Myritol® PC

Naturechem® MAR; Neobee® 18; Neobee® 1054; Neobee® M-5; Neobee® M-20; Neobee® O; Neosolv™ 2; Neosolv™ 4; Neosolv™ 6; Neosolv™ 8; Neosolv™ 24; Neosolv™ 150; Nikkol BWA-5; Nikkol BWA-10

Olepal ISO

Palatinol® DBP; PD-23; PD-25; PD-28; Pecosil® DCU; Pelemol® G7A; Penreco 2251 Oil; Penreco 2257 Oil; Penreco 2260 Oil; Penreco 2263 Oil; Penreco Snow; Permethyl® 99A; Pluracol® E200; Pluracol® E300; Pluracol® E400; Pluracol® E600; Pluracol® E600 NF; Pluracol® E1450 NF; Pluracol® E4000; Polychol 15; Poly-Solv® DM; Poly-Solv® DPM; Poly-Solv® EM; Poly-Solv® MPM; Poly-Solv® TM; Poly-Solv® TPM; Primarol 1006; Primarol 1107; Procetyl 10; Procetyl 30; Procetyl 50; Promyr; Propal; Propylene Glycol USP/FCC Ultra Grade; Prox-onic MG-020 P; Punctilious® SDA 1-1

Renex® 648; Renex® PEG 300; Renex® PEG 400; Renex® PEG 600; Renex® PEG 1000; Renex® PEG 1500; Renex® PEG 4000; Renex® PEG 6000; Renex® PEG 8000; Rhodacal® 330; Rhodasurf® PEG 300; Ricinion; Ritawax AEO; Ritawax ALA

Schercemol 1818; Schercemol CO; Schercemol DIA; Schercemol

Cosmetic and Personal Care Products *(cont'd.)*

DIS; Schercemol NGDC; Schercemol NGDO; Schercemol OLO; Schercemol PGDP; Schercemol PGML; Secoster® DO 600; Secoster® MA 300; Solimate; Solulan® 5; Solulan® 16; Solulan® 75; Solulan® C-24; Solulan® L-575; Solvent APV Spec.; Standamul® G; Standamul® G-16; Stearate 400 WL 817; Stearate 1500; Stearate 6000 WL 1644; Sucro Ester 15

Tebol™; Tebol™ 99; Tegosoft® CI; Tegosoft® CT; Tegosoft® M; Tegosoft® P; Teric PPG 1650; Transcutol; Trivent DIA; Trivent NP-13

Unichem BZBN; Unimate® BYS; Uniplex 110; Uniplex 150

Varonic® DM55; Vista LPA

Wickenol® 136; Wickenol® 151; Wickenol® 152; Wickenol® 153; Wickenol® 158; Wickenol® 171; Witcamide® CMEA; Witcohol™ 66; Witconol™ APEM; Witconol™ APM; Witconol™ APS

York Krystal Kleer Castor Oil; York USP Castor Oil

Chemicals:

Acetamide MEA; Ammonium xylenesulfonate; n-Amyl alcohol

Benzyl benzoate; Benzyl laurate; s-Butyl acetate; t-Butyl alcohol; Butyl benzoate; n-Butyl propionate; Butyl stearate

C12-15 alkyl benzoate; Caprylic/capric/lauric triglyceride; Caprylic/capric/linoleic triglyceride; Caprylic/capric triglyceride; Caprylic/capric triglyceride PEG-4 esters; Carbomer; Cetearyl alcohol; Cetyl lactate; Cetyl octanoate; C10-11 isoparaffin; C11-12 isoparaffin; C11-13 isoparaffin; C13-14 isoparaffin; Cocoa butter

Decyl oleate; Dibutyl phthalate; Diethylene glycol dioctanoate/diisononanoate; Diethyl phthalate; Diisocetyl adipate; Diisopropyl adipate; Diisopropyl sebacate; Dimethicone copolyol undecylenate; Dimethyl caprylamide-capramide; Dimethyl isosorbide; Dipropylene glycol; Dipropylene glycol butyl ether

Ethoxydiglycol; Ethylene glycol propyl ether

Formic acid

Glycereth-7 triacetate; Glyceryl formal

Hexane; Hybrid safflower oil

Isocetyl alcohol; Isocetyl salicylate; Isodecyl isononanoate; Isodecyl oleate; Isododecane; Isononyl isononanoate; Isooctyl stearate; Isopropyl laurate; Isopropyl myristate; Isopropyl palmitate; Isopropyl stearate; Isostearyl alcohol; Isostearyl isostearate; Isostearyl lactate; Isotridecyl isononanoate

Lactamide DGA; Laneth-5; Laneth-10; Laneth-15; Lanolin; Lanolin oil

Methoxyethanol; Methoxyisopropanol; Methyl benzoate; 2-Methyl-1,3-propanediol; Mineral oil; Mixed terpenes; Myristyl propionate

Neopentyl glycol dicaprate; Neopentyl glycol dioctanoate

Octyl hydroxystearate; Octyl isononanoate; Oleth-2; Oleyl alcohol; Oleyl oleate

PEG-4 dilaurate; PEG-8 dilaurate; PEG-20 dilaurate; PEG-4 dioleate; PEG-8 dioleate; PEG-20 dioleate; PEG-2 lactamide; PEG-5 lanolate; PEG-10 lanolate; PEG-20 lanolate; PEG-75 lanolin; PEG-6 methyl ether; PEG-10 methyl ether; PEG-16

Cosmetic and Personal Care Products *(cont'd.)*

methyl ether; PEG-8 oleate; PEG-20 oleate; Petrolatum; Poloxamer 181; Polysorbate 60; Potassium hydroxide; PPG-9; PPG-17; PPG-20; PPG-26; PPG-30; PPG-10 butanediol; PPG-40 diethylmonium chloride; PPG-24-glycereth-24; PPG-66-glycereth-12; PPG-10 glyceryl ether; PPG-55 glyceryl ether; PPG-2 methyl ether; PPG-3 methyl ether; PPG-2 myristyl ether propionate; PPG-15 stearyl ether; Propylene glycol caprylate; Propylene glycol dicaprylate; Propylene glycol dicaprylate/dicaprate; Propylene glycol dioctanoate; Propylene glycol dipelargonate; Propylene glycol ethyl ether; Propylene glycol ethyl ether acetate; Propylene glycol n-propyl ether
Shea butter; Sucrose palmitate
Tallow alcohol; α-Terpineol; Tridecyl neopentanoate
Walnut oil; Wheat germ oil

Dry-Cleaning Chemicals

Trade names:
DF-2000™
Emcol® CC-42; Emcol® CC-57
Nopalcol Series
PPG Perchlor
Rhodacal® 330
Shell TS-28; Shell VM&P Naphtha EC; Shell Cyclo Sol 53; Shell Cyclo Sol 63; Shell Mineral Spirits 135; Shell Mineral Spirits 145-EC; Shell Mineral Spirits 146 HT; Shell Mineral Spirits 200 HT; Shell Sol 70; Shell Sol 71; Shell Sol 71 (H), 71 (WR); Shell Sol 72; Shell Sol 140; Shell Sol 142 HT; Shell Sol 340; Shell Sol 340 HT; Shell Sol B; Shell Super VM&P Naphtha EC; Shell Toluene; Shell Tolu-Sol 5; Shell Tolu-Sol 6-EC; Shell Tolu-Sol 19-EC; Shell Xylene; Sovatex C Series
T-Mulz® 426
Vista LPA; Vista LPA-110

Chemicals:
Butyl lactate
1,4-Dioxane
Mineral spirits
Naphtha
Perchloroethylene
Stoddard solvent
1,1,1-Trichloroethane; Trichloroethylene; Trichlorofluoromethane; Trichlorotrifluoroethane

Electronics Manufacturing

Trade names:
Arcosolv® DPM; Arcosolv® DPM Acetate; Arcosolv® DPNB; Arcosolv® PE; Arcosolv® PE Acetate; Arcosolv® PM; Arcosolv® PNB; Arcosolv® PNP; Arcosolv® PTB; Arcosolv® TPM
CHP®
Decap; DMSO; Dynasolve 180; Dynasolve 185; Dynasolve 185 SG; Dynasolve 190; Dynasolve 210; Dynasolve 220; Dynasolve 225; Dynasolve 230; Dynasolve 700; Dynasolve 710; Dynasolve 711;

Electronics Manufacturing *(cont'd.)*

Dynasolve 750; Dynasolve XD 27-2
Formula 111
GBL
HEP®
MS-195/CO2 Safezone® Cleaning Solvent & Flux Remover
NMP.; NMP-EL; Nusolv™ ABP-62; Nusolv™ ABP-74; Nusolv™ ABP-87; Nusolv™ ABP-103; Nusolv™ ABP-164
Oxsol® 100
PPG Trichlor; Purasolv® BL; Purasolv® EL; Purasolv® ELECT; Purasolv® ELS; Purasolv® IPL; Purasolv® ML
TF Solvent; THF; TP-35 Solvent
Uresolve 411; Uresolve Blue.; Uresolve HF.; Uresolve Plus.; Uresolve Plus 500.; Uresolve Plus SG.

Chemicals: Bis(methylethyl)-1,1´-biphenyl; Butoxypropanol; Butyl lactate
p-Chlorobenzotrifluoride; N-Cyclohexyl pyrrolidone
1,4-Dioxane; Dipropylene glycol butyl ether
Ethyl lactate
Isopropyl lactate
Methoxyisopropanol; Methyl lactate; N-Methyl-2-pyrrolidone
PPG-2 methyl ether; PPG-3 methyl ether; PPG-2 methyl ether acetate; Propylene glycol ethyl ether; Propylene glycol ethyl ether acetate; Propylene glycol methyl ether acetate; Propylene glycol n-propyl ether
1,1,2-Trichloroethane; Trichlorotrifluoroethane

Food Processing

Trade names: Carbowax® PEG 300; Carbowax® Sentry® PEG 300; Carbowax® Sentry® PEG 400; Carbowax® Sentry® PEG 540 Blend; Carbowax® Sentry® PEG 600; Carbowax® Sentry® PEG 900; Carbowax® Sentry® PEG 1000; Carbowax® Sentry® PEG 1450; Carbowax® Sentry® PEG 3350; Carbowax® Sentry® PEG 4600; Carbowax® Sentry® PEG 8000; Crill 1; Crill 2
PD-23; PD-25; PD-28; Penreco 2251 Oil; Penreco 2257 Oil; Penreco 2259 Oil; Penreco 2260 Oil; Penreco 2263 Oil

Chemicals: Ethylene dichloride
Glycerin
Hexane
Methyl alcohol; Methyl bromide; Methylene chloride; Methyl ethyl ketone
2-Nitropropane
PEG-8 dioleate; Petroleum distillates
Trichloroethylene

Foods

Trade names: Aldo® MCT KFG; Axol® E 61; Axol® E 66
Capmul® MCM; Captex® 200; Captex® 300; Captex® 350; Captex® 810A; Captex® 810D; Captex® 8227; Crill 1; Crill 2; Crodamol GTCC
Delios® C; Delios® S; Delios® V; Dow E400 NF; Dow E600 NF;

Foods *(cont'd.)*

Dow E900 NF; Dow E1000 NF; Dow E1450 NF; Dow E3350 NF; Dow E4500 NF; Dow E8000 NF
Emery® 912; Emery® 916; Emery® 917; Emery® 918; Estasan GT 8-40 3578; Estasan GT 8-60 3575; Estasan GT 8-60 3580; Estasan GT 8-65 3577; Estasan GT 8-65 3581; Estasan GT 8-70 3579
Isoamyl Alcohol 95%; Isoamyl Alcohol 99%
Lexol® PG-800; Lexol® PG-865; Liponate GC; Liponate PC; Lipovol SES
Miglyol® 810; Miglyol® 812; Miglyol® 840
Neobee® 18; Neobee® 1054; Neobee® M-5; Neobee® M-20; Neobee® O
Polarin® Range; Propylene Glycol USP/FCC Ultra Grade; Purasolv® EL; Purasolv® ELECT; Purasolv® ELS
Unichem BZBN
Vista LPA-142

Chemicals: Acetal; Acetic acid; Acetone; Acetylated hydrogenated lard glyceride; Acetylated hydrogenated soybean oil glyceride; Acetylated mono- and diglycerides of fatty acids; Alcohol; Ammonium hydroxide
Benzyl benzoate; Bornyl acetate; Butylene glycol; n-Butyl propionate
Capric triglyceride; Caprylic alcohol; Caprylic/capric/lauric triglyceride; Caprylic/capric/linoleic triglyceride; Caprylic/capric/oleic triglyceride
Ethyl acetate
Glycerin; Glyceryl oleate; Glyceryl stearate; Glyceryl tricaprylate/caprate
Lactic acid
Methyl alcohol; Methylene chloride; Mono- and diglycerides of fatty acids
Naphtha
Oleyl alcohol
PEG-4 dilaurate; PEG-20 dilaurate; PEG-4 dioleate; PEG-8 dioleate; PEG-20 dioleate; PEG-8 oleate; PEG-20 oleate; Propylene glycol; Propylene glycol dicaprylate/dicaprate; Propylene glycol dioctanoate
Triundecanoin

Fuel Additives

Trade names: Arcosolv® TPM
Butyl Diglyme; BVA ND-205 Solv.
Diglyme; Dytek® EP
Eastman® EB; Ethyl Diglyme
Halso® 99; High Purity MTBE
Jeffsol™ EC
Monoglyme
Norpar® 12; Norpar® 13; Norpar® 15
PD-23; PD-25; PD-28; Penreco 2251 Oil; Penreco 2257 Oil; Penreco 2259 Oil; Penreco 2260 Oil; Penreco 2263 Oil
Shell VM&P Naphtha HT; Shell Mineral Spirits 146 HT; Shell Mineral Spirits 200 HT; Shell Sol 71; Shell Sol 142 HT; Shell Sol

Fuel Additives *(cont'd.)*

340 HT; Shell Sol B HT
Tetraglyme; Tolu-Sol® A HT Solv.; Tolu-Sol® W HT Solv.; Triglyme
Vista LPA; Vista LPA-110; Vista LPA-142

Chemicals: Alcohol denatured
n-Decane; Diethylenetriamine; Dimethylamine; 2,3-Dimethylpentane; 2,4-Dimethylpentane; 3,3-Dimethylpentane; n-Dodecane
Ethylbenzene
Gasoline
Heptane
Isopentane; Isopropyl alcohol
Kerosene
Methoxyethanol; Methoxyisopropanol; Mineral spirits; Monochlorotoluene
Nitromethane; 2-Nitropropane
Octane
PEG-3 dimethyl ether; Pentylamine; PPG-3 methyl ether
2,2,3-Trimethylbutane
n-Undecane
Xylene; m-Xylene

Gas Scrubbing

Trade names: Adol® 42; Adol® 80
HEP®
Polyglyme

Chemicals: Monochlorotoluene

Industrial, Institutional, and Household Cleaning

Trade names: Acintene® AC-P; Acintene® DP 738; Acintene® LS 160 A; Acintene® LS 160 B; Acintene® LS 160 D; Actrel® 1111L; Actrel® 1138L; Actrel® 1160L; Actrel® 1960L; Actrel® 1971L; Actrel® 3338L; Actrel® 3349L; Actrel® 3360L; Actrel® 4493L; Adol® 80; Alkasurf® OD-410; Arcosolv® DPM; Arcosolv® DPM Acetate; Arcosolv® DPNB; Arcosolv® DPNP; Arcosolv® PM; Arcosolv® PNB; Arcosolv® PNP; Arcosolv® PTB; Arcosolv® TPM; Aromatic 150; Aromatic 150; Arosolve 570-HF; Arosolve 570 Special; Arosolve B-950; Arosolve MRC-A; Arosolve MRC-HF; Arosolve SH-180 LF; Arosolve XNF-1
Berol 02; BLO®; Butyl Di-Icinol; Butyl Icinol; Butyl Oxitol®; BVA ND-205 Solv.
Cardolite® NC-507; Cardolite® NC-510; Chemax E-200 MO; Chemax E-1000 MO; Citroflex® 1; Citroflex® 2; Citroflex® 4; Citroflex® 6; Citroflex® A-1; Citroflex® A-4; CNC Sol BD; Corexit CL8594; Crill 1; Crill 2
Dalpad A; DB; Dba; DBE; DBE-2, -2SPG; DBE-3; DBE-4; DBE-5; DBE-6; DBE-9; DBE-IB; DE; DEA; Degreez; Dizene; DM; Dowanol® DB; Dowanol® DM; Dowanol® DPM; Dowanol® DPMA; Dowanol® DPnB; Dowanol® EB; Dowanol® EPh;

Industrial, Institutional, and Household Cleaning *(cont'd.)*

Industrial, Institutional, and Household Cleaning *(cont'd.)*

Renex® 648; Rexoclean; Rexoclean 200N; Rexoclean GA; Rexoclean JA; Rexoclean NB Conc.; Rexosolve 150; Rexosolve BCT; Rexosolve EP, EPN; Rexosolve GA, GAX; Rexosolve HP; Rexosolve MS; Rexosolve OA, OAC; Rexosolve RF-2; Rexosolve SL; Rhodacal® 330

Santosol™ DMA; Santosol™ DME; Santosol™ DME-2; Santosol™ DME-3; Santosol™ DMG; Santosol™ DMS; Scour NFP; Secoster® A; Secoster® DO 600; Secoster® MA 300; Secoster® MO 400; Shell VM&P Naphtha HT; Shell Cyclo Sol 53; Shell Mineral Spirits 135; Shell Mineral Spirits 146 HT; Shell Mineral Spirits 200 HT; Shell Sol 71; Shell Sol 142 HT; Shell Sol 340; Shell Sol 340 HT; Ship Shape® Resin Cleaner; Solvenol® No. 2; Solvenon® DIP; Solvenon® DPM; Solvenon® IPP; Solvenon® PM; Solvenon® PP; Surfadone® LP-100; Surfadone® LP-300; Syn Fac® TEA-97

T-Det® N-4; T-Det® N-6; Tebol™; Tebol™ 99; TF Solvent; T-Mulz® 426; TP-35 Solvent; Tri-Ethane

Ureresolve; Uresolve 411; Uresolve Blue.; Uresolve HF.; Uresolve Plus.; Uresolve Plus 500.

Varonic® DM55; Vista LPA; Vista LPA-110; Vista LPA-142

Chemicals: Acetyl tributyl citrate; Acetyl triethyl citrate; Acetyl trimethyl citrate; Alcohol denatured

Bis(methylethyl)-1,1′-biphenyl; 2-Butanol; Butoxydiglycol; 1-Butoxyethoxy-2-propanol; Butoxypropanol; t-Butyl alcohol; Butyl lactate

C9 alkyl acetate; C10 alkyl acetate; C13 alkyl acetate; Caprylyl pyrrolidone; Cetoleth-2; Cetoleth-3; Cetoleth-6; p-Chlorobenzotrifluoride; C13-14 isoparaffin; Cyclohexanone

Diacetone alcohol; 3,4-Dichlorobenzotrifluoride; Diethylene glycol propyl ether; Dimethyl caprylamide-capramide; Dimethyl glutarate; Dimethyl succinate; Dioctyl maleate; 1,4-Dioxane; Dipropylene glycol butyl ether; Dipropylene glycol dimethyl ether; Dipropylene glycol ethyl ether; Dodoxynol-7

Ethoxyethanol; Ethylene glycol diethyl ether; Ethylene glycol 2-ethylhexyl ether; Ethyl lactate

N-Hydroxyethylpyrrolidone

Isobutyl alcohol

Kerosene

Lauryl pyrrolidone; d-Limonene

Methoxyisopropanol; Methylcyclohexanol; Methylene chloride; N-Methyl-2-pyrrolidone; Mineral spirits; Mixed terpenes; Monochlorotoluene

Nonoxynol-4; Nonoxynol-5; Nonoxynol-6; Nonoxynol-9 phosphate

Octyl hydroxystearate; Oleth-2; Oxo-heptyl acetate; Oxo-hexyl acetate

PEG-75 lanolin; PEG-4 oleate; PEG-5 oleate; 3-(n-Pentadecyl) phenol; Perchloroethylene; Perfluoroheptane; Perfluorohexane; Perfluoro-N-methylmorpholine; Perfluoropentane; Petroleum distillates; Phenoxydiglycol; PPG-40 diethylmonium chloride; PPG-2 methyl ether; PPG-3 methyl ether; PPG-2 methyl ether acetate; n-Propyl alcohol; Propylene glycol

Industrial, Institutional, and Household Cleaning *(cont'd.)*

t-butyl ether; Propylene glycol ethyl ether; Propylene glycol ethyl ether acetate; Propylene glycol phenyl ether; Propylene glycol n-propyl ether
Sodium xylenesulfonate
α-Terpineol; 1,1,2,2-Tetrachloro-1,2-difluoroethane; Tetrahydronaphthalene; 1,3,5-Trichlorobenzene; 1,1,1-Trichloroethane; Trichloroethylene; Trichlorofluoromethane; Trichlorotrifluoroethane; Triglycol monomethyl ether; Trihexyl citrate; Trimethyl citrate; Tripropylene glycol n-butyl ether; Tripropylene glycol ethyl ether; Turpentine oil
Xylene

Inks

Trade names:
Acintene® AC-P; Acintene® LS 160 B; Acintene® LS 160 D; Adol® 42; Adol® 52 NF; Adol® 80; Agrisynth® BLO; Agsol Ex BLO; Albrite® TBP; Alkasurf® OD-410; Arcosolv® DPM Acetate; Arcosolv® DPNB; Arcosolv® PE; Arcosolv® PE Acetate; Arcosolv® PM; Arcosolv® PNB; Arcosolv® PNP; Arcosolv® PTB; Arcosolv® TPM; Aromatic 100; Aromatic 150
BLO®; Butyl Di-Icinol; Butyl Dioxitol®; Butyl Icinol
Citroflex® 1; Citroflex® 2; Citroflex® 4; Citroflex® 6; Citroflex® A-1; Citroflex® A-2; Citroflex® A-4; Crill 1; Crill 2
Dalpad A; DBE; DBE-2, -2SPG; DBE-4; DBE-6; DBE-IB; DMSO; Dowanol® DB; Dowanol® DM; Dowanol® DPMA; Dowanol® EB; Dowanol® EPh; Dowanol® PM; Dowanol® TPM; Dowanol® TPnB; Dynasolve 200
Eastman® C-11 Ketone; Eastman® CDA-19; Eastman® DB; Eastman® DB Acetate; Eastman® DE Acetate; Eastman® DM; Eastman® EB Acetate; Eastman® PM Acetate; Eccoflex 090; Ecco Strip LS-W; Ektasolve® DM; Ethyl Icinol; Exxate® 600; Exxate® 700; Exxate® 800; Exxate® 900; Exxate® 1000; Exxate® 1300; Exx-Print® 176 D; Exx-Print® 283 D; Exx-Print® 287 D; Exx-Print® 314 A; Exx-Print® 392 D; Exx-Print® 493 D; Exx-Print® 588 D; Exxsol® D 40; Exxsol® D 60; Exxsol® Heptane
Fancol OA-95; Filmex® A-2; Filmex® B; Filmex® C; Filmex® D-1; Filmex® D-2
HEP®; Hisolve DM; Hisolve MC
Icinol BE33; Icinol DPM; Icinol EE22; Icinol ME33; Icinol ME45; Icinol PM; Igepal® OD-410; Isopar® V
Jeffox™ PPG-400; Jeffox™ PPG-2000; Jeffox™ WL-440; Jeffox™ WL-660; Jeffox™ WL-1400; Jeffox™ WL-5000
Kemstrene® 96.0% USP; Kemstrene® 99.0%; Kemstrene® 99.7% USP; Kemstrene® High Gravity; Kodaflex® DMP; Kronitex® TBP
Marlican®; Methyl Di-Icinol; Monolan® PPG440, PPG1100, PPG2200
NE™ 101; NEP®; Nirez® 9007; Nirez® 9011; NMP; Norpar® 12; Norpar® 13; Norpar® 15; Nusolv™ ABP-62; Nusolv™ ABP-74; Nusolv™ ABP-87; Nusolv™ ABP-103; Nusolv™ ABP-164
#15 Oil; #30 Oil; #40 Oil; Oxsol® 73; Oxsol® 100; Oxsol® 253;

Inks *(cont'd.)*

Oxsol® 325; Oxsol® 550; Oxsol® 1000

Palatinol® DBP; Pale 4; Pale 16; Pale 170; Pale 1000; Penreco 2251 Oil; Penreco 2257 Oil; Penreco 2259 Oil; Penreco 2260 Oil; Penreco 2263 Oil; Pluracol® E400 NF; Pluracol® E600 NF; Pluracol® E1450 NF; Poly-G® WI 285; Poly-G® WI 625; Poly-G® WI 1715; Poly-G® WS 100; Poly-G® WS 170; Poly-G® WS 260; Poly-G® WS 660; Poly-G® WS 2000; Poly-G® WS 3520; Poly-G® WS 5100; Poly-G® WT 9150; Poly-G® WT 90,000; Poly-Solv® DM; Poly-Solv® DPM; Poly-Solv® EM; Poly-Solv® MPM; Poly-Solv® TM; Polysolvan® O; Printsolve™ Ink Remover; Propylene Glycol USP/FCC Ultra Grade; Purasolv® BL; Purasolv® ELECT; Purasolv® ELS; 2-Pyrol®

Rhodasurf® PEG 300

Santosol™ DMA; Santosol™ DME; Santosol™ DME-2; Shell VM&P Naphtha HT; Shell Cyclo Sol 53; Shell Cyclo Sol 63; Shell Mineral Spirits 135; Shell Mineral Spirits 146 HT; Shell Mineral Spirits 200 HT; Shell Sol B HT; Solvenon® DIP; Solvenon® DPM; Solvenon® PM; SR-256; Surfadone® LP-100; Surfadone® LP-300; Sylfat® DX; Sylfat® MM

Tecsol® A; Tecsol® A-2; Tecsol® B; Tecsol® C; Tecsol® D; Tecsol® D-2; Tecsol® H; Tolu-Sol® A HT Solv.; Tolu-Sol® W HT Solv.; Tri-Ethane

Ucon® 50-HB-55; Ucon® 50-HB-100; Ucon® 50-HB-170; Ucon® 50-HB-260; Ucon® 50-HB-400; Ucon® 50-HB-660; Ucon® 50-HB-2000; Ucon® 50-HB-3520; Ucon® 50-HB-5100; Ucon® 75-H-450; Ucon® LB-65; Ucon® LB-135; Ucon® LB-165; Ucon® LB-285; Ucon® LB-385; Ucon® LB-525; Ucon® LB-625; Ucon® LB-1145; Ucon® LB-1715; Ucon® LB-3000

Varsol® 1; Varsol® 18; Vista 47; Vista LPA; Vista LPA-110; Vista LPA-142; Vista LPA-170; Vista LPA-210; Vorite 105; Vorite 110; Vorite 115; Vorite 120; Vorite 125

Chemicals: Acetone; Acetyl tributyl citrate; Acetyl triethyl citrate; Acetyl trimethyl citrate; Alcohol denatured

Benzyl alcohol; Bis(methylethyl)-1,1′-biphenyl; Butoxydiglycol; 1-Butoxyethoxy-2-propanol; Butoxypropanol; Butyl lactate; Butyrolactone

C9 alkyl acetate; C10 alkyl acetate; C13 alkyl acetate; Caprylyl pyrrolidone; Castor oil, polymerized; p-Chlorobenzotrifluoride; Cumylphenyl benzoate; Cumylphenyl neodecanoate

3,4-Dichlorobenzotrifluoride; Diethylene glycol butyl ether acetate; Dimethyl adipate; Dimethyl glutarate; Dimethyl succinate; Dipropylene glycol; Dipropylene glycol butyl ether; Dipropylene glycol ethyl ether; n-Dodecane

Ethyl acetate; Ethyl hexanediol; Ethyl lactate

Glycol

Heptane; Hexylene glycol

Isopropyl acetate; Isostearyl alcohol

Lauryl pyrrolidone

Methoxydiglycol; Methoxyethanol; Methoxyisopropanol; Methyl ethyl ketone; Mineral spirits

Nitroethane; Nitropropane; 2-Nitropropane

Oleyl alcohol; Oxo-heptyl acetate; Oxo-hexyl acetate

Inks (*cont'd.*)

Phenoxydiglycol; Phenoxyethanol; PPG-400; PPG-2000; PPG-2-buteth-3; PPG-3-buteth-5; PPG-5-buteth-7; PPG-7-buteth-10; PPG-9-buteth-12; PPG-12-buteth-16; PPG-20-buteth-30; PPG-28-buteth-35; PPG-33-buteth-45; PPG-5 butyl ether; PPG-9 butyl ether; PPG-15 butyl ether; PPG-18 butyl ether; PPG-22 butyl ether; PPG-24 butyl ether; PPG-33 butyl ether; PPG-40 butyl ether; PPG-53 butyl ether; PPG-3 methyl ether; PPG-2 methyl ether acetate; Propyl acetate; Propylene glycol t-butyl ether; Propylene glycol ethyl ether; Propylene glycol ethyl ether acetate; Propylene glycol methyl ether acetate; Propylene glycol phenyl ether; Propylene glycol n-propyl ether

Tetrahydrofuran; 1,1,1-Trichloroethane; Triglycol monomethyl ether; 2,2,4-Trimethyl-1,3-pentanediol monoisobutyrate; Tripropylene glycol n-butyl ether; Tripropylene glycol ethyl ether; Turpentine oil

Leather Treatment

Trade names:

Corial®; Crill 1; Crill 2
Eccoflex 090; Ecco Strip LS-W
Kemester® 143
Nopalcol Series
#15 Oil; #30 Oil; #40 Oil
Pale 4; Pale 16; Pale 170; Pale 1000; Poly-Solv® DM; Poly-Solv® DPM; Poly-Solv® EM; Poly-Solv® MPM; Poly-Solv® TM
QO® Tetrahydrofurfuryl Alcohol (THFA®)
Sovatex C Series
Vorite 105; Vorite 110; Vorite 120; Vorite 125

Chemicals:

n-Butyl acetate; s-Butyl acetate
Diacetone alcohol
Ethoxyethanol acetate; Ethyl acetate
Glyceryl acetate
Oleth-2
Tetrahydrofurfuryl alcohol; Triglycol monomethyl ether
Xylene

Metal Treating and Processing

Trade names:

Actrel® 1111L; Actrel® 1138L; Actrel® 1160L; Actrel® 1171L; Actrel® 1960L; Actrel® 1971L; Actrel® 3338L; Actrel® 3349L; Actrel® 3360L; Actrel® 4493L; Adol® 42; Adol® 52 NF; Adol® 66; Adol® 80
Butyl Oxitol®
Crill 1; Crill 2
Emerest® 2624; Emery® Methyl Oleate; Exxate® 600; Exxate® 700; Exxate® 800; Exxate® 900; Exxate® 1000; Exxate® 1300
Fancol OA-95
Hisolve DM
Jeffsol™ PC
Kemester® 115; Kemester® 143; Kemester® 205; Kemester® 213;

Metal Treating and Processing *(cont'd.)*

Kemester® 226; Kemester® 2050; Kemester® 4516; Kemester® 9018; Kemester® 9022; Kemester® EGDS; Kemester® EGMS
Mackazoline™ T
Norpar® 12; Norpar® 13; Norpar® 15
PD-23; PD-25; PD-28; Poly-G® WI 285; Poly-G® WI 625; Poly-G® WI 1715; Poly-G® WS 100; Poly-G® WS 170; Poly-G® WS 260; Poly-G® WS 660; Poly-G® WS 2000; Poly-G® WS 3520; Poly-G® WS 5100; Poly-G® WT 9150; Poly-G® WT 90,000; Purasolv® EL; Purasolv® IPL; Purasolv® ML
Rhodacal® 330
Secoster® MA 300; Sylfat® DX; Sylfat® MM
Vista LPA; Vista LPA-110; Vista LPA-170; Vista LPA-210
Witconol™ APEM; Witconol™ APM

Chemicals: C8 alkyl acetate; C9 alkyl acetate; C10 alkyl acetate; C13 alkyl acetate
Dipropylene glycol
Ethyl lactate
Isopropyl lactate; Isostearyl alcohol
Methyl lactate; 2-Methyl-1-pentanol; Monochlorotoluene
Oleyl alcohol; Oxo-heptyl acetate; Oxo-hexyl acetate
PEG-4 dilaurate; PEG-8 dioleate; Propylene glycol t-butyl ether
Tall oil hydroxyethyl imidazoline; 1,1,1-Trichloroethane

Mineral Processing

Trade names: Adol® 42; Adol® 52 NF; Adol® 66; Adol® 80
Emcol® CC-42
Glidsol® 140
Jeffox™ WL-440; Jeffox™ WL-660; Jeffox™ WL-1400; Jeffox™ WL-5000
Shell Sol 142 HT
Tolu-Sol® A HT Solv.

Chemicals: Hydrobromic acid
Isostearyl alcohol

Mining

Trade names: Arcosolv® DPM; Arcosolv® DPM Acetate; Arcosolv® PM; Arcosolv® TPM
Isoamyl Alcohol 95%; Isoamyl Alcohol 99%
Shell Sol 142 HT
Vista LPA; Vista LPA-110; Vista LPA-142; Vista LPA-170; Vista LPA-210

Chemicals: PPG-2 methyl ether; PPG-3 methyl ether; PPG-2 methyl ether acetate

Paints and Coatings

Trade names: Acintene® AC-P; Acintene® DP 738; Acintene® LS 160 A; Acintene® LS 160 B; Acintene® LS 160 D; Agrisynth® BLO;

Paints and Coatings *(cont'd.)*

Arcal™ AAP; Arconate® 1000; Arconate® HP; Arconate® PC; Arcosolv® DPM; Arcosolv® DPM Acetate; Arcosolv® DPNB; Arcosolv® DPTB; Arcosolv® PE; Arcosolv® PE Acetate; Arcosolv® PM; Arcosolv® PM Acetate; Arcosolv® PNP; Arcosolv® PTB; Arcosolv® TPM; Arcosolv® TPNB; Aromatic 100; Aromatic 150; Aromatic 150; Aromatic 200

Butoxyethoxy Propasol; Butoxyl®; Butyl Carbitol®; Butyl Carbitol® Acetate; Butyl Cellosolve®; Butyl Cellosolve® Acetate; Butyl Di-Icinol; Butyl Dioxitol®; Butyl Dipropasol; Butyl Icinol; Butyl Oxitol®; Butyl Propasol

Capmul® MCM; Carbitol® PM-600; Carbitol® Low Gravity; Carbopol® 940; Carbowax® PEG 300; Cellosolve®; Cellosolve® Acetate; Citroflex® 1; Citroflex® 2; Citroflex® 4; Citroflex® 6; Citroflex® A-1; Citroflex® A-2; Citroflex® A-4; Crill 1; Crill 2

DB; Dba; DBE; DBE-2, -2SPG; DBE-3; DBE-4; DBE-5; DBE-6; DBE-9; DBE-IB; DE; DEA; Decalin; Diglyme; Dowanol® DPM; Dowanol® DPMA; Dowanol® DPnB; Dowanol® DPnP; Dowanol® EB; Dowanol® PM; Dowanol® PMA; Dowanol® PnB; Dowanol® PnP; Dowanol® TPnB; DP; Dynasolve 150

Eastman® C-11 Ketone; Eastman® CDA-19; Eastman® DB; Eastman® DB Acetate; Eastman® DBP; Eastman® DE Acetate; Eastman® DM; Eastman® DP; Eastman® EB Acetate; Eastman® EEH; Eastman® EEP; Eastman® EP; Eastman® PM; Eastman® Triacetin; EB; EBA; Eccoflex 090; Ecco Strip LS-W; EE; EEA; EEH; EH; Ektapro® EEP Solvent; Ektasolve® DM; EM; EMA; Emcol® CC-57; EP; Ethyl Di-Icinol; Ethyl Icinol; Exxate® 600; Exxate® 700; Exxate® 800; Exxate® 900; Exxate® 1000; Exxate® 1300; Exxsol® D 40; Exxsol® D 60; Exxsol® D 75/100 Naphtha; Exxsol® D 80; Exxsol® D 110; Exxsol® D 115/145 Naphtha; Exxsol® D-130; Exxsol® D 3135

Fancol OA-95; Filmer EHC; Filmex® A-2; FoamFlush® Urethane Remover

Hallcomid® M-18-OL; Halso® 99; Hetoxide MPC; Hexyl Carbitol®; Hexyl Cellosolve®; Hisolve DM; Hisolve MC

Icinol BE33; Icinol DPM; Icinol EE22; Icinol ME33; Icinol ME45; Icinol PM; Isoamyl Alcohol 95%; Isoamyl Alcohol 99%; Isopar® C; Isopar® E; Isopar® G; Isopar® H; Isopar® K; Isopar® L; Isopar® M

Jeffamine® D-230; Jeffsol™ PC

Kenplast® ES-2HP; Kodaflex® DBP; Kodaflex® HS-3; Kodaflex® HS-4

Methyl Carbitol®; Methyl Cellosolve®; Methyl Cellosolve® Acetate; Methyl Di-Icinol; Methyl Dipropasol; Methyl Icinol; Methyl Propasol®; Methyl Propasol® Acetate

NE™ 101; NiPar S-10™; NMP; NMP-EL; Nusolv™ ABP-62; Nusolv™ ABP-74; Nusolv™ ABP-87; Nusolv™ ABP-103; Nusolv™ ABP-164

#30 Oil; #40 Oil; Oxsol® 73; Oxsol® 100; Oxsol® 253; Oxsol® 325; Oxsol® 550; Oxsol® 1000

Pale 4; Pale 16; Pale 170; Pale 1000; Panasol™ Solvents; Penreco 2251 Oil; Penreco 2257 Oil; Penreco 2259 Oil; Penreco 2260 Oil;

Paints and Coatings *(cont'd.)*

Penreco 2263 Oil; Poly-Solv® DM; Poly-Solv® DPM; Poly-Solv® EM; Poly-Solv® MPM; Poly-Solv® TM; Poly-Solv® TPM; Proglyde DMM; Propsolv I-1; Propsolv I-2; Propsolv III-1; Propsolv III-2; Polysolvan® O; Propyl Carbitol®; Propyl Cellosolve®; Propyl Dipropasol; Propyl Propasol; Purasolv® BL; Purasolv® EL; Purasolv® ELECT; Purasolv® ELS; Purasolv® IPL; Purasolv® ML; 2-Pyrol®

QO® Tetrahydrofurfuryl Alcohol (THFA®)

Rhodacal® 330

Santosol™ DMA; Santosol™ DME; Santosol™ DME-2; Santosol™ DME-3; Santosol™ DMG; Santosol™ DMS; Shell TS-28; Shell VM&P Naphtha EC; Shell VM&P Naphtha HT; Shell Cyclo Sol 53; Shell Cyclo Sol 63; Shell Mineral Spirits 135; Shell Mineral Spirits 145-EC; Shell Mineral Spirits 146 HT; Shell Mineral Spirits 200 HT; Shell Sol 70; Shell Sol 71; Shell Sol 71 (H), 71 (WR); Shell Sol 72; Shell Sol 140; Shell Sol 142 HT; Shell Sol 340; Shell Sol 340 HT; Shell Sol B; Shell Sol B HT; Shell Super VM&P Naphtha EC; Shell Toluene; Shell Tolu-Sol 5; Shell Tolu-Sol 6-EC; Shell Tolu-Sol 19-EC; Shell Xylene; Ship Shape® Resin Cleaner; Solvenon® DIP; Solvenon® IPP; Solvenon® PM; Solvenon® PP; SR-256; Sulfolane W; Surfadone® LP-100; Surfadone® LP-300

Tebol™; Tebol™ 99; Tecsol® 1; Tecsol® 3; Tecsol® A; Tecsol® A-2; Tecsol® B; Tecsol® C; Tecsol® D; Tecsol® D-2; Tecsol® H; Teric PPG 400; Texanol® Ester-Alcohol; THF; Tolu-Sol® A HT Solv.; Tolu-Sol® W HT Solv.

Ucar® n-Butyl Propionate; Ucar® Ester EEP; Ucar® Filmer IBT; Ucar® n-Pentyl Propionate; Ucon® 50-HB-55; Ucon® 50-HB-100; Ucon® 50-HB-170; Ucon® 50-HB-260; Ucon® 50-HB-400; Ucon® 50-HB-660; Ucon® 50-HB-2000; Ucon® 50-HB-3520; Ucon® 50-HB-5100; Ucon® 75-H-450; Ucon® LB-65; Ucon® LB-135; Ucon® LB-165; Ucon® LB-285; Ucon® LB-385; Ucon® LB-525; Ucon® LB-625; Ucon® LB-1145; Ucon® LB-1715; Ucon® LB-3000; Uniplex 150

Varsol® 1; Varsol® 18; Vista LPA; Vista LPA-110; Vista LPA-142; Vista LPA-170; Vista LPA-210; Vorite 105; Vorite 110; Vorite 115; Vorite 120; Vorite 125

Chemicals: Acetone; Acetophenone; Acetyl triethyl citrate; Acetyl trimethyl citrate; Alcohol denatured; Amyl acetate; s-Amyl acetate; s-Amyl alcohol; t-Amyl alcohol

Benzene; Benzonitrile; Benzyl alcohol; Bis(methylethyl)-1,1′-biphenyl; 2-Butanol; Butoxydiglycol; Butoxyethanol; Butoxyethanol acetate; 1-Butoxyethoxy-2-propanol; Butoxypropanol; n-Butyl acetate; s-Butyl acetate; Butyl alcohol; t-Butyl alcohol; Butyl lactate; n-Butyl propionate; Butyrolactone

C8 alkyl acetate; C9 alkyl acetate; C10 alkyl acetate; C13 alkyl acetate; Caprylic alcohol; Caprylyl pyrrolidone; Castor oil, polymerized; p-Chlorobenzotrifluoride; C7-8 isoparaffin; C8-9 isoparaffin; C10-11 isoparaffin; C11-12 isoparaffin; C11-13 isoparaffin; C13-14 isoparaffin; Cyclohexane; Cyclohexanol; Cyclohexanone; Cyclohexyl acetate

Decahydronaphthalene; Diacetone alcohol; Dibutyl ether; Dibu-

Paints and Coatings *(cont'd.)*

Paper and Pulp Processing

Trade names:
Adol® 52 NF; Adol® 66; Adol® 80
Carbowax® PEG 300
Emery® Methyl Oleate
Fancol OA-95
Marlican®
Nopalcol Series; Norpar® 12; Norpar® 13; Norpar® 15; Nusolv™ ABP-62; Nusolv™ ABP-74; Nusolv™ ABP-87; Nusolv™ ABP-103; Nusolv™ ABP-164
PD-23; PD-25; PD-28; Penreco 2251 Oil; Penreco 2257 Oil; Penreco 2259 Oil; Penreco 2260 Oil; Penreco 2263 Oil; Pilot SXS-40; Pluracol® E400 NF; Pluracol® E1450 NF; Printsolve™ Ink Remover
QO® Tetrahydrofurfuryl Alcohol (THFA®)
Shell Sol B HT
T-Det® N-6; Tolu-Sol® W HT Solv.
Ucon® 50-HB-55; Ucon® 50-HB-100; Ucon® 50-HB-170; Ucon® 50-HB-260; Ucon® 50-HB-400; Ucon® 50-HB-660; Ucon® 50-HB-2000; Ucon® 50-HB-3520; Ucon® 50-HB-5100; Ucon® 75-H-450; Ucon® LB-65; Ucon® LB-135; Ucon® LB-165; Ucon® LB-285; Ucon® LB-385; Ucon® LB-525; Ucon® LB-625; Ucon® LB-1145; Ucon® LB-1715; Ucon® LB-3000
Vista LPA; Vista LPA-110; Vista LPA-142; Vista LPA-170; Vista LPA-210

Chemicals:
Acetyl tributyl citrate
Bis(methylethyl)-1,1´-biphenyl
n-Decane
Isostearyl alcohol
Mineral oil
Oleth-2; Oleyl alcohol
Polysorbate 60
Turpentine oil

Petroleum and Oil Processing

Trade names:
Acintene® AC-P; Acintene® DP 738; Adol® 42; Adol® 52 NF; Adol® 66; Adol® 80; Agrisynth® BLO; Agsol Ex BLO; Arconate® PC; Arcosolv® DPM; Arcosolv® DPM Acetate; Arcosolv® PM; Arcosolv® TPM; Arlox FE
BLO®
Cardolite® NC-511
Dytek® EP
Fancol OA-95
GBL
M-Pyrol®
NMP
2-Pyrol®
QO® Furfural
Secoster® A; Secoster® DO 600; Secoster® MO 400; Shell Cyclo Sol 63; Shell Sol 340 HT; Sulfolane W; Sylfat® DX; Sylfat® MM
Tolu-Sol® A HT Solv.; Tolu-Sol® W HT Solv.

Chemicals:
2-(2-Aminoethoxy) ethanol

Petroleum and Oil Processing *(cont'd.)*

Furfural
Heptane; Hexylene glycol
Isostearyl alcohol
N-Methyl-2-pyrrolidone
Oleyl alcohol
PPG-2 methyl ether; PPG-3 methyl ether; PPG-2 methyl ether acetate; Propionitrile
Sulfolane
n-Undecane

Pharmaceuticals

Trade names:

Antarox® P-400; Antarox® P-700; Argobase 125T; Arlasolve® DMI
Butyl Diglyme
Capmul® MCM; Capmul® MCM-90; Captex® 200; Captex® 300; Captex® 350; Captex® 810A; Captex® 810D; Captex® 8227; Carbopol® 940; Carbopol® 980; Carbowax® PEG 300; Carbowax® Sentry® PEG 300; Carbowax® Sentry® PEG 400; Carbowax® Sentry® PEG 540 Blend; Carbowax® Sentry® PEG 600; Carbowax® Sentry® PEG 900; Carbowax® Sentry® PEG 1000; Carbowax® Sentry® PEG 1450; Carbowax® Sentry® PEG 3350; Carbowax® Sentry® PEG 4600; Carbowax® Sentry® PEG 8000; Ceraphyl® 140; Ceraphyl® 140-A; Ceraphyl® ICA; Comperlan® COD; Corona Lanolin; Coronet Lanolin; Cremophor® EL; Crill 1; Crill 2; Crillet 6; Crodamol CAP; Crodamol GTCC; Crodamol PMP
Diglyme; DMSO; Dow E300 NF; Dow E400 NF; Dow E600 NF; Dow E900 NF; Dow E1000 NF; Dow E1450 NF; Dow E3350 NF; Dow E4500 NF; Dow E8000 NF; Dynacerin® 660
Eastman® Triacetin; EmCon Rice Bran; EmCon W; EmCon Walnut; Emery® 912; Emery® 916; Emery® 917; Emery® 918; Emery® Methyl Oleate; Estasan GT 8-40 3578; Estasan GT 8-60 3575; Estasan GT 8-60 3580; Estasan GT 8-65 3577; Estasan GT 8-65 3581; Estasan GT 8-70 3579; Estol IPM 1509 (BIO-IPM); Ethyl Diglyme
Fancol ALA; Fancol CB; Fancol CB Extra; Fancol Karite Butter; Fancol OA-95; Fancorsil A; Filmex® A-2; Filmex® B; Filmex® C; Filmex® D-1; Filmex® D-2
GBL; Glucam® E-10; Glucam® E-20; Glucam® P-10; Glucam® P-20
HEP®; High Purity MTBE
Imwitor® 742; Incrocas 35
Kemstrene® 96.0% USP; Kemstrene® 99.0%
Labrafac® CM10; Labrafac® Hydro; Labrafil® Isostearique; Labrasol; Lexol® 3975; Lexol® GT-865; Lexol® PG-800; Lexol® PG-865; Lipovol SES
Miglyol® 810; Miglyol® 812; Miglyol® 840; Miglyol® 8810; Monoglyme; Myritol® PC
Neobee® 1054; Neobee® M-5; Neobee® M-20; Neobee® O; NEP®; Novol NF
Olepal ISO

Pharmaceuticals *(cont'd.)*

PD-23; PD-25; PD-28; Pelemol® G7A; Penreco Snow; Pharmasolve™; Pluracol® E300; Pluracol® E400; Pluracol® E400 NF; Pluracol® E600; Pluracol® E600 NF; Pluracol® E1450 NF; Pluracol® E4000; Polychol 15; Propylene Glycol USP/FCC Ultra Grade; Punctilious® SDA 1-1

Renex® PEG 300; Renex® PEG 400; Renex® PEG 600; Renex® PEG 1000; Renex® PEG 1500; Renex® PEG 4000; Renex® PEG 6000; Renex® PEG 8000; Rhodasurf® PEG 600; Ricinion; Ritawax AEO; Ritawax ALA

Shell VM&P Naphtha HT; Shell Mineral Spirits 135; Shell Mineral Spirits 146 HT; Shell Mineral Spirits 200 HT; Shell Sol B HT; Solimate; Solulan® 16; Solulan® 75; Solulan® C-24; Solulan® L-575; Soluphor® P; Solutol® HS 15; Stearate 400 WL 817; Stearate 1500; Stearate 6000 WL 1644; Sucro Ester 15; Super Refined® Almond Oil; Super Refined® Corn Oil; Super Refined® Cottonseed Oil; Super Refined® Olive Oil; Super Refined® Peanut Oil; Super Refined® Safflower Oil; Super Refined® Sesame Oil; Super Refined® Soybean Oil USP

Tebol™; Tebol™ 99; Transcutol; Triglyme

Unichem BZBN

Wickenol® 158; Wickenol® 171

York Krystal Kleer Castor Oil; York USP Castor Oil

Chemicals:

Acetal; Acetic acid; Acetic acid, glacial; Acetophenone; t-Amyl alcohol; Apricot kernel oil PEG-6 esters

Benzyl benzoate; Bornyl acetate; Butyl alcohol; t-Butyl alcohol; Butylene glycol; n-Butyl propionate; Butyrolactone

Caprylic/capric glycerides; Caprylic/capric/lauric triglyceride; Caprylic/capric/linoleic triglyceride; Caprylic/capric triglyceride PEG-4 esters; Carbomer; Castor oil; Cetearyl isononanoate; Chloroform; Cocoa butter; Corn oil; Cottonseed oil

Decyl oleate; Diacetylated monoglycerides; Diethylene glycol dibutyl ether; Diethylene glycol dimethyl ether; Diisopropyl sebacate; Dimethicone copolyol undecylenate; Dimethyl isosorbide; Dimethyl sulfoxide; Dioctyl adipate; Dioctyl phthalate; 1,4-Dioxane

Ethoxydiglycol; Ethyl acetate; Ethylacetoacetate; Ethylene glycol dimethyl ether; Ethyl hexanediol; Ethyl oleate

Glycerin; Glyceryl caprylate; Glyceryl caprylate/caprate; Glyceryl formal; Glycol

Hexane; Hybrid safflower oil; Hydrogenated castor oil

Isodecyl oleate; Isopropyl laurate; Isopropyl myristate; Isopropyl palmitate; Isopropyl stearate; Isostearyl lactate

Laneth-5; Laneth-10; Laneth-15

Methyl alcohol; Methyl t-butyl ether; Methylene chloride; Methyl gluceth-10; Methyl gluceth-20; Methyl isobutyl ketone; Mineral oil; Mineral spirits

Octyl hydroxystearate; Octyl isononanoate; Oleth-2; Oleyl alcohol; Oleyl erucate; Olive oil

Peanut oil; PEG-4; PEG-8 caprylic/capric glycerides; PEG-33 castor oil; PEG-35 castor oil; PEG-4 dilaurate; PEG-8 dilaurate; PEG-20 dilaurate; PEG-3 dimethyl ether; PEG-4 dimethyl ether; PEG-4 dioleate; PEG-8 dioleate; PEG-20 dioleate; PEG-15

Pharmaceuticals *(cont'd.)*

hydroxystearate; PEG-6 isostearate; PEG-8 oleate; PEG-20 oleate; PEG-8 palmitostearate; PEG-150 palmitostearate; PEG-20 sorbitan isostearate; Petrolatum; Phenethyl alcohol; β-Picoline; γ-Picoline; Polyethylene glycol monomethyl ether; Polyglyceryl-3 diisostearate; Polypropylene glycol; Polysorbate 60; PPG-10 methyl glucose ether; PPG-20 methyl glucose ether; PPG-15 stearyl ether; Propylene glycol; Propylene glycol alginate; Propylene glycol caprylate; Propylene glycol dicaprylate; Propylene glycol dicaprylate/dicaprate; Propylene glycol dioctanoate; Propylene glycol dipelargonate; Propylene glycol laurate; 2-Pyrrolidone

Rice bran oil

Safflower oil; SD alcohol 3A; Sesame oil; Sweet almond oil

Tetrahydrofuran; Tetrahydroxypropyl ethylenediamine; 1,1,2-Trichloroethane; Triisostearin PEG-6 esters; Turpentine oil

Walnut oil

Plastics and Resins

Trade names:

Agrisynth® BLO; Agsol Ex BLO; Alkasurf® OD-410; Arconate® 1000; Arconate® HP; Arconate® PC; Arcosolv® DPM; Arcosolv® DPM Acetate; Arcosolv® PM; Arcosolv® PM Acetate; Arcosolv® PTB; Arcosolv® TPM; Aromatic 100; Aromatic 150; Aromatic 150; Aromatic 200

Berol 02; BLO®; Butoxyl®; Butyl Dioxitol®; Butyl Oxitol®; BVA ND-205 Solv.

CHP®; Citroflex® 2; Citroflex® 4; Citroflex® A-2; Citroflex® A-4; Crill 1; Crill 2

DBE; DBE-2, -2SPG; DBE-3; DBE-4; DBE-5; DBE-6; DBE-9; Decalin; Dermol ICSA; Diglycolamine® Agent (DGA®); DMSO; Dynasolve 200

Eastman® DE; Eastman® DP; Eastman® EB; Eastman® PM Acetate; Ecco Tipping Solution #3; Ektapro® EEP Solvent; Emcol® CC-42; Emcol® CC-57; Eposolve 299-R; Ethylan® HB1; Exxate® 600; Exxate® 700; Exxate® 800; Exxate® 900; Exxate® 1000; Exxate® 1300

Fancol OA-95; FoamFlush® Urethane Remover

GAFsolver™ TMP; GBL

High Purity MTBE; Hisolve MC; Hisolve MDM

Igepal® Cephene Lave; Igepal® LAVE; Igepal® OD-410; Isoamyl Alcohol 95%; Isoamyl Alcohol 99%

Jeffsol™ EC; Jeffsol™ PC

Kenflex® N1; Kenplast® A-450; Kenplast® ES-2HP; Kenplast® G; Kodaflex® DBP; Kodaflex® DEP; Kodaflex® DMP; Kodaflex® HS-3; Kodaflex® HS-4; Kronitex® TOF

M-1421-7B, 12B Nylon Tipping Solutions; Macol® 85; Mod Acid; Monolan® PPG440, PPG1100, PPG2200; M-Pyrol®

Naturechem® GTR; Naxol®; Naxonac™ 510; NMP; Nusolv™ ABP-62; Nusolv™ ABP-74; Nusolv™ ABP-87; Nusolv™ ABP-103; Nusolv™ ABP-164

PentoXone; Pilot SXS-40; Polystab RB1483; 2-Pyrol®

QO® Furfuryl Alcohol (FA®)

Plastics and Resins *(cont'd.)*

Rhodacal® 330; Rhodasurf® PEG 300

Shell Mineral Spirits 146 HT; Shell Mineral Spirits 200 HT; Shell Sol 71; Shell Sol 142 HT; Shell Sol 340 HT; Shell Sol B HT; Ship Shape® Resin Cleaner; Solvenon® DIP; Solvenon® IPP; Solvent B-100; Solvent B-200 (B-2); Sulfolane W

TBP; Tebol™; Tebol™ 99; THF; THF; Tolu-Sol® A HT Solv.; Tolu-Sol® W HT Solv.; Tri-Ethane

Ucon® 50-HB-55; Ucon® 50-HB-100; Ucon® 50-HB-170; Ucon® 50-HB-260; Ucon® 50-HB-400; Ucon® 50-HB-660; Ucon® 50-HB-2000; Ucon® 50-HB-3520; Ucon® 50-HB-5100; Ucon® 75-H-450; Ucon® LB-65; Ucon® LB-135; Ucon® LB-165; Ucon® LB-285; Ucon® LB-385; Ucon® LB-525; Ucon® LB-625; Ucon® LB-1145; Ucon® LB-1715; Ucon® LB-3000; Uniplex 110; Uniplex 155

Uresolve 411

Vista LPA-170

York Krystal Kleer Castor Oil; York USP Castor Oil

Chemicals: Acetone; Acetophenone; Acetylacetone; s-Amyl acetate

Benzaldehyde; Benzene; Benzonitrile; Benzyl acetate; Benzyl alcohol; Benzyltriethyl ammonium chloride; Benzyl trimethyl ammonium chloride; Bis(methylethyl)-1,1′-biphenyl; 2-Butanol; Butoxydiglycol; Butoxyethanol; 1-Butoxyethoxy-2-propanol; Butyl alcohol; t-Butyl alcohol; Butyl benzoate; Butyl lactate; n-Butyl propionate; Butyl stearate; Butyrolactone

C8 alkyl acetate; C9 alkyl acetate; C10 alkyl acetate; C13 alkyl acetate; Carbon disulfide; Cumylphenyl benzoate; Cumylphenyl neodecanoate; Cyclohexane; Cyclohexanol; Cyclohexanone; Cyclohexyl acetate; N-Cyclohexyl pyrrolidone; Cyclopentane

Decahydronaphthalene; Diacetone alcohol; Dibutyl ether; Dibutyl phthalate; Diethylene glycol; Diethylene glycol diethyl ether; Diethylene glycol propyl ether; Diethylenetriamine; Diethyl phthalate; Diethyl toluamide; Diisobutyl ketone; Diisobutyl phthalate; Dimethyl adipate; Dimethyl formamide; Dimethyl glutarate; Dimethyl phthalate; Dimethyl sebacate; Dimethyl succinate; Dimethyl sulfoxide; Dipropylene glycol; Dipropylene glycol butyl ether; Dipropylene glycol dimethyl ether; Dipropylene glycol ethyl ether

Erucamide; Ethoxydiglycol; Ethoxyethanol; Ethyl acetate; Ethylbenzene; 2-Ethyl-1-butanol; Ethylene carbonate; Ethylene glycol acetate; Ethylene glycol diacetate; Ethylene glycol propyl ether; Ethyl 3-ethoxypropionate; Ethyl formate; 2-Ethylhexanol; 2-Ethylhexyl acetate; Ethyl propionate

Furfural; Furfuryl alcohol

Glycol

2-Heptanol

Isoamyl acetate; Isoamyl alcohol; Isobutyl acetate; Isobutyl heptyl ketone; Isocetyl salicylate; Isophorone; 1-Isopropoxy-2-propanol; 2-Isopropoxy-1-propanol; Isopropyl acetate; Isopropyl alcohol; Isopropyl ether

dl-Limonene

Methoxyethanol; Methoxyethanol acetate; Methoxyisopropanol;

Plastics and Resins *(cont'd.)*

Methyl acetate; Methyl alcohol; Methyl amyl alcohol; Methyl benzoate; Methyl t-butyl ether; Methyl butynol; Methylcyclohexanol; Methyl ethyl ketone; N-Methyl-2-pyrrolidone; Mineral oil; Morpholine

Nitroethane; Nitromethane; Nitropropane; 2-Nitropropane

Oxo-heptyl acetate; Oxo-hexyl acetate

PEG-4 dilaurate; PEG-20 dilaurate; PEG-4 dioleate; PEG-20 dioleate; PEG-8 oleate; PEG-20 oleate; Phenoxydiglycol; Phenoxyethanol; β-Picoline; γ-Picoline; Polypropylene glycol; Polysorbate 60; PPG-40 diethylmonium chloride; PPG-2 methyl ether; PPG-3 methyl ether; PPG-2 methyl ether acetate; Propyl acetate; n-Propyl alcohol; Propylbenzene; Propylene carbonate; Propylene glycol ethyl ether; Propylene glycol methyl ether acetate; Propylene glycol phenyl ether

Sulfolane

α-Terpineol; Tetraethylenepentamine; Tetrahydrofuran; Tetrahydrofurfuryl alcohol; Toluene; Tributyl citrate; Trichloroacetic acid; Triethyl citrate; 2,2,4-Trimethyl-1,3-pentanediol monoisobutyrate; Tripropylene glycol ethyl ether

Xylene

Rubber Compounding and Processing

Trade names:

Acintene® AC-P; Acintene® DP 738

Carbowax® PEG 300

Decalin

Kronitex® TOF

Monolan® PPG440, PPG1100, PPG2200; #15 Oil; #30 Oil; #40 Oil

Pale 4; Pale 16; Pale 170; Pale 1000; Pluracol® E200; Pluracol® E300; Pluracol® E400; Pluracol® E400 NF; Pluracol® E600; Pluracol® E600 NF; Pluracol® E1450 NF

QO® Furfural

Shell Sol B HT; Shell Toluene; Shell Xylene; Ship Shape® Resin Cleaner

TBP; Tetralin; Tolu-Sol® A HT Solv.; Tolu-Sol® W HT Solv.

Vorite 105; Vorite 110; Vorite 115; Vorite 120; Vorite 125

Chemicals:

Benzonitrile; Butyl stearate

Carbon disulfide; 1-Chloro-1-nitropropane; Cyclohexanol; Cyclohexanone; Cyclohexyl acetate

Decahydronaphthalene; n-Decane; Diethylene glycol propyl ether; Diisobutyl ketone; 1,4-Dioxane

Ethylamine

Heptane; Hexane

Methyl benzoate

Naphtha; Nitropropane; 2-Nitropropane

β-Picoline; γ-Picoline; Propyl acetate

Tetrahydrofurfuryl alcohol; Tetrahydronaphthalene; 1,1,2-Trichloroethane; Turpentine oil

Xylene

Textiles and Fibers

Trade names: Adol® 42; Adol® 52 NF; Adol® 66; Adol® 80; Agrisynth® BLO; Agsol Ex BLO; Arconate® 1000; Arconate® HP; Arconate® PC; Arcosolv® DPNB; Arcosolv® PE; Arcosolv® PE Acetate; Arcosolv® PNB; Arcosolv® PNP; Arosolve 570-HF; Arosolve 570 Special; Arosolve B-950; Arosolve CON; Arosolve DBM; Arosolve DMI-F; Arosolve FX-6; Arosolve IWS; Arosolve MN-LF; Arosolve RCB; Arosolve TMB; Arosolve VNO-LF; Arosolve XNF-1

BLO®; Butyl Diglyme; BVA ND-205 Solv.

Carbowax® PEG 300; CNC Sol BD; CNC Sol UE; CNC Sol XN; CNC Sol XNN #11; Consoscour P; Consoscour M; Consoscour TEK; Crill 1; Crill 2

Dalpad A; DBE; DBE-2, -2SPG; DBE-3; DBE-4; DBE-5; DBE-6; DBE-9; DBE-IB; DF-2000™; Diglyme; Dowanol® DB; Dowanol® DM; Dowanol® EPh; Dowanol® PPh; Dowanol® TPnB; Dytek® EP

Eastman® DB; Eastman® DE Acetate; Eastman® TEP; Ecco Cleaner 930 Conc.; Ektasolve® DM; Emcol® CC-42; Emcol® CC-55; Emcol® CC-57; Emkapon DAC-500; Emkapon KW; Emulsifier Q; Ethyl Diglyme; Exxate® 600; Exxate® 700; Exxate® 800; Exxate® 900; Exxate® 1000; Exxate® 1300

Filmex® A-2; Filmex® B; Filmex® C; Filmex® D-1; Filmex® D-2; Flexricin® 9

GBL; Glyezin®

Herco® Pine Oil; Hipochem MS-BW

Intravon® SOL; Intravon® SOL-N; Intravon® SOL-W

Jeffsol™ EC; Jeffsol™ PC

Marvanol® Solvent Scour 34 ; Monoglyme; MPDiol® Glycol; M-Pyrol®

Neosolv™ 2; Neosolv™ 4; Neosolv™ 6; Neosolv™ 8; Neosolv™ 24; Neosolv™ 150; NEP®

Pluracol® E200; Pluracol® E300; Pluracol® E400; Pluracol® E400 NF; Pluracol® E600; Pluracol® E600 NF; Pluracol® E1450 NF; Poly-Solv® DM; Poly-Solv® DPM; Poly-Solv® EM; Poly-Solv® MPM; Poly-Solv® TM; Poly-Solv® TPM

QO® Tetrahydrofurfuryl Alcohol (THFA®)

Rexosolve; Rexosolve CR; Rhodasurf® PEG 300

Scour NFP; Secoster® DO 600; Sesolvan® L; Shell Cyclo Sol 63; Shell Mineral Spirits 135; Shell Mineral Spirits 146 HT; Shell Mineral Spirits 200 HT; Shell Sol 142 HT; Solvent Scour 25/27; Solvent-Scour 263; Solvent-Scour 880; Sovatex C Series; Surfadone® LP-100; Surfadone® LP-300

Tolu-Sol® A HT Solv.; Tri-Ethane; Triglyme

Ucon® 50-HB-55; Ucon® 50-HB-100; Ucon® 50-HB-170; Ucon® 50-HB-260; Ucon® 50-HB-400; Ucon® 50-HB-660; Ucon® 50-HB-2000; Ucon® 50-HB-5100; Ucon® 75-H-450; Ucon® LB-65; Ucon® LB-135; Ucon® LB-165; Ucon® LB-285; Ucon® LB-385; Ucon® LB-525; Ucon® LB-625; Ucon® LB-1145; Ucon® LB-1715; Ucon® LB-3000

Varonic® DM55

Chemicals: Benzyl alcohol; Butyl stearate

C9 alkyl acetate; C10 alkyl acetate; C13 alkyl acetate; Caprylyl

Textiles and Fibers *(cont'd.)*

pyrrolidone; Cyclohexanol
Diacetone alcohol; Dimethyl adipate; Dimethyl glutarate; Dimethyl succinate; Dipropylene glycol; Dipropylene glycol butyl ether
Ethoxyethanol
Hexylene glycol
Isostearyl alcohol
Methoxydiglycol; Methoxyisopropanol; Mineral oil; Mineral spirits
Oleyl alcohol; Oxo-heptyl acetate; Oxo-hexyl acetate
PEG-4 dilaurate; PEG-3 dimethyl ether; PEG-8 dioleate; Polysorbate 60; PPG-40 diethylmonium chloride; Propylene carbonate; Propylene glycol ethyl ether; Propylene glycol ethyl ether acetate; Propylene glycol n-propyl ether
Sodium thiocyanate; Stoddard solvent
Thiodiglycol; 1,1,1-Trichloroethane; Triethylene glycol; Triglycol monomethyl ether

Waxes and Polishes

Trade names:
Arcosolv® DPM Acetate; Arcosolv® PM; Arcosolv® TPM
Crill 1; Crill 2
Decalin; Dermol ICSA
Gaftex® HI-8; Gaftex® HI-12
Isopar® M; Isopar® V
M-Pyrol®
Naturechem® GTR; Neosolv™ 2; Neosolv™ 4; Neosolv™ 6; Neosolv™ 8; Neosolv™ 24; Neosolv™ 150; Norpar® 13; Norpar® 15
#15 Oil; #30 Oil; #40 Oil
Pale 4; Pale 16; Pale 170; Pale 1000; PD-23; PD-25; PD-28; Penreco 2251 Oil; Penreco 2257 Oil; Penreco 2260 Oil; Penreco 2263 Oil; Pluracol® E400 NF; Pluracol® E600 NF; Pluracol® E1450 NF; Poly-Solv® DM; Poly-Solv® DPM; Poly-Solv® EM; Poly-Solv® MPM; Poly-Solv® TM; Poly-Solv® TPM
Shell VM&P Naphtha HT; Shell Mineral Spirits 135; Shell Mineral Spirits 146 HT; Shell Mineral Spirits 200 HT; Shell Sol 142 HT; Shell Sol 340 HT
Tetralin; Tolu-Sol® A HT Solv.; Tolu-Sol® W HT Solv.
Vorite 105; Vorite 110; Vorite 115; Vorite 120; Vorite 125

Chemicals:
Acetone
Benzene; Benzyl alcohol; Butyl alcohol
Carbon disulfide; Castor oil, polymerized; Cyclohexane; Cyclohexanone
Decahydronaphthalene; Diacetone alcohol; Dioctyl maleate; 1,4-Dioxane; 1,3-Dioxolane
Erucamide; 2-Ethyl-1-butanol; Ethyl ether; 2-Ethylhexyl acetate
Isoamyl acetate; Isocetyl salicylate; Isopropyl ether
dl-Limonene
Methoxyethanol acetate; Methoxyisopropanol; Methyl amyl alcohol; Methylcyclohexanol; Morpholine
Nitroethane; Nitromethane

Waxes and Polishes *(cont'd.)*

2-Octanol

Petroleum distillates; Polypropylene glycol; PPG-2 methyl ether; PPG-3 methyl ether; PPG-2 methyl ether acetate; n-Propyl alcohol

Tetrahydronaphthalene; 1,1,1-Trichloroethane; Trichloroethylene; Tridecyl neopentanoate; Triglycol monomethyl ether; Turpentine oil

Part IV
Manufacturers Directory

Manufacturers Directory

Aarhus Oliefabrik A/S

Postboks 50, Bruunsgade 27, DK-8100 Aarhus C Denmark (Tel.: 86-12 60 00; Telefax: 86-183839; Telex: 64341)

Aarhus Inc.

131 Marsh St., PO Box 4240, Newark, NJ 07114 USA (Tel.: 201-344-1300; Telefax: 201-344-9049)

ABITEC Corp., Subsid. of Associated British Foods (ABF)

525 W. First Ave., PO Box 569, Columbus, OH 43216-0569 USA (Tel.: 614-299-3131; 800-848-1340; Telefax: 614-299-8279; Telex: 245494 capctyprdcol)

AB Technology Ltd., Div. of ABITEC

Salthouse Rd, Brackmills, Northampton, Northamptonshire, NN4 0EX UK (Tel.: 44 1604 768999; Telefax: 44 1604 701503; Telex: 311848)

Aceto Corporation

1 Hollow Lane, Suite 201, Lake Success, NY 11042-1215 USA (Tel.: 516-627-6000; Telefax: 516-627-6093; Telex: 62662)

Pfaltz & Bauer Inc., Research chemicals subsid. of Aceto Corp., 172 E. Aurora St., Waterbury, CT 06708 USA (Tel.: 203-574-0075; 800-225-5172; Telefax: 203-574-3181; Telex: 996471)

AC Industries, Inc., Sattva Chemical Co. Div.

2 Landmark Sq., Suite 214, Stamford, CT 06901-2410 USA (Tel.: 203-348-8002; 800-736-7893; Telefax: 203-348-3666; Telex: 6819048 SATVA UW)

Actrachem L.P.

7201 W. 65th St., Bedford Park, IL 60501 USA (Tel.: 708-458-8450; 800-323-3231; Telefax: 708-458-0286)

ADM, Archer Daniels Midland Co.

Box 1470, Decatur, IL 62525 USA

Advance Research Chemicals Inc.

5005 N. Skiatook Rd., Catossa, OK 74015 USA (Tel.: 918-266-6789; Telefax: 918-266-6796; Telex: 74-8086)

Air Liquide

Air Liquide Hellas S.A., 26-28 Asklipiou St, GR-10679 Athens Greece (Tel.: 30 1-360 84 11; Telefax: 30 1-362-54 31; Telex: 216444)

Air Liquide America Corp., 2121 N. California Blvd., Suite 350, Walnut Creek, CA 94596 USA (Tel.: 510-977-6500; Telefax: 510-746-6306; Telex: (ITT) 470020)

Air Products

Air Products and Chemicals, Inc., 7201 Hamilton Blvd., Allentown, PA 18195-1501 USA (Tel.: 610-481-4911; 800-345-3148; Telefax: 610-481-5900; Telex: 847416)

Air Products Nederland B.V., Kanaalweg 15, PO Box 3193, 3502 GD Utrecht The Netherlands (Tel.: 31-30-857100; Telefax: 31-30 857111)

Air Products Asia, Inc., Room 1901, Peregrine Tower, Lippo Centre, 89 Queensway, Central Hong Kong (Tel.: (852)524-7110; Telefax: (852) 877-0094)

Air Products Japan, Inc., Shuwa No. 2 Kamiyacho Bldg., 3-18-19, Toranomon, Minato-ku, Tokyo, 105 Japan (Tel.: (81)(3)3432-7046; Telefax: (81)(3)3432-7048)

Air-Scent Int'l.

PO Box 1000, 215 Eighth Ave., Braddock, PA 15104 USA (Tel.: 412-351-2100; 800-247-0770; Telefax: 412-351-7701)

Ajay Chemicals, Inc.

1400 Industry Rd., Powder Springs, GA 30073 USA (Tel.: 404-943-6202; Telefax: 404-439-0369; Telex: 804-468)

Akcros

Akcros Chemicals, Eccles Site, PO Box 1, Eccles, Manchester, M30 0BH UK (Tel.: 44 161 785 1111; Telefax: 44 161 78 87 886)

Akcros Chemicals France S.A., Zone Industrielle BP 40, 441220 St Laurent-Nouan France (Tel.: 33 54 87 73 89; Telefax: 33 54 87 79 40; Telex: 750679 Tinstab)

Akcros Chemicals GmbH & Co. KG, Chemiewerk Greiz-Dolau, Liebigstrasse 7, 07973 Greiz Germany (Tel.: 49 36 61 780; Telefax: 49 36 61 782 02; Telex: 331141)

Akcros Chemicals Iberica S.L., Autovia Castelldefels Km. 4, 65, 08820 El Prat De Llobregat, Barcelona Spain (Tel.: 34 3 4785 755; Telefax: 34 3 4780 888; Telex: 16152 Akrcos dk)

Akcros Chemicals Italia Srl, Via Cristina Belgioioso 13/15, 20021 Baranzate DiBollate (MI) Italy (Tel.: 39 2 38200484; Telefax: 39 2 38200443)

Akcros Chemicals Nordic ApS, Naverland 2, 8. floor, DK-2600 Glostrup-Copenhagen Denmark (Tel.: 45 4344 4201; Telefax: 45 4344 4203; Telex: 16152 Akrcos dk)

Akcros Chemicals v.o.f., Roermond Site, PO Box 44, 6040 AA Roermond The Netherlands (Tel.: 0475-356446; Telefax: 0475-356300)

Akcros Chemicals America, 500 Jersey Ave., PO Box 638, New Brunswick, NJ 08903 USA (Tel.: 908-247-2202; 800-500-7890; Telefax: 908-247-2287)

Akcros Chemicals, Taiwan Branch, Room 1408, 145h Fl. No. 96, Chung Shan N. Rd., Sec. 2, Taipei 10449, Taiwan, R.O.C. (Tel.: 886 2 562 6922; Telefax: 886 2 536 5287; Telex: 15185)

Akcros Chemicals (Asia Pacific) Pte. Ltd., 7500A Beach Rd., 15-309 The Plaza, Singapore 0719 (Tel.: 65 292 1966; Telefax: 65 292 9665; Telex: RS 23241 AKCROS)

Akzo Nobel

Akzo Nobel België N.V./S.A., 13 Marnix Ave., 1050 Bruxelles Belgium (Tel.: 2-518 04 07; Telefax: 02 518 05 05; Telex: 62 664)

Akzo Nobel Chemicals Inc., 300 S. Riverside Plaza, Chicago, IL 60606 USA (Tel.: 312-906-7500; 800-227-7070; Telefax: 312-906-7680; Telex: 25-3233)

Akzo Nobel Surface Chemistry, Inc., Meritt 8 Corporate Park, 99 Hawley Lane, Stratford, CT 06497 USA (Tel.: 203-378-0500; Telefax: 203-378-5960)

Akzo Nobel Chemicals Ltd., 1-5, Queens Road, Hersham, Walton-on-Thames, Surrey, KT12 5NL UK (Tel.: 44 1932 247891; Telefax: 44 1932-231204; Telex: 21997)

Akzo Nobel Chemicals bv, Postbus 247, Stationsstraat 48, 3800 AE Amersfoort The Netherlands (Tel.: 31-33-467 67 67; Telefax: 31-33-467 61 00; Telex: 79322)

Akzo Nobel Surface Chemistry AB, S-444 85 Stenungsund Sweden (Tel.: 46 303 850 00; Telefax: 46 303 845 58)

Akzo Nobel K.K., Godo Kaikan Bldg., 3-27 Kioi-cho, Chiyoda-ku, Tokyo, 102 Japan (Tel.: (03)5275-6300; Telefax: (03)3222-0465; Telex: J24262)

Alba International Inc.

508 Clearwater Dr., N. Aurora, IL 60542 USA (Tel.: 708-897-4200; 800-669-9333; Telefax: 708-377-5330)

Albemarle

Albemarle Corp., 451 Florida St., Baton Rouge, LA 70801 USA (Tel.: 504-388-7040; 800-535-3030; Telefax: 504-388-7686; Telex: 586441, 586431)

Albemarle S.A., Div. of Albemarle Corp., 523 Ave. Louise, Box 19, B-1050 Brussels Belgium (Tel.: 32-2-642-4411; Telefax: 32-2-648-0560; Telex: 22549)

Albemarle Asia Pacific Co., #13-06 PUB Bldg., Devonshire Wing, 111 Somerset Rd., Singapore 0923 Singapore (Tel.: 65-732-6286; Telefax: 65-737-4123)

Albright & Wilson

Albright & Wilson Ltd., European & Corporate Headquarters, PO Box 3, 210-222 Hagley Rd. West, Oldbury, Warley, West Midlands, B68 0NN UK (Tel.: 44 121 420 5297; Telefax: 44 121-420-5462; Telex: 336291)

Albright & Wilson Americas Inc., PO Box 26229, Richmond, VA 23260-6229 USA (Tel.: 804-550-4300; 800-446-3700; Telefax: 804-550-4385)

Albright & Wilson Saint Mihiel S.A., B.P. 19 Han-Sur-Meuse, 55300 Saint-Mihiel France (Tel.: 33 29 91 73 00; Telefax: 33 29 91 73 99; Telex: 961058)

Albright & Wilson Am. (Canada), PO Box 2220, St. John's, Newfoundland, A1C 6E6 Canada (Tel.: 709-753-0838; Telefax: 709-753-5353)

Albright & Wilson (Australia) Ltd., PO Box 20, Yarraville, Victoria, 3013 Australia (Tel.: 61 3-688-7777; Telefax: 61 3-688-7788)

Albright & Wilson Asia Pacific Pte. Ltd., 6 Jalan Besut, Jurong Industrial Estate, Singapore 2261 (Tel.: 011-65-261-2151; Telefax: 011-65-265-1941; Telex: 37007)

Albright & Wilson Asia Pacific Pte. Ltd., Rm. 4051, Hua Ting Sheraton Hotel, 1200 Cao Xi Bei Lu, Shanghai, 200030 China (Tel.: 86 21 439100 x4051; Telefax: 86 21 4397856)

Albright & Wilson Ltd. Japan, No. 2 Okamotoya Bldg. 6 Fl., 1-24, Toranomon 1-chome, Minato-ku, Tokyo, 105 Japan (Tel.: (03) 3508 9461; Telefax: (03) 3591 0733; Telex: 2226721)

Aldrich

Aldrich Chemical Co., Inc., 1001 W. St. Paul Ave., Milwaukee, WI 53233 USA (Tel.: 414-273-3850; 800-558-9160; Telefax: 414-273-4979; Telex: 26843 ALDRICH MI)

Aldrich Chemical Co. Ltd., The Old Brickyard, New Rd., Gillingham, Dorset, SP8 4JL UK (Tel.: 44 1747 822211, 0800-71 71 81; Telefax: 44 1747 823779; Telex: 417238)

Sigma-Aldrich N.V./S.A., K. Cardijnplei 8, B-2880 Bornem Belgium (Tel.: 038991301; Telefax: 038991311)

Aldrich-Chimie S.a.r.l., BP 701, 38297 Saint Quentin Fallavier, , Cedex France (Tel.: 74822800; Telefax: 74956808; Telex: 308215 Aldrich F)

Aldrich Chemical, Unit 2, 10 Anella Ave., Castle Hill, NSW, 2154 Australia (Tel.: 028999977; Telefax: 028999742)

Aldrich Japan Inc., Kyodo Bldg., Shinkanda, 10 Kanda-Mikura-cho, Chiyoda-ku, Tokyo, 101 Japan (Tel.: (03)3258 0155; Telefax: (03)3258 0157)

Alemark Chemicals

1177 High Ridge Rd., Stamford, CT 06905 USA (Tel.: 203-966-7410; Telefax: 203-966-4276)

Alexander Chemical Corp.

One Corporate Lakes, 2525 Cabot Drive, Lisle, IL 60532 USA (Tel.: 708-955-6050; 800-445-9458; Telefax: 708-955-6065)

Alfa Aesar Johnson Matthey

30 Bond St., Ward Hill, MA 01835 USA (Tel.: 508-521-6300; 800-322-4757; Telefax: 508-521-6366)

Allan Chemical Corp.

PO Box 1837, Fort Lee, NJ 07024-8337 USA (Tel.: 201-592-8122; 800-633-0274; Telefax: 201-592-9298; Telex: 170991)

Allchem Industries Inc.

4001 Newberry Rd, Suite E-3, Gainesville, FL 32607 USA (Tel.: 904-378-9696; Telefax: 904-338-0400; Telex: 509540 ALLCHEM UD)

AlliedSignal

AlliedSignal Inc., PO Box 1053, 101 Columbia Rd., Morristown, NJ 07960 USA (Tel.: 201-455-2000; 800-526-0717; Telefax: 201-707-4555; Telex: 136410)

AlliedSignal Europe N.V., Haasrode Research Park, Grauwmeer 1, B-3001 Heverlee (Leuven) Belgium (Tel.: 32-16-39 12 33; Telefax: 32-16-40 03 77)

Alnor Oil Company, Inc.

70 E. Sunrise Hwy., Suite 418, Valley Stream, NY 11581 USA (Tel.: 516-561-6146; Telefax: 516-561-6123; Telex: 221512 ALNOR UR)

Alzo Inc.

6 Gulfstream Blvd., Matawan, NJ 07747 USA (Tel.: 908-254-1901; Telefax: 908-254-4423)

Amber Synthetics, Amsyn Inc.

1177 High Ridge Rd., Stamford, CT 06905 USA (Tel.: 203-972-7401; Telefax: 203-966-4276)

Amerchol

Amerchol Corp., Subsid. of Union Carbide, PO Box 4051, 136 Talmadge Rd., Edison, NJ 08818-4051 USA (Tel.: 908-248-6000; 800-367-3534; Telefax: 908-287-4186; Telex: 833472)

Amerchol Europe, Havenstraat 86, B-1800 Vilvoorde Belgium (Tel.: 32-2-252-4012; Telefax: 32-2-252-4909; Telex: 846-69105 AMRCHL B)

AmeriBrom, Inc., Member of the Dead Sea Bromine Group

52 Vanderbilt Ave., New York, NY 10017 USA (Tel.: 212-286-4000; 800-280-2766; Telefax: 212-286-4475; Telex: RCA 220531)

American Biorganics, Inc.

2236 Liberty Dr, Niagara Falls, NY 14304 USA (Tel.: 716-283-1434; 800-648-6689; Telefax: 716-283-1570; Telex: 926074)

American Chemet Corp.

400 Lake Cook Rd., Deerfield, IL 60015 USA (Tel.: 708-948-0800; Telefax: 708-948-0811; Telex: 72-4301)

American Ingredients

American Ingredients Co., 3947 Broadway, Kansas City, MO 64111 USA (Tel.: 816-561-9050; 800-669-4092; Telefax: 816-561-1132)

American Ingredients Co./Patco Polymer Additives Div., 3947 Broadway, Kansas City, MO 64111 USA (Tel.: 816-561-9050; 800-669-2250; Telefax: 816-561-0422)

American International Chemical, Inc.

17 Strathmore Rd, Natick, MA 01760 USA (Tel.: 508-655-5805; 800-238-0001; Telefax: 508-655-0927; Telex: 948342)

American Oil & Supply Co.

238 Wilson Ave., Newark, NJ 07105 USA (Tel.: 201-589-0250; 800-582-3267; Telefax: 201-465-5083)

Amoco

Amoco Chemical Co., 200 East Randolph Dr., Mail code 4106, Chicago, IL 60601 USA (Tel.: 312-856-3200; 800-621-4567; Telefax: 312-856-4151; Telex: 25-3731)

Amoco Chemical (Europe) S.A., Div. of Amoco Corp., 15, Rue Rothschild, 1211 Geneva 21 Switzerland (Tel.: 41-22-715-0701; Telex: 422787)

Amoco Chemical Asia Pacific Ltd., 16th Floor, Great Eagle Centre, 23 Harbour Rd. Hong Kong (Tel.: (852) 2586-8899)

Amoco Oil Co., Lubricants Div., 2021 Spring Rd., Oak Brook, IL 60562-1857 USA (Tel.: 708-571-7100; Telefax: 708-571-7174)

AMRESCO

30175 Solon Ind. Pkwy., Solon, OH 44139 USA (Tel.: 216-349-1199; 800-829-2802; Telefax: 216-349-1182; Telex: 985582)

Anar Chemical Co.

1765F Cortland Court, Addison, IL 60101 USA (Tel.: 708-953-1660; 800-344-1660; Telefax: 708-953-1698)

ANGUS

ANGUS Chemical Co., 1500 E. Lake Cook Rd., Buffalo Grove, IL 60089-6556 USA (Tel.: 847-215-8600; 800-362-2580; Telefax: 847-215-8626; Telex: 275422 ANGUS UR)

ANGUS Chemie GmbH, Unit 7, Rotunda Business Centre, Thorncliffe Park Estate, Chapeltown, Sheffield, S30 4PH UK (Tel.: 44 114 2571322; Telefax: 44 114 2571336)

ANGUS Chemie GmbH, Huyssenallee 5, 45128 Essen Germany (Tel.: (49) 201-233531; Telefax: (49)201-238661; Telex: 8571563 ANGE D)

ANGUS Chemie GmbH, Le Bonaparte, Centre d'Affaires, Paris Nord, F-93153 Le Blanc Mesnil France (Tel.: (33)1-48-65-73-40; Telefax: (33)1-48-65-73-20; Telex: ANGUS 232089 F)

ANGUS Chemical (Singapore) Pte. Ltd., 150 Beach Rd., #17-01 Gateway West, Singapore 18972 (Tel.: (65) 293-1738; Telefax: (65) 293-3307)

Aquaness Chemical, A Div. of Baker Performance Chemicals, Inc.

3900 Essex Lane, Suite 1120, PO Box 27714, Houston, TX 77227-7714 USA (Tel.: 713-599-7400; Telefax: 713-599-7460)

9101 W. 21st St., PO Box 370, Sand Springs, OK 74063 USA (Tel.: 918-241-4976)

Aquatec Quimica S.A.

Av. Paulista no. 37-12° andar, 01311-000 Sao Paulo-SP Brazil (Tel.: 55 11-284-4188; Telefax: 55 11-287 7841; Telex: 1121312)

Arakawa

Arakawa Chemical Industries Ltd., 3-7, Hiranomachi 1-chome, Chuo-ku, Osaka, 541 Japan (Tel.: (06) 209-8581; Telefax: (06) 209-8542; Telex: 5222296 ARKOSA J)

Arakawa Chemical (USA) Inc., 625 N. Michigan Ave., Suite 1700, Chicago, IL 60611 USA (Tel.: 312-642-1750; Telefax: 312-642-0089; Telex: 26-5514)

ARC Chemical, Div. of Balchem Corp.

P.O. Box 175, Slate Hill, NY 10973 USA (Tel.: 914-355-2891; 800-431-5641; Telefax: 914-355-6314)

ARCO

ARCO Chemical/Headquarters, Research & Engineering Center, 3801 West Chester Pike, Newtown Sq., PA 19073-2387 USA (Tel.: 610-359-2000; 800-345-0252; Telefax: 610-359-2841)

ARCO Chemical Canada Inc., 100 Consilium Pl., Suite 306, Scarborough, Ontario, M1H 3E3 Canada (Tel.: 416-296-9864; Telefax: 416-296-9834)

ARCO Chemical Pan American, Inc., Paseo de la Reforma, 390 Decimo Piso, 06600 Mexico City Mexico (Tel.: 905-514-6833)

ARCO Chemical Europe Inc., Bridge Ave., Maidenhead, Berkshire, SL6 1YP UK (Tel.: 44-1628-77-5000; Telefax: 44-1628-77-5180; Telex: 847436)

ARCO Quimica do Brazil Ltda., Av. Roque Petroni jr., 999 cj 123, Sao Paulo SP 04708-000 Brazil (Tel.: 55-1-535-5673; Telefax: 55-11535-3321)

ARCO Chemical Japan, Inc., Hamacho Center Bldg., 2-31-1 Nihonbashi, Hama-cho, Chuo-ku, Tokyo, 103 Japan (Tel.: (03) 5641-4500; Telefax: (03) 5641-4550)

ARCO Chemical Asia Pacific Ltd., 38th Floor East Tower, Bond Centre, 89 Queensway Hong Kong (Tel.: (852) 846 9200; Telefax: (852) 840 1690; Telex: 691 30)

ARCO Chemical (Singapore) Pte. Ltd., 37th Floor, 152 Beach Rd., #37-00 The Gateway East, Singapore 0718 (Tel.: 65-291-2011; Telefax: 65-297-0517)

Arista Industries, Inc.

1082 Post Rd., Darien, CT 06820 USA (Tel.: 203-655-0881; 800-255-6457; Telefax: 203-656-0328; Telex: 996493)

Aristech Chemical Corp.

600 Grant St., Room 1028, Pittsburgh, PA 15230-0250 USA (Tel.: 412-433-7873; Telefax: 412-433-1816; Telex: 6503608865)

Arizona Chemical Co., Div. of International Paper

1001 E. Business Hwy. 98, Panama City, FL 32401-3633 USA (Tel.: 904-785-6700; 800-526-5294; Telefax: 904-785-2203; Telex: 514411)

Arol Chemical Products Co.

649 Ferry St., Newark, NJ 07105 USA (Tel.: 201-344-1510; Telefax: 201-344-7127)

Asahi Chemical Industry Co., Ltd.

Hibiya Mitsui Bldg., 1-2, Yuraku-cho 1-chome, Chiyoda-ku, Tokyo, 100 Japan (Tel.: (03) 3507-2730; Telefax: (03) 3507-2495; Telex: 222-3518 BEMBRGJ)

Asahi Denka Kogyo K.K.

Furukawa Bldg. 2-8, Nihonbashi Muro-machi 2-chome Chuo-ku, Tokyo, 103 Japan (Tel.: (03) 5255-9002; Telefax: (03) 3270-2463; Telex: 222-2407 TOKADK)

Asahi-Penn Chemical Co. Ltd., Joint venture of Asahi Glass Co. Ltd./PPG Industries Inc.

13-3, Nihonbashi Kobuna-cho, Chuo-ku, Tokyo, 103 Japan (Tel.: (03) 3662-0520; Telefax: (03) 3662-0596)

Ashland

Ashland Chemical Co./Industrial Chemicals & Solvents,Div. of Ashland Inc., PO Box 2219, Columbus, OH 43216 USA (Tel.: 614-790-3333; 800-526-4032; Telefax: 614-889-3465)

Ashland Chemical Co./General Polymers Div., Subsid. of Ashland Oil Inc., 2463 Royal Windsor Drive, Mississauga, Ontario, L5J 1K9 Canada (Tel.: 905-823-7975)

Ashland Chemical China, Business Bldg., Rm. 408A, Holiday Inn Lido Beijing, Jichang Rd., Jiangtai Rd., Beijing, 10004 China (Tel.: 86 10 437 6620; Telefax: 86 10 437 6623)

ASHTA Chemicals, Inc.

P.O. Box 858, 3509 Middle Rd., Ashtabula, OH 44005-0858 USA (Tel.: 216-997-5221; 800-492-5082; Telefax: 216-992-0151)

The Associated Octel Co. Ltd.

PO Box 17, Oil Site Rd., Ellesmere Port, S. Wirral, L65 4HF UK (Tel.: 44 151 3553611, FAx 44 151 3566106)

Atomergic Chemetals Corp.

222 Sherwood Ave., Farmingdale, NY 11735-1718 USA (Tel.: 516-694-9090; Telefax: 516-694-9177; Telex: 6852289)

J.T. Baker, Inc.

600 North Broad St., Phillipsburg, NJ 08865 USA (Tel.: 908-859-2151)

Balchem Corp.

PO Box 175, Slate Hill, NY 10973 USA (Tel.: 914-355-2861; Telefax: 914-355-6314)

Barco Chemical Products Inc.

PO Box 268448, 6900 N. Glenwood Ave., Chicago, IL 60626 USA (Tel.: 312-465-3827)

Barker Industries, Inc.

2841 Old Steele Creek Rd., Charlotte, NC 28208 USA (Tel.: 704-391-1023; Telefax: 704-393-0464)

BASF

BASF AG, Carl-Bosch Str. 38, 67056 Ludwigshafen Germany (Tel.: 49-621-60-99739; Telefax: 49-621-60-93344; Telex: 62157120BASF)

BASF Corp., 3000 Continental Dr. North, Mount Olive, NJ 07828-1234 USA (Tel.: 201-426-2600; 800-367-9861, 201-426-4999)

BASF plc, PO Box 4, Earl Road, Cheadle Hulme, Cheadle, Cheshire, SK8 6QG UK (Tel.: 44 161 485-6222; Telefax: 44 161-486-0891; Telex: 669211 BASFCH G)

BASF Japan Ltd., 3-3, Kioi-cho, Chiyoda-ku, Tokyo, 102 Japan (Tel.: (03) 3238-2300; Telex: 222-2130 BASFTK)

Baychem, Inc.

3200 Moon Station Rd., Kennesaw, GA 30144 USA (Tel.: 404-429-1405; Telefax: 404-425-9141)

Bayer

Bayer AG, Bayerwerk, D-51368 Leverkusen Germany (Tel.: (0214)30-1; Telefax: 0214-30-6 51 36; Telex: 85103-0 byd)

Bayer plc, Stoke Court, Stoke Poges, Slough, Berkshire, SL2 4LY UK (Tel.: 44 1753 645151)

Bayer Antwerpen N.V., Div. of Bayer AG, Haven 507, Scheldelaan 420, B-2040 Antwerp Belgium (Tel.: 32-3540 3011; Telefax: 32-3541 6936; Telex: 71175 BAYANT B)

Bayer Corp./Fibers, Organics & Rubber Div., Bayer Rd., Pittsburgh, PA 15205-9741 USA (Tel.: 412-777-2000; 800-662-2927; Telefax: 412-777-7840; Telex: 1561261)

Bayer Corp./Fibers, Organics & Rubber Div., 2603 W. Market St., Akron, OH 44313 USA (Tel.: 216-836-0451; 800-321-0997; Telefax: 216-838-0200; Telex: 1561261)

Berje Inc.

5 Lawrence St., Bloomfield, NJ 07003 USA (Tel.: 201-748-8980; Telefax: 201-680-9618; Telex: 475-4165A)

Bernel Chemical Co., Inc.

174 Grand Ave., Englewood, NJ 07631 USA (Tel.: 201-569-8934; Telefax: 201-569-1741)

B.I. Chemicals, Inc.

2820 N. Normandy Dr., Petersburg, VA 23805 USA (Tel.: 804-862-1990; 800-BICHEM6; Telefax: 804-862-3246)

Biddle Sawyer Corp.

2 Penn Plaza, New York, NY 10121 USA (Tel.: 212-736-1580; Telefax: 212-239-1089; Telex: 427471BSCE)

BOC Gases

The Priestley Centre, The Surrey Research Park, Guildford, Surrey, GU2 5XY UK (Tel.: 44-1483-579857; Telefax: 44-1483-505211; Telex: 858078)

Boehle Chemicals, Inc.

19306 W. 10 Mile Rd., PO Box 2001, Southfield, MI 48037 USA (Tel.: 313-255-2210; Telefax: 313-255-4474)

Boliden Intertrade Inc.

3379 Peachtree Rd. NE, Suite 300, Atlanta, GA 30326 USA (Tel.: 404-239-6700; 800-241-1912; Telefax: 404-239-6701; Telex: 981036)

BP

BP Chemicals Ltd., Subsid. of The British Petroleum Co. plc, Britannic House, 6th Flr., 1 Finsbury Circus, London, EC2M 7BA UK (Tel.: 44 171 496-4867; Telefax: 44 171-496-4898; Telex: 266883 BPCLBHG)

BP Oil UK Ltd., Div. of The British Petroleum Co. plc, BP House, Breakspear Way, Hemel Hempstead, Hertsfordshire, HP2 4UL UK (Tel.: 44 1442 232323; Telefax: 44 1442-225225; Telex: 827711)

BP Chemicals Inc., 4440 Warrensville Center Rd., Warrensville Hts., OH 44128 USA (Tel.: 216-586-6455; 800-272-4367; Telefax: 216-586-5588; Telex: 6873120, 6873119)

Brand-Nu Laboratories, Inc.

PO Box 895, 30 Maynard St., Meriden, CT 06450 USA (Tel.: 203-235-7989; 800-243-3768; Telefax: 203-235-7163)

Brown Chemical Co., Inc./Industrial and Fine Chems.

302 W. Oakland Ave., Oakland, NJ 07436-0785 USA (Tel.: 201-337-0900)

Browning Chemical Corp.

707 Westchester Ave., White Plains, NY 10604 USA (Tel.: 914-686-0300; 800-771-0008; Telefax: 914-686-0310; Telex: 235039)

Buffalo Color Corp.

959 Rte. 46 East, Suite 403, Parsippany, NJ 07054 USA (Tel.: 201-316-5600; 800-631-0171; Telefax: 201-316-5828; Telex: 7109885924)

Burdick & Jackson, Div. of Baxter Diagnostics Inc.

1953 S. Harvey St., Muskegon, MI 49442 USA (Tel.: 616-726-3171; 800-368-0050; Telefax: 616-728-8226)

Burlington Chemical Co., Inc.

PO Box 111, Burlington, NC 27216 USA (Tel.: 910-584-0111; 800-672-5888; Telefax: 910-584-3548; Telex: 9102502503)

Calgene Chemical Inc.

7247 North Central Park Ave., Skokie, IL 60076-4093 USA (Tel.: 708-675-3950; 800-432-7187; Telefax: 708-675-3013; Telex: 72-4417)

Carbochem, Inc.

551 Lancaster Ave., Haverford, PA 19041 USA (Tel.: 610-527-0808; Telefax: 610-527-6422; Telex: 272746)

Carbonic Industries Corp.

3700 Crestwood Pkwy., Suite 200, Deluth, GA 30136-5583 USA (Tel.: 800-241-5882; Telefax: 404-717-2222)

Cardolite Corp.

500 Doremus Ave., Newark, NJ 07105-4805 USA (Tel.: 201-344-5015; 800-322-7365; Telefax: 201-344-1197; Telex: 325446)

R.E. Carroll, Inc.

1570 N. Olden Ave., PO Box 139, Trenton, NJ 08638 USA (Tel.: 609-695-6211; 800-257-9365; Telefax: 609-695-0102)

CasChem Inc., A Cambrex Co.

40 Ave. A, Bayonne, NJ 07002 USA (Tel.: 201-858-7900; 800-CASCHEM; Telefax: 201-437-2728; Telex: 710-729-4466)

Catawba-Charlab, Inc., Affiliate of CHT

5046 Old Pineville Rd., PO Box 240497, Charlotte, NC 28224 USA (Tel.: 704-523-4242; Telefax: 704-522-8142)

Cedar Concepts Corp.

4342 S. Wolcott Ave., Chicago, IL 60609 USA (Tel.: 312-890-5790; Telex: 297-220 CLTD)

Central Soya

Central Soya Co., Inc./Chemurgy Div., PO Box 2507, 1946 West Cook Rd., Fort Wayne, IN 46801-2507 USA (Tel.: 219-425-5100; 800-348-0960; Telefax: 219-425-5301; Telex: 49609682)

Central Soya Aarhus A/S, Skansevej 2, PO Box 380, DK-8100 Aarhus C Denmark (Tel.: 45 89 31 21 11; Telefax: 45 89 31 21 12; Telex: 64348)

Cerac, Inc.

PO Box 1178, 407 N. 13th St., Milwaukee, WI 53201 USA (Tel.: 414-289-9800; Telefax: 414-289-9805; Telex: RCA 286122)

Charkit Chemical Corp.

PO Box 1725, 330 Post Rd., Darien, CT 06820 USA (Tel.: 203-655-3400; Telefax: 203-655-8643; Telex: 6819184)

Chemax, Inc.

PO Box 6067, Greenville, SC 29606 USA (Tel.: 864-277-7000; 800-334-6234; Telefax: 864-277-7807; Telex: 570412 IPM15SC)

Chemcentral Corp.

7050 W. 71 St., PO Box 730, Bedford Park, IL 60499-0730 USA (Tel.: 708-594-7000; 800-331-6174; Telefax: 708-594-6328)

The Chemical Co.

PO Box 436, 19B Narragansett Ave, Jamestown, RI 02835 USA (Tel.: 401-423-3100; Telefax: 401-423-3102)

Chemie Linz

Chemie Linz GmbH, Div. of Chemie Holding AG, Postfach 296, St Peter-Strasse 25, A-4021 Linz Austria (Tel.: 43-732-59160; Telefax: 43-732-3800; Telex: 221324)

Chemie Linz UK Ltd., Div. of Chemie Linz GmbH, 12 The Green, Richmond, Surrey, TW9 1PX UK (Tel.: 44 181 948 6966; Telefax: 44 181-332 2516; Telex: 924941)

Chemie Linz North America, Inc., 65 Challenger Rd, Ridgefield Park, NJ 07660 USA (Tel.: 201-641-6410; Telefax: 201-641-2323; Telex: 853211 Chemie Linz)

Chemisphere Corp.

2101 Clifton Ave., St. Louis, MO 63139 USA (Tel.: 314-644-1300; Telefax: 314-644-1425)

Chemisphere Ltd.

38 King St., Chester, Cheshire, CH1 2AH UK (Tel.: 44 1244 320878; Telefax: 44 1244 320858; Telex: 61398 CHEMSPR G)

ChemMark Development Inc.

70 Tyler Place, South Plainfield, NJ 07080 USA (Tel.: 908-561-5200; Telefax: 908-561-9174)

Chemol Co.

2410 Randolph Ave., PO Drawer 20687, Greensboro, NC 27406 USA (Tel.: 910-333-3060; 800-849-3050; Telefax: 910-273-4645)

Chemoxy International plc, Div. of Suter plc

All Saints Refinery, Cargo Fleet Rd, Middlesbrough, Cleveland, TS3 6AF UK (Tel.: 44 1642 248555; Telefax: 44 1642 244340; Telex: 587185)

Chemron Corp.

PO Box 2299, Paso Robles, CA 93447 USA (Tel.: 805-239-1550; Telefax: 805-239-8551; Telex: 501532)

Chemtronics Inc.

8125 Cobb Center Dr., Kennesaw, GA 30144 USA (Tel.: 404-424-4888; 800-645-5244; Telefax: 404-423-0748; Telex: 968567)

Chevron

Chevron Chemical Co., 1301 McKinney, PO Box 3766, Houston, TX 77253-3766 USA (Tel.: 713-754-2000; 800-231-3260; Telex: 762799)

Chevron International Oil Co., Inc., PO Box 7146, San Francisco, CA 94120-7146 USA (Tel.: 415-894-5341; 800-344-5650; Telefax: 415-894-1083; Telex: RCA278755 ITT 470074)

Chisso

Chisso Corp., Tokyo Bldg., 7-3, Marunouchi 2-chome, Chiyoda-ku, Tokyo, 100 Japan (Tel.: (03) 3284-8411; Telefax: (03) 3284-8412; Telex: 02225212 CHISSO J)

Chisso America Inc., 1185 Ave of the Americas, New York, NY 10036 USA (Tel.: 212-302-0500; Telefax: 212-302-0643; Telex: WU 147029 CHISSO NYK)

Climax Performance Materials Corp. *See* Actrachem L.P.

CM Chemical Products Inc.

50 Valley Rd., Berkeley Heights, NJ 07922 USA (Tel.: 908-665-8593; 800-654-8059; Telefax: 908-665-8543)

CNC International, L.P.

PO Box 3000, 20 Privilege St., Woonsocket, RI 02895 USA (Tel.: 401-769-6100; Telefax: 401-769-4509)

Coalite Chemicals Div.

PO Box 152, Buttermilk Lane, Bolsover, Chesterfield, Derbyshire, S44 6AZ UK (Tel.: 44 1246 826816; Telefax: 44 1246 240309; Telex: 547624)

Condea

Condea Chemie GmbH, Überseering 40, D-22297 Hamburg Germany (Tel.: 49-40-6375-0; Telefax: 49-40 6375 3588; Telex: 215166 conh d)

Condea Vista Company, 900 Threadneedle, PO Box 19029, Houston, TX 77224-9029 USA (Tel.: 713-588-3000; 800-231-8216; Telefax: 713-588-3236; Telex: 794557)

Condea Augusta SpA, Via Medici del Vascello 26, I-20138 Milano Italy (Tel.: 39-2-520-1; Telefax: 39-2-520-29558)

Condea Vista Japan Inc., PO Box 110, Kasumigaseki Bldg., 3-2-5 Kasamugaseki, Chiyoda-ku, Tokyo, 100 Japan (Tel.: 3593-0611; Telefax: 3593-0615; Telex: J 29368 VISTACHM)

Condea Vista Singapore, 152 Beach Rd. #02-08, Gateway East, Singapore 189721)

Consos, Inc.

PO Box 34186, Charlotte, NC 28234 USA (Tel.: 704-596-2813; Telefax: 704-596-4861)

Cookson Specialty Additives

1000 Wayside Rd., Cleveland, OH 44110 USA (Tel.: 216-531-6010; 800-321-4236; Telefax: 216-486-6638)

Coulton Chemical Corp.

6600 Sylvania Ave., Sylvania, OH 43560 USA (Tel.: 419-885-4661)

Coyne Chemical

3015 State Rd., Croydon, PA 19021 USA (Tel.: 215-785-3000; Telefax: 215-785-1585)

Coz Corp., Allied Products Corp.

Providence Rd., Northbridge, MA 01534 USA (Tel.: 508-234-6151; Telefax: 508-234-4087)

CPS Chemical Co. Inc.

PO Box 162, Old Water Works Rd., Old Bridge, NJ 08857 USA (Tel.: 908-607-2700; Telefax: 908-607-2562; Telex: 844532-CPSOLDB)

Croda

Croda Chemicals Ltd., Div. of Croda International plc, Cowick Hall, Snaith, Goole, North Humberside, DN14 9AA UK (Tel.: 44 1405 860551; Telefax: 44 1405 862253; Telex: 57601)

Croda Food Products Ltd., Div. of Croda International plc, Cowick Hall, Snaith, Goole, North Humberside, DN14 9AA UK (Tel.: 44 1405 860551; Telefax: 44 1405 860205; Telex: 57601)

Croda Surfactants Ltd., Cowick Hall, Snaith, Goole, North Humberside, DN14 9AA UK (Tel.: 44 1405 860551; Telefax: 44 1405 860205; Telex: 57601)

Croda Universal Ltd., Div. of Croda International plc, Cowick Hall, Snaith, Goole, North Humberside, DN14 9AA UK (Tel.: 44 1405 860551; Telefax: 44 1405 860205; Telex: 57601)

Croda Inc., 7 Century Dr., Parsippany, NJ 07054-4698 USA (Tel.: 201-644-4900; Telefax: 201-644-9222; Telex: 57601)

Croda Japan K.K., Aceman Bldg., 1-10, Tokui-cho 1-chome, Chuo-ku, Osaka, 540 Japan (Tel.: (06) 942-1791; Telefax: (06) 942-1790; Telex: 5233117)

Crompton & Knowles Corp./Dyes & Chems. Div.

PO Box 33157, Charlotte, NC 28233 USA (Tel.: 704-372-5890; 800-323-4383; Telefax: 704-332-8785)

Crosfield

Crosfield Co., Member of the Unilever Specialty Chemicals group, 101 Ingalls Ave., Joliet, IL 60435 USA (Tel.: 815-727-3651; 800-727-3651; Telefax: 815-727-5312)

Crosfield Brazil, Ave Maria Coehlo de Aguiar 215, Bloco C 3 andar, CEP 05804 Sao Paulo Brazil (Tel.: (55) 11 545 1481; Telefax: (55) 11 545 2226; Telex: 1155026 GELE BR)

Crosfield Group, Div. of Unilever plc, PO Box 26, Warrington, Cheshire, WA5 1AB UK (Tel.: 44 1925 416100; Telefax: 44 1925 59828; Telex: 627067)

Crosfield SpA, Via dei Cipressi, 10, 37033, Montorio Verona Italy (Tel.: (39) 45 557159; Telefax: (39) 45 884 0099)

Crosfield B.V., Ir. Rocourstraat 28, 6245 ZG Eijsden The Netherlands (Tel.: 31 4409-9333; Telefax: 31 4409-3995; Telex: 56884 CROSF NL)

Crosfield Singapore, 320 Jalan Boon Lay, 2261 Singapore (Tel.: (65) 261 3490; Telefax: (65) 261 3254; Telex: LEVSIN RS 24757)

Cross World Sales Corp.

1891 Main St., Mohegan Lake, NY 10547 USA (Tel.: 914-528-9142; Telefax: 914-528-9205)

Crowley

Crowley Chemical Co., 261 Madison Ave., New York, NY 10016 USA (Tel.: 212-682-1200; Telefax: 212-953-3487; Telex: 12-7662)

Crowley Tar Products Co., Inc., 261 Madison Ave., New York, NY 10016 USA (Tel.: 212-682-1200)

CTC Organics

792 Windsor St. SW, Atlanta, GA 30315 USA (Tel.: 404-524-6744; Telefax: 404-577-1651)

Cytec

Cytec Industries Inc., A Business Unit of American Cyanamid Co., Five Garret Mountain Plaza, West Paterson, NJ 07424 USA (Tel.: 201-357-3100; 800-438-5615; Telefax: 201-357-3065; Telex: 130400)

Cytec Industries UK Ltd., Bowling Park Dr., Bradford, West Yorkshire, BD4 7TT UK (Tel.: 44 1274 733891; Telefax: 44-1274 734770; Telex: 51295)

Cytec Japan Ltd., No. 25 Mori Bldg., 23rd Floor, 1-4-30 Roppongi, Minato-ku, Tokyo, 106 Japan (Tel.: 81-3-3586-9716; Telefax: 81-3-3586-9732; Telex: 22439 CYANAMID J)

Daicel

Daicel Chemical Industries, Ltd., Toranomon Mitsui Bldg., 8-1 Kasumigaseki, 3-Chome, Chiyoda-ku, Tokyo, 100 Japan (Tel.: (03)507-3203; Telefax: (03)507-3198; Telex: 2224632 DAICEL J)

Daicel (USA) Inc., Subsid. of Diacel Chem. Industries, Ltd., One Parker Plaza, 400 Kelby St., Fort Lee, NJ 07024 USA (Tel.: 201-461-4466; Telefax: 201-461-2776)

Daicel (Europa) GmbH, Subsid. of Daicel Chem. Industries Ltd., Ost St. 22, 40211 Düsseldorf Germany (Tel.: (211)369848; Telefax: (211)364429; Telex: (41)8588042 DCELD)

Daihachi Chemical Industry Co., Ltd.

Sanyo Nissei Kawaramachi Bldg., 2-7, Kawaramachi 2-chome, Chuo-ku, Osaka, 541 Japan (Tel.: (06) 201-1455; Telefax: (06) 201-1458)

The Degen Co.

200 Kellogg St., PO Box 5240, Jersey City, NJ 07305 USA (Tel.: 201-432-1192; Telefax: 201-432-8483; Telex: 325999 DEGEN CO)

Degussa

Degussa AG, Postfach 110533, Weossfrauenstrasse 9, D-60287 Frankfurt am Main Germany (Tel.: +49 69-218-01; Telefax: +49 69-218-3218; Telex: 415200-25 dwd)

Degussa Corp., Wholly owned subsid. of Degussa AG, 65 Challenger Rd., Ridgefield Park, NJ 07660 USA (Tel.: 201-807-3224; 800-237-6745; Telefax: 201-807-3111; Telex: 221420 degus ur)

Jan Dekker B.V.

Postbus 10, NL-1520 AA Wormerveer The Netherlands (Tel.: 31-75-2782 78; Telefax: 31-75-21 38 83; Telex: 19273)

The Dial Corp., A Greyhound Dial Co.

2000 Aucutt Rd., Montgomery, IL 60538 USA (Tel.: 708-892-4381; 800-323-5385)

Diamine and Chemicals Ltd.

1 National Chambers, 2nd Flr., Near Dena Bank, Ashram Rd., Ahmedabad, 380 009 India (Tel.: 91-79-409626; Telefax: 91-79-409552)

Dixie Chem Co., Inc.

PO Box 130410, 300 Jackson Hill, Houston, TX 77219 USA (Tel.: 713-863-1947; Telefax: 713-863-8316; Telex: 910-18815089)

Dover Chemical Corp., Subsid. of ICC Industries Inc.

3676 Davis Rd. N.W., PO Box 40, Dover, OH 44622 USA (Tel.: 216-343-7711; 800-321-8805/6; Telefax: 216-364-1579; Telex: 983466)

Dow

Dow Chemical North America, 2040 Willard H. Dow Center, Midland, MI 48674 USA (Tel.: 517-636-1000; 800-441-4DOW; Telefax: 517-636-9752; Telex: 227455)

Dow Chemical Canada Inc., PO Box 1012, 1086 Modeland Rd., Sarnia, Ontario, N7T 7K7 Canada (Tel.: 519-339-3131; 800-441-4369)

Dow Quimica Mexicana, S.A. de C.V., Av. Paseo de las Palmas 555-3, Lomas de Chapultepec Mexico (11000 D.F. Mexico (Tel.: 525-227-1900)

Dow Europe S.A., Bachtobelstrasse 3, CH-8810 Horgen Switzerland (Tel.: 41-1-728-2095; Telefax: 41-1-728-3081; Telex: 826940)

Dow Corning Corp.

PO Box 0994, Midland, MI 48686-0994 USA (Tel.: 517-496-4000; 800-248-2481; Telefax: 517-496-4586; Telex: 227450)

DSM Engineering Plastics, N. Am. Headquarters

PO Box 3333, 2267 West Mill Road, Evansville, IN 47732-3333 USA (Tel.: 812-435-7500; 800-333-4237; Telefax: 812-435-7702)

DuPont

DuPont Chemicals, 1007 Market St., Wilmington, DE 19898 USA (Tel.: 800-441-7515)

DuPont Nylon, Barley Mill Plaza, PO Box 80025, Wilmington, DE 19880-0025 USA (Tel.: 302-999-3213; 800-231-0998; Telefax: 302-999-3441)

DuPont Canada Inc., PO Box 2200, Streetsville Postal Station, Mississauga, Ontario, L5M 2H3 Canada (Tel.: 416-821-3300; 800-6686942)

DuPont France, 137 rue de L'Université, F-75334 Paris Cedex 07 France (Tel.: 33-45 50 63 32; Telefax: 33-47 53 09 65; Telex: 206772)

DuPont (Germany), DuPont Strasse 1, D-61343 Bad Hamburg v.d.H. Germany (Tel.: 49 61 72 87 1313; Telefax: 49 61 72 87 1314)

DuPont de Nemours International S.A., 2, chemin du Pavillon, PO Box 50, CH-1218 Le Grand-Saconnex Geneva Switzerland (Tel.: 41-22-717 51 11; Telex: 415 777 DUP CH)

DuPont (Australia) Ltd., Northside Gardens, 168 Walker St., PO Box 930, North Sydney, NSW, 2060 Australia (Tel.: 011-612 923-6111)

DuPont Asia Pacific, Ltd., 1122 New World Office Bldg., East Wing, Salisbury Rd., Kowloon Hong Kong (Tel.: 852-734-5345; Telefax: 852-724-4458)

DuPont Singapore Pte., Ltd., Maritime Square, #07-01World Trade Center, Singapore 0409 (Tel.: 011-65-273-2244)

DuPont China Ltd., 1122 New World Office Bldg., E. Wing, Salisbury Rd., Kowloon Hong Kong (Tel.: 011-852-734-5345)

Dynaloy, Inc.

7 Great Meadow Lane, Hanover, NJ 07936 USA (Tel.: 201-887-9270; Telefax: 201-887-3678; Telex: 642033)

Eastern Chemical, Div. of Admiral Spec. Prods., Subsid. of United Guardian

PO Box 2500, Smithtown, NY 11787 USA (Tel.: 516-273-0900; 800-645-5566; Telefax: 516-273-0858; Telex: 4974275 GCC HAUP)

Eastern Color & Chemical Co.

35 Livingston St., PO Box 6161, Providence, RI 02904 USA (Tel.: 401-331-9000; Telefax: 401-331-2155)

Eastman Chemical Products, Inc.

PO Box 431, Kingsport, TN 37662-5280 USA (Tel.: 423-229-2000; 800-EASTMAN; Telefax: 423-229-1196; Telex: 6715569)

Electron Microscopy Sciences Corp.

P.O. Box 251, 321 Morris Rd., Ft. Washington, PA 19034 USA (Tel.: 215-646-1566; 800-523-5874; Telefax: 215-646-8931)

Elf Atochem

Elf Atochem S.A., 4, cours Michelet, La Défense 10, F-92091 Paris Cedex 42 France (Tel.: 49-00-8080; Telefax: 49-00-7447; Telex: 611922 ATO F)

Elf Atochem UK Ltd., Colthrop Lane, Thatcham, Newbury, Berkshire, RG13 4LW UK (Tel.: 44 1635 870-000; Telefax: 44 1635-861212; Telex: 847689 ATOKEM G)

Elf Atochem North America Inc., Headquarters, 2000 Market St., Philadelphia, PA 19103-3222 USA (Tel.: 215-419-7000; 800-628-4453; Telefax: 215-419-7875)

Elf Atochem Canada, PO Box 278, Oakville, Ontario, L6J 5A3 Canada (Tel.: 905-827-9841; Telefax: 905-827-7913)

EM Industries, Inc./Fine Chems. Div., An Associate of E. Merck

5 Skyline Dr., Hawthorne, NY 10532 USA (Tel.: 914-592-4660; Telefax: 914-592-9469)

Emkay Chemical Co.

319-325 Second St., PO Box 6537, Elizabeth, NJ 07206 USA (Tel.: 908-352-7053; 800-EMKAY CHEM; Telefax: 908-352-6398)

EniChem America, Inc.

2000 West Loop South, Suite 2010, Houston, TX 77027 USA (Tel.: 713-940-0700; 800-441-3646; Telefax: 713-940-0724)

Esprit Chemical Co.

Palm Tower Office Bldg., 1343 Main St., Suite 700, Sarasota, FL 34236 USA (Tel.: 813-953-2000; 800-608-2888; Telefax: 813-954-0871)

Exxon

Exxon Chemical Co., PO Box 3272, Houston, TX 77253-3272 USA (Tel.: 713-870-6000; 800-526-0749; Telefax: 713-870-6661; Telex: 794588)

Exxon Chemical Geopolymers Ltd., Div. of Exxon Corp., PO Box 122, 4600 Parkway, Solent Business Park, Whiteley, Fareham, Hampshire, PO15 7AZ UK (Tel.: 44 1489 884400; Telefax: 44 1489 884403; Telex: 47437)

Exxon Chemical Europe Inc., Blvd du Souverain 280, B-1160 Bruxelles Belgium (Tel.: 32-2-76931; Telefax: 32-2-7693225; Telex: 24733)

Imperial Oil, Chemicals Div., Sales Office for Exxon, 111 St. Clair Ave. West, Toronto, Ontario, M5W 1K3 Canada (Tel.: 416-968-4046; Telex: 06986899)

Fabrichem Inc.

211 Sigwin Dr., Fairfield, CT 06430 USA (Tel.: 203-259-5512; Telefax: 203-254-7886)

Fabriquimica S.R.L.

Calle 32 No. 3313, San Martin 1650, Argentina (Tel.: 1-755-7290; Telefax: 54-1-755-7290)

The Fanning Corp.

2450 W. Hubbard St., Chicago, IL 60612-1408 USA (Tel.: 312-563-1234; Telefax: 312-563-0087)

Farleyway Chemicals Ltd.

Ham Lane, , Kingswinford, West Midlands, DY6 7JU UK (Tel.: 44 1384 400 222; Telefax: 44 1384 400 020; Telex: 339528 FAR G)

Ferro

Ferro Corp./World Headquarters, 1000 Lakeside Ave., Cleveland, OH 44114-1183 USA (Tel.: 216-641-8580; Telefax: 216-696-6958; Telex: 98-0165)

Ferro Corp./Bedford Chemical Div., 7050 Krick Rd., Bedford, OH 44146 USA (Tel.: 216-641-8580; 800-321-9946; Telefax: 216-439-7686; Telex: 98-165)

Ferro Corp./Grant Chemical Div., PO Box 263, Baton Rouge, LA 70821 USA (Tel.: 504-654-6801; Telefax: 504-654-3268; Telex: 980165)

Ferro Corp./Keil Chemical Div., 3000 Sheffield Ave., Hammond, IN 46320 USA (Tel.: 219-931-2630; 800-628-9079; Telefax: 219-931-0895; Telex: 725484)

Fina Chemicals, Div. of Petrofina S.A.

52 Rue de l'Industrie, B-1040 Brussels Belgium (Tel.: 32 2-288 9132; Telefax: 32-2-288-3322; Telex: 21 556 PFINA B)

Finetex Inc.

418 Falmouth Ave., PO Box 216, Elmwood Park, NJ 07407 USA (Tel.: 201-797-4686; Telefax: 201-797-6558; Telex: 710-988-2239)

Florida Chemical Co., Inc.

351 Winter Haven Blvd., N.E., Winter Haven, FL 33881 USA (Tel.: 813-294-8483; Telefax: 813-294-7783)

Fluka

Fluka Chemical Corp., 980 South Second St., Ronkonkoma, NY 11779 USA (Tel.: 516-467-0980; 800-FLUKA-US; Telefax: 800-441-8841; Telex: 96-7807)

Fluka Sarl, L'isle D'Abeau Che, snes BP 705, St. Quentin Fallavier, Cedex 38297 France (Tel.: 74 82 28 00; Telefax: 74 95 68 08)

Fluka Chemie AG, Industriestrasse 25, CH-9470 Buchs Switzerland (Tel.: 081 755 2511; Telefax: 081 756 5449; Telex: 855 282)

Fluka Chemical Ltd., The Old Brickyard, New Road, Gillingham, Dorset, SP8 4JL UK (Tel.: 44 1747 823097; Telefax: 44 1747 824596; Telex: 417238)

FMC

FMC Corp./Chemical Products Group, 1735 Market St., Philadelphia, PA 19103 USA (Tel.: 215-299-6000; 800-346-5101; Telefax: 215-299-5999; Telex: 685-1326)

FMC Corp. Canada, 625 Howe St., Suite 1200, Vancouver, B.C., V6C 2T6 Canada (Tel.: 604-685-6508; Telefax: 604-689-5917)

FMC Corp. N.V., Ave. Louise 480-B9, 1050 Brussels Belgium (Tel.: 32-2-645 9211; Telefax: 32-2-640 0564; Telex: 23063 FMCEUR B)

Focus Chemical Corp.

875 Greenland Rd., Orchard Park, Suite B9, Portsmouth, NH 03801 USA (Tel.: 603-430-9802; Telefax: 603-431-5390)

Freedom Textile Chemicals Co.

801 Washington St., Conshohocken, PA 19428 USA (Tel.: 610-828-3800; 800-533-4514; Telefax: 610-834-7855; Telex: 5106608845)

Garratt Callahan Company

111 Rollins Rd., Millbrae, CA 94030, 415-697-5811)

Gattefosse

Gattefosse S.A., 36 Chemin de Genas, BP 603, 69804 Saint Priest, Cedex France (Tel.: 33 78 90 63 11; Telefax: 33 78 90 45 67; Telex: 340 240 F)

Gattefosse Corp., 372 Kinderkamack Rd., Westwood, NJ 07675 USA (Tel.: 201-358-1700; Telefax: 201-358-4050)

Gaylord Chemical Co.

PO Box 1209, 106 Galeria Blvd., Slidell, LA 70459-1209 USA (Tel.: 504-649-5464; 800-426-6620; Telefax: 504-649-0068; Telex: 901 4748663)

Gelest Inc.

612 William Leigh Dr., Tullytown, PA 19007-6308 USA (Tel.: 215-547-1015; Telefax: 215-547-2484)

General Chemical

General Chemical Corp., 90 East Halsey Rd., Parsippany, NJ 07054-0373 USA (Tel.: 201-515-0900; 800-631-8050; Telefax: 201-515-2468; Telex: 139-450 GEN-CHEMPAPY)

General Chemical Canada Ltd., 201 City Centre Dr., Mississauga, Ontario, L5B 3A3 Canada (Tel.: 416-896-9595; 800-668-0433; Telefax: 416-276-6594)

Geoliquids, Inc.

15 E. Palatine Rd., Suite 109, Prospect Heights, IL 60070 USA (Tel.: 708-215-0938; 800-827-2411; Telefax: 708-215-9821)

Georgia Gulf Corp.

PO Box 105197, Atlanta, GA 30348 USA

Georgia-Pacific

Georgia-Pacific Chemical Div., 1754 Thorne Rd., Tacoma, WA 98421 USA (Tel.: 206-572-8181; Telefax: 206-572-4721)

Georgia-Pacific Resins Inc., Subsid. of Georgia-Pacific Corp., 2883 Miller Rd, Decatur, GA 30035 USA (Tel.: 770-593-6859; 800-765-7374; Telefax: 800-395-6885; Telex: 804600)

Gist-brocades

Gist-brocades, 8720 Red Oak Blvd., Suite 401, Charlotte, NC 28217 USA (Tel.: 704-527-9000; 800-438-1361; Telefax: 704-527-8184)

Gist-brocades SpA, Via Milano 42, I-27045 Casteggio Italy (Tel.: 39-383-8931; Telefax: 39-383-805397; Telex: 321197 VINAL I)

Giulini Corp.

105 East Union Ave., Bound Brook, NJ 08805 USA (Tel.: 908-469-6504; Telefax: 908-469-8418; Telex: 700179)

Givaudan

Givaudan-Roure Corp., 100 Delawanna Ave., Clifton, NJ 07014 USA (Tel.: 201-365-8277; Telefax: 201-777-9304; Telex: 219259 givc ur)

Givaudan Iberica S.A., Plá d'en Batlle, Sant Celoni, E-08470 Barcelona Spain (Tel.: 34 3-867 06 00; Telefax: 34 3-867 03 19; Telex: 94000)

Goldschmidt

Goldschmidt AG, Th., Goldschmidtstrasse 100, Postfach 101461, D-45127 Essen Germany (Tel.: 49 201-173-01; Telefax: 49 201-224916; Telex: 857170)

Goldschmidt Ltd., Th., Subsid. of Goldschmidt AG, Tego House, Victoria Road, Ruislip, Middlesex, HA4 0YL UK (Tel.: 44 181 422 7788; Telefax: 44 181-864 8159; Telex: 923146 tego g)

Goldschmidt Chemical Corp., 914 E. Randolph Rd., PO Box 1299, Hopewell, VA 23860 USA (Tel.: 804-541-8658; 800-446-1809; Telefax: 804-541-2783; Telex: 710-958-1350)

Goldschmidt Canada, 2150 Winston Park Dr., Unit 201, Oakville, Ontario, L6H 5V1 Canada (Tel.: 905-829-2233; 800-799-2483; Telefax: 905-829-2575)

Goldschmidt Japan Co. Ltd., Th., Rm. 1113, Shuwa Kioi-cho TBR Bldg. 5-7, Koji-machi, Chiyoda-ku, Tokyo, 102 Japan (Tel.: (033) 234-2831; Telefax: (033) 234-3019; Telex: 2322587 tego j)

Goldschmidt Industrial Chemicals Corp., PO Box 279, McDonald, PA 15057 USA (Tel.: 412-796-1511; 800-426-7273; Telefax: 412-922-6657)

Goodrich

BFGoodrich Specialty Chemicals, 9911 Brecksville Rd., Brecksville, OH 44141-3247 USA (Tel.: 216-447-5000; 800-331-1144; Telefax: 216-447-5770; Telex: 4996831)

Goodrich Canada, 125 Northfield Dr. West, Waterloo, Ontario, N2L 6K4 Canada (Tel.: 519-888-3330; Telefax: 519-888-3337)

Goodrich Holding Corp., Subsid. of BFGoodrich, Av. La Paseo de Las Palmas No. 735-103, Col. Lomas de Barrilaco, CP 11010 Mexico, D.F. Mexico (Tel.: 52-5-520-3199; Telefax: 52-5-520-8542)

BFGoodrich Chemical (UK) Ltd., The Lawn, 100 Lampton Road, Hounslow, Middlesex, TW3 4EB UK (Tel.: 44 181 570 4700; Telefax: 44 181-570 0850)

BFGoodrich Chemical Europe, 742 Rue de Verdun, B-1130 Brussels Belgium (Tel.: 32-2-247-1911; Telefax: 32-2-247-1991)

Grain Processing Corp.

1600 Oregon St., Muscatine, IA 52761 USA (Tel.: 319-264-4265; Telefax: 319-264-4289; Telex: 46-8497)

Great Lakes Chemical Corp.

PO Box 2200, One Great Lakes Blvd., W. Lafayette, IN 47906-0200 USA (Tel.: 317-497-6100; 800-621-9521; Telefax: 317-497-6123)

Great Western Chemical Co.

808 SW 15th Ave., Portland, OR 97205 USA (Tel.: 503-228-2600; Telefax: 503-221-5752; Telex: 910-464-4733)

R.W. Greeff & Co. Inc.

777 West Putnam Ave, Greenwich, CT 06830 USA (Tel.: 203-532-2900; Telefax: 203-532-2980; Telex: 996609)

Grindsted Products

Grindsted Products A/S, Edwin Rahrs Vej 38, DK-8220 Brabrand Denmark (Tel.: 45 86-25-3366; Telefax: 45 86-25-1077; Telex: 64177 gvdan dk)

Grindsted Products, Inc., 201 New Century Pkwy., PO Box 26, Industrial Airport, KS 66031 USA (Tel.: 913-764-8100; 800-255-6837; Telefax: 913-764-5407; Telex: 4-37295)

Grünau GmbH, Chemische Fabrik, A Henkel Group Co.

Postfach 1063, Robert-Hansen-Strasse 1, D-89251 Jllertissen Germany (Tel.: (07303)13-706; Telefax: (07303)13203; Telex: 719114 gruea-d)

Haarmann & Reimer Corp.

PO Box 175, 70 Diamond Road, Springfield, NJ 07081 USA (Tel.: 201-467-5600; 800-422-1559; Telefax: 201-467-3514; Telex: 219134 HAR UR)

C.P. Hall Co.

311 South Wacker Dr., Suite 4700, Chicago, IL 60606-6622 USA (Tel.: 312-554-7400; 800-881-5851; Telefax: 312-554-7499)

7300 South Central Ave., Chicago, IL 60638-0428 USA (Tel.: 708-594-6000; 800-321-8242; Telefax: 708-458-0428)

Howard Hall, Div. of R.W. Greeff & Co. Inc.

777 West Putnam Ave., Greenwich, CT 06830 USA (Tel.: 203-532-2900; Telefax: 203-532-2980; Telex: 681 9012)

Halocarbon Products Corp.

PO Box 661, 887 Kinderkamack Rd., River Edge, NJ 07661 USA (Tel.: 201-262-8899; Telefax: 201-262-0019)

Hampshire

Hampshire Chemical Corp., 55 Hayden Ave., Lexington, MA 02173 USA (Tel.: 617-861-9700; Telefax: 617-863-8043; Telex: 200076 GRLX UR)

Hampshire Chemical Ltd., Northdale House, North Circular Road, London, NW10 7UH UK (Tel.: 44 181 961-9366; Telefax: 44 181 963 0928; Telex: 25139)

Harcros Chemicals Inc./Harcros Organics

PO Box 2930, 5200 Speaker Rd., Kansas City, KS 66110-2930 USA (Tel.: 913-621-7749; Telefax: 913-621-7746; Telex: 477266)

Hardman Inc., A Harcros Chemical Group Co.

600 Cortlandt St., Belleville, NJ 07109 USA (Tel.: 201-751-3000; Telefax: 201-751-8407; Telex: TWX: 710-995-4940)

Harris Specialty Chemicals, PCR, Inc.

P.O. Box 1466, Gainesville, FL 32602 USA (Tel.: 904-376-8246; 800-331-6313; Telefax: 904-371-6246)

Hart Products Corp.

173 Sussex St., Jersey City, NJ 07302 USA (Tel.: 201-433-6632; Telefax: 201-435-7268)

Harwick Chemical Corp.

60 S. Seiberling St., PO Box 9360, Akron, OH 44305-0360 USA (Tel.: 216-798-9300; Telefax: 216-798-0214; Telex: TWX: 810-431-2126)

Hatco Corp.

1020 King George Post Rd., Fords, NJ 08863-0601 USA (Tel.: 908-738-1000; Telefax: 908-738-9385; Telex: 84-4545)

Henkel

Henkel KGaA, Henkelstrasse 67, D-40191 Düsseldorf Germany (Tel.: 49-211-797-3300; Telefax: 49-211-798-9638; Telex: 858170)

Henkel KGaA/Cospha, Postfach 101100, D-40191 Düsseldorf Germany (Tel.: 49-211-797-0; Telefax: 49-211-798-7696; Telex: 85817-129)

Henkel Ltd., Div. of Henkel KG, Henkel House, 292-308 Southbury Road, Enfield, Middlesex, EN1 1TS UK (Tel.: 44 181 804 3343; Telefax: 44 181 443 2777; Telex: 922708 HENKEL G)

Henkel (Ireland) Ltd., Western Industrial Estate, Naas Road, Dublin, 12 Ireland (Tel.: 35 31 4 505 622; Telefax: 35 31 4 503 649)

Henkel Belgium S.A., Div. of Henkel KG, 66 ave du Port, Havenlaan 65, 1210 Brussels Belgium (Tel.: 32-423 17 11; Telefax: 32-428 34 67; Telex: 21294 HENKEL B)

Henkel Iberica s.a., Div. Pulcra, Sector E C/42, Zona Franca, 08040 Barcelona Spain (Tel.: 34 3 290 48 50; Telefax: 34 3 290 48 78; Telex: 98301)

Henkel Corp./Cospha, 300 Brookside Ave., Ambler, PA 19002 USA (Tel.: 215-628-1476; 800-531-0815; Telefax: 215-628-1450; Telex: 125854)

Henkel Corp./Functional Products, 300 Brookside Ave., Ambler, PA 19002 USA (Tel.: 215-628-1583; 800-654-7588; Telefax: 215-628-1155; Telex: 215-685-1092)

Henkel Corp./Organic Products Div., 300 Brookside Ave., Ambler, PA 19002 USA (Tel.: 215-628-1441; 800-634-2436; Telefax: 215-628-1200; Telex: 6851092 amchm uw)

Henkel Corp./Textile Chemicals, 11709 Fruehauf Dr., Charlotte, NC 28241 USA (Tel.: 800-634-2436; Telefax: 704-587-3800)

Henkel Canada Ltd., 2290 Argentia Rd., Mississauga, Ontario, L5N 6H9 Canada (Tel.: 416-542-7588; 800-668-6023; Telefax: 416-542-7566)

Henkel Argentina S.A., Carabelas 2398, 1870 Avellaneda, Casilla de Correo 3496, AR-1000 Buenos Aires Argentina (Tel.: 541 2042056; Telefax: 541 205 33 60; Telex: 22475 hen ar)

Henkel S.A. Indústrias Químicas, Avenida das Nacoes, Unidas 10.989, CEP 04578 Sao Paulo, SP Brazil (Tel.: 55 11 828 2340; Telefax: 55 11 828 2326; Telex: 038-1138417 hbiq br)

Henkel Mexicana S.A. de C.V., Calz. de la Viga S/N, Fracc. Los Laureles en Tulpetlac, Ecatepec de Morelos, C.P. 55090 Mexico (Tel.: 525 787-1899; Telefax: 525 729-9804; Telex: 1762295 henk me)

Henkel Australia Pty. Ltd., 1 Clyde St., Silverwater, NSW, 2141 Australia (Tel.: 61 2 748 4355; Telefax: 61 2 748 3863; Telex: AA 25058)

Henkel Corp./Emery Group, 11501 Northlake Dr., Cincinnati, OH 45249 USA (Tel.: 513-530-7300; 800-543-7370; Telefax: 513-530-7581; Telex: 4333016)

Hercules

Hercules Inc., Hercules Plaza-6205SW, Wilmington, DE 19894-0001 USA (Tel.: 302-594-5000; 800-247-4372; Telefax: 302-594-5400; Telex: 835-479)

Hercules Ltd., Div. of Hercules Inc., 31 London Road, Reigate, Surrey, RH2 9YA UK (Tel.: 44 1737 242434; Telefax: 44 1737-224288; Telex: 25803)

Hercules B.V., 8 Veraartlaan, NL-2288GM Rijswijk The Netherlands (Tel.: 31-70-150-000; Telefax: 31-70-3989893; Telex: 31172)

Heterene Chemical Co., Inc.

PO Box 247, 295 Vreeland Ave., Paterson, NJ 07543 USA (Tel.: 201-278-2000; Telefax: 201-278-7512; Telex: 883358)

Hexcel Corp./Chemical Products Div.

215 N. Centennial St., Zeeland, MI 49464 USA (Tel.: 616-772-2193; Telefax: 616-772-7344; Telex: 226375)

High Point Chemical Corp.

PO Box 2316, 243 Woodbine St., High Point, NC 27261 USA (Tel.: 910-884-2214; 800-727-2214; Telefax: 910-884-5039)

Hoechst

Hoechst AG, Postfach 800320, D-65926 Frankfurt am Main Germany (Tel.: 49 69 305-5753; Telefax: 49 69 316700; Telex: 41234-0 HOD)

Hoechst Celanese/Int'l. Headquarters, Rt. 202-206 North, Somerville, NJ 08876 USA (Tel.: 201-635-2600; 800-235-2637; Telefax: 201-635-4330; Telex: 136346)

Hoechst Japan Ltd./Chemicals Dept., New Hoechst Bldg., 10-16, Akasaka 8-chome, Minato-ku, Tokyo, 107 Japan (Tel.: (03) 3479-5118; Telefax: (03) 3479-6715)

Honeywill & Stein Ltd., Div. of BP Chemicals Ltd.

Times House, Throwley Way, Sutton, Surrey, SM1 4AF UK (Tel.: 44 181 770 7090; Telefax: 44 181-770 7295; Telex: 946560 BPCLGH G)

Houghton Chemical Corp.

PO Box 307, 52 Cambridge St., Allston, MA 02134 USA (Tel.: 617-254-1010; 800-777-2466; Telefax: 617-254-2713)

J.M. Huber Corp./Chemicals Div.

PO Box 310, 907 Revolution St., Havre de Grace, MD 21078 USA (Tel.: 410-939-3500; Telefax: 410-939-7313)

Hüls

Hüls AG, Postfach 1320, D-45764 Marl Germany (Tel.: 49 2365-49-1; Telefax: 49-2365-49-2000; Telex: 829211 HSD)

Hüls (UK) Ltd., Featherstone Rd., Wolverton Mill South, Milton Keynes, Buckinghamshire, MK12 5TB UK (Tel.: 44 1908 226 444; Telefax: 44 1908 224950; Telex: 826500)

Hüls America Inc., 220 Davidson Ave., Somerset, NJ 08873 USA (Tel.: 908-560-6345; 800-631-5275)

Hüls Canada, Inc., 235 Orenda Rd., Brampton, Ontario, L6T 1E6 Canada (Tel.: 905-451-3810; Telefax: 905-451-4469; Telex: 0697557)

Hüls Japan Ltd., Mita Kokusai Bldg., 4-28, Mita 1-cho, Minato-ku, Tokyo, 108 Japan (Tel.: (03) 3455 1981; Telefax: (03) 3453 3233; Telex: 2422288 huels jp j)

The Humphrey Chemical Co. Inc., A Cambrex Co.

45 Devine St., North Haven, CT 06473-0325 USA (Tel.: 203-281-0021; 800-652-3456; Telefax: 203-287-9197; Telex: 994487)

Huntsman

Huntsman Corp., 3040 Post Oak Blvd., Houston, TX 77056 USA (Tel.: 713-235-6000; 800-826-0868; Telefax: 800-831-1782)

Huntsman Chemical Corp., 2000 Eagle Gate Tower, Salt Lake City, UT 84111 USA (Tel.: 801-532-5200; 800-421-2411; Telefax: 801-536-1581)

Huntsman Corp. Canada Inc., 150 Research Lane, Suite 307, Guelph, Ontario, N1G 2T2 Canada (Tel.: 519-824-3280; Telefax: 519-824-4979)

Huntsman de Brasil Participacoes Ltda., R. Heloisa Pamplona, 628, Sao Caetano Do Sul, SP 09520, Sao Paulo Brazil (Tel.: 55 11-442-9264; Telefax: 55 11-441-7216)

Huntsman Corp. Belgium N.V., Woluwe Office Garden, Woluwedal 26, B-1932 Zaventem Belgium (Tel.: 32-2-718-01-20; Telefax: 32-2-718-02-11)

Huntsman International Trading Corp., 350 Orchard Rd., #11-07/10 Shaw House, Singapore 0922 (Tel.: 65 730-0288; Telefax: 65 730-0222)

ICC Industries Inc.

720 Fifth Ave., New York, NY 10019 USA (Tel.: 212-903-1732; Telefax: 212-903-1726; Telex: CCI 7607944)

ICI

ICI plc, Imperial Chemical House, 9 Millbank, London, SW11 3JS UK (Tel.: 44 171 834 4444; Telefax: 44 171 834 2040; Telex: 21324)

ICI Surfactants Ltd. (UK), PO Box 90, Wilton Centre, Middlesbrough, Cleveland, TS90 8JE UK (Tel.: 44 1642 454144; Telefax: 44 1642 437374; Telex: 587461)

ICI Americas, Inc., Subsid. of ICI plc, PO Box 15391, 3411 Silverside Rd., Wilmington, DE 19850 USA (Tel.: 302-887-3000; 800-441-7780; Telefax: 302-887-4320; Telex: 62032112)

ICI Specialty Chemicals, Concord Pike & New Murphy Rd., Wilmington, DE 19897 USA (Tel.: 302-886-3000; 800-822-8215; Telefax: 302-886-2972)

ICI Surfactants Americas, Concord Plaza, 3411 Silverside Rd., PO Box 15391, Wilmington, DE 19850-5391 USA (Tel.: 302-887-3000; 800-822-8215; Telefax: 302-887-3525; Telex: 4945649)

ICI Surfactants, Everslaan 45, B-3078 Everberg Belgium (Tel.: 02-758-9361; Telefax: 02-758-9686; Telex: 26151 ICIEVB B)

ICI Australia Operations Pty. Ltd./ICI Surfactants Australia, Newsom St., Ascot Vale, 3032 Australia (Tel.: 61 3 272 5343; Telefax: 61 3 272 5353; Telex: 30192)

ICI Japan Ltd., Palace Bldg., 1-1-1, Marunouchi, Chiyoda-ku, Tokyo, 100 Japan (Tel.: (03) 3211-3115)

Inolex Chemical Co.

Jackson & Swanson Sts., Philadelphia, PA 19148-3497 USA (Tel.: 215-271-0800; 800-521-9891; Telefax: 215-289-9065; Telex: 834617)

Inspec Group plc

Charleston Ind. Estate, Hythe, Southampton, SO45 3ZG UK (Tel.: 44-1703-894666; Telefax: 44-1703-243113)

Integra Chemical Co.

710 Thomas Ave. SW, Renton, WA 98055 USA (Tel.: 206-277-9244; Telefax: 206-277-9246)

International Chemical Group

PO Box 43800, San Ysidro, CA 92173 USA (Tel.: 619-441-9108; Telefax: 619-441-2739; Telex: 650-552-1658)

International Chemical Inc.

One World Trade Center, Suite 8665, New York, NY 10048 USA (Tel.: 212-912-0655; 800-914-2436; Telefax: 212-912-0669; Telex: 420001 ITT UI)

International Flavors & Fragrances (IFF)

1515 Hwy 36, Union Beach, NJ 07735 USA (Tel.: 908-264-4500; Telefax: 908-888-2595; Telex: 275284)

Ishihara Corp. (USA)

600 Montgomery St., San Francisco, CA 94111 USA (Tel.: 415-421-8207; 800-477-2833; Telefax: 415-397-5403)

ISP

ISP, International Specialty Products, World Headquarters, 1361 Alps Rd., Wayne, NJ 07470-3688 USA (Tel.: 201-628-4000; 800-522-4423; Telefax: 201-628-4117; Telex: 219264)

ISP (Canada) Inc., 1075 The Queensway East, Box 1740, Station B, Mississauga, Ontario, L4Y 4C1 Canada (Tel.: 905-277-0381; Telefax: 905-272-0552; Telex: 06961186)

ISP Europe, 40 Alan Turing Rd., Surrey Research Park, Guildford, Surrey, GU2 5YF UK (Tel.: 44 1483 301757; Telefax: 44 1483 302175; Telex: 859142)

ISP Global Technologies Deutschland GmbH, Rudolf-Diesel-Strasse 25, Postfach 1380, 50203 Frechen Germany (Tel.: 02234 105-0; Telefax: 02234 105-211; Telex: 889931)

ISP (Österreich) GmbH, Belvederegasse 18/1, A-1040 Wien Vienna Austria (Tel.: 43 1 504-76-21; Telefax: 43 1 505-89-44; Telex: 133990)

ISP (Australasia) Pty. Ltd., 73-75 Derby St., Silverwater, N.S.W., 2141 Australia (Tel.: Sydney (02) 648-5177; Telefax: (02) 647-1608; Telex: 73711)

ISP Asia Pacific Pte. Ltd., 200 Cantonment Rd., Hex 06-07 Southpoint, 0208 Singapore (Tel.: (65) 2249406; Telefax: (65) 2260853; Telex: 25071)

ISP (Japan) Ltd., Shinkawa Iwade Bldg. 8F, 26-9, Shinkawa 1-Chome, Chuo-Ku, Tokyo, 104 Japan (Tel.: (03) 3555-1571; Telefax: (03) 3555-1660; Telex: J23568)

ISP Van Dyk, Inc., Member of the ISP Group, 11 William St., Belleville, NJ 07109 USA (Tel.: 201-450-7722; Telefax: 201-751-2047; Telex: 710-995-4928)

Itochu Specialty Chemicals Inc.

50 Main St., White Plains, NY 10606 USA (Tel.: 914-448-7700; Telefax: 9124-448-7748

Ivax Corp.

8800 Northwest 36th St., Miami, FL 33178-2404 USA (Tel.: 305-590-2200; Telefax: 305-590-2252)

Janssen Chimica

Janssen Chimica, Div. of Janssen Pharmaceutica, Janssen Pharmaceuticalaan 3, B-2440 Geel Belgium (Tel.: 14-60 42 00; Telefax: 14-60 42 20; Telex: 34103)

Janssen Chimica, Div. of Janssen Pharmaceutica, 755 Jersey Ave., New Brunswick, NJ 08901 USA (Tel.: 908-214-1300; 800-772-8786; Telefax: 908-220-6553; Telex: 182395 BIOSPECT)

JLM Marketing, Inc.

8675 Hidden River Pkwy., Tampa, FL 33637 USA (Tel.: 813-632-3300; Telefax: 813-632-3301; Telex: 666886 JLM UW)

K3 Corp.

10115 Walker Lake Dr., Great Falls, VA 22066 USA (Tel.: 703-759-9731; Telefax: 703-759-3799)

Kalama Chemical, Inc.

1110 Bank of California Center, Seattle, WA 98164 USA (Tel.: 206-682-7890; 800-742-6147; Telefax: 206-682-1907; Telex: 910-444-2294)

Kanto Chemical Co., Inc.

2-8, Nihonbashi Hon-cho 3-chome, Chuo-ku, Tokyo, 103 Japan (Tel.: (03) 3279-1751; Telefax: (03) 3279-5560; Telex: 2223446 CICA J)

Kao Corp. S.A.

Puig dels Tudons, 10, 08210 Barbera Del Valles, Barcelona Spain (Tel.: 81 3-729-0000; Telefax: 81 3-719 0534; Telex: 59749)

Kaufholz & Co., Inc.

P.O. Box 99, Lansdowne, PA 19050-0099 USA (Tel.: 610-622-0170; Telefax: 610-622-0171)

Kelco International Ltd.

Tadworth Surrey, Waterfield, KT20 5HQ UK (Tel.: 44-1737-377000; Telefax: 44-1737-377100)

Kenrich Petrochemicals, Inc.

140 E. 22nd St., PO Box 32, Bayonne, NJ 07002-0032 USA (Tel.: 201-823-9000; 800-LICA KPI; Telefax: 201-823-0691; Telex: 125023)

Kerr-McGee Chemical Corp.

Kerr-McGee Ctr., PO Box 25861, Oklahoma City, OK 73125 USA (Tel.: 405-270-1313; 800-654-3911; Telefax: 405-270-3123; Telex: 747-128)

Koch Chemicals Ltd.

2 Marshgate Drive, Hertford, Hertfordshire, SG13 7JY UK (Tel.: 44 1992 553781; Telefax: 44 1992 586961; Telex: 817136 KOCHEM G)

Koei Chemical Co., Ltd., Subsid. of Sumitomo Chem. Co. Ltd.

Sumkia Fudosan Yokobori Bldg., 6-17, Koraibashi 4-chome, Chuo-ku, Osaka, 541 Japan (Tel.: (06) 204-1515; Telefax: (06) 204-1510; Telex: 522-7415 KOEI J)

Koster Keunen, Inc.

1021 Echo Lake Rd., PO Box 69, Watertown, CT 06795-0069 USA (Tel.: 860-945-3333; Telefax: 860-945-0330)

Koyo Chemical Co., Ltd.

Iidabashi, Hitown-Bldg., 2-28, Shimomiyabi-cho, Shinjuku-ku, Tokyo, 162 Japan (Tel.: (03) 3268-1717; Telefax: (03) 3268-1723)

Kraft Chemical Co.

1975 N. Hawthorne, Melrose Park, IL 60160 USA (Tel.: 708-345-5200; Telefax: 708-345-4005; Telex: 754268 KRAFT)

Kyowa Hakko USA Inc.

599 Lexington Ave., Suite 2780, New York, NY 10022 USA (Tel.: 212-319-5353; Telefax: 212-421-1283; Telex: 6973541 KHKNYK)

Lanaetex Products, Inc.

151-157 Third Ave., PO Box 52 Station A, Elizabeth, NJ 07206 USA (Tel.: 908-351-9700; Telefax: 908-351-8753; Telex: 3792268 TLAP1)

Langley Smith & Co. Ltd.

36 Spital Square, London, E1 6DY UK (Tel.: 44 171 247 7473; Telefax: 44 171-375 1470; Telex: 883013)

LaRoche Chemicals Inc.

PO Box 1031, Airline Hwy., Baton Rouge, LA 70821-1031 USA (Tel.: 504-356-8406; 800-548-6336; Telefax: 504-356-8405; Telex: 784581)

LaRoche Industries Inc.

1100 Johnson Ferry Rd. NE, Atlanta, GA 30342 USA (Tel.: 404-851-0300; Telefax: 404-851-0476)

Lebermuth Co. Inc.

PO Box 4103, South Bend, IN 46634 USA (Tel.: 219-259-7000; 800-648-1123; Telefax: 219-258-7450; Telex: 276424 BMEN)

Lindau Chems, Inc.

731 Rosewood Dr., Columbia, SC 29201 USA (Tel.: 803-799-6863; Telefax: 803-256-3639)

Lipo Chemicals Inc.

207 19th Ave., Paterson, NJ 07504 USA (Tel.: 201-345-8600; Telefax: 201-345-8365; Telex: 130117)

Liquid Carbonic

810 Jorie Blvd., Oak Brook, IL 60521-2216 USA (Tel.: 708-572-7500

LNP Engineering Plastics Inc.

475 Creamery Way, Exton, PA 19341 USA (Tel.: 610-363-4500; 800-854-8774; Telefax: 610-363-4749; Telex: 4973041)

Lohmann Chemicals, Dr. Paul Lohmann GmbH KG

PO Box 1220, D-31857 Emmerthal Germany (Tel.: 49 5155 63-0; Telefax: 49 5155 63-118; Telex: 92858 lohma d)

Lomas International

7705 N.E. Industrial Blvd., PO Box 10225, Macon, GA 31297 USA (Tel.: 912-784-1900; 800-652-9297; Telefax: 912-784-1745)

Lonza

Lonza Ltd./Fine Chemicals, A member of the A-L Alusuisse-Lonza Group, Münchensteinerstrasse 38, CH-4002 Basle Switzerland (Tel.: 41 61-316 81 11; Telefax: 41 61-316 83 01; Telex: 965960 Ion ch)

Lonza SpA, Via Vittor Pisani, 31, I-20124 Milan Italy (Tel.: (02) 66 99 91; Telefax: (02) 66 98 76 30; Telex: 312431 ftalmi i)

Lonza UK Ltd., Imperial House, Lypiatt Road, Cheltenham, Gloucestershire, GL50 2QJ UK (Tel.: 44 1242 513211; Telefax: 44 1242 222294; Telex: 43152)

Lonza Inc., 17-17 Route 208, Fair Lawn, NJ 07410 USA (Tel.: 201-794-2400; 800-777-1875 (tech.); Telefax: 201-794-2695; Telex: 4754539 LONZAF)

Lonza Japan Ltd., Kyowa Shinkawa Bldg., 8th Fl., 20-8, Shinkawa 2-chome, Chuo-ku, Tokyo, 104 Japan (Tel.: (03) 5566-0612; Telefax: (03) 5566-0619)

Lowenstein Dyes & Cosmetics Inc.

420 Morgan Ave., Brooklyn, NY 11222 USA (Tel.: 718-388-5410; Telefax: 718-387-3806)

Magie Bros. Oil, Div. Pennzoil Products Co.

9101 Fullerton Ave., Franklin Park, IL 60131 USA (Tel.: 708-455-4500; 800-MAGIE 47; Telefax: 708-455-0383)

Mallinckrodt, Inc./Drug and Fine Chemicals Div.

Mallinckrodt & 2nd Street, PO Box 5439, St Louis, MO 63147 USA (Tel.: 314-895-2000; 800-325-8888; Telefax: 314-539-1251)

Marathon Oil Co.

539 S. Main St., Findlay, OH 45840 USA (Tel.: 419-421-2973; Telefax: 419-421-2264; Telex: TRT 196230)

Marcus International, Inc.

616 FM 1960 West, Suite 210, Houston, TX 77090 USA (Tel.: 713-893-8970; 800-487-8970; Telefax: 713-893-2502)

Marine Bioproducts Inc.

333 W 1 St., Boston, MA 02127 USA (Tel.: 617-268-0758)

Marlowe-Van Loan Corp.

PO Box 1851, High Point, NC 27261 USA (Tel.: 910-886-7126; 800-422-4MVL; Telefax: 910-889-6663; Telex: TWX: 510-926-1589)

McIntyre Group Ltd.

24601 Governors Hwy., University Park, IL 60466 USA (Tel.: 708-534-6200; 800-645-6457; Telefax: 708-534-6216)

Meer Corp.

PO Box 9006, 9500 Railroad Ave., N. Bergen, NJ 07047-1206 USA (Tel.: 201-861-9500; Telefax: 201-861-9267; Telex: 219130)

E. Merck

Postfach 4119, Frankfurter Strasse 250, D-64293 Darmstadt Germany (Tel.: 49-6151-72-0; Telefax: 49-6151-72-2000; Telex: 419328-0 em d)

Merichem Co.

4800 Texas Commerce Tower, Houston, TX 77002 USA (Tel.: 713-224-3030; 800-231-3030; Telefax: 713-224-4403)

M. Michel & Co., Inc.

90 Broad St., New York, NY 10004 USA (Tel.: 212-344-3878; Telefax: 212-344-3880; Telex: 421468)

Midwest Grain Products Inc.

1300 Main St., PO Box 130, Atchison, KS 66002 USA (Tel.: 913-367-1480; 800-255-0302; Telefax: 913-367-0192)

Miller-Stephenson Chem. Co., Inc.

George Washington Hwy., Danbury, CT 06810 USA (Tel.: 203-743-4447; 800-992-2424; Telefax: 203-791-8702)

514 Carlingview Dr., Rexdale, Ontario, M9W 5R3 Canada (Tel.: 416-675-3204; 800-323-4621; Telefax: 416-674-2987)

Milliken

Milliken Chemical, A Div. of Milliken & Co., PO Box 1927, M-400, Spartanburg, SC 29304-1927 USA (Tel.: 803-503-2200; 800-345-0372; Telefax: 803-503-2430; Telex: 810-282-2580)

Milliken Europe N.V., 18 Ham, B-9000 Gent Belgium (Tel.: 32 9 265 1084; Telefax: 32 9 265 11 95)

Mitsubishi

Mitsubishi Chemical Corp., Mitsubishi Bldg., 5-2, Marunouchi 2-chome, Chiyoda-ku, Tokyo, 100 Japan (Tel.: (03) 3283-6531; Telefax: (03) 3283-6658; Telex: BISICH J 24901)

Mitsubishi Chemical America, Inc., 81 Main St., Suite 401, White Plains, NY 10601 USA (Tel.: 914-286-3600; Telefax: 914-681-0995; Telex: 233570 MCI UR)

Mitsubishi Chemical Europe GmbH, Niederkasseler Lohweg 8, D-40547 Duesseldorf Germany (Tel.: (0211) 52392-0; Telefax: (0211) 591272; Telex: 8587716 MCI D)

Mitsubishi Gas

Mitsubishi Gas Chemical Co., Inc., Mitsubishi Bldg., 2-5-2, Marunouchi, Chiyoda-ku, Tokyo, 100 Japan (Tel.: (03) 3283-5000; Telefax: (03) 3287-0844; Telex: 222-2624 MGCHO J)

Mitsubishi Gas Chemical Co., Inc., 520 Madison Ave., 9th Floor, New York, NY 10022 USA (Tel.: 212-752-4620; Telefax: 212-758-4012; Telex: 649545 MGC UR NYK)

Mitsubishi Petrochemical Co., Ltd.

Mitsubishi Bldg., 5-2, Marunouchi 2-chome, Chiyoda-ku, Tokyo, 100 Japan (Tel.: (03) 3283-5700; Telefax: (03) 3283-5472; Telex: 222-3172)

Mitsui Petrochemical

Mitsui Petrochemical Industries, Ltd., Kasumigaseki Bldg., 2-5, Kasumigaseki 3-chome, Chiyoda-ku, Tokyo, 100 Japan (Tel.: (03) 3580-3616; Telefax: (03) 3593-0028; Telex: J22984 MIPECA)

Mitsui Petrochemicals (America) Ltd., 250 Park Ave., Suite 950, New York, NY 10017 USA (Tel.: 212-682-2366; Telefax: 212-490-6694; Telex: 7105814089)

Mitsui Toatsu

Mitsui Toatsu Chemicals, Inc., Kasumigaseki Bldg., 2-5, Kasumigaseki 3-chome, Chiyoda-ku, Tokyo, 100 Japan (Tel.: (03) 3592-4111; Telefax: (03) 3592-4267; Telex: 2223622 MTCHEM J)

Mitsui Toatsu Chemicals, Inc., MTC America Inc., 2500 Westchester Ave., Suite 110, Purchase, NY 10577 USA (Tel.: 914-253-0777; Telefax: 914-253-0790)

Mobil Chemical Co.

PO Box 3029, Edison, NJ 08818-3029 USA (Tel.: 908-321-6000)

Mona Industries Inc.

PO Box 425, 76 E. 24th St., Paterson, NJ 07544 USA (Tel.: 201-345-8220; 800-553-6662; Telefax: 201-345-3527; Telex: 130308)

Monomer-Polymer & Dajac Labs, Inc.

1675 Bustleton Pike, Feasterville, PA 19053 USA (Tel.: 215-364-1155; Telefax: 215-364-1583)

Monsanto

Monsanto Chemical Co., A Unit of Monsanto Company, 800 N. Lindbergh Blvd., St. Louis, MO 63167 USA (Tel.: 314-694-1000; 800-325-4330; Telefax: 314-694-7625; Telex: 650 397 7820)

Monsanto Canada Inc., 2330 Argentia Rd., Box 787, Streetsville Postal Station, Mississauga, Ontario, L5M 2G4 Canada (Tel.: 905-826-9222; Telefax: 905-826-3119)

Monsanto Commercial S.A. de C.V., Bosques de Durazno 61, 3R Piso, Bosques de las Lomas, Mexico City, DF 11700 Mexico (Tel.: 52-5-251-2715; Telefax: 52-5-251-7923)

Monsanto Europe S.A., Ave. de Tervuren 270-272, PO Box 1, B-1150 Brussels Belgium (Tel.: 32-2-761-41-11; Telefax: 32-2-776-40-40; Telex: 62927 Mesab)

Monsanto Australia Ltd., 600 St. Kilda Rd., 12th Floor, Melbourne, Victoria, 3004 Australia (Tel.: 61-3-522-7122; Telefax: 61-3-525-2253)

Monsanto Japan Ltd., 31F, ARK Mori Bldg., 1-12-32, Akasaka, Minato-ku, Tokyo, 107 Japan (Tel.: 81-3 5562-2600; Telefax: 81-3 5562-2601; Telex: J22614)

Montedipe SpA

Via Rosellini 15-17, I-20124 Milan Italy (Tel.: 39 2-63331; Telefax: 39 2-63338943; Telex: 310679)

Morflex

Morflex, Inc., 2110 High Point Rd., Greensboro, NC 27403 USA (Tel.: 910-292-1781; Telefax: 910-292-9961; Telex: 910240 7846)

Reilly Chemicals S.A., Rue Defacqz 115, Bte 19, B-1060 Bruxelles Belgium (Tel.: 32-2-543-7500; Telefax: 32-2 537 1208)

Morton International

Morton International Inc., 100 North Riverside Plaza, Chicago, IL 60606-1598 USA (Tel.: 312-807-2000; Telefax: 312-807-3150; Telex: 353-645)

Morton International, Inc./Specialty Chemicals Group, 2000 West St., Cincinnati, OH 45215 USA (Tel.: 513-733-2100; Telefax: 513-733-2133)

Morton International Ltd./Specialty Chem., Ind. Chem. & Addit., 7900-A Taschereau Blvd., Suite 106, Brossard, Quebec, J4X 1C2 Canada (Tel.: 514-466-7764; Telefax: 514-466-7771)

Morton International Ltd., Westward House, 155-157 Staines Rd., Hounslow, Middlesex, TW3 3JB UK (Tel.: 44 181 570 7766; Telefax: 44 181-570 6943; Telex: 262002)

Morton International S.A., Chaussee de la Hulpe 130, Boite 5, B-1050 Brussels Belgium (Tel.: 32-2-790211; Telefax: 32-2-790250; Telex: 23708)

Mosselman N.V./S.A.

Blvd. Industriel 80, 1070 Brussels Belgium (Tel.: 32-2-524 18 78; Telefax: 32-2-5200158; Telex: 23533 MOSS B)

National Diagnostics, Inc.

305 Patton Dr., Atlanta, GA 30336 USA (Tel.: 404-699-2121; 800-526-3867; Telefax: 404-699-2077)

Natural Gas Odorizing, Inc.

PO Box 1429, Baytown, TX 77522-1429 USA (Tel.: 713-424-5568; Telefax: 713-424-3681)

Nayler Chemicals Ltd.

Old Rolling Mill, Edge Green Rd, Ashton in Makerfield, Wigan, Lancashire, WN4 8YA UK (Tel.: 44 1942 720988; Telefax: 44 1942 271251; Telex: 669367 BREEZE G)

Nepera, Inc., A Cambrex Co.

Route 17, Harriman, NY 10926 USA (Tel.: 914-782-1202; Telefax: 914-782-2418; Telex: 510-249-4847)

New Japan Chemical Co., Ltd.

1-8, Bingo-machi 2-chome, Chuo-ku, Osaka, 541 Japan (Tel.: (06) 202-0621; Telefax: (06) 222-0062; Telex: 522-4703)

Nihon Kagaku Sangyo Co., Ltd.

20-5, Shitaya 2-chome, Taito-ku, Tokyo, 110 Japan (Tel.: (03) 3876-3131; Telefax: (03) 3876-3278; Telex: J 28318 NIKKASAN)

Nikko Chemicals

Nikko Chemicals Co., Ltd., 4-8, Nihonbashi, Bakurocho 1-chome, Chuo-ku, Tokyo, 103 Japan (Tel.: 81 3 3661-1677; Telefax: 81 3 3664-8620; Telex: 2522744 NIKKOL J)

Barnet Products Corp., Distributor for Nikko Chemicals Co., Ltd., 560 Sylvan Ave., Englewood Cliffs, NJ 07632 USA (Tel.: 201-569-6622; Telefax: 201-569-8847)

Nippon Nyukazai Co., Ltd., Subsid. of Sankyo Co., Ltd.

Yoshizawa Bldg., 9-19, Ginza 3-chome, Chuo-ku, Tokyo, 104 Japan (Tel.: 81-3 3543-8571; Telefax: 81 3 3546-3174)

Nissan Chemical

Nissan Chemical Industries, Ltd., Kowa-Hitotsubashi Bldg., 3-7-1, Kanda-Nishiki-cho, Chiyoda-ku, Tokyo, 101 Japan (Tel.: (03) 3296-8065; Telefax: (03) 3296-8360; Telex: 222-3071)

Nissan Chemical America Corp., 303 South Broadway, Tarrytown, NY 10591 USA (Tel.: 914-332-4745; Telefax: 914-332-4808)

Noah Chemical, Div. of Noah Technologies Corp.

7001 Fairgrounds Pkwy, San Antonio, TX 78238 USA (Tel.: 512-680-9000; Telefax: 512-521-3323)

Norman, Fox & Co.

5511 S. Boyle Ave., PO Box 58727, Vernon, CA 90058 USA (Tel.: 213-583-0016; 800-632-1777; Telefax: 213-583-9769)

Norsk Hydro AS

Bygdoyalle 2, N 0240 Oslo 2 Norway (Tel.: 472 243 2100; Telefax: 472 243 2725; Telex: 78350 HYDRO N)

Northwestern Flavors, Inc.

120 N. Aurora St., W. Chicago, IL 60185 USA (Tel.: 708-231-6111; Telefax: 708-231-1302)

Novachem Corp.

PO Box 6379, High Point, NC 27262 USA (Tel.: 910-885-0041; Telefax: 910-885-4964)

NOVA Molecular Technologies, Inc.

1 Parker Place, Suite 495, Janesville, WI 53545 USA (Tel.: 608-754-NOVA; Telefax: 608-754-6878)

O'Brien Industries, Inc., A Zinkan Enterprises Co.

10574 Ravenna Rd., Twinsburg, OH 44087 USA (Tel.: 216-487-1500; Telefax: 216-425-8202)

Occidental

Occidental Chemical Corp., 5005 LBJ Freeway, Dallas, TX 75244 USA (Tel.: 214-404-3800; 800-752-5151; Telefax: 214-404-3669; Telex: 229835)

Occidental Chemical Corp., PO Box 344, Niagara Falls, NY 14302-0344 USA (Tel.: 800-733-1165; Telefax: 716-278-7297)

Occidental Chemical Europe S.A., Holidaystraat 5, B-1831 Diegem Belgium (Tel.: 32 725 44 50; Telefax: 32 725 46 76; Telex: 23046)

Occidental Chemical Corp. (Australia Pty. Ltd.), Suite 2, 16th Floor, 275 Alfred St., N. Sidney, NSW, 2060 Australia (Tel.: 61-29-957-6722; Telefax: 61-29-957-1548)

Occidental Chemical Asia Ltd., Toranomon 34 Mori Bldg., 9th Floor, 25-5 Toranomon 1-Chome, Minato-ku, Tokyo, 105 Japan (Tel.: (03) 3502-4651; Telefax: (03) 3502-4640)

Olin

Olin Corp., 120 Long Ridge Rd., PO Box 1355, Stamford, CT 06904 USA (Tel.: 203-356-2000; 800-462-6546; Telefax: 203-271-4060; Telex: 420202)

Olin Chemicals, 501 Merritt 7, PO Box 4500, Norwalk, CT 06856-4500 USA (Tel.: 203-750-3429; 800-462-6546; Telefax: 203-270-3752)

Olin Quimica, S.A. de C.V., Campos Eliseos No. 385, Piso 9, Torre A, Col. Polanco, Delg. Miguel Hidalgo, 11560 Mexico (Tel.: (52-5)281-2045; Telefax: (52-5)281-2037)

Olin Brasil Ltda., Nacoes Unidas, 11.857, 12th Flr., Sao Paulo, 04578-000 Brazil (Tel.: 55-11-505-0382; Telefax: 55-11 505-1950; Telex: 11-25034)

Olin S.A., ZAC Paris Nord II, 209 Ave. des Nations, B.P. 60019, F-95970 Roissy CDG Cedex France (Tel.: (33-1) 48 63 21 66; Telefax: (33-1) 48 63 22 25; Telex: 650769)

Olin Australia Ltd., PO Box 141, Level 2, Suite 1, 601 Pacific Hwy., St. Leonards, N.S.W., 2065 Australia (Tel.: 61-2-906-4455; Telefax: 61-2-439-4198; Telex: 26328)

Olin Far East Ltd., Korea, 80-6 Soosong-dong, Suktan Bldg., Chongro-ku, Seoul, 110-140, Korea (Tel.: (82-2)737-2840; Telefax: (82-2) 730-7387)

Olin Far East Ltd., Taiwan, 7F-2 No. 137, Fu Shing S. Road Sec. 1, Taipei, Taiwan, R.O.C. (Tel.: (886-2) 752-4413; Telefax: (886-2) 741-2113)

Olin Industrial H.K. Ltd., 111 Peninsula Centre, 67 Mody Rd., Tsim Sha Tsui-East, Kowloon Hong Kong (Tel.: (852) 366-8303; Telefax: (852) 367-1309)

Olin Japan Inc., Shiozaki Bldg., 7-1 Hirakawa-Cho 2-Chome, Chiyoda-ku, Tokyo, 102 Japan (Tel.: (813) 3263-4615; Telex: (81-3) 3264-2750)

Olin Pte., Ltd., 501 Orchard Rd., 08-03 Lane Crawford Rd., Singapore 0923 (Tel.: (65) 735-1268; Telefax: (65) 735-1298)

Olin Pty. Ltd., PO Box 114, Bergvlei, 2012 Rep. of S. Africa (Tel.: (27-11)444-2244; Telefax: (27-11)444-2241)

OM Group

2301 Scranton Rd., Cleveland, OH 44113 USA (Tel.: 216-781-8383; 800-321-9696; Telefax: 216-781-5919; Telex: 810-421-8322)

Osaka Organic Chemical Ind. Co., Ltd.

Shin Toyama Bldg., 1-7-20, Azuchi-machi, Chuo-ku, Osaka, 541 Japan (Tel.: (06) 264-5071; Telefax: (06) 264-1675)

Outokumpu Oy

PO Box 280, SF-02101 Espoo Finland (Tel.: 358-4211; Telefax: 358-4213888; Telex: 124441 OKHI SF)

Oxiteno S/A Industria E Comercio

Av. Brigedeiro Luiz Antonio 1343, 7 andar, Sao Paulo SP, 01350-900 Brazil (Tel.: (55-11)283 6118; Telefax: (55-11)2893533; Telex: (55-11) TLX 31727)

Patco Products Div. *See under* American Ingredients

PCR

PCR, Inc., PO Box 1466, Gainesville, FL 32602 USA (Tel.: 904-376-8246; 800-331-6313; Telefax: 904-371-6246)

PCR, Inc./Harris Specialty Chemical Div., 8570 Phillips Hwy., Suite 101, Jacksonville, FL 32256-8208 USA (Tel.: 904-376-8246; 800-331-6313; Telefax: 904-371-6246; Telex: 810-825-6342)

Penreco, Div. of Pennzoil Prods. Co.

138 Petrolia St., Karns City, PA 16041 USA (Tel.: 412-756-0110; 800-245-3952; Telefax: 412-756-1050; Telex: 1561596)

Pentagon Chemicals Ltd.

Northside, Workington, Cumbria, CA14 1JJ UK (Tel.: 44 1900 604371; Telefax: 44 1900 66943; Telex: 64353 PENTA G)

Penta Manufacturing Co., A Division of Penta International Corp.

PO Box 1448, Fairfield, NJ 07007-1448 USA (Tel.: 201-740-2300; Telefax: 201-740-1839; Telex: 219472 PENT UR)

Peridot Chemicals (NJ) Inc.

1680 Rt. 23 North, Wayne, NJ 07470 USA (Tel.: 201-696-9000; 800-222-0121; Telefax: 201-696-2501)

Pfaltz & Bauer Inc. *See under* Aceto

Pfanstiehl Laboratories, Inc.

1219 Glen Rock Ave., PO Box 439, Waukegan, IL 60085 USA (Tel.: 708-623-0370; 800-383-0126; Telefax: 708-623-9173; Telex: 25 3672 PFANLAB)

Pfizer International

235 East 42nd Street, New York, NY 10017 USA (Tel.: 212-573-2323; 800-336-9008; Telefax: 212-573-2273; Telex: 420440)

Phibro Energy USA Inc.

500 Dallas, Suite 3200, Houston, TX 77002 USA (Tel.: 713-646-5042; Telefax: 713-646-5293; Telex: 3736083)

Phillips

Phillips Chemical Co., Div. of Phillips Petroleum Co., PO Box 968, Borger, TX 79008 USA (Tel.: 806-274-5236; 800-858-4327; Telefax: 806-274-5230)

Phillips Chemical Co., Div. of Phillips Petroleum Co., 101 ARB Plastics Tech. Center, Bartlesville, OK 74004 USA (Tel.: 918-661-9845; Telefax: 918-662-2929)

Phillips Petroleum Chemicals N.V., Steenweg op Brussels 355, B-3090 Overijse Belgium (Tel.: 32-2-689 1211; Telefax: 32-2-689 1472; Telex: 22197)

Phillips Petroleum International, Ltd., Shin-Tokyo Bldg., 3-1, Marunouchi 3-chome, Chioyda-ku, Tokyo, 100 Japan (Tel.: (03) 3216-6951; Telefax: (03) 3216-6960; Telex: J24641 PHILPET)

Phoenix Chemical, Inc.

322 Courtyard Dr., Somerville, NJ 08876 USA (Tel.: 908-707-0232; Telefax: 908-707-0186)

Piedmont Chem. Industries, Inc.

PO Box 2728, High Point, NC 27261 USA (Tel.: 910-885-5131; Telefax: 910-887-5563)

Pilot Chemical Co.

11756 Burke St., Santa Fe Springs, CA 90670 USA (Tel.: 310-723-0036; 800-707-4568; Telefax: 310-945-1877; Telex: 4991200 PILOT)

PMC Specialties Group, Inc., Div. of PMC, Inc.

20525 Center Ridge Road, Rocky River, OH 44116 USA (Tel.: 216-356-0700; 800-543-2466; Telefax: 216-356-2787; Telex: 4332035)

501 Murray Rd., Cincinnati, OH 45217 USA (Tel.: 513-242-3300; 800-543-2466; Telefax: 513-482-7353; Telex: 5106000948)

Pokonobe Industries Inc.

Box 1756, Santa Monica, CA 90406 USA (Tel.: 310-392-1259; Telefax: 310-392-3659)

Polarome Mfg. Co. Inc.

200 Theodore Conrad Dr., Jersey City, NJ 07305 USA (Tel.: 201-309-4500; Telefax: 201-433-0638; Telex: RCA 233 176)

Polymer Research Corp. of America

2186 Mill Ave., Brooklyn, NY 11234 USA (Tel.: 718-444-4300; Telefax: 718-241-3930)

PPG

PPG Industries, Inc., One PPG Place, Pittsburgh, PA 15272 USA (Tel.: 412-434-3131; 800-CHEM-PPG; Telefax: 412-434-2891; Telex: 86 6570)

PPG Industries, Inc./Specialty Chemicals Business Unit, 3938 Porett Dr., Gurnee, IL 60031 USA (Tel.: 847-244-3410; 800-323-0856; Telefax: 847-244-9633; Telex: 25-3310)

PPG Industrial do Brazil Ltda., Edificio Grande Avenida, Paulista Ave. 1754, Suite 153, 01310 Sao Paulo Brazil (Tel.: 55-011-284-0433; Telefax: 55-011-289-2105; Telex: 011-39104 PPGB BR)

PPG Mexico, S.A. de C.V., Av. Presidente Juarez No. 1978, Col. San Jeronimo Tepetlacalco, Tlalnepantla, Edo., C.P. 54090 Mexico (Tel.: 525-397-8222; Telefax: 525-398-5133)

PPG Industries (UK) Ltd./Specialty Chems., Carrington Business Park, Carrington, Urmston, Manchester, M31 4DD UK (Tel.: 44 161 777-9203; Telefax: 44 161-777-9064; Telex: 851-94014896 mazu g)

PPG Ouvrie S.A., 64, rue Faldherbe, B.P. 127, 59811 Lesquin Cedex France (Tel.: 33-2087-0510; Telefax: 33-2087-5631; Telex: 131419F)

PPG Industries Taiwan, Ltd., Suite 601, Worldwide House, No. 131, Ming East Rd., Sec. 3, Taipei, 105, Taiwan, R.O.C. (Tel.: 886-2-514-8052; Telefax: 886-2-514-7957; Telex: 10985 PPGTWN)

PPG Industries-Asia/Pacific Ltd., Takanawa Court, 5th floor, 13-1 Takanawa 3-Chome, Minato-Ku, Tokyo, 108 Japan (Tel.: 81-3-3280-2911; Telefax: 81-3-3280-2920; Telex: 02-42719 PPGPACJ)

PQ

PQ Corp., PO Box 840, Valley Forge, PA 19482-0840 USA (Tel.: 610-651-4200; 800-944-7411; Telefax: 610-251-9118; Telex: 476 1129 PQCO VAF)

Vitro-PQ S.A., Apdo. 382, Tlalnepantla, Edo. de Mexico Mexico (Tel.: 5-227-6800; Telefax: 5-227-6898)

Akzo-PQ Silica VOF, Amersfoort The Netherlands (Tel.: 033-676-777; Telefax: 033-676-169)

PQ Australia Pty. Ltd., Melbourne, Victoria Australia (Tel.: 03-791-4066; Telefax: 03-791-6511)

PQ Asia Inc., Tokyo Japan (Tel.: 03-5275-7051; Telefax: 03-5275-6670)

Presperse Inc.

601 Hadley Rd., PO Box 735, South Plainfield, NJ 07080 USA (Tel.: 908-756-2023; Telefax: 908-756-8754; Telex: (910)2409388)

Primachem Inc.

12 Greenwoods Rd., Old Tappan, NJ 07675 USA (Tel.: 201-784-3434; Telefax: 201-784-7997; Telex: 175094 PRIMA)

Procter & Gamble

Procter & Gamble Co./Chemicals Div., Rm C2N09A, 11530 Reed Hartman Hwy., Cincinnati, OH 45241 USA (Tel.: 513-626-3701; 800-477-8899; Telefax: 513-626-3145; Telex: 21-4185, P&GCIN)

Procter & Gamble Inc. Canada, 4711 Yonge St., PO Box 355, Station A, Toronto, Ontario, M5W 1C5 Canada (Tel.: 416-730-4064; Telefax: 416-730-4122)

Procter & Gamble Ltd./Europe Div., PO Box 9, Hayes Gate House, 27 Uxbridge Rd., Hayes, Middlesex, UB4 0JD UK (Tel.: 44 181 242 2303; Telefax: 44 181 242 2294)

Procter & Gamble Inc. Japan, 1-17 Koyocho-naka 1-chome, Higashinaka-ku, Kobe, 658 Japan (Tel.: 81-78-845-5328; Telefax: 81-78-845-6912)

Protameen Chemicals, Inc.

375 Minnisink Rd., PO Box 166, Totowa, NJ 07511 USA (Tel.: 201-256-4374; Telefax: 201-256-6764; Telex: 130125)

Protex

Protex S.A., B.P. 177, 6 rue Barbès, 92305 Levallois-Paris France (Tel.: 33 1-41 34 14 00; Telefax: 33 1-41 34 14 16; Telex: 620987)

Protex Chemicals Ltd., Astley Lane Industrial Estate, Astley Way, Swillington, Leeds, L228 8XT UK (Tel.: (0113) 2876002; Telefax: (0113) 2875003)

Protex Nederland, Broekhovenseweg 130 R, 502 LJ Tilburg The Netherlands (Tel.: 13/36 74 77; Telefax: 13/36 65 92)

Protex Chemie Basel AG, Thannerstrasse 72, 4054 Basel (CH) Switzerland (Tel.: 061-302 0066; Telefax: 061-302 0076)

Protex Extrosa, Postfach 1415, 79504 Lorrach Germany (Tel.: 07621-84772; Telefax: 07621-12429)

Protex Korea Co. Ltd., PO Box 1193, Seoul, 1351010, Korea (Tel.: 02/548 69 93; Telefax: 02/548 65 64; Telex: Panwool K 28817)

Provital

Centro Industrial Santiaga, Talleres 6, no. 15, Apartado Correos 78, Barcelona Spain (Tel.: 93-718-80-12; Telefax: 93-718-38-30; Telex: 98476 DITT E)

Punda Mercantile Inc.

310 Victoria Ave., Montreal, Quebec, H3Z 2M9 Canada (Tel.: 514-489-7278)

Purac

Purac Biochem, Postbus 21, Arkelsedijk 46, NL-4200 AA Gorinchem The Netherlands (Tel.: 31 1830-41799; Telefax: 31 1830-22741; Telex: 23615 purac nl)

Purac America, Inc., 111 Barclay Blvd., Suite 280, Lincolnshire Corporate Center, Lincolnshire, IL 60069 USA (Tel.: 708-634-6330; Telefax: 708-634-1992; Telex: 280231 PURACINC ARHT)

QO Chemicals

QO Chemicals, Inc., Subsid. of Great Lakes Chem. Corp., PO Box 2500, One Great Lakes Blvd., West Lafayette, IN 47906 USA (Tel.: 317-497-6100; 800-621-9521; Telefax: 317-497-7930; Telex: 446968 QOC UD)

QO Chemicals, Inc., Industrieweg 12, Haven 391, 2030 Antwerp Belgium (Tel.: 32-3-541-21-65; Telefax: 32-3541-65-03; Telex: 32338)

QO Chemicals Europe, S.A.R.L., 8, rue Bellini, 75782 Paris Cedex 16 France (Tel.: 33-1-45031450; Telefax: 33-1-47043604)

QO Chemicals (Australia), Inc., 16 Princess St., Kew, Victoria, 3101 Australia (Tel.: 61-3-853-6464; Telefax: 61-3-853-5929)

Quantum

Quantum Chemical Co., 11500 Northlake Dr., PO Box 429550, Cincinnati, OH 45249 USA (Tel.: 513-530-6500; 800-323-4905; Telefax: 513-530-6119; Telex: 155116)

Quantum Chemical Co., 1 Pierce Place, Suite 250C, Itasca, IL 60143 USA (Tel.: 708-285-1111; 800-323-7659; Telefax: 708-285-3316)

Quantum Chemical Europe B.V., Lange Bunder 7, 4854 MB Bavel The Netherlands (Tel.: 1613-6600; Telefax: 1613-3500)

Quest International

Quest International, Postbus 2, 1400 CA Bussum The Netherlands (Tel.: 31 2159-99111; Telefax: 31 2159-46067; Telex: 43050 QSTN NL)

Quest International, Lindtsedijk 8, 3336 le Zwijndrecht The Netherlands (Tel.: 31 78-128511; Telefax: 31 78-195279; Telex: 29477)

Quest International (UK) Ltd., Bromborough Port, Wirral, Merseyside, L62 4SU UK (Tel.: 44 151 645 2060; Telefax: 44 151 645 6975; Telex: 627173)

Quest International Inc., PO Box 630, Woods Corners, Norwich, NY 13815 USA (Tel.: 607-334-9951; Telefax: 607-334-5022; Telex: 646056)

Rasa Industries, Ltd.

Yaesu Dai Bldg., 1-1, Kyobashi 1-chome, Chuo-ku, Tokyo, 104 Japan (Tel.: (03) 3278-3801; Telefax: (03) 3281-6697; Telex: 2225818 RASAKOJ)

Raschig Corp.

5000 Old Osborne Tpke., PO Box 7656, Richmond, VA 23231 USA (Tel.: 804-222-9516; Telefax: 804-226-1569)

Reichhold Chemicals, Inc./Corporate Headquarters

2400 Ellis Rd., Durham, NC 27703-5543 USA (Tel.: 919-990-7500; 800-448-3482; Telefax: 919-990-7711)

Reilly Chemicals S.A. *See under* Morflex

Reilly Industries Inc.

1500 S. Tibbs Ave., Indianapolis, IN 46241 USA (Tel.: 317-248-6411; Telefax: 317-248-6402; Telex: 27 404)

Research Organics Inc.

4353 E. 49th St., Cleveland, OH 44125 USA (Tel.: 216-883-8025; 800-321-0570; Telefax: 216-883-1576)

Rhône-Poulenc

Rhône-Poulenc S.A., 25 quai Paul Doumer, 92408 Courbevoie Cedex France (Tel.: 33 1 47 68 12 34; Telex: 610500 F RHONE)

Rhône-Poulenc Chimie (France)/Secteur Specialites Chimiques, Cedex 29, F-92097 Paris La Défense France (Tel.: 47-68 12 34; Telefax: 47-68 09 00)

Rhône-Poulenc, Inc., CN7500, Prospect Plain Rd., Cranbury, NJ 08512-7500 USA (Tel.: 609-860-4000; 800-922-2189; Telefax: 609-860-0269)

Rhône-Poulenc North America Inc./North American Sales, One Corporate Dr., Shelton, CT 06484 USA (Tel.: 203-925-8164; Telefax: 203-925-3670)

Rhône-Poulenc Basic Chemical Co., One Corporate Dr., Box 881, Shelton, CT 06484 USA (Tel.: 800-642-4200; Telefax: 203-925-3627)

Rhône-Poulenc, Inc./Detergents and Personal Care, CN 7500, Prospect Plains Rd., Cranbury, NJ 08512 USA (Tel.: 609-860-4600; 800-253-5052)

Rhône-Poulenc, Inc./Surfactants & Specialty Chemicals, CN 7500, Prospect Plains Rd., Cranberry, NJ 08512-7500 USA (Tel.: 609-860-4000; 800-922-2189; Telefax: 609-860-0459)

Rhône-Poulenc, Inc./Textile Div., 1000 Hurricane Shoals Rd., Lawrenceville, GA 30243 USA

Richman Chemical, Inc.

768 N. Bethlehem Pike, Lower Gwynedd, PA 19002 USA (Tel.: 215-628-2946; Telefax: 215-628-4262)

Ridge Technologies, Inc.

117 Lyons Rd., Basking Ridge, NJ 07920 USA (Tel.: 908-766-1915; Telefax: 908-204-0312)

R.I.T.A. Corp.

1725 Kilkenny Court, PO Box 585, Woodstock, IL 60098 USA (Tel.: 815-337-2500; 800-426-7759; Telefax: 815-337-2522; Telex: 72-2438)

Rit-Chem Co. Inc.

109 Wheeler Ave., PO Box 435, Pleasantville, NY 10570-0435 USA (Tel.: 914-769-9110; Telefax: 914-769-1408; Telex: 229 639 RTCH)

Ronsheim & Moore Ltd., Div. of Hickson & Welch Ltd.

Ings Lane, Castleford, Yorkshire, WF10 2JT UK (Tel.: 44 1977 556565; Telefax: 44 1977-518058; Telex: 55378)

R S A Corp.

36 Old Sherman Tpke., Danbury, CT 06810 USA (Tel.: 203-790-8100; Telefax: 203-790-1709; Telex: 131148)

Ruetgers-Nease Chemical Co., Inc., Subsid. of Rütgerswerke AG

201 Struble Rd., State College, PA 16801 USA (Tel.: 814-238-2424; Telefax: 814-238-1567)

Ruger Chemical Co. Inc.

85 Cordier St., Irvington, NJ 07111 USA (Tel.: 201-926-0331; 800-274-RUGER; Telefax: 201-926-4921)

Samson Chem. Co.

8915 Sorenson Ave., PO Box 2163, Santa Fe Springs, CA 90670 USA (Tel.: 310-945-9188; Telefax: 310-698-7571)

Sanko Kagaku K.K.

2-18-29, Jyuhachijyo, Yodogawa-ku, Osaka, 532 Japan (Tel.: (06) 391-1971)

San Yuan Chemical Co., Ltd.

PO Box 26-1134, Taipei, Taiwan, R.O.C. (Tel.: 886-2-3946002; Telefax: 886-2-3579803)

Sartomer

Sartomer Co., Oaklands Corp. Center, 502 Thomas Jones Way, Exton, PA 19341 USA (Tel.: 610-363-4100; 800-SARTOMER; Telefax: 610-363-4140; Telex: 173071)

Sartomer International, Inc., 331 North Bridge Rd., #23-06 Odeon Towers, Singapore 188720 (Tel.: 65-334-2645/334-2646; Telefax: 65-334-2647)

Schenectady Chemicals, Inc.

PO Box 1046, Schenectady, NY 12301 USA (Tel.: 518-370-4200; Telefax: 518-346-3111; Telex: 145457)

Scher Chemicals, Inc.

Industrial West & Styertowne Rd., PO Box 4317, Clifton, NJ 07012 USA (Tel.: 201-471-1300; Telefax: 201-471-3783; Telex: 642643 Scherclif)

Schumacher

1969 Palomar Oaks Way, Carlsbad, CA 92009 USA (Tel.: 619-931-9555; 800-545-9242; Telefax: 619-931-7819; Telex: 910 322 1382)

Schweizerhall Inc.

10 Corporate Place South, Piscataway, NJ 08854 USA (Tel.: 908-981-8200; 800-243-6564; Telefax: 908-981-8282; Telex: 4754581 SUSA)

SCM

SCM Glidco Organics, PO Box 389, Jacksonville, FL 32201 USA (Tel.: 904-768-5800; 800-231-6728; Telefax: 904-768-2200; Telex: 441763)

SCM Glidco Organics, 210 Summit Ave., Montvale, NJ 07645-1526 USA (Tel.: 201-391-3040; Telefax: 201-391-3284)

N.V. SCM Europe S.A., 141 Rue Saint Lambert, 1200 Brussels Belgium (Tel.: 32-2-771-2110; Telefax: 32-2-771-4902)

SCM Metal Products, Inc., 2601 Weck Dr., PO Box 12166, Research Triangle Park, NC 27709-2166 USA (Tel.: 919-544-8090; Telefax: 919-544-7996; Telex: 196 072)

Scott Specialty Gases

6141 Easton Rd., Plumsteadville, PA 18949 USA (Tel.: 215-766-8861; Telefax: 215-766-2476)

Sea-Land Chemical Co.

902 Westpoint Pkwy., Suite 350, Westlake, OH 44145 USA (Tel.: 216-871-7887; Telefax: 216-871-7949)

Seeler Industries Inc.

2000 N. Broadway, Joliet, IL 60435 USA (Tel.: 815-740-2640; 800-336-2422; Telefax: 815-740-6469)

Shance Chemical Corp.

1 Depot Plaza, Mamaroneck, NY 10543 USA (Tel.: 914-698-1493; Telefax: 914-698-1520)

Shell

Shell Chemical Co., A Div. of Shell Oil Co., One Shell Plaza, PO Box 2463, Houston, TX 77252 USA (Tel.: 713-241-6161; 800-872-7435; Telefax: 713-241-4043; Telex: 762248)

Shell Chemical Co., 3200 Southwest Freeway, Suite 1230, Houston, TX 77027 USA (Tel.: 713-241-8101; Telefax: 713-241-8107)

Pecten Chemicals, Inc., Representing for Int'l. sales, PO Box 4407, Houston, TX 77210 USA (Tel.: 713-241-6161; Telefax: 713-241-5194)

Showa Denko K.K.

13-9, Shiba-Daimon 1-chome, Minato-ku, Tokyo, 105 Japan (Tel.: (03) 5470-3533; Telefax: (03) 3436-2625; Telex: J26232)

Sigma. *See under* Aldrich

SKW Chemicals Inc.

1509 Johnson Ferry Rd., Suite 150, Marietta, GA 30062 USA (Tel.: 404-971-1317; Telefax: 404-971-4306)

Werner G. Smith, Inc.

1730 Train Ave., Cleveland, OH 44113 USA (Tel.: 216-861-3676; 800-535-8343; Telefax: 216-861-3680)

SNIA (UK) Ltd.

36 Broadway, St. James's, London, SW1H 0BH UK (Tel.: 44 171 222 8696; Telefax: 44 171 222 8705; Telex: 23377)

Soda Aromatic Co., Ltd.

15-9, Nihonbashi Honcho 4-chome, Chuo-ku, Tokyo, 103 Japan (Tel.: (03) 3666-0553; Telefax: (03) 3666-2913; Telex: SODAROMA J27930)

Solvay Deutschland GmbH

Postfach 220, 30002 Hannover Germany (Tel.: 511-857-0; Telefax: 511-282126; Telex: 922-755)

Southern Clay Products, Inc.

1212 Church St., Gonzales, TX 78629 USA (Tel.: 210-672-2891; 800-324-2891; Telefax: 210-672-3081)

Spectrum Chemical Mfg. Corp.

14422 S. San Pedro St., Gardena, CA 90248 USA (Tel.: 310-516-8000; 800-772-8786; Telefax: 800-525-2299; Telex: 182395)

A.E. Staley Manufacturing Co., Subsid. of Tate & Lyle PLC

2200 E. Eldorado St., PO Box 151, Decatur, IL 62525 USA (Tel.: 217-423-4411; 800-258-7536; Telefax: 217-421-2881)

Stanchem, A Business Unit of HCl Inc.

43 Jutland Rd., Etobicoke, Ontario, M8Z 2G3 Canada (Tel.: 416-259-8231; Telefax: 416-259-5333)

Stan Chem International Ltd.

4 Kings Rd, Reading, Berkshire, RG1 3AA UK (Tel.: 44 1734 580247; Telefax: 44 1734 589580; Telex: 847746)

Standard Chemical Co.

Mill Lane, Cheadle, Cheshire, SK8 2NX UK (Tel.: 44 161 428-5225; Telefax: 44 161-428-0890; Telex: 666413)

Stepan

Stepan Co., 22 West Frontage Rd., Northfield, IL 60093 USA (Tel.: 708-446-7500; 800-745-7837; Telefax: 708-501-2443; Telex: 910-992-1437)

Stepan/Food Ingredients Dept., 100 West Hunter Ave., Maywood, NJ 07607 USA (Tel.: 201-845-3030; 800-523-3614; Telefax: 201-845-6754; Telex: 710-990-5170)

Stepan Canada, PO Box 307, Orillia, Ontario, L3V 6J6 Canada (Tel.: 705-326-7329; Telefax: 705-326-4523)

Stepan Europe, BP127, 38340 Voreppe France (Tel.: 33-76-50-51-00; Telefax: 33-7656-7165; Telex: 320511 F)

Stevenson Cooper

PO Box 38349, 1039 West Venango St., Philadelphia, PA 19140 USA (Tel.: 215-223-2600; Telefax: 215-223-3597)

Stockhausen, Inc., A Company of the Hüls Group

2401 Doyle St., Greensboro, NC 27406 USA (Tel.: 910-333-3500; Telefax: 910-333-3545; Telex: 574405)

Suchema AG

Haupstrasse 15, CH-8251 Kaltenbach Switzerland (Tel.: 41 54-41 12 65; Telefax: 41 54-41 32 34; Telex: 897170 BGK CH)

Süd-Chemie

Süd-Chemie AG, Postfach 830953, D-81709 Munchen Germany (Tel.: 49 89-5110 0; Telefax: 49 89 51 10 444; Telex: 523821 SCMU D)

Süd-Chemie Rheologicals, Lenbachplatz 6, D-80333 München Germany (Tel.: 49 89-5110 0; Telefax: 49 89 51 10-412)

Sunnyside

225 Carpenter Ave., Wheeling, IL 60090 USA (Tel.: 312-541-5700; 800-323-8611; Telefax: 708-541-9043)

Surco Products, Inc.

PO Box 777, Eighth & Pine Aves., Braddock, PA 15104-0777 USA (Tel.: 412-351-7700; 800-556-0111; Telefax: 412-351-7701)

Sybron

Sybron Chemicals Inc., PO Box 125, Hwy. 29, Wellford, SC 29385 USA (Tel.: 803-439-6333; 800-677-3500; Telefax: 803-439-1612; Telex: 570326)

Sybron Chemie Nederland B.V., Postbus 46, NL-6710 BA Ede The Netherlands (Tel.: 31-8380-70911; Telefax: 31-8380-30236; Telex: 37249)

Synthetic Chemicals Ltd., Div. of Shell Chemicals UK Ltd.

Four Ashes, Wolverhampton, West Midlands, WV10 7BP UK (Tel.: 44 1902 794000; Telefax: 44 1902 794300; Telex: 337306)

Synthetic Products Co., Subsid. of Cookson America Inc.

1000 Wayside Rd., Cleveland, OH 44110 USA (Tel.: 216-531-6010; 800-321-4236; Telefax: 216-486-6638)

Takasago International Corp.

19-22, Takanawa 3-chome, Minato-ku, Tokyo, 108 Japan (Tel.: (03) 3442-1211; Telefax: (03) 3442-1285; Telex: TAKAS A J32508)

Texaco

Texaco Chemical Co., PO Box 27707, Houston, TX 77227-7707 USA (Tel.: 713-961-3711; 800-231-3107; Telefax: 713-235-6437; Telex: 227-031 TEX UR)

Texaco Ltd., Div. of Texaco Chemical Europe, 195 Knightsbridge Green, London, SW7 1RU UK (Tel.: 44 171 581 5500; Telefax: 44 171-581 9163; Telex: 8956681 TEXACO G)

Texas Petrochemicals Corp.

8707 Katy Frwy., Suite 300, Houston, TX 77024 USA (Tel.: 713-461-3322; Telefax: 713-461-1029; Telex: TWX 510-891-7831)

Thatcher Co.

1900 Fortune Rd., PO Box 27407, Salt Lake City, UT 84127 USA (Tel.: 801-972-4587; Telefax: 801-972-4606)

3M Co./Specialty Chemicals Div.

3M Center Bldg. 223-6S-04, St. Paul, MN 55144-1000 USA (Tel.: 612-733-3064; Telefax: 612-737-7635)

Toho Chemical Industry Co., Ltd.

No. 1-2-5, Ningyo-cho, Nihonbashi, Chuo-ku, Tokyo, 103 Japan (Tel.: (81-3) 3668-2279; Telefax: (81-3) 3668-2278; Telex: 252-2332 TOHO K J)

Toray Fine Chemicals Co., Ltd.

273, Katsube-cho, Moriyama-shi, Shiga, 524 Japan (Tel.: (0775) 83-2570; Telefax: (0775) 83-3312)

Toray Thiokol Co., Ltd.

Toray Bldg., 1-8-1, Mihama, Urayasu-shi, Chiba, 279 Japan (Tel.: (0473) 50-6151; Telefax: (0473) 50-6091)

Tosoh

Tosoh Corp., 7-7 Akasaka 1-chome, Minato-ku, Tokyo, 107 Japan (Tel.: (03) 3585-9891; Telefax: (03) 3582-8120; Telex: J24475tosoh)

Tosoh USA Inc., 1100 Circle 75 Pkwy., Suite 600, Atlanta, GA 30339 USA (Tel.: 770-956-1100; Telefax: 770-956-7368; Telex: 542272 tosoh atl)

Total Petroleum Inc.

East Superior St., Alma, MI 48801 USA (Tel.: 517-463-9630; 800-292-9033; Telefax: 517-463-9623)

Toyo Kasei Kogyo Co., Ltd.

Shindai Bldg., 1-2-6, Dojimahama, Kita-ku, Osaka, 530 Japan (Tel.: (06) 346-6707; Telefax: (06) 341-6715)

Toyotama Perfumery Co., Ltd.

Kyodo Bldg. (Showa), 3-8, Nihonbashi Honcho 1-chome, Chuo-ku, Tokyo, 103 Japan (Tel.: (03) 3270-5511; Telefax: (03) 3279-6947; Telex: J 23403)

Traco Labs, Inc.

205 S. Main St., Seymour, IL 61875 USA (Tel.: 217-687-2800; 800-798-7226; Telefax: 217-687-2700)

TR-AMC Chemicals, Div. of TR-Metro Chemicals Inc.

PO Box 296, Hudson Ave., Ridgefield, NJ 07657-0296 USA (Tel.: 201-941-7706; Telefax: 201-941-7702; Telex: 130594)

T&R Chemicals, Inc.

700 Celum Rd., Box 330, Clint, TX 79836 USA (Tel.: 915-851-2761)

Tri-Iso, Inc.

480 N. Indian Hill Blvd., Ste. 2A, Claremont, CA 91711 USA (Tel.: 909-621-4635; Telefax: 909-621-9119)

Tri-K Industries, Inc.

27 Bland St., PO Box 312, Emerson, NJ 07630 USA (Tel.: 201-261-2800; 800-526-0372; Telefax: 201-261-1432; Telex: 215085 TRIK UR)

Triple Crown America, Inc.

13 N. Seventh St., Perkasie, PA 18944 USA (Tel.: 215-453-2500; Telefax: 215-453-2508)

Trivent Chemical Co., Inc.

4266 U.S. Rt. 1, PO Box 194, Monmouth Junction, NJ 08852 USA (Tel.: 908-438-0995; Telefax: 908-438-1020)

UCB

UCB S.A./Chemical Sector, 33 rue d'Anderlecht, B-1620 Drogenbos Belgium (Tel.: 32-2-3714923; Telefax: 32-2-3714924; Telex: 22342 UCBOS B)

UCB S.A./Chemical Sector, 326 avenue Louise, B-1050 Brussels Belgium (Tel.: 32(2)-641-1411; Telefax: 32(2)-640 9860; Telex: 21280)

UCB (Chem) Ltd., Div. of UCB S.A., Star House, 69 Clarendon Road, Watford, Hertfordshire, WD1 1DJ UK (Tel.: 44 1923 248011; Telefax: 44 1923-250225; Telex: 23958)

Ungerer & Co.

4 Bridgewater Lane, PO Box U, Lincoln Park, NJ 07035 USA (Tel.: 201-628-0600; Telefax: 201-628-0251; Telex: 4754267)

Unichema

Unichema International, Part of the Unilever Speciality Chemicals Group, Postbus 2, 2800 AA Gouda The Netherlands (Tel.: 31-0-1820-42911; Telefax: 31-0-1820-42250; Telex: 20661)

Unichema Chemicals Ltd., Div. of Unichema International, Bebington, Wirral, Merseyside, L62 4UF UK (Tel.: 44 151 645-2020; Telefax: 44 151-645-9197; Telex: 629408)

Unichema Chemie GmbH, Postfach 100963, D-46249 Emmerich Germany (Tel.: 49 0 2822-720; Telefax: 49 0 2822-72276; Telex: 8125113)

Unichema France S.A., 148 Boulevard Haussemann, 75008 Paris France (Tel.: 33 1 44 95 08 40; Telefax: 33 1 42563188; Telex: 643217)

Unichema North America, 4650 S. Racine Ave., Chicago, IL 60609 USA (Tel.: 312-376-9000; 800-833-2864; Telefax: 312-376-0095; Telex: 176068)

Unichema Japan, Sankei Bldg. 7F 708, 4-9, Umeda 2-chome, Kita-ku, Osaka, 530 Japan (Tel.: 81 6341-7221; Telefax: 81 6341-7725)

Unichema Australia Pty. Ltd., 164 Ingles St., Port Melbourne, Victoria, 3207 Australia (Tel.: 61-3 647-9311; Telefax: 61-3 645 3001; Telex: 30130)

Union Camp Corp./Chem. Prods. Div.

1600 Valley Rd., Wayne, NJ 07470 USA (Tel.: 201-628-2290; 800-733-1374; Telefax: 201-628-2840; Telex: 130735)

Union Carbide

Union Carbide Corp., 39 Old Ridgebury Rd., Danbury, CT 06817-0001 USA (Tel.: 203-794-2000; 800-335-8550; Telefax: 203-794-3170)

Union Carbide Canada Ltd., 7400 Blvd des Galleries, d'Anjou, Quebec, H1M 3M2 Canada (Tel.: 514-493-2610; Telefax: 514-493-2619)

Union Carbide Argentina S.A.I.C.S., Av. Leandro N. Alem 1051, Piso 13, (1001) Buenos Aires, Argentina (Tel.: 541-313-8883; Telefax: 541-313-8907; Telex: 25854 UCARB AR)

Union Carbide Produtos Quimicos Ltda., Rua Dr. Eduardo De Souza Aranha, 153, Sao Paulo, 04530 Brazil (Tel.: 55 11 828 1168; Telefax: 55 11 828 1144)

Union Carbide (Europe) S.A., 7, rue du Pre-Bouvier, CH-1217 Meyrin (Geneve) Switzerland (Tel.: 41-22-989 66 53; Telefax: 41-22-989 6545; Telex: 419207 UNC CH)

Union Carbide Austria GmbH, Storchengasse 1, A-1150 Wien Austria (Tel.: 43-1-891500; Telefax: 43-1-8915030; Telex: 131436 UCAA)

Union Derivan S.A.

Av. Meridiana 133, Barcelona E-08026 Spain (Tel.: 343-2322113; Telefax: 343-2323951; Telex: 98204 UNDER E)

United Catalysts Inc., Süd-Chemie Rheologicals Group

PO Box 32370, Louisville, KY 40232 USA (Tel.: 502-634-7200; Telefax: 502-637-3732; Telex: 204190, 204239)

United-Guardian, Inc.

230 Marcus Blvd, PO Box 2500, Smithtown, NY 11787 USA (Tel.: 516-273-0900; 800-645-5566; Telefax: 516-273-0858)

United Mineral & Chemical Corp., UMC

1100 Valley Brook Ave., Lyndhurst, NJ 07071-3608 USA (Tel.: 201-507-3300; 800-777-0505; Telefax: 201-507-1506; Telex: 6505113226)

Unitex Chemical Corp.

PO Box 16344, 520 Broome Rd., Greenboro, NC 27406 USA (Tel.: 910-378-0965; Telefax: 910-272-4312)

Universal Preserv-A-Chem Inc./UPI

297 North 7th St., Brooklyn, NY 11211 USA (Tel.: 718-782-7429; Telefax: 718-782-8109)

UOP

25 E. Algonquin Rd., Des Plaines, IL 60017-5017 USA (Tel.: 708-391-2395; 800-348-0832; Telex: 25-3285)

Van Waters & Rogers Inc., Subsid. of Univar Corp.

6100 Carillon Point, Kirkland, WA 98033 USA (Tel.: 206-889-3400; 800-234-4588; Telefax: 206-889-4133)

Veckridge Chemical Co. Inc.

60-70 Central Ave., South Kearny, NJ 07032 USA (Tel.: 201-344-1818; Telefax: 201-690-5936)

Veitsiluoto Oy/Forest Chemicals Industry

PO Box 196, SF-90101 Oulu 10 Finland (Tel.: 358-81-316 3111; Telefax: 358-81-378 5755; Telex: 32125 oulpk sf)

Velsicol Chemical Corp.

10400 W. Higgins Road, Rosemont, IL 60018 USA (Tel.: 708-298-9000; 800-843-7759; Telefax: 708-298-9014; Telex: 3730755)

Vigoro Industries, Inc., Vigoro Corp.

2017 W. Hwy. 50, P.O. Box 4139, Fairview Heights, IL 62208 USA (Tel.: 618-624-5522; Telefax: 618-624-8148; Telex: 54-6419)

Vitamins, Inc.

200 E. Randolph Dr., Chicago, IL 60601-7799 USA (Tel.: 312-861-0700; Telefax: 312-861-0708; Telex: 25 4717)

Vrymeer Commodities, A Ronstadt Group Co.

PO Box 545, 36W171 Indian Mound Rd., St. Charles, IL 60174-0545 USA (Tel.: 708-377-2584; Telefax: 708-377-5521)

Vulcan Chemicals, Div. of Vulcan Materials Co.

One Metroplex Dr., PO Box 530390, Birmingham, AL 35253-0390 USA (Tel.: 205-877-3000; 800-633-8280; Telefax: 205-877-3802; Telex: 59-6108)

Wacker

Wacker-Chemie GmbH, Hanns-Seidel-Platz 4, D-81737 München Germany (Tel.: (089) 6279 01; Telefax: (089) 6279-1770; Telex: 5291210)

Wacker Chemicals Ltd., Div. of Wacker-Chemie GmbH, The Clock Tower, Mount Felix, Bridge Street, Walton-on-Thames, Surrey, KT12 1AS UK (Tel.: 44 1932 246111; Telefax: 44 1932-240141; Telex: 28 391 wacker g)

Wacker Chemie Danmark A/S, Hovedvejen 91, DK-2600 Glostrop Denmark (Tel.: (43) 43 03 00; Telefax: (43) 43 03 16)

Wacker Chemicals (USA) Inc., 535 Connecticut Ave., Norwalk, CT 06855 USA (Tel.: 203-866-9400; 800-634-6752; Telefax: 203-866-9427; Telex: 643 444)

Wacker Méxicana, S.A. de C.V., Apartado Postal 99-070, Av. Periférico Sur 3343/PH, San Jeronimo, 10200, Mexico, D.F. (Tel.: 525-595-7599; Telefax: 525-683-8434; Telex: 17 72 920 waxk me)

Wacker Química do Brasil Ltda., Avenida Adolfo Pinheiro, 2056-9° Andar, Santo Amaro, 04734-003 Sao Paulo-SP Brazil (Tel.: (011) 548 81 33; Telefax: (011) 2 46 99 75; Telex: 1 156 946 wack br)

Wacker Chemicals Australia Pty. Ltd., Level 1, 566 St. Kilda Rd., Melbourne, Victoria, 3004 Australia (Tel.: (03) 525 16 00; Telefax: (03) 521 27 43; Telex: 30 367 hoechst aa)

Wacker Chemicals East Asia Ltd., TOHMA Nishi-Waseda Bldg., 2-14-1, Nishi Waseda, Shinjuku-ku, Tokyo, 169 Japan (Tel.: (03)52 72 31 21; Telefax: (03) 52 72 31 40)

Welch, Holme & Clark Co. Inc.

7 Ave. L, Newark, NJ 07105 USA (Tel.: 201-465-1200; Telefax: 201-465-7332)

West Agro, Inc.

11100 N. Congress, Kansas City, MO 64153 USA (Tel.: 312-298-5505; 800-421-1905; Telefax: 816-891-1595; Telex: 437012)

Westbrook Lanolin Co., A Div. of Woolcombers (Holdings) Ltd.

Argonaut Works, Laisterdyke, Bradford, West Yorkshire, BD4 8AU UK (Tel.: 44 1274 663331; Telefax: 44 1274-667665; Telex: 51502)

Westco Chemicals, Inc.

11312 Hartland St., North Hollywood, CA 91605 USA (Tel.: 213-877-0077; Telefax: 818-766-7170; Telex: 673150)

Westlake Plastics Co.

PO Box 127, W. Lenni Rd., Lenni, PA 19052 USA (Tel.: 215-459-1000; Telefax: 215-459-1084; Telex: 83-5406)

Wilke International Inc.

1375 N. Winchester, Olathe, KS 66061 USA (Tel.: 913-780-5544; 800-779-5545; Telefax: 913-780-5574)

Witco

Witco Corp., One American Lane, Greenwich, CT 06831-2559 USA (Tel.: 203-552-2000; 800-779-4826 x6400; Telefax: 203-552-2010)

Witco Corp./Lubricants Group Golden Bear Prods., 10100 Santa Monica Blvd., Suite 1470, Los Angeles, CA 90067-4183 USA (Tel.: 310-277-4511; 800-429-4826; Telefax: 310-201-0383)

Witco Corp./Oleochemicals and Surfactants Group, One American Lane, Greenwich, CT 06831-2559 USA (Tel.: 203-552-2000; 800-779-4826; Telefax: 203-552-2010)

Witco Corp./Organics Div., One American Lane, Greenwich, CT 06831-2559 USA (Tel.: 203-552-2000; Telefax: 203-552-2010)

Witco Corp./Petroleum Specialties Group, One American Lane, Greenwich, CT 06831-2559 USA (Tel.: 203-861-6280; 800-494-8673; Telefax: 203-552-2448)

Witco Corp./Polymer Additives Group, One American Lane, Greenwich, CT 06831-2559 USA (Tel.: 203-861-6279; 800-494-8737; Telefax: 203-552-2449)

Witco Corp./Polymer Additives Group, Bussey Rd., PO Box 1439, Marshall, TX 75671-1439 USA (Tel.: 903-938-5141; Telefax: 903-938-2647)

Witco Canada Ltd., 2 Lansing Sq., Suite 1200, Willowdale, Ontario, M2J 4Z4 Canada (Tel.: 416-497-9991; Telefax: 416-497-7110)

Witco GmbH, Industriegebiet West, Max-Wolf Strasse 7, D-35396 Steinau an der Strasse Germany (Tel.: 49 66 63 54-0; Telefax: 49 66 63 54 10-5)

Witco Surfactants GmbH, Postfach 1160, 36392 Steinau an der Strasse Germany (Tel.: 49 6663-54-0; Telefax: 49 6663-54-129)

Witco UK, Union Lane, Droitwich, Worcestershire, WR9 9BB UK (Tel.: 44 190 579 4795; Telefax: 44 190 579 4002)

Witco S.A., 20/22 rue de la Ville l'Evêque, 75008 Paris France (Tel.: 33 1 44 51 05 05; Telefax: 33 1 42 65 67 61; Telex: 842 290233)

Witco B.V., St. Canisiussingel 26, 6511 TJ Nijmegen The Netherlands (Tel.: 31-80-231534; Telefax: 31-80-224562)

Witco Italiana S.R.L., Viale Lunigiana 40, 20125 Milano Italy (Tel.: 39 2 676 011; Telefax: 39 2 6698 4203; Telex: 39 33 21 71)

Witco Ltd., PO Box 10245, 26112 Haifa Bay Israel (Tel.: 972 4-469-111; Telefax: 972 4-455778; Telex: 45198)

Witco Australia, 53 Restwell St., Bankstown, NSW, 2200 Australia (Tel.: 61 2 709 4699; Telefax: 61 2 709 3637)

Witco Asia Pacific Pte. Ltd., 396 Alexandra Rd., #06-04 Tower, Singapore 0511 (Tel.: 65 274 3878; Telefax: 65 274 8693)

Wright Corp.

PO Box 9009, 102 Orange St., Wilmington, NC 28402 USA (Tel.: 910-251-8952; Telefax: 910-762-9223)

Yorkshire Pat-Chem Inc.

11 Worley Rd., PO Box 1926, Greenville, SC 29602 USA (Tel.: 803-233-3941; 800-443-9358; Telefax: 803-232-3542)

Yuki Gosei Kogyo Co., Ltd.

Hirakawa-cho CH Bldg., 2-3-24, Hirakawa-cho, Chiyoda-ku, Tokyo, 102 Japan (Tel.: (03) 5275-5067; Telefax: (03) 5275-5079)

Zeeland Chemicals, Inc., A Cambrex Co.

215 N. Centennial St., Zeeland, MI 49464 USA (Tel.: 616-772-2193; 800-223-0453; Telefax: 616-772-7344; Telex: 226375)

Zeneca Inc.

1800 Concord Pike, Wilmington, DE 19897 USA

Zeochem

1314 S. 12 St., PO Box 35940, Louisville, KY 40232 USA (Tel.: 502-634-7600; 800-626-5355; Telefax: 502-634-8133; Telex: 204190, 204239)

Zinkan Enterprises, Inc.

10574 Ravenna Rd., Twinsbury, OH 44087 USA (Tel.: 216-487-1500; Telefax: 216-425-8202)

Appendices

CAS Number-to-Trade Name Cross-Reference

CAS	Trade name
56-81-5	Emery® 912
56-81-5	Emery® 916
56-81-5	Emery® 917
56-81-5	Emery® 918
56-81-5	Kemstrene® 96.0% USP
56-81-5	Kemstrene® 99.0%
56-81-5	Kemstrene® 99.7% USP
56-81-5	Kemstrene® High Gravity
57-55-6	Propylene Glycol USP/FCC Ultra Grade
64-17-5	Eastman® CDA-19
64-17-5	Punctilious® Ethyl Alcohol
64-17-5	Punctilious® SDA 1-1
64-17-5	Tecsol® D-1
67-68-5	Decap
67-68-5	DMSO
68-12-2	Dynasolve 100
75-09-2	Dynasolve 200
75-09-2	Dynasolve 210
75-65-0	Tebol™
75-65-0	Tebol™ 99
77-89-4	Citroflex® A-2
77-90-7	Citroflex® A-4
77-93-0	Citroflex® 2
77-94-1	Citroflex® 4
78-40-0	Eastman® TEP
78-42-2	Kronitex® TOF
78-42-2	Reomol® TOP
78-78-4	Exxsol® Isopentane
79-01-6	PPG Trichlor
79-24-3	NE™ 101
84-66-2	Kodaflex® DEP
84-69-5	Uniplex 155
84-74-2	Eastman® DBP
84-74-2	Kodaflex® DBP
84-74-2	Palatinol® DBP
84-74-2	Uniplex 150
84-78-6	Kodaflex® HS-3
91-17-8	Decalin
96-48-0	Agrisynth® BLO
96-48-0	Agsol Ex BLO
96-48-0	BLO®
96-48-0	Dynasolve 699
96-48-0	Gaftex® HI-B
96-48-0	GBL
96-49-1	Jeffsol™ EC
96-49-1	Texacar® EC
97-64-3	Purasolv® EL
97-64-3	Purasolv® ELECT
97-64-3	Purasolv® ELS
97-99-4	QO® Tetrahydrofurfuryl Alcohol (THFA®)
98-00-0	QO® Furfuryl Alcohol (FA®)
98-01-1	QO® Furfural
98-56-6	Oxsol® 100
101-34-8	Naturechem® GTR
102-76-1	Eastman® Triacetin
103-23-1	Wickenol® 158
106-65-0	DBE-4
106-65-0	Santosol™ DMS
107-70-0	PentoXone
107-98-2	Arcosolv® PM
107-98-2	Dowanol® PM
107-98-2	Eastman® PM
107-98-2	Icinol PM
107-98-2	Methyl Propasol®
107-98-2	Poly-Solv® MPM
107-98-2	Solvenon® PM
108-03-2	NiPar S-10™
108-32-7	Arconate® 1000
108-32-7	Arconate® HP
108-32-7	Arconate® PC
108-32-7	Jeffsol™ PC
108-32-7	Solvenon® PC
108-32-7	Texacar® PC
108-65-6	Arcosolv® PM Acetate
108-65-6	Dowanol® PMA
108-65-6	Eastman® PM Acetate
108-65-6	Ektasolve® PM Acetate
108-65-6	Methyl Propasol® Acetate
108-88-3	Shell Toluene
108-93-0	Naxol®
109-86-4	EM
109-86-4	Hisolve MC
109-86-4	Methyl Cellosolve®
109-86-4	Methyl Icinol
109-86-4	Poly-Solv® EM
109-99-9	Dynasolve 150
109-99-9	QO® Tetrahydrofuran (THF)
109-99-9	THF
110-27-0	Crodamol IPM
110-27-0	Estol IPM 1509 (BIO-IPM)

CAS	Trade name
110-27-0	Kessco® IPM
110-27-0	Lexol® IPM
110-27-0	Lexol® IPM-NF
110-27-0	Promyr
110-27-0	Tegosoft® M
110-49-6	EMA
110-49-6	Methyl Cellosolve® Acetate
110-71-4	Monoglyme
110-80-5	Cellosolve®
110-80-5	EE
110-80-5	Ethyl Icinol
110-90-0	Ethyl Di-Icinol
111-15-9	Cellosolve® Acetate
111-15-9	EEA
111-60-4	Kemester® EGMS
111-76-1	Butyl Icinol
111-76-2	Butyl Cellosolve®
111-76-2	Butyl Oxitol®
111-76-2	Dowanol® EB
111-76-2	Eastman® EB
111-76-2	EB
111-77-3	DM
111-77-3	Dowanol® DM
111-77-3	Eastman® DM
111-77-3	Ektasolve® DM
111-77-3	Hisolve DM
111-77-3	Methyl Carbitol®
111-77-3	Methyl Di-Icinol
111-77-3	Poly-Solv® DM
111-82-0	Emery® 2270
111-82-0	Emery® 2290
111-82-0	Emery® 2296
111-90-0	Carbitol®
111-90-0	Carbitol® Low Gravity
111-90-0	DE
111-90-0	Eastman® DE
111-90-0	Solvent APV Spec.
111-90-0	Transcutol
111-96-6	Diglyme
111-96-6	Hisolve MDM
112-07-2	Butyl Cellosolve® Acetate
112-07-2	Eastman® EB Acetate
112-07-2	EBA
112-10-7	Wickenol® 127
112-15-2	Carbitol® Acetate
112-15-2	DEA
112-15-2	Eastman® DE Acetate
112-25-4	EH
112-25-4	Hexyl Cellosolve®
112-34-5	Butyl Carbitol®
112-34-5	Butyl Di-Icinol
112-34-5	Butyl Dioxitol®
112-34-5	DB
112-34-5	Dowanol® DB
112-34-5	Eastman® DB
112-35-6	Hisolve TM
112-35-6	Poly-Solv® TM
112-39-0	Emery® 2216
112-49-2	Triglyme
112-50-5	Dowanol® TMH
112-59-4	Hexyl Carbitol®
112-61-8	Emery® 2218
112-62-9	Emery® 2219
112-73-2	Butyl Diglyme
112-73-2	Ethyl Diglyme
119-64-2	Tetralin
120-51-4	Unichem BZBN
122-99-6	Dowanol® EPh
122-99-6	Rewopal® MPG 10
122-99-6	Rewopal® MPG 40
123-51-3	Isoamyl Alcohol 95%
123-51-3	Isoamyl Alcohol 99%
123-95-5	Dermol EB
123-95-5	Unimate® BYS
124-10-7	Emery® 2214
124-17-4	Butyl Carbitol® Acetate
124-17-4	Dba
124-17-4	DB Acetate
124-17-4	Eastman® DB Acetate
126-33-0	Sulfolane W
126-73-8	Albrite® TBP
126-73-8	Amgard® TBPO^4
126-73-8	Kronitex® TBP
126-73-8	Reomol® TBP
126-73-8	TBP
127-18-4	PPG Perchlor
131-11-3	Kodaflex® DMP
131-11-3	Uniplex 110
138-22-7	Purasolv® BL
138-86-3	Dipentene No. 122
140-03-4	Naturechem® MAR
140-25-0	Mazon® EE-1
142-26-7	Acetamide MEA
142-26-7	Foamid AME-70
142-26-7	Foamid AME-75
142-26-7	Foamid AME-100
142-26-7	Incromectant AMEA-70
142-26-7	Upamide ACMEA
142-26-7	Witcamide® CMEA
142-82-5	Exxsol® Heptane
142-91-6	Crodamol IPP
142-91-6	Kessco® IPP
142-91-6	Propal
142-91-6	Tegosoft® P
143-22-6	Dowanol® TBH
143-28-2	Adol® 80
143-28-2	Adol® 85 NF
143-28-2	Fancol OA-95
143-28-2	Novol NF
143-28-2	Witcohol™ 350
328-84-7	Oxsol® 1000
335-57-9	PF-5070
355-42-0	PF-5060
501-24-6	Cardolite® NC-510
501-24-6	Cardolite® NC-507
547-64-8	Purasolv® ML
589-37-7	Dytek® EP
590-01-2	Ucar® n-Butyl Propionate
616-45-5	2-Pyrol®
616-45-5	Soluphor® P

CAS	Trade name
617-51-6	Purasolv® IPL
624-54-4	Ucar® n-Pentyl Propionate
627-83-8	Kemester® EGDS
627-93-0	DBE-6
627-93-0	Santosol™ DMA
629-14-1	Ethyl Glyme
678-26-2	PF-5050
756-79-6	Antiblaze® DMMP
763-69-9	Ektapro® EEP Solvent
763-69-9	Ucar® Ester EEP
770-35-4	Dowanol® PPh
872-50-4	GAFsolver™ TMP
872-50-4	M-Pyrol®
872-50-4	NMP
872-50-4	NMP-EL
872-50-4	Pharmasolve™
929-06-6	Diglycolamine® Agent (DGA®)
929-77-1	Kemester® 9022
1119-40-0	DBE-5
1119-40-0	Santosol™ DMG
1120-28-1	Kemester® 2050
1300-72-7	Pilot SXS-40
1321-94-4	Arosolve MN-LF
1330-20-7	Shell Xylene
1338-39-2	Crill 1
1344-00-9	EZA®
1559-35-9	EEH
1569-01-3	Arcosolv® PNP
1569-01-3	Dowanol® PnP
1587-20-8	Citroflex® 1
1634-04-4	High Purity MTBE
2163-42-0	MPDiol® Glycol
2687-91-4	NEP®
2687-96-9	Gaftex® HI-12
2687-96-9	Surfadone® LP-300
2807-30-9	Eastman® EP
2807-30-9	EP
2807-30-9	Propyl Cellosolve®
3158-56-3	Cardolite® NC-507
3445-11-2	HEP®
3687-45-4	Schercemol OLO
4435-53-4	Butoxyl®
5131-66-8	Arcosolv® PNB
5131-66-8	Dowanol® PnB
5306-85-4	Arlasolve® DMI
5333-42-6	Standamul® G
5881-94-3	Eastman® DP
6221-95-0	Lonzest® 143-S
6837-24-7	CHP®
6881-94-3	DP
6938-94-9	Crodamol DA
6938-94-9	Schercemol DIA
6938-94-9	Trivent DIA
7384-98-7	Nikkol Sefsol 228
7397-62-8	Polysolvan® O
7491-02-3	Schercemol DIS
8001-22-7	Super Refined® Soybean Oil USP
8001-23-8	Neobee® 18
8001-23-8	Super Refined® Safflower Oil
8001-25-0	EmCon Olive
8001-25-0	Super Refined® Olive Oil
8001-29-4	Super Refined® Cottonseed Oil
8001-30-7	Super Refined® Corn Oil
8001-79-4	Cosmetol® X
8001-79-4	Crystal® O
8001-79-4	Crystal® Crown
8001-79-4	EmCon CO
8001-79-4	York Krystal Kleer Castor Oil
8001-79-4	York USP Castor Oil
8002-03-7	Super Refined® Peanut Oil
8002-09-3	Herco® Pine Oil
8002-31-1	Fancol CB
8002-31-1	Fancol CB Extra
8006-54-0	Corona Lanolin
8006-54-0	Coronet Lanolin
8006-95-9	EmCon W
8007-24-7	Cardolite® NC-511
8007-69-0	Super Refined® Almond Oil
8008-74-0	Lipovol SES
8008-74-0	Super Refined® Sesame Oil
8009-03-8	Penreco Snow
8024-09-7	EmCon Walnut
8029-91-2	Axol® E 61
8042-47-5	Drakeol® 7
8042-47-5	Penreco 2259 Oil
8042-47-5	Penreco 2260 Oil
8052-41-3	3135 Naphtha
8052-41-3	Varsol® 1
8052-41-3	Varsol® 18
9003-11-6	Ucon® 75-H-450
9003-13-8	Dow L910
9003-13-8	Dow L1150
9003-13-8	Ucon® Fluid AP
9003-13-8	Ucon® LB-65
9003-13-8	Ucon® LB-135
9003-13-8	Ucon® LB-285
9003-13-8	Ucon® LB-385
9003-13-8	Ucon® LB-525
9003-13-8	Ucon® LB-625
9003-13-8	Ucon® LB-1145
9003-13-8	Ucon® LB-1715
9003-13-8	Ucon® LB-3000
9004-74-4	Dow MPEG350
9004-74-4	Dow MPEG550
9004-74-4	Dow MPEG750
9004-74-4	Upiwax 350
9004-78-8	Marlophen® P 7
9004-96-0	Chemax E-200 MO
9004-96-0	Chemax E-1000 MO
9004-96-0	Emerest® 2624
9004-96-0	Olepal 1
9004-96-0	Secoster® MO 400
9004-98-2	Eumulgin® PWM2
9005-02-1	Cithrol 2DL
9005-02-1	Cithrol 4DL

CAS	Trade name
9005-02-1	Cithrol 6DL
9005-02-1	Cithrol 10DL
9005-07-6	Cithrol 2DO
9005-07-6	Cithrol 4DO
9005-07-6	Cithrol 6DO
9005-07-6	Cithrol 10DO
9005-07-6	Secoster® DO 600
9016-45-9	Renex® 648
9016-45-9	T-Det® N-4
9016-45-9	T-Det® N-6
9035-85-2	Procetyl 10
9035-85-2	Procetyl 30
9035-85-2	Procetyl 50
9042-19-7	Arcal™ AAP
9064-13-5	Macol® 85
9076-43-1	Emcol® CC-42
9082-00-2	Dow 15-200
9082-00-2	Dow 112-2
10233-13-3	Crodamol IPL
13475-82-6	Permethyl® 99A
13552-80-2	Captex® 8227
16544-70-0	Citroflex® 6
17673-56-2	Dynacerin® 660
24991-55-7	Polyglyme
24991-55-7	Tetraglyme
25168-05-2	Halso® 99
25231-21-4	Arlamol® E
25231-21-4	Witconol™ APS
25265-77-4	Texanol® Ester-Alcohol
25265-77-4	Ucar® Filmer IBT
25322-68-3	Carbowax® PEG 300
25322-68-3	Carbowax® Sentry® PEG 300
25322-68-3	Carbowax® Sentry® PEG 400
25322-68-3	Carbowax® Sentry® PEG 600
25322-68-3	Carbowax® Sentry® PEG 900
25322-68-3	Carbowax® Sentry® PEG 1000
25322-68-3	Carbowax® Sentry® PEG 1450
25322-68-3	Carbowax® Sentry® PEG 3350
25322-68-3	Carbowax® Sentry® PEG 4600
25322-68-3	Carbowax® Sentry® PEG 8000
25322-68-3	Dow E200
25322-68-3	Dow E300
25322-68-3	Dow E300 NF
25322-68-3	Dow E400
25322-68-3	Dow E400 NF
25322-68-3	Dow E600
25322-68-3	Dow E600 NF
25322-68-3	Dow E900
25322-68-3	Dow E900 NF
25322-68-3	Dow E1000
25322-68-3	Dow E1000 NF
25322-68-3	Dow E1450
25322-68-3	Dow E1450 NF
25322-68-3	Dow E3350
25322-68-3	Dow E3350 NF
25322-68-3	Dow E4500
25322-68-3	Dow E4500 NF
25322-68-3	Dow E8000
25322-68-3	Dow E8000 NF
25322-68-3	Lutrol® E 300
25322-68-3	Lutrol® E 400
25322-68-3	Lutrol® E 1500
25322-68-3	Lutrol® E 4000
25322-68-3	Lutrol® E 6000
25322-68-3	Pluracol® E200
25322-68-3	Pluracol® E300
25322-68-3	Pluracol® E400
25322-68-3	Pluracol® E400 NF
25322-68-3	Pluracol® E600
25322-68-3	Pluracol® E600 NF
25322-68-3	Pluracol® E1450 NF
25322-68-3	Pluracol® E4000
25322-68-3	Renex® PEG 300
25322-68-3	Renex® PEG 400
25322-68-3	Renex® PEG 600
25322-68-3	Renex® PEG 1000
25322-68-3	Rhodasurf® PEG 300
25322-68-3	Rhodasurf® PEG 600
25322-68-3	Teric PEG 200
25322-68-3	Teric PEG 300
25322-68-3	Teric PEG 400
25322-69-4	Dow P425
25322-69-4	Dow P1000TB
25322-69-4	Dow P1200
25322-69-4	Dow P2000
25322-69-4	Dow P4000
25498-49-1	Arcosolv® TPM
25498-49-1	Dowanol® TPM
25640-78-2	Nusolv™ ABP-62
25791-96-2	Dow PT250
25791-96-2	Dow PT700
25791-96-2	Dow PT3000
26264-05-1	Rhodacal® 330
26266-57-9	Crill 2
26266-58-0	Nikkol SO-30
26266-58-0	Nikkol SO-30R
26402-26-6	Imwitor® 742
26402-31-3	Flexricin® 9
26446-38-8	Sucro Ester 15
26447-10-9	Stepanate® AXS
27194-74-7	Schercemol PGML
27458-93-1	Adol® 66
27458-93-1	Witcohol™ 66
27841-06-1	Schercemol NGDC
28510-23-8	Schercemol NGDO
29383-26-4	Wickenol® 171
29911-27-1	Arcosolv® DPNP
29911-27-1	Dowanol® DPnP
29911-28-2	Arcosolv® DPNB
29911-28-2	Dowanol® DPnB
31565-12-5	Nikkol Sefsol 218

CAS	Trade name
34590-94-8	Arcosolv® DPM
34590-94-8	Dowanol® DPM
34590-94-8	Icinol DPM
35274-05-6	Ceraphyl® 28
35545-57-4	Hetoxide BN-13
36311-34-9	Ceraphyl® ICA
36311-34-9	Standamul® G-16
36653-82-4	Adol® 52 NF
37311-04-9	Witconol™ APEM
41395-83-9	Schercemol PGDP
41395-89-5	Wickenol® 152
41669-30-1	Schercemol 1818
42131-25-9	Wickenol® 151
52125-53-8	Arcosolv® PE
53637-25-5	Dow EP530
55934-93-5	Dowanol® TPnB
56002-14-3	Olepal ISO
56519-71-2	Lexol® PG-800
57018-52-7	Arcosolv® PTB
57018-52-7	Butyl Propasol
57569-76-3	Pelemol® G7A
59130-69-7	Schercemol CO
59231-34-4	Ceraphyl® 140
59231-34-4	Ceraphyl® 140-A
61788-59-8	Emery® 2253
61788-59-8	Emery® 2254
61788-61-2	Emery® 2203
61788-61-2	Kemester® 143
61790-81-6	Solulan® 75
61790-81-6	Solulan® L-575
61791-12-6	Cremophor® EL
61791-12-6	Incrocas 35
61791-12-6	Ricinion
61791-20-6	Nikkol BWA-5
61791-20-6	Nikkol BWA-10
61791-20-6	Polychol 5
61791-20-6	Polychol 15
61791-31-9	Comperlan® COD
61849-72-7	Glucam® P-10
61849-72-7	Glucam® P-20
61849-72-7	Prox-onic MG-020 P
63793-60-2	Witconol™ APM
64742-14-9	Penreco 2251 Oil
64742-46-7	Isopar® V
64742-46-7	Penreco 2257 Oil
64742-47-8	DX 3641 Naphtha
64742-47-8	Exxsol® D 40
64742-47-8	Exxsol® D 60
64742-47-8	Exxsol® D 80
64742-47-8	Exxsol® D 110
64742-47-8	Exxsol® D-130
64742-47-8	Exxsol® D 3135
64742-47-8	Isopar® M
64742-47-8	140 Naphtha
64742-47-8	Penreco 2263 Oil
64742-48-9	Isopar® C
64742-48-9	Isopar® E
64742-48-9	Isopar® G
64742-48-9	Isopar® G Naphtha
64742-48-9	Isopar® H
64742-48-9	Isopar® K
64742-48-9	Isopar® K Naphtha
64742-48-9	Isopar® L
64742-48-9	2024 Naphtha
64742-49-0	Exxsol® D 75/100 Naphtha
64742-49-0	Exxsol® D 115/145 Naphtha
64742-49-0	Exxsol® Hexane
64742-49-0	Exxsol® Methylpentane Naphtha
64742-49-0	1520 Naphtha
64742-49-0	RS Naphtha
64742-89-8	2429 Naphtha
64742-89-8	Shell Tolu-Sol® A HT Solv.
64742-89-8	Shell Tolu-Sol® W HT Solv.
64742-94-5	Aromatic 150
64742-94-5	Aromatic 200
64742-94-5	BVA ND-205 Solv.
64742-94-5	Hi-Sol® 15
64742-95-6	Aromatic 100
64742-95-6	Hi-Sol® 10
64771-72-8	Norpar® 12
64771-72-8	Norpar® 13
64771-72-8	Norpar® 15
65381-09-1	Captex® 300
65381-09-1	Delios® S
65381-09-1	Delios® V
65381-09-1	Estasan GT 8-40 3578
65381-09-1	Estasan GT 8-60 3575
65381-09-1	Estasan GT 8-60 3580
65381-09-1	Estasan GT 8-65 3577
65381-09-1	Estasan GT 8-65 3581
65381-09-1	Estasan GT 8-70 3579
65381-09-1	Labrafac® LIPO
65381-09-1	Lexol® GT-865
65381-09-1	Liponate GC
65381-09-1	Miglyol® 810
65381-09-1	Miglyol® 812
65381-09-1	Neobee® M-5
65381-09-1	Neobee® O
65381-09-1	Tegosoft® CT
66082-42-6	Emerest® 2452
66794-58-9	Crillet 6
67701-28-4	Captex® 810A
67701-28-4	Captex® 810D
67774-74-7	Marlican®
68239-42-9	Glucam® E-10
68239-43-0	Glucam® E-20
68411-27-8	Crodamol AB
68412-54-4	Berol 02
68424-60-2	Fancol Karite Butter
68459-50-7	Agnosol 5
68459-50-7	Lanpol 5
68553-81-1	EmCon Rice Bran
68583-51-7	Captex® 200
68583-51-7	Myritol® PC
68583-51-7	Neobee® M-20
68937-83-7	Emery® 2209
68990-55-6	Axol® E 66
68991-68-4	Captex® 350

CAS Number-to-Chemical Cross-Reference

CAS	Chemical
50-21-5	Lactic acid
56-23-5	Carbon tetrachloride
56-37-1	Benzyltriethyl ammonium chloride
56-81-5	Glycerin
56-93-9	Benzyl trimethyl ammonium chloride
57-55-6	Propylene glycol
60-12-8	Phenethyl alcohol
60-24-2	2-Mercaptoethanol
60-29-7	Ethyl ether
64-17-5	Alcohol
64-17-5	Alcohol denatured
64-18-6	Formic acid
64-19-7	Acetic acid
64-19-7	Acetic acid, glacial
67-56-1	Methyl alcohol
67-63-0	Isopropyl alcohol
67-64-1	Acetone
67-66-3	Chloroform
67-68-5	Dimethyl sulfoxide
68-12-2	Dimethyl formamide
71-23-8	n-Propyl alcohol
71-36-3	Butyl alcohol
71-41-0	n-Amyl alcohol
71-43-2	Benzene
71-55-6	1,1,1-Trichloroethane
74-83-9	Methyl bromide
74-87-3	Methyl chloride
74-95-3	Methylene bromide
74-98-6	Propane
75-00-3	Ethyl chloride
75-04-7	Ethylamine
75-05-8	Acetonitrile
75-09-2	Methylene chloride
75-12-7	Formamide
75-15-0	Carbon disulfide
75-25-2	Bromoform
75-31-0	Isopropylamine
75-45-6	Chlorodifluoromethane
75-52-5	Nitromethane
75-56-9	Propylene oxide
75-61-6	Dibromodifluoromethane
75-65-0	t-Butyl alcohol
75-68-3	1,1,1-Chlorodifluoroethane
75-69-4	Trichlorofluoromethane
75-71-8	Dichlorodifluoromethane
75-84-3	2,2-Dimethylpropanol
75-85-4	t-Amyl alcohol
76-03-9	Trichloroacetic acid
76-05-1	Trifluoroacetic acid
76-12-0	1,1,2,2-Tetrachloro-1,2-difluoroethane
76-13-1	Trichlorotrifluoroethane
76-14-2	Dichlorotetrafluoroethane
76-49-3	Bornyl acetate
77-75-8	Methyl pentynol
77-89-4	Acetyl triethyl citrate
77-90-7	Acetyl tributyl citrate
77-93-0	Triethyl citrate
77-94-1	Tributyl citrate
78-06-8	Dibutyltin β-mercaptopropionate
78-40-0	Triethyl phosphate
78-42-2	Trioctyl phosphate
78-59-1	Isophorone
78-78-4	Isopentane
78-83-1	Isobutyl alcohol
78-87-5	Propylene dichloride
78-92-2	2-Butanol
78-93-3	Methyl ethyl ketone
79-00-5	1,1,2-Trichloroethane
79-01-6	Trichloroethylene
79-20-9	Methyl acetate
79-24-3	Nitroethane
79-33-4	L-Lactic acid
79-34-5	1,1,2,2-Tetrachloroethane
79-46-9	2-Nitropropane
79-92-5	Camphene
80-56-8	α-Pinene
84-66-2	Diethyl phthalate
84-69-5	Diisobutyl phthalate
84-74-2	Dibutyl phthalate
84-78-6	Butyl octyl phthalate
91-17-8	Decahydronaphthalene
93-58-3	Methyl benzoate
94-96-2	Ethyl hexanediol
95-47-6	o-Xylene
95-50-1	o-Dichlorobenzene
96-22-0	Diethyl ketone
96-37-7	Methylcyclopentane
96-48-0	Butyrolactone
96-49-1	Ethylene carbonate
97-64-3	Ethyl lactate

CAS	Chemical
97-85-8	Isobutyl isobutyrate
97-95-0	2-Ethyl-1-butanol
97-99-4	Tetrahydrofurfuryl alcohol
98-00-0	Furfuryl alcohol
98-01-1	Furfural
98-51-1	p-t-Butyl toluene
98-55-5	α-Terpineol
98-56-6	p-Chlorobenzotrifluoride
98-82-8	Cumene
98-86-2	Acetophenone
99-62-7	1,3-Diisopropyl benzene
99-86-5	α-Terpinene
100-18-5	1,4-Diisopropyl benzene
100-41-4	Ethylbenzene
100-47-0	Benzonitrile
100-51-6	Benzyl alcohol
100-52-7	Benzaldehyde
101-34-8	Glyceryl triacetyl ricinoleate
102-60-3	Tetrahydroxypropyl ethylenediamine
102-76-1	Triacetin
102-82-9	Tributylamine
103-09-0	2-Ethylhexyl acetate
103-23-1	Dioctyl adipate
103-65-1	Propylbenzene
104-76-7	2-Ethylhexanol
105-30-6	2-Methyl-1-pentanol
105-37-3	Ethyl propionate
105-45-3	Methyl acetoacetate
105-46-4	s-Butyl acetate
105-57-7	Acetal
105-99-7	Dibutyl adipate
106-36-5	Propyl propionate
106-42-3	p-Xylene
106-61-6	Glyceryl acetate
106-65-0	Dimethyl succinate
106-79-6	Dimethyl sebacate
106-97-8	Butane
107-06-2	Ethylene dichloride
107-12-0	Propionitrile
107-15-3	Ethylenediamine (anhyd.)
107-21-1	Glycol
107-31-3	Methyl formate
107-41-5	Hexylene glycol
107-70-0	4-Methoxy-4-methyl-pentanone-2
107-83-5	Isohexane
107-87-9	Methyl propyl ketone
107-88-0	Butylene glycol
107-98-2	Methoxyisopropanol
108-03-2	Nitropropane
108-08-7	2,4-Dimethylpentane
108-10-1	Methyl isobutyl ketone
108-11-2	Methyl amyl alcohol
108-20-3	Isopropyl ether
108-21-4	Isopropyl acetate
108-32-7	Propylene carbonate
108-38-3	m-Xylene
108-65-6	Propylene glycol methyl ether acetate
108-67-8	1,3,5-Trimethylbenzene
108-70-3	1,3,5-Trichlorobenzene
108-83-8	Diisobutyl ketone
108-84-9	Methyl amyl acetate
108-87-2	Methyl cyclohexane
108-88-3	Toluene
108-89-4	γ-Picoline
108-93-0	Cyclohexanol
108-94-1	Cyclohexanone
108-99-6	β-Picoline
109-02-4	p-Methyl morpholine
109-06-8	α-Picoline
109-60-4	Propyl acetate
109-66-0	n-Pentane
109-79-5	n-Butyl mercaptan
109-86-4	Methoxyethanol
109-89-7	Diethylamine
109-94-4	Ethyl formate
109-99-9	Tetrahydrofuran
110-01-0	Tetrahydrothiophene
110-02-1	Thiophene
110-12-3	Methyl isoamyl ketone
110-19-0	Isobutyl acetate
110-27-0	Isopropyl myristate
110-34-9	Isobutyl palmitate
110-43-0	Methyl n-amyl ketone
110-49-6	Methoxyethanol acetate
110-54-3	Hexane
110-58-7	Pentylamine
110-71-4	Ethylene glycol dimethyl ether
110-80-5	Ethoxyethanol
110-82-7	Cyclohexane
110-86-1	Pyridine
110-89-4	Piperidine
110-91-8	Morpholine
110-98-5	Dipropylene glycol
111-03-5	Glyceryl oleate
111-13-7	Methyl hexyl ketone
111-15-9	Ethoxyethanol acetate
111-27-3	Hexyl alcohol
111-40-0	Diethylenetriamine
111-46-6	Diethylene glycol
111-48-8	Thiodiglycol
111-55-7	Ethylene glycol diacetate
111-60-4	Glycol stearate
111-62-6	Ethyl oleate
111-65-9	Octane
111-69-3	Adiponitrile
111-70-6	Heptyl alcohol
111-76-2	Butoxyethanol
111-77-3	Methoxydiglycol
111-82-0	Methyl laurate
111-84-2	n-Nonane
111-87-5	Caprylic alcohol
111-90-0	Ethoxydiglycol
111-92-2	Dibutylamine
111-96-6	Diethylene glycol dimethyl ether
112-07-2	Butoxyethanol acetate

CAS	Chemical
112-10-7	Isopropyl stearate
112-14-1	Octyl acetate
112-15-2	Ethoxydiglycol acetate
112-25-4	Ethylene glycol hexyl ether
112-27-6	Triethylene glycol
112-30-1	n-Decyl alcohol
112-34-5	Butoxydiglycol
112-35-6	Triglycol monomethyl ether
112-36-7	Diethylene glycol diethyl ether
112-39-0	Methyl palmitate
112-40-3	n-Dodecane
112-49-2	PEG-3 dimethyl ether
112-50-5	Ethoxytriglycol
112-57-2	Tetraethylenepentamine
112-59-4	Diethylene glycol n-hexyl ether
112-60-7	PEG-4
112-61-8	Methyl stearate
112-62-9	Methyl oleate
112-70-9	Tridecyl alcohol
112-73-2	Diethylene glycol dibutyl ether
112-80-1	Oleic acid
112-84-5	Erucamide
112-92-5	Stearyl alcohol
115-19-5	Methyl butynol
117-81-7	Dioctyl phthalate
117-82-8	Dimethoxyethyl phthalate
117-84-0	Dioctyl phthalate
119-64-2	Tetrahydronaphthalene
120-51-4	Benzyl benzoate
120-82-1	1,2,4-Trichlorobenzene
121-44-8	Triethylamine
121-69-7	n,n-Dimethylaniline
122-62-3	Dioctyl sebacate
122-99-6	Phenoxyethanol
123-18-2	Isobutyl heptyl ketone
123-42-2	Diacetone alcohol
123-51-3	Isoamyl alcohol
123-54-6	Acetylacetone
123-86-4	n-Butyl acetate
123-91-1	1,4-Dioxane
123-92-2	Isoamyl acetate
123-94-4	Glyceryl stearate
123-95-5	Butyl stearate
123-96-6	2-Octanol
124-10-7	Methyl myristate
124-16-3	1-Butoxyethoxy-2-propanol
124-17-4	Diethylene glycol butyl ether acetate
124-18-5	n-Decane
124-29-8	Cetyl alcohol
124-38-9	Carbon dioxide
124-40-3	Dimethylamine
126-33-0	Sulfolane
126-73-8	Tributyl phosphate
127-18-4	Perchloroethylene
127-19-5	Dimethyl acetamide
127-91-3	β-Pinene

CAS	Chemical
131-11-3	Dimethyl phthalate
134-62-3	Diethyl toluamide
136-60-7	Butyl benzoate
137-32-6	2-Methyl-1-butanol
138-22-7	Butyl lactate
138-86-3	dl-Limonene
139-87-7	N-Ethyldiethanolamine
140-03-4	Methyl acetyl ricinoleate
140-11-4	Benzyl acetate
140-25-0	Benzyl laurate
141-04-8	Diisobutyl adipate
141-70-8	Isododecane
141-78-6	Ethyl acetate
141-79-7	Mesityl oxide
141-97-9	Ethylacetoacetate
142-16-5	Dioctyl maleate
142-26-7	Acetamide MEA
142-55-2	Propylene glycol laurate
142-82-5	Heptane
142-91-6	Isopropyl palmitate
142-92-7	Hexyl acetate
142-96-1	Dibutyl ether
143-22-6	Butoxytriglycol
143-28-2	Oleyl alcohol
287-92-3	Cyclopentane
328-84-7	3,4-Dichlorobenzotrifluoride
335-57-9	Perfluoroheptane
355-42-0	Perfluorohexane
464-06-2	2,2,3-Trimethylbutane
470-67-7	1,4-Cineole
470-82-6	Eucalyptol
501-24-6	3-(n-Pentadecyl) phenol
540-72-7	Sodium thiocyanate
540-84-1	Isooctane
540-88-5	t-Butyl acetate
542-55-2	Isobutyl formate
542-59-6	Ethylene glycol acetate
543-49-7	2-Heptanol
547-64-8	Methyl lactate
562-49-2	3,3-Dimethylpentane
563-80-4	Methyl isopropyl ketone
565-59-3	2,3-Dimethylpentane
583-57-3	1,2-Dimethylcyclohexane
584-02-1	3-Pentanol
589-37-7	1,3-Pentanediamine
589-82-2	3-Heptanol
589-90-2	1,4-Dimethylcyclohexane
590-01-2	n-Butyl propionate
590-35-2	Neoheptane
590-66-9	1,1-Dimethylcyclohexane
591-21-9	1,3-Dimethylcyclohexane
591-76-4	Isoheptane
598-82-3	DL-Lactic acid
599-64-4	Cumyl phenol
600-25-9	1-Chloro-1-nitropropane
616-45-5	2-Pyrrolidone
617-51-6	Isopropyl lactate
622-45-7	Cyclohexyl acetate
624-29-3	cis-1,4-Dimethylcyclo-

CAS	Chemical
	hexane
624-54-4	n-Pentyl propionate
624-92-0	Methyl disulfide
626-38-0	s-Amyl acetate
627-83-8	Glycol distearate
627-93-0	Dimethyl adipate
628-63-7	Amyl acetate
629-14-1	Ethylene glycol diethyl ether
629-70-9	Cetyl acetate
638-04-0	cis-1,3-Dimethylcyclohexane
646-06-0	1,3-Dioxolane
657-84-1	Sodium toluenesulfonate
661-19-8	Behenyl alcohol
678-26-2	Perfluoropentane
696-29-7	Isopropylcyclohexane
756-79-6	Dimethyl methylphosphonate
763-69-9	Ethyl 3-ethoxypropionate
770-35-4	Propylene glycol phenyl ether
822-50-4	1,2-Dimethylcyclopentane
872-50-4	N-Methyl-2-pyrrolidone
929-06-6	2-(2-Aminoethoxy) ethanol
929-77-1	Methyl behenate
1119-40-0	Dimethyl glutarate
1120-21-4	n-Undecane
1120-28-1	Methyl eicosenate
1192-18-3	1,2-Dimethylcyclopentane
1300-71-6	Xylenol
1300-72-7	Sodium xylenesulfonate
1310-58-3	Potassium hydroxide
1310-73-2	Sodium hydroxide
1317-38-0	Copper oxide (ic)
1318-02-1	Sodium silicoaluminate
1321-67-1	Naphthol spirits
1323-38-2	Castor oil
1323-65-5	Dinonyl phenol
1330-20-7	Xylene
1336-21-6	Ammonium hydroxide
1338-24-5	Naphthenic acid
1338-39-2	Sorbitan laurate
1344-00-9	Sodium silicoaluminate
1559-35-9	Ethylene glycol 2-ethylhexyl ether
1559-36-0	Diethylene glycol 2-ethylhexyl ether
1569-01-3	Propylene glycol n-propyl ether
1569-02-4	Propylene glycol ethyl ether
1587-20-8	Trimethyl citrate
1634-04-4	Methyl t-butyl ether
1638-26-2	1,1-Dimethylcyclopentane
1640-89-7	Ethylcyclopentane
1678-91-7	Ethylcyclohexane
1678-92-8	Propylcyclohexane
1678-93-9	Butylcyclohexane
1678-98-4	Isobutylcyclohexane
2040-95-1	Butylcyclopentane
2040-96-2	Propylcyclopentane
2050-92-2	Di-n-amylamine
2116-84-9	Phenyl trimethicone
2163-42-0	2-Methyl-1,3-propanediol
2207-01-4	cis-1,2-Dimethylcyclohexane
2207-03-6	trans-1,3-Dimethylcyclohexane
2207-04-7	trans-1,4-Dimethylcyclohexane
2425-77-6	2-Hexyl-1-decanol
2432-87-3	Dioctyl sebacate
2452-99-5	1,2-Dimethylcyclopentane
2453-00-1	1,3-Dimethylcyclopentane
2517-43-3	3-Methoxybutanol
2532-58-3	1,3-Dimethylcyclopentane
2615-15-8	PEG-6
2687-91-4	N-Ethyl-2-pyrrolidone
2687-96-9	Lauryl pyrrolidone
2807-30-9	Ethylene glycol propyl ether
2915-53-9	Dioctyl maleate
3158-56-3	3-(n-Pentadecyl) phenol
3178-22-1	t-Butylcyclohexane
3386-18-3	PEG-9
3445-11-2	N-Hydroxyethylpyrrolidone
3687-45-4	Oleyl oleate
3687-46-5	Decyl oleate
3788-32-7	Isobutylcyclopentane
3875-51-2	Isopropylcyclopentane
3875-52-3	t-Butylcyclopentane
4131-74-2	Dimethyl thiodipropionate
4169-04-4	Propylene glycol phenyl ether
4254-14-2	(-)-Propylene glycol
4254-15-3	(+)-Propylene glycol
4254-16-4	(±)-Propylene glycol
4259-00-1	1,1,2-Trimethylcyclopentane
4390-04-9	Isohexadecane
4435-53-4	Methoxybutyl acetate
4478-97-1	Ceteth-5
4516-69-2	1,1,3-Trimethylcyclopentane
4740-78-7	Glyceryl formal
4850-32-2	s-Butylcyclopentane
5117-19-1	PEG-8
5131-66-8	Butoxypropanol
5274-65-7	Oleth-2
5306-85-4	Dimethyl isosorbide
5333-42-6	Octyldodecanol
5353-27-5	Oleth-5
5650-20-4	Tetraethylene glycol ethyl ether
5959-89-7	Sorbitan laurate
5989-27-5	d-Limonene
5989-54-8	l-Limonene
6032-29-7	s-Amyl alcohol
6221-95-0	Myristyl propionate
6780-13-8	Ethylenediamine (monohydrate)
6790-09-6	PEG-12

CAS	Chemical
6837-24-7	N-Cyclohexyl pyrrolidone
6876-23-9	trans-1,2-Dimethylcyclohexane
6881-94-3	Diethylene glycol propyl ether
6938-94-9	Diisopropyl adipate
7311-27-5	Nonoxynol-4
7384-98-7	Propylene glycol dicaprylate
7397-62-8	Glycolic acid-n-butyl ester
7425-87-8	N-Octyl pyrrolidone
7446-09-5	Sulfur dioxide
7491-02-3	Diisopropyl sebacate
7681-82-5	Sodium iodide
7785-26-4	(L, -)-α-Pinene
7785-70-8	(+)-α-Pinene
8000-41-7	α-Terpineol (mixt.)
8001-22-7	Soybean oil
8001-23-8	Safflower oil
8001-25-0	Olive oil
8001-29-4	Cottonseed oil
8001-30-7	Corn oil
8001-78-3	Hydrogenated castor oil
8001-79-4	Castor oil
8002-03-7	Peanut oil
8002-05-9	Petroleum distillates
8002-09-3	Pine oil
8002-31-1	Cocoa butter
8005-44-5	Cetearyl alcohol
8006-54-0	Lanolin (anhyd.)
8006-61-9	Gasoline
8006-64-2	Turpentine oil
8006-95-9	Wheat germ oil
8007-24-7	3-(n-Penta-8′-decenyl) phenol
8007-69-0	Sweet almond oil
8008-20-6	Kerosene
8008-74-0	Sesame oil
8009-03-8	Petrolatum (NF)
8012-95-1	Mineral oil
8020-83-5	Mineral oil (white)
8020-84-6	Lanolin (hyd.)
8024-09-7	Walnut oil
8027-32-5	Petrolatum (USP)
8027-33-6	Lanolin alcohol
8029-91-2	Acetylated hydrogenated lard glyceride
8029-92-3	Acetylated lard glyceride
8029-92-3	Diacetylated monoglycerides
8030-30-6	Naphtha
8032-32-4	Mineral spirits
8038-43-5	Lanolin oil
8039-09-6	PEG-75 lanolin
8042-47-5	Mineral oil
8051-30-7	Cocamide DEA
8052-41-3	Stoddard solvent
9003-01-4	Carbomer
9003-11-6	PEG/PPG-17/6 copolymer
9003-11-6	Poloxamer 181
9003-13-8	PPG-5 butyl ether
9003-13-8	PPG-9 butyl ether
9003-13-8	PPG-14 butyl ether
9003-13-8	PPG-15 butyl ether
9003-13-8	PPG-18 butyl ether
9003-13-8	PPG-22 butyl ether
9003-13-8	PPG-24 butyl ether
9003-13-8	PPG-33 butyl ether
9003-13-8	PPG-40 butyl ether
9003-13-8	PPG-53 butyl ether
9004-74-4	PEG-6 methyl ether
9004-74-4	PEG-10 methyl ether
9004-74-4	PEG-16 methyl ether
9004-74-4	Polyethylene glycol monomethyl ether
9004-95-9	Ceteth-5
9004-95-9	Ceteth-16
9004-95-9	Ceteth-24
9004-96-0	PEG-4 oleate
9004-96-0	PEG-5 oleate
9004-96-0	PEG-6 oleate
9004-96-0	PEG-8 oleate
9004-96-0	PEG-20 oleate
9004-98-2	Oleth-2
9004-98-2	Oleth-5
9004-98-2	Oleth-16
9005-00-9	Steareth-5
9005-00-9	Steareth-16
9005-02-1	PEG-4 dilaurate
9005-02-1	PEG-8 dilaurate
9005-02-1	PEG-12 dilaurate
9005-02-1	PEG-20 dilaurate
9005-07-6	PEG-4 dioleate
9005-07-6	PEG-8 dioleate
9005-07-6	PEG-12 dioleate
9005-07-6	PEG-20 dioleate
9005-37-2	Propylene glycol alginate
9005-65-6	Polysorbate 80
9005-67-8	Polysorbate 60
9006-65-9	Dimethicone
9007-16-3	Carbomer
9007-17-4	Carbomer
9007-20-9	Carbomer
9014-92-0	Dodoxynol-7
9016-00-6	Dimethicone
9016-45-9	Nonoxynol-4
9016-45-9	Nonoxynol-5
9016-45-9	Nonoxynol-6
9035-85-2	PPG-10 cetyl ether
9035-85-2	PPG-30 cetyl ether
9035-85-2	PPG-50 cetyl ether
9038-95-3	PPG-2-buteth-3
9038-95-3	PPG-3-buteth-5
9038-95-3	PPG-5-buteth-7
9038-95-3	PPG-7-buteth-10
9038-95-3	PPG-9-buteth-12
9038-95-3	PPG-12-buteth-16
9038-95-3	PPG-20-buteth-30
9038-95-3	PPG-28-buteth-35
9038-95-3	PPG-33-buteth-45

CAS	Chemical
9062-04-8	Propylene glycol dicaprylate/dicaprate
9065-63-8	PPG-2-buteth-3
9065-63-8	PPG-3-buteth-5
9065-63-8	PPG-5-buteth-7
9065-63-8	PPG-7-buteth-10
9065-63-8	PPG-9-buteth-12
9065-63-8	PPG-12-buteth-16
9065-63-8	PPG-20-buteth-30
9065-63-8	PPG-28-buteth-35
9065-63-8	PPG-33-buteth-45
9076-43-1	PPG-40 diethylmonium chloride
9082-00-2	PPG-24-glycereth-24
10035-10-6	Hydrobromic acid
10108-25-5	PEG-4 oleate
10143-53-0	Diethylene glycol ethylvinyl ether
10143-56-3	Diethylene glycol methylpentyl ether
10213-77-1	PPG-3 methyl ether
10233-13-3	Isopropyl laurate
10326-41-7	D-Lactic acid
11099-07-3	Glyceryl stearate
12068-03-0	Sodium toluenesulfonate
13429-07-7	PPG-2 methyl ether
13475-82-6	Isododecane
13552-80-2	Triundecanoin
13952-84-6	s-Butylamine
15764-24-6	Dipropylene glycol ethyl ether
16544-70-0	Trihexyl citrate
17673-56-2	Oleyl erucate
17673-59-3	Oleyl linoleate
18172-67-3	(-)-β-Pinene
18912-80-6	Diethylene glycol isobutyl ether
20178-34-1	Tripropylene glycol ethyl ether
20324-33-8	PPG-3 methyl ether
20636-48-0	Nonoxynol-5
23336-36-9	PEG-5 oleate
23783-42-8	Tetraethylene glycol methyl ether
24133-73-1	Cumylphenyl acetate
24991-55-7	PEG-3 dimethyl ether
24991-55-7	PEG-4 dimethyl ether
25168-05-2	Monochlorotoluene
25190-05-0	Oleth-16
25231-21-4	PPG-11 stearyl ether
25231-21-4	PPG-15 stearyl ether
25265-71-8	Dipropylene glycol
25265-77-4	2,2,4-Trimethyl-1,3-pentanediol monoisobutyrate
25322-68-3	PEG-4
25322-68-3	PEG-6
25322-68-3	PEG-8
25322-68-3	PEG-9
25322-68-3	PEG-12
25322-68-3	PEG-14
25322-68-3	PEG-16
25322-68-3	PEG-20
25322-68-3	PEG-32
25322-68-3	PEG-75
25322-68-3	PEG-100
25322-68-3	PEG-150
25322-68-3	Polyethylene glycol
25322-69-4	Polypropylene glycol
25322-69-4	PPG-9
25322-69-4	PPG-17
25322-69-4	PPG-20
25322-69-4	PPG-26
25322-69-4	PPG-30
25496-72-4	Glyceryl oleate
25498-49-1	PPG-3 methyl ether
25637-84-7	Glyceryl dioleate
25639-42-3	Methylcyclohexanol
25640-78-2	1,1′-Biphenyl (1-methylethyl)
25791-96-2	PPG-10 glyceryl ether
25791-96-2	PPG-55 glyceryl ether
26027-38-3	Nonoxynol-4
26027-38-3	Nonoxynol-5
26027-38-3	Nonoxynol-6
26183-52-8	Deceth-6
26264-02-8	Nonoxynol-5
26264-05-1	Isopropylamine dodecylbenzenesulfonate
26266-57-9	Sorbitan palmitate
26266-58-0	Sorbitan trioleate
26401-47-8	Dodoxynol-7
26402-26-6	Caprylic/capric glycerides
26402-26-6	Glyceryl caprylate
26402-31-3	Propylene glycol ricinoleate
26446-35-5	Glyceryl acetate
26446-38-8	Sucrose palmitate
26447-10-9	Ammonium xylenesulfonate
26952-21-6	Isooctyl alcohol
27176-97-2	Nonoxynol-4
27177-01-1	Nonoxynol-6
27177-05-5	Nonoxynol-6
27194-74-7	Propylene glycol laurate
27321-96-6	Choleth-24
27458-93-1	Isostearyl alcohol
27841-06-1	Neopentyl glycol dicaprate
28510-23-8	Neopentyl glycol dioctanoate
29383-26-4	Octyl hydroxystearate
29710-25-6	Octyl hydroxystearate
29911-27-1	Dipropylene glycol n-propyl ether
29911-28-2	Dipropylene glycol butyl ether
31565-12-5	Propylene glycol caprylate
31566-31-1	Glyceryl stearate
31807-55-3	Isododecane
34590-94-8	PPG-2 methyl ether
35274-05-6	Cetyl lactate

CAS	Chemical
35545-57-4	PEG-13 betanaphthol ether
36311-34-9	Isocetyl alcohol
36653-82-4	Cetyl alcohol
36936-60-4	PEG-3 triethanolamine
37200-49-0	Polysorbate 80
37205-87-1	Nonoxynol-4
37205-87-1	Nonoxynol-5
37205-87-1	Nonoxynol-6
37220-82-9	Glyceryl oleate
37286-64-9	PPG-3 methyl ether
37311-04-9	PPG-3-myreth-3
41395-83-9	Propylene glycol dipelargonate
41395-89-5	Isodecyl isononanoate
41669-30-1	Isostearyl isostearate
42131-25-9	Isononyl isononanoate
42131-27-1	Isotridecyl isononanoate
42131-28-2	Isostearyl lactate
51258-15-2	PPG-24-glycereth-24
51258-15-2	PPG-66-glycereth-12
51609-41-7	Nonoxynol-9 phosphate
51811-79-1	Nonoxynol-9 phosphate
52125-53-8	Propylene glycol ethyl ether
52581-71-2	PPG-23 oleyl ether
52581-71-2	PPG-37 oleyl ether
52688-97-0	PEG-4 dioleate
52688-97-0	PEG-8 dioleate
52688-97-0	PEG-12 dioleate
52688-97-0	PEG-20 dioleate
52993-54-1	s-Butylcyclohexane
53637-25-5	Poloxamer 181
55934-93-5	Tripropylene glycol n-butyl ether
56002-14-3	PEG-6 isostearate
56519-71-1	Propylene glycol dioctanoate
57018-52-7	Propylene glycol t-butyl ether
57307-99-0	PEG-8 caprylic/capric glycerides
57533-90-1	Diisocetyl adipate
57569-76-3	Glycereth-7 triacetate
58262-41-2	Diisocetyl adipate
58748-27-9	Propylene glycol dicaprylate/dicaprate
59130-69-7	Cetyl octanoate
59130-70-7	Cetearyl octanoate
59231-34-4	Isodecyl oleate
59231-37-7	Isotridecyl isononanoate
60344-26-5	PEG-6 oleate
61788-49-6	Acetylated lanolin alcohol
61788-59-8	Methyl cocoate
61788-61-2	Methyl tallowate
61790-81-6	PEG-75 lanolin
61790-86-1	Polysorbate 80
61791-12-6	PEG-12 castor oil
61791-12-6	PEG-33 castor oil
61791-12-6	PEG-35 castor oil
61791-20-6	Laneth-5
61791-20-6	Laneth-10
61791-20-6	Laneth-15
61791-20-6	Laneth-16
61791-31-9	Cocamide DEA
61849-72-7	PPG-10 methyl glucose ether
61849-72-7	PPG-20 methyl glucose ether
63148-62-9	Dimethicone
63449-39-8	Paraffin, chlorinated
63793-60-2	PPG-3 myristyl ether
64475-85-0	Mineral spirits
64742-14-9	Petroleum distillates
64742-47-8	Petroleum distillates
64742-48-9	C7-8 isoparaffin
64742-48-9	C8-9 isoparaffin
64742-48-9	C10-11 isoparaffin
64742-48-9	C11-12 isoparaffin
64742-48-9	C11-13 isoparaffin
64742-48-9	C13-14 isoparaffin
64742-49-0	Hexane
64742-89-8	Heptane
64742-95-6	Naphtha (light aromatic)
65381-09-1	Caprylic/capric triglyceride
66082-42-6	Polyglyceryl-3 diisostearate
66197-78-2	Nonoxynol-9 phosphate
66794-58-9	PEG-20 sorbitan isostearate
67701-28-4	Caprylic/capric/linoleic triglyceride
67701-28-4	Caprylic/capric/oleic triglyceride
67701-32-0	Mono- and diglycerides of fatty acids
67701-33-1	Mono- and diglycerides of fatty acids
67762-27-0	Cetearyl alcohol
67762-38-3	Methyl oleate
67762-39-4	Methyl caprylate/caprate
67762-40-7	Methyl laurate
68037-74-1	Dimethicone (branched)
68239-42-9	Methyl gluceth-10
68239-43-0	Methyl gluceth-20
68411-27-8	C12-15 alkyl benzoate
68412-53-3	Nonoxynol-9 phosphate
68424-60-2	Shea butter
68425-31-0	Gasoline
68459-50-7	PEG-3 lanolate
68459-50-7	PEG-5 lanolate
68459-50-7	PEG-10 lanolate
68459-50-7	PEG-20 lanolate
68514-15-8	Gasoline
68526-79-4	Hexyl alcohol
68526-83-0	Isooctyl alcohol
68526-85-2	n-Decyl alcohol
68526-86-3	Tridecyl alcohol
68551-17-7	C10-13 isoparaffin
68553-81-1	Rice bran oil
68583-51-7	Propylene glycol dicaprylate/dicaprate
68584-24-7	Isopropylamine dodecylbenzenesulfonate

CAS	Chemical
68603-42-9	Cocamide DEA
68920-06-9	Naphtha
68956-56-9	Terpinolene
68988-72-7	Propylene glycol dicapry- late/dicaprate
68990-53-4	Mono- and diglycerides of fatty acids
68990-55-6	Acetylated mono- and diglycerides of fatty acids
68990-58-9	Acetylated mono- and diglycerides of fatty acids
68991-68-4	Caprylic/capric/lauric triglyceride
69009-90-1	Bis(methylethyl)-1,1´-biphenyl
69430-24-6	Cyclomethicone
70321-63-0	Lanolin oil
70693-04-8	Isostearyl alcohol
71093-13-5	Steareth-5
71566-49-9	Octyl isononanoate
73398-61-5	Caprylic/capric triglyceride
74623-31-7	PPG-5-buteth-7
74623-31-7	PPG-12-buteth-16
76050-42-5	Carbomer
84540-57-8	Propylene glycol methyl ether acetate
84604-00-2	Walnut oil
84650-19-1	Laneth-15
84696-37-7	Rice bran oil
85049-36-1	Ethyl oleate
85186-88-5	Sorbitan trioleate
85251-77-0	Glyceryl stearate
85404-84-8	Polyglyceryl-3 diisostearate
85409-09-2	Caprylic/capric triglyceride
85536-08-9	PEG-8 caprylic/capric glycerides
85586-21-6	Methyl stearate
85666-92-8	Glyceryl stearate
85736-49-8	PEG-12 dioleate
88230-35-7	Oxo-hexyl acetate
88917-22-0	PPG-2 methyl ether acetate
90438-79-2	Oxo-heptyl acetate
91031-48-0	Isooctyl stearate
97281-23-7	Glycol stearate
97488-91-0	Apricot kernel oil PEG-6 esters
98516-30-4	Propylene glycol ethyl ether acetate
106436-39-9	Tridecyl neopentanoate
108419-32-5	C8 alkyl acetate
108419-33-6	C9 alkyl acetate
108419-34-7	C10 alkyl acetate
108419-35-8	C13 alkyl acetate
111109-77-4	Dipropylene glycol dimeth- yl ether
132739-31-2	Dipropylene glycol t-butyl ether
134141-38-1	PEG-4 dioleate
138208-68-1	Isocetyl salicylate
146340-22-9	Acetyl trimethyl citrate
159100-33-1	Dimethicone copolyol undecylenate
977026-99-5	Shea butter

EINECS Number-to-Trade Name Cross-Reference

EINECS	Trade name
200-289-5	Emery® 912
200-289-5	Emery® 916
200-289-5	Emery® 917
200-289-5	Emery® 918
200-289-5	Kemstrene® 96.0% USP
200-289-5	Kemstrene® 99.0%
200-289-5	Kemstrene® 99.7% USP
200-289-5	Kemstrene® High Gravity
200-338-0	Propylene Glycol USP/FCC Ultra Grade
200-578-6	Eastman® CDA-19
200-578-6	Punctilious® Ethyl Alcohol
200-578-6	Punctilious® SDA 1-1
200-578-6	Tecsol® D-1
200-664-3	Decap
200-664-3	DMSO
200-679-5	Dynasolve 100
200-838-9	Dynasolve 200
200-838-9	Dynasolve 210
200-889-7	Tebol™
200-889-7	Tebol™ 99
201-066-5	Citroflex® A-2
201-067-0	Citroflex® A-4
201-070-7	Citroflex® 2
201-071-2	Citroflex® 4
201-114-5	Eastman® TEP
201-116-6	Kronitex® TOF
201-116-6	Reomol® TOP
201-142-8	Exxsol® Isopentane
201-167-4	PPG Trichlor
201-188-9	NE™ 101
201-550-6	Kodaflex® DEP
201-557-4	Eastman® DBP
201-557-4	Kodaflex® DBP
201-557-4	Palatinol® DBP
201-557-4	Uniplex 150
201-562-1	Kodaflex® HS-3
202-046-9	Decalin
202-509-5	Agrisynth® BLO
202-509-5	Agsol Ex BLO
202-509-5	BLO®
202-509-5	Dynasolve 699
202-509-5	Gaftex® HI-B
202-509-5	GBL
202-510-0	Jeffsol™ EC
202-510-0	Texacar® EC
202-598-0	Purasolv® EL
202-598-0	Purasolv® ELECT
202-598-0	Purasolv® ELS
202-625-6	QO® Tetrahydrofurfuryl Alcohol (THFA®)
202-626-1	QO® Furfuryl Alcohol (FA®)
202-627-7	QO® Furfural
202-935-1	Naturechem® GTR
203-051-9	Eastman® Triacetin
203-090-1	Wickenol® 158
203-419-9	DBE-4
203-419-9	Santosol™ DMS
203-512-4	PentoXone
203-539-1	Arcosolv® PM
203-539-1	Dowanol® PM
203-539-1	Eastman® PM
203-539-1	Icinol PM
203-539-1	Methyl Propasol®
203-539-1	Poly-Solv® MPM
203-539-1	Solvenon® PM
203-544-9	NiPar S-10™
203-572-1	Arconate® 1000
203-572-1	Arconate® HP
203-572-1	Arconate® PC
203-572-1	Jeffsol™ PC
203-572-1	Solvenon® PC
203-572-1	Texacar® PC
203-603-9	Arcosolv® PM Acetate
203-603-9	Dowanol® PMA
203-603-9	Eastman® PM Acetate
203-603-9	Ektasolve® PM Acetate
203-603-9	Methyl Propasol® Acetate
203-625-9	Shell Toluene
203-630-6	Naxol®
203-713-7	EM
203-713-7	EM
203-713-7	Hisolve MC
203-713-7	Methyl Cellosolve®
203-713-7	Methyl Icinol
203-713-7	Poly-Solv® EM
203-726-8	Dynasolve 150
203-726-8	QO® Tetrahydrofuran (THF)
203-726-8	THF
203-726-8	THF
203-751-4	Crodamol IPM
203-751-4	Estol IPM 1509 (BIO-IPM)

EINECS	Trade name
203-751-4	Kessco® IPM
203-751-4	Lexol® IPM
203-751-4	Lexol® IPM-NF
203-751-4	Promyr
203-751-4	Tegosoft® M
203-772-9	EMA
203-772-9	Methyl Cellosolve® Acetate
203-777-6	Exxsol® Hexane
203-794-9	Monoglyme
203-804-1	Cellosolve®
203-804-1	EE
203-804-1	Ethyl Icinol
203-839-2	Cellosolve® Acetate
203-839-2	EEA
203-886-9	Kemester® EGMS
203-905-0	Butyl Cellosolve®
203-905-0	Butyl Icinol
203-905-0	Butyl Oxitol®
203-905-0	Dowanol® EB
203-905-0	Eastman® EB
203-905-0	EB
203-906-6	DM
203-906-6	Dowanol® DM
203-906-6	Eastman® DM
203-906-6	Ektasolve® DM
203-906-6	Hisolve DM
203-906-6	Methyl Carbitol®
203-906-6	Methyl Di-Icinol
203-906-6	Poly-Solv® DM
203-911-3	Emery® 2270
203-911-3	Emery® 2290
203-911-3	Emery® 2296
203-919-7	Carbitol®
203-919-7	Carbitol® Low Gravity
203-919-7	DE
203-919-7	Eastman® DE
203-919-7	Ethyl Di-Icinol
203-919-7	Solvent APV Spec.
203-919-7	Transcutol
203-924-4	Diglyme
203-924-4	Hisolve MDM
203-934-9	Eastman® EB Acetate
203-934-9	Wickenol® 127
203-940-1	Carbitol® Acetate
203-940-1	DEA
203-940-1	Eastman® DE Acetate
203-951-1	EH
203-951-1	Hexyl Cellosolve®
203-961-6	Butyl Carbitol®
203-961-6	Butyl Di-Icinol
203-961-6	Butyl Dioxitol®
203-961-6	DB
203-961-6	Dowanol® DB
203-961-6	Eastman® DB
203-966-3	Emery® 2216
203-978-9	Dowanol® TMH
203-988-3	Hexyl Carbitol®
203-989-9	Carbowax® Sentry® PEG 900
203-989-9	Carbowax® Sentry® PEG 1000
203-989-9	Carbowax® Sentry® PEG 1450
203-989-9	Carbowax® Sentry® PEG 3350
203-989-9	Carbowax® Sentry® PEG 4600
203-989-9	Carbowax® Sentry® PEG 8000
203-989-9	Dow E200
203-989-9	Dow E300
203-989-9	Dow E400
203-989-9	Dow E600
203-989-9	Dow E900
203-989-9	Dow E1000
203-989-9	Dow E1000 NF
203-989-9	Dow E1450
203-989-9	Dow E1450 NF
203-989-9	Dow E3350
203-989-9	Dow E3350 NF
203-989-9	Dow E4500
203-989-9	Dow E4500 NF
203-989-9	Dow E8000
203-989-9	Dow E8000 NF
203-989-9	Lutrol® E 1500
203-989-9	Lutrol® E 4000
203-989-9	Lutrol® E 6000
203-989-9	Pluracol® E200
203-989-9	Pluracol® E1450 NF
203-989-9	Pluracol® E4000
203-989-9	Renex® PEG 1000
203-989-9	Teric PEG 200
204-001-9	Butyl Diglyme
204-001-9	Ethyl Diglyme
204-340-2	Tetralin
204-402-9	Unichem BZBN
204-405-8	Mazon® EE-1
204-589-7	Dowanol® EPh
204-589-7	Rewopal® MPG 10
204-589-7	Rewopal® MPG 40
204-633-5	Isoamyl Alcohol 95%
204-633-5	Isoamyl Alcohol 99%
204-648-7	2-Pyrol®
204-648-7	Soluphor® P
204-666-5	Dermol EB
204-666-5	Unimate® BYS
204-680-1	Emery® 2214
204-685-9	Butyl Carbitol® Acetate
204-685-9	Dba
204-685-9	DB Acetate
204-685-9	Eastman® DB Acetate
204-783-1	Sulfolane W
204-800-2	Albrite® TBP
204-800-2	Amgard® TBPO^4
204-800-2	Kronitex® TBP
204-800-2	Reomol® TBP
204-800-2	TBP
204-825-9	PPG Perchlor
205-011-6	Kodaflex® DMP
205-011-6	Uniplex 110

EINECS	Trade name
205-341-0	Dipentene No. 122
205-392-9	Naturechem® MAR
205-495-9	Permethyl® 99A
205-530-8	Acetamide MEA
205-530-8	Foamid AME-70
205-530-8	Foamid AME-75
205-530-8	Foamid AME-100
205-530-8	Incromectant AMEA-70
205-530-8	Upamide ACMEA
205-530-8	Witcamide® CMEA
205-542-3	Schercemol PGML
205-563-8	Exxsol® Heptane
205-563-8	Shell Tolu-Sol® A HT Solv.
205-571-1	Crodamol IPP
205-571-1	Kessco® IPP
205-571-1	Propal
205-571-1	Tegosoft® P
205-592-6	Dowanol® TBH
205-597-3	Adol® 80
205-597-3	Adol® 85 NF
205-597-3	Fancol OA-95
205-597-3	Novol NF
205-597-3	Witcohol™ 350
206-392-1	PF-5070
206-585-0	PF-5060
208-930-0	Purasolv® ML
211-014-3	Kemester® EGDS
211-020-6	DBE-6
211-020-6	Santosol™ DMA
211-076-1	Ethyl Glyme
212-052-3	Antiblaze® DMMP
212-828-1	GAFsolver™ TMP
212-828-1	M-Pyrol®
212-828-1	NMP
212-828-1	NMP-EL
212-828-1	Pharmasolve™
213-195-4	Diglycolamine® Agent (DGA®)
213-207-8	Kemester® 9022
214-277-2	DBE-5
214-277-2	Santosol™ DMG
214-304-8	Kemester® 2050
215-090-9	Pilot SXS-40
215-535-7	Shell Xylene
215-684-8	EZA®
216-653-1	High Purity MTBE
219-370-1	Primarol 1006
220-045-1	Carbowax® PEG 300
220-045-1	Carbowax® Sentry® PEG 300
220-045-1	Dow E300 NF
220-045-1	Lutrol® E 300
220-045-1	Pluracol® E300
220-045-1	Renex® PEG 300
220-045-1	Rhodasurf® PEG 300
220-045-1	Teric PEG 300
220-548-6	Eastman® EP
220-548-6	EP
220-548-6	Propyl Cellosolve®
222-980-4	Schercemol OLO
222-981-6	Ceraphyl® 140
224-644-9	Butoxyl®
225-856-4	Carbowax® Sentry® PEG 400
225-856-4	Dow E400 NF
225-856-4	Lutrol® E 400
225-856-4	Pluracol® E400
225-856-4	Pluracol® E400 NF
225-856-4	Renex® PEG 400
225-856-4	Teric PEG 400
225-878-4	Arcosolv® PNB
225-878-4	Dowanol® PnB
226-159-8	Arlasolve® DMI
226-242-9	Standamul® G
226-300-9	Lonzest® 143-S
229-859-1	Carbowax® Sentry® PEG 600
229-859-1	Dow E600 NF
229-859-1	Pluracol® E600
229-859-1	Pluracol® E600 NF
229-859-1	Renex® PEG 600
229-919-7	CHP®
229-985-7	Eastman® DP
230-770-5	T-Det® N-4
230-962-9	Nikkol Sefsol 228
230-991-7	Polysolvan® O
231-306-4	Schercemol DIS
232-274-4	Super Refined® Soybean Oil USP
232-276-5	Neobee® 18
232-276-5	Super Refined® Safflower Oil
232-277-0	EmCon Olive
232-277-0	Super Refined® Olive Oil
232-280-7	Super Refined® Cottonseed Oil
232-281-2	Super Refined® Corn Oil
232-293-8	Cosmetol® X
232-293-8	Crystal® O
232-293-8	Crystal® Crown
232-293-8	EmCon CO
232-293-8	York Krystal Kleer Castor Oil
232-293-8	York USP Castor Oil
232-296-4	Super Refined® Peanut Oil
232-298-5	Penreco 2263 Oil
232-348-6	Corona Lanolin
232-348-6	Coronet Lanolin
232-370-6	Lipovol SES
232-370-6	Super Refined® Sesame Oil
232-373-2	Penreco Snow
232-453-7	Shell Mineral Spirits 135
232-453-7	Shell Mineral Spirits 145-EC
232-453-7	Shell Mineral Spirits 146 HT
232-453-7	Shell Mineral Spirits 200 HT
232-453-7	Shell Sol 71
232-453-7	Shell Sol 142 HT
232-453-7	Shell Sol 340 HT
232-489-3	Varsol® 1
232-489-3	Varsol® 18

EINECS Number-to-Chemical Cross-Reference

EINECS	Chemical
200-018-0	Lactic acid
200-262-8	Carbon tetrachloride
200-270-1	Benzyltriethyl ammonium chloride
200-289-5	Glycerin
200-300-3	Benzyl trimethyl ammonium chloride
200-338-0	Polypropylene glycol
200-338-0	Propylene glycol
200-456-2	Phenethyl alcohol
200-464-6	2-Mercaptoethanol
200-467-2	Ethyl ether
200-578-6	Alcohol
200-578-6	Alcohol denatured
200-579-1	Formic acid
200-580-7	Acetic acid
200-580-7	Acetic acid, glacial
200-659-6	Methyl alcohol
200-661-7	Isopropyl alcohol
200-662-2	Acetone
200-663-8	Chloroform
200-664-3	Dimethyl sulfoxide
200-679-5	Dimethyl formamide
200-746-9	n-Propyl alcohol
200-751-6	Butyl alcohol
200-752-1	n-Amyl alcohol
200-753-7	Benzene
200-756-3	1,1,1-Trichloroethane
200-827-9	Propane
200-835-2	Acetonitrile
200-838-9	Methylene chloride
200-842-0	Formamide
200-843-6	Carbon disulfide
200-849-9	Polysorbate 80
200-876-6	Nitromethane
200-879-2	Propylene oxide
200-889-7	t-Butyl alcohol
200-892-3	Trichlorofluoromethane
200-893-9	Dichlorodifluoromethane
200-908-9	t-Amyl alcohol
200-929-3	Trifluoroacetic acid
200-936-1	Trichlorotrifluoroethane
201-055-5	Methyl pentynol
201-066-5	Acetyl triethyl citrate
201-067-0	Acetyl tributyl citrate
201-070-7	Triethyl citrate
201-071-2	Tributyl citrate

EINECS	Chemical
201-114-5	Triethyl phosphate
201-116-6	Trioctyl phosphate
201-126-0	Isophorone
201-142-8	Isopentane
201-148-0	Isobutyl alcohol
201-152-2	Propylene dichloride
201-158-5	2-Butanol
201-159-0	Methyl ethyl ketone
201-166-9	1,1,2-Trichloroethane
201-167-4	Trichloroethylene
201-185-2	Methyl acetate
201-188-9	Nitroethane
201-209-1	2-Nitropropane
201-296-2	L-Lactic acid
201-550-6	Diethyl phthalate
201-557-4	Dibutyl phthalate
201-562-1	Butyl octyl phthalate
202-046-9	Decahydronaphthalene
202-259-7	Methyl benzoate
202-377-9	Ethyl hexanediol
202-422-2	o-Xylene
202-425-9	o-Dichlorobenzene
202-490-3	Diethyl ketone
202-509-5	Butyrolactone
202-510-0	Ethylene carbonate
202-598-0	Ethyl lactate
202-612-5	Isobutyl isobutyrate
202-625-6	Tetrahydrofurfuryl alcohol
202-626-1	Furfuryl alcohol
202-627-7	Furfural
202-708-7	Acetophenone
202-773-1	1,3-Diisopropyl benzene
202-826-9	1,4-Diisopropyl benzene
202-849-4	Ethylbenzene
202-855-7	Benzonitrile
202-859-9	Benzyl alcohol
202-860-4	Benzaldehyde
202-935-1	Glyceryl triacetyl ricinoleate
203-041-4	Tetrahydroxypropyl ethylenediamine
203-051-9	Triacetin
203-090-1	Dioctyl adipate
203-132-9	Propylbenzene
203-234-3	2-Ethylhexanol
203-291-4	Ethyl propionate
203-299-8	Methyl acetoacetate
203-300-1	s-Butyl acetate

EINECS	Chemical
203-310-6	Acetal
203-350-4	Dibutyl adipate
203-389-7	Propyl propionate
203-396-5	p-Xylene
203-419-9	Dimethyl succinate
203-431-4	Dimethyl sebacate
203-448-7	Butane
203-458-1	Ethylene dichloride
203-468-6	Ethylenediamine
203-473-3	Glycol
203-473-3	Polyethylene glycol
203-489-0	Hexylene glycol
203-512-4	4-Methoxy-4-methyl-pentanone-2
203-523-4	Isohexane
203-528-1	Methyl propyl ketone
203-529-7	Butylene glycol
203-539-1	Methoxyisopropanol
203-544-9	Nitropropane
203-550-1	Methyl isobutyl ketone
203-551-7	Methyl amyl alcohol
203-560-6	Isopropyl ether
203-561-1	Isopropyl acetate
203-572-1	Propylene carbonate
203-603-9	Propylene glycol methyl ether acetate
203-604-4	1,3,5-Trimethylbenzene
203-608-6	1,3,5-Trichlorobenzene
203-620-1	Diisobutyl ketone
203-624-3	Methyl cyclohexane
203-625-9	Toluene
203-626-4	γ-Picoline
203-630-6	Cyclohexanol
203-631-1	Cyclohexanone
203-636-9	β-Picoline
203-640-0	p-Methyl morpholine
203-643-7	α-Picoline
203-686-1	Propyl acetate
203-692-4	n-Pentane
203-705-3	n-Butyl mercaptan
203-713-7	Methoxyethanol
203-716-3	Diethylamine
203-721-0	Ethyl formate
203-726-8	Tetrahydrofuran
203-728-9	Tetrahydrothiophene
203-729-4	Thiophene
203-737-8	Methyl isoamyl ketone
203-745-1	Isobutyl acetate
203-751-4	Isopropyl myristate
203-767-1	Methyl n-amyl ketone
203-772-9	Methoxyethanol acetate
203-777-6	Hexane
203-780-2	Pentylamine
203-794-9	Ethylene glycol dimethyl ether
203-804-1	Ethoxyethanol
203-806-2	Cyclohexane
203-809-9	Pyridine
203-813-0	Piperidine
203-815-1	Morpholine
203-821-4	Dipropylene glycol
203-827-7	Glyceryl oleate
203-837-1	Methyl hexyl ketone
203-839-2	Ethoxyethanol acetate
203-852-3	Hexyl alcohol
203-865-4	Diethylenetriamine
203-872-2	Diethylene glycol
203-874-3	Thiodiglycol
203-881-1	Ethylene glycol diacetate
203-886-9	Glycol stearate
203-889-5	Ethyl oleate
203-892-1	Octane
203-897-9	Heptyl alcohol
203-905-0	Butoxyethanol
203-906-6	Methoxydiglycol
203-911-3	Methyl laurate
203-917-6	Caprylic alcohol
203-919-7	Ethoxydiglycol
203-924-4	Diethylene glycol dimethyl ether
203-933-3	Butoxyethanol acetate
203-934-9	Isopropyl stearate
203-940-1	Ethoxydiglycol acetate
203-951-1	Ethylene glycol hexyl ether
203-953-2	Triethylene glycol
203-956-9	n-Decyl alcohol
203-961-6	Butoxydiglycol
203-963-7	Diethylene glycol diethyl ether
203-966-3	Methyl palmitate
203-978-9	Ethoxytriglycol
203-986-2	Tetraethylenepentamine
203-988-3	Diethylene glycol n-hexyl ether
203-989-9	PEG-4
203-989-9	PEG-20
203-989-9	PEG-32
203-989-9	PEG-75
203-989-9	PEG-100
203-989-9	PEG-150
203-990-4	Methyl stearate
203-992-5	Methyl oleate
203-998-8	Tridecyl alcohol
204-001-9	Diethylene glycol dibutyl ether
204-007-1	Oleic acid
204-009-2	Erucamide
204-017-6	Stearyl alcohol
204-070-5	Methyl butynol
204-211-0	Dioctyl phthalate
204-340-2	Tetrahydronaphthalene
204-402-9	Benzyl benzoate
204-405-8	Benzyl laurate
204-428-0	1,2,4-Trichlorobenzene
204-469-4	Triethylamine
204-493-5	n,n-Dimethylaniline
204-558-8	Dioctyl sebacate
204-589-7	Phenoxyethanol
204-626-7	Diacetone alcohol
204-633-5	Isoamyl alcohol

EINECS	Chemical
204-634-0	Acetylacetone
204-648-7	2-Pyrrolidone
204-658-1	n-Butyl acetate
204-661-8	1,4-Dioxane
204-662-3	Isoamyl acetate
204-664-4	Glyceryl stearate
204-666-5	Butyl stearate
204-680-1	Methyl myristate
204-685-9	Diethylene glycol butyl ether acetate
204-696-9	Carbon dioxide
204-783-1	Sulfolane
204-800-2	Tributyl phosphate
204-825-9	Perchloroethylene
204-826-4	Dimethyl acetamide
205-011-6	Dimethyl phthalate
205-149-7	Diethyl toluamide
205-252-7	Butyl benzoate
205-341-0	dl-Limonene
205-392-9	Methyl acetyl ricinoleate
205-399-7	Benzyl acetate
205-450-3	Diisobutyl adipate
205-495-9	Isododecane
205-500-4	Ethyl acetate
205-502-5	Mesityl oxide
205-516-1	Ethylacetoacetate
205-524-5	Dioctyl maleate
205-530-8	Acetamide MEA
205-542-3	Propylene glycol laurate
205-563-8	Heptane
205-571-1	Isopropyl palmitate
205-572-7	Hexyl acetate
205-592-6	Butoxytriglycol
205-597-3	Oleyl alcohol
206-016-6	Cyclopentane
206-392-1	Perfluoroheptane
206-585-0	Perfluorohexane
207-431-5	Eucalyptol
208-754-4	Sodium thiocyanate
208-760-7	t-Butyl acetate
208-930-0	Methyl lactate
209-526-7	3-Pentanol
209-661-1	3-Heptanol
209-954-4	DL-Lactic acid
210-736-6	Cyclohexyl acetate
210-840-1	cis-1,4-Dimethylcyclohexane
210-871-0	Methyl disulfide
211-014-3	Glycol distearate
211-020-6	Dimethyl adipate
211-047-3	Amyl acetate
211-076-1	Ethylene glycol diethyl ether
211-103-7	Cetyl acetate
211-135-1	1,1,2,2-Tetrachloroethane
211-546-6	Behenyl alcohol
212-052-3	Dimethyl methylphosphonate
212-828-1	N-Methyl-2-pyrrolidone
213-195-4	2-(2-Aminoethoxy) ethanol
213-207-8	Methyl behenate
214-277-2	Dimethyl glutarate
214-304-8	Methyl eicosenate
215-090-9	Sodium xylenesulfonate
215-181-3	Potassium hydroxide
215-185-5	Sodium hydroxide
215-269-1	Copper oxide (ic)
215-356-4	Dinonyl phenol
215-535-7	Xylene
215-647-6	Ammonium hydroxide
215-662-8	Naphthenic acid
215-663-3	Sorbitan laurate
215-665-4	Polysorbate 80
215-684-8	Sodium silicoaluminate
216-653-1	Methyl t-butyl ether
218-320-6	Phenyl trimethicone
219-370-1	2-Hexyl-1-decanol
220-045-1	PEG-6
220-548-6	Ethylene glycol propyl ether
222-206-1	PEG-9
222-980-4	Oleyl oleate
222-981-6	Decyl oleate
224-506-8	Isohexadecane
224-644-9	Methoxybutyl acetate
225-248-9	Glyceryl formal
225-856-4	PEG-8
225-878-4	Butoxypropanol
226-159-8	Dimethyl isosorbide
226-242-9	Octyldodecanol
226-300-9	Myristyl propionate
227-729-9	Sorbitan laurate
227-813-5	d-Limonene
227-815-6	l-Limonene
229-859-1	PEG-12
229-919-7	N-Cyclohexyl pyrrolidone
229-985-7	Diethylene glycol propyl ether
230-072-0	Diisopropyl adipate
230-770-5	Nonoxynol-4
230-962-9	Propylene glycol dicaprylate
230-991-7	Glycolic acid-n-butyl ester
231-195-2	Sulfur dioxide
231-306-4	Diisopropyl sebacate
231-679-3	Sodium iodide
232-077-3	(-)-α-Pinene
232-087-8	(+)-α-Pinene
232-274-4	Soybean oil
232-276-5	Safflower oil
232-277-0	Olive oil
232-280-7	Cottonseed oil
232-281-2	Corn oil
232-292-2	Hydrogenated castor oil
232-293-8	Castor oil
232-296-4	Peanut oil
232-298-5	Petroleum distillates
232-348-6	Lanolin
232-366-4	Kerosene
232-370-6	Sesame oil
232-373-2	Petrolatum

EINECS	Chemical
232-384-2	Mineral oil
232-430-1	Lanolin alcohol
232-453-7	Mineral spirits
232-455-8	Mineral oil
232-489-3	Stoddard solvent
233-113-0	Hydrobromic acid
233-293-0	PEG-4 oleate
233-560-1	Isopropyl laurate
233-713-2	D-Lactic acid
234-325-6	Glyceryl stearate
235-088-1	Sodium toluenesulfonate
236-547-9	PPG-2 methyl ether
236-935-8	Triundecanoin
241-654-9	Oleyl erucate
242-060-2	(-)-b-Pinene
246-770-3	Dipropylene glycol
246-771-9	2,2,4-Trimethyl-1,3-pentanediol monoisobutyrate
247-144-2	Glyceryl dioleate
247-152-6	Methylcyclohexanol
247-555-7	Nonoxynol-5
247-556-2	Isopropylamine dodecylbenzenesulfonate
247-568-8	Sorbitan palmitate
247-569-3	Sorbitan trioleate
247-668-1	Glyceryl caprylate
247-669-7	Propylene glycol ricinoleate
247-706-7	Sucrose palmitate
247-710-9	Ammonium xylenesulfonate
248-133-5	Isooctyl alcohol
248-470-8	Isostearyl alcohol
250-705-4	Glyceryl stearate
252-104-2	PPG-2 methyl ether
252-478-7	Cetyl lactate

EINECS	Chemical
252-964-9	Isocetyl alcohol
253-149-0	Cetyl alcohol
253-407-2	Glyceryl oleate
255-350-9	Propylene glycol dipelargonate
255-485-3	Isostearyl isostearate
255-674-0	Isostearyl lactate
260-799-9	Diisocetyl adipate
261-619-1	Cetyl octanoate
261-673-6	Isodecyl oleate
262-980-8	Acetylated lanolin alcohol
262-988-1	Methyl cocoate
262-989-7	Methyl tallowate
263-163-9	Cocamide DEA
265-724-3	Caprylic/capric triglyceride
266-231-6	Nonoxynol-9 phosphate
267-008-6	Cetearyl alcohol
267-015-4	Methyl oleate
270-112-4	C12-15 alkyl benzoate
271-366-9	C10-13 isoparaffin
271-397-8	Rice bran oil
271-657-0	Cocamide DEA
274-273-1	C7-8 isoparaffin
274-559-6	Lanolin oil
275-637-2	Octyl isononanoate
285-206-0	Ethyl oleate
286-074-7	Sorbitan trioleate
286-490-9	Glyceryl stearate
287-824-6	Methyl stearate
288-459-5	PEG-12 dioleate
291-548-1	Polyglyceryl-3 diisostearate
292-951-5	Isooctyl stearate
306-522-8	Glycol stearate
308-980-4	Tallow alcohol
404-640-5	Dipropylene glycol dimethyl ether

Glossary

aprotic solvent. A solvent that cannot act as a proton acceptor or donor.

aromatic hydrocarbons. A major group of unsaturated cyclic hydrocarbons, containing one or more rings. They are derived chiefly from petroleum and coal tar, e.g., benzene.

CAA (Clean Air Act). These 1990 amendments of the United States Federal Government require the EPA to issue approximately 400 new regulations and a number of guidelines and policy statements covering such areas as attainment of NAAQs (National Ambient Air Quality Standards) in all areas, acid rain, protection of the stratospheric ozone layer, hazardous air pollutants, and operating permits.

cosolvent. *See* latent solvent.

CWA (Clean Water Act). Regulates the discharge of nontoxic and toxic pollutants into surface waters. Its objective is to restore and maintain the chemical, physical, and biological integrity of the United States' waters.

diluent. A liquid that by itself is not necessarily a true or active solvent for a resin in a coating, but which aids dissolution when used in conjunction with the active solvent.

drying solvent. A solvent used to remove water from metal surfaces by displacing the water, e.g., for precision equipment and electronic components.

exempt solvent. Specified organic compounds that are not subject to the requirements of a regulation. The solvents that have been deemed of negligible photochemical reactivity by the EPA are methane, ethane, 1,1,1-trichloroethane (methyl chloroform), methylene chloride, trichlorofluoromethane (CFC-11), dichlorodifluoromethane (CFC-12), chlorodifluoromethane (CFC-22), trifluoromethane (FC-23), trichlorotrifluoroethane (CFC-114), and chloropentafluoroethane (CFC-115). However, many chlorinated and fluorinated compounds are being phased out rapidly under the Montreal Protocol on Substances that Deplete the Ozone Layer.

flash point. The lowest temperature at which a solvent will flash under conditions of the test. It is a critical test for volatile and flammable solvents.

HAPs (Hazardous Air Pollutants). A list of 189 compounds and classes of compounds in the Clean Air Act Amendments which are singled out for special emission reduction measures. Compounds were placed on the list on the basis of a combination of factors: the total amount produced; the number of people exposed to the compound; and the inherent level of toxicity displayed.

33/50 Initiative(Air Toxics Initiative). A voluntary EPA program in which over 500 companies were asked to reduce emissions of 17 compounds at a faster rate than would occur under the Air Toxics provisions of the Clean Air Act Amendments. The objective was a 33% reduction by the end of 1992 compared to 1988 emission levels, and a 50% reduction by the end of 1995.

Kauri-Butanol value (KB value). A measure of the aromatic content and hence the solvent power of a hydrocarbon liquid. Kauri gum is readily soluble in butanol but insoluble in hydrocarbons. The KB value is a measure of the volume of the solvent required to produce turbidity in a standard solution containing kauri gum dissolved in butanol. Naphtha fractions have a KB value of about 30 and toluene about 105.

ketone. A class of liquid organic compounds in which the carbonyl group, is attached to two alkyl groups. The simplest member of the series is acetone; others include diethyl ketone and methyl ethyl ketone.

latent solvent. An organic liquid that dissolves nitrocellulose when in combination with an active solvent—usually alcohols and is used widely in nitrocellulose lacquers in a ratio of one part alcohol to two parts active solvent; a solvent that in conjunction with another solvent can dissolve a solute.

LCLo. The lowest lethal airborne concentration tested.

LD50. The amount of a material, given all at once, which causes the death of 50% (one half) of a group of test animals. The LD50 can be determined for any route of entry, but dermal (applied to skin) and oral (given by mouth) LD50's are most common. The LD50 is one measure of the short-term poisoning potential of a material.

mineral spirits. The common name for several types of hydrocarbon solvents used primarily in the coating and dry-cleaning industry. They are divided into four types: Stoddard solvent; high flash point; odorless; and low dry point.

miscible. Two liquids are said to be miscible if they are partially or completely soluble in each other. Commonly, the term miscible is understood to mean that the two liquids are completely soluble in each other.

Montreal Protocol on Substances that Deplete the Ozone Layer. An international agreement signed by almost all industrialized nations which sets a phaseout for the production of ozone depleting substances (ODSs). The compounds to be phased out include chlorofluorocarbons (CFCs), halons, carbon tetrachloride, methyl chloroform (1,1,1-trichloroethane), and hydrochlorofluorocarbons (HCFCs). The original phaseout dates were revised in the 1992 modification of the Montreal Protocol.

NA (North American) number. Transportation identification number used only for shipments within Canada and the United States.

nonpolar solvent. Solvent having a low dielectric constant, e.g., n-hexane.

ozone depleting potential (ODP). A factor established to reflect the ozone-depletion potential of a substance, on a mass per kilogram basis, as compared to chlorofluorocarbon-11 (CFC-11). Such factor is based on the substance's atmospheric lifetime, the molecular weight of bromine and chlorine, and the substance's ability to be photolytically disassociated, as well as other factors.

photochemically reactive. Any solvent which exceeds the levels permitted in any of the three restricted categories (*see under* Rule 66). The photochemically reactive concept has been largely supplanted by the VOC reduction concept in regulations. However, some states in the United States still apply it to any source not governed by a specific VOC emission reduction regulation.

polar solvent. Solvent having a high dielectric constant such as water. It orients around ions such that its dipole moments provide electrostatic attraction to both anions and cations. The attraction of ions and dipoles result in high dielectric constant.

Proposition 65. Also known as the California Safe Drinking Water and Toxic Enforcement Act, it requires that the state establishes two lists, one of which contains compounds known to the state to cause cancer, and the other list contains compounds known to be fetotoxic.

reactive compound. A volatile reactive compound that is a precursor, i.e., reacts in the atmosphere to form photochemical oxidants

Rule 66. The regulation originally adopted in Los Angeles, California in 1967 which was the first air pollution regulation in the United States to control solvent blend composition and specifies three categories of solvents: Category I, comprised of olefins (may not exceed 5% by volume); Category II contains C8+ aromatics, excluding ethyl benzene, (may not exceed 8% by volume); and Category III, composed of ethylbenzene, toluene, branched ketones, and trichloroethylene (the limit is 20% by volume).

SARA Title III. This is the Emergency Planning and Community Right-to-Know Act of 1986 of the Superfund Amendments and Reauthorization Act which requires local and state emergency planning, and provides citizens and local governments with information about potential chemical hazards in their communities.

solubility. The ability of a material to dissolve in water or another liquid and is in equilibrium with an excess of the substance within which it is dissolved. Solubility may be expressed as a ratio or may be described using words such as insoluble, very soluble or miscible.

solubilizer. An agent that increases the solubility of a substance.

solute. One or more substances dissolved in another substance, called the solvent; the solute is uniformly dispersed in the solvent in the form of either molecules or ions; the resulting mixture is a solution.

solution. A uniformly dispersed mixture at the molecular or ionic level, of one or more substances (the solute) in one or more other substances (the solvent).

solvation. The adsorption of a microlayer or film of water or other solvent on individual dispersed particles of a solution or dispersion.

solvent. A material, usually a liquid, which is capable of dissolving another chemical. Chemicals commonly called solvents can dissolve many different chemicals. Examples of common solvents are water, ethanol, acetone, hexane, and toluene.

solvent extraction. A process using solvents to separate a solid from a solid, a solid from a liquid, or a liquid from a liquid.

SNAP (Significant New Alternatives Policy). Program which implements section 612 of the Clean Air Act. EPA evaluates substitutes that companies want to use instead of the ozone-depleting chemicals, and makes sure that they won't cause greater damage to human health and the environment than the ozone depleters they are replacing or than other available substitutes. Based on this evaluation, EPA then adds these substitutes to its lists of acceptable or unacceptable substitutes.

UN (United Nations) number. The UN number is a four-digit number assigned to a potentially hazardous material (such as gasoline, UN1203) or class of materials (such as corrosive liquids, UN 1760). These numbers are used by firefighters and other emergency response personnel for identification of materials during transportation emergencies. UN (United Nations) numbers are internationally recognized.

vapor degreasing. Method of cleaning metal parts whereby the part is suspended in the vapors of the cleaning solvent.

vapor pressure. The pressure exerted when a solid or liquid is in equilibrium with its own vapor. The vapor pressure is a function of the substance and of the temperature.

VOC (volatile organic compounds). All compounds of carbon having a vapor pressure exceeding 0.5 mm Hg at 104 C; they participate in atmospheric photochemical reactions in sunlight to produce ozone and nitrogen oxides.

volatile. Ability of a material to evaporate. The term volatile is commonly understood to mean that a material evaporates easily.